		IIIA	IVA	VA	VIA	VIIA	0
							4.003 He 2
		10.81 B 5	12.01 C 6	14.01 N 7	16.00 O 8	19.00 F 9	20.18 Ne 10
IB	IIB	26.98 Al 13	28.09 Si 14	30.97 P 15	32.06 S 16	35.45 Cl 17	39.95 Ar 18
63.54 Cu 29	65.37 Zn 30	69.72 Ga 31	72.59 Ge 32	74.92 As 33	78.96 Se 34	79.90 Br 35	83.80 Kr 36
107.9 Ag 47	112.4 Cd 48	114.8 In 49	118.7 Sn 50	121.8 Sb 51	127.6 Te 52	126.9 I 53	131.3 Xe 54
197.0 Au 79	200.6 Hg 80	204.4 Tl 81	207.2 Pb 82	209.0 Bi 83	[209] Po 84	[210] At 85	[222] Rn 86

158.9 Tb 65	162.5 Dy 66	164.9 Ho 67	167.3 Er 68	168.9 Tm 69	173.0 Yb 70	175.0 Lu 71
[247] Bk 97	[251] Cf 98	[254] Es 99	[257] Fm 100	[258] Md 101	[255] No 102	[256] Lr 103

CHEMISTRY

CHEMISTRY EXPERIMENT AND THEORY
SECOND EDITION

THOMAS L. ALLEN
RAYMOND M. KEEFER
UNIVERSITY OF CALIFORNIA, DAVIS

HARPER & ROW, PUBLISHERS, New York
Cambridge, Philadelphia, San Francisco,
London, Mexico City, São Paulo, Sydney

Sponsoring Editor: Malvina Wasserman
Project Editor: Nora Helfgott
Designer: Michel Craig
Senior Production Manager: Kewal K. Sharma
Compositor: York Graphic Services, Inc.
Printer and Binder: R. R. Donnelley & Sons Company
Art Studio: Vantage Art, Inc.
Cover: Micrographics of potassium ferrocyanide

CHEMISTRY: Experiment and Theory, Second Edition

Copyright © 1982 by Thomas L. Allen and Raymond M. Keefer.

All rights reserved. Printed in the United States of America. No part of this book may be used or reproduced in any manner whatsoever without written permission, except in the case of brief quotations embodied in critical articles and reviews. For information address Harper & Row, Publishers, Inc., 10 East 53d Street, New York, NY 10022.

Library of Congress Cataloging in Publication Data

Allen, Thomas L.
  Chemistry, experiment and theory.

  Includes index.
  1. Chemistry. I. Keefer, Raymond M. II. Title.
QD31.2.A44   1982   540   81-7078
ISBN 0-06-040209-1   AACR2

To Pat and Hilda

CONTENTS

Preface xvii

CHAPTER 1 CHEMISTRY AS A SCIENCE 1

    1.1 The Study of Chemistry 2
    1.2 Heterogeneous Systems and Homogeneous Systems 3
    1.3 Substances and Solutions 5
    1.4 Compounds and Elements 9
    1.5 Isotopes 10
    1.6 Macroscopic Forms of Matter 11
    1.7 Observation and Interpretation 11
    1.8 The Law of Conservation of Mass and the Law of Definite Proportions 12
    1.9 The Atomic-Molecular Theory 13
    1.10 Induction and Deduction 14
    1.11 The Dynamic Character of Chemistry 15
    1.12 Measurement and Units 15

CHAPTER 2 ATOMS AND MOLECULES 21

    2.1 Atoms and Atomic Masses 22
    2.2 The Mole and Avogadro's Number 23
    2.3 Molecules, Molecular Masses, and Moles 26
    2.4 Composition, Atomic Masses, and Simplest Formulas 30
    2.5 Simultaneous Determination of the Simplest Formula and Atomic Masses 32

2.6 Chemical Reactions and Chemical Equations 32
2.7 Nonequivalent Amounts in Chemical Reactions 34
2.8 Solutions and Molarity 37

CHAPTER 3 EXPERIMENTAL STUDIES OF GASES 43

3.1 Physical Properties of Gases 44
3.2 Dependence of Volume on Amount at Constant Temperature and Pressure 45
3.3 Pressure Measurements 46
3.4 Dependence of Volume on Pressure at Constant Temperature and Amount 48
3.5 Gaseous Solutions 50
3.6 Dependence of Volume on Temperature at Constant Pressure and Amount 52
3.7 Volume Relations of Reacting Gases 53
3.8 Simultaneous Determination of Atomic Masses and Molecular Formulas 55
3.9 Experimental Determination of Molecular Masses from Gas Densities 57
3.10 The Ideal or Perfect Gas Equation 58
3.11 Real Gases 61
3.12 Equations of State for Real Gases 61

CHAPTER 4 THE KINETIC MOLECULAR THEORY OF GASES 69

4.1 Pressure Exerted by Molecules Having Speed u 70
4.2 Correction for Molecules Having Different Speeds 73
4.3 Temperature and Average Translational Kinetic Energy 74
4.4 RMS Speed and Graham's Law of Diffusion 76
4.5 Distribution of Molecular Speeds 78
4.6 Real Gases 81

CHAPTER 5 ENERGY CHANGES IN CHEMICAL SYSTEMS 87

5.1 The Concept of Energy 88
5.2 The First Law of Thermodynamics 91
5.3 Heat Capacity 96
5.4 Heat Capacities of Pure Substances 98

5.5 Explanation of Thermodynamic Properties   100
5.6 Enthalpies of Fusion and Enthalpies of Vaporization   103
5.7 Enthalpies of Reaction   104
5.8 Standard Enthalpies of Formation   107
5.9 Calorimetry   110

CHAPTER 6   ELECTRONS AND NUCLEI   117

6.1 General Features of Atomic Structure   118
6.2 Discovery of the Electron   119
6.3 Measurement of the Electronic Charge   124
6.4 Discovery of the Nucleus   127
6.5 Nuclear Charge   130
6.6 Isotopic Mass   134

CHAPTER 7   PARTICLES AND WAVES   141

7.1 Wave Motion   142
7.2 Light Waves   144
7.3 Energy Quanta   147
7.4 Photons   152
7.5 Wave Properties of Electrons   154
7.6 Standing de Broglie Waves for a Particle in a Box   156
7.7 The Heisenberg Uncertainty Principle   158
7.8 Wave Functions and Probability Densities   161

CHAPTER 8   ATOMIC STRUCTURE   165

8.1 Atomic Spectra   166
8.2 The Spectrum of Atomic Hydrogen   168
8.3 Quantum Numbers for the Hydrogen Atom   171
8.4 Description of Hydrogen Atom Orbitals   174
8.5 Electron Spin   178
8.6 The Pauli Exclusion Principle   180
8.7 The Shielding Effect for Many-Electron Atoms   181
8.8 Electron Configurations   184
8.9 The Building-up Principle and the Periodic Law   185
8.10 Electron Correlation   190
8.11 Electron Densities and Atomic Radii   192
8.12 Ionization Potentials and Electron Affinities   194
8.13 Electron-Dot Symbols   196

CHAPTER 9 MOLECULAR STRUCTURE AND CHEMICAL BONDING 201

- 9.1   Molecular Properties   202
- 9.2   Simple Ionic Bonding Theory   204
- 9.3   Internuclear Distances in Ionic Crystals   206
- 9.4   Bond Energies of Ionic Molecules   207
- 9.5   Covalent Bonding—Lewis Structures   210
- 9.6   Violations of the Octet Rule   213
- 9.7   Donor–Acceptor Bonds and Formal Charge   214
- 9.8   Covalent Bond Angles and the VSEPR Theory   216
- 9.9   Covalent Bond Radii   222
- 9.10 Resonance   223
- 9.11 Polar Bonds and Electronegativity   224
- 9.12 Molecules in Electric Fields   230
- 9.13 Dipole Moments   231
- 9.14 Molecular Orbitals and Chemical Bonding   235
- 9.15 Molecular Orbitals in Diatomic Molecules Formed by Like Atoms   238
- 9.16 Molecular Orbital Description of Diatomic Molecules Formed by Like Atoms   240
- 9.17 Molecular Orbitals in Diatomic Molecules Formed by Unlike Atoms   244
- 9.18 Valence Bond Theory   246
- 9.19 Linear Molecules and sp Hybrid Orbitals   246
- 9.20 Triangular Molecules and sp^2 Hybrid Orbitals   249
- 9.21 Tetrahedral Molecules and sp^3 Hybrid Orbitals   251

CHAPTER 10 LIQUIDS AND LIQUID SOLUTIONS 259

- 10.1   Vapor Pressure   260
- 10.2   Phase Diagrams   263
- 10.3   Kinetic Molecular Theory of Liquids   265
- 10.4   Intermolecular Forces   267
- 10.5   Concentration   271
- 10.6   Vapor Pressure Lowering   273
- 10.7   Boiling Point Elevation and Freezing Point Depression   279
- 10.8   Osmotic Pressure   282
- 10.9   Solubility of Nonelectrolytes   285
- 10.10 Electrolytes   286
- 10.11 Solubility of Electrolytes in Water   290
- 10.12 Chemical Reactions and Net Reactions   290

CHAPTER 11 THE CHEMICAL ELEMENTS AND THEIR COMPOUNDS 299

- 11.1 Metals, Metalloids, and Nonmetals 300
- 11.2 Hydrogen 301
- 11.3 Acids, Bases, and Salts 306
- 11.4 The Noble Gases 309
- 11.5 The Alkali Metals 310
- 11.6 The Alkaline Earth Metals 311
- 11.7 The Group IIIA Elements 312
- 11.8 Carbon—Organic Chemistry 313
- 11.9 Isomerism 317
- 11.10 Silicon, Germanium, Tin, and Lead 319
- 11.11 The Group VA Elements 320
- 11.12 The Group VIA Elements 322
- 11.13 The Halogens 324
- 11.14 The Transition Metals 326

CHAPTER 12 CHEMICAL EQUILIBRIA 333

- 12.1 Gas Phase Equilibria 334
- 12.2 Calculation of Equilibrium Concentrations 339
- 12.3 Conventions Used for Equilibrium Constants 340
- 12.4 Le Châtelier's Principle 342
- 12.5 Dissociation of Weak Electrolytes 345
- 12.6 Heterogeneous Equilibria 348

CHAPTER 13 SIMULTANEOUS EQUILIBRIA 359

- 13.1 Generalized Systems of Acids and Bases 360
- 13.2 The Dissociation of Water; Solutions of Strong Acids and Strong Bases 363
- 13.3 The Concept of pH 365
- 13.4 Relative Strengths of Acids and Bases 366
- 13.5 Weak Acids and Bases; Exact Solutions of Equilibrium Problems 368
- 13.6 Approximations 370
- 13.7 Salt Solutions 374
- 13.8 Buffers 377
- 13.9 Acid–Base Indicators 380
- 13.10 Acid–Base Titrations 381
- 13.11 Polyprotic Acids 386
- 13.12 Sulfide Precipitations 391

13.13 Complex Ions 393
13.14 Use of Acids or Ligands to Dissolve Slightly Soluble Electrolytes 396

CHAPTER 14 ELECTRON-TRANSFER REACTIONS 405

14.1 Oxidation and Reduction 406
14.2 Oxidation States 407
14.3 The Concept of Half-Reactions 410
14.4 Balancing Electron-Transfer Reactions 413
14.5 The Standard Potential of an Electrochemical Cell 418
14.6 Standard Electrode Potentials 421
14.7 The Relative Strength of Oxidizing Agents and Reducing Agents 425
14.8 Effect of Concentration and Pressure on Cell Voltage; Determination of Equilibrium Constants 426
14.9 Primary Cells, Storage Cells, and Fuel Cells 430

CHAPTER 15 THERMODYNAMICS AND CHEMICAL EQUILIBRIA 439

15.1 Temperature Dependence of Vapor Pressures and Equilibrium Constants 440
15.2 Entropy 446
15.3 Entropies of Chemical Substances 447
15.4 Entropy Changes in Spontaneous Irreversible Processes 451
15.5 Heat Engines 454
15.6 Entropy and Probability 456
15.7 The Gibbs Free Energy 458
15.8 Standard Free-Energy Changes 462

CHAPTER 16 CHEMICAL KINETICS 471

16.1 Rate of Reaction 472
16.2 Rate Laws 475
16.3 Method of Initial Rates 477
16.4 Integrated Form of the First-Order Rate Law 478
16.5 Integrated Form of the Second-Order Rate Law 480
16.6 Reaction Rates and Equilibrium 483
16.7 Effect of Temperature on the Rate Constant 484
16.8 The Collision Theory of Reaction Rates 488
16.9 The Transition-State Theory 493

16.10 Reaction Mechanisms 494
16.11 Catalysts and Enzymes 501

CHAPTER 17 THE SOLID STATE 509

17.1 Macroscopic Properties of Solids 510
17.2 The Seven Systems of Crystals 513
17.3 X-ray Diffraction 513
17.4 Metallic Crystal Structures 518
17.5 The Metallic Bond 520
17.6 Ionic Crystal Structures 521
17.7 The Born-Haber Cycle 529
17.8 Covalent Network Crystals 531
17.9 Molecular Crystals 531
17.10 Hydrogen-Bonded Crystals 533
17.11 The Free-Electron Theory of Metals 534
17.12 The Electronic Band Theory 539
17.13 Magnetic Properties 540
17.14 Crystal Defects 543

CHAPTER 18 THE ELEMENTS OF GROUPS IA–IVA 549

18.1 The Group IA Elements 551
18.2 The Group IIA Elements 557
18.3 The Group IIIA Elements 562
18.4 The Group IVA Elements 567

CHAPTER 19 THE ELEMENTS OF GROUPS VA–VIIA AND THE NOBLE GASES 577

19.1 The Group VA Elements 578
19.2 The Group VIA Elements 588
19.3 The Group VIIA Elements 598
19.4 The Noble Gases 602

CHAPTER 20 THE TRANSITION METALS—GENERAL ASPECTS 607

20.1 General Features 608
20.2 Natural Occurrence and Separation 610
20.3 Metals 613

20.4 Oxidation States 615
20.5 Standard Electrode Potentials 617
20.6 Oxides, Hydroxides, Hydrous Oxides, and Oxyacids 619

CHAPTER 21 THE TRANSITION METALS—COORDINATION CHEMISTRY 625

21.1 Octahedral Complexes 627
21.2 The d Orbitals 629
21.3 Ligand-Field Splitting 631
21.4 Magnetic Properties 636
21.5 Visible Absorption Spectra 638
21.6 Ligand-Field Stabilization Energies 640
21.7 Linear, Tetrahedral, and Square Planar Complexes 644
21.8 The Jahn-Teller Effect 649
21.9 Molecular Orbitals of an Octahedral Complex 649
21.10 Chelates 651

CHAPTER 22 ORGANIC CHEMISTRY 657

22.1 Saturated Hydrocarbons (Alkanes) 658
22.2 Unsaturated Hydrocarbons 663
22.3 Aromatic Hydrocarbons 665
22.4 Alkyl Halides 668
22.5 Oxygen Derivatives—Alcohols, Phenols, and Ethers 670
22.6 Oxygen Derivatives—Aldehydes, Ketones, Carboxylic Acids, and Esters 672
22.7 Nitrogen Derivatives 676
22.8 Formation of Carbon–Carbon Bonds 678
22.9 Structure Determination 680

CHAPTER 23 BIOLOGICAL CHEMISTRY 689

23.1 Amino Acids, Polypeptides, and Proteins 690
23.2 Structures of Polypeptides and Proteins 694
23.3 Carbohydrates 696
23.4 Nucleic Acids 702
23.5 Enzymes 707
23.6 Thermodynamics of Chemical Reactions in Biological Systems 712

CHAPTER 24 NUCLEAR CHEMISTRY 719

24.1 Natural Radioactivity 720
24.2 Artificial Nuclear Reactions and Radioactivity 728
24.3 Nuclear Energy 732
24.4 Nuclear Structure 735
24.5 Applications of Isotopes 737

APPENDIXES A-1

A THE INTERNATIONAL SYSTEM OF UNITS (SI) A-3

B FUNDAMENTAL CONSTANTS AND CONVERSION FACTORS A-5

B-1 Fundamental Constants to Four Significant Figures A-5
B-2 Conversion Factors A-5

C MATHEMATICAL OPERATIONS A-7

C-1 Powers and Exponents A-7
C-2 Significant Figures A-8
C-3 Functions and Graphs A-11
C-4 Solving Problems A-13
C-5 Logarithms A-14

D CONCEPTS FROM PHYSICS A-18

D-1 Velocity and Acceleration A-18
D-2 Mass, Force, and Weight A-19
D-3 Linear Momentum and Angular Momentum A-19
D-4 Energy, Work, and Heat A-20
D-5 Electric Charge and Current A-21
D-6 Electric Potential, Field Strength, and Resistance A-22

E ELECTRON CONFIGURATIONS AND FIRST IONIZATION POTENTIALS OF ATOMS A-23

F THERMODYNAMIC PROPERTIES OF SUBSTANCES AT 1 atm PRESSURE AND 25°C   A-24

G SUPPLEMENTAL DERIVATIONS   A-30

    G-1 Kinetic Theory Including Molecular Motion in Any Direction A-30

    G-2 Relation Between Coulombic Energy of Attraction and Internuclear Distance A-32

H BRIEF ANSWERS TO PROBLEMS   A-34

INDEX   I-1

PREFACE

The world about us presents to our senses an endless variety of natural phenomena, and there is little to suggest to the casual observer any sort of underlying unity. Careful observations, combined with the processes of rational thought, have shown that all these phenomena result from the behavior of different combinations of a few kinds of very small particles, which are present in enormous numbers and which obey certain natural laws. The science of chemistry includes the study of these particles and their combinations with one another, the laws governing their behavior, and the natural phenomena resulting therefrom. This is, indeed, a large order, for at one extreme it includes the phenomena occurring within our own bodies and at another extreme the phenomena occurring in the most distant celestial bodies.

Experiment and theory are the twin pillars on which modern science is founded. In the ensuing pages we shall stress the importance of each in reaching an understanding of chemical phenomena. It is within an experimental setting that the meaning of a theoretical concept is revealed, and it is within the context of theory that an experiment acquires purpose and significance. Thus each illuminates the other, and together they show the path toward understanding.

We have written this book to present the science of chemistry as a full-year course at the beginning university or college level. The approach taken here assumes that the student is interested in one of the physical or biological sciences or in an applied field such as agriculture, engineering, or medicine, but is not necessarily interested in a career in chemistry itself. Most students who study chemistry at this level are still assessing and reassessing their interests and goals. This process will be enlightened by acquiring a good overview of chemistry. The student who is already inclined toward a career in chemistry will also benefit from this approach.

In principle and practice, chemistry is a quantitative science. At various points its expression requires a certain amount of mathematics, mainly arithmetic and elementary algebra. Nevertheless, in discussing quantitative relationships the emphasis here is on their physical significance rather than on their mathematical aspects. To the practicing chemist, even one engaged in areas seemingly dominated by sophisticated mathematics, the chemistry comes first—mathematics is only a tool, albeit an essential one.

Subject to the limitation that more advanced mathematics (including calculus) is avoided, we have followed a course designed to present chemistry in a form acceptable to the serious student—chemistry the way it really is. In our view, this approach is the most practical in the long run, for it will most quickly bring the student to an understanding of modern chemistry.

The introductory chapters of this book emphasize macroscopic phenomena. Topics introduced in this section include the different forms of matter (Chapter 1), stoichiometry (Chapter 2), and the physical properties of gases (Chapter 3). The kinetic theory of gases is then developed (Chapter 4) both as a model explaining the behavior of gases and as an introduction to the study of individual molecules. Energy and related topics, including the first law of thermodynamics and thermochemistry, are introduced in Chapter 5. In addition to the inherent importance of all these subjects, they provide the basis for understanding the behavior of individual particles, and they are a necessary foundation for beginning laboratory work.

We then turn to the microscopic world. Following introductory chapters on electrons and nuclei (Chapter 6) and the modern interpretation of their behavior (Chapter 7), we take up atomic structure (Chapter 8), molecular structure (Chapter 9), and the interactions of molecules in liquids and solutions (Chapter 10).

The remaining chapters include a study of chemical equilibria (Chapters 12 and 13), equilibria in redox systems (Chapter 14), and the second and third laws of thermodynamics (Chapter 15). The topic of chemical kinetics is explored in Chapter 16; the solid state is covered in Chapter 17.

Descriptive chemistry is extensively developed in six chapters including a general introduction (Chapter 11), which provides a change of pace partway through the chapters on chemical principles, the inorganic chemistry of the main group elements (Chapters 18 and 19) and the transition metals (Chapters 20 and 21), and organic chemistry (Chapter 22). It should also be mentioned that a substantial amount of descriptive chemistry is woven into the preceding chapters on chemical principles in examples of their applications. We conclude with biological chemistry (Chapter 23) and nuclear chemistry (Chapter 24). Appendixes include reference materials, basic mathematics, and concepts from elementary physics.

A word of explanation is in order regarding the treatment of thermodynamics. We believe that a solid grounding in the concept of energy is an important prelude to the study of atomic and molecular structure, and therefore we have placed the relatively easy first-law topics in an early chapter. The concept of entropy, however, is more difficult for most students, and we find that second- and third-law topics are most easily grasped when the student has first developed an understanding of chemical equilibria.

NEW TO THIS EDITION

Our principal goals in revising this text have been to maintain its accuracy and clarity of exposition, and to bring the material up to date. Although the second edition is comprehensive, covering all important topics in general chemistry in sufficient depth for the student to acquire a working understanding, the book has been kept to a reasonable length so that all the material can be covered at a comfortable pace in a single academic year.

For the most part, only SI units are used, but we also include certain other units that are still widely used in chemistry: the liter and milliliter are used for volume, the atmosphere and torr are most often used as pressure units, and length at the molecular level is typically expressed in ångströms.

In general, the organization of the first edition, which we have found to be a successful one, has been retained. Material on solution stoichiometry has been added to Chapter 2 as an aid to the early use of solutions in laboratory work. The discussion of hybrid orbitals in Chapter 9 has been expanded by including material on hydrocarbons formerly in Chapter 22. The materials of Chapters 10 and 11 have been interchanged so that the introduction to descriptive chemistry can make use of the discussion of liquids and liquid solutions, particularly the concept of electrolytes. In Chapter 16, the sections on rate laws and mechanisms have been expanded substantially. An extensive table of thermodynamic properties of chemical substances is included as a new Appendix F. The more important equations in each chapter are marked in color for emphasis.

The number of problems has been increased by over 60 percent, and the more difficult problems are now marked in color. Brief answers to all problems not requiring a discussion or other extensive material are listed in Appendix H.

SUPPLEMENTARY MATERIALS

A complete *Solutions Manual* is available, showing in detail how to work each problem and answer each question. Of course, students are encouraged to attempt each problem on their own, but this supplement is designed to provide step-by-step assistance where it is needed. An *Instructor's Manual* is also provided, with discussions of the pedagogy of each chapter and sample examination problems.

ACKNOWLEDGMENTS

The authors wish to acknowledge their indebtedness to the many persons who have assisted in various ways in shaping this book. The ideas on which the book is based and their manner of presentation have been developed through many discussions with our colleagues and with the students whom we have been privileged to teach.

Russell J. Boyd, Dalhousie University; Louis deHayes, California State Polytechnic University at Pomona; Bruce E. Eichinger, University of Washington; James H. Espenson, Iowa State University; and Lawrence S. Weiler, University of British Columbia, have all made many helpful comments and suggestions concerning the manuscript. Our editors, Malvina Wasserman and Lieselotte Hofmann, have been of invaluable assistance in improving the finished product. Marilee Urban typed the entire manuscript with care and efficiency. The production has been under the able direction of Nora Helfgott, and the design is by Michel Craig. Many illustrations have been generously contributed by various scientists and publishers; individual acknowledgments appear in the figure captions.

Thomas L. Allen
Raymond M. Keefer

CHEMISTRY

CHAPTER 1
CHEMISTRY AS A SCIENCE

1

1.1 THE STUDY OF CHEMISTRY

Since the earliest days of recorded history humans have sought to understand the nature of their environment. The greatest thinkers have asked questions and suggested answers concerning various aspects of this all-pervading theme. Methods for asking questions and testing answers by means of experiments have been developed and refined during the past few centuries, and these methods and the answers they provide about the nature of matter constitute the science of chemistry. More specifically, chemistry is the study of the composition and structure of the various forms of matter, the transformations or processes which they undergo, and the phenomena they exhibit under various conditions. Therefore all tangible forms of matter, including ourselves, lie within the domain of chemistry.

There are several important reasons for studying chemistry:

1. Most simply and perhaps most importantly, because of intellectual curiosity. When you puzzle about the constitution of a raindrop or a grain of salt, or what happens when the two meet, you join a distinguished group of predecessors in trying to understand the nature of matter.
2. As an introduction to scientific methods. In the chemical laboratory you can learn how to test hypotheses by means of controlled experiments. Chemistry has played a central role in the development of scientific methods.
3. Because of its many beneficial uses. Chemistry has provided us with many new products as well as methods of predicting the effects that a particular substance will have under given conditions. Our understanding of how matter behaves is still incomplete, but what is known gives us the opportunity for control over our environment to an ever-increasing degree.
4. Because undesirable side effects such as air pollution and water pollution may result from indiscriminate use of otherwise beneficial products. Chemistry can be and is being used to identify and eliminate sources of pollution, although it is becoming progressively clearer that determining all the significant effects of a given substance is a complex and challenging problem.

In the ensuing pages we shall outline the methods by which chemists investigate the nature of matter and some of the more important results of these investigations. The interplay of experiment and theory in chemistry will be illustrated by examining both the development of laws and theories to interpret experimental observations, and the design of experiments to test laws and theories. We shall also see how chemistry extends, on one hand, to the synthesis of new substances not found in nature and, on the other, to new insights into the seemingly ordinary phenomena of everyday occurrence.

1.2 HETEROGENEOUS SYSTEMS AND HOMOGENEOUS SYSTEMS

The materials that form our universe, as well as the reaction products formed in the chemical laboratory, typically consist of complex mixtures in the solid, liquid, or gaseous form. How, then, do we proceed to resolve these mixtures into their component parts and determine their composition? We first note that materials are recognized by their characteristics, or **properties,** such as color, hardness, boiling point, and density (mass per unit volume). These properties are sometimes classified as physical properties, such as boiling point and density, which do not depend on the composition of measuring instruments; chemical properties, such as solubility and reactivity, which depend on the composition of other materials; and biological properties, such as odor, taste, and the efficacy or toxicity of drugs, which involve effects on living organisms. Numerous other properties of materials exist, and they will be defined and utilized later.

Next, we choose a region of interest and designate it as a **system.** (Everything else is called the **surroundings** of the system.) In examining different systems, we generally find that they have regions in which samples drawn from different parts are identical in all their properties—such regions are said to be **homogeneous.** (The sample size cannot be too small, or one encounters the heterogeneity associated with the very small particles that constitute all matter. Samples large enough to avoid this difficulty are called macroscopic.) Each homogeneous region is called a **phase.** A system containing two or more phases is called a **heterogeneous system.** If there is only one phase, the system is a **homogeneous system.**

For example, in Figure 1.1 we may choose our system of interest as everything inside the inner surface of the bottle. This is then a heterogeneous system consisting of four phases: air, ice, the solution of sugar and water, and the sugar crystals. (The different parts of a phase need not be contiguous. Thus we have only one ice phase, made up of two ice cubes.) If we choose the outer surface of the bottle as the boundary of our system, the system then contains a fifth phase—the glass of the bottle itself. Each phase has uniform properties throughout; samples removed from, say, different parts of the liquid are identical in all their properties.

Note that a determination of heterogeneity requires more than a casual examination of the properties of the system. A mixture of ordinary table salt and table sugar is heterogeneous, although the distinction between the salt and sugar phases may become apparent only on close examination of the densities or other properties of the individual crystals. As another example, cast iron is a heteroge-

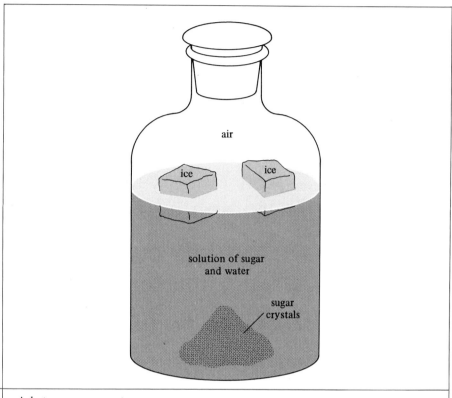

FIGURE 1.1 A heterogeneous system.

neous mixture of carbon and iron, but the distinction between the two phases becomes apparent only by examination with a microscope. Also note that different phases may be identical in some (but not all) of their properties. Thus a mixture of ice and liquid water is heterogeneous; although the two phases have the same composition, they differ in other properties such as density and resistance to flow.

Different phases can be separated from each other by mechanical manipulation. For example, two liquid phases may be separated by placing the mixture in a separatory funnel (Figure 1.2) and withdrawing the denser phase through the stopcock at the bottom. Separation of a finely divided solid from the liquid in which it is suspended is a fairly typical laboratory and industrial problem. It is usually accomplished by a process known as filtration, in which the mixture is allowed to pass through a solid porous material, such as filter paper, sintered glass, or asbestos, which retains the solid materials and allows the liquid to pass through. Denser or coarser solids can often be removed by decantation, that is, by allowing the solid to settle to the bottom of the container and then carefully pouring off the liquid. A centrifuge is often used to hasten the settling process.

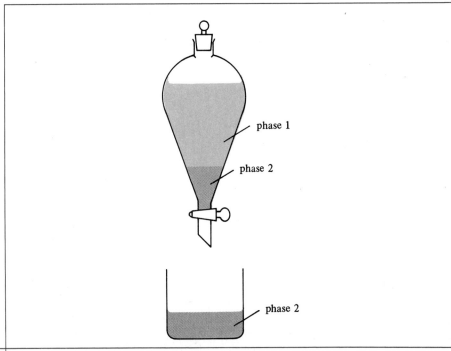

FIGURE 1.2 Separation of liquid phases by means of a separatory funnel. The density of phase 2 is greater than that of phase 1.

Mixtures containing two solid phases may be separated by treating the solid with a liquid in which one solid phase is soluble and the other is not soluble to any significant extent. For example, addition of water to a mixture of sugar and sand will dissolve the sugar. The water-sugar solution can then be removed from the sand by decantation.

1.3 SUBSTANCES AND SOLUTIONS

It has been shown experimentally that homogeneous systems may be divided into two categories depending on their behavior while undergoing any change of phase between the solid, liquid, or gaseous phases. These two types of homogeneous systems are called **substances** and **solutions.** (For emphasis, a substance is often called a **pure substance.**) As we shall see, solutions are homogeneous mixtures of substances.

To illustrate the difference in behavior between a substance and a solution, let us examine the behavior of two homogeneous liquids as they are converted to gases by heating and then reliquefied by cooling. This process is called **distillation,** and it can be accomplished using the apparatus shown in Figure 1.3. A thermometer is used to monitor the temperature of the vapor during the distillation. For

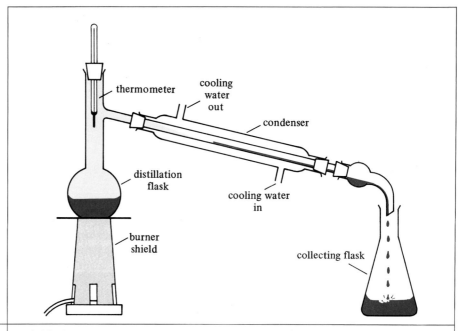

FIGURE 1.3

Distillation apparatus. The substance vaporizes upon heating and escapes from the liquid phase. When the vapor reaches the water-cooled condenser, it condenses back to the liquid phase and drains into the collecting flask.

our two liquids we shall choose water and a 50% (by weight) mixture of ethyl alcohol and water.

When water is placed in the distillation flask and the flask is heated, we note that the temperature rises until the water begins to boil. As the water boils, the vapor enters the condenser, where it is cooled and liquefied and the liquid drains into the collecting flask. Throughout the distillation, the temperature registered by the thermometer remains constant at 100°C (Figure 1.4). It may also be observed that the composition of the water is unaffected by the distillation. This behavior of water is typical of that found for any substance. In general, *a substance is defined as any homogeneous system that undergoes all changes of phase at constant pressure without a change in temperature or composition.*

If now we place the 50% (by weight) mixture of alcohol and water in the distillation flask and heat it, we again find that the temperature increases until the liquid boils. However, the temperature now continues to rise slowly as the liquid is distilled (Figure 1.4). We note that the liquid begins to boil at 82°C and that the first liquid that passes into the collecting flask contains 77% alcohol. As evaporation proceeds, the greater loss of alcohol causes the boiling liquid in the distillation flask to become richer in water. As a result, the temperature rises, and it reaches 92°C before all of the liquid is vaporized. (Of course, after *all* of the alcohol-water mixture has been distilled, the liquid in the collecting flask is again

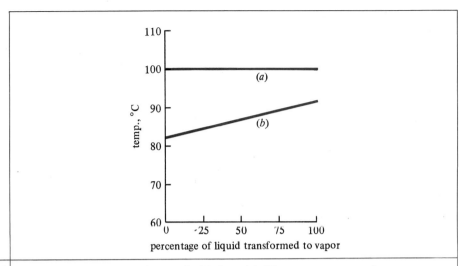

FIGURE 1.4

Temperature changes during distillation. (a) Water—temperature is constant at 100°C during vaporization; (b) 50% (by weight) mixture of ethyl alcohol and water—temperature varies from 82 to 92°C during vaporization.

50% by weight of each component.) This behavior of the alcohol-water mixture is typical of that found for any solution. In general, *a solution is defined as any homogeneous system that undergoes a change of phase at constant pressure with a change in temperature and composition.*

Any solution can be separated into two or more substances. For example, by collecting different fractions of the distillate from the alcohol-water mixture, it is possible to effect a partial separation of the components. Repeated distillation of the fractions leads to a more complete separation. This process may be done in one step using a **fractionating column** between the distillation flask and the condenser. One type of column is simply a long vertical tube packed with glass beads or other inert material. A more efficient type of column, called a bubble-cap column, is shown in Figure 1.5. As the vapor passes through the fractionating column, it condenses to a liquid that may then revaporize. Condensation and vaporization are repeated many times before the vapor reaches the condenser, resulting in much greater separation of the components of the solution. This process, called **fractional distillation,** is of great importance in the oil industry, where it is used to separate petroleum into commercial products such as gasoline (boiling range 85°C to 200°C) and kerosene (boiling range 200°C to 275°C). Gasoline and kerosene are themselves solutions, and further separation can be achieved by further distillation.

Both substances and solutions may be in the liquid, solid, or gaseous phase. For example, air is a solution. Alloys of gold and silver are solid solutions; 18-karat gold is a solid solution in the proportions $\frac{18}{24}$ of gold and $\frac{6}{24}$ of another metal, usually silver. Carbon steels are solid solutions of iron and carbon.

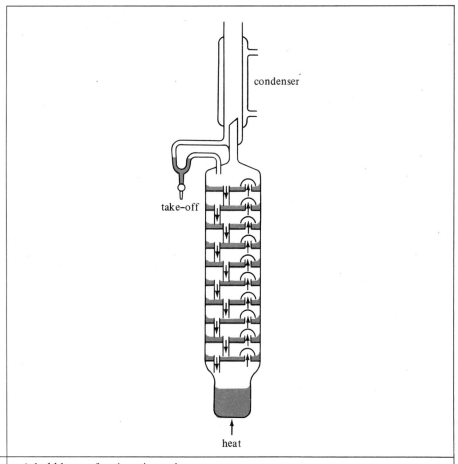

FIGURE 1.5

A bubble-cap fractionating column.

There are a few solutions that behave like a pure substance for a single change of phase. For example, a mixture of 96% ethyl alcohol and 4% water boils at 78.2°C with no change in temperature or composition. However, the same solution freezes over a range of temperatures to form a heterogeneous mixture of ice and solid ethyl alcohol. Solutions of this type are called constant-boiling solutions, or **azeotropes.** There also exist certain solutions that freeze at constant temperature to form heterogeneous mixtures; these are called constant-freezing solutions, or **eutectics.**

Changes of phase, the separation of one phase from another, and the separation of a solution into its component substances are called **physical processes.**

It should be noted that the distinction between a substance and a solution is sometimes based on the idea that a substance contains only one kind of elementary unit (called a molecule), whereas a solution contains more than one kind of molecule. For example, water is made up of water molecules, whereas the water-alcohol solution contains both water molecules and alcohol molecules. Although this distinction fits these examples, it does not provide a rigorous basis for the distinction between a solution and a substance for the following reasons:

1. Although rare, there are some substances that contain more than one kind of molecule.
2. It is very difficult to study individual molecules. Given a homogeneous liquid, how could you determine if it contained more than one kind of molecule? On the other hand, the distinction based on change of temperature and composition is easy to apply.

Still another distinction between a substance and a solution is based on the idea that the composition of a substance is fixed, whereas the composition of a solution is variable. For example, water has a definite composition (as will be discussed in Section 1.8), whereas water and alcohol may be mixed in any proportions. Although this method of distinguishing between a solution and a substance is applicable to these examples, it has two deficiencies:

1. It is difficult to apply. Given a homogeneous liquid, is its composition fixed or variable? One must first effect a separation into components, and then see if these components can be combined in variable proportions.
2. The theory of atoms and molecules was developed to explain the discovery that substances have definite proportions. It would be difficult to understand why this was an important discovery if it were true by definition. By using operational definitions of a substance and a solution based on temperature and composition behavior observed during a change of phase, one has a logical basis on which to proceed toward the development of the atomic-molecular theory.

1.4 COMPOUNDS AND ELEMENTS

Although substances do not change composition during changes of phase, many of them can be decomposed into two or more substances by more drastic treatment. A substance that can be shown to be composed of two or more substances is called a **compound.** For example, when a direct electric current is passed through water, the water decomposes. Oxygen and hydrogen, two gases having very different properties, are produced in the decomposition, one forming at each electrode (Figure 1.6). Therefore, water is a compound. Other familiar examples of compounds are ordinary salt, sugar, limestone, penicillin, aspirin, and DDT. About 8 million different compounds are now known, and the number is increasing by about 350,000 each year.

Originally, a substance that could not be decomposed into two or more different substances was called an **element.** This definition was necessarily somewhat difficult to apply. If all efforts to decompose a substance had failed, was it an

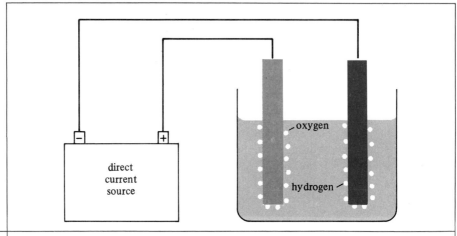

FIGURE 1.6

Decomposition of water by a direct electric current: Oxygen gas forms at one electrode, hydrogen gas at the other.

element? In many cases it was, but occasionally further efforts succeeded. For example, a compound of vanadium (vanadium monoxide) was for many years mistakenly believed to be the element vanadium. But now that the elements have been investigated with sufficient care, we do not expect any similar discoveries in the future. In any case, with the discovery of isotopes, it became necessary to modify the definition of an element given above, as described in the next section.

There are now 106 known elements. Hydrogen and oxygen, the products of the decomposition of water, are elements. Other familiar examples are carbon, copper, aluminum, iron, and mercury. Most of the elements occur naturally. A few have been created as products of certain processes called nuclear reactions. It is expected that future investigations will lead to the production of additional new elements.

The decomposition of a compound into its elements and the conversion of one set of substances into another are called chemical processes, or **chemical reactions.**

1.5 ISOTOPES

During the twentieth century it has been found that most elements are separable, usually with great difficulty, into two or more substances having *very similar* chemical properties. For example, the element silver has been separated into two substances, one slightly denser than the other. Because of their almost identical chemical properties, they are considered to be *different forms of the same element,* and the term **isotope** is used whenever it is necessary to specify which form of the element is being used. Thus the element silver as it occurs in nature consists of two isotopes. Several other isotopes of silver have been made by nuclear reactions.

Some elements, such as sodium, occur in nature as single isotopes. Tin has the largest number of naturally occurring isotopes—10. The separation of an element into its isotopes is an allowed exception to the definition of the term element given in the preceding section.

Both physical processes and chemical reactions are used to separate an element into its isotopes. At the macroscopic level further separation has not been possible.

1.6 MACROSCOPIC FORMS OF MATTER

The way in which different forms of matter have been classified above may be outlined as follows:

I. Heterogeneous systems (two or more phases)
II. Homogeneous systems (one phase)
    A. Solutions (homogeneous systems containing two or more substances)
    B. Substances
        1. Compounds (substances composed of two or more elements)
        2. Elements
            a. Mixtures of isotopes
            b. Pure isotopes

In trying to understand matter in its rich complexity, chemists have found that this separation into simpler forms is an essential first step. It was the study of elements and compounds that led to the most important theoretical development in chemistry—the atomic-molecular theory—and it is to this development that we now turn. At the same time we must consider in some detail the complementary roles played by theory and experiment in chemistry.

1.7 OBSERVATION AND INTERPRETATION

Chemists today feel that they have a highly detailed understanding of the nature of matter. As these ideas are presented, it is important to develop an understanding of the experimental basis of our knowledge. Prior to the modern scientific revolution, there were two important systems of understanding—one based on authority and one based on pure reason. Under the first of these, one simply accepted the ideas of some person or group; under the second, contemplation led to beliefs about the way the universe was constructed. Neither was satisfactory, because there was no way to resolve conflicts between different authorities or different thinkers. In the scientific approach, a combination of reason and observation is used. All ideas are subject to question; all meaningful ideas can be tested by suitable observations.

Observations are the direct perceptions obtained with the senses of sight, touch, smell, hearing, and taste. They necessarily relate to highly specific events, often to seemingly unimportant ones. It is notoriously difficult for beginners to appreciate the crucial importance of their observations, to trust them, and to distinguish between these observations and the interpretations thereof.

Observations are subject to interpretation and generalization at various levels as facts, laws, and theories. A fact summarizes a number of observations of one particular type. A law also summarizes the results of experience but is of much more general applicability. A theory provides a conceptual framework to explain observations, facts, and laws.

To facilitate interpretation and generalization, observations are usually made under controlled conditions such as constant temperature and pressure. A procedure designed to obtain observations of a particular type under a desired set of conditions is called an **experiment.**

1.8 THE LAW OF CONSERVATION OF MASS AND THE LAW OF DEFINITE PROPORTIONS

The development of the atomic-molecular theory is one of the most important and interesting examples of the scientific approach to the study of matter. During the latter part of the eighteenth century, following the introduction of the modern concept of the chemical elements by Antoine Laurent Lavoisier, chemists became interested in determining the elemental composition of different substances. By decomposing a sample of a compound into its elements and carefully weighing the elements, they could determine the mass of each element in the sample of compound. (See Appendix D-2 for a discussion of the weighing operation and the distinction between mass and weight.) Studies of this type form the basis of an important branch of chemistry called **stoichiometry** (from the Greek *stoicheion,* element, and *metrein,* to measure).

Implicit in the early work of Lavoisier and others was the **law of conservation of mass.** When a substance is decomposed or one set of substances is changed into another, *the total mass of the substances produced is identical with the total mass of the substances consumed.* The law of conservation of mass has since been verified to a very high degree of accuracy. For example, the decomposition of exactly 100 grams (g) of water produces hydrogen and oxygen having a total mass of exactly 100 g.

Each determination of the composition of a sample of water involves a large number of observations of the decomposition, observations of the analytical balance at various times during the weighing operations, and so on. It is a remarkable experimental fact based on these observations that samples of water from different sources (naturally occurring water from rain, rivers, lakes, wells, and oceans, and water made in the laboratory by burning hydrogen) all have the same composition, 11.19% hydrogen and 88.81% oxygen. (It is, of course, necessary to purify the water by distillation or some other method prior to the analysis.) Similar results were obtained during the eighteenth century for many substances; each had a characteristic composition. This work culminated in the **law of definite proportions,** stated by Joseph Louis Proust in 1799: *Different samples of a substance contain elements in the same proportion.*

There ensued a celebrated controversy between Proust and Claude Louis Berthollet, who maintained that the composition of many substances depends on the conditions under which they are prepared. However, Proust and others were able to show that many of the examples cited by Berthollet were heterogeneous mixtures or solutions and that accurate analyses of carefully purified substances supported the law of definite proportions. Certain kinds of solution, particularly solid solutions whose composition varies over a limited range depending on the conditions of preparation, are now called berthollides, or **nonstoichiometric compounds.**

The law of definite proportions is essentially descriptive—it summarizes the results of many chemical analyses and it predicts that future analyses of new substances will show that each has a definite composition. It does not explain why substances exhibit this interesting property. But that explanation came quickly, with the atomic theory proposed by John Dalton in 1803.

1.9 THE ATOMIC-MOLECULAR THEORY

Dalton proposed that the different elements are composed of small indivisible particles called **atoms**, that all atoms of an element are identical, and that atoms of different elements have different properties.[1] Compounds are formed when atoms of one element combine with atoms of another to form particular combinations. These combinations are now called **molecules.** Thus a molecule of water is formed when a certain number of atoms of hydrogen combine with a certain number of atoms of oxygen. (Atoms of the same element may also combine to form molecules. For example, oxygen gas consists of molecules, each molecule containing two oxygen atoms.)

This provides a simple explanation of the law of definite proportions. Because all samples of water are made up of water molecules and because each water molecule contains a certain fixed number of atoms of each type, it necessarily follows that all samples have the same composition.

But the atomic-molecular theory, like any good theory, provided a great deal more than an explanation of something that was already known. It provided a useful way to think about the structure of matter, with innumerable consequences that could be tested by experiment. The first of these, deduced by Dalton, was the **law of multiple proportions.**

Consider the case of two elements that can combine to form two or more substances. For example, carbon and oxygen form two very important compounds, now called carbon monoxide and carbon dioxide. Dalton hypothesized that each molecule of the first compound contains one atom of carbon and one atom of oxygen, and that a molecule of the second compound has one atom of carbon combined with two atoms of oxygen. Because twice as many atoms of

[1] That matter is composed of small particles had been suggested by many persons prior to Dalton. In ancient Greece an atomic theory was proposed by Democritus, and later refuted by Aristotle. Dalton's unique contribution was to develop an atomic theory connected to experimental observations.

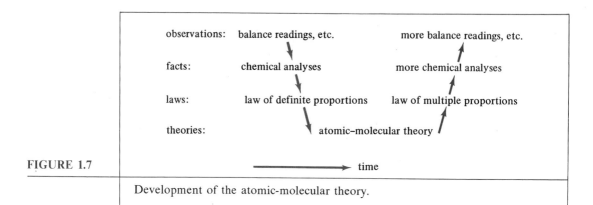

FIGURE 1.7 Development of the atomic-molecular theory.

oxygen are combined with an atom of carbon in the latter molecule, the *mass* of oxygen that is combined with a *given mass* of carbon should be twice as large. Here is something that can be tested experimentally. And, in fact, each gram of carbon is combined with 1.33 g of oxygen in carbon monoxide and with twice as much oxygen, 2.66 g, in carbon dioxide.

Dalton recognized that he had no way of knowing exactly how many atoms there were in each molecule. The masses of oxygen listed above show only that there must be twice as many atoms of oxygen per atom of carbon in the second compound. He always chose the simplest combinations of atoms consistent with the experimental data, and as it turned out he was correct in these two cases. (He even represented the carbon dioxide molecule with what eventually was shown to be the correct structure—the three atoms in a row, with the carbon atom in the middle.) Nevertheless, the number of atoms in every molecule must be a whole number (integer), according to the atomic theory. Thus, if we take the same mass of one element in each case, the masses of another element should always be in a ratio of integers (such as 1:1, 2:1, 3:1, or 3:2) for different compounds of the same elements. To state the **law of multiple proportions** somewhat more formally: *When two or more elements unite to form more than one compound, the masses of one element that combine with a given mass of another are in a ratio of small integers.* Careful analyses of many substances soon placed the law of multiple proportions on a sound experimental basis, and this, in turn, provided important evidence to support the atomic-molecular theory.

A schematic diagram of the development of the atomic-molecular theory is shown in Figure 1.7.

1.10 INDUCTION AND DEDUCTION

As we proceed from the specific to the general, from observations to facts to laws to theories, we are engaged in a logical process called **induction.** It is exciting but dangerous to generalize, for induction does not lead to unique results. There are usually several possible interpretations of any set of observations, as every reader of murder mysteries knows. For this reason interpretations are usually proposed in tentative form, pending experimental tests—at this point they are called **hypotheses.**

By contrast, the path from the general to the specific, called **deduction,** is unique. Each hypothesis has certain definite consequences that can be tested experimentally. If the predictions are not borne out by appropriate observations, it is necessary to revise the hypothesis or, perhaps, construct a completely new one. Hypotheses that survive experimental tests become facts, laws, or theories, and the experimental tests are regarded as supporting evidence.

1.11 THE DYNAMIC CHARACTER OF CHEMISTRY

How, then, does one *prove* a fact or a law or a theory? How many experimental tests must it survive before it is established with complete certainty? One cannot foresee what new observations will be made, and therefore it is always possible that existing interpretations may have to be revised. (This is an aspect of science that helps to make it such a lively affair for those deeply involved in it.)

As we shall see in more detail later, a series of important experimental discoveries beginning in about 1890 opened up an exciting new era of chemistry that apparently has not yet reached its peak. One of the casualties of these discoveries was the atomic theory in its original form. It was found that under certain conditions atoms are divisible, for they are composed of yet smaller particles (electrons and nuclei). Even the law of definite proportions had to be revised to take into account the existence of isotopes (atoms of the same element with different masses).

And yet with these new developments came much more direct confirmation of the essentials of the atomic-molecular theory, so that with modern instruments we can "see" atoms and molecules in the sense of determining their structures and their motions with a high degree of accuracy.

1.12 MEASUREMENT AND UNITS

Many of our investigations in chemistry involve measuring various quantities and expressing the results in terms of a standardized set of units. The set of units now agreed upon by international convention is called the International System of Units (abbreviated SI, from the French name *Le Système International d'Unités*). A summary of SI units is given in Appendix A. There is no point in now going over units for quantities that will not be meaningful to the reader until later, but a few general observations as well as a discussion of certain fundamental units are in order here.

For conciseness the result of a measurement is often indicated as a number between 1 and 10, multiplied by a positive or negative power of 10. This notation is especially useful for very large and very small numbers. For example, the length of time required for the radioactive decay of 50% of a sample of the isotope uranium-238 is written as 4.51×10^9 years rather than as 4,510,000,000 years. (A more detailed discussion of exponential notation is given in Appendix C-1.) Alternatively, this length of time could be written as 4.51 gigayears (abbreviated Gy). (Names and abbreviations of terms used for multiples and submultiples of units are listed in Appendix A.)

An important advantage of exponential notation is that the accuracy of a result can be indicated in a simple way. For example, if a distance has been found to be roughly 800 meters (m), that information is conveyed by the form 8×10^2 m. On the other hand, if the distance has been carefully measured and is known to be almost exactly 800 m, the appropriate notation is 8.00×10^2 m. We say that the first result, 8×10^2, has only one significant figure, while the second result, 8.00×10^2, has three significant figures. For a more detailed discussion of significant figures, see Appendix C-2.

Measurements of length, mass, and time are of fundamental importance in the physical sciences. The units used to express these measurements are as follows.

Length

In scientific work, length is expressed in terms of the meter. (The spelling of this unit is meter in the United States, and metre in most other countries.) For many years the meter was defined as the length at 0°C of a certain bar made of a platinum-iridium alloy, which is kept at the International Bureau of Weights and Measures in Sèvres, France. In 1960 the meter was redefined in terms of the wavelength of light emitted by the krypton-86 isotope under certain specified conditions. (The concept of wavelength will be discussed in Chapter 7.) One advantage of the new definition is that the standard of length occurs in nature, so that the existence of a particular object is no longer essential to its definition.

When the meter was first established (in 1791 in France), it was intended that it be one 10-millionth part of the distance along the earth's surface from the pole to the equator along the meridian passing through Paris. As a result, the distance from the pole to the equator is very close to 10,000 kilometers (km). (Modern measurements have shown the meter to be about 0.2 millimeters (mm) shorter than intended, which is still remarkable accuracy for a geodetic survey made in the 1790s.)

Mass

The fundamental standard of mass is the kilogram (kg), defined as a mass equal to that of a particular cylinder of a platinum-iridium alloy kept at the International Bureau of Weights and Measures. (The terms mass and weight are often confused; for a discussion of the distinction between mass and weight, see Appendix D-2.) In the chemical laboratory one usually works with masses that are much smaller than a kilogram, and the common unit of mass is the gram, where 1000 g = 1 kg.

When the kilogram was first established, it was made to have a mass equal to the mass of exactly 1 cubic decimeter (dm^3) of water at its temperature of maximum density, 3.98°C, and a pressure of 1 atmosphere (atm). Thus it was intended that the kilogram, like the meter, would be based on properties found in nature. The attempt was remarkably accurate, but 1 kg of water under the conditions specified above has a volume slightly larger than 1 dm^3—its volume is 1.000028 dm^3.

Time

The fundamental standard of time is the second, abbreviated s. It was originally based on the average period of rotation of the earth, called a mean solar day, with 1 mean solar day equaling 86,400 s. (There are 24 hours (h) in a day, 60 minutes (min) in an hour, and 60 s in a minute; hence $24 \times 60 \times 60 = 86,400$ s in a day.) Because of small variations in the rotation of the earth, the second was redefined in terms of the year (the period of the orbital motion of the earth about the sun). Then in 1967 the second was redefined in terms of an "atomic clock": A second is the duration of a certain number of periods of the light absorbed by the cesium-133 isotope under certain specified conditions. (A period is the time required for light to travel one wavelength.)

Other Units

Many other units are based on the meter, kilogram, and second. For example, the fundamental unit of volume is the cubic meter, or m^3. Both for historical reasons and because this unit is too large for convenient use, the unit of volume usually encountered in chemistry is the cubic decimeter, or dm^3, commonly called a **liter** (L). (In most countries this unit is spelled litre; the usual spelling in the United States is liter.)

When a derived unit is the product of two or more fundamental units, that fact may be indicated by a space or a dot, as in kg m or kg · m. Division may be indicated by a horizontal line, an oblique line, or a negative power. For example, a speed of 2.5 m per second may be written

$$2.5 \frac{m}{s}, \; 2.5 \text{ m/s, or } 2.5 \text{ m s}^{-1}$$

An important advantage of negative powers is that confusion is avoided in complicated expressions.

As an example of a derived unit, consider mass per unit volume, called **density**.[2]

$$\text{density} = \frac{\text{mass}}{\text{volume}} \tag{1.1}$$

---

[2] Throughout this book the more important equations are indicated with colored equation numbers.

It is usually expressed in grams per cubic centimeter—that is, $g \cdot cm^{-3}$ (or in grams per milliliter, $g\ mL^{-1}$, which has the same units). Because of the original attempt to define a kilogram as the mass of a cubic decimeter of water, the density of water is very close to $1.00\ g\ cm^{-3}$.

Many units are named after scientists whose work was closely connected with the properties measured in those units. For example, the SI unit of force is the newton, abbreviated N, named after Isaac Newton, who first formulated the laws governing the relationships between force, mass, and acceleration. Note that when named after an individual the unit starts with a lowercase letter, but the abbreviation is capitalized.

Consideration of units for other quantities, such as temperature, energy, and electricity, will be postponed until the need arises.

PROBLEMS

Note: Appendix H lists brief answers to those problems that do not require a discussion or other extensive material. In each chapter, the more difficult problems appear at the end of the problem section and are indicated by colored numbers.

1.1 From materials with which you are familiar suggest examples of
   (a) heterogeneous systems
   (b) solutions
   (c) compounds
   (d) elements

1.2 Suppose that the bottle of Figure 1.1 were filled to the top with water and that the ice melted and all of the solid sugar dissolved. Is the system inside the bottle now heterogeneous or homogeneous? Is it a solution or a substance?

1.3 Compare several properties of a substance in the liquid and solid forms at the freezing point.

1.4 Acetic acid melts at 16.6°C. State whether acetic acid is homogeneous or heterogeneous
   (a) at 10°C
   (b) at 16.6°C when part of the acetic acid has melted
   (c) at 20°C

1.5 What experimental test could you apply to determine whether a given sample of some material is
   (a) heterogeneous or homogeneous?
   (b) a solution or a substance?
   (c) a compound or an element?

1.6 To what kind of system is each of the following types of separation apparatus applicable?
(a) fractionating column
(b) filter
(c) separatory funnel

1.7 A mixture of 62 g of tin and 38 g of lead is melted. On cooling, the liquid freezes at a constant temperature of 183 °C to form a heterogeneous mixture of tin and lead crystals. What kind of system is this? What property distinguishes it from a true substance?

1.8 Show that Dalton's atomic-molecular theory is consistent with the law of conservation of mass.

1.9 Two compounds of nitrogen and oxygen contain 36.4% oxygen and 69.6% oxygen (by mass), respectively. Use these data to calculate the mass of one element (say, oxygen) combined with a given mass of the other (say, exactly 1 g of nitrogen). Do your results obey the law of multiple proportions?

1.10 Which of the steps shown in Figure 1.7 are inductive? Which are deductive?

1.11 Using Appendix Table A-2, express a length of 7.5 centimeters (cm) in meters, kilometers, decimeters, millimeters, and nanometers (nm).

1.12 Using the conversion factors in Appendix B-2, convert
(a) a height of 5 ft, 2 in., to meters
(b) 25 cm to inches
(c) 125 lb to kilograms
(d) 4.0 qt (U.S., liquid) to liters
(e) 0.50 qt (Imperial) to liters

1.13 An automobile requires 7.4 gal (U.S., liquid) of gasoline for a trip of 259 mi. Using the conversion factors in Appendix B-2, calculate the average distance traveled per unit volume of gasoline in
(a) miles per gallon (U.S., liquid)
(b) kilometers per liter
(c) miles per gallon (Imperial)

1.14 A gold cylinder has a diameter of 1.08 cm, a length of 2.75 cm, and a mass of 48.6 g. Calculate the density of gold in grams per cubic centimeter and in kilograms per cubic meter. (The formula for the volume of a cylinder is $\pi d^2 l/4$, where $\pi = 3.14$, d is the diameter, and l is the length.)

1.15 An aluminum cylinder has a mass of exactly 1 kg. Its length is 8.00 cm. Using 2.70 g cm^{-3} for the density of aluminum, calculate the diameter of the cylinder. (The formula for the volume of a cylinder is given in the preceding problem.)

1.16 One compound of elements X and Y contains 35.2% X (by mass); another compound contains 24.6% X.
(a) Show that these data illustrate the law of multiple proportions.
(b) If molecules of the first compound contain three atoms of Y per atom of X, how many atoms of Y per atom of X are there in molecules of the second compound?

CHAPTER 2
ATOMS AND MOLECULES

2

2.1 ATOMS AND ATOMIC MASSES

It is convenient to use symbols for the different elements and their atoms. A table inside the back cover lists the known elements and the symbols in common use. Most of the symbols consist of the first letter, or the first and one additional letter, taken from the common name of the element. For example, Cl represents the element chlorine, or one atom of chlorine, or some collection of chlorine atoms—the usage is usually clear from the context. In some cases the symbol is derived from the Latin name: antimony, Sb (*stibium*); copper, Cu (*cuprum*); gold, Au (*aurum*); iron, Fe (*ferrum*); lead, Pb (*plumbum*); mercury, Hg (*hydrargyrum*); potassium, K (*kalium*); silver, Ag (*argentum*); sodium, Na (*natrium*); tin, Sn (*stannum*). The symbol for tungsten, W, comes from *wolfram,* another name for the element.

The masses of individual atoms are so small that they are inconvenient numbers to use. Indeed, it was not possible to measure them for many years following Dalton's discoveries. It was, however, possible for Dalton and other chemists in the nineteenth century to determine the *relative masses* of different atoms. These are called **atomic masses** or **atomic weights.**[1] They still provide the most convenient way to calculate stoichiometric relations.

A scale of relative values requires that one value be chosen as a standard; all other values are then compared to it. In the past both hydrogen and oxygen atoms have been used as standards at various times. The scale adopted by international agreement in 1960 assigns exactly 12 to the mass of the most abundant carbon isotope (carbon-12). All other atomic masses are measured relative to this.

The original basis for the assignment of atomic masses was the analysis of compounds of an assumed formula. For example, 0.220 g of carbon dioxide contains 0.060 g of carbon and 0.160 g of oxygen. Let us assume that carbon dioxide is composed of molecules containing one atom of carbon and two atoms of oxygen as implied by the formula (CO_2). If we let n represent the number of molecules of CO_2, m_c represent the mass of a carbon atom, and m_o represent the mass of an oxygen atom, then we may write that 0.220 g of CO_2 contains a

$$\text{mass of carbon} = nm_c = 0.060 \text{ g}$$

---

[1] Although the latter term is the one in common usage, the term atomic mass is more precise, and it is slowly being accepted, perhaps in anticipation of chemistry in outer space, where all objects are practically weightless.

and a

$$\text{mass of oxygen} = 2nm_o = 0.160 \text{ g}$$

We can now eliminate n by dividing the second equation by the first to obtain

$$\frac{2nm_o}{nm_c} = \frac{0.160 \text{ g}}{0.060 \text{ g}} \quad \text{or} \quad \frac{m_o}{m_c} = \frac{0.160}{2(0.060)} = 1.33 \quad (2.1)$$

Thus an oxygen atom has a mass 1.33 times as large as that of a carbon atom. If the atomic mass of carbon were 12.0, then the relative mass of oxygen would be (1.33)(12.0), or 16.0.

In chemistry, we usually work with naturally occurring elements that may contain two or more isotopes (atoms of the same element with different masses). Since most chemical methods involve the use of large numbers of atoms, we obtain by the methods of the preceding paragraph an average atomic mass for each element. Fortunately the isotopic content of naturally occurring elements is remarkably constant. However, it is possible by physical methods (Chapter 6) to determine the mass of individual atoms and, hence, the masses of each isotope of an element. By such methods we find that the element carbon as it occurs in nature consists of 98.89% of a light isotope to which we assign an atomic mass of exactly 12, 1.11% of a heavier isotope of atomic mass 13.00335, and extremely small amounts of a still heavier isotope of atomic mass 14.00324.

The different isotopes of an element are most conveniently designated by the atomic mass rounded off to the nearest integer, called the **mass number.** The mass number either follows the name of the element (as in carbon-12) or appears as a superscript to the left of the symbol for the element (as in ^{12}C). Neglecting the contribution of ^{14}C, we can calculate the **average atomic mass** for carbon to be $(0.9889 \times 12) + (0.0111 \times 13.00335)$, or 12.011. (For simplicity the word "average" is usually dropped, and average atomic mass is called **atomic mass.**) The atomic masses of the elements, expressed to four significant figures, are listed in the table inside the back cover. In this table we find, for example, that the atomic mass of hydrogen is 1.008, that of carbon is 12.01, and that of oxygen is 16.00.

2.2 THE MOLE AND AVOGADRO'S NUMBER

In the preceding section we have shown that the mass of n oxygen atoms relative to the mass of n carbon atoms is 1.33 regardless of how large n may be (Eq. 2.1). Conversely, if we take samples whose relative masses are the same as the ratio of the masses of the atoms (or the ratio of atomic masses), each sample must contain the same number of atoms. For example, the number of oxygen atoms in 16.00 g of oxygen is the same as the number of carbon atoms in 12.01 g of carbon. This type of reasoning has led to the extremely important **mole concept** of chemistry.

By international agreement the **mole** (abbreviated **mol**) is defined as *the amount of substance that contains as many elementary entities as there are atoms in* *exactly 0.012 kg (or 12 g) of carbon-12.* (It is necessary to specify the nature of the elementary entities—for the moment we shall restrict ourselves to atoms.) Because the relative masses of atoms of different elements are expressed by the atomic masses, and because the mole is defined in terms of the atomic mass of carbon-12 in grams, it follows that *a mole of atoms of any element has a mass equal to its atomic mass in grams.* For example, a mole of hydrogen atoms has a mass of 1.008 g, a mole of carbon atoms has a mass of 12.01 g, and a mole of oxygen atoms has a mass of 16.00 g. (At first sight it may seem peculiar that a mole of carbon atoms has a slightly higher mass than a mole of carbon-12 atoms. The reason for this is, of course, that by "carbon" we mean the naturally occurring isotopic mixture, which contains over 1% of the heavier isotope carbon-13.)

The number of atoms contained in exactly 12 g of carbon-12, and therefore the number of elementary entities in a mole of any substance, is called **Avogadro's number**, named after Amedeo Avogadro. Avogadro's number is denoted by N_0. It existed as a concept long before it was possible to measure it. Some of the experiments in which Avogadro's number has been determined will be described in later chapters. The most accurate value now available is $6.022045(31) \times 10^{23}$, where the last two digits are uncertain. It will be sufficiently accurate for problems in this book to use only three significant figures, 6.02×10^{23}. (The exponential notation used for very large and very small numbers is described in Appendix C-1, and significant figures are discussed in Appendix C-2.)

We are now in a position to consider some useful calculations involving atoms and moles. Given the number of grams of an element, or the number of moles of atoms, or the number of atoms, the other two quantities can be determined as shown in the following diagram:

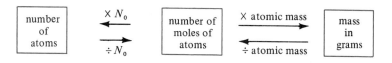

Examples of such calculations are given next. If any difficulty is encountered with these examples, see Appendix C-4 for a discussion of problem solving.

EXAMPLE 1

A sample of oxygen atoms (O) contains 2.50×10^{22} atoms. How many moles of oxygen atoms does the sample contain? What is the mass of the sample?

As there are 6.02×10^{23} atoms in 1 mol of O, the number of moles of oxygen atoms in this sample is

$$\frac{2.50 \times 10^{22} \text{ atoms}}{6.02 \times 10^{23} \text{ atoms (mol of O)}^{-1}} = 0.0415 \text{ mol of O}$$

Note that we must divide by atoms (mol of O)$^{-1}$ to obtain the correct units for the answer. Always include units in your calculations and in your answers.

2.2 THE MOLE AND AVOGADRO'S NUMBER

The atomic mass of oxygen (from the table inside the back cover) is 16.00, and therefore we have

0.0415 mol of O × 16.00 g of O (mol of O)$^{-1}$ = 0.664 g of O

Again note that including units in the calculation gives the answer in the desired units.

Alternatively, we can use the diagram above to obtain the same answers. To go from number of atoms to number of moles of atoms we must divide by Avogadro's number, N_0. Then we must multiply by the atomic mass to obtain the mass of the sample in grams.

EXAMPLE 2

A block of aluminum (Al) has a mass of 75.0 g. How many moles of aluminum atoms does the block contain? How many atoms of aluminum are there in the block?

From the table inside the back cover we find that aluminum has an atomic mass of 26.98. Therefore 1 mol of aluminum atoms has a mass of 26.98 g, and the 75.0-g block contains

$$\frac{75.0 \text{ g of Al}}{26.98 \text{ g of Al (mol of Al)}^{-1}} = 2.78 \text{ mol of Al}$$

Again we note that some units cancel and that the answer is obtained in the correct units.

Each mole of aluminum atoms contains 6.02×10^{23} atoms, and therefore 2.78 mol of aluminum atoms contains

2.78 mol of Al × 6.02 × 10^{23} atoms of Al (mol of Al)$^{-1}$
$$= 1.67 \times 10^{24} \text{ atoms of Al}$$

Note that the use of the diagram preceding Example 1 leads to the same answers.

From the atomic masses and Avogadro's number, the average mass of one atom of any element can be calculated.

EXAMPLE 3

Calculate the average mass of an atom of gold (Au).

The atomic mass of gold is 197.0. Thus 1 mol of gold atoms (which contains Avogadro's number of atoms) has a mass of 197.0 g, and the average mass of a gold atom is

$$\frac{197.0 \text{ g (mol of Au)}^{-1}}{6.02 \times 10^{23} \text{ atoms of Au (mol of Au)}^{-1}} = 3.27 \times 10^{-22} \text{ g (atom of Au)}^{-1}$$

In comparing the masses of individual atoms, the basic standard is one-twelfth of the mass of a carbon-12 atom. This unit is called the atomic mass unit (abbreviated u), or dalton, and it is sometimes attached as a dimension to atomic masses. (For example, "the atomic mass of gold is 197.0 u.") In this book we shall follow the definition given previously and treat atomic mass as a dimensionless number.

Because 1 mol of u has a mass of exactly 1 g, and contains Avogadro's number of u, the mass of 1 u is therefore

$$\frac{1 \text{ g } \cancel{\text{mol}^{-1}}}{6.02 \times 10^{23} \text{ u } \cancel{\text{mol}^{-1}}} = 1.66 \times 10^{-24} \text{ g u}^{-1}$$

2.3 MOLECULES, MOLECULAR MASSES, AND MOLES

Combinations of atoms, or molecules, are represented by *molecular formulas* constructed from the symbols for their constituent atoms. Each molecule of hydrogen chloride, a gas at ordinary pressures and temperatures, contains one atom of hydrogen and one atom of chlorine, and the molecular formula for the substance hydrogen chloride or its individual molecules is accordingly HCl. If a molecule contains more than one atom of any element, the number is indicated by a subscript to the right of the symbol for the element. Thus ammonia, another gas, has the molecular formula NH_3; that is, each ammonia molecule is a combination of one nitrogen atom and three hydrogen atoms. The element hydrogen exists as a gas at room temperature and its molecular formula is H_2.

In some cases the structure of a molecule is roughly indicated by the molecular formula. Thus ethyl alcohol, a liquid, could be represented by the symbol C_2H_6O, which indicates that it is composed of two carbon atoms, six hydrogens, and one oxygen, but the usual symbol is C_2H_5OH, indicating that one of the six hydrogen atoms is attached to the oxygen atom. This symbol also serves to distinguish it from another compound in which the molecules have the same number of atoms of each kind in a different arrangement—methyl ether (also called dimethyl ether), CH_3OCH_3, a gas at room temperature. In a molecule of methyl ether, all the hydrogen atoms are attached to the carbon atoms, none to the oxygen atom. (Such substances, which have the same number of atoms of each kind but different structures and different properties, are called **structural isomers**.)

A better representation of the molecular structure is given by **structural formulas**:

$$\text{H—Cl} \qquad \begin{array}{c} \text{H—N—H} \\ | \\ \text{H} \end{array} \qquad \begin{array}{c} \text{H} \quad \text{H} \\ | \quad | \\ \text{H—C—C—O—H} \\ | \quad | \\ \text{H} \quad \text{H} \end{array} \qquad \begin{array}{c} \text{H} \quad \text{H} \\ | \quad | \\ \text{H—C—O—C—H} \\ | \quad | \\ \text{H} \quad \text{H} \end{array}$$

hydrogen chloride   ammonia   ethyl alcohol   methyl ether

FIGURE 2.1

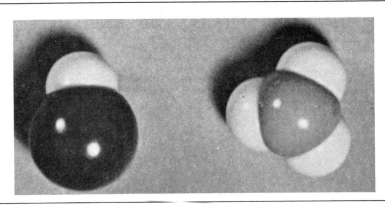

Molecular models of hydrogen chloride, HCl (left), and ammonia, NH_3 (right).

The lines connecting the atomic symbols represent the chemical bonds holding the atoms together. The nature of these bonds and the rules for writing structural formulas will be discussed in Chapter 9.

Still better representations of molecular structures are given by **molecular models,** which do not suffer from the restrictions imposed by trying to represent a three-dimensional object on a sheet of paper with only two dimensions. Molecular model representations of HCl and NH_3 are shown in Figure 2.1, whereas Figure 2.2 shows similar representations of ethyl alcohol and methyl ether.

The experimental methods used to determine molecular formulas and molecular structures will be discussed later. A first step in these investigations is often the experimental determination of the **simplest formula** (sometimes called the **empirical formula**), which gives the ratio of the number of atoms of each element. For example, hydrogen peroxide molecules contain one hydrogen atom for each

FIGURE 2.2

Molecular models of ethyl alcohol, C_2H_5OH (left), and methyl ether, CH_3OCH_3 (right).

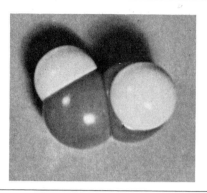

FIGURE 2.3

Molecular model of hydrogen peroxide, H_2O_2.

oxygen atom, and the simplest formula is therefore HO. The molecular formula is H_2O_2, and the structural formula is

$$\begin{array}{c} H \\ | \\ O-O \\ | \\ H \end{array}$$

A molecular model of hydrogen peroxide is shown in Figure 2.3.

The mass of any molecule is simply the sum of the masses of its constituent atoms, and, therefore, the relative **molecular mass** on the carbon-12 scale is the sum of the atomic masses.[2] For example, the molecular mass of hydrogen chloride is 1.008 (the atomic mass of hydrogen) + 35.45 (the atomic mass of chlorine), or 36.46. Ethyl alcohol and methyl ether have the same molecular mass, (2 × 12.01) + (6 × 1.008) + 16.00 (for the two carbon atoms, the six hydrogen atoms, and the oxygen atom, respectively), or a total of 46.07.

Using the basic definition of the mole given in the preceding section, we see that *a mole of a compound contains Avogadro's number of molecules*. Reasoning similar to that used previously for atoms shows that *its mass in grams is equal to its molecular mass*. Thus a mole of hydrogen chloride has a mass of 36.46 g, and it contains 6.02×10^{23} molecules of HCl.

It is very useful to think about moles in terms of the *number* of atoms or molecules they contain, even though Avogadro's number is so enormous as to strain the imagination. A good way of cutting it down to size is to realize that it is simply a numerical unit, similar to more familiar units such as the dozen, gross, case, and ream. The crucial step in calculations involving elements and compounds is the development of a mental picture of the relative numbers of atoms and molecules, and this can easily be done by thinking about moles in this way.

[2] The molecular mass is often called the molecular weight.

2.3 MOLECULES, MOLECULAR MASSES, AND MOLES

The term mole is applied to any collection of Avogadro's number of entities, whether atoms, molecules, ions, or electrons. Difficulty arises in practice unless the entity is specified. For example, for elements that form molecules containing more than one atom, such as H_2, O_2, N_2, Cl_2, P_4, and S_8, it is necessary to specify moles of atoms or moles of molecules. Thus 1 mol of oxygen atoms (1 mol of O) has a mass of 16.00 g while 1 mol of oxygen molecules (1 mol of O_2) has a mass of 32.00 g. Because it is not clear which of the above is meant by "a mole of oxygen," it is essential to state whether one means a mole of oxygen atoms (O) or a mole of oxygen molecules (O_2).

When one is working with moles of molecules, the following diagram provides a useful summary of the various relationships involved. Note that it is very similar to the diagram in the preceding section, the only differences being that "atoms" have been changed to "molecules," and "atomic mass" has been changed to "molecular mass."

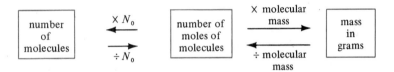

We can now consider a simple problem illustrating the use of moles of molecules.

EXAMPLE 4

A certain sample of carbon dioxide, CO_2, has a mass of 9.24 g. How many moles of CO_2 is this? How many molecules of CO_2 are contained in this sample? How many atoms of each element?

To convert from grams of CO_2 to moles of CO_2 we must use the mass of 1 mol of CO_2, that is, the molecular mass. The molecular mass of CO_2 is 12.01 + (2 × 16.00), or 44.01. Therefore the number of moles of CO_2 in 9.24 g of CO_2 is

$$\frac{9.24 \text{ g}}{44.01 \text{ g (mol of } CO_2)^{-1}} = 0.210 \text{ mol of } CO_2$$

As each mole of CO_2 contains 6.02×10^{23} molecules of CO_2, there are

$$0.210 \text{ mol} \times 6.02 \times 10^{23} \text{ molecules mol}^{-1} = 1.26 \times 10^{23} \text{ molecules of } CO_2$$

Since there are 1.26×10^{23} CO_2 molecules, each containing one carbon atom and two oxygen atoms, there are 1.26×10^{23} carbon atoms and 2.52×10^{23} oxygen atoms in the sample.

Alternatively one may solve this problem using the diagram above. We are moving from right to left, so we must divide the mass by the molecular mass to obtain the number of moles of CO_2, and then we must multiply by Avogadro's number to obtain the number of molecules of CO_2.

2.4 COMPOSITION, ATOMIC MASSES, AND SIMPLEST FORMULAS

Each compound is characterized by a certain molecular formula (and simplest formula) and also by a certain definite elemental composition by mass, as discussed in the preceding chapter. These two properties and the relative atomic masses of the different elements are closely interconnected—given any two, we can calculate the third. For instance, from the composition of water and the atomic masses of hydrogen and oxygen, we can calculate the simplest formula of water, as shown in Example 5.

EXAMPLE 5

Experiments on the decomposition of water show that it is 11.2% hydrogen and 88.8% oxygen by mass. From these data and the atomic masses, calculate the simplest formula of water.

Problems of this type are most easily handled by visualizing a specific amount, say, 100.0 g.

We wish to determine the simplest formula, which is the ratio of the number of hydrogen atoms to the number of oxygen atoms in a molecule. This is the same as the ratio of the number of moles of hydrogen atoms to the number of moles of oxygen atoms in a mole of water, or in a sample of any size.

The sample chosen contains

$$\frac{11.2 \text{ g}}{1.008 \text{ g (mol of H)}^{-1}} = 11.1 \text{ mol of H}$$

and

$$\frac{88.8 \text{ g}}{16.00 \text{ g (mol of O)}^{-1}} = 5.55 \text{ mol of O}$$

These numbers are in a ratio of 11.1:5.55 or 2.00:1.00. Thus the simplest formula for water is H_2O.

If we wish also to determine the molecular formula, additional experimental data are required. Molecular formulas consistent with the simplest formula are H_2O, H_4O_2, H_6O_3, and so on. The corresponding molecular masses are 18.0, 36.0, 54.0, and so on. Experimental determinations of the molecular mass (by methods to be described later) give the result 18.0, and therefore H_2O is also the molecular formula.

Given the simplest formula and the elemental composition of a compound, we can determine the ratio of the atomic masses or, if one atomic mass is also given, then the other or others can be determined. This type of calculation is illustrated in Example 6.

The third variation on this theme, namely, calculating the elemental composition of a compound from the atomic masses and the simplest formula or molecular formula, is relatively easy; it is illustrated by Example 7.

2.4 COMPOSITION, ATOMIC MASSES, AND SIMPLEST FORMULAS

EXAMPLE 6

The simplest formula of hematite or ferric oxide, an important iron ore, is Fe_2O_3, and its composition is 69.9% iron and 30.1% oxygen. Using 16.00 for the atomic mass of oxygen, calculate the atomic mass of iron.

Again it is convenient to consider a 100.0-g sample of the compound, containing in this case 69.9 g of iron combined with 30.1 g of oxygen. The atomic mass of iron is then 69.9 g of iron divided by the number of moles of iron atoms. Therefore we look for information that will give us the number of moles of iron atoms. By using the simplest formula, we can calculate the number of moles of iron atoms from the number of moles of oxygen atoms. Thus our first task is to calculate the latter quantity.

To obtain the number of moles of oxygen atoms, we divide the mass of oxygen by the atomic mass of oxygen:

$$\frac{30.1 \text{ g}}{16.00 \text{ g (mol of O)}^{-1}} = 1.88 \text{ mol of O}$$

From the simplest formula we know that the compound contains 2 mol of iron atoms combined with 3 mol of oxygen atoms, or $\frac{2}{3}$ mol of iron atoms combined with 1 mol of oxygen atoms. This sample then must contain

$$1.88 \text{ mol of O} \times \tfrac{2}{3} \text{ mol of Fe (mol of O)}^{-1} = 1.25 \text{ mol of Fe}$$

Thus we know that 1.25 mol of iron atoms has a mass of 69.9 g, and therefore the mass of 1 mol of iron atoms is

$$\frac{69.9 \text{ g}}{1.25 \text{ mol of Fe}} = 55.9 \text{ g (mol of Fe)}^{-1}$$

and the atomic mass of iron is 55.9. (More accurate data are required to obtain the value of 55.85 listed in the table of atomic masses.)

EXAMPLE 7

Calculate the elemental composition (in percentage by mass) of nitrous oxide, N_2O.

For a problem of this type it is most convenient to choose 1 mol of the compound as a sample. The molecular mass of N_2O is $(2 \times 14.01) + 16.00 = 44.02$. Therefore the 1-mol sample has a mass of 44.02 g, and it contains 28.02 g of nitrogen and 16.00 g of oxygen. Thus the elemental composition of N_2O is

$$\frac{28.02 \text{ g of N}}{44.02 \text{ g of N}_2\text{O}} \times 100\% = 63.65\% \text{ N in N}_2\text{O}$$

$$\frac{16.00 \text{ g of O}}{44.02 \text{ g of N}_2\text{O}} \times 100\% = 36.35\% \text{ O in N}_2\text{O}$$

2.5 SIMULTANEOUS DETERMINATION OF THE SIMPLEST FORMULA AND ATOMIC MASSES

What if we begin with only a set of experimental determinations of elemental composition? How do we then determine at the same time the simplest formulas *and* the atomic masses? This was the position in which Dalton found himself, and it is indeed a difficult one. Either we must have additional information of a type not yet considered or else we must make some arbitrary assumptions. The latter course was the only one open to Dalton.

He chose to assume that each compound had the simplest possible formula unless there was experimental evidence to the contrary. For example, he assumed that a molecule of water contained one atom of hydrogen and one atom of oxygen. As it turned out, he was correct in some cases and wrong in others. Because of the errors in formulas, there were corresponding errors in his scale of relative atomic masses. Still, this procedure served to start the atomic-molecular theory on its way, and the errors were corrected by later generations of chemists. The experimental evidence that eventually solved the problem came mainly from studies of the properties of gases, which will be considered in the next chapter.

2.6 CHEMICAL REACTIONS AND CHEMICAL EQUATIONS

When substances interact to produce different substances, a chemical process, or **chemical reaction,** is said to occur. An example is the combustion of methane (molecular formula CH_4). It combines with oxygen to form carbon dioxide and water vapor. (This is the principal reaction in the flame of a Bunsen burner, a gas stove, or a furnace burning natural gas.) A chemical reaction is conveniently represented by a **chemical equation,** with the substances that change, called **reactants,** on the left side, and the substances that are formed, called **products,** on the right.

$$CH_4 + 2O_2 \longrightarrow CO_2 + 2H_2O$$

This reaction requires two oxygen molecules, each containing two atoms of oxygen, to provide the four atoms of oxygen required to form one molecule of carbon dioxide and two molecules of water. The number of molecules is indicated by a coefficient; for example, the symbol $2O_2$ represents two molecules of oxygen. When there is only one molecule, the number 1 is omitted; for example, the symbol CH_4 represents one molecule of methane. When the correct coefficients are shown, then the same number of atoms of an element appears on each side of the equation. In the example given above there are one atom of carbon, four atoms of hydrogen, and four atoms of oxygen on each side. This is then said to be a **balanced equation.** When an equation is balanced an equality sign is often used instead of an arrow.

$$CH_4 + 2O_2 = CO_2 + 2H_2O \tag{2.2}$$

The equality sign does not imply that the reactants are identical with the products.

Determining the correct coefficients (called balancing the equation) is done, as in the above example, by considering how many atoms of each type are required to form the products. For complex reactions this procedure is too tedious; standard methods of balancing complicated equations will be considered in Chapter 14.

In addition to the interpretation in terms of molecules, the above equation may also be interpreted in terms of moles. Thus 1 mol of methane (CH_4) and 2 mol of oxygen (O_2) react to form 1 mol of carbon dioxide (CO_2) and 2 mol of water (H_2O). A third alternative is to interpret the equation in terms of the masses corresponding to these numbers of moles. The molecular masses of the substances appearing in Eq. 2.2 are (to three significant figures): 16.0 (CH_4), 32.0 (O_2), 44.0 (CO_2), and 18.0 (H_2O). Therefore 16.0 g of methane and 64.0 g of oxygen react to form 44.0 g of carbon dioxide and 36.0 g of water. Finally, we can interpret the equation in terms of any other masses in the same ratio. For example, on dividing each of the above numbers by 16.0, we find that 1.00 g of methane and 4.00 g of oxygen react to form 2.75 g of carbon dioxide and 2.25 g of water. These various interpretations are summarized in Table 2.1.

Although balanced equations are useful in considering chemical reactions, it should be realized that there is a great deal of important information not conveyed by an equation. Does the reaction actually occur to form the products shown? If so, under what conditions, and how fast? Does the system also react in the opposite direction to a significant extent, so that it approaches some intermediate composition in which both reactants and products are present in appreciable amounts? These are some of the questions to be considered later on. At this point, however, it should be mentioned that a pair of arrows is often used to emphasize that the reverse reaction also occurs to a significant extent, as in the synthesis of ammonia from its elements,

$$N_2 + 3H_2 \rightleftharpoons 2NH_3$$

A single arrow is sometimes used to emphasize the forward direction or to indicate that the reverse reaction is unimportant, as in the decomposition of hydrogen peroxide:

$$2H_2O_2 \longrightarrow 2H_2O + O_2$$

TABLE 2.1

INTERPRETATIONS OF A CHEMICAL EQUATION

Methane CH_4	+	Oxygen $2O_2$	=	Carbon dioxide CO_2	+	Water $2H_2O$
1 molecule		2 molecules		1 molecule		2 molecules
1 mol		2 mol		1 mol		2 mol
16.0 g		64.0 g		44.0 g		36.0 g
1.00 g		4.00 g		2.75 g		2.25 g

2.7 NONEQUIVALENT AMOUNTS IN CHEMICAL REACTIONS

When the relative amounts of reactants used in a chemical reaction are equal to the relative amounts of reactants required by the chemical equation, we say that the reactants are present in **equivalent amounts.** As an example, if 8.0 g of CH_4 and 32.0 g of oxygen are allowed to react according to Eq. 2.2, we say that the reactants are present in equivalent amounts since we have utilized 4.0 g of oxygen for each gram of methane (Table 2.1). In practice it often happens that the reactants are not present in equivalent amounts. As we shall see, the amount of product formed is then determined by the reactant present in *less than* the equivalent amount. We say that this reactant is present in *limiting amount* since it determines the amount of product that can be formed. (Such a reactant is often called the *limiting reagent*.) For example, the combustion of methane occurs according to Eq. 2.2 even if excess oxygen is present. If we pass 1 mol of CH_4 and 5 mol of O_2 through a burner, only 2 mol of O_2 is required for reaction, and $5 - 2 = 3$ mol of O_2 will remain unreacted. Therefore the hot gases leaving the burner contain 1 mol of CO_2, 2 mol of H_2O, and 3 mol of O_2, as shown in Table 2.2. The methane is said to be present in limiting amount since it controls the amounts of carbon dioxide and water that can be produced.

To obtain a general procedure for determining which substance in a chemical reaction is present in limiting amount, consider the general equation for a reaction of two substances:

$$a\text{A} + b\text{B} = c\text{C} + d\text{D}$$

where A, B, C, and D represent the formulas of chemical substances, and a, b, c, and d represent the coefficients in the balanced equation. Further, let n_A and n_B be the numbers of moles of A and B, respectively. Then B is the limiting reagent if

$$\frac{n_A}{n_B} > \frac{a}{b}$$

(Of course, if the ratio n_A/n_B is less than a/b, then A is the limiting reagent.) For a reaction involving three or more substances, this rule must be applied in turn to each pair of substances.

When the masses of reactants used in a chemical reaction are given, it is a simple matter to convert them to moles of reactants, and then to determine which

TABLE 2.2

THE COMBUSTION OF METHANE AND EXCESS OXYGEN

Substance	Initial amount (moles)	Final amount (moles)
CH_4	1	—
O_2	5	3
CO_2	—	1
H_2O	—	2

2.7 NONEQUIVALENT AMOUNTS IN CHEMICAL REACTIONS

reactant is present in limiting amount. Then the amount of the various products can be determined. The following example shows this type of calculation.

EXAMPLE 8

Iron (Fe) may be formed by the reaction of carbon monoxide (CO) and an iron ore (Fe_2O_3) as shown in the following equation:

$$Fe_2O_3 + 3CO = 2Fe + 3CO_2$$

How many grams of iron could be produced by the reaction of 10.0 g of Fe_2O_3 and 4.00 g of CO?

$$\frac{10.0 \text{ g}}{159.7 \text{ g (mol of } Fe_2O_3)^{-1}} = 0.0626 \text{ mol of } Fe_2O_3$$

$$\frac{4.00 \text{ g}}{28.01 \text{ g (mol of CO)}^{-1}} = 0.143 \text{ mol of CO}$$

Because 3 mol of CO reacts with only 1 mol of Fe_2O_3 (see the previous equation for the chemical reaction), it follows that 0.143 mol of CO can react with only

$$\frac{0.143 \text{ mol of CO}}{3 \text{ mol of CO (mol of } Fe_2O_3)^{-1}} = 0.0477 \text{ mol of } Fe_2O_3$$

Therefore excess Fe_2O_3 is present.

Thus carbon monoxide is the reactant present in limiting amount, and the number of moles of iron which will be produced is

$$\frac{2 \text{ mol of Fe}}{3 \text{ mol of CO}} \times 0.143 \text{ mol of CO} = 0.0953 \text{ mol of Fe}$$

This amount of iron has a mass of 0.0953 mol of Fe $\times$ 55.85 g (mol of Fe)$^{-1}$ = 5.32 g of Fe. Thus 5.32 g of iron would be produced.

In **combustion analysis** a sample of a substance is burned in a stream of gaseous oxygen and the various products are trapped and weighed. For example, a compound containing carbon and hydrogen (or carbon, hydrogen, and oxygen) burns in excess oxygen to produce water and carbon dioxide. The masses of hydrogen and carbon in the sample can be determined from the masses of water and carbon dioxide produced. The mass of oxygen can then be obtained by subtracting the sum of the masses of hydrogen and carbon from the mass of the sample. The following example illustrates this method of analyzing compounds.

EXAMPLE 9

A 0.230-g sample of a substance, known to contain C, H, and O only, was burned in a stream of excess oxygen and the products of the combustion were passed through a weighed tube containing magnesium perchlorate, which absorbed the water vapor, and then through a weighed tube containing sodium hydroxide and calcium oxide to absorb the carbon dioxide. When the first tube was weighed after combustion, it had increased in mass by 0.270 g because of the water produced in the reaction. The second tube increased in mass by 0.440 g because of the carbon dioxide formed in the reaction. What, then, is the mass of carbon, hydrogen, and oxygen in the sample and what is the simplest formula of the substance?

The mass of hydrogen in the sample is equal to the mass of hydrogen in the water produced. As the molecular mass of water (H_2O) is 18.02, the amount of water produced is

$$\frac{0.270 \text{ g of } H_2O}{18.02 \text{ g of } H_2O \text{ (mol of } H_2O)^{-1}} = 0.0150 \text{ mol of } H_2O$$

Each mole of water contains 2 mol of hydrogen atoms, and therefore this amount of water contains 0.0300 mol of hydrogen atoms, or

$$0.0300 \text{ mol of } H \times 1.008 \text{ g of H (mol of H)}^{-1} = 0.0302 \text{ g of H}$$

The mass of carbon in the sample is equal to the mass of carbon in the carbon dioxide produced. As the molecular mass of carbon dioxide (CO_2) is 44.01, the amount of carbon dioxide produced is

$$\frac{0.440 \text{ g of } CO_2}{44.01 \text{ g of } CO_2 \text{ (mol of } CO_2)^{-1}} = 0.0100 \text{ mol of } CO_2$$

This amount of CO_2 contains 0.0100 mol of carbon atoms, which have a mass of

$$0.0100 \text{ mol of } C \times 12.01 \text{ g of C (mol of C)}^{-1} = 0.120 \text{ g of C}$$

The mass of oxygen can now be obtained by subtracting the masses of hydrogen and carbon from the mass of the sample. Thus we have $0.230 - 0.0302 - 0.120 = 0.080$ g of oxygen. (Note that the mass of oxygen in the sample cannot be obtained by simply adding the mass of oxygen in the water to the mass of oxygen in the carbon dioxide, because an unknown part of the oxygen in the products comes from the gaseous oxygen used to burn the sample.) The amount of oxygen is therefore

$$\frac{0.080 \text{ g of O}}{16.00 \text{ g of O (mol of O)}^{-1}} = 0.0050 \text{ mol of O}$$

To obtain the simplest formula we need to know the relative number of moles of atoms of each element. The sample contains

0.0300 mol of hydrogen atoms
0.0100 mol of carbon atoms
0.0050 mol of oxygen atoms

If we divide the larger numbers of moles by the smallest one, we obtain

$$\frac{0.0300 \text{ mol of H}}{0.0050 \text{ mol of O}} = 6.0 \text{ mol of H per mol of O}$$

$$\frac{0.0100 \text{ mol of C}}{0.0050 \text{ mol of O}} = 2.0 \text{ mol of C per mol of O}$$

Thus the simplest formula is C_2H_6O.

2.8 SOLUTIONS AND MOLARITY

Chemical reactions are typically carried out in liquid solutions. Such solutions are usually prepared by dissolving one or more substances (which may be in the solid, liquid, or gaseous state) in a liquid. The substance present in largest amount is called the **solvent,** whereas the substances present in smaller amounts are called **solutes.** Water is often used as a solvent, in which case the solution is called an **aqueous** solution.

The composition of the solution is called its **concentration,** which can be expressed in many ways. One of the most useful concentration units is the molar concentration, or molarity. (Other concentration units, as well as the physical properties of solutions, will be taken up in Chapter 10.) The **molarity** (symbol c_M) is defined as the *number of moles of solute contained in 1 L (or 1 dm³) of solution.* Therefore, in general,

$$c_M = \frac{n}{V} \tag{2.3}$$

where n is the number of moles and V is the volume in liters. The unit of molarity, mol L^{-1}, is abbreviated M. Thus a $0.50 M$ H$_2$SO$_4$ solution is one that contains 0.50 mol of H$_2$SO$_4$ (sulfuric acid) in 1 L of solution.

The essential feature of concentration is that it depends on the *relative* amounts of solvent and solutes. By itself concentration tells us nothing about the total quantity present. Thus the quantity of a $0.50 M$ H$_2$SO$_4$ solution might be a drop, a liter, or a tank car.

The following three examples indicate some of the common ways in which chemists use solutions. Preparation of a solution of known concentration is illustrated by the first example.

EXAMPLE 10

An aqueous solution is prepared by dissolving 10.0 g of solid sodium hydroxide (NaOH) in sufficient water to prepare 200 mL of solution. What is the molarity of NaOH in the solution?

By definition the molarity is the number of moles of NaOH per liter of solution, that is,

$$c_M = \frac{n}{V}$$

To obtain the number of moles of NaOH we divide the mass of NaOH by its molecular mass, 40.00.

$$n = \frac{10.0 \text{ g of NaOH}}{40.00 \text{ g of NaOH (mol of NaOH)}^{-1}} = 0.250 \text{ mol of NaOH}$$

The volume of the solution is 200 mL or 0.200 L, and therefore the molarity is

$$c_M = \frac{0.250 \text{ mol of NaOH}}{0.200 \text{ L}} = 1.25 M \text{ NaOH}$$

The next example illustrates the determination of the amount of solute in a given volume of solution.

EXAMPLE 11

How many moles of H_2SO_4 are there in 25.0 mL of $0.200M$ H_2SO_4?

$$c_M = \frac{n}{V} \quad \text{or} \quad n = c_M V$$

Therefore

$$n = (0.200 \text{ mol of } H_2SO_4 \text{ L}^{-1})(0.0250 \text{ L})$$
$$= 0.00500 \text{ mol of } H_2SO_4$$

Finally, we consider what volume of solution is required to measure out a given amount of the solute.

EXAMPLE 12

What volume of $2.00M$ HCl (hydrochloric acid) must be used to obtain 0.0400 mol of HCl?

$$c_M = \frac{n}{V} \quad \text{or} \quad V = \frac{n}{c_M}$$

Therefore

$$V = \frac{0.0400 \text{ mol of HCl}}{2.00 \text{ mol of HCl L}^{-1}} = 0.0200 \text{ L} = 20.0 \text{ mL}$$

PROBLEMS

2.1 Distinguish between
  (a) an atom and a mole of atoms
  (b) atomic mass and a mole of atoms
  (c) an atom and a molecule
  (d) a molecule and a mole
  (e) the mass of an atom and the atomic mass

2.2 The element oxygen consists of three isotopes with atomic masses of 15.99491, 16.99914, and 17.99916. The natural abundances of these isotopes in oxygen are 99.759% ^{16}O, 0.037% ^{17}O, and 0.204% ^{18}O. Calculate the average atomic mass of oxygen.

2.3 The element magnesium consists of 78.70% ^{24}Mg of atomic mass 23.98504, 10.13% ^{25}Mg of atomic mass 24.98584, and 11.17% ^{26}Mg of atomic mass 25.98259. Calculate the average atomic mass of magnesium.

PROBLEMS

2.4 What is the molecular mass of each of the following?
 (a) S_2
 (b) S_8
 (c) C_6H_6
 (d) HNO_3
 (e) H_2SO_4
 (f) $NaHCO_3$

2.5 Calculate the number of atoms in
 (a) 8.00 g of sulfur
 (b) 10 molecules of benzene (C_6H_6)
 (c) 0.200 mol of oxygen molecules (O_2)

2.6 The atomic mass of nitrogen is 14.01.
 (a) What is the average mass in grams of 1 atom of nitrogen?
 (b) How many moles of nitrogen atoms are there in 7.0 g of nitrogen? How many moles of nitrogen molecules (N_2) are there?
 (c) How many atoms of nitrogen are there in 7.0 g of nitrogen? How many N_2 molecules?

2.7 How many moles of sulfur atoms are there in
 (a) 6.47 g of sulfur?
 (b) 125 g of sulfuric acid (H_2SO_4)?
 (c) 100 atoms of sulfur?
 (d) 6.70 mol of sodium sulfate (Na_2SO_4)?
 (e) 10 molecules of SO_2?

2.8 Common sugar is $C_{12}H_{22}O_{11}$. Calculate the mass percentage of carbon, hydrogen, and oxygen in sugar.

2.9 The molecular formula of nitric acid is HNO_3. Calculate the mass percentage of hydrogen, nitrogen, and oxygen in nitric acid.

2.10 Table salt contains 39.4% sodium (Na) and 60.6% chlorine (Cl) by mass. What is the simplest formula of table salt (sodium chloride)?

2.11 A compound was found to contain only sulfur and oxygen, which were present in equal masses. What is the simplest formula of the compound?

2.12 Nicotine (a compound obtained from tobacco) consists of 74.0% C, 8.70% H, and 17.3% N by mass. What is the simplest formula of nicotine?

2.13 A metal oxide of formula MO_2 contains 60% of the metal and 40% oxygen by mass. Calculate the atomic mass of the metal.

2.14 Element X forms a compound with oxygen having the molecular formula X_4O_{10}. It contains 56.3% oxygen by mass. Calculate the atomic mass of X.

2.15 Dalton's scale of atomic weights was based on hydrogen taken as exactly 1. For the composition of water he used Lavoisier's analysis of 85% oxygen and 15% hydrogen. From these data and his formula for water (HO), calculate his atomic weight for oxygen.

2.16 Balance the following equations by inserting appropriate coefficients. (All reactants and products are shown.)
  (a) $K + Cl_2 \rightarrow KCl$
  (b) $SO_2 + O_2 \rightarrow SO_3$
  (c) $KClO_3 \rightarrow KCl + O_2$
  (d) $C_2H_2 + O_2 \rightarrow CO_2 + H_2O$
  (e) $CH_4 + Cl_2 \rightarrow CCl_4 + HCl$

2.17 How many moles of carbon dioxide (CO_2)
  (a) are there in 100 g of carbon dioxide?
  (b) could be obtained by heating 100 g of $CaCO_3$ if the $CaCO_3$ decomposed on heating as shown by the equation
  $$CaCO_3 = CaO + CO_2$$
  (c) are there in 1 million molecules of CO_2?
  (d) could be produced from the burning of 10 g of sugar ($C_{12}H_{22}O_{11}$) in excess oxygen?

2.18 (a) Tartaric acid is composed of carbon, hydrogen, and oxygen. A sample weighing 0.400 g is subjected to combustion analysis. The masses of water and carbon dioxide produced are found to be 0.144 g and 0.469 g, respectively. Calculate the simplest formula of tartaric acid.
  (b) The molecular mass of tartaric acid (determined by methods to be described in later chapters) is 150. From this result and the simplest formula, calculate the molecular formula.

2.19 Ethyl alcohol can be burned in the presence of oxygen to produce water and carbon dioxide as shown by the following balanced equation:
$$C_2H_5OH + 3O_2 = 2CO_2 + 3H_2O$$
  (a) How many moles of C_2H_5OH will react with 1 mol of O_2?
  (b) If 0.20 mol of C_2H_5OH and 0.80 mol of O_2 are heated in a closed container until the above reaction has gone to completion, what substances would be present and in what amounts?
  (c) What mass of oxygen is required to react with 1.0 g of C_2H_5OH?
  (d) How much water would be produced if 1.0 g of C_2H_5OH is burned in the presence of 2.5 g of oxygen?

PROBLEMS

2.20 (a) How many moles of potassium dichromate, $K_2Cr_2O_7$, can be made from 0.50 mol of potassium atoms (K), 0.75 mol of chromium atoms (Cr), and 3.5 mol of oxygen atoms (O)?
(b) How many moles of potassium dichromate can be made from 2.65 g of potassium, 4.00 g of chromium, and 3.35 g of oxygen?

2.21 How many grams of sodium chloride (NaCl) are required to prepare 200 mL of $0.20M$ NaCl?

2.22 How many moles of potassium nitrate (KNO_3) are there in 2.20 L of $1.2M$ KNO_3?

2.23 If 45.0 mL of $0.280M$ NaOH are required for a certain chemical reaction, how many moles of NaOH were used?

2.24 If 12.0 mL of $6.0M$ HCl are diluted with water to a final volume of 60.0 mL, what is the concentration of HCl?

2.25 The following two solutions of sulfuric acid are mixed together: 125 mL of $0.750M$ H_2SO_4, and 285 mL of $0.500M$ H_2SO_4. Assuming that the volume of the mixture is the sum of the volumes of the individual solutions (which is a good approximation), what is the molarity of H_2SO_4 in the mixture?

2.26 A solution of hydrochloric acid (HCl) has a density of 1.18 g mL^{-1}, and it contains 37.1% HCl (by mass). Calculate the molarity of HCl.

2.27 (a) The hemoglobin in the blood of mammals contains about 0.335% iron. Calculate a minimum value for the molecular mass of hemoglobin.
(b) Actually there are four atoms of iron in each molecule of hemoglobin. What then is the actual molecular mass of hemoglobin?

CHAPTER 3
EXPERIMENTAL STUDIES OF GASES

3

Experimental investigations of gases, together with hypotheses to explain the observations, were very important to the development of the atomic theory and the concept of molecules. In this chapter we shall examine some of the physical properties of gases, along with the laws that summarize the experimental data. These laws led Avogadro to propose the important relationship between the volume of a gas and the number of molecules of the gas. This, in turn, led to a rational system of atomic masses of the elements and to a rapid development of our knowledge of the atomic composition of substances. An understanding of the behavior of gases is fundamental to chemistry and provides a valuable insight into the scientific methods used in developing our knowledge of chemical substances.

3.1 PHYSICAL PROPERTIES OF GASES

The results of a large number of experimental investigations indicate that the volume (V) of a gaseous, liquid, or solid substance is dependent on the temperature (T), the pressure (P), and the amount of the substance expressed as the number of moles (n). As we shall see, if three of the above properties of a gas are known, the value of the fourth property is fixed. The relationship between these properties may be expressed mathematically as

$$V = f(T, P, n) \qquad (3.1)$$

where f stands for some function of temperature, pressure, and amount. This equation is known as an **equation of state** of the gas. In ensuing sections we shall show how the exact form of the equation of state has been determined.

In marked contrast to liquids and solids, gases show much larger changes in volume with changes in temperature or pressure. Furthermore, these changes are quite uniform from one gas to another. To establish the mathematical relationship (f) between the volume of a gas and the three variables (T, P, n), we employ a method often used by scientists, namely, evaluation of the relationship by considering changes in only one variable at a time. Thus a complex problem can be separated into two or more much simpler problems.

3.2 DEPENDENCE OF VOLUME ON AMOUNT AT CONSTANT TEMPERATURE AND PRESSURE

On an intuitive basis, one might expect that doubling the mass of a gas will double the volume (if all other variables are fixed), and this is indeed the case. In fact, the volume is directly proportional to the mass (assuming constant temperature and pressure). In a similar way, one might expect a direct proportionality between volume and amount (still holding the temperature and pressure constant), and this is also true.

A more formal argument in support of these relationships begins by noting that one of the properties of a homogeneous system that is constant throughout the phase at any given temperature and pressure is the density, d, defined as the mass per unit volume. For a sample of any size it may be calculated by dividing the mass, m, by the volume, V, of the sample.

$$d = \frac{m}{V} \quad \text{or} \quad V = \frac{m}{d} \tag{3.2}$$

Thus the volume, V, is directly proportional to the mass, m. (Properties such as the volume that are dependent on the quantity of material used are known as **extensive properties**. Properties such as temperature and pressure that are independent of the quantity of material are known as **intensive properties**.)

The mass is equal to the number of moles, n, multiplied by the molecular mass, M. Thus Eq. 3.2 may be rewritten

$$V = \frac{m}{d} = \frac{nM}{d} = n\left(\frac{M}{d}\right) = n\overline{V} \tag{3.3}$$

where $\overline{V}$, called the molar volume, is obtained (in liters per mole) by dividing the molecular mass (in grams per mole) by the density (in grams per liter). Thus the volume, V, is directly proportional to the amount, n.

EXAMPLE 1

The density of oxygen gas at 0°C and 1 atmosphere (atm) pressure is 1.429 g L^{-1}. Calculate the following quantities at 0°C and 1 atm pressure:

(a) volume occupied by 1.00 g of oxygen
(b) molar volume of oxygen assuming that 1 mol of oxygen has a mass of 32.0 g
(c) volume occupied by 0.0200 mol of oxygen

(a) Applying Eq. 3.2,

$$V = \frac{m}{d} = \frac{1.00 \text{ g}}{1.429 \text{ g L}^{-1}} = 0.700 \text{ L}$$

(b) Applying the definition of the molar volume, $\bar{V}$, in Eq. 3.3,

$$\bar{V} = \frac{M}{d} = \frac{32.0 \text{ g mol}^{-1}}{1.429 \text{ g L}^{-1}} = 22.4 \text{ L mol}^{-1}$$

(c) Applying Eq. 3.3,

$$V = n\bar{V} = (0.0200 \text{ mol})(22.4 \text{ L mol}^{-1}) = 0.448 \text{ L}$$

3.3 PRESSURE MEASUREMENTS

The fact that gases can exert an outward force on the surface of a container is evident to anyone who has inflated an automobile tire or a balloon. The pressure of a gas is defined as the force per unit area. For a surface of any size the pressure, P, is the force, F, exerted on the surface divided by the area of the surface, A.

$$P = \frac{F}{A} \tag{3.4}$$

Force is defined as the product of mass, m, and its acceleration, a (see Appendix D-2).

$$F = ma \tag{3.5}$$

The unit of force is the newton (N), which is the force required to increase the velocity of a 1-kg mass by 1 m s^{-1} in a time interval of 1 s (1 N = 1 kg m s^{-2}).

There are several units of pressure in common use. The SI unit of pressure is the pascal (Pa), which is the pressure exerted on an area of 1 m^2 by a force of 1 N (1 Pa = 1 kg m^{-1} s^{-2}).

The pressure exerted by a gas is ordinarily measured with a device known as a **manometer**, which consists of a transparent glass U-tube partly filled with a liquid and having one end of the tube sealed, as shown in Figure 3.1. The pressure due to a gas in the closed tube of the manometer, P_2, will be exerted at Y. It is balanced by the pressure at X, which is equal to the sum of P_1 and the pressure (P_L) exerted by the column of liquid above X. That is, $P_2 = P_1 + P_L$.

The pressure, P_L, is equal to the force, F, divided by the cross-sectional area of the column of liquid, A. The force exerted by this column of liquid is simply its weight, W; and its weight is the product of its mass, m, and the local acceleration of gravity, g. The mass of liquid is given by Vd, where V is its volume and d is its density. Finally, the volume, V, is equal to hA, where h is the height of the cylindrical column of liquid and A is the cross-sectional area. Applying these relations in succession, we have

$$P_L = \frac{F}{A} = \frac{W}{A} = \frac{mg}{A} = \frac{Vdg}{A} = \frac{hAdg}{A}$$

3.3 PRESSURE MEASUREMENTS

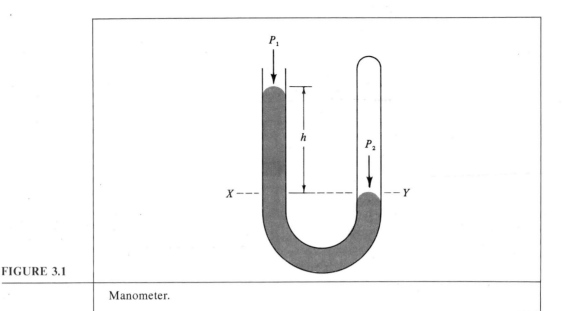

FIGURE 3.1 Manometer.

Canceling the A's in the last term, we have an important formula for the pressure exerted by a liquid column of height, h,

$$P_L = hdg \tag{3.6}$$

Thus

$$P_2 = P_1 + P_L = P_1 + hdg \tag{3.7}$$

Atmospheric pressure, which varies with time and place, is measured with a **barometer**, invented by Evangelista Torricelli in 1643. A glass tube, closed at one end, is filled with mercury (Hg), and then inverted over a dish of mercury (Figure 3.2). If this is done carefully so as to exclude air, water, and other foreign substances, then the only gas in the space above the column of mercury is mercury vapor. The pressure exerted by the atmosphere (P_A) is then equal to the pressure due to the mercury vapor plus that of the mercury column itself, hdg (Eq. 3.7).

The pressure due to the mercury vapor is so small (2.34×10^{-7} atm at $0°C$) compared to the atmospheric pressure that it may be neglected, and the pressure exerted by the atmosphere is simply hdg. At any given location, g is a constant, and d is a constant at constant temperature, so that the atmospheric pressure is therefore proportional to the height of the mercury column.

The commonly used pressure units for gases are the standard atmosphere and the height of a column of mercury. A standard atmosphere (usually called an atmosphere and abbreviated atm) is defined as exactly 1.01325×10^5 Pa. The standard acceleration of gravity is 9.80665 m s^{-2} (see Appendix D-2), and at $0°C$

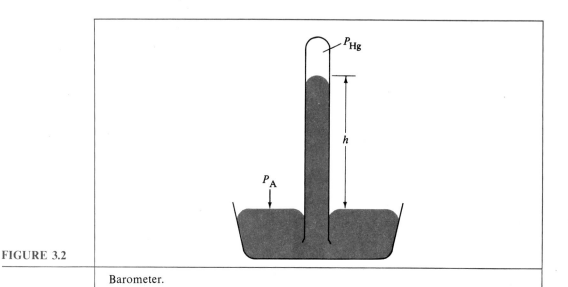

FIGURE 3.2 Barometer.

the density of mercury is 13.5951 g cm^{-3}, or 1.35951 × 10^4 kg m^{-3}. From these values and Eq. 3.6, we can calculate the height of a mercury column equivalent to 1 atm:

$$h = \frac{P_L}{dg} = \frac{1.01325 \times 10^5 \text{ kg m}^{-1} \text{s}^{-2}}{(1.35951 \times 10^4 \text{ kg m}^{-3})(9.80665 \text{ m s}^{-2})}$$

$$= 0.760000 \text{ m} \quad \text{or} \quad 760.000 \text{ mm}$$

Thus 1 atm = 1.01325 × 10^5 Pa or 760 mm Hg at 0°C. For this reason the unit of 1/760 atm has special significance. Its name is the torr (symbol Torr), after Torricelli.

3.4 DEPENDENCE OF VOLUME ON PRESSURE AT CONSTANT TEMPERATURE AND AMOUNT

The original determination of the effect of pressure on the volume of a gas was made by Robert Boyle in 1662. He used a manometer like the one shown in Figure 3.1. In the closed tube he trapped a small amount of air and adjusted the mercury columns until both were at the same height. The portion of the tube containing the air was divided into 48 equal parts so that the volume at the particular prevailing atmospheric pressure ($29\frac{2}{16}$ in. Hg according to Boyle's report) was recorded as 48 units. He then added mercury to the open end and recorded the volume and the difference in heights of the two mercury columns. The total pressure (P) on his enclosed sample of air may be calculated using Eq. 3.7. Boyle's data are plotted in Figure 3.3(a), and it is evident that the volume decreased (but not linearly) as the pressure was increased. Although it is possible

3.4 DEPENDENCE OF VOLUME ON PRESSURE AT CONSTANT TEMPERATURE AND AMOUNT

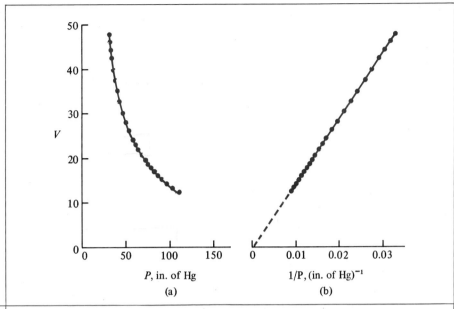

FIGURE 3.3 (a) Effect of pressure on volume of air (constant temperature and amount). (b) The same data as in (a), plotted with the reciprocal of pressure as the horizontal coordinate.

to correlate data with nonlinear plots, it is much simpler to plot the data so that a straight line is obtained. Deviations from the straight line due to experimental errors or other causes are then easy to observe. The relationship noted in Figure 3.3(a) suggests that the volume could possibly be linearly related to the reciprocal of the pressure ($1/P$). To test this idea, Boyle's data are replotted in Figure 3.3(b), and it is evident that a straight line can be drawn through the experimental points and that this line passes through the origin. The equation for this straight line is

$$V = C\left(\frac{1}{P}\right) \quad \text{or} \quad PV = C \tag{3.8}$$

where C is a constant whose numerical value depends on the temperature and amount. (See Appendix C-3 for a discussion of the equation for a straight line.) The relationship expressed by Eq. 3.8 is known as **Boyle's law,** which may be expressed as follows: *At constant temperature the volume of a gas is inversely proportional to its pressure.* Thus, if the pressure on the gas is increased, the volume will decrease—and vice versa.

It should be noted that these results were obtained with crude equipment by present-day standards and that, although the quantity of air was kept constant, the temperature was subject to slight changes. However, Boyle was a very clever investigator, and to demonstrate that small temperature changes would not markedly influence his results, he noted that cooling the gas by placing a wet cloth over the tube or warming the tube with a candle flame produced only slight changes in the volume.

EXAMPLE 2

If a certain quantity of a gas occupies 50 mL at 0.80 atm, at what pressure would it occupy 20 mL if the temperature remains unchanged?

Because PV is constant (Boyle's law), $P_i V_i = P_f V_f$ where i and f refer to the initial and final conditions, respectively. Thus

$$P_f = P_i \left(\frac{V_i}{V_f}\right) = (0.80 \text{ atm}) \left(\frac{50 \text{ mL}}{20 \text{ mL}}\right) = 2.0 \text{ atm}$$

The development of scientific knowledge has been strongly influenced by the formation of laws, such as Boyle's law, and by the subsequent testing of these laws by more precise measurements and over a wider range of conditions. Frequently this leads to cases or conditions for which the law is not in agreement with experiment. For example, precise data show deviations from Eq. 3.8, particularly at high pressure. In Section 3.11 we shall consider deviations from Boyle's law in more detail.

3.5 GASEOUS SOLUTIONS

If a flask containing oxygen at 1 atm pressure is connected by a valve to a second flask containing nitrogen at 1 atm pressure, and the valve connecting the two flasks is opened allowing the gases to mix, what will be the resulting pressure? Experimentally we find that the final pressure is 1 atm. This result holds for mixtures of other gases (providing no chemical reaction occurs) and is independent of the relative sizes of the two flasks. How can we explain such a result? Let us simplify the problem by assuming that the flasks are identical in size. If only one of the flasks contained gas and the valve was opened, the volume would then increase by a factor of 2 and the final pressure would be $\frac{1}{2}$ atm according to Eq. 3.8. If each gas in the two-gas system behaves the same as it would behave if it were present alone, the total pressure in this system would be $\frac{1}{2}$ atm $+ \frac{1}{2}$ atm $=$ 1 atm, which was the experimental result.

The above results can be generalized to give the law of partial pressures, discovered by John Dalton in 1801. This law states: *If the partial pressure of each gas* (P_i) *in a mixture of gases is defined as the pressure that the gas would exert if it were present in the container alone, then the total pressure* (P_T) *is the sum of the partial pressures,* that is,

$$P_T = P_1 + P_2 + P_3 + \cdots = \sum_i P_i \qquad (3.9)$$

3.5 GASEOUS SOLUTIONS

TABLE 3.1 — VAPOR PRESSURE OF WATER

Temperature (°C)	Pressure (Torr)	Temperature (°C)	Pressure (Torr)
0	4.6	50	92.5
10	9.2	60	149.4
20	17.5	70	233.7
22	19.8	80	355.1
24	22.4	90	525.8
26	25.2	98	707.3
28	28.3	99	733.2
30	31.8	100	760.0
40	55.3	101	787.6

The symbol $\sum_i P_i$ represents the sum of the pressures exerted by the individual gases.

Because most laboratory experiments on gases involve measuring the volume of gas collected over a liquid (often water), we are usually dealing with a mixture of gases—the gas collected and the vaporized liquid. For example, if gaseous oxygen is collected over liquid water, the total pressure is the sum of the partial pressure of oxygen and the partial pressure of water vapor. Fortunately, in a short period of time, a gas in contact with a liquid in a closed container becomes saturated with respect to the liquid substance—the partial pressure of the vaporized liquid reaches a constant value, called the **vapor pressure,** which depends only on the temperature. Experimental data on the vapor pressure of water at various temperatures are shown in Table 3.1. The following problem indicates how gas volumes may be corrected for the effect of vaporized liquid.

EXAMPLE 3

Oxygen gas was collected over water at 24°C. The volume of gas collected was 20.0 L when the pressure was 750 Torr. What volume would the oxygen have occupied if it had been dry at 760 Torr and 24°C?

Frequently the solution of a problem may be simplified by arranging the data so that it is readily apparent what is known and what answer is desired. In the present case this can be done as follows (n and T are constant):

	P_{oxygen}	V_{oxygen}
initial	$(750 - P_w)$ Torr	20.0 L
final	760 Torr	V_f L

The vapor pressure of water (P_w) at 24°C is 22.4 Torr (Table 3.1); therefore initially $P_{oxygen} = 750 - 22 = 728$ Torr (only three significant figures are needed). Because PV is constant (Boyle's law), $P_i V_i = P_f V_f$, where the subscripts i and f refer to the initial and final conditions, respectively. Therefore

$$V_f = V_i \left(\frac{P_i}{P_f}\right) = (20.0 \text{ L}) \left(\frac{728 \text{ Torr}}{760 \text{ Torr}}\right) = 19.2 \text{ L}$$

3.6 DEPENDENCE OF VOLUME ON TEMPERATURE AT CONSTANT PRESSURE AND AMOUNT

All of us have a qualitative concept of temperature in that we assign a higher temperature to hot objects than we do to cool objects. One of the most common quantitative temperature scales is the Celsius scale. At 1 atm pressure an ice-water mixture has a temperature of zero degrees Celsius (0°C), and the vapor of boiling water has a temperature of 100°C. To determine temperatures other than 0° and 100°C, we may use any property of a system that can be easily measured and that changes with temperature. The most commonly used property is the volume of a fixed amount of mercury, which increases with increasing temperature. The common thermometer consists of sufficient mercury to fill a glass bulb and to partially fill a capillary tube connected to the bulb. The Celsius scale is marked on the capillary tube by noting the height of the mercury column when the bulb is placed in an ice-water mixture and then in the vapor of boiling water. The intervening distance is then divided into 100 equal parts so that each part represents 1°C.

We can use the mercury thermometer to investigate the volume of a given amount of a gas at several temperatures. In Figure 3.4 the volume of 1-g samples of oxygen, hydrogen, and helium at 1 atm pressure are plotted against the temperature. It is evident that for each gas a straight line can be drawn through the experimental points. If these straight lines are extrapolated (dotted lines) to zero volume, then they all intersect the temperature axis at about -273°C. More precise data show that the intersection occurs at -273.15°C. Because the existence of a lower temperature would seem to imply negative volumes, -273.15°C was considered to be the lowest possible temperature, and it proved useful to define a new temperature scale using the same size of degree as on the Celsius scale but with the zero point at -273.15°C. This is known as the absolute temperature scale. The absolute scale is divided into kelvin units (K), where 1 K has the same size as 1°C. Thus

$$T = t + 273.15 \tag{3.10}$$

where T is the absolute temperature in kelvins and t is the temperature in degrees Celsius. With this temperature scale, the relation of volume of gas to absolute temperature is

$$V = C'T \tag{3.11}$$

where C' is a constant depending on n and P. This relation was discovered by Jacques Charles in 1787 and Joseph Gay-Lussac in 1802.

Although the zero point of the absolute scale corresponds to zero volume of gas (Eq. 3.11), it is impossible to verify this idea experimentally, because all known gases liquefy or solidify before reaching the absolute zero. We shall see later (Chapter 15) that the temperature scale defined above is identical with the thermodynamic temperature scale and that one of the consequences of the laws of thermodynamics is that absolute zero cannot be reached.

3.7 VOLUME RELATIONS OF REACTING GASES

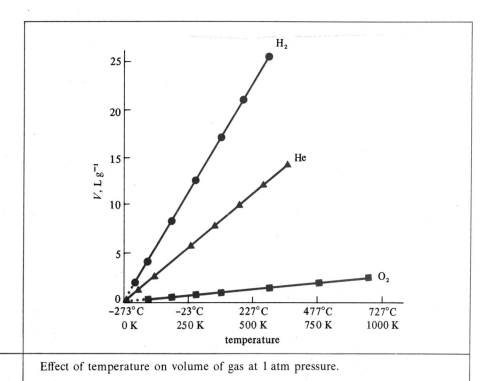

FIGURE 3.4 Effect of temperature on volume of gas at 1 atm pressure.

EXAMPLE 4

If a given amount of a gas occupies 80 mL at 27°C and 1 atm pressure, what volume would the same quantity of gas occupy at 57°C and 1 atm pressure?

Given that the amount and pressure are constant, $V = C'T$; that is, volume increases with an increase in absolute temperature. The initial absolute temperature is $273 + 27 = 300$ K; the final absolute temperature is $273 + 57 = 330$ K. Because $V/T = C'$,

$$\boxed{\frac{V_i}{T_i} = \frac{V_f}{T_f}}$$

where i and f refer to the initial and final states, respectively. Therefore

$$V_f = V_i \left(\frac{T_f}{T_i}\right) = (80 \text{ mL})\left(\frac{330 \, K}{300 \, K}\right) = 88 \text{ mL}$$

3.7 VOLUME RELATIONS OF REACTING GASES

In 1809, Gay-Lussac summarized the available information on the chemical reactions of various gases and noted that the volumes of gases that react chemically to form a new substance, when measured at the same temperature and pressure, are related by simple integers, as illustrated in the following examples:

$$\text{hydrogen(g)} + \text{oxygen(g)} = \text{water(g)} \quad (3.12)$$
$$\text{2 L} \quad\quad\quad \text{1 L} \quad\quad\quad \text{2 L}$$

$$\text{ammonia(g)} + \text{hydrogen chloride(g)} = \text{ammonium chloride(s)} \quad (3.13)$$
$$\text{1 L} \quad\quad\quad\quad \text{1 L}$$

$$\text{ammonia(g)} = \text{nitrogen(g)} + \text{hydrogen(g)} \quad (3.14)$$
$$\text{2 L} \quad\quad\quad \text{1 L} \quad\quad\quad \text{3 L}$$

$$\text{hydrogen(g)} + \text{chlorine(g)} = \text{hydrogen chloride(g)} \quad (3.15)$$
$$\text{1 L} \quad\quad\quad \text{1 L} \quad\quad\quad \text{2 L}$$

In these equations g designates the gaseous state, and s the solid state. (Later on l will be used for the liquid state where appropriate.) As ammonium chloride is a solid, its volume is not relevant to this discussion.

On the basis of such observations, Gay-Lussac formulated the law of combining volumes: *The volumes of gases that react with each other or that are produced in a chemical reaction are in the ratio of simple integers.*

Shortly thereafter (1811), Amedeo Avogadro proposed a simple theoretical interpretation of the law of combining volumes. He hypothesized that *equal numbers of molecules are contained in equal volumes of gases at the same temperature and pressure.* This hypothesis may be stated mathematically using Eq. 3.3

$$V = n\overline{V}$$

by noting that all gases have the same molar volume at the same temperature and pressure. Dalton had earlier rejected such a concept because he believed that the molecules of gaseous elements were all monatomic. Thus he argued that for the reaction of hydrogen and oxygen (Eq. 3.12), one molecule of oxygen could not give rise to two molecules of water (each containing oxygen) since this would require the oxygen atom to split into two parts. However, Avogadro pointed out that there was no experimental evidence to support Dalton's assumption that molecules of gaseous elements were monatomic and suggested that groups of atoms of the same element could be joined in a single molecule. Thus, if the oxygen molecule were diatomic (O_2), then it could split to form two molecules of water (Eq. 3.12) each containing one atom of oxygen.

Other evidence (such as Eq. 3.15) suggested that the hydrogen molecule is also diatomic. Thus when hydrogen and oxygen react, the balanced equation is

$$2H_2(g) + O_2(g) = 2H_2O(g)$$

The coefficients of 2, 1 (implied), and 2 in this equation are the same as the relative volumes shown in Eq. 3.12, a result that follows from Avogadro's hypothesis whenever the reactant or product is gaseous. Equations 3.13–3.15 are interpreted similarly:

$$NH_3(g) + HCl(g) = NH_4Cl(s)$$

$$2NH_3(g) = N_2(g) + 3H_2(g)$$

$$H_2(g) + Cl_2(g) = 2HCl(g)$$

For many years Avogadro's hypothesis was not generally accepted, primarily because of objections raised as to the nature of the forces that could hold two atoms of the same type together in a single molecule [a subject which was not satisfactorily explained until 1916-1927 (see Chapter 9)]. Then, in 1858, Stanislao Cannizzaro revived Avogadro's hypothesis and showed how it could be used to solve the riddle of atomic masses, as explained in the next section. Avogadro's hypothesis is now called **Avogadro's law.**

The following example shows how Avogadro's law may be used to calculate a molecular formula.

EXAMPLE 5

A certain gaseous compound contains only carbon and hydrogen. It reacts with gaseous fluorine, F_2, to produce gaseous carbon tetrafluoride, CF_4, and hydrogen fluoride, HF. If 1.00 L of the compound reacts with 5.00 L of fluorine to produce 2.00 L of CF_4 and 4.00 L of HF, what is the molecular formula of the compound? (All volumes are measured at the same temperature and pressure.)

To apply Avogadro's law, let N be the number of molecules in 1.00 L of any gas under these conditions of temperature and pressure. Then N molecules of the compound produce $2N$ molecules of CF_4 and $4N$ molecules of HF. Thus each molecule of the compound contains two carbon atoms and four hydrogen atoms, and the molecular formula is C_2H_4.

3.8 SIMULTANEOUS DETERMINATION OF ATOMIC MASSES AND MOLECULAR FORMULAS

If one combines Avogadro's law with experimental data on gases and with the concept that a molecule of a substance must contain an integral number of atoms of each of its constituent elements, one has a basis for determining the atomic masses of the elements. This procedure was developed by Cannizzaro in 1858 and led to the establishment of a rational system of atomic masses as well as providing confirming evidence for the validity of Avogadro's hypothesis. The method used was essentially that shown next.

The mass of oxygen (M_O) in 1 L of a gaseous substance can be evaluated from the density, d (in grams per liter), and the mass percent of oxygen:

$$M_O = \frac{\text{density} \times \% \text{ oxygen}}{100}$$

(Division by 100 is required to convert percentage of oxygen to the fraction of oxygen in the gas.)

If N is the number of molecules of the gas in 1 L, and i is the number of oxygen atoms in each molecule, then the total number of oxygen atoms in 1 L is given by the product Ni. Denoting the mass of an oxygen atom by m_O, the total mass of these oxygen atoms is Nim_O, and therefore

$$M_O = Nim_O \tag{3.16}$$

By the same line of reasoning, for a gaseous substance containing hydrogen we have the analogous equations

$$M_H = \frac{\text{density} \times \% \text{ hydrogen}}{100}$$

and

$$M_H = Njm_H \tag{3.17}$$

where M_H is the mass of hydrogen in 1 L of the substance, N is the number of molecules per liter, j is the number of hydrogen atoms per molecule, and m_H is the mass of a hydrogen atom.

Densities measured at 0°C and 1 atm pressure (commonly designated as standard temperature and pressure, abbreviated STP) are given in Table 3.2 for a series of gases containing oxygen as well as a series containing hydrogen. The values of M_O and M_H in column 4 are obtained from the experimental data in columns 2 and 3. Thus 1 L of carbon monoxide contains $(1.250)(57.1)/100 = 0.714$ g of oxygen, and 1 L of hydrogen chloride contains $(1.639)(2.76)/100 = 0.0452$ g of hydrogen.

Since the temperature and pressure are the same, N must be a constant (Avogadro's law), and of course m_O is a constant. Therefore the values of M_O must be integral multiples of Nm_O, that is, $i \cdot Nm_O$. Two values of M_O are about 0.72 g, and the other three values are about twice as large, 1.44 g. Thus we may assume that i is 1 for those substances with $M_O \approx 0.72$ g, and i is 2 for those substances with $M_O \approx 1.44$ g. These values of i are shown in the last column of Table 3.2. In the series of hydrogen compounds the smallest value is about 0.045 g and again the other values are approximately equal to simple multiples of this value (0.090, 0.135, 0.18, and 0.27 g). If we assume that 0.045 g corresponds to one atom of hydrogen per molecule, we obtain the number of atoms per molecule shown in the last column of Table 3.2.

The results of the two series of values can be combined to obtain the relative masses of the atoms of oxygen and hydrogen because, at the same temperature and pressure, N is the same for both series:

$$\frac{(M_O \text{ at } i = 1)}{(M_H \text{ at } j = 1)} = \frac{Nm_O}{Nm_H} = \frac{m_O}{m_H} = \frac{0.72 \text{ g}}{0.045 \text{ g}} = 16$$

By similar calculations Cannizzaro was able to determine the number of atoms

TABLE 3.2 — CANNIZZARO'S METHOD

Substance	d at STP (g L^{-1})	Mass % oxygen	M_O (g L^{-1})	i(atoms O/ molecule)
Carbon monoxide	1.250	57.1	0.714	1
Carbon dioxide	1.977	72.7	1.431	2
Methyl ether	2.091	34.8	0.728	1
Sulfur dioxide	2.927	50.0	1.463	2
Oxygen	1.429	100.0	1.429	2

Substance	d at STP (g L^{-1})	Mass % hydrogen	M_H (g L^{-1})	j(atoms H/ molecule)
Hydrogen chloride	1.639	2.76	0.0452	1
Acetylene	1.173	7.74	0.0908	2
Ammonia	0.7710	17.76	0.1369	3
Methane	0.7168	25.13	0.1801	4
Ethane	1.357	20.11	0.2729	6
Hydrogen	0.0899	100.00	0.0899	2

per molecule of other elements that form gaseous compounds and, hence, their relative atomic masses.

Logically one must consider the possibility that all of these values for i and j are too small by some integral factor. However, no gaseous substances containing oxygen or hydrogen are known for which M_O or M_H is not an integral multiple of 0.72 g or 0.045 g, respectively. Furthermore, direct evidence that these values of i and j are correct has been obtained in recent years by spectroscopic and diffraction methods.

3.9 EXPERIMENTAL DETERMINATION OF MOLECULAR MASSES FROM GAS DENSITIES

The results of the preceding section provide evidence that the formula of the oxygen molecule is O_2 and that for hydrogen is H_2. If we assume these values to be correct, the volume of 1 mol of each gas at STP (0°C and 1 atm) may be evaluated from the molecular masses and the densities given in Table 3.2 as follows: Because 1.429 g of oxygen occupies 1 L, 32.00 g of oxygen or 1 mol of oxygen would occupy $1 \text{ L} \times (32.00 \text{ g}/1.429 \text{ g})$, or 22.39 L. Therefore $\bar{V}_{O_2}$ = 22.39 L mol^{-1}. By a similar calculation we find that $\bar{V}_{H_2}$ = 22.42 L mol^{-1}.

In general, we will take the value of 22.4 L mol^{-1} for the molar volume[1] of any gas at STP and assume that Boyle's law is sufficiently precise for our calculations. The following problem illustrates how this result, together with the gas laws, may be utilized to estimate the molecular mass of a substance.

[1] Even at 1 atm pressure many gases show slight deviations from Boyle's law. If one corrects for these deviations by extrapolating to zero pressure (as discussed in Section 3.11), the best value for $P\bar{V}$ is 22.414 L atm mol^{-1}.

EXAMPLE 6

If 1.10 g of a gaseous substance occupies a volume of 632 mL at 27°C and a pressure of 740 Torr, calculate the approximate molecular mass of the substance.

We know that 1 mol of a gas occupies 22.4 L at STP. Thus we have the following sets of conditions:

	P (Torr)	T (K)	V (L)	n (mol)
STP	760	273	22.4	1.00
measured	740	300	0.632	?

In order to evaluate n, we find it convenient to calculate first the volume of 1 mol of gas (V_f) at the experimental conditions (740 Torr and 300 K):

	P (Torr)	T (K)	V (L)	n (mol)
STP	760	273	22.4	1.00
measured	740	300	V_f	1.00

But V_f must be 22.4 L multiplied by a factor to correct for the change in pressure and a factor to correct for the change in temperature. Thus V_f = 22.4 L × (pressure factor) × (temperature factor). We know that a decrease in pressure will result in an increase in volume, so that the pressure factor must be greater than 1, that is, 760 Torr/740 Torr. An increase in temperature will also cause an increase in volume so that the temperature factor must be 300 K/273 K. Thus

$$V_f = (22.4 \text{ L mol}^{-1}) \left(\frac{760 \text{ Torr}}{740 \text{ Torr}}\right)\left(\frac{300 \text{ K}}{273 \text{ K}}\right) = 25.3 \text{ L mol}^{-1}$$

Because a volume of 25.3 L contains 1 mol at 740 Torr and 300 K, 0.632 L must contain 0.632 L/25.3 L mol^{-1} = 0.0250 mol under the same conditions. From the data given in the problem we know that this many moles have a mass of 1.10 g, and therefore 1 mol has a mass of 1.10 g/0.0250 mol = 44.0 g mol^{-1}. Thus the molecular mass is 44.0.

3.10 THE IDEAL OR PERFECT GAS EQUATION

Throughout the development of chemistry as a science it has repeatedly been very useful to postulate the concept of ideal behavior, that is, a situation in which certain laws are assumed to predict exactly the behavior of a given system. For example, we may define an **ideal gas,** or **perfect gas,** as one that precisely obeys the gas laws described above under all conditions. Such a gas is to be contrasted with **real gases,** such as oxygen, hydrogen, and carbon dioxide, which do not precisely obey the gas laws. However, as the pressure of any real gas is made lower and lower, its behavior approaches that described by the gas laws, and we can say that all real gases approach the ideal or perfect gas at very low pressure. Thus the ideal or perfect gas represents the limiting behavior of real gases as the pressure approaches zero.

To obtain the equation of state of an ideal gas, we can combine Boyle's law, Avogadro's law, and the law of Charles and Gay-Lussac. The application of Avogadro's law to Eq. 3.3 ($\bar{V} = n\bar{V}$) shows that $\bar{V}$ is constant for all gases at the same temperature and pressure and that it is equal to 22.414 L mol^{-1} at 0°C and 1 atm pressure. Therefore, employing Boyle's law and the law of Charles and Gay-Lussac, we conclude that at any temperature T and pressure P,

$$V = n(\bar{V})(\text{pressure factor})(\text{temperature factor})$$
$$= n(22.414 \text{ L mol}^{-1})\left(\frac{1 \text{ atm}}{P}\right)\left(\frac{T}{273.15 \text{ K}}\right)$$

or

$$PV = nRT \tag{3.18}$$

where

$$R = \frac{22.414 \text{ L mol}^{-1} \text{ atm}}{273.15 \text{ K}} = 0.082057 \text{ L atm mol}^{-1} \text{ K}^{-1}$$

R is called the **gas constant**.

The numerical value of R depends on the units used for pressure and volume. In SI units, the value of R is

$$R = \frac{(0.082057 \text{ L atm mol}^{-1} \text{ K}^{-1})(1.01325 \times 10^5 \text{ Pa atm}^{-1})}{1 \times 10^3 \text{ L m}^{-3}}$$
$$= 8.3144 \text{ m}^3 \text{ Pa mol}^{-1} \text{ K}^{-1}$$

To illustrate the usefulness of Eq. 3.18, called the **ideal** or **perfect gas equation**, we return to Example 6. Since the pressure, temperature, and volume are known, it is possible to solve for n using Eq. 3.18 if one expresses the variables in the correct units:

$$n = \frac{PV}{RT} = \frac{(740/760 \text{ atm})(632/1000 \text{ L})}{(0.08206 \text{ L atm mol}^{-1} \text{ K}^{-1})(300 \text{ K})} = 0.0250 \text{ mol}$$

From the number of moles and the given mass, the molecular mass is calculated as before.

As another example of the use of the ideal gas equation, consider the following problem.

EXAMPLE 7

Calculate the volume occupied by 1.60 g of oxygen gas (O_2) at 50°C and 0.10 atm pressure.

Since $PV = nRT$,

$$V = \frac{nRT}{P}$$

$$n = \frac{1.60 \text{ g}}{32.00 \text{ g(mol of } O_2)^{-1}} = 0.0500 \text{ mol of } O_2$$

$$V = \frac{(0.0500 \text{ mol})(0.08206 \text{ L atm mol}^{-1} \text{ K}^{-1})(323 \text{ K})}{0.100 \text{ atm}}$$

$$= 13.3 \text{ L}$$

Use of the gas constant in SI units is illustrated by the following example.

EXAMPLE 8

Calculate the pressure (in pascals) exerted by 0.237 mol of gaseous chlorine in a volume of 19.4 L at 18°C.

To work this problem it is most convenient to use the value of R in SI units, and therefore one must convert the volume from liters to cubic meters.

$$V = \frac{19.4 \text{ L}}{1 \times 10^3 \text{ L m}^{-3}} = 1.94 \times 10^{-2} \text{ m}^3$$

Then, as $PV = nRT$, we have

$$P = \frac{nRT}{V} = \frac{(0.237 \text{ mol})(8.314 \text{ m}^3 \text{ Pa mol}^{-1} \text{ K}^{-1})(291 \text{ K})}{1.94 \times 10^{-2} \text{ m}^3}$$

$$= 2.96 \times 10^4 \text{ Pa}$$

The ideal gas equation can be used to obtain an equation that is useful in calculating the value of P, V, n, or T under one set of conditions from the value of that property under a different set of conditions. Since PV/nT is always equal to a constant (namely R),

$$\frac{P_1 V_1}{n_1 T_1} = \frac{P_2 V_2}{n_2 T_2} \tag{3.19}$$

where the subscripts 1 and 2 refer to different conditions. If a property does not change in going from one set of conditions to another, then of course it cancels

3.12 EQUATIONS OF STATE FOR REAL GASES

out. For example, if n does not change, then $n_1 = n_2$, and Eq. 3.19 reduces to the simpler form

$$\frac{P_1 V_1}{T_1} = \frac{P_2 V_2}{T_2} \tag{3.20}$$

EXAMPLE 9

What volume would a gas occupy at 17°C and 200 Torr if the same amount of gas occupied a volume of 6.20 L at 50°C and 600 Torr?

Since n is constant, we can apply Eq. 3.20, and

$$V_2 = \frac{P_1 V_1}{T_1} \times \frac{T_2}{P_2}$$

$$= \frac{(600 \text{ Torr})(6.20 \text{ L})}{323 \text{ K}} \times \frac{290 \text{ K}}{200 \text{ Torr}}$$

$$= 16.7 \text{ L}$$

3.11 REAL GASES

Although the various gas laws described above are good approximations to the behavior of real gases observed in experiments at ordinary pressures, significant deviations occur at high pressures. To describe these deviations, it is convenient to consider a property called the **compressibility factor**, Z, defined as PV/nRT. An ideal or perfect gas would of course have a compressibility factor of 1 under all conditions. This is shown as the horizontal line in Figure 3.5. The experimental data plotted in Fig. 3.5 were measured on samples of oxygen gas at the temperatures and pressures shown. (Other gases show similar behavior.) At low temperatures there is an initial decrease in Z with increasing pressure, followed by a steady increase in Z as the pressure is increased further. At high temperatures, there is an increase in Z as the pressure is increased even at low pressures (see Figure 3.5 for O_2 gas at 200°C). For all gases there is an intermediate temperature, known as the Boyle temperature, at which there is a considerable range in pressure over which Boyle's law or the ideal gas law predicts the behavior of the gas quite precisely.

3.12 EQUATIONS OF STATE FOR REAL GASES

We have seen that the simplest equation of state, the ideal gas equation, approximates the behavior of gases at low pressures. To represent the behavior of gases with greater accuracy, particularly at high pressures, more complicated equations

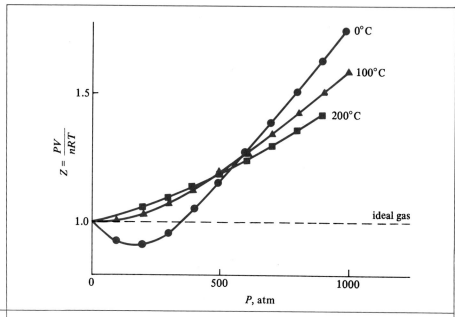

FIGURE 3.5 Deviations from ideal gas behavior of oxygen gas at 0°C, 100°C, and 200°C.

of state are required. One of the more important equations of this sort was proposed by Johannes D. van der Waals in 1873:

$$\left(P + \frac{an^2}{V^2}\right)(V - nb) = nRT \tag{3.21}$$

In terms of the molar volume ($\bar{V} = V/n$) this takes the simpler form

$$\left(P + \frac{a}{\bar{V}^2}\right)(\bar{V} - b) = RT \tag{3.22}$$

TABLE 3.3

VAN DER WAALS CONSTANTS

Gas	a (L² atm mol⁻²)	b (L mol⁻¹)
Helium	0.03412	0.02370
Hydrogen	0.2444	0.02661
Oxygen	1.360	0.03183
Carbon dioxide	3.592	0.04267
Methyl ether	8.073	0.07246
n-Butane	14.47	0.1226

3.12 EQUATIONS OF STATE FOR REAL GASES

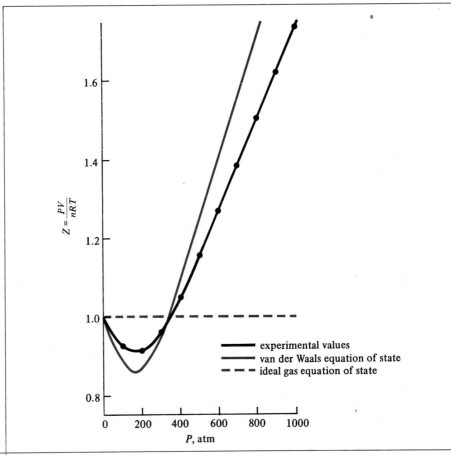

FIGURE 3.6

Comparison of experimental and calculated compressibility factors at pressures up to 1000 atm for oxygen gas at 0°C.

The van der Waals constants a and b are chosen to provide the best fit of the equation to the experimental data. Table 3.3 lists van der Waals constants for several gases. (The physical significance of these constants will be discussed in the next chapter.) Figure 3.6 compares the actual behavior of oxygen at 0°C and at various pressures with that predicted by the ideal gas and van der Waals equations of state.

Other equations of state of greater complexity give even better agreement with the experimental data, and, in fact, it is possible to set up equations of state that will reproduce the experimental behavior of any gas to the desired accuracy. Because of the mathematical complexities of such equations, this topic will not be pursued further, but it should be noted that the study of real gases is an important aspect of chemistry. Not only do reliable equations of state have practical applications in the design of engines and compressors, but they are also the main source of information on the nature of forces between molecules. The interpretation of the ideal gas and van der Waals equations in this latter context will be considered in the next chapter.

PROBLEMS

3.1 What volume does 1 mol of an ideal gas occupy under standard conditions (0°C and 1 atm pressure)? If we accept this value as a good approximation for real gases, what is
   (a) the mass of 10.0 L of CO_2 at standard conditions?
   (b) the volume occupied at standard conditions by 1 million molecules of hydrogen?
   (c) the volume occupied at standard conditions by 5 g of ethylene (C_2H_4)?

3.2 If a barometer was constructed using water instead of mercury, to what height would the water rise if the atmospheric pressure was 1 atm and the water temperature was 25°C where the density of water is 0.997 g cm^{-3}?

3.3 If a certain quantity of gas occupies a volume of 200 mL at 750 Torr, at what pressure would it occupy 50 mL if the temperature remains the same?

3.4 A 2.0-L tank containing oxygen at 0.50 atm pressure is connected by means of a tube and a valve to a 3.0-L tank containing nitrogen at 2.0 atm pressure at the same temperature. If the valve is opened, what will be the total pressure if the temperature remains the same?

3.5 One liter each of O_2, N_2, and CO_2 at 1 atm pressure and room temperature are forced into a 0.50-L container. What is the resulting pressure if the temperature remains the same?

3.6 A 10.0-L container is filled with a gas mixture consisting of 1.00 g of O_2, 2.00 g of H_2, and 0.50 g of CO_2. What is the total pressure if the temperature is 27°C? What is the partial pressure of each substance?

3.7 Nitrogen gas is collected over water at 22°C, and the volume is measured as 284 mL when the total pressure of the gas is 764 Torr. What volume would the nitrogen occupy at the same temperature if it were dry and the pressure were 760 Torr?

3.8 If a gas occupies a volume of 20 mL at 27°C, at what temperature will its volume be 10 mL if the pressure remains the same?

3.9 A sample of a gas has a pressure of 1.00 atm. When its temperature is reduced by 50°C (the volume being held constant), the pressure drops to 0.800 atm. What is the initial temperature of the gas?

3.10 What volume of oxygen is required to burn 2.0 L of methane if both are measured at the same temperature and pressure?
$$CH_4(g) + 2O_2(g) = CO_2(g) + 2H_2O(g)$$
What volume of CO_2 would be produced?

3.11 Two volumes of hydrogen react with one volume of oxygen to produce two volumes of gaseous water if all volumes are measured at the same temperature and pressure. From this result and Avogadro's law, show that the oxygen molecule must contain an even number of atoms.

3.12 A certain gaseous compound contains carbon, hydrogen, and oxygen. It reacts with oxygen to produce gaseous carbon dioxide and water vapor. If 0.50 L of the compound reacts with 1.25 L of oxygen, O_2, to produce 1.00 L of CO_2 and 1.00 L of water vapor, what is the molecular formula of the compound? (All volumes are measured at the same temperature and pressure.)

3.13 The molecular masses and compositions of several gaseous substances containing element X are listed in the next table. From these data what value would you choose for the atomic mass of the element? How many atoms of X are there in a molecule of each gas?

	Molecular mass	% X (by mass)
Gas 1	38	100.0
Gas 2	88	86.3
Gas 3	121	31.4
Gas 4	20	95.0

3.14 The densities at 0°C and 1 atm pressure of several gaseous substances containing element X are listed in the next table, together with the composition as % X (by mass). Based on these data, how many atoms of X are there in a molecule of each gas?

	Density at STP (g L^{-1})	% X (by mass)
Gas 1	1.260	85.6
Gas 2	2.009	81.7
Gas 3	1.250	42.9
Gas 4	2.091	52.1

3.15 Ammonia (NH_3) is a gas at room temperature. Calculate the density of ammonia gas at 27°C and 1.00 atm pressure.

3.16 The density of a gas is 0.357 g L^{-1} at 0°C and 0.50 atm. If the gas is 25% hydrogen and 75% carbon by mass, what is its molecular formula?

3.17 (a) A 0.195-g sample of a liquid substance was vaporized, and the gas occupied a volume of 144 mL at 77°C and 380 Torr. What is the molecular mass of the gas?
(b) The substance was found to contain 7.75% hydrogen and 92.25% carbon. What is the molecular formula of the gas?

3.18 A good vacuum pump is required to produce a pressure as low as 10^{-4} Pa. How many molecules are there in 1 L of a gas at this pressure and 0°C?

3.19 (a) Oxygen is carried by blood as a compound of oxygen and hemoglobin, in which each hemoglobin molecule combines with four O_2 molecules. How many hemoglobin molecules are required to combine with all of the O_2 molecules in 100 mL of oxygen gas (measured at 27°C and 1.00 atm pressure)?
(b) Calculate the mass of these hemoglobin molecules. (The molecular mass of hemoglobin is 6.8×10^4.)

3.20 A gaseous mixture contains oxygen, O_2, and ozone, O_3. The partial pressure of oxygen is 0.347 atm, and the partial pressure of ozone is 0.653 atm. Calculate the pressure of oxygen after all of the ozone has decomposed,
$$2O_3(g) = 3O_2(g)$$
(The temperature and volume are held constant.)

3.21 If an automobile tire is inflated to a gauge pressure of 30 lb in.$^{-2}$ at 25°C, what will the gauge pressure be if the tire, after running at high speed, reaches a temperature of 50°C? Assume that the volume is constant and note that gauge pressure measures pressure above the atmospheric pressure of 14.7 lb in.$^{-2}$.

3.22 What pressure in atmospheres corresponds to each of the following:
(a) 700 Torr?
(b) 365 mm Hg?
(c) 437 Pa?
(d) 20 lb in.$^{-2}$? (*Note:* 1 atm = 14.7 lb in.$^{-2}$.)
(e) gauge pressure of 30 lb in.$^{-2}$? (See Problem 3.21 for the definition of gauge pressure.)

3.23 When an open flask containing air is heated from 27° to 87°C, what fraction of the air in the flask is expelled? Assume that the volume of the flask and the atmospheric pressure are constant.

3.24 Calculate the volume of CO_2 (measured at 1 atm and 27°C) produced by the combustion of 1.0 L of gasoline. The density of gasoline is 0.68 g mL^{-1}. Assume that the chemical reaction is

$$C_7H_{16}(l) + 11O_2(g) = 7CO_2(g) + 8H_2O(g)$$

3.25 The data in the next table indicate that PV for the listed gases varies slightly with pressure in the range 0.01 to 10 atm. Plot the values of PV against P and from your graph determine the value of PV at $P = 0$, designated $(PV)_0$, at 0°C. Will one value of $(PV)_0$ suffice for the three gases? What value of $(PV)_0$ would you choose for an ideal gas?

P(atm)	PV (L atm mol^{-1} at 0°C)		
	CO_2	H_2	O_2
0.01	22.412	22.414	22.413
0.10	22.398	22.416	22.411
0.40	22.353	—	22.405
0.70	22.306	—	22.398
1.00	22.260	22.428	22.392
4.00	21.793	22.468	22.328
7.00	21.313	22.510	22.263
10.00	20.818	22.555	22.198

Up to what pressure can each of the three gases be represented by the ideal gas law without introducing an error of over 1%?

3.26 Using the van der Waals constants in Table 3.3, calculate the pressure at which carbon dioxide has a molar volume of exactly 2.00 L at a temperature of 0°C. Compare your answer with the pressure exerted by an ideal gas at the same temperature and molar volume.

CHAPTER 4
THE KINETIC MOLECULAR THEORY OF GASES

4

The discovery of laws governing natural phenomena, such as the gas laws, challenges scientists to find a reasonable explanation for these laws. In general this is done by developing a model, in terms of familiar concepts, from which laws may be deduced and experimental behavior predicted. Often a close analysis of the ensuing model will lead to predictions of new facts or relationships that can then be subjected to experimental confirmation, thus providing further evidence as to the validity and usefulness of the proposed model. The kinetic theory of gases is a good example of such a model.

4.1 PRESSURE EXERTED BY MOLECULES HAVING SPEED u

The kinetic theory assumes that a gas is composed of molecules in random motion and that the collisions of the molecules with the walls of the containing vessel produce an outward pressure on the walls. Thus, if the number of molecules of gas in a given container were increased, one would expect more collisions with the walls per second and hence an increase in pressure. On the other hand, if the number of molecules were kept constant and the gas were compressed to a smaller volume, the molecules would travel shorter distances to reach the walls, there would be more collisions with the walls per second, and hence there would again be an increase in pressure. Both of these conclusions are in qualitative agreement with the gas laws described in the preceding chapter.

It is necessary to formulate our model more precisely to predict quantitative relationships and to understand how pressure is exerted by a gas. The kinetic theory model of an ideal gas will be developed using the following assumptions.

1. A gas is composed of molecules that can be treated as point masses (i.e., we neglect the size of the molecule and therefore disregard molecular collisions).
2. The molecules move in straight lines in random directions.
3. There are no forces between the molecules.
4. Collisions of the molecules with the walls of the container are elastic (i.e., if a molecule with speed u strikes a wall perpendicularly, it will rebound with a speed u but will move in the opposite direction. Thus the velocity would change from $+u$ to $-u$).

Let us consider the motion of a molecule of mass m in a rectangular box of length l. (For background information on the concepts used in the following

4.1 PRESSURE EXERTED BY MOLECULES HAVING SPEED u

paragraphs, see Appendix D.) The motion of the molecule can be expressed in terms of both its **velocity,** which is a measure of its **speed** u (distance traveled in unit time), and the direction in which it is going. Let us assume that the molecule is moving in a direction parallel to the length of the box with a velocity $+u$ as shown in Figure 4.1. When the molecule strikes the right-hand end of the box, it will rebound with a velocity of $-u$, that is, with the same speed in meters per second but moving in the opposite direction.

The force, F, exerted on the molecule due to a collision with the right wall may be calculated by Newton's second law of motion,

$$F = ma \tag{4.1}$$

The **acceleration,** a, is merely the change in velocity of the molecule per unit of time $\Delta u/\Delta t$, where the symbol Δ is used to signify change in a quantity. Thus, if at time t_2 the velocity is u_2 and at time t_1 the velocity is u_1, then

$$\frac{\Delta u}{\Delta t} = \frac{u_2 - u_1}{t_2 - t_1}$$

Substituting this expression for acceleration into Eq. 4.1, we have

$$F = m\frac{\Delta u}{\Delta t} = \frac{\Delta(mu)}{\Delta t}$$

since m is constant. The product of mass times velocity is known as **momentum,** and we see from the above equation that force is equal to the rate of change of momentum with time.

Although the change in velocity ($+u$ to $-u$) of the molecule is known, the time required for the change is not known. Nevertheless, since the molecule experiences a force, the wall will be subjected to an equal force in the opposite direction and hence an outward pressure. This will be a rapidly fluctuating pressure as it will be exerted only during the very short time interval of a collision. Our usual measuring devices do not respond to such rapid fluctuations, nor are they sensitive enough to measure the effect of a single collision. However, for gases at

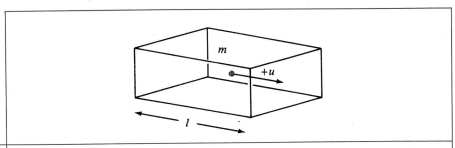

FIGURE 4.1 The motion of a molecule of mass m and velocity u in a rectangular box of length l.

appreciable pressures, we are dealing with an enormous number of molecules, and what we actually measure is the average pressure exerted on the wall due to an enormous number of collisions over a short period of time. This measurement allows us to utilize the *average* force exerted on the wall, which we obtain from the change in momentum per collision with the wall, multiplied by the number of collisions per second.

$$F = \begin{pmatrix} \text{change in momentum} \\ \text{per collision} \end{pmatrix} \times \begin{pmatrix} \text{number of collisions} \\ \text{per second} \end{pmatrix} \quad (4.2)$$

The change in momentum experienced by a molecule striking the right wall is the momentum after the collision ($-mu$) minus the momentum before the collision (mu) or $-2mu$. Because the total momentum of the system (wall and molecule) must remain constant during the elastic collision, the change in momentum experienced by the wall during each collision is the same quantity with opposite sign, or $+2mu$. Thus the first term on the right-hand side of Eq. 4.2 is $2mu$.

To determine the number of collisions with the right-hand wall in 1 s, we note that the molecule travels u meters (m) in 1 s and that a collision with this wall will occur every time the molecule travels $2l$ m. Therefore, the number of collisions per second with the right-hand wall will be

$$\frac{u(\text{m s}^{-1})}{2l(\text{m collision}^{-1})} = \frac{u}{2l} \text{ (collisions s}^{-1}\text{)}$$

Thus the second term on the right-hand side of Eq. 4.2 is $u/2l$. Substituting these results into Eq. 4.2, we have

$$F = (2mu)\left(\frac{u}{2l}\right) = \frac{mu^2}{l} \quad (4.3a)$$

If the number of molecules in the box is increased from one molecule to N molecules, all having the same speed u, but moving in various directions, then on the average there will be $N/3$ molecules moving parallel to the length of the box, $N/3$ molecules moving parallel to the width of the box, and $N/3$ molecules moving parallel to the height of the box. Because only the $N/3$ molecules moving parallel to the length of the box will ever strike the right-hand wall, the average force exerted on this wall will be larger by a factor of $N/3$ than the force due to a single molecule, and we have

$$F = \frac{N}{3} \times \frac{mu^2}{l} \quad (4.3b)$$

(To simplify the derivation we have assumed that all of the molecules are moving parallel to the sides of the box. This rather artificial assumption leads to the same results as a more complicated treatment of the problem, described in

Appendix G-1, which allows for molecular motion in all directions.) Because pressure is merely force per unit area, the pressure is given by

$$P = \frac{F}{A} = \frac{Nmu^2}{3lA} = \frac{Nmu^2}{3V} \qquad (4.4)$$

where A is the area of the right-hand wall and V is the volume of the box ($V = lA$). At this point in the derivation we can note some interesting results. Not only has the proposed model provided us with an explanation of how a gas can exert a pressure, but Eq. 4.4 shows that the pressure is proportional to the concentration of molecules (N/V), to the mass of a molecule (m), and to the square of the speed of the molecules (u^2).

Up to this point we have assumed that the molecules all have the same speed. As the next step in the derivation of the results of the kinetic theory, we shall consider a gas containing molecules with different speeds.

4.2 CORRECTION FOR MOLECULES HAVING DIFFERENT SPEEDS

Although a rectangular container was chosen to simplify the development in the preceding section, Eq. 4.4 can be obtained regardless of the shape of the container chosen. However, the assumption that all molecules have the same speed is an oversimplification, and we now ask what type of an average speed we must use in Eq. 4.4 to allow for molecules of different speeds.

In any sample of a gas the total pressure of the gas may be expressed as the sum of the contributions to the pressure due to the collisions of each of the n_i molecules of speed u_i with the wall; that is,

$$P = \frac{1}{3V}n_1 m u_1^2 + \frac{1}{3V}n_2 m u_2^2 + \frac{1}{3V}n_3 m u_3^2 + \cdots$$

or

$$P = \frac{m}{3V}(n_1 u_1^2 + n_2 u_2^2 + n_3 u_3^2 + \cdots)$$

or

$$P = \frac{m}{3V}\sum_i n_i u_i^2 \qquad (4.5)$$

The symbol $\sum_i$ represents the sum over all molecules present. If we multiply both the numerator and denominator of the right side of Eq. 4.5 by the total number of molecules, N, we obtain

$$P = \frac{Nm}{3V} \frac{\sum_i n_i u_i^2}{N} \tag{4.6}$$

It is now convenient to introduce the notation $\overline{u^2}$ to represent the arithmetic mean of the squared speeds of all the molecules present.

$$\overline{u^2} = \frac{\sum_i n_i u_i^2}{N} \tag{4.7}$$

Equation 4.6 can now be rewritten in the simpler form,

$$P = \frac{Nm\overline{u^2}}{3V} \tag{4.8}$$

Note that Eq. 4.8 is the same as Eq. 4.4, except that u^2 has now been replaced by the mean of the squares of the various molecular speeds, $\overline{u^2}$.

On multiplying both sides of Eq. 4.8 by V, we obtain

$$PV = \tfrac{1}{3}Nm\overline{u^2} \tag{4.9}$$

4.3 TEMPERATURE AND AVERAGE TRANSLATIONAL KINETIC ENERGY

Our next objective is to determine how to fit temperature into our model. We can do this by equating the right sides of Eq. 4.9 and the perfect gas equation, $PV = nRT$, because the left side of each is PV.

$$\tfrac{1}{3}Nm\overline{u^2} = nRT$$

or

$$m\overline{u^2} = \frac{3nRT}{N}$$

Because the total number of molecules (N) is equal to the number of moles (n) times Avogadro's number (N_0), we obtain

$$m\overline{u^2} = \frac{3nRT}{nN_0} = \frac{3RT}{N_0}$$

The quantity R/N_0 appearing on the right-hand side of this equation is called the Boltzmann constant, k_B. Its numerical value is given in Table 4.1. Substituting k_B for R/N_0, we have

$$m\overline{u^2} = 3k_B T \tag{4.10}$$

4.3 TEMPERATURE AND AVERAGE TRANSLATIONAL KINETIC ENERGY

TABLE 4.1 VALUES OF THE GAS CONSTANT R AND THE BOLTZMANN CONSTANT k_B

$R = 0.08206$ L atm K^{-1} mol^{-1}
$R = 8.314$ J K^{-1} mol^{-1} [a]
$k_B = 1.381 \times 10^{-23}$ J K^{-1} molecule^{-1} [a]

[a] 1 J = 1 kg m^2 s^{-2} = 1 Pa m^3.

At this point it is useful to introduce the concept of kinetic energy. Like any moving object, a molecule has kinetic energy because of its motion through space (called **translational kinetic energy**), which is equal to one-half its mass multiplied by the square of its speed, or

$$\tfrac{1}{2}mu^2$$

Using ε_k to represent the *average* translational kinetic energy of the gas molecules, we obtain

$$\varepsilon_k = \tfrac{1}{2}m\overline{u^2} \tag{4.11a}$$

where we have used the average value of the squares of the molecular speeds. Combining this result with Eq. 4.10, we obtain

$$\varepsilon_k = \tfrac{3}{2}k_B T \tag{4.11b}$$

This simple equation is the main result of our long derivation, and it is of major importance in chemistry. It provides us with a mechanical interpretation of the concept of temperature, namely, that *the absolute temperature of a gas is directly proportional to the average translational kinetic energy of the gas molecules*. Furthermore, this equation enables one to calculate the average translational kinetic energy of the gas molecules from the Boltzmann constant and the absolute temperature.

In considering the implications of Eq. 4.11b, one should note the absence of such properties as P, V, and m. Thus the average translational kinetic energy of the gas molecules does *not* depend on the pressure of the gas, nor on its volume, nor on the mass of the molecules—it depends only on the absolute temperature.

Equations 4.11a and 4.11b are often encountered on a molar basis rather than on a molecular basis. To convert to a molar basis, let $E_k = N_0 \varepsilon_k$ be the average translational kinetic energy *per mole*, and let $M = N_0 m$ be the molecular mass. Then on multiplying both sides of Eq. 4.11a by N_0 we obtain

$$E_k = \tfrac{1}{2}M\overline{u^2} \tag{4.12a}$$

On multiplying both sides of Eq. 4.11b by N_0, and noting that $N_0 k_B = R$, we get the result

$$E_k = \tfrac{3}{2}RT \tag{4.12b}$$

The gas constant R or, on a molecular basis, the Boltzmann constant k_B, appears in many different equations, and it is necessary to consider the units (liter atmospheres, joules, etc.) in selecting the numerical value to be used. Some of the more commonly used values of R and k_B are given in Table 4.1.

The SI unit of energy is the joule (abbreviated J). When a force of 1 N acts through a distance of 1 m, 1 J of energy is expended; that is,

$$1 \text{ J} = 1 \text{ N m} = 1 \text{ kg m}^2 \text{ s}^{-2}$$

To convert from liter atmospheres to joules,

$$\begin{aligned} 1 \text{ L atm} &= (1 \times 10^{-3} \text{ m}^3)(1.0132 \times 10^5 \text{ Pa}) \\ &= 101.32 \text{ m}^3 \text{ kg m}^{-1} \text{ s}^{-2} \\ &= 101.32 \text{ kg m}^2 \text{ s}^{-2} = 101.32 \text{ J} \end{aligned}$$

(A summary of units and conversion factors is given in Appendix B.)

4.4 RMS SPEED AND GRAHAM'S LAW OF DIFFUSION

The role played by $\overline{u^2}$ in Eqs. 4.11a and 4.12a naturally leads to a particular kind of average speed, called the **root mean square (rms) speed.** It is defined as the square root of $\overline{u^2}$.

$$u_{\text{rms}} = \sqrt{\overline{u^2}} = \sqrt{\frac{1}{N} \sum_i n_i u_i^2} \tag{4.13}$$

Note that the operations occur from right to left in the phrase "root mean square": First one takes the *square* of the individual speeds, then one computes the arithmetic *mean* of the squares (by adding together the individual squares and dividing the sum by the number of molecules N), and finally one takes the square *root* of this mean. Although this procedure is necessary to define the rms speed, it is more easily obtained from Eqs. 4.12a and 4.12b. Because the right-hand sides of these equations equal E_k,

$$\tfrac{1}{2} M \overline{u^2} = \tfrac{3}{2} RT$$

$$\overline{u^2} = \frac{3RT}{M}$$

$$u_{\text{rms}} = \sqrt{\overline{u^2}} = \sqrt{\frac{3RT}{M}} \tag{4.14}$$

EXAMPLE 1

Calculate the rms speed of nitrogen molecules, N_2, at 25°C.

Since we wish to obtain a speed, we choose a value of R involving meters and seconds, that is, joules. (The units of R also include the kilogram, and therefore it is necessary to express the molecular mass M in kilograms per mole.) Substituting R in joules per kelvin per mole from Table 4.1 into Eq. 4.14, along with the absolute temperature and molecular mass, and noting that $1\text{ J} = 1\text{ kg m}^2\text{ s}^{-2}$, we obtain

$$u_{\text{rms}} = \sqrt{\frac{(3)(8.314\text{ kg m}^2\text{ s}^{-2}\text{ K}^{-1}\text{ mol}^{-1})(298\text{ K})}{28.0 \times 10^{-3}\text{ kg mol}^{-1}}}$$
$$= 515\text{ m s}^{-1}$$

To convert from meters per second to the more familiar units of miles per hour, we use the conversion factor in Appendix B:

$$u_{\text{rms}} = 515\text{ m s}^{-1} \times 2.237\text{ mi h}^{-1}\text{ (m s}^{-1})^{-1}$$
$$= 1.15 \times 10^3\text{ mi h}^{-1}$$

It is thus seen that nitrogen molecules (which constitute 78% of air molecules) are moving very rapidly, even at ordinary temperatures.

To compare the rms speeds $u_{\text{rms, A}}$ and $u_{\text{rms, B}}$ in two different gases A and B having different molecular masses (and possibly different temperatures), Eq. 4.14 is applied to the ratio of the rms speeds,

$$\frac{u_{\text{rms, A}}}{u_{\text{rms, B}}} = \frac{\sqrt{3RT_A/M_A}}{\sqrt{3RT_B/M_B}} = \sqrt{\frac{T_A M_B}{T_B M_A}} \tag{4.15}$$

where the factors of $3R$ have canceled out.

Note that the ratio of speeds is *directly* proportional to the square root of the temperatures, and *inversely* proportional to the square root of the molecular masses. Both results are qualitatively reasonable. Because gas molecules have higher average translational kinetic energy at higher temperature, one would expect the molecules to have higher speeds at higher temperatures. Because the molecules of different gases have the same average translational kinetic energy at the same temperature, it follows that the molecules having the smaller mass must have the higher speed. It may also seem intuitively reasonable that the heavier molecules move more slowly, while the lighter molecules dart about more rapidly.

In the special case that the temperatures are the same, then Eq. 4.15 takes the simpler form

$$\frac{u_{\text{rms, A}}}{u_{\text{rms, B}}} = \sqrt{\frac{M_B}{M_A}} \tag{4.16}$$

This equation was first discovered in 1830 by Thomas Graham, on the basis of his experimental studies of the diffusion of gases. It is called **Graham's law of diffusion**: the rates at which gases diffuse are inversely proportional to the square roots of their molecular masses. This law provides another example of the success of the kinetic theory in explaining the observed behavior of gases.

EXAMPLE 2

In a comparison of samples of oxygen, O_2, and hydrogen, H_2, at the same temperature, which molecules have the higher rms speed? By what factor do the two rms speeds differ?

Applying Eq. 4.16,

$$\frac{u_{rms, H_2}}{u_{rms, O_2}} = \sqrt{\frac{M_{O_2}}{M_{H_2}}} = \sqrt{\frac{32.0}{2.0}} = \sqrt{16} = 4.0$$

Thus the hydrogen molecules have the higher rms speed by a factor of 4.0.

The following simple experiment illustrates Graham's law of diffusion. Two gases, ammonia (NH_3) and hydrogen chloride (HCl), are introduced at opposite ends of a long glass tube filled with air (see Figure 4.2). After diffusing through the tube, the gases meet, at which time they react to produce solid ammonium chloride (NH_4Cl) in the form of a white smoke.

$$NH_3(g) + HCl(g) = NH_4Cl(s)$$

Because the ammonia has the smaller molecular mass, it diffuses more rapidly than the hydrogen chloride. Therefore the smoke appears not at the middle of the tube, but rather at a point closer to the end where the HCl was injected.

4.5 DISTRIBUTION OF MOLECULAR SPEEDS

We have noted in the preceding sections that gaseous molecules of a substance do not all have the same energy or speed. However, it was possible to treat such gases in terms of their average energies or speeds. We now turn our attention to the problem of how these molecular speeds are distributed about some average value and how this distribution may vary with temperature. The distribution problem was first solved by James Clerk Maxwell (1860) in terms of the speeds of the molecules, and later Ludwig Boltzmann (1871) solved it in terms of the energies of the molecules. Thus the term **Maxwell-Boltzmann distribution** is frequently used to signify the distribution of speeds or energies. The distribution of speeds at 300 K and 600 K is shown in Figure 4.3, where the fraction of molecules having a given speed is plotted against the molecular speed.

Note that the fraction of molecules having a particular speed increases rap-

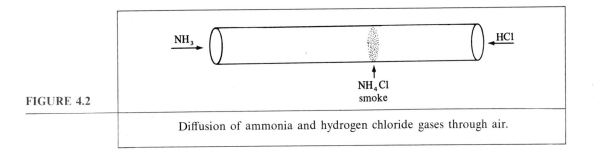

FIGURE 4.2

Diffusion of ammonia and hydrogen chloride gases through air.

idly with increasing speed until it reaches a maximum. Beyond the maximum the fraction decreases and approaches the horizontal axis asymptotically. A rise in temperature from 300 K to 600 K results in a decrease in the height of the maximum and also shifts the maximum to higher speeds. Therefore, *an increase in temperature markedly increases the fraction of the molecules with high speeds.* (This fact will prove very useful in explaining the effect of temperature on many systems of chemical interest.)

The Maxwell-Boltzmann distribution has been verified experimentally by a number of investigators. One type of experimental investigation uses a beam of molecules at a known temperature, a velocity selector, and a method of detecting the number of molecules in the selected range of velocity. A simplified diagram of an apparatus that has been used to measure the velocity distribution of gaseous thallium (Tl) or potassium (K) molecules is shown in Figure 4.4.

Thallium metal is vaporized in an oven whose temperature can be main-

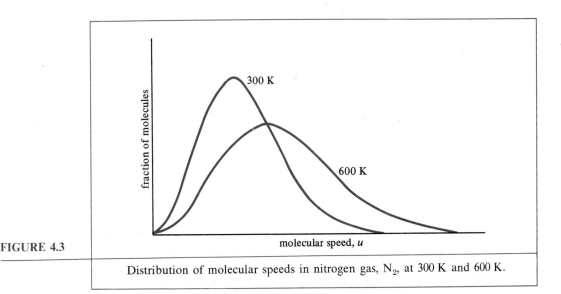

FIGURE 4.3

Distribution of molecular speeds in nitrogen gas, N_2, at 300 K and 600 K.

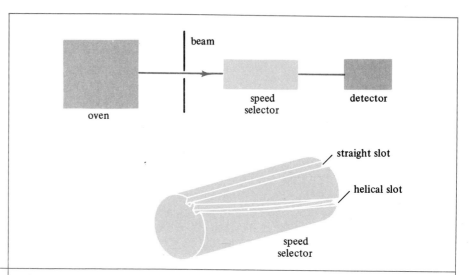

FIGURE 4.4 Apparatus for measuring the distribution of speeds in a beam of molecules. [SOURCE: R. C. Miller and P. Kusch, *Physical Review*, **99**, *1314* (*1955*).]

tained constant to within ±0.25°C, and the gaseous molecules (which are monatomic under these conditions) emerge through a slit in the oven. The emerging molecules then pass through a hole in a metallic plate to give a beam of molecules. The speed selector consists of a solid rotating cylinder on whose surface is cut a helical slot (Figure 4.4). (A straight slot is also provided to allow adjustment of the beam of molecules between the source and the detector so that it is parallel to the cylinder axis.) As the cylinder rotates at a certain speed the beam of molecules enters the helical slot. The only molecules that pass through the speed selector are those that are traveling just fast enough to keep up with the motion of the helical slot. Molecules that are moving too slowly or too rapidly strike the edges of the slot and stick. The speed of the molecules selected for observation is controlled by the speed of rotation of the cylinder.

Robert C. Miller and Polykarp Kusch obtained the speed distributions for thallium vapor shown in Figure 4.5. The number of thallium molecules reaching the detector in unit time (intensity) is plotted against the reduced speed of the molecules. The reduced speed is the measured speed divided by the speed at which maximum intensity is obtained. (The speed at the intensity maximum is 376 m s^{-1} at 870 K and 392 m s^{-1} at 944 K.) The solid line of Figure 4.5 represents the calculated theoretical curve, which is in good agreement with the experimental points obtained for the two temperatures. Furthermore, the measured values of the speed at the intensity maximum agree within 0.5% with the values calculated from the temperature of the oven.

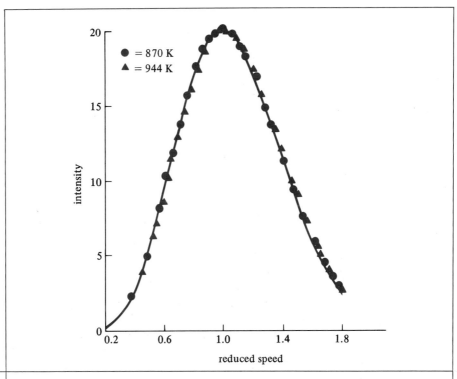

FIGURE 4.5 Typical thallium speed distributions at 870 K and 944 K. The reduced speed is the actual speed divided by the speed at the intensity maximum. [SOURCE: *R. C. Miller and P. Kusch,* Physical Review, **99**, *1314* (*1955*).]

4.6 REAL GASES

Since the kinetic molecular theory in its simplest form provides us with a good explanation for the behavior of an ideal gas, it is reasonable to reexamine the assumptions of the theory (see Section 4.1) in order to ascertain why real gases deviate from ideal behavior at higher pressures, as described in the preceding chapter. That two of the assumptions are oversimplifications is indicated by the fact that all real gases condense to the liquid state at sufficiently low temperature and high pressure. Because gases condense to liquids, there must be attractive forces between the molecules, and therefore assumption 3 is not strictly valid. Because liquids occupy space, the molecules are not point masses, and therefore assumption 1 is not strictly valid.

In 1873 van der Waals modified the kinetic molecular theory by considering each molecule to be a sphere of finite size. Owing to the size of the molecules, there will be a certain volume, known as the excluded volume, that is not available for molecules to move around in. If b represents the volume of 1 mol of these spheres, then the excluded volume is nb, and the volume available for the molecules to move around in (called the *ideal volume*) is smaller than the measured volume, V, by nb; that is

$$V_{ideal} = V - nb$$

It may be shown that the excluded volume per mole, b, is actually four times the volume of the molecules in 1 mol. Using v_0 to represent the volume of a single molecule, then $b = 4N_0 v_0$, where N_0 is Avogadro's number. If the diameter of a molecule is d, then, from the formula for the volume of a sphere, $v_0 = \pi d^3/6$. Therefore

$$b = (4N_0)\left(\frac{\pi d^3}{6}\right) = \frac{2\pi N_0 d^3}{3} \tag{4.17}$$

The second correction that van der Waals made was to include the effect of the attractive forces between the molecules. He suggested that as a molecule approaches a wall of the container, attraction to the other molecules in the body of the gas slows it down, so that it strikes the wall at a lower speed and exerts a smaller pressure. This effect is largest when the molecules are close to one another, a situation that occurs when the concentration n/V is large. Through a line of reasoning that we shall omit here, van der Waals concluded that the attractive forces were proportional to $(n/V)^2$, hence equal to $a(n/V)^2$, where a is a proportionality constant. This term must be added to the measured pressure, P, to obtain the pressure that the gas would exert if it behaved ideally (called the *ideal pressure*). Thus

$$P_{ideal} = P + a\left(\frac{n}{V}\right)^2$$

When the above values of V_{ideal} and P_{ideal} were substituted into the perfect gas equation in the form $P_{ideal} V_{ideal} = nRT$, van der Waals obtained his equation of state:

$$\left(P + \frac{an^2}{V^2}\right)(V - nb) = nRT$$

It is interesting to test the above interpretations of the van der Waals constants a and b by comparing these constants with other parameters that reflect the intermolecular attraction and molecular size of various substances. For example, in Table 4.2, the values of b for several gases are tabulated, together with the molar volumes of the corresponding liquids at their boiling points. Because liquids are not easily compressed, one may assume that the molecules in a liquid are very close together; therefore the approximate agreement between the molar volumes of the liquids and the excluded volume b tends to provide confirmation of the finite size of the gaseous molecules. The value of a is a measure of the attrac-

TABLE 4.2

COMPARISON OF GAS AND LIQUID PROPERTIES

	Molar volume at boiling point (mL mol^{-1})	b (mL mol^{-1})	Normal boiling point (K)	a $\left(\dfrac{L^2\,atm}{mol^2}\right)$
Hydrogen	28.4	26.6	20.4	0.244
Oxygen	27.9	31.8	90.2	1.36
Methyl ether	60	72.5	248.3	8.07
n-Butane	96.5	122.6	272.7	14.5
Water	18.8	30.5	373.2	5.46

tive forces between gaseous molecules, and one might expect that it would, therefore, be proportional to the boiling point of the liquid. It is apparent from the data of Table 4.2 that, with the exception of water, the value of a increases as the boiling point increases. As we shall see in Section 10.4, the discrepancy for water provides a useful clue to understanding the interaction between water molecules in liquid water.

PROBLEMS

4.1 A mixture of 1 mol of oxygen, O_2, and 1 mol of carbon dioxide, CO_2, is placed in a 22.4-L container at 25°C. Which type of molecule will undergo the greater number of collisions per second with the walls of the container?

4.2 Make a qualitative comparison of N_2, O_2, Ar, H_2O, and CO_2 molecules in air at room temperature with respect to
(a) average translational kinetic energy per molecule
(b) rms speed of molecules

4.3 (a) What is the average translational kinetic energy of nitrogen molecules, N_2, at 0°C and 100°C?
(b) What is the average translational kinetic energy per mole of nitrogen at the same temperatures?

4.4 Calculate the average speed and the rms speed of two molecules having speeds of 1×10^2 and 7×10^2 m s^{-1}. Which is greater?

4.5 The rms speed of oxygen molecules, O_2, at 25°C is 4.82×10^2 m s^{-1}. At what temperature do hydrogen molecules, H_2, have the same rms speed?

4.6 In Example 1 we found that the rms speed of nitrogen molecules at 25°C was 1.15×10^3 mi h^{-1}. Using this result, calculate the rms speed at 25°C (in miles per hour) of
(a) helium atoms, He
(b) carbon dioxide molecules, CO_2

4.7 A gas diffuses four times as fast as oxygen, O_2. What is its molecular mass?

4.8 A typical speed for a bullet leaving the muzzle of a rifle is about 1.0×10^3 m s^{-1}. At what temperature do hydrogen molecules, H_2, have an rms speed equal to this value?

4.9 The density of gas A is higher than that of gas B by a factor of 2.89 (when measured at the same temperature and pressure). Which gas diffuses more rapidly? By what factor?

4.10 (a) Calculate the ratio of rms speeds for ammonia, NH_3, and hydrogen chloride, HCl, at the same temperature.
(b) If the glass tube in Figure 4.2 is 1.00 m long, at what point will the ammonium chloride smoke form?

4.11 Given a sample of methane gas, CH_4, at STP, what could you do to double the rms speed of the molecules? What could you do to double the average translational kinetic energy of the molecules?

4.12 One container is filled with oxygen gas, O_2, at 25°C and 1 atm pressure while a second container of equal volume is filled with hydrogen gas, H_2, at 25°C and 1 atm pressure. Compare quantitatively the following properties in each container:
(a) number of molecules
(b) average translational kinetic energy per molecule
(c) rms speed of molecules

4.13 Given two containers of the same volume, one of which contains methane gas, CH_4, at 0°C and 1.0 atm pressure and the other sulfur dioxide gas, SO_2, at 273°C and 2.0 atm pressure, compare quantitatively the following properties in each container:
(a) number of molecules
(b) average translational kinetic energy of the molecules
(c) rms speed of molecules

4.14 Each of the curves in Figure 4.3 has a maximum. What is the physical meaning of these maxima?

4.15 Using the van der Waals constant b for hydrogen in Table 4.2 and Eq. 4.17, calculate the diameter of a hydrogen molecule (in meters).

4.16 For gases A and B under the conditions indicated at the beginning of each part, how are the properties listed at the end related? Use $>$ for greater than, $<$ for less than, $=$ for equals, and ? when no conclusion can be drawn. For example, in part (a), is $n_A > n_B$, $n_A < n_B$, $n_A = n_B$, or $n_A ? n_B$?

(a) Same P, V, T. n_A _____ n_B.
(b) Same P, V, T; $M_A > M_B$. $u_{rms, A}$ _____ $u_{rms, B}$. (M represents the molecular mass.)
(c) $T_A > T_B$, $u_{rms, A} < u_{rms, B}$. M_A _____ M_B.
(d) Same n, V, u_{rms}. P_A _____ P_B.
(e) Same u_{rms}; $M_A > M_B$. T_A _____ T_B.

4.17 The mean molecular speed, $\bar{u}$, is defined as

$$\bar{u} = \frac{n_1 u_1 + n_2 u_2 + n_3 u_3 + \cdots}{N} = \frac{\sum_i n_i u_i}{N}$$

where n_1 molecules have speed u_1, n_2 molecules have speed u_2, and so on, and N is the total number of molecules. It can be shown that for gas molecules having a Maxwell-Boltzmann distribution of speeds,

$$\bar{u} = \sqrt{\frac{8RT}{\pi M}}$$

(a) Calculate $\bar{u}$ for nitrogen molecules, N_2, at 25°C.
(b) Which is larger, $\bar{u}$ or u_{rms} (see Example 1) for nitrogen molecules at 25°C?
(c) Show that in general,

$$\bar{u} = \sqrt{\frac{8}{3\pi}} u_{rms} = 0.921 \, u_{rms}$$

4.18 It can be shown that the number of collisions per second, z, of gas molecules with a wall of area A is

$$z = \frac{N \bar{u} A}{4V}$$

where N is the number of molecules contained in volume V, and $\bar{u}$ is the mean molecular speed (see preceding problem).
(a) Calculate the number of collisions per second with a wall area of 1.00 m² for 1.00 mol of N_2 at 25°C and 1.00 atm pressure.
(b) Compare z for samples of hydrogen gas, H_2, and oxygen gas, O_2, at STP, and wall areas of the same size.
(c) Explain why the pressures exerted by the two gases in part (b) are the same, even though z is larger for hydrogen.

CHAPTER 5
ENERGY CHANGES IN CHEMICAL SYSTEMS

5

Although we have developed methods for describing by chemical equations the chemical reactions that may occur in any system, we have not examined the reasons why the reactions occur. If we consider a large number of reactions, we find that many (but not all) occur with the evolution of energy. Indeed, the energy released in the burning of wood, coal, or fuel oil is a common method for heating buildings and for generating electrical energy. In this chapter we shall study the various types of energy that are of interest in chemistry and the conversion of one type of energy to another. In addition we shall investigate the energy changes associated with chemical reactions as a first step in our effort to understand why chemical reactions occur. This is a part of a subject known as chemical thermodynamics, which enables one to predict the extent to which a chemical reaction will occur from a knowledge of the macroscopic properties of the reactants and the products.

5.1 THE CONCEPT OF ENERGY

One of the basic concepts of science is the **law of conservation of energy.** There is no known exception to this law, which states that *there is a certain quantity, which we call energy, that remains constant in any isolated system.* Energy may exist in many different forms, such as kinetic energy, potential energy, thermal energy, electrical energy, chemical energy, radiant energy, nuclear energy, and mass energy. In fact, the law of conservation of energy proved to be such a useful concept that apparent exceptions to the law led to the postulation of new forms of energy. To make the subject of energy less abstract, let us consider some of the forms of energy of primary interest to chemists and some of the processes whereby energy may be changed from one form to another.

As noted in Chapter 4, we define the **translational kinetic energy,** E_k, of an object of mass m moving with a speed u by:

$$E_k = \tfrac{1}{2}mu^2 \tag{5.1}$$

An object may also have various kinds of **potential energy,** E_p, owing to its location. In general a system acquires potential energy from other forms of energy, or by means of *work, w,* done on the system. Now mechanical work, as the term is used in the sciences, is defined as follows: When an object is moved against a resisting force, F, through a distance, d, then the mechanical work required is equal to the product, Fd.

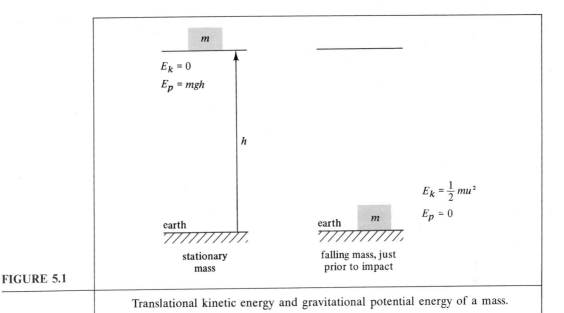

FIGURE 5.1 Translational kinetic energy and gravitational potential energy of a mass.

$$w = Fd \tag{5.2}$$

Just as work may increase the potential energy (or kinetic energy) of a system, so a system possessing potential energy (or kinetic energy) may do an equal quantity of work if it is suitably harnessed.[1]

Potential energy measures the difference in energy between two locations. This may be illustrated by considering the **gravitational potential energy** possessed by a mass m at a height h above the earth (Figure 5.1). The gravitational force is mg, where g is the acceleration of free fall, so that by using Eq. 5.2 we find that the work required to raise the mass to a height h is mgh. Therefore the gravitational potential energy at a height h is

$$E_p = mgh \tag{5.3}$$

If the mass were allowed to fall freely until it struck the earth, the potential energy would decrease as the distance of the mass from the earth decreased, and the kinetic energy of the mass would increase as the velocity of the mass increased. Just before the mass struck the earth the potential energy would be zero since h would be zero. All the energy would now be in the form of kinetic energy, as shown in the following example.

---

[1] A common definition of energy is the capacity for doing work. This definition perhaps places too much emphasis on work, which is only one manifestation of energy. It also neglects the restriction on the amount of work which can be obtained from certain kinds of energy, a restriction expressed by the second law of thermodynamics (Chapter 15).

EXAMPLE 1

What would be the maximum kinetic energy and the maximum speed attained by a 2.50-kg mass if it were dropped from a height of 12.0 m above the earth? (Assume that the resistance of air can be neglected.)

The acceleration of free fall, g, varies from place to place on the earth's surface (see Appendix D-2). For problems of this type it is conventional to use the standard acceleration of free fall, defined as 9.80665 m s^{-2}. At a height of 12.0 m the potential energy of the 2.50-kg mass would be

$$E_p = mgh = (2.50 \text{ kg})(9.80665 \text{ m s}^{-2})(12.0 \text{ m})$$
$$= 294 \text{ kg m}^2 \text{ s}^{-2} = 294 \text{ J}$$

Just before the mass strikes the earth all the potential energy will be converted to kinetic energy. Therefore the maximum kinetic energy is

$$E_k = 294 \text{ J}$$

Since $E_k = \frac{1}{2}mu^2$,

$$u = \sqrt{\frac{2E_k}{m}} = \sqrt{\frac{(2)(294 \text{ kg m}^2 \text{ s}^{-2})}{2.50 \text{ kg}}} = 15.3 \text{ m s}^{-1}$$

What happens to the kinetic energy after the mass strikes the earth? The potential energy relative to the earth is now zero, and, because the velocity of the mass is zero, its kinetic energy is also zero. Experimentally one could provide a qualitative answer by noting that the temperature of the mass increases as a result of its collision with the earth. Thus, as a result of the collision, kinetic energy has been converted to *thermal energy* as evidenced by the rise in temperature of the mass.

We have seen that work (w) is an important method of transferring energy. The other common mode of energy transfer is **heat** (q). When two objects at different temperatures are in close proximity, there is a transfer of thermal energy from one to the other. Ultimately, an equilibrium condition is reached, at which time the temperature of the two objects will be the same. We say that heat flows from the hotter object to the cooler object in this process. Thus heat is a *measure of the energy transferred because of a temperature difference*. (At one time thermal energy, the energy an object possesses because of its temperature, was commonly called heat, and this usage still persists to some extent. The modern scientific definition of the term heat is that given above.)

The first investigator to show the relationship between mechanical work and thermal energy was James P. Joule. A diagram of Joule's apparatus is shown in Figure 5.2. The two masses are connected by cords to the central axle in such a way that, as the masses fall, they turn the axle. Paddles attached to the axle stir water in the container shown in a cutaway view at the bottom of the figure. This stirring action increases the temperature of the water. The mechanical work done

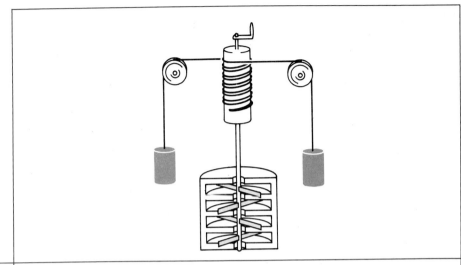

FIGURE 5.2 Joule's apparatus for determining the thermal energy produced by mechanical work.

could be calculated from the known masses and the distance they fell (i.e., mgh), and the rise in temperature of the known mass of water could be measured. Joule found that a mass of 772 lb falling a distance of 1 ft increased the temperature of 1 lb of water by 1° Fahrenheit. Joule was not content with this one method, and he proceeded to demonstrate that the production of thermal energy by friction (rubbing iron rings together) and by electrical work gave the same value for the "mechanical equivalent of heat" (thermal energy produced by work done on the system) within his experimental error of 5%.

Many chemical reactions, such as the burning of coal, proceed with the evolution of considerable quantities of heat. We may satisfy the law of conservation of energy in this system by saying that we are converting the *chemical energy* stored in the coal and oxygen to thermal energy. The above examples point out the usefulness of the concept of energy, and we will defer consideration of other types of energy until the need for them arises.

5.2 THE FIRST LAW OF THERMODYNAMICS

Thermodynamics is that area of science that is concerned primarily with temperature, with energy changes, and with the extent to which a system acquires an orderly arrangement. It is somewhat abstract and is based on three general laws that summarize a large number of experimental observations. These laws cannot be derived, but there are no known exceptions to them.

In this section we focus our attention on the energy of a system and on the first law of thermodynamics.

The laws of thermodynamics are based on experimental observations of macroscopic amounts of substances and therefore are independent of any assumptions about the molecular structures of the substances. Far from being a weakness, this independent character is one of the great strengths of thermodynamics. Often nothing is known about the molecular structure of a compound, yet, from its macroscopic properties, highly accurate predictions may be made about chemical equilibria involving the compound. Furthermore, thermodynamic studies often produce important information about molecular structure. Conversely, when sufficient information about molecular structure is available, it can be used to calculate the macroscopic properties of a substance because, after all, these properties reflect the behavior of the individual molecules of which the substance is composed.

An understanding of thermodynamics requires precision in the use of certain terms. A **system** is any portion of the universe with definite boundaries that we choose to investigate, and we designate the other portions of the universe as the **surroundings** of the system. Thermodynamics deals mainly with the equilibrium states of systems. In an **equilibrium state** the properties of the system are fixed and do not vary with time, nor do the properties depend on the previous history of the system. These equilibrium states can be identified by specifying the values of certain properties of the system that are called **state functions.** It is most convenient to use pressure, volume, and temperature, in addition to other properties that will be identified later, as state functions.

Two important features of state functions require emphasis. First, we can specify the equilibrium state of a system by fixing the values of a sufficiently large number of state functions. Second, the change in any state function (X) in going from state A with a value of X_A to state B with a value of X_B is simply $\Delta X = X_B - X_A$ regardless of the path taken in going from state A to state B. This is analogous to the difference in elevation between two mountain peaks, which is also independent of the path taken in going from one peak to the other.

We can illustrate the above features by using an ideal gas as a model system. If we specify the values of two state functions, namely, the pressure and temperature of a given amount of a gas, then the volume is fixed, as are the values of all other state functions of the system. The second feature is illustrated by the following example.

EXAMPLE 2

Calculate ΔP, ΔT, and ΔV when 2.00 mol of an ideal gas at 1.00 atm and 300 K is compressed to 2.00 atm and 400 K.

$$\Delta P = P_B - P_A = 2.00 \text{ atm} - 1.00 \text{ atm} = 1.00 \text{ atm}$$
$$\Delta T = T_B - T_A = 400 \text{ K} - 300 \text{ K} = 100 \text{ K}$$

To calculate ΔV, we first calculate the initial and final volumes.

$$V_A = \frac{nRT_A}{P_A} = \frac{(2.00 \text{ mol})(0.08216 \text{ L atm mol}^{-1} \text{ K}^{-1})(300 \text{ K})}{1.00 \text{ atm}}$$
$$= 49.3 \text{ L}$$

5.2 THE FIRST LAW OF THERMODYNAMICS

$$V_B = \frac{nRT_B}{P_B} = \frac{(2.00 \text{ mol})(0.08216 \text{ L atm mol}^{-1}\text{K}^{-1})(400 \text{ K})}{2.00 \text{ atm}}$$

$$= 32.9 \text{ L}$$

$$\Delta V = V_B - V_A = 32.9 \text{ L} - 49.3 \text{ L} = -16.4 \text{ L}$$

Note that these results do not depend on the path taken in going from state A to state B. For example, we could first increase the temperature from 300 K to 400 K (at constant pressure) and then increase the pressure from 1.00 atm to 2.00 atm (at constant temperature); alternatively, we could first increase the pressure, and then increase the temperature.

We noted that, if a system is linked mechanically to its surroundings, it may gain energy by having work done on it. It may also gain energy by absorbing heat if its surroundings are at a higher temperature. Neither the work done on the system nor the heat absorbed by the system is independent of the path taken in going from an initial state to a final state, so that neither w nor q is a state function. However, the **first law of thermodynamics** states that there exists a state function, the **internal energy,** E, such that *the increase in internal energy of a system,* ΔE, *is the sum of q, the heat absorbed by the system from its surroundings, plus w, the total of all types of work done on the system by its surroundings*. The first law of thermodynamics may be expressed symbolically by the equation[2]

$$\Delta E = q + w \tag{5.4}$$

The most common type of work done on or by chemical systems is work resulting from changes in volume. A simple example is the work involved when a gas is allowed to expand against a constant external pressure. An experimental device for demonstrating this consists of a cylinder (to contain the gas) that is fitted with a movable piston of cross-sectional area A, as shown in Figure 5.3. (Actually, we shall neglect frictional effects due to the motion of the piston in the cylinder and consider an ideal situation that can be approached experimentally, as we construct pistons with less and less frictional resistance, but that can never be equaled.) The external pressure P exerted on the gas is equal to the atmospheric pressure plus the pressure due to the weight of the piston. If we increase the temperature of the system, the volume of the gas will increase from V_A to V_B as the piston rises a distance d. The change in volume, $\Delta V = V_B - V_A$, is equal to the volume of the cylindrical region swept out by the piston as it moves through distance d. Since the volume of a cylinder is equal to the product of its cross-sectional area and its height, $\Delta V = Ad$. Because pressure is force divided by area, the product of pressure and volume gives force times distance, which is defined as work:

$$P\Delta V = (FA^{-1})(Ad) = Fd = \text{work} \tag{5.5}$$

---

[2] In some texts, work is defined as the work done on the surroundings by the system, in which case the sign of w in Eq. 5.4 would be the opposite of that used here.

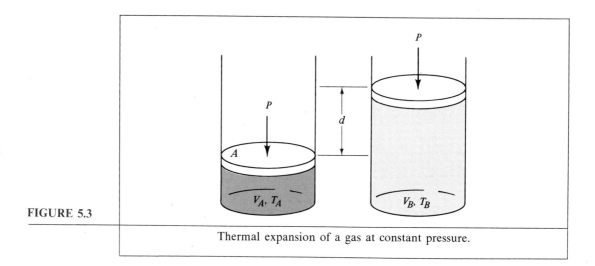

FIGURE 5.3 Thermal expansion of a gas at constant pressure.

In this process, because the surroundings have been pushed back, ΔV is positive, and the work done *by the system* on the surroundings is $P\Delta V$. By convention the w used in the first law is defined as work done *on the system* by its surroundings, and therefore we must change sign,

$$w = -P\Delta V \tag{5.6}$$

If the gas is compressed, then the surroundings do work on the system, ΔV is negative, and w is positive, as would be obtained by using Eq. 5.6.

The fact that w and q are not state functions can be demonstrated by calculating the work done for each of two possible paths that can be used to convert 1 mol of an ideal gas at 0°C from a pressure of 1.00 atm and a volume of 22.4 L to a pressure of 0.500 atm and a volume of 44.8 L. These two paths are illustrated in Figure 5.4.

PATH 1

We shall keep the pressure constant at 1.00 atm and increase the volume to 44.8 L by increasing the temperature from 273 K to 546 K (line AC of Figure 5.4). The work done on the system is

$$w = -P\Delta V = -1.00 \text{ atm} \times (44.8 - 22.4) \text{ L}$$
$$= -22.4 \text{ L atm}$$

(This work is represented by the shaded area under line AC in Figure 5.4.) Now we can maintain the volume constant at 44.8 L and decrease the temperature to 273 K, thus decreasing the pressure to 0.500 atm, as shown by line CB. Since $\Delta V = 0$ for this latter change, no work is done.

5.2 THE FIRST LAW OF THERMODYNAMICS

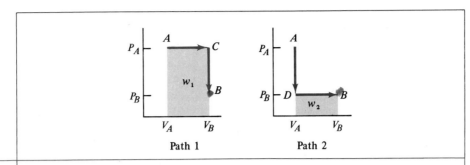

FIGURE 5.4

Two possible paths to change a gas from state A to state B.

The total work in going from A to B by Path 1 is therefore

$$w_1 = -22.4 \text{ L atm}$$

PATH 2

We shall keep the volume constant at 22.4 L and decrease the pressure to 0.500 atm by decreasing the temperature from 273 K to 136.5 K (line AD in Figure 5.4). Because $\Delta V = 0$, $w = 0$ for this change. Now, while the pressure is kept constant at 0.500 atm, let us raise the temperature from 136.5 K to 273 K, thus causing the volume to increase from 22.4 L to 44.8 L. For this step the work done on the system is

$$w = -P\Delta V = -0.500 \text{ atm} \times (44.8 - 22.4) \text{ L}$$
$$= -11.2 \text{ L atm}$$

(This work is represented by the shaded area under line DB in Figure 5.4.) The total work in going from A to B by Path 2 is therefore

$$w_2 = -11.2 \text{ L atm}$$

Since the work done is dependent on the path taken from A to B, w is not a state function. We can also conclude that q is not a state function because ΔE must be the same for both paths:

$$\Delta E = w_1 + q_1 = -22.4 \text{ L atm} + q_1$$
$$\Delta E = w_2 + q_2 = -11.2 \text{ L atm} + q_2$$

Therefore, $q_1 = q_2 + 11.2$ L atm and q is not a state function.

We have noted above that in a constant-volume process there is no pressure-volume work involved and, hence, the energy change in a constant-volume process is equal to the heat absorbed by the system:

$$\Delta E = q_V + w = q_V \tag{5.7}$$

For q_V the subscript V is used as a reminder that the volume has been held constant. It is also understood that only pressure-volume work is being considered. If other types of work are involved (e.g., electrical work), they must be allowed for in Eq. 5.7.

Most chemical reactions are conducted at constant pressure (usually atmospheric pressure). If we again consider processes in which no work is involved except pressure-volume work, we note that $w = -P\Delta V$, and therefore $\Delta E = q_P - P\Delta V$, or

$$q_P = \Delta E + P\Delta V \tag{5.8}$$

where q_P indicates that the pressure is held constant during the process being considered.

At this point it is convenient to introduce a very important thermodynamic property called the heat content or **enthalpy** (H). It is defined as *the sum of the internal energy and the pressure-volume product of the system.* This definition is expressed symbolically by the equation

$$H = E + PV \tag{5.9}$$

As the enthalpy is defined in terms of state functions, it is also a state function.

For any change between two equilibrium states A and B having the same pressure, the change in enthalpy is

$$\begin{aligned} \Delta H = H_B - H_A &= (E_B + PV_B) - (E_A + PV_A) \\ &= E_B - E_A + P(V_B - V_A) \end{aligned}$$

or

$$\Delta H = \Delta E + P\Delta V \tag{5.10}$$

Comparing Eq. 5.8 and Eq. 5.10, we see that

$$\Delta H = q_P \tag{5.11}$$

The two results contained in Eqs. 5.7 and 5.11 express parallel relations involving ΔE and ΔH. For any process at constant *volume*, the heat absorbed by the system is equal to ΔE. For any process at constant *pressure*, the heat absorbed by the system is equal to ΔH. It is this latter relation that makes enthalpy such an important thermodynamic property.

5.3 HEAT CAPACITY

If one wishes to increase the temperature of a substance, this is usually accomplished by adding heat. The quantity of heat required to increase the temperature

5.3 HEAT CAPACITY

of 1 mol of substance by 1°C is known as the **molar heat capacity** (C) of the substance. Because the heat absorbed is dependent on the path taken in going from the initial to the final state, there are many possible values of the heat capacity; however, two are commonly used. The heat capacity at constant volume is denoted C_V, and the heat capacity at constant pressure is denoted C_P.

It has been found experimentally that the quantity of heat required to increase the temperature of a substance by 1°C is proportional to the number of moles. Further, if the temperature of a given amount of a substance is increased from T_A to T_B, the quantity of heat required is proportional to ΔT. Combining these results with the definition of molar heat capacity, we obtain two important equations:

$$\Delta E = q_V = nC_V\Delta T \qquad (5.12)$$

$$\Delta H = q_P = nC_P\Delta T \qquad (5.13)$$

(It is assumed that C_V and C_P are constant over the temperature interval from T_A to T_B, and this means that caution should be used in applying these equations to large temperature intervals.) The subscripts in Eqs. 5.12 and 5.13 serve as reminders that the volume or pressure is held constant. Note that because the size of the degree is the same on the Celsius and Kelvin scales, ΔT may be in either °C or K.

The following examples illustrate the use of Eq. 5.13.

EXAMPLE 3

When 0.15 mol of liquid iodine, I_2, is heated from 400 K to 425 K at 1 atm pressure, 297 J of heat is absorbed. Calculate the molar heat capacity of liquid iodine.

Solving Eq. 5.13 for C_P, we obtain

$$C_P = \frac{q_P}{n\Delta T} = \frac{297 \text{ J}}{(0.15 \text{ mol})(25 \text{ K})}$$

$$= 79 \text{ J K}^{-1} \text{ mol}^{-1}$$

EXAMPLE 4

How many joules of heat are required to raise the temperature of 20.0 g of aluminum (Al) from 10.0°C to 40.0°C at 1 atm pressure? For aluminum, C_P is 24.3 J K^{-1} mol^{-1}.

The heat required may be calculated using Eq. 5.13

$$q_P = nC_P\Delta T$$

$$= \frac{20.0 \text{ g}}{26.98 \text{ g (mol of Al)}^{-1}} \times 24.3 \text{ J K}^{-1}\text{(mol of Al)}^{-1} \times (40.0 - 10.0) \text{ K}$$

$$= 540 \text{ J}$$

EXAMPLE 5

If 2.40 mol of gold atoms (Au) at 100°C is added to 10.0 mol of water at 25.0°C, what will be the final temperature if no heat is lost to the surroundings? (The pressure is constant at 1 atm. For gold, C_P is 25.4 J K^{-1} mol^{-1}. For water, C_P is 75.3 J K^{-1} mol^{-1}.)

If we let t be the final temperature in °C, and note that ΔT is always $T_{\text{final}} - T_{\text{initial}}$, then

$$q_P(\text{water}) = nC_P\Delta T$$
$$= 10.0 \text{ mol of H}_2\text{O} \times 75.3 \text{ J K}^{-1}(\text{mol of H}_2\text{O})^{-1} \times (t - 25)\text{K}$$
$$= [753\,t - (18.8 \times 10^3)]\text{J}$$

$$q_P(\text{gold}) = nC_P\Delta T$$
$$= 2.40 \text{ mol of Au} \times 25.4 \text{ J K}^{-1}(\text{mol of Au})^{-1} \times (t - 100)\text{K}$$
$$= [61.0t - (6.10 \times 10^3)]\text{J}$$

Because the system does not lose or gain heat from the surroundings,

$$q_P(\text{system}) = 0 = q_P(\text{gold}) + q_P(\text{water})$$

or

$$q_P(\text{water}) = -q_P(\text{gold})$$
$$[753t - (18.8 \times 10^3)]\text{J} = -[61.0t - (6.10 \times 10^3)]\text{J}$$
$$814t = 24.9 \times 10^3$$
$$t = 30.6$$

Thus the final temperature is 30.6°C.

5.4 HEAT CAPACITIES OF PURE SUBSTANCES

Most elements in the solid state have heat capacities at constant volume, C_V, which approach zero at very low temperatures and which approach $3R \approx$ 25 J K^{-1} (mol of atoms)$^{-1}$ at higher temperatures. Data for several elements are shown in Figure 5.5. At ordinary temperatures, most elements of higher atomic mass have the full value of $3R$ for C_V.

For solids, values of C_P are only slightly greater than C_V. As shown in Table 5.1, *for most elements $C_P \approx 26$ J K^{-1} (mol of atoms)$^{-1}$ at ordinary temperatures*. This generalization was discovered in 1819 by Alexis T. Petit and Pierre L. Dulong. The **law of Petit and Dulong** was helpful in establishing the correct values for the atomic masses of the elements, particularly of those whose compounds could not be easily volatilized and determined by the method of Cannizzaro. For example, Petit and Dulong showed that the atomic masses of several elements of this type (iron, copper, gold, lead, etc.) were only half as large as the values previously reported.

To a fairly good approximation, the heat capacities of solid compounds are equal to the sum of the heat capacities of the constituent elements—**Kopp's law**

5.4 HEAT CAPACITIES OF PURE SUBSTANCES

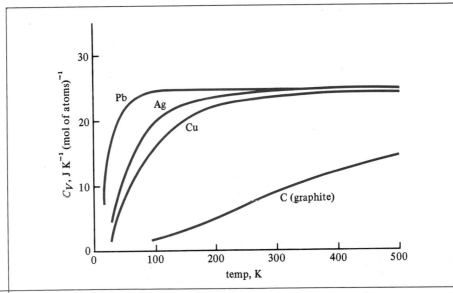

FIGURE 5.5 Molar heat capacities at constant volume of lead (Pb), silver (Ag), copper (Cu), and the graphite form of carbon (C).

(formulated in 1864 by Hermann Kopp). For example, C_P of calcium sulfide, CaS, is 47.4 J K^{-1} mol^{-1} at 25°C, which is about the same as the sum of the heat capacities of elemental calcium and sulfur at the same temperature, $26.3 + 22.6 = 48.9$ J K^{-1} mol^{-1}. If there are n atoms in a molecule of the compound, the molar heat capacity at constant volume, C_V, approaches $3Rn$ at higher temperatures.

The heat capacity of a liquid substance near the melting point is in general about the same or slightly greater than the heat capacity of the solid near the same temperature (see Table 5.2). Water is unusual in that the heat capacity of

TABLE 5.1

HEAT CAPACITIES OF SOLID ELEMENTS AT 25°C AND 1 atm PRESSURE

Element	C_P [J K^{-1} (mol of atoms)$^{-1}$]	Element	C_P [J K^{-1} (mol of atoms)$^{-1}$]
Lithium	24.6	Molybdenum	23.8
Carbon (diamond)	6.1	Silver	25.5
Carbon (graphite)	8.5	Tin	26.4
Aluminum	24.3	Iodine	27.2
Silicon	20.0	Cesium	31.4
Sulfur (rhombic)	22.6	Barium	26.4
Calcium	26.3	Gold	25.4
Chromium	23.4	Lead	26.8
Iron	25.1	Uranium	27.8

TABLE 5.2

MOLAR HEAT CAPACITIES OF SOLIDS AND LIQUIDS NEAR THE MELTING POINT

Substance	Temperature (K)	C_P (solid) (J K^{-1} mol^{-1})	C_P (liquid) (J K^{-1} mol^{-1})
Carbon monoxide	68	51.9	60.2
Propane	85	52.7	84.5
Chlorine	172	55.6	66.9
Mercury	234	28.5	28.5
Water	273	38.9	76.0
Tin	505	30.5	30.5
Zinc	693	29.7	31.4
Sodium chloride	1073	63.6	66.9

liquid water is much larger than that of the solid (ice). An energy unit that is often used in chemistry is the **calorie** (abbreviated cal). It was originally defined in terms of the heat capacity of liquid water—a calorie was the heat required to raise the temperature of 1 g of liquid water by 1°C at 1 atm pressure. (The molar heat capacity of water was therefore 18 cal K^{-1} mol^{-1}.) The calorie is now defined in terms of the SI unit of energy, namely, 1 cal = 4.184 J (exactly). In the field of nutrition a kilocalorie is called a Calorie (abbreviated Cal—note the capital C), and this is the term one often hears in connection with diets. Thus the statement that "an average slice of whole-wheat bread provides 55 Cal" means that the metabolism of this amount of food releases 55 Cal = 55 kcal = 230 kilojoules (kJ).

Figure 5.6 shows the temperature variation of C_P for several gases. For all gases in which each molecule is a single atom (mercury and the noble gases—helium, neon, argon, etc.), $C_P = \frac{5}{2}R \approx 21$ J K^{-1} mol^{-1} independent of temperature. Diatomic gases (hydrogen, hydrogen chloride, chlorine, etc.) have values intermediate between $\frac{7}{2}R \approx 29$ J K^{-1} mol^{-1} and $\frac{9}{2}R \approx 37$ J K^{-1} mol^{-1} at most temperatures. (At low temperatures the heat capacity of hydrogen falls below $\frac{7}{2}R$, approaching $\frac{5}{2}R$ at very low temperatures.) Polyatomic gases have even higher heat capacities.

For all gases at reasonably low pressures, C_P is greater than C_V by $R \approx 8.3$ J K^{-1} mol^{-1}. This difference may be expressed by the equation

$$C_P = C_V + R \tag{5.14}$$

The reason for this difference between the two heat capacities of a gas is explained in the next section.

5.5 EXPLANATION OF THERMODYNAMIC PROPERTIES

What happens to the energy absorbed by a substance when its temperature is increased? Why do different substances have different heat capacities? The answers to these questions lie in the behavior of the atoms and molecules composing

5.5 EXPLANATION OF THERMODYNAMIC PROPERTIES

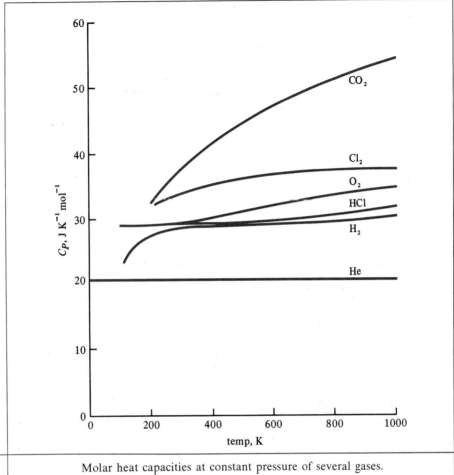

FIGURE 5.6

Molar heat capacities at constant pressure of several gases.

the substances. It is customary to preface the answers by a disclaimer, pointing out that thermodynamics is a science of macroscopic states and, as such, is unaffected by the truth or error of explanations of thermodynamic properties in terms of microscopic behavior. Nevertheless, thermodynamic data provide important information concerning the structure of matter.

In the preceding chapter we saw that the translational kinetic energy of 1 mol of a perfect gas is $E_k = \frac{3}{2}RT$. Raising the temperature from T to $(T + 1)$K increases the translational kinetic energy by

$$\Delta E_k = \tfrac{3}{2}R(T + 1) - \tfrac{3}{2}RT = \tfrac{3}{2}R$$

Thus the molar heat capacity at constant volume is $C_V = \Delta E/\Delta T = \frac{3}{2}R$. As we have seen above, this result is exactly what is found experimentally for the monatomic gases, and thus it provides additional evidence for the kinetic molecular theory of gases. The heat absorbed when we increase the temperature of a monatomic gas at constant volume simply increases the translational kinetic energy of the molecules.

If, instead of constant volume, the temperature of 1 mol of a monatomic gas is increased by 1 K at constant pressure, then the heat absorbed is $C_P = C_V + R = \frac{5}{2}R$. What happens to this extra R of energy? When the gas is heated at constant pressure it expands, doing work against the surroundings, and the energy transmitted to the surroundings as work must be replaced by absorbing additional energy as heat. The additional energy needed is R, as may be seen by the following calculation. To calculate the work done by the system, we use Eq. 5.6, $w = -P\Delta V$. For 1 mol of gas heated from T_A to T_B, $w = -P\Delta V = -(PV_B - PV_A) = -(RT_B - RT_A) = -R\Delta T$. If ΔT is 1 K, then $w = -R$. The sign convention for w tells us that the system does work on the surroundings equal to R; therefore additional energy equal to R must be added as heat. To summarize, of the total of $\frac{5}{2}R$ added as heat in this constant pressure process, $\frac{3}{2}R$ is used to increase the translational kinetic energy of the molecules, and R is used to do work on the surroundings. (The above argument for the work done when a gas is heated at constant pressure is applicable to any gas, thus explaining why Eq. 5.14 applies to all gases.)

The situation with respect to gases containing diatomic and polyatomic molecules is more complicated, as they have not only translational kinetic energy but also rotational energy (the molecules rotate like spinning tops) and vibrational energy (the atoms in a molecule vibrate back and forth). The classical theory of such gases (which is analogous to that described in Chapter 4 for translational motion) was able to explain their heat capacities at high temperatures but not why the heat capacities were lower at lower temperatures. For example, in a gas containing diatomic molecules the theory predicted that rotation and vibration each contribute R to the heat capacity, so that

$$C_V = \tfrac{3}{2}R(\text{translation}) + R(\text{rotation}) + R(\text{vibration}) = \tfrac{7}{2}R$$

and

$$C_P = C_V + R = \tfrac{9}{2}R$$

as found experimentally at high temperatures (Figure 5.6).

The reason heat capacities decrease at lower temperatures was not known until the quantum theory was discovered on the basis of other evidence (see Chapter 7). According to the quantum theory of heat capacities, that part of the heat capacity of a diatomic gas due to vibration decreases from R at high temperatures to zero at low temperatures; and, at very low temperatures, there is a similar falloff in the rotational contribution to the heat capacity. Most gases condense to the liquid state before the latter phenomenon occurs, but it has been observed for hydrogen, the diatomic gas having the lowest boiling point (Figure 5.6).

In a solid the atoms vibrate in three-dimensional space. According to the classical theory, vibration in each dimension contributes R to C_V, and thus C_V of the solid elements should be $3R$ J K^{-1} (mol of atoms)$^{-1}$, as found experimentally at high temperatures (Figure 5.5). The quantum theory is again required to explain the data at low temperatures, where for all elements C_V approaches zero as the absolute temperature approaches zero. (It should be noted that the atoms in a solid continue to vibrate even at temperatures near the absolute zero, but the vibrational energy is essentially temperature-independent in this region and hence C_V is practically zero.)

5.6 ENTHALPIES OF FUSION AND ENTHALPIES OF VAPORIZATION

Energy is required to melt a solid or vaporize a liquid, even though for a pure substance the temperature does not change during either process. The amounts of energy required are called the **enthalpy of fusion** and **enthalpy of vaporization**, respectively. (In the past, the term heat was often used in place of enthalpy, and thus these properties were called heat of fusion and heat of vaporization.) Some melting and boiling points and molar enthalpies of fusion and vaporization at a pressure of 1 atm are shown in Table 5.3.

As shown by these data, *substances that boil at higher temperatures have larger enthalpies of vaporization, such that the molar enthalpy of vaporization divided by the absolute temperature of the boiling point (called the molar entropy of vaporization) is roughly constant (about 88 J K^{-1} mol^{-1}).* This result, discovered in 1884 by Frederick Trouton, is called **Trouton's rule.**

Enthalpies of fusion and enthalpies of vaporization are interpreted in terms of the structural changes that accompany the processes of melting and evaporation. The atoms and molecules in a solid are arranged in a regular pattern or structure; energy is required to break up this structure when the solid melts. When a liquid evaporates, the molecules are no longer in close contact, and energy is required to overcome the attractive forces holding the molecules together.

To calculate the heat required to melt (or vaporize) a particular sample of a

TABLE 5.3 MELTING AND BOILING POINTS, AND MOLAR ENTHALPIES OF FUSION AND VAPORIZATION AT 1 atm PRESSURE

Substance	T_{mp} (K)	ΔH_{fus} (kJ mol^{-1})	T_{bp} (K)	ΔH_{vap} (kJ mol^{-1})	$\Delta H_{vap}/T_{bp}$ (J K^{-1} mol^{-1})
Hydrogen chloride	158.9	2.01	188.1	16.2	80.8
Chlorine	172.2	6.40	239.1	20.4	85.3
Ammonia	195.4	5.65	239.7	23.3	97.2
Methyl ether	131.7	4.94	248.3	21.5	86.6
Water	273.15	6.008	373.15	40.66	109.0
Zinc	692.7	6.69	1180	115	97.5
Sodium chloride	1073	28.4	1738	171	98.4

substance at constant pressure, one simply multiplies the molar enthalpy of fusion (or vaporization) by the number of moles.

Melting: $\quad \Delta H = q_P = n\Delta H_{fus}$ (5.15)

Vaporization: $\quad \Delta H = q_P = n\Delta H_{vap}$ (5.16)

Use of these equations is illustrated by the following example.

EXAMPLE 6

How many joules of heat are required to convert 2.0 mol of ice at $-10\,°C$ to steam (gaseous water) at $100\,°C$? (The pressure is held constant at 1 atm.)

This problem may be divided into four parts: ΔH_1 to heat the ice to $0\,°C$; ΔH_2 to melt the ice to liquid water at $0\,°C$; ΔH_3 to raise the water temperature from $0\,°C$ to $100\,°C$; and ΔH_4 to convert the water to steam at $100\,°C$.
The molar heat capacities of ice and liquid water are taken from Table 5.2, and the molar enthalpies of fusion and vaporization are taken from Table 5.3.

$$\Delta H_1 = nC_p\Delta T = 2.0 \text{ mol} \times 38.9 \text{ J K}^{-1}\text{ mol}^{-1} \times [0 - (-10)]\text{K}$$
$$= 0.78 \text{ kJ}$$

$$\Delta H_2 = n\Delta H_{fus} = 2.0 \text{ mol} \times 6.008 \text{ kJ mol}^{-1}$$
$$= 12 \text{ kJ}$$

$$\Delta H_3 = nC_p\Delta T = 2.0 \text{ mol} \times 76.0 \text{ J K}^{-1}\text{ mol}^{-1} \times [100 - 0]\text{K}$$
$$= 15 \text{ kJ}$$

$$\Delta H_4 = n\Delta H_{vap} = 2.0 \text{ mol} \times 40.66 \text{ kJ mol}^{-1}$$
$$= 81 \text{ kJ}$$

$$q_P = \Delta H_{total} = (0.78 + 12 + 15 + 81)\text{kJ} = 109 \text{ kJ}$$

5.7 ENTHALPIES OF REACTION

The quantity of heat involved in a chemical reaction is as essential a feature of the reaction as one of the reactants or products, and it may be included in the equation for the reaction, as shown for the combustion of hydrogen gas at 1 atm pressure to form liquid water:

$$H_2(g) + \tfrac{1}{2}O_2(g) = H_2O(l) + 285.8 \text{ kJ at } 25\,°C \quad (5.17)$$

As might be inferred from Eq. 5.17, 285.8 kJ is evolved for each mole of water produced. Because of the sign convention for heat, $q_P = -285.8$ kJ, and

$$\Delta H = H_{products} - H_{reactants} = -285.8 \text{ kJ at } 25\,°C$$

(Note that when measuring or calculating ΔH for a reaction, the reactants and products are taken at the same temperature—in this case $25\,°C$.)

5.7 ENTHALPIES OF REACTION

It is customary to associate the enthalpy change with the amount of reaction (in moles) indicated by the equation for the reaction. Thus for

$$H_2(g) + \tfrac{1}{2}O_2(g) = H_2O(l) \qquad \Delta H = -285.8 \text{ kJ}$$

and for

$$2H_2(g) + O_2(g) = 2H_2O(l) \qquad \Delta H = -571.6 \text{ kJ}$$

In the second case 2 mol of water is produced, whereas the first equation indicates that only 1 mol of water is produced.

Reactions that occur with the evolution of heat ($\Delta H < 0$) are classified as exothermic, whereas those that occur with absorption of heat ($\Delta H > 0$) are called endothermic. The combustion of hydrogen gas is an exothermic reaction, but the decomposition of water to form hydrogen and oxygen is endothermic,

$$H_2O(l) = H_2(g) + \tfrac{1}{2}O_2(g) \qquad \Delta H = 285.8 \text{ kJ at } 25°C$$

(Note that reversing a chemical equation simply changes the sign of ΔH.)

In the laboratory most reactions are done at constant pressure (atmospheric pressure) so that ΔH values are most useful. However, reactions involving high temperatures or large volume changes are usually studied under conditions of constant volume using the bomb calorimeter described in Section 5.9. The ΔE values obtained under conditions of constant volume may be converted to values of ΔH using the following method.

For any chemical reaction represented by

$$aA + bB = cC + dD \tag{5.18}$$

the change in enthalpy is given by

$$\Delta H = \Delta E + \Delta(PV) \tag{5.19}$$

where

$$\Delta(PV) = cP_C \bar{V}_C + dP_D \bar{V}_D - aP_A \bar{V}_A - bP_B \bar{V}_B \tag{5.20}$$

where $\bar{V}_A$ is the molar volume of A, $\bar{V}_B$ is the molar volume of B, and so on. If all the reactants and products are liquids or solids, $\Delta(PV)$ is small[3] and may be neglected in comparison to ΔE. However, for reactions involving gaseous reactants or products, $\Delta(PV)$ may be quite large. If we assume that all the reactants

---

[3] Because 1 J = 0.00987 L atm, a change in volume of 0.987 L (much larger than in most reactions involving only liquids or solids) at 1 atm pressure means that the $\Delta(PV)$ term is equivalent to 100 J. Values of ΔE for chemical reactions usually involve many kilojoules.

and products in Eq. 5.18 are gases whose behavior can be closely approximated by the ideal gas law, we can write $\Delta(PV)$ of Eq. 5.20 in the form

$$\Delta(PV) = cRT + dRT - aRT - bRT = (c + d - a - b)RT$$
$$= (\Delta n)RT$$

where $\Delta n = c + d - a - b$, and where the reactants and products are measured at the same temperature T. In general, Δn is the number of moles of *gaseous* products less the number of moles of *gaseous* reactants. (If any of the reactants or products are liquids or solids, they are neglected in calculating Δn.) Substituting this value of $\Delta(PV)$ in Eq. 5.19, we obtain

$$\Delta H = \Delta E + (\Delta n)RT \tag{5.21}$$

EXAMPLE 7

The combustion of 1 mol of hydrogen was carried out in a bomb calorimeter at 25°C, and ΔE was found to be -282.1 kJ. Calculate ΔH.

$$H_2(g) + \tfrac{1}{2}O_2(g) = H_2O(l) \qquad \Delta E = -282.1 \text{ kJ at } 25°C$$

$\Delta n =$ moles of gaseous products $-$ moles of gaseous reactants
$= (0 - 1 - \tfrac{1}{2})$ mol
$= -\tfrac{3}{2}$ mol

$\Delta H = \Delta E + (\Delta n)RT$
$= -282.1 \text{ kJ} - (\tfrac{3}{2} \text{ mol})(8.314 \text{ J K}^{-1} \text{ mol}^{-1})(298.2 \text{ K})$
$= -285.8$ kJ

Because both H and E are state functions, the enthalpy or energy change for any reaction must be independent of the path in going from reactants to products. This principle was originally established experimentally by Germain H. Hess and is called Hess's law or the **law of constant heat summation.**[4] It enables us to calculate values of the enthalpy change for reactions that cannot be determined directly. For example, consider the reaction of carbon with oxygen to form carbon monoxide:

$$C(s) + \tfrac{1}{2}O_2(g) = CO(g) \qquad \Delta H_1 = ? \tag{5.22}$$

Unless excess oxygen is used, some carbon remains unreacted. If excess oxygen is used, some carbon dioxide is formed in addition to the carbon monoxide. Therefore, it is not feasible to measure ΔH_1 directly. However, the enthalpy changes for the following two reactions have been determined experimentally:

$$C(s) + O_2(g) = CO_2(g) \qquad \Delta H_2 = -393.5 \text{ kJ} \tag{5.23}$$
$$CO(g) + \tfrac{1}{2}O_2(g) = CO_2(g) \qquad \Delta H_3 = -283.0 \text{ kJ} \tag{5.24}$$

[4] Discovered before Joule's work on the mechanical equivalent of heat, Hess's law was later recognized as simply a special case of the first law of thermodynamics.

5.8 STANDARD ENTHALPIES OF FORMATION

To apply the law of constant heat summation, we recall that *chemical equations may be added to give new equations*. By Hess's law, the ΔH values are also additive. In this case, reaction 5.22 is obtained by adding reaction 5.23 to the reverse of reaction 5.24. On reversing reaction 5.24, the sign of ΔH changes. Therefore ΔH_1 is equal to $\Delta H_2 - \Delta H_3$.

$$C(s) + O_2(g) = CO_2(g) \qquad \Delta H_2 = -393.5 \text{ kJ}$$
$$CO_2(g) = CO(g) + \tfrac{1}{2}O_2(g) \qquad -\Delta H_3 = +283.0 \text{ kJ}$$

$$C(s) + \tfrac{1}{2}O_2(g) = CO(g) \qquad \Delta H_1 = \Delta H_2 - \Delta H_3 = -110.5 \text{ kJ}$$

The following example provides another illustration of the use of the law of constant heat summation.

EXAMPLE 8

From the enthalpies of combustion of reactions 2, 3, and 4, calculate ΔH_1 for reaction 1.

(1) $\quad C(s) + 2H_2(g) = CH_4(g) \qquad \Delta H_1 = ?$
(2) $\quad H_2(g) + \tfrac{1}{2}O_2(g) = H_2O(l) \qquad \Delta H_2 = -285.8 \text{ kJ}$
(3) $\quad C(s) + O_2(g) = CO_2(g) \qquad \Delta H_3 = -393.5 \text{ kJ}$
(4) $\quad CH_4(g) + 2O_2(g) = CO_2(g) + 2H_2O(l) \qquad \Delta H_4 = -890.3 \text{ kJ}$

(All data are for a temperature of 25°C.)

To obtain reaction 1, we need C on the left, so we start with reaction 3. We need $2H_2$ on the left, so we add twice reaction 2. We need CH_4 on the right, so we reverse reaction 4.

$$C(s) + O_2(g) = CO_2(g) \qquad \Delta H_3$$
$$2H_2(g) + O_2(g) = 2H_2O(l) \qquad 2\Delta H_2$$
$$CO_2(g) + 2H_2O(l) = CH_4(g) + 2O_2(g) \qquad -\Delta H_4$$

$$C(s) + 2H_2(g) = CH_4(g) \qquad \Delta H_1 = \Delta H_3 + 2\Delta H_2 - \Delta H_4$$

On adding the first three reactions, we get the desired reaction. (The other substances all cancel out. For example, the first reaction has CO_2 on the right, and the third reaction has CO_2 on the left.) Adding the corresponding ΔH values gives the desired ΔH.

$$\Delta H_1 = \Delta H_3 + 2\Delta H_2 - \Delta H_4$$
$$= [-393.5 + (2)(-285.8) - (-890.3)] \text{kJ}$$
$$= -74.8 \text{ kJ}$$

5.8 STANDARD ENTHALPIES OF FORMATION

It has proved convenient to tabulate information on enthalpies of reactions in terms of the formation of 1 mol of each compound from its elements. Thus we define the **standard enthalpy of formation,** or **standard heat of formation,** of a compound (ΔH_f°) as *the enthalpy change when* 1 *mol of the compound in its standard state is formed from its elements in their standard states.* We choose as our *standard state* the most stable form of each substance at the specified temperature and 1 atm pressure. Thus the standard enthalpy of formation of water from the elements at 25°C would be

$$H_2(g) + \tfrac{1}{2}O_2(g) = H_2O(l) \qquad \Delta H_f^\circ = -285.84 \text{ kJ mol}^{-1}$$

where each substance is at a pressure of 1 atm and a temperature of 25°C. From the above definition of standard enthalpies of formation, it is evident that by convention *the standard enthalpy of formation of an element in its standard state is zero.* The fact that enthalpies of reaction involve standard states is designated by using the superscript °, as in $H°$.

Standard enthalpies of formation of some compounds are given in Table 5.4. (There is a more extensive table of enthalpies of formation in Appendix F.) It is interesting to note that these enthalpies of formation can be combined to obtain the enthalpy changes of reactions between compounds. This can be shown graphically for the formation of CO_2 by either Eq. 5.23 or Eq. 5.24 (Figure 5.7). If we take as our reference point the reactant elements all at zero enthalpy, we can place our products relative to the reference point. It is easily seen that for the reaction

$$C(s) + \tfrac{1}{2}O_2(g) = CO(g) \qquad \Delta H° = -110.5 \text{ kJ}$$

as we found earlier. If we extend this reasoning to other compounds, we find that the standard enthalpy change ($\Delta H°$) for any reaction is simply the difference

TABLE 5.4

STANDARD ENTHALPIES OF FORMATION AT 25°C

Substance	Formula	Form	ΔH_f° (kJ mol^{-1})
Water	H_2O	l	−285.84
Hydrogen chloride	HCl	g	−92.312
Hydrogen bromide	HBr	g	−36.4
Carbon dioxide	CO_2	g	−393.51
Methane	CH_4	g	−74.848
Ethane	C_2H_6	g	−84.667
Ethylene	C_2H_4	g	52.47
Acetylene	C_2H_2	g	226.75
Benzene	C_6H_6	l	49.03
Ethyl alcohol	C_2H_5OH	l	−277.63
Acetic acid	CH_3CO_2H	l	−487.0
Ammonia	NH_3	g	−45.90
Aluminum oxide	Al_2O_3	s	−1675
Ferric oxide	Fe_2O_3	s	−825.5
Calcium oxide	CaO	s	−635.1
Calcium hydroxide	$Ca(OH)_2$	s	−986.1
Calcium carbonate	$CaCO_3$	s	−1206.9
Cupric oxide	CuO	s	−155.9

5.8 STANDARD ENTHALPIES OF FORMATION

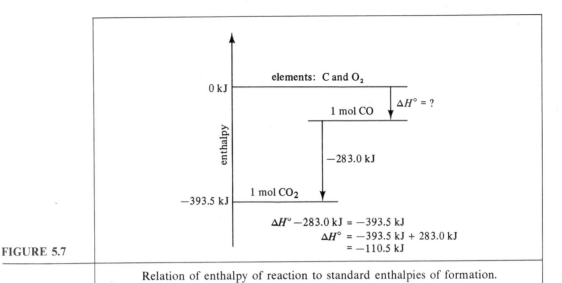

FIGURE 5.7 Relation of enthalpy of reaction to standard enthalpies of formation.

between the sum of the standard enthalpies of formation of the products and the corresponding sum for the reactants.

$$aA + bB = cC + dD \qquad \Delta H° = ?$$

$$\Delta H° = c[\Delta H_f° \text{ of C}] + d[\Delta H_f° \text{ of D}] - a[\Delta H_f° \text{ of A}] - b[\Delta H_f° \text{ of B}]$$
$$= \Sigma \Delta H_f° \text{ (products)} - \Sigma \Delta H_f° \text{ (reactants)} \qquad (5.25)$$

When using Eq. 5.25, one must multiply the enthalpy of formation per mole of each substance involved in the reaction by the coefficient of that substance in the balanced equation for the reaction. Then the sum of these terms for the reactants is subtracted from the sum of terms for the products.

EXAMPLE 9

Compute the standard enthalpy change for the combustion of acetic acid, CH_3CO_2H, to produce carbon dioxide and water at 25°C.

$$CH_3CO_2H(l) + 2\,O_2(g) = 2\,CO_2(g) + 2\,H_2O(l)$$

By convention the standard enthalpy of formation of gaseous oxygen is zero. The other standard enthalpies of formation are listed in Table 5.4. With these data and Eq. 5.25, we obtain

$$\Delta H° = 2\text{ mol} \times [\Delta H_f° \text{ of } CO_2(g)] + 2\text{ mol} \times [\Delta H_f° \text{ of } H_2O(l)] - 1\text{ mol} \times [\Delta H_f° \text{ of } CH_3CO_2H(l)] - 2\text{ mol} \times [\Delta H_f° \text{ of } O_2(g)]$$
$$= (2\text{ mol})(-393.51\text{ kJ mol}^{-1}) + (2\text{ mol})(-285.84\text{ kJ mol}^{-1})$$
$$\quad - (1\text{ mol})(-487.0\text{ kJ mol}^{-1}) - (2\text{ mol})(0\text{ kJ mol}^{-1})$$
$$= -871.7\text{ kJ}$$

5.9 CALORIMETRY

The experimental evaluation of heat changes associated with chemical or physical processes comprises the field of calorimetry. Let us first consider methods of determining experimentally the amounts of energy associated with increases of temperature of a substance or a change of state of a substance.

A simple calorimeter for determining the amount of energy absorbed by a liquid is shown in Figure 5.8. This calorimeter is designed to minimize the flow of heat between the calorimeter and its surroundings. Processes that involve no exchange of heat ($q = 0$) are called **adiabatic processes;** hence the above calorimeter could be designated an adiabatic calorimeter. The calorimeter consists of a double-walled container with a vacuum between the walls to minimize heat transfer to the surroundings, a thermometer, a stirrer, and a heater. To reduce heat transfer even further, the container may be placed in a liquid bath whose temperature may be adjusted to correspond to the temperature within the calorimeter. The liquid to be investigated is brought to the desired temperature and placed in the calorimeter. After the calorimeter has reached a constant temperature, the temperature is recorded. The calorimeter and its contents are then heated by passing an electric current through the heater for a known period of time, after which the temperature is again recorded.

For the above experiment the calorimeter and its contents form the system that is being investigated. We note that only two types of work are involved, namely, pressure-volume work, w_{PV}, and electrical work, w_{el}. If we apply the first law to the change produced by the electrical current, we may write

$$\Delta E = q + w_{PV} + w_{el} = w_{PV} + w_{el}$$

or

$$w_{el} = \Delta E - w_{PV} \tag{5.26}$$

as $q = 0$ for an adiabatic process. The electrical work (w_{el}) done on the system in joules[5] due to the electric current is

$$w_{el} = VQ = VIt \tag{5.27}$$

where V is the electrical potential in volts, Q is the electrical charge transported through the heater in coulombs, I is the current in amperes, and t is the time in seconds.

Experimentally, we can show that the increase in temperature of the calorimeter and its contents is directly proportional to the time the electric current flows and, therefore, to the electrical work done. However, the energy produced by this work is utilized in increasing the temperature of both the liquid and the calorimeter components. We may determine the energy required to increase the temperature of the calorimeter components by repeating the experiment using

---

[5] $1 \text{ J} = 1$ volt coulomb (V C) $= 1$ volt ampere second (V A s).

5.9 CALORIMETRY

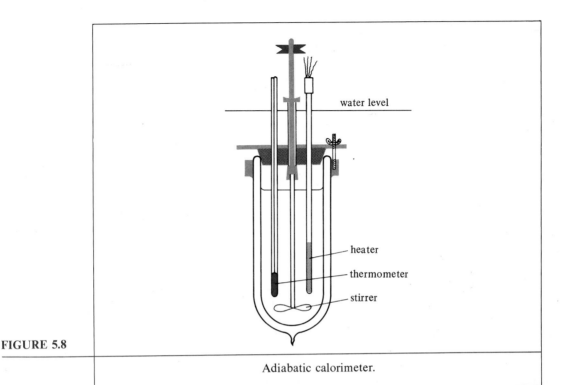

FIGURE 5.8 Adiabatic calorimeter.

different amounts of liquid in the same calorimeter. The results of such experiments indicate that the electrical energy required to raise the temperature of the system by 1°C is a linear function of the number of moles of liquid (n) placed in the calorimeter,

$$\frac{w_{el}}{\Delta T} = nC + C_{cal} \tag{5.28}$$

where C is the molar heat capacity of the substance in the calorimeter, and C_{cal}, the calorimeter constant, is the heat required to raise the temperature of the calorimeter 1°C.

If we substitute two sets of experimental data for different amounts (n_1 and n_2 mol) of substance in Eq. 5.28 and then subtract the resulting equations, we can eliminate C_{cal} and obtain C:

$$\left(\frac{w_{el}}{\Delta T}\right)_2 - \left(\frac{w_{el}}{\Delta T}\right)_1 = n_2 C - n_1 C + C_{cal} - C_{cal}$$

$$C = \frac{(w_{el}/\Delta T)_2 - (w_{el}/\Delta T)_1}{n_2 - n_1} \tag{5.29}$$

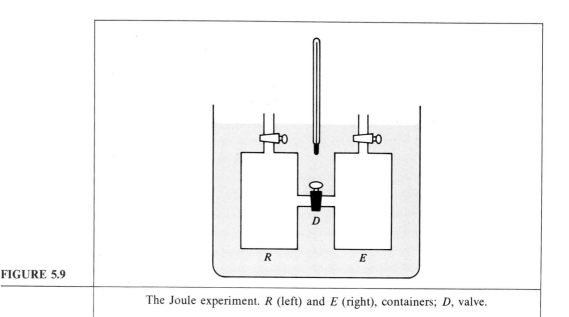

FIGURE 5.9 The Joule experiment. R (left) and E (right), containers; D, valve.

If both measurements were done in a closed container so that the volume is constant, we obtain C_V. If both measurements are at constant pressure, then C_P is obtained.

The heat required to produce a change of phase, either the enthalpy of fusion (solid to liquid) or the enthalpy of vaporization (liquid to gas), may be determined by using suitable calorimeters. The calorimeter shown in Figure 5.8 may be utilized to determine the enthalpy of fusion of a substance; for this purpose, all that is required is to determine the amount of liquid produced when a given quantity of heat is added.

An experiment of great significance in the study of gases was performed by Joule in 1844. He constructed a calorimeter as shown in Figure 5.9. The left container (R) was filled with air to a pressure of 22 atm and the right container (E) was evacuated. The two containers were immersed in a water bath. After the system had reached thermal equilibrium, he recorded the temperature of the water and opened the valve (D), allowing the air to rush into the right container. No work can be done on or by the surroundings, therefore $w = 0$. Joule noted that the temperature of the water did not change, indicating that there was no heat involved in the expansion of the gas; thus $q = 0$. Since w and q are both zero, it is evident that $\Delta E = 0$ and that E is not a function of the volume. Later experiments, using more sensitive methods of detecting small heat changes, indicate that real gases do show detectable changes in temperature with change in volume. These changes in temperature are attributed to the forces between the molecules of real gases. For an ideal gas the energy is a function of temperature alone and is independent of pressure or volume (see Eq. 4.10b).

Many types of calorimeter have been used to investigate the heat involved

5.9 CALORIMETRY

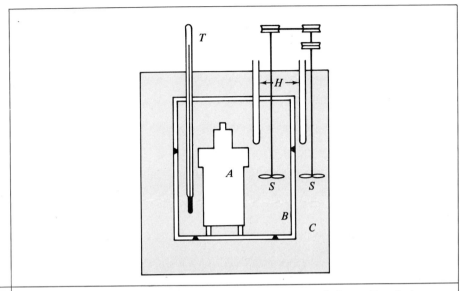

FIGURE 5.10

Bomb calorimeter. *A*, bomb in which the reaction is carried out; *B*, calorimeter vessel; *H*, electrical heaters; *T*, thermometer; *S*, stirrers; *C*, surrounding bath.

in chemical reactions. A typical apparatus is shown in Figure 5.10. It consists of a reaction chamber (*A*) surrounded by a known amount of liquid in an insulated container (*B*) to minimize heat transfer with the surrounding bath (*C*), which is maintained at the same temperature as the liquid in *B*. For reactions that involve high temperatures, a strong metal container of fixed volume, called a **bomb**, is used as the reaction chamber. For example, in combustion reactions, such as the reaction of carbon with oxygen to produce carbon dioxide,

$$C(s) + O_2(g) = CO_2(g)$$

a weighed amount of carbon is introduced into the bomb and then the bomb is filled with oxygen gas at high pressure and sealed off. The bomb is placed in the calorimeter and allowed to come to a uniform temperature, which is recorded. The combustion is then initiated by electrically igniting the mixture. After the system has again reached an equilibrium temperature, the temperature of the calorimeter is measured. The heat evolved during the chemical reaction at constant volume is equal to the increase in thermal energy[6] of the calorimeter and its contents. Another very precise way of determining the heat given off in a reaction is to replace the fluids in *B* and *C* with a mixture of ice and water and then

---

[6] The increase in thermal energy of the calorimeter may be determined by measuring the electrical heat required to produce the same temperature change as noted for the reaction; or it may be calculated from the amount of fluid, the heat capacity of the fluid, the temperature rise, and the calorimeter constant.

measure the amount of ice melted as a result of the reaction. With the known enthalpy of fusion of ice, one may easily calculate the heat evolved in the reaction by multiplying the molar enthalpy of fusion by the number of moles of ice converted to water.

PROBLEMS

5.1 For each of the following changes state whether the energy of the system will decrease, increase, or remain constant:
   (a) The system expands against an external pressure, while $q = 0$.
   (b) The system absorbs heat from the surroundings, while $w = 0$.
   (c) The surroundings do work on the system, while $q = 0$.
   (d) The system is cooled by placing it in an ice bath, while $w = 0$.

5.2 A system is changed from state A to state B by the following sequence of steps:
   Step 1. The system absorbs 300 J of heat while doing 100 J of work on the surroundings.
   Step 2. The system loses 50 J of heat to the surroundings at constant volume.
   Step 3. The surroundings do 500 J of work on the system with no heat change.
   What is the change in energy, ΔE, for each step and for the overall change from state A to state B?

5.3 (a) A mass of 1.00 g falls from a height of 100 m. Calculate its kinetic energy (in joules and calories) just prior to striking the ground. (Use $g = 9.81$ m s^{-2}.)
   (b) By using your answer to part (a), calculate the temperature increase (in °C) of a 100-m waterfall. (Assume that all of the kinetic energy of the falling water is converted to thermal energy of the water, and that C_P for water is 75.3 J K^{-1} mol^{-1}.)

5.4 If 12.0 g of silver metal at 100°C were placed in 10.0 g of water at 20.0°C, what would the final temperature be if no heat were lost to the surroundings? (The pressure is constant at 1 atm. For silver, C_P is 25.5 J K^{-1} mol^{-1}. For water, C_P is 75.3 J K^{-1} mol^{-1}.)

5.5 At 1 atm pressure, how many joules of heat are required to
   (a) raise the temperature of 25.0 g of water from 50 to 60°C?
   (b) convert 25.0 g of ice at 0°C to liquid water at 0°C?
   (c) convert 25.0 g of ice at -10°C to liquid water at 25°C?
   (d) convert 25.0 g of liquid water at 100°C to water vapor at 100°C?
   Note: For C_P of liquid water, use 75.3 J K^{-1} mol^{-1}; for C_P of ice, use 38.9 J K^{-1} mol^{-1}. See Table 5.3 for ΔH_{fus} and ΔH_{vap} of water.

5.6 The C_P value of $H_2O(g)$ at 110°C is 36.3 J K^{-1} mol^{-1}. When exactly 1 mol of steam is heated from 110.00 to 111.00°C at 1 atm pressure,
   (a) What are q, w, ΔE, and ΔH?
   (b) What is the increase in the translational kinetic energy (in joules) of the water molecules?
   (c) Subtract the answer to question (b) from ΔE. In what form is this part of the energy?

5.7 A man of mass 75 kg wants to climb a mountain with an elevation of 1.0 km above his position on the valley floor. His metabolism permits the conversion of approximately one-quarter of the heat of combustion of his food into work. Given that the enthalpy of combustion of cane sugar (sucrose) is 16.5 kJ g^{-1}, calculate the minimum quantity of sugar (in grams) that he would need to consume to furnish the energy for his climb.

5.8 Calculate $\Delta H°$ at 25°C for the reaction
$$FeCl_3(s) + \tfrac{3}{2}H_2(g) = Fe(s) + 3HCl(g)$$
from the law of constant heat summation and the following data:
$Fe(s) + \tfrac{3}{2}Cl_2(g) = FeCl_3(s)$ $\Delta H° = -399.4$ kJ
$H_2(g) + Cl_2(g) = 2\,HCl(g)$ $\Delta H° = -184.6$ kJ
(All data are for a temperature of 25°C.)

5.9 Using the data in Table 5.4, calculate $\Delta H°$ and $\Delta E°$ for each of the following reactions at 25°C:
   (a) $CaCO_3(s) = CaO(s) + CO_2(g)$
   (b) $2Al(s) + Fe_2O_3(s) = 2Fe(s) + Al_2O_3(s)$
   (c) $C_2H_2(g) + 2H_2(g) = C_2H_6(g)$
   (d) $C_2H_6(g) + \tfrac{7}{2}O_2(g) = 2CO_2(g) + 3H_2O(l)$
   (e) $3C_2H_2(g) = C_6H_6(l)$

5.10 Use the data in Appendix F to calculate $\Delta H°$ for each of the following reactions at 25°C and 1 atm pressure:
   (a) $BaCO_3(s) + CaSO_4(s) = BaSO_4(s) + CaCO_3(s)$
   (b) $C_6H_6(l) + 7\tfrac{1}{2}O_2(g) = 6\,CO_2(g) + 3H_2O(l)$
   (c) $SO_2(g) + 2H_2S(g) = 3S(s) + 2H_2O(l)$
   (d) $CaSO_4(s) + 2H_2O(g) = CaSO_4 \cdot 2H_2O(s)$
   (e) $Hg_2Cl_2(s) = Hg(l) + HgCl_2(s)$
   (f) $CH_4(g) + 2Cl_2(g) = CCl_4(l) + 4HCl(g)$

5.11 The combustion of ethyl alcohol at 25°C and 1 atm pressure proceeds according to the equation
$$C_2H_5OH(l) + 3O_2(g) = 2CO_2(g) + 3H_2O(l)$$
   (a) What are $\Delta H°$ and $\Delta E°$ for the above reaction?
   (b) The standard enthalpy of formation of $C_2H_5OH(g)$ is -235 kJ mol^{-1} at

25°C and 1 atm pressure. Calculate the values of $\Delta H°$ and $\Delta E°$ for the vaporization of 1 mol of ethyl alcohol at 25°C.

5.12 A sample of sucrose (common sugar), $C_{12}H_{22}O_{11}$, weighing 0.411 g, is burned in gaseous oxygen in a constant-volume calorimeter to give gaseous CO_2 and liquid H_2O. The temperature of the calorimeter and its contents rises from 25.00 to 27.12°C. In a separate experiment, 3.20 kJ are required to raise the temperature of the calorimeter and its contents 1.00°C.
(a) How much heat was produced by the combustion?
(b) Write the equation for the reaction.
(c) Calculate ΔE and ΔH for the reaction.
(d) Assuming that the ΔH value calculated in part (c) is equal to $\Delta H°$ of the reaction, what is the standard enthalpy of formation ($\Delta H_f°$) of sucrose?

5.13 As discussed in Section 5.9, the energy of a perfect gas does not depend on pressure or volume, but is a function of the temperature only. Show that the enthalpy of a perfect gas is also a function of the temperature only.

5.14 Exactly one mole of an ideal monatomic gas is put through Path 1 and Path 2 of Figure 5.4. The pressure and temperature in state A are 4.00 atm and 546 K. The pressure and temperature of state B are 2.00 atm and 546 K. For each path, evaluate q, w, and ΔE for each step and for the sum of the steps in going from A to B.

5.15 Analysis of a metal oxide shows that it contains 60.0% (by mass) of the metal. The amount of heat required to raise the temperature of 1.00 g of the metal by 1.00°C (called the specific heat) is 0.523 J.
(a) Using the law of Petit and Dulong, calculate an approximate value for the atomic mass of the metal.
(b) From this value and the percent composition, calculate the simplest formula of the oxide.
(c) From the simplest formula and the percent composition, calculate a precise value for the atomic mass.

5.16 When a particular chemical reaction occurs in a sealed container of fixed volume in an ice calorimeter, the total volume of ice and water decreases by 0.169 mL. Calculate q, w, and ΔE for the reaction. (The density of ice is 0.917 g mL^{-1}, and the density of liquid water is 1.000 g mL^{-1}. The enthalpy of fusion of ice is 6.008 kJ mol^{-1}.)

CHAPTER 6
ELECTRONS AND NUCLEI

6

6.1 GENERAL FEATURES OF ATOMIC STRUCTURE

Discoveries in the late nineteenth and early twentieth centuries showed that an atom is not an indivisible entity as Dalton had supposed but rather is composed of very small particles called **electrons, protons, and neutrons.** Various experiments show that electrons and protons are electrically charged particles—electrons are negative and protons are positive. (For a discussion of electric charge, see Appendix D-5.) A neutron is electrically neutral as its name implies. Protons and neutrons have about the same mass; electrons are much lighter. The masses and charges of these particles are summarized in Table 6.1.

An extremely small nucleus, containing all of the protons and neutrons, is located at the center of each atom. Very strong forces, called nuclear forces, bind the protons and neutrons together. The protons give rise to a positive charge on the nucleus of $+Ze$, where Z is the number of protons (called the **atomic number**) and e is the elementary unit of charge. All atoms of an element have the same number of protons, and thus each element is characterized by its atomic number. For example, the nucleus of each carbon atom contains six protons, so the atomic number of carbon is 6. The atomic number is sometimes indicated as a subscript at the left of the symbol for the element (e.g., $_6$C).

Because protons and neutrons are so much heavier than electrons, practically all of the mass of the atom is concentrated in the nucleus. Each proton and each neutron contributes about one unit on the atomic mass scale to the mass of the atom, and hence the total mass of the atom is approximately equal to $Z + N$, where N is the number of neutrons. In Chapter 2 it was pointed out that the atomic mass of an isotope rounded off to the nearest integer is called the mass number. Designating the mass number by A, we see that $A = Z + N$, and therefore $N = A - Z$.

The various isotopes of an element have the same number of protons but different numbers of neutrons and, therefore, different mass numbers. For example, there are three important carbon isotopes of mass numbers 12, 13, and 14, designated $^{12}_6$C, $^{13}_6$C, and $^{14}_6$C, respectively. There are six neutrons in the nucleus of a $^{12}_6$C atom, seven neutrons in a $^{13}_6$C nucleus, and eight neutrons in a $^{14}_6$C nucleus.

Nuclei are very small, ranging in diameter from about 10^{-15} to 10^{-14} m. For example, various experiments indicate that a $^{12}_6$C nucleus has a spherical shape approximately 3×10^{-15} m in diameter. Electrons are known to be much smaller than nuclei. Exactly how small is not known—at present they are considered to be structureless points.

TABLE 6.1	PROPERTIES OF ATOMIC PARTICLES		
	Particle	Charge[a]	Mass[b]
	Electron	-1	0.0005486
	Proton	$+1$	1.007277
	Neutron	0	1.008665

[a] In units of elementary charge ($e = 1.602 \times 10^{-19}$ coulomb).
[b] On atomic mass scale ($^{12}C = 12$ exactly).

The remaining space of an atom is occupied by the electrons, equal in number to the protons in the nucleus. They move about the nucleus with very high kinetic energies, attracted to the nucleus by the coulombic force of attraction that exists between electric charges of opposite sign. Although atoms do not have sharp boundaries, the electrons are mainly within about 10^{-10} m of the nucleus of the atom. [The unit of distance 10^{-10} m is called an ångström (Å), named after Anders J. Ångström.]

For example, each carbon atom has six electrons. Experimental studies of diamond, a solid form of carbon, show that the distance between the nuclei of adjacent carbon atoms is 1.54 Å. If this result is interpreted as the distance between the nuclei of two spherical atoms that "touch" each other, then the radius of a carbon atom is $\frac{1}{2} \times 1.54$ Å $= 0.77$ Å.

If one or more electrons are removed, an atom (or molecule) acquires a net positive charge and is called a **positive ion.** An atom or molecule that has acquired one or more extra electrons has a net negative charge and is called a **negative ion.** The net electric charge (in units of e, the elementary charge) is indicated by a right superscript. For example, F^- is a fluorine atom with an extra electron, and Al^{3+} is an aluminum atom that has lost three electrons. Because the atomic numbers of fluorine and aluminum are 9 and 13, respectively, F^- has $9 + 1 = 10$ electrons, and Al^{3+} has $13 - 3 = 10$ electrons also.

This, then, is a brief description of the general features of an atom. Before beginning a more detailed discussion of atomic structure, let us consider some of the experimental evidence on which this picture is based.

6.2 DISCOVERY OF THE ELECTRON

The atomic character of electricity and the intimate connection between electricity and chemistry were first suggested by the experimental investigations of Michael Faraday in the first half of the nineteenth century. Faraday found that the mass of an element formed during electrolysis was proportional to the quantity of electricity passed through the electrolytic cell (see Figure 1.6). For a given quantity of electricity the relative mass of each element formed was not only characteristic of that element, it was the same as that involved in ordinary chemical reactions. Thus Faraday found that, when hydrogen, oxygen, chlorine, iodine, lead, and tin were produced by the same quantity of electricity in a series of electrolysis experiments, their relative masses were approximately 1, 8, 36, 125, 104, and 58, respec-

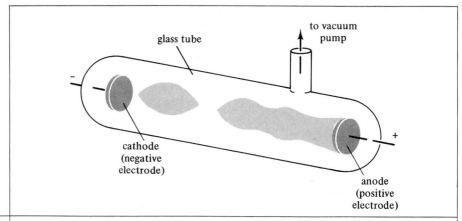

FIGURE 6.1

Electric discharge through air at a pressure of about 0.1 Torr. Luminescent regions are shown in color.

tively. [We now recognize these numbers as approximately equal to the atomic mass (for hydrogen, chlorine, and iodine) or one-half the atomic mass (for oxygen, lead, and tin). More precise data show that the relative amounts of the various elements are exactly equal to their atomic masses or to their atomic masses divided by a small integer.]

Faraday referred to these results as the doctrine of definite electrochemical action; they are now called **Faraday's laws of electrolysis.** He wrote:

> I think I cannot deceive myself in considering the doctrine of definite electrochemical action as of the utmost importance. It touches by its facts more directly and closely than any former fact, or set of facts, have done, upon the beautiful idea, that ordinary chemical affinity is a mere consequence of the electrical attractions of the particles of different kinds of matter; and it will probably lead us to the means by which we may enlighten that which is at present so obscure, and either fully demonstrate the truth of the idea, or develope that which ought to replace it.[1]

Succeeding generations of scientists were to find the means for demonstrating the truth of Faraday's idea.

On considering Faraday's results in connection with the atomic theory, G. Johnstone Stoney in the latter part of the nineteenth century suggested the existence of an elementary unit of electric charge, for which he proposed the name electron.

During the same period many investigators were studying the conduction or discharge of electricity through gases. A simple discharge tube is made from a glass tube with two pieces of metal (called electrodes) sealed in the ends (see Figure 6.1). When the electrodes are connected to a battery or an electric genera-

---

[1] M. Faraday, *Philosophical Transactions of the Royal Society* (*London*), **124**, 115 (1834).

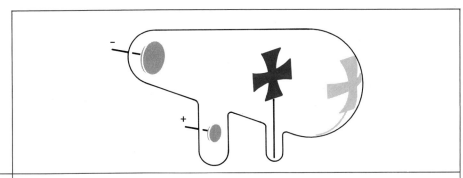

FIGURE 6.2 Experiment showing that rays emitted by the cathode cause luminescence of the glass.

tor and most of the air is removed from the tube, an electric current flows through the tube, accompanied by emission of light (**luminescence**) from various regions within the tube. Neon lights are familiar examples of this phenomenon.

In 1859 Julius Plücker discovered that, at very low pressures, a greenish-yellow light is emitted by the glass walls of the tube itself. Other investigators found that an object placed in the tube casts a shadow on the glass (Figure 6.2). The position of the shadow showed that rays emitted by the cathode (negative electrode) caused the luminescence of the glass. Such an experiment was described in 1879 by William Crookes:

> To obtain this action in a more striking manner, a tube was made having a metal cross on a hinge opposite the negative pole. The sharp image of the cross was projected on the phosphorescent end of the bulb, where it appeared black on a green ground. After the coil had been playing for some time a sudden blow caused the cross to fall down, when immediately there appeared on the glass a bright green cross on a darker background. The part of the glass formerly occupied by the shadow, having been protected from bombardment, now shone out with full intensity, whilst the adjacent parts of the glass had lost some of their sensitiveness, owing to previous bombardment.[2]

These **cathode rays** were extensively studied during the last decade of the nineteenth century by J. J. Thomson and others. Thomson's apparatus is shown in Figure 6.3. By using an anode (positive electrode) with a narrow slit in its center, he was able to obtain a beam of cathode rays that produced a bright spot at the end of the tube. (Modern television tubes operate on the same principle, and Thomson's apparatus may be viewed as the world's first TV tube.) Thomson described his cathode ray tube as follows:

> The rays from the cathode C pass through a slit in the anode A, which is a metal plug fitting tightly into the tube and connected with the earth; after pass-

[2] W. Crookes, *Philosophical Transactions of the Royal Society* (*London*), **170**, 645 (1879).

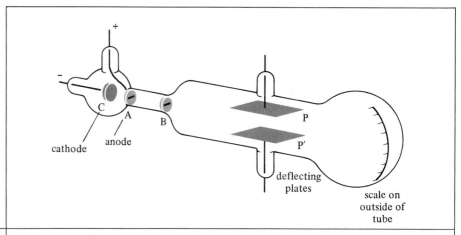

FIGURE 6.3

J. J. Thomson's cathode ray tube. [SOURCE: *Adapted from J. J. Thomson*, Philosophical Magazine and Journal of Science, [5] **44**, *296 (1897)*.]

ing through a second slit in another earth-connected metal plug B, they travel between two parallel aluminium plates about 5 cm long by 2 broad and at a distance of 1.5 cm apart; they then fall on the end of the tube and produce a narrow well-defined phosphorescent patch. A scale pasted on the outside of the tube serves to measure the deflexion of this patch. At high exhaustions the rays were deflected when the two aluminium plates were connected with the terminals of a battery of small storage-cells; the rays were depressed when the upper plate was connected with the negative pole of the battery, the lower with the positive, and raised when the upper plate was connected with the positive, the lower with the negative pole. The deflexion was proportional to the difference of potential between the plates, and I could detect the deflexion when the potential-difference was as small as two volts.[3]

The deflection of the cathode rays is shown in Figure 6.4.

Thomson interpreted his experiments as showing that cathode rays are negatively electrified particles. (Thomson did not call these particles electrons, a name he apparently disliked; nevertheless, the term became universally used. Cathode rays are electrons.) By measuring the effects of electric and magnetic fields, he was able to measure m/e, the ratio of the mass (m) of each of the particles to the electric charge (e) carried by it. As the particles pass between plates P and P' (Figure 6.4) they are attracted by the positive plate and repelled by the negative plate. It is known from studies of the behavior of electrically charged particles in electric fields that the force on the particles is eV/d, where V is the voltage difference between the two plates and d is the distance between them. The angle (θ) through which the rays were deflected as a result of the voltage on the deflection plates was found to agree with the equation

---

[3] J. J. Thomson, *Philosophical Magazine and Journal of Science*, [5] **44**, 296 (1897).

6.2 DISCOVERY OF THE ELECTRON

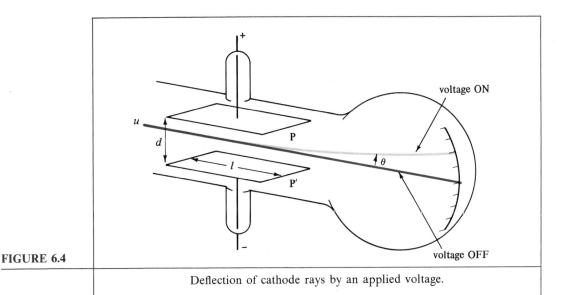

FIGURE 6.4 Deflection of cathode rays by an applied voltage.

$$\theta = \frac{Vel}{dmu^2} \tag{6.1}$$

where l is the length of the plates and u is the velocity of the electrons. Thus *the angle of deflection increases with an increase in the voltage, the charge on the particle, and the length of the deflecting plates. The angle is decreased if the deflecting plates are placed farther apart, if the mass of the particle is increased, or if the velocity of the particle is increased.* (The greater the mass or velocity of an object, the more difficult it is to deflect it.)

To determine the velocity of the electrons, Thomson disconnected the voltage to the deflection plates and applied a magnetic field (H) perpendicular to the flow of electrons. Studies of the behavior of electrically charged particles in magnetic fields show that the deflecting force on the electrons caused by their motion through the magnetic field is equal to the product of H, e, and u. Thomson adjusted the magnetic field so that he obtained the same deflection as with the electric field. Under these conditions the deflecting forces must be the same; therefore

$$Heu = \frac{eV}{d} \tag{6.2}$$

and

$$u = \frac{V}{dH} \tag{6.3}$$

If this result for the velocity is substituted in Eq. 6.1, we find that

$$\theta = \frac{Vel}{dm(V/dH)^2} = \frac{eldH^2}{mV} \qquad (6.4)$$

or

$$\frac{m}{e} = \frac{ldH^2}{\theta V} \qquad (6.5)$$

Thomson measured l, d, H, θ, and V in a series of seven experiments and obtained results for m/e ranging from 1.1×10^{-8} to 1.5×10^{-8} g C^{-1}. He found that the results did not depend on the nature of the cathode (he used aluminum and platinum) nor on the residual gas in the tube (air, hydrogen, or carbon dioxide).

He also determined m/e by a completely independent method involving calorimetric measurements of the kinetic energy of the particles, their radius of curvature in a magnetic field, and the total electric charge carried in a given time. In 26 experiments, he obtained results ranging from 0.32×10^{-8} to 1.0×10^{-8} g C^{-1}. The "scatter" in these experimental values reflects inaccuracies in the methods of determination available at the time these measurements were made. The best modern value is 0.56857×10^{-8} g C^{-1}.

The results for m/e were about 1000 times smaller than the mass-to-charge ratio of the hydrogen ion as determined by electrolysis experiments. Either the mass of the electron was very much smaller than the mass of the smallest ion or else it carried a much larger electric charge, or perhaps it had both a smaller mass and a larger charge. This question was settled in favor of the first possibility by precise measurements of e.

6.3 MEASUREMENT OF THE ELECTRONIC CHARGE

The first accurate measurements of e, the fundamental unit of electric charge, were made during the period 1909–1912 by Robert A. Millikan. By investigating the velocity of fall of neutral and electrically charged oil drops, he was able to show that the electric charge could be varied only by e or an integral multiple of e, and his measurements gave a very accurate determination of the magnitude of e.

A beam of X rays (see Section 6.4), which create positive and negative ions as they pass through air, was used to vary the electric charge on a tiny drop of oil. Millikan described his apparatus, shown in Figure 6.5, as follows:

> A droplet from an oil spray blown by an atomizer over plate C drifts through a minute pinhole in C into the space between C and D, where it is successively moved up and down at will by throwing on or off the 10,000-volt electric field between C and D. The droplet's charge may be changed at will when the field is on by shooting a narrow beam of X-rays through L and L', thus producing ions along its path. When C is negative this exposes the droplet to a rain of

6.3 MEASUREMENT OF THE ELECTRONIC CHARGE

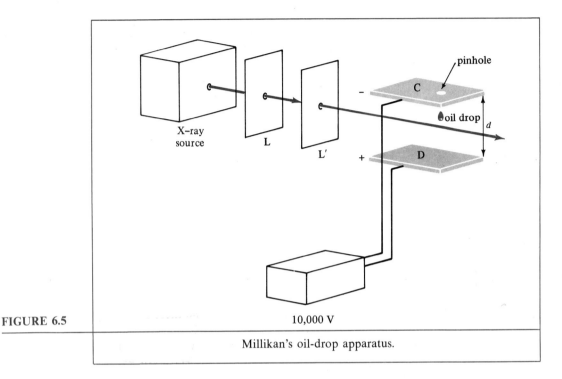

FIGURE 6.5 Millikan's oil-drop apparatus.

positive ions, one or more of which may be caught by the drop lying in wait for them.[4]

By means of the switch the direction of the electric field could be reversed, making the upper plate positive and the lower plate negative. Thus the charge on the droplet could be changed from positive to negative or to zero as desired.

As an oil drop falls under the influence of gravity or rises under the influence of the applied voltage, its velocity is determined by measuring the time required to move a certain distance. Some results obtained by Millikan with a single oil drop are shown in Table 6.2. Other oil drops show similar behavior.

We find by experiment that a small droplet falls through air at a constant velocity, which indicates that the total force on the droplet must be zero. Thus the weight (W) tending to produce a downward motion (which we will take as negative) is balanced by the frictional force (F_f) of the air opposing the motion of the drop,

$$W + F_f = 0$$

In 1849 George Stokes found that

[4] R. A. Millikan, *Autobiography,* Prentice-Hall, Englewood Cliffs, N.J., 1950, p. 79.

TABLE 6.2 **MILLIKAN'S MEASUREMENTS ON VELOCITIES OF AN OIL DROP**

Velocity of fall under gravity $u_0 = -1.083 \times 10^{-4}$ m s^{-1} [a]
Velocity of rise, u, with a constant electric voltage:

u (m s^{-1})	$u - u_0$ (m s^{-1})	Q
1.924×10^{-4}	3.007×10^{-4}	e
4.936×10^{-4}	6.019×10^{-4}	$2e$
7.897×10^{-4}	8.980×10^{-4}	$3e$
10.95×10^{-4}	12.03×10^{-4}	$4e$

[a] The upward direction is taken as positive; thus u_0 is negative.

$$F_f = -6\pi\eta r u_0 \qquad (6.6)$$

where η is the coefficient of viscosity of air, r is the radius of the drop, and u_0 is the velocity of the drop. The negative sign in Eq. 6.6 indicates that the frictional force is in the direction opposite to the motion of the drop. Combining these two equations, we obtain

$$W = 6\pi\eta r u_0 \qquad (6.7)$$

If the drop has an electric charge Q, then, when a voltage V is applied between plates C and D, the drop is subject to an additional force QV/d, where d is the distance between C and D. If the applied voltage results in an attractive force of sufficient magnitude between the drop and the upper plate C, the drop will rise at a constant velocity, u. (Note that u is the velocity of the drop when the voltage is on, and u_0 is the velocity of free fall when the voltage is off.) Because the total force on the drop must again be zero, we find that

$$W + F_f + \frac{QV}{d} = 0$$

or

$$W - 6\pi\eta r u + \frac{QV}{d} = 0 \qquad (6.8)$$

By combining Eqs. 6.7 and 6.8, we see that

$$\frac{QV}{d} = -6\pi\eta r u_0 + 6\pi\eta r u \qquad (6.9)$$

or

$$Q = \frac{6\pi\eta rd}{V}(u - u_0) \tag{6.10}$$

For given values of r, d, and V obtained by studying the motion of a single drop both with and without an applied voltage, Q is proportional to $u - u_0$. As shown in Table 6.2, the results indicate that the values for $u - u_0$ are in a ratio of $1:2:3:4$ and therefore the net charges on the oil drop are also in the ratio of $1:2:3:4$.

Although it was possible to discharge the oil drop completely, so that, when the voltage was applied, the drop fell in exactly the same time as when there was no voltage, no values intermediate between zero and 3.007×10^{-4} m s^{-1} were obtained for $u - u_0$ of this oil drop. Furthermore, the higher velocities were always such that $u - u_0$ was an integral multiple of this value (within the experimental error). Therefore when rising at the lowest velocity, the net charge on the oil drop was e, the electronic charge. When moving at higher velocities the charge was $2e$, $3e$, $4e$, and so on.

By determining η, r, d, and V, it was possible to calculate the electronic charge e; Millikan obtained a result of 1.59×10^{-19} C. This was later recalculated (using an improved value for η, the coefficient of viscosity of air) as 1.6018×10^{-19} C. The best modern value is $\underline{1.60219 \times 10^{-19}}$ C.

6.4 DISCOVERY OF THE NUCLEUS

The experimental evidence for the ubiquitous electrons led to questions concerning their place in atoms. How many electrons are there in an atom, and how are they arranged? As atoms are electrically neutral, they must have regions of positive charge to balance the negative electons. One model of the atom (the "plum pudding" or "raisin muffin" model) postulated a uniform sphere of positive charge in which the electrons were embedded.

Like the discovery of the electron, the discovery of the nucleus in 1911 also originated from experiments with cathode ray tubes, although here the path was somewhat less direct. In 1895 Wilhelm Röntgen discovered a highly penetrating kind of radiation (called **X rays**) emitted from the bright luminescent spot where cathode rays strike the glass tube. X rays cause photographic plates to darken. Their high penetration is evidenced by the fact that they darken plates even when the plates are wrapped in black paper. They cause various substances to luminesce, and, as mentioned earlier, they are strongly ionizing radiations. (X rays were subsequently shown to be light of very short wavelength.)

The discovery of X rays led Henri Becquerel to study the luminescence of uranium compounds. In the course of his research he made the startling discovery that uranium and its compounds *spontaneously* emit very penetrating radiations. These radiations result from occurrences in the materials themselves, and do not require the sort of external stimuli that generate X rays. This phenomenon is called radioactivity. Other radioactive elements were discovered by Marie and

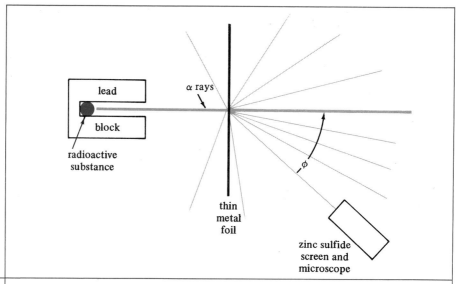

FIGURE 6.6 Measurement of number of alpha particles scattered through an angle φ.

Pierre Curie. The radiations from radioactive substances were shown to be of three types: positively charged particles called **alpha rays** (eventually shown to be rapidly moving helium nuclei); negatively charged particles called **beta rays** (these turned out to be high-speed electrons); and uncharged radiations called **gamma rays** (light of even shorter wavelength than X rays).

It was work with alpha rays that led to the discovery of the nucleus. Ernest Rutherford and Hans Geiger found that when a beam of alpha rays is directed at a thin metal foil, most of the rays pass through it without deflection, and a few are deflected (scattered) through a few degrees of angle so that they emerge from the foil traveling in a slightly different direction. The alpha rays could be observed by allowing them to strike a screen coated with zinc sulfide. On hitting the screen, each alpha particle gave off a flash of light (scintillation), observable through a microscope. The angular dependence of the scattering could then be measured by placing the screen at different angles (Figure 6.6). The rest of the story, involving a student named Ernest Marsden, was told many years later by Rutherford:

> One day Geiger came to me and said, "Don't you think that young Marsden, whom I am training in radioactive methods, ought to begin a small research?" Now I had thought that too, so I said, "Why not let him see if any alpha particles can be scattered through a large angle?" I may tell you in confidence that I did not believe that they would be, since we knew that the alpha particle was a very fast massive particle, with a great deal of energy, and you could show that if the scattering was due to the accumulated effect of a number of small scatterings the chance of an alpha particle's being scattered backwards was very small. Then I remember two or three days later Geiger coming to me in great excitement and saying, "We have been able to get some of the alpha particles coming

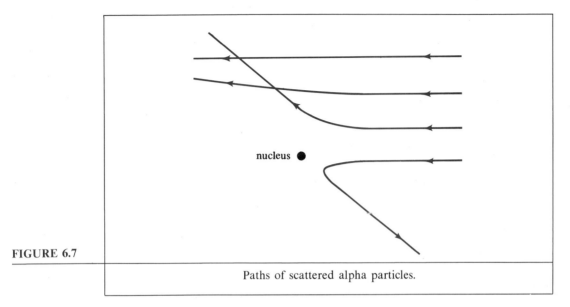

FIGURE 6.7 Paths of scattered alpha particles.

backwards. . . ." It was quite the most incredible event that has ever happened to me in my life. It was almost as incredible as if you fired a 15-inch shell at a piece of tissue paper and it came back and hit you.[5]

Although only about 1 in 10^4 alpha particles was scattered through more than 90° (emerging from the foil on the same side from which it had entered), Rutherford realized that this phenomenon could be accounted for only by assuming that the positive charge and most of the mass of the atom are concentrated in a very small scattering center or nucleus. The positive alpha particles are repelled by the positive nuclei, following paths of the type shown in Figure 6.7.

Most of the alpha particles miss the tiny nuclei by a wide margin, and they are not scattered appreciably. A few pass near the nucleus and are deflected through large angles. Still fewer score direct hits on a nucleus and bounce backward in the direction from which they came. Rutherford used his nuclear model to derive a scattering formula for the number of alpha particles (N) scattered in unit time through an angle ϕ, and he predicted that it should be directly proportional to the thickness of the foil, t, and to the square of the charge on the nucleus, Ze, and inversely proportional to the fourth power of sin ($\phi/2$) and to the fourth power of the alpha-particle velocity, u.

These points were confirmed by Geiger and Marsden in a series of careful experiments. Some of their results for the angular dependence of the scattering are listed in Table 6.3. As shown in the last column, the product of the number of scintillations and the fourth power of sin ($\phi/2$) is approximately constant as predicted by Rutherford from the nuclear theory of the atom.

[5] E. Rutherford, in *Background to Modern Science,* J. Needham and W. Pagel, eds., Cambridge University Press, London, 1938, p. 68.

TABLE 6.3	ANGULAR DEPENDENCE OF SCATTERING FROM A GOLD FOIL		
	Angle of deflection, ϕ	No. of scintillations per unit time, N	$N \times \sin^4\left(\dfrac{\phi}{2}\right)$
	15°	132,000	38.4
	22.5°	27,300	39.6
	30°	7,800	35.0
	37.5°	3,300	35.3
	45°	1,435	30.8
	60°	477	29.8
	75°	211	29.1
	105°	69.5	27.5
	120°	51.9	29.0
	135°	43.0	31.2
	150°	33.1	28.8

6.5 NUCLEAR CHARGE

The variation with nuclear charge predicted by Rutherford could not be confirmed directly, as the nuclear charge was unknown at this time, but the scattering experiments showed that Z was roughly one-half the atomic mass. It was then suggested by Antonius van der Broek that Z is identical with the ordinal position (atomic number) of the element in the periodic table.

Following Cannizzaro's solution to the problem of atomic masses in 1858, it had been found that a periodic relationship between the properties of the elements was evident when they were arranged in a certain order, which was approximately the order of increasing atomic mass. Periodic tables exhibiting this relationship were developed independently by Dmitriĭ I. Mendeleev and J. Lothar Meyer, beginning in about 1869.[6] Mendeleev's second periodic table, first published in 1871, is shown in Table 6.4, and a modern periodic table is shown on the inside front cover. Elements in the same vertical column, called a **group** and identified by a roman numeral, resemble each other in various chemical and physical properties. The observed periodicity of the properties of the elements is called the **periodic law,** and the horizontal rows of elements are known as **periods.**

In some cases it was necessary to place a heavier element ahead of a lighter element so as to have elements of similar properties in the same group. Most of the apparent inversions were caused by errors in the atomic masses, but three were not. As shown in the periodic table on the inside front cover, for elements 18 and 19, 27 and 28, and 52 and 53, the heavier element has the lower atomic number.

---

[6] In 1866 John A. R. Newlands had suggested a "law of octaves," with the first element (hydrogen) similar to the eighth (fluorine), the second (lithium) similar to the ninth (sodium), and so on. (The noble gases were unknown at this time.) Because of its musical connotations the idea was ridiculed.

6.5 NUCLEAR CHARGE

TABLE 6.4   MENDELEEV'S SECOND PERIODIC TABLE

	Группа I.	Группа II.	Группа III.	Группа IV.	Группа V.	Группа VI.	Группа VII.	Группа VIII. Переходъ къ групп. I.
Типическіе элементы.	H=1 Li=7	Be=9,4	B=11	C=12	N=14	O=16	F=19	
Первый періодъ { Рядъ 1-й	Na=23	Mg=24	Al=27,3	Si=28	P=31	S=32	Cl=35,5	
— 2-й	K=39	Ca=40	—=44	Ti=50?	V=51	Cr=52	Mn=55	Fe=56, Co=59, Ni=59, Cu=63
Второй періодъ { — 3-й	(Cu=63)	Zn=65	—=68	—=72	As=75	Se=78	Br=80	
— 4-й	Rb=85	Sr=87	(?Yt=88?)	Zr=90	Nb=94	Mo=96	—=100	Ru=104, Rh=104, Pd=104, Ag=108
Третій періодъ { — 5-й	(Ag=108)	Cd=112	In=113	Sn=118	Sb=122	Te=128?	J=127	
— 6-й	Cs=133	Ba=137	—=137	Ce=138?	—	—	—	—
Четвер. періодъ { — 7-й	—	—	—	—	—	—	—	
— 8-й	—	—	—	—	Ta=182	W=184	—	Os=199?, Ir=198?, Pt=197, Au=197
Пятый періодъ { — 9-й	(Au=197)	Hg=200	Tl=204	Pb=207	Bi=208	—	—	
— 10-й	—	—	—	Th=232	—	Ur=240	—	
Высшая соляная окись.	R²O	R²O² или RO	R²O³	R²O⁴ или RO²	R²O⁵	R²O⁶ или RO³	R²O⁷	R²O⁸ или RO⁴
Высшее водородное соединеніе.			(RH³?)	RH⁴	RH³	RH²	RH	

Source: D. I. Mendeleev, *Journal of the Russian Physico-Chemical Society*, **3**, 25 (1871).

Mendeleev found that the periodic relationships required him to leave blank spaces in the table for unknown elements. The elements directly below boron, aluminum, and silicon (which he called "eka boron," "eka aluminum," and "eka silicon") were estimated to have atomic masses of 44, 68, and 72, as shown in Table 6.4. He also predicted many of the other properties of these elements and their compounds. When the elements (named scandium, gallium, and germanium) were discovered shortly thereafter, his predictions were found to be highly accurate, and this of course strengthened confidence in the validity and usefulness of the periodic table. By the time the nucleus was discovered the periodic law was a well-developed concept.

Direct evidence for van der Broek's hypothesis that the ordinal position of the element in the periodic table is identical with Z, the number of fundamental units of electric charge on the nucleus, was obtained in 1913 by Henry G. J. Moseley from X-ray spectra. Moseley bombarded different elements with cathode rays and studied the X rays emitted. He found that each element gave several X rays, and that the X rays changed in a regular way with increasing atomic number, as shown in Figure 6.8. Moseley's results are summarized in Figure 6.9. In this figure the square root of a certain property of the X rays (the frequency) is plotted versus the atomic number for elements having atomic numbers between 13 (aluminum) and 79 (gold). (The concept of frequency will be discussed in detail in the next chapter; for the purpose of this section one may simply regard X-ray frequencies as characteristic properties of the different elements.) To obtain a

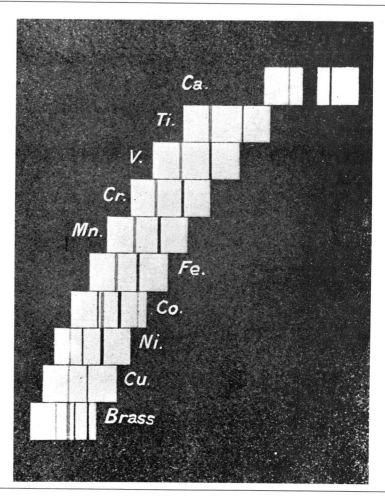

FIGURE 6.8

X-ray lines as photographed by Moseley for several neighboring elements. These are high-frequency X rays called the K series. There are extra lines in the cobalt spectrum caused by iron and nickel impurities. Brass is an alloy of copper and zinc; its spectrum includes the X-ray lines of both metals. [SOURCE: *H. G. J. Moseley, Philosophical Magazine,* [6] **26**, *1024 (1913)*.]

regular change in the square root of frequency with increasing atomic number, it was necessary to leave vacant lines for unknown elements at 43, 61, and 75. All three elements have been discovered since.[7] Moseley wrote:

> We have here a proof that there is in the atom a fundamental quantity, which

---

[7] Later work showed some errors in the assignment of atomic numbers between 66 and 72 shown in Figure 6.9. Moseley's method was used to determine the correct atomic numbers. Hafnium (atomic number 72) was discovered a few years later with the aid of the X-ray method.

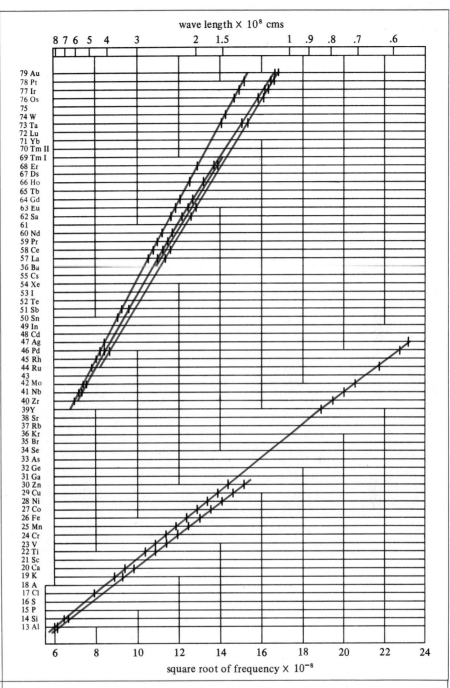

FIGURE 6.9 Moseley's diagram of X-ray frequencies. Each element emits X rays of several frequencies. There are typically two high-frequency X rays, designated Kα and Kβ. They are shown at the bottom of the figure. There are also several low-frequency X rays, of which only four have been plotted. They are designated α, β, φ, and γ of the L series, and they lie in the upper part of the figure. Some frequencies are missing because they had not been measured at the time Moseley published this work. [SOURCE: *H. G. J. Moseley,* Philosophical Magazine, [6] **27**, *703 (1914)*.]

increases by regular steps as we pass from one element to the next. This quantity can only be the charge on the central positive nucleus, of the existence of which we already have definite proof. Rutherford has shown, from the magnitude of the scattering of α [alpha] particles by matter, that this nucleus carries a + charge approximately equal to that of $A/2$ electrons, where A is the atomic weight. Barkla, from the scattering of X-rays by matter, has shown that the number of electrons in an atom is roughly $A/2$, which for an electrically neutral atom comes to the same thing. Now atomic weights increase on the average by about 2 units at a time, and this strongly suggests the view that N increases from atom to atom always by a single electronic unit.[8] We are therefore led by experiment to the view that N is the same as the number of the place occupied by the element in the periodic system.[9]

6.6 ISOTOPIC MASS

At about the time that Moseley was determining the nuclear charge, J. J. Thomson developed a method for measuring nuclear mass. Once again the method originated in experiments with cathode ray tubes. In the region between the electrodes the high-speed electrons collide with gas molecules, knocking off electrons and creating positive ions. With a perforated cathode the positive ions pass through the holes in the cathode and emerge as luminescent positive rays, as was discovered by Eugen Goldstein in 1886. The nature of the positive rays depends on the gas in the discharge tube, different gases giving different colors. Thomson studied their mass-to-charge ratios by deflecting them with electric and magnetic fields. In principle his method was similar to the deflection of cathode rays described earlier. He showed that the element neon consists of atoms of two different masses, 20 and 22, designated ^{20}Ne and ^{22}Ne. Later work showed the presence of a relatively small number of atoms of intermediate mass, ^{21}Ne.

Because of their similarity in chemical and physical properties, atoms of the same nuclear charge but different mass belong in the same place in the periodic table, hence they are called **isotopes** (Greek *isos*, same; *topos*, place). In general, isotopes have different nuclear properties. For example, ^{12}C and ^{13}C are stable, but ^{14}C is radioactive, emitting beta rays. Isotopes had been discovered somewhat earlier in studies of the radioactive elements, but Thomson's work was the first indication that stable elements contained different isotopes.

A modern analytical instrument called a mass spectrometer (Figure 6.10) operates on the same principle as Thomson's experiment. Positive ions are accelerated to high velocities and then passed through a magnetic field that deflects them so that they move along curved paths. Each curved path is an arc of a circle, which has a certain radius called the **radius of curvature** (see Figure 6.11). This radius depends on the mass (m) and charge (Q) of the ions:

1. Ions of small mass are more easily deflected than ions of large mass and

---

[8] As used here N is the same as Z.
[9] H. G. J. Moseley, *Philosophical Magazine*, [6] **26**, 1031 (1913).

6.6 ISOTOPIC MASS

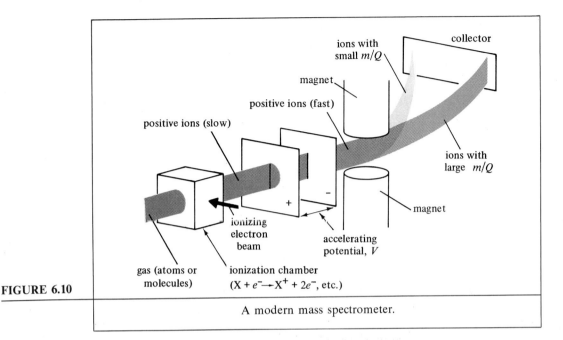

FIGURE 6.10 A modern mass spectrometer.

the same charge. For example, ^{20}Ne$^+$ has a smaller radius of curvature than ^{22}Ne$^+$.

2. The deflecting force on an ion of large Q is greater than the deflecting force on an ion of the same mass and small Q. For example, the doubly charged ion ^{20}Ne^{2+} has a smaller radius of curvature than the singly charged ion of the same mass, ^{20}Ne$^+$.

The relationship between the various quantities involved is

$$\frac{m}{Q} = \frac{H^2 R^2}{2V} \qquad (6.11)$$

where H is the magnetic field strength, R is the radius of curvature, and V is the accelerating potential. Thus the magnetic field separates the ions into rays of different m/Q. To study the relative numbers of ions having different mass-to-charge ratios, either the intensity of the magnetic field is varied or, as is more often done in practice, the accelerating potential is varied, sending the ions of different m/Q successively into the collector. The electric current due to positive ions of different m/Q is measured and recorded.

The plot of current against mass of the ions is called a **mass spectrum.** The mass spectrum of mercury is shown in Figure 6.12. The highest peak is due to the most abundant isotope, ^{202}Hg. There are smaller peaks from the isotopes of mass numbers 198, 199, 200, 201, and 204, and a very small peak at 196. (In this spectrum all ions have a single positive charge.)

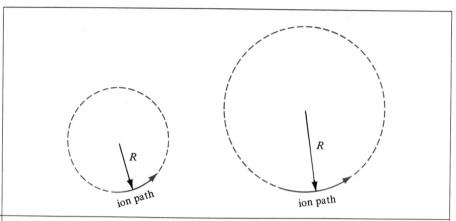

FIGURE 6.11 Radii of curvature. Left: An ion with small m/Q and small R. Right: An ion with large m/Q and large R.

The atomic masses and relative abundances of the mercury isotopes obtained from the mass spectrum are listed in Table 6.5. From these data the average atomic mass of mercury may be calculated by weighting each mass according to its abundance. Most of the atomic masses in the table on the inside back cover have been obtained by this method.

EXAMPLE 1

A beam of positive ions in a mass spectrometer contains $^{12}C^+$, $^{13}C^+$, $^{12}C^{2+}$, $^{13}C^{2+}$, $^{24}Mg^+$, and $^{24}Mg^{2+}$ ions. Which ion has the smallest radius of curvature? Which ion has the largest? Which ions have the same radius of curvature (assuming that the atomic masses are equal to the mass numbers)?

First, calculate m/Q for the different ions. Expressing m on the atomic mass scale, and Q in electronic charge units, we obtain the m/Q value written below each ion.

$^{12}C^+$	$^{13}C^+$	$^{12}C^{2+}$	$^{13}C^{2+}$	$^{24}Mg^+$	$^{24}Mg^{2+}$
12	13	6	6.5	24	12

The $^{12}C^{2+}$ ions have the smallest mass-to-charge ratio and the smallest radius of curvature. The $^{24}Mg^+$ ions have the largest mass-to-charge ratio and the largest radius of curvature. To this accuracy, the $^{12}C^+$ and $^{24}Mg^{2+}$ ions have the same m/Q and the same radius of curvature.

Molecules as well as atoms have characteristic mass spectra. Typically the spectrum has an important peak (called the "parent peak") owing to the singly charged positive ion of the molecule. In addition there are various peaks because of the fragments formed when the molecule dissociates under the impact of high-energy electrons in the ionization chamber. For example, the mass spectrum of methane, CH_4, includes peaks owing to CH_4^+, CH_3^+, CH_2^+, and CH^+. Each

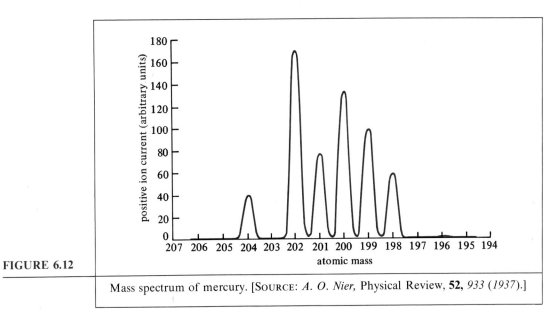

FIGURE 6.12 Mass spectrum of mercury. [SOURCE: *A. O. Nier,* Physical Review, **52**, *933 (1937)*.]

chemical substance has a characteristic mass spectrum that is highly useful in analyzing mixtures of gases for the relative amounts of the various components. Mass spectra are also useful in determining certain physical properties of molecules, such as the energy required to break a particular chemical bond. For these reasons the mass spectrometer is an important instrument in the modern chemical laboratory.

TABLE 6.5

COMPOSITION OF NATURALLY OCCURRING MERCURY		
Mass number	Relative abundance (%)	Atomic mass
196	0.146	195.96582
198	10.02	197.96677
199	16.84	198.96826
200	23.13	199.96834
201	13.22	200.97031
202	29.80	201.97063
204	6.85	203.97348

PROBLEMS

6.1 How many neutrons, protons, and electrons are there in each of the following?
(a) $^{16}_{8}O$
(b) $^{18}_{8}O^{2-}$
(c) $^{35}_{17}Cl^{-}$
(d) $^{3}_{1}H^{+}$
(e) $^{226}_{88}Ra^{2+}$

6.2 Which atom or ion has the following numbers of particles?
   (a) 15 protons, 16 neutrons, and 15 electrons
   (b) 26 protons, 30 neutrons, and 23 electrons
   (c) 35 protons, 44 neutrons, and 36 electrons

6.3 Fill in the blanks in the following table:

Atom	Protons	Neutrons	Electrons
$^{4}_{2}He$			
$^{23}_{11}Na^+$			
	22	26	22
	22	28	22
	22	26	18

6.4 (a) Compare the modern values of the atomic masses of hydrogen, chlorine, and iodine, and one-half the atomic masses of oxygen, lead, and tin, with Faraday's results for the relative masses produced by the same quantity of electricity in an electrolysis experiment.
   (b) Suggest an explanation for the fact that a given quantity of electricity produces 1 mol of atoms of some elements and only $\frac{1}{2}$ mol of atoms of some other elements.

6.5 How would each of the following changes alter the deflection of a beam of electrons by an electric field (Figure 6.4)? (State whether the deflection would remain the same, increase, or decrease, and by what factor.)
   (a) doubling the voltage applied to P and P'
   (b) halving the length of the deflecting plates P and P'
   (c) halving the distance between plates P and P'
   (d) doubling the velocity of the electrons

6.6 With a magnetic field in place of an electric field, how would each of the following changes affect the force on a moving electron:
   (a) doubling the magnetic field
   (b) halving the velocity of the electron

6.7 If, in a Millikan-type experiment, you observed the following velocities of an oil drop, what multiples of e would you assign for the charge on the oil drop for each velocity?

u (m s^{-1})	Applied voltage	n (where $Q = ne$)
3.00×10^{-4}	On	
-1.10×10^{-4}	Off	
5.05×10^{-4}	On	
9.15×10^{-4}	On	

6.8 Explain qualitatively why the number of alpha particles scattered at a particular angle ϕ in unit time should increase when
(a) the thickness of the foil is increased
(b) the charge on the nucleus is increased
(c) the alpha particles are moving more slowly

6.9 Compare the modern values of the atomic masses of scandium, gallium, and germanium with Mendeleev's estimates.

6.10 The highest frequency (in hertz) of X rays emitted by elements upon bombardment with cathode rays are as follows:

Sulfur (Z = 16), 0.56×10^{18}
Chromium (Z = 24), 1.31×10^{18}
Germanium (Z = 32), 2.38×10^{18}
Cadmium (Z = 48), 5.57×10^{18}
Gadolinium (Z = 64), 10.2×10^{18}

Prepare the following plots:
(a) frequency against atomic number
(b) square root of frequency against atomic number
What frequency would you predict for the element of atomic number 43? (In working this problem, consider that X-ray frequency is simply a characteristic property of each element. The concept of frequency will be explained in the next chapter.)

6.11 In a mass spectrometer, predict the effect on the radius of curvature of an ion of each of the following:
(a) doubling the magnetic field strength
(b) doubling the accelerating potential
(c) doubling the mass of the ion
(d) doubling the charge on the ion
(e) changing the sign of the charge on the ion but not its magnitude

6.12 A beam of positive ions in a mass spectrometer contains $^6Li^+$, $^6Li^{2+}$, $^7Li^+$, $^7Li^{2+}$, $^{12}C^+$, and $^{12}C^{2+}$ ions. Arrange these ions in order of increasing radius of curvature. (Assume that the atomic masses are equal to the mass numbers.)

6.13 How many electrons must be removed from a ^{48}Ti atom to give a positive ion having the same radius of curvature as (a) a $^{16}O^+$ ion, (b) a $^{24}Mg^+$ ion, (c) a $^{24}Mg^{2+}$ ion?

6.14 Magnesium contains 78.70% of an isotope of atomic mass 23.98504, 10.13% of an isotope of atomic mass 24.98584, and 11.17% of an isotope of atomic mass 25.98259. Calculate to the correct number of significant figures the average atomic mass of a sample of magnesium.

6.15 Calculate the average atomic mass of mercury from the data given in Table 6.5.

6.16 The juvenile hormone (JH), secreted by insects, is a substance that controls their growth and development. Its simplest formula (determined by combustion analysis) is $C_6H_{10}O$. A sample of it is placed in a mass spectrometer, and its positive ion (JH$^+$) is found to have an m/Q value 21.0 times larger than that of a ^{14}N$^+$ ion. What are the molecular mass and molecular formula of the juvenile hormone?

6.17 (a) From the data in Section 6.1, calculate the ratio of the atomic radius to the nuclear radius of a ^{12}C atom.
  (b) Calculate the ratio of the atomic volume to the nuclear volume of a ^{12}C atom. (The volume of a sphere of radius R is $\frac{4}{3}\pi R^3$.)

6.18 (a) From the mass of an electron (Table 6.1) and the mass of a ^{12}C atom (exactly 12), calculate the mass of a ^{12}C nucleus on the atomic mass scale.
  (b) Calculate the percentage of the mass of a ^{12}C atom that is contained in the nucleus.
  (c) From this result and the answers to Problem 6.17, calculate the ratio of the nuclear density to the atomic density of a ^{12}C atom.

CHAPTER 7
PARTICLES AND WAVES

7

To describe the structure of atoms and molecules requires that we consider the behavior of electrons. How do they move? How do they interact with nuclei? The laws governing the motion of objects of ordinary size (i.e., visible to the naked eye or with an optical microscope) were discovered in the seventeenth century by Isaac Newton. It was naturally assumed that very small objects—atoms, molecules, and the particles of which they are composed—obey the same laws. This assumption was rudely shaken in 1900 by the investigations of Max Planck. To show how Planck was led to his revolutionary discovery we must consider the wave theory of light and how it developed. But first a review of the general concept of wave motion may be helpful.

7.1 WAVE MOTION

The most familiar waves are those that travel across the surface of a body of water disturbed by the wind, tide, and ships. If the disturbance is moderate, the waves have the general shape shown in Figure 7.1. The section of a wave from one crest to the next is called a **cycle.** The distance from one crest to the next is known as the **wavelength,** λ (lowercase Greek letter lambda). It is measured in meters per cycle. (The "per cycle" is often omitted.) The height of a crest above the average water level is called the **amplitude.**

Figure 7.1 represents an instantaneous view of a wave that is traveling in the x direction. If we let the speed of the wave motion be u, then each crest and each trough moves u m in the x direction per second. The number of crests passing a fixed point in one second is known as the **frequency,** ν (lowercase Greek letter nu). It is measured in hertz (abbreviated Hz), with dimensions of cycles per second (often abbreviated as s^{-1}). Speed, frequency, and wavelength are related, because the distance traveled per second is equal to the distance per cycle times the number of cycles per second.

$$u(\text{m s}^{-1}) = \lambda(\text{m cycle}^{-1}) \, \nu(\text{cycles s}^{-1}) \tag{7.1}$$

Another term used to describe wave motion is the **period** of a wave—the interval of time between passage of two successive crests of the wave. The period of a wave is the reciprocal of its frequency.

The speed of wave motion depends on the type of wave and the medium

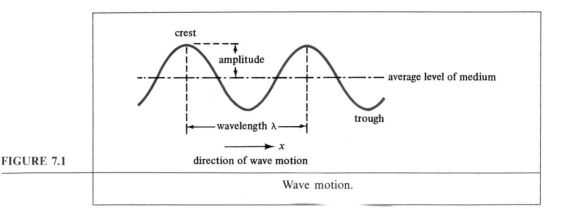

FIGURE 7.1 Wave motion.

through which the wave travels. Thus ocean waves, sound waves in air, and light waves all move with different speeds.

EXAMPLE 1

The range of sound frequencies audible to human beings varies with age and the individual but is approximately 15 Hz to 20,000 Hz. Given that the speed of sound in air is 340 m s^{-1}, calculate the wavelength of a sound wave whose frequency is 20,000 Hz.

From Eq. 7.1 we have

$$u = \lambda \nu \quad \text{or} \quad \lambda = \frac{u}{\nu}$$

Therefore

$$\lambda = \frac{340 \text{ m s}^{-1}}{20,000 \text{ s}^{-1}} = 1.70 \times 10^{-2} \text{ m} \quad \text{or} \quad 1.70 \text{ cm}$$

One of the most interesting and important phenomena observed with waves is that of **interference,** which occurs when waves generated by two or more different sources come together. Figure 7.2 shows the interference pattern formed by waves from two wave sources in a water tank.

Studies of wave interference have shown that the height of the surface at a particular point where two waves meet may be calculated in a simple way from their heights—one just adds them together. At those points where each wave taken separately would produce a crest of height a at the same time, the two waves together produce a crest of height $2a$. Similarly, where each wave would produce a trough of depth $-a$ at the same time, they generate an even deeper trough of depth $-2a$. At these points we say that the waves are in step or in phase and that there is **constructive interference** (Figure 7.3).

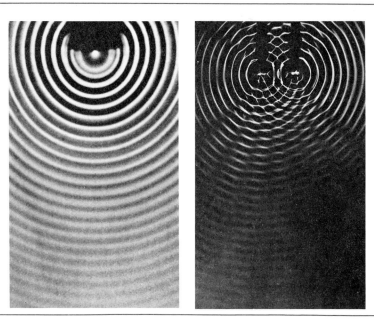

FIGURE 7.2

Left: Photograph of water waves spreading out from a vibrating point source near the top of the figure. Wave crests appear white, troughs black. Right: Photograph of interference pattern produced by two point sources (near the top of the figure) vibrating in phase. (SOURCE: The Project Physics Course, F. J. Rutherford, G. Holton, and F. G. Watson, directors, Holt, Rinehart and Winston, Publishers, New York, © 1970, Unit 3, p. 115.)

When the waves are out of step or out of phase they cancel each other—this is called **destructive interference.** At a point where one wave would produce a crest of height a at the same time the other wave produced a trough of depth $-a$, the net result is zero (Figure 7.3). Constructive interference at some points and destructive interference at others produce the interference pattern or diffraction pattern shown in Figure 7.2. (Diffraction and interference are essentially synonymous. The term **interference** is commonly used when there are only two wave sources, as in the experiment described here; the term **diffraction** is used when there are more than two wave sources.)

7.2 LIGHT WAVES

Two rival theories of light originated in the seventeenth century to explain the properties of light that was known to travel in straight lines. Isaac Newton advanced the concept that light consisted of a stream of particles or "corpuscles." A contemporary, Christiaan Huygens, proposed a wave theory to explain the same phenomena that Newton explained using his corpuscular theory. Huygens's the-

7.2 LIGHT WAVES

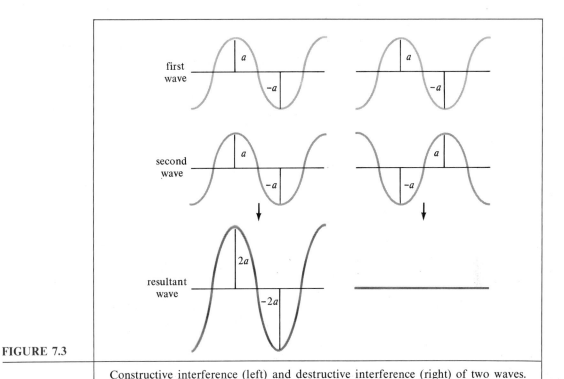

FIGURE 7.3 Constructive interference (left) and destructive interference (right) of two waves.

ory pictured light as a stream of waves similar to the water waves described in the preceding section.

During the nineteenth century numerous experiments on light led to a decision in favor of the wave theory. One of these experiments was done by Thomas Young, who found that when a light beam passes simultaneously through two adjacent pinholes, a band of alternating light and dark regions is produced on a screen behind the pinholes, that is, an interference pattern is formed (see Figure 7.4).

The explanation of the interference pattern by the wave theory of light is as follows (see Figure 7.5). Point B in the center of the screen is the same distance from either pinhole, P or P', and therefore light waves striking the edges of the pinholes and scattered toward B reach it in step or in phase (e.g., both waves at a crest). Two waves in phase reinforce each other (constructive interference), resulting in a region of brightness at the center of the screen. On either side of B, one wave must travel farther than the other. At points D and D', where the extra distance traveled is just one-half of a wavelength, the two waves are exactly out of step or out of phase (e.g., one at a crest and the other at a trough). Two waves out of phase cancel each other (destructive interference), resulting in darkness at D and D'. Farther from the center, at points B' and B'', the extra distance traveled is

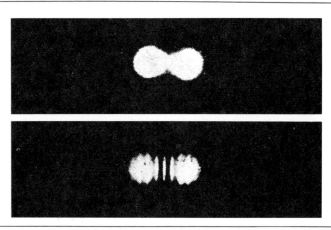

FIGURE 7.4 Interference experiment. The lower photograph shows the interference pattern produced by light passing through two adjacent pinholes simultaneously. The upper photograph shows only two light spots when the two beams passed through the pinholes at different times. (One pinhole was opened, then covered, and the other pinhole was opened.) (SOURCE: *A. Einstein and L. Infeld,* The Evolution of Physics, *Simon & Schuster, New York, 1938, Plate II; photographed by V. Arkadiev.*)

one wavelength, so that the waves are in phase again, illuminating B' and B''. The argument is readily extended to points still farther from the center, and thus the observed interference pattern is explained. Furthermore, the pattern changes when the pinholes are placed closer together or farther apart, or when the screen is moved closer or farther from the pinholes, and these changes agree quantitatively with the wave theory. From the observed interference pattern and the distance between the holes, Young was able to make the first determination of the wavelength of light.

It should be emphasized that, although the wave theory provides a good explanation for the interference patterns first observed by Young, the waves themselves were not observed. (Figure 7.4 is an observation, and Figure 7.5 is a drawing that explains the observation.) At the same time, the observed interference pattern is regarded as experimental evidence supporting the wave theory.

The nature of light waves was a mystery until, in the latter part of the nineteenth century, James Clerk Maxwell proposed the electromagnetic theory of light, namely, that *light waves are alternating electric and magnetic fields*. Experimental confirmation of Maxwell's theory was obtained by Heinrich R. Hertz, who found a method for generating electromagnetic waves, a discovery on which radio, television, and radar transmission is based. (The unit of frequency is named after Hertz.) Visible light is only a small portion of the spectrum of electromagnetic radiation, which includes radio and microwaves, infrared, visible, and ultraviolet light as well as X and gamma rays, as shown in Figure 7.6.

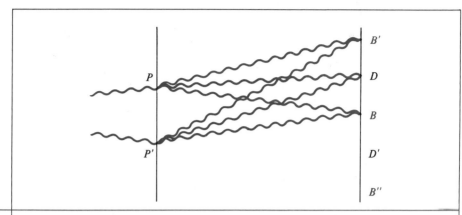

FIGURE 7.5

Wave theory of Young's interference experiment. Light falling on the left screen passes through pinholes P and P', producing an interference pattern on the right screen. Bright bands appear at B, B', and B''; at D and D' there is darkness.

The speed of light (c) of any frequency in a vacuum is 3.00×10^8 m s^{-1}. Equation 7.1 may be rewritten for light as

$$c = \lambda \nu \tag{7.2}$$

EXAMPLE 2

Calculate the wavelength (in ångströms) of blue light of frequency 6.5×10^{14} Hz. Note that $1\ \text{Å} = 10^{-10}$ m.

From Eq. 7.2, $c = \lambda \nu$; therefore

$$\lambda = \frac{c}{\nu} = \frac{3.00 \times 10^8\ \text{m s}^{-1}}{6.5 \times 10^{14}\ \text{s}^{-1}}$$

$$= 4.6 \times 10^{-7}\ \text{m} \quad \text{or} \quad \frac{4.6 \times 10^{-7}\ \text{m}}{10^{-10}\ \text{m (Å)}^{-1}} = 4.6 \times 10^3\ \text{Å}$$

7.3 ENERGY QUANTA

The discoveries about light, described in the preceding section, provided a highly successful theory of light. It was in complete agreement with experiment except for one seemingly minor point—the theory could not explain the experimental data on the light emitted by a hot furnace.

Solid objects absorb and emit light of various frequencies. We will be concerned with the **intensity** of the light, defined as the radiant energy emitted per unit area per second. (Light intensity is conveniently measured with an exposure meter or light meter of the type used in photography.) Studies of light emitted by an object show that the intensity of light emitted at various frequencies (called the

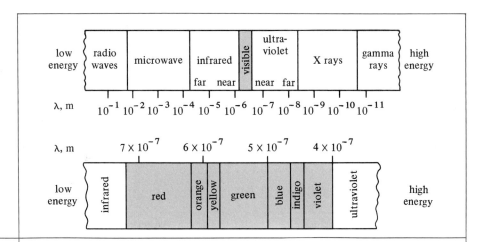

FIGURE 7.6 Wavelengths of electromagnetic radiation. The upper diagram is the entire spectrum; in the lower diagram the visible region has been magnified to show wavelengths of different colors.

intensity distribution) depends markedly on the temperature. At ordinary temperatures, practically all of the emitted light is in the infrared region, but as the temperature is increased the object glows, first with a dull red color (e.g., coals in a fireplace), changing at still higher temperature to a brilliant bluish-white (e.g., the carbon anode of a searchlight).

A furnace with a large central cavity is convenient for studying this radiation (Figure 7.7). The interior surface of the furnace is composed of vibrating atoms that emit and absorb light, so that the cavity becomes filled with light of various frequencies. This light is called cavity radiation or **blackbody radiation.** A small hole in the furnace permits the escape of some of the light, which can be resolved into its various frequencies by passing it through a prism. The light meter can then be placed in a position to determine the intensity at various frequencies.

Some results obtained in the latter part of the nineteenth century are shown in Figure 7.8. Unlike the intensity distribution of light emitted by flames and gas-discharge tubes, the intensity distribution of blackbody radiation does not depend on the material of which the furnace is constructed, but only on its temperature. This invariant quality indicates a phenomenon of fundamental importance. For a given temperature the intensity increases with increasing frequency up to a maximum value, and then drops off as the frequency is further increased. At a higher temperature the intensity is greater at all frequencies, and the point of maximum intensity is shifted toward higher frequency. (This is why the color of a hot object shifts toward the blue end of the spectrum as the temperature rises.)

These intensity distribution curves are somewhat reminiscent of the distribution of molecular energies in the kinetic theory of gases, but application of methods similar to those that had so successfully explained gas properties met with only partial success. At low frequencies the agreement between experiment

7.3 ENERGY QUANTA

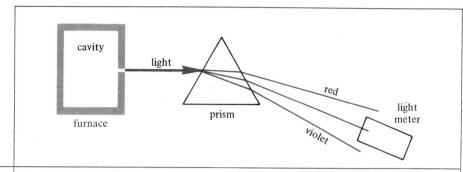

FIGURE 7.7 Experimental determination of the intensity of blackbody radiation at various frequencies.

and theory was good, but, instead of reaching a maximum, the theoretical curves continued to increase with increasing frequency into the ultraviolet region. It was to this conflict between theory and experiment, called the "ultraviolet catastrophe," that Planck directed his attention.

Planck was able to obtain theoretical curves in good agreement with experiment but only with the aid of a radically new hypothesis. He assumed that the atoms in the solid walls of the furnace vibrate with various frequencies and that *an atom oscillating with a particular frequency, ν, can emit or absorb light of that same frequency only.* Furthermore, he assumed that it does so *only* in a discrete unit or *quantum* of energy, and that the amount of energy associated with the quantum is directly proportional to the vibrational frequency. Thus *the change in energy of the vibrating atom is given by the equation*

$$\Delta E = h\nu \tag{7.3}$$

where h is the proportionality constant, now called Planck's constant. The most accurate measurements give a value for h of $6.626176(36) \times 10^{-34}$ J s, where the last two digits are uncertain.

Planck's assumptions require that only certain energies of oscillation are allowed (see Figure 7.9). From its lowest energy state the vibrating atom can change to the next quantum state by absorbing energy $h\nu$, either as light or by collision with other atoms. It can then either undergo a transition to the next higher state or else emit light of energy $h\nu$, and thereby return to the state of lowest energy. Thus the oscillator is restricted to certain characteristic energies, called **energy levels**, as indicated in Figure 7.9. An atom in one of these energy levels is said to be in a **stationary state**. The lowest state is called the **ground state**; the others are **excited states**. Transitions from one state to another are indicated by arrows. The different states can be labeled by a vibrational quantum number, v. By convention, $v = 0$ for the ground state, $v = 1$ for the first excited state, and so on. (Do not confuse the quantum number v with the vibrational frequency ν.)

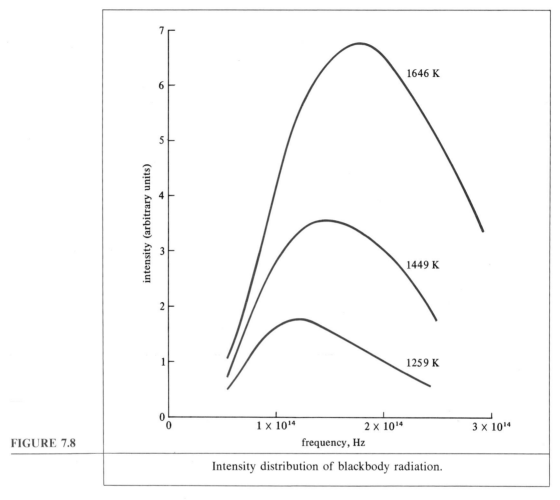

FIGURE 7.8

Intensity distribution of blackbody radiation.

It was later found that the energy of the ground state is $\frac{1}{2}h\nu$, and thus the allowed energies are

$$E = (v + \tfrac{1}{2})h\nu \quad \text{where } v = 0 \text{ or } 1 \text{ or } 2 \text{ or } 3, \text{ etc.} \tag{7.4}$$

This equation is applicable to the vibrational motion of atoms in molecules as well as in solids.

The lowest possible energy of a vibrating atom, $\frac{1}{2}h\nu$, is called its **zero-point energy,** because the atoms in a solid continue to vibrate with this much energy even at very low temperatures (approaching the zero point on the absolute temperature scale).

For vibrating systems of ordinary size (e.g., a mass attached to a spring) the energy levels given by Eq. 7.4 are so close together that in effect all energies are possible, and quantization of the energy cannot be detected. It is only for systems

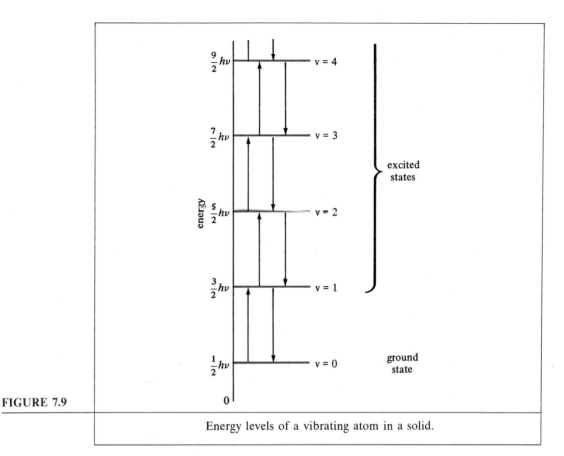

FIGURE 7.9 Energy levels of a vibrating atom in a solid.

of very small mass (atoms and molecules) that quantization of the energy is important.

EXAMPLE 3

The atoms in a certain diatomic molecule vibrate with a frequency of 1.0×10^{14} Hz. When this molecule drops from the first excited state to the ground state, what is the frequency and energy (per molecule and per mole) of the light emitted? What is the energy of the second excited state?

By Planck's fundamental postulate, the frequency of the light emitted is the same as the vibrational frequency, 1.0×10^{14} Hz. The energy of the quantum of light emitted is given by Eq. 7.3.

$$\Delta E = h\nu = (6.626 \times 10^{-34} \text{ J s})(1.0 \times 10^{14} \text{ s}^{-1})$$
$$= 6.6 \times 10^{-20} \text{ J}$$

This is the energy emitted when a single molecule changes its state. To obtain the energy change per mole, we must multiply by Avogadro's number.

$$\Delta E = (6.6 \times 10^{-20} \text{ J molecule}^{-1})(6.02 \times 10^{23} \text{ molecules mol}^{-1})$$
$$= 40 \text{ kJ mol}^{-1}$$

The energy of the second excited state of the molecule may be obtained using Eq. 7.4 with $v = 2$.

$$E = (v + \tfrac{1}{2})h\nu = (2.5)(6.626 \times 10^{-34} \text{ J s})(1.0 \times 10^{14} \text{ s}^{-1})$$
$$= 1.7 \times 10^{-19} \text{ J}$$

The connection between energy and frequency expressed by Planck's equation explained why the intensity curves in Figure 7.8 did not continue to increase with increasing frequency into the ultraviolet region. The explanation is as follows. For an atom vibrating with very high frequency the energy change given by Eq. 7.3 is very large. As a result it is extremely difficult to excite a high-frequency oscillation. Because an atom vibrating with high frequency is seldom excited to the $v = 1$ state, it seldom drops back down to the ground state (emitting high-frequency light). In fact this event occurs so rarely that only a small fraction of blackbody radiation is in the ultraviolet region, as found experimentally (Figure 7.8). Thus Planck's hypothesis averted the ultraviolet catastrophe. Calculated curves agreed with the experimental data at all frequencies. It should be emphasized, however, that this result was achieved only by assuming that atoms obey laws different from classical mechanics (the laws discovered by Newton and others from studies of the behavior of large objects). The century of the quantum had begun.

7.4 PHOTONS

Planck's hypothesis of energy quantization extended only to the act of emission or absorption of light. A few years later (1905), Albert Einstein suggested that light itself is quantized, so that a change from a higher to a lower energy level results in a bulletlike unit or particle of light, called a **photon**. The energy of a photon is

$$E = h\nu \tag{7.5}$$

Transition from a lower to a higher energy level occurs when a photon is absorbed.

Einstein arrived at the photon concept of light through his analysis of experimental results on the **photoelectric effect**. When a light shines on a metal surface, electrons are ejected from the metal, provided that the frequency of the light exceeds some minimum or threshold frequency, ν_0, characteristic of the metal. As the frequency of the light is increased above the threshold, the kinetic energy of the electrons increases linearly (see Figure 7.10). Thus electrons acquire more kinetic energy from blue light than from green light. For a given frequency the

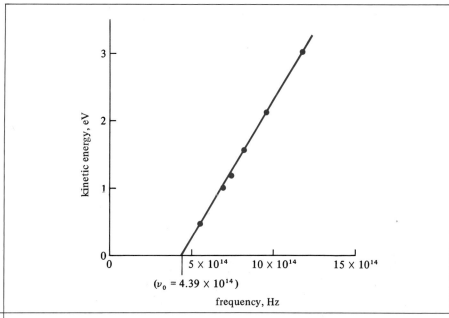

FIGURE 7.10 The kinetic energy of electrons ejected from metallic sodium by light of different frequencies. (1 eV = 1.60×10^{-19} J.)

number of electrons emitted per second (i.e., the photoelectric current) increases with increasing light intensity, but the light intensity does not affect the kinetic energy of the electrons. Thus a bright blue light gives a larger electric current than a dim blue light, but the electrons are emitted with the same kinetic energy.

These results are readily understood if it is assumed that each electron is ejected from the metal by a photon of energy $h\nu$. Light of higher frequency consists of photons of higher energy and therefore it ejects electrons of higher kinetic energy. A more intense light of the same frequency consists of more photons of the same energy and therefore it ejects more electrons having the same kinetic energy.

Part of the photon energy is required to overcome the attractive forces of the metal for the electron; the remainder goes into the kinetic energy of the electron. The principle of conservation of energy requires that

$$h\nu = w + E_k \tag{7.6}$$

where w is the energy required to remove an electron from the metal surface, and E_k is the kinetic energy of the electron. Therefore

$$E_k = h\nu - w$$

At the threshold frequency, ν_0, the energy of the photon is just sufficient to overcome the attractive forces of the metal, so that $h\nu_0 = w$. Thus

$$E_k = h\nu - h\nu_0 = h(\nu - \nu_0) \tag{7.7}$$

This equation fits the experimental data in Figure 7.10, and the slope of the line gives Planck's constant, h. The first accurate measurements of h were obtained in this way.[1] The numerical value of h obtained in photoelectric studies was in good agreement with the value obtained from experimental data on blackbody radiation, considering the limited accuracy of the latter data.

The photoelectric effect is not limited to metals but also occurs with other solids, liquids, and gases. It is an important source of information on the energies required to remove electrons from various substances. It also has many other applications (e.g., sound motion pictures, television, light meters, and door openers).

Another phenomenon that can be explained only in terms of photons was discovered in 1923 by Arthur H. Compton. He found that when X rays collide with electrons the frequency of the X rays drops, and the decrease in frequency is exactly what would be expected from a collision between a photon and an electron. Part of the momentum and energy of the photon is transferred to the electron, just like a collision between two billiard balls. Because the energy of the photon is decreased by the collision and because for a photon $E = h\nu$, the frequency is also decreased.

According to Einstein's theory of relativity, the momentum of a photon, p, is given by

$$p = \frac{h}{\lambda} \tag{7.8}$$

where λ is the wavelength of the light.[2] Equation 7.8, together with the laws of conservation of energy and conservation of momentum, led to a quantitative explanation of the experimental results on the frequency changes accompanying the scattering of X rays (now called the **Compton effect**).

7.5 WAVE PROPERTIES OF ELECTRONS

Newton's corpuscles had reappeared as Einstein's photons, but this time with strong experimental evidence supporting their claim to existence. And yet other

---

[1] It should be mentioned that at the time of Einstein's analysis of the photoelectric effect only relatively crude experimental data were available. More accurate data (such as those shown in Figure 7.10) confirming his interpretation were obtained by Robert A. Millikan in 1914–1916.

[2] From Planck's equation the energy of the photon is $E = h\nu$. Einstein's equation relating mass and energy is $E = mc^2$, where m is the mass of the photon and c is the velocity of light. Thus $mc^2 = h\nu$ or $mc = h\nu/c$. But mc is the momentum of the photon, p, and thus $p = h\nu/c$. The frequency, wavelength, and velocity of light are related by Eq. 7.2 ($c = \lambda\nu$). Therefore $p = h\nu/c = h/\lambda$.

7.5 WAVE PROPERTIES OF ELECTRONS

experimental results, such as interference patterns, could be explained only by the wave theory. Furthermore, there seemed to be an inherent contradiction in the theory of photons. To obtain the energy and momentum of a photon (particle properties), one calculates them using Eqs. 7.5 and 7.8 from frequency and wavelength (wave properties).

What, then, is light—a beam of particles or a kind of wave motion?

This strange state of affairs suggested to Louis de Broglie that wave properties might be associated with what had heretofore been regarded as a kind of particle—the electron. An electron of mass m and velocity u has a momentum $p = mu$. If Eq. 7.8 applies to electrons as well as to photons, then $mu = h/\lambda$ or

$$\lambda = \frac{h}{mu} \tag{7.9}$$

This relationship, proposed in 1924, gives what is now called the **de Broglie wavelength** of an electron.

EXAMPLE 4

What is the kinetic energy and the de Broglie wavelength of an electron with a speed of 1.20×10^6 m s^{-1}?

The translational kinetic energy of any object is given by

$$E_k = \tfrac{1}{2}mu^2$$

Thus for an electron of mass 9.110×10^{-31} kg (Appendix B) and a speed of 1.20×10^6 m s^{-1}, the kinetic energy is

$$E_k = (\tfrac{1}{2})(9.110 \times 10^{-31} \text{ kg})(1.20 \times 10^6 \text{ m s}^{-1})^2$$
$$= 6.56 \times 10^{-19} \text{ kg m}^2 \text{ s}^{-2} \quad \text{or} \quad 6.56 \times 10^{-19} \text{ J}$$

The de Broglie wavelength of the electron may be obtained from Eq. 7.9:

$$\lambda = \frac{h}{mu} = \frac{6.626 \times 10^{-34} \text{ J s}}{(9.110 \times 10^{-31} \text{ kg})(1.20 \times 10^6 \text{ m s}^{-1})}$$
$$= \frac{6.06 \times 10^{-10} \text{ kg m}^2 \text{ s}^{-2} \text{ s}}{\text{kg m s}^{-1}} = 6.06 \times 10^{-10} \text{ m} \quad \text{or} \quad 6.06 \text{ Å}$$

Experimental confirmation of Eq. 7.9 was obtained shortly after de Broglie proposed it by Clinton J. Davisson and Lester H. Germer and, independently, by George P. Thomson, J. J. Thomson's son. They obtained interference or diffraction patterns of electrons scattered from various metals, and the wavelengths calculated from the observed diffraction patterns agreed closely with the values calculated from Eq. 7.9. Electron diffraction and the electron microscope are today important experimental tools for structural investigations making direct use of the wave properties of electrons. Neutron diffraction has also become important in recent years. (It should be emphasized that, although interference patterns of de Broglie waves have been observed many times, the de Broglie waves themselves are not observable. The situation here is similar to that pointed out for light waves in Section 7.2.)

Thus not only light but electrons and neutrons and, in fact, all forms of matter have both wave properties and particle properties. This characteristic of nature is called **wave-particle duality.** It can be interpreted in various ways. One way is to note that both particle and wave concepts are borrowed from our experience with objects of ordinary size (baseballs, ocean waves, etc.). To deal with matter on a very small scale neither concept is sufficient by itself, but an understanding of the nature of matter can be obtained if both concepts are used, the wave properties being manifested in some experiments, and the particle properties in others.

A closely related view holds that light, electrons, neutrons, and the like are particles whose motion is controlled by wavelike laws, somewhat analogous to the influence of an ocean wave on the motion of the water molecules. In the remainder of this chapter we consider a very simple system—a single particle confined to a box—and see how the behavior of the particle is related to its de Broglie wavelength. Besides having some important applications, the particle in a box illustrates the essential features of the quantum theory in a relatively simple way.

7.6 STANDING DE BROGLIE WAVES FOR A PARTICLE IN A BOX

The waves associated with a beam of light or a stream of electrons are traveling waves, analogous to ocean waves. *When a particle is confined to a particular region of space, the appropriate waves are* **standing waves** analogous to the vibrations of a plucked string in a guitar or other stringed instrument. The simplest vibrational mode of a string stretched between two fixed points is called the **fundamental** or **first harmonic** (see Figure 7.11). All parts of the string move up and down in phase. After the string reaches a maximum distance, or amplitude (the solid curve), it collapses through a position where it is straight (dashed line) to the opposite configuration (dashed curve) and then it rises and repeats the sequence. The more complicated wave patterns of the second and third harmonics are also shown in Figure 7.11. In the second harmonic, the left half of the string moves up while the right half moves down, and vice versa. In the third harmonic, the left and right thirds move in the same direction, while the middle third moves in the opposite direction.

The points along the string that do not move at all are called **nodes.** The fixed ends of the string are of course nodes in every vibrational mode. In the second harmonic, there is one node halfway between the ends. In the third harmonic, there are two nodes between the ends, and in the nth harmonic there are $n - 1$ nodes between the ends of the string. The distance between adjacent nodes is just one-half of a wavelength ($\lambda/2$). Thus the condition for setting up standing waves is that there must be an integral number of half wavelengths in the length of the string, a, or

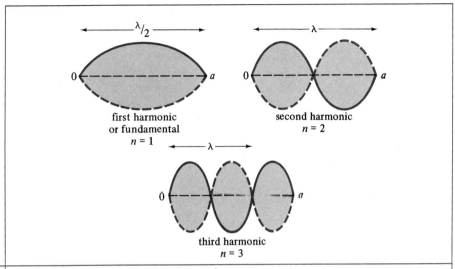

FIGURE 7.11 Standing waves in a string or standing de Broglie waves of a particle in a box.

$$\frac{n\lambda}{2} = a \quad \text{where } n = 1, 2, 3, \text{ etc.} \tag{7.10}$$

Let us consider the behavior of a particle of mass m confined to a box of length a. To simplify the problem, we first consider motion in one dimension only. If we assume that the standing de Broglie waves associated with the motion of the particle are analogous to the various vibrational modes of a string,[3] we may combine Eqs. 7.8 and 7.10. Solving each equation for λ, we obtain

$$\lambda = \frac{h}{p} = \frac{2a}{n}$$

or

$$p = \pm \frac{nh}{2a} \quad n = 1, 2, 3, \text{ etc.} \tag{7.11}$$

The momentum is, therefore, $\pm h/2a$ or $\pm 2h/2a$ or $\pm 3h/2a$, and so on. (The de Broglie wavelength depends only on the magnitude of the momentum, not on its direction. Both positive and negative signs are included to indicate that motion in either direction is possible.)

From the momentum we can obtain the kinetic energy, because

$$E_k = \frac{1}{2} mu^2 = \frac{(mu)^2}{2m} = \frac{p^2}{2m} \tag{7.12}$$

---

[3] A rigorous quantum-mechanical treatment (which requires rather sophisticated mathematics) shows that this analogy is an appropriate one and that the results presented here are correct.

If we combine Eqs. 7.11 and 7.12, then

$$E_k = \frac{n^2 h^2}{8ma^2} \qquad n = 1, 2, 3, \text{etc.} \tag{7.13}$$

According to Eq. 7.13 the kinetic energy of the particle may have only certain discrete values. The allowed energy levels obtained from this equation for the four states of lowest energy are shown in Figure 7.12. *It may be seen that the requirement of standing de Broglie waves has led directly to a quantum number n and to quantization of the energy.* Note that the spacing between adjacent levels increases as n increases. In units of $h^2/8ma^2$, the differences between adjacent levels are 3, 5, 7, and so on. This pattern of energy levels may be compared with that of a vibrating atom (Figure 7.9), where the levels are equally spaced.

For a particle in a three-dimensional box there are de Broglie waves extending in each direction, and the corresponding energy expression is the sum of three terms such as Eq. 7.13:

$$E_k = \frac{h^2}{8m}\left(\frac{n_x^2}{a^2} + \frac{n_y^2}{b^2} + \frac{n_z^2}{c^2}\right) \qquad \text{where} \quad \begin{cases} n_x = 1, 2, 3, \ldots \\ n_y = 1, 2, 3, \ldots \\ n_z = 1, 2, 3, \ldots \end{cases} \tag{7.14}$$

Here a, b, and c are the dimensions of the box, and n_x, n_y, and n_z are three quantum numbers, one for each dimension. Each quantum number is an integer. Note that the lowest value of each quantum number is one; for a vibrating atom, the lowest value of the quantum number is zero.

The results obtained in this section form part of the basis of the quantum theory of gases. For many properties of a gas, such as the pressure, the quantum theory gives results identical with the kinetic theory of gases described in Chapter 4, but certain properties (in particular, entropy, discussed in Chapter 15) can be calculated only with the aid of Eq. 7.14. The agreement between calculated values and experimental data for the entropies of various gases is evidence for the validity of this equation. The quantum theory of electrons in metals (discussed in Chapter 17) is also based on the model of a particle in a box.

7.7 THE HEISENBERG UNCERTAINTY PRINCIPLE

Although we have expressions for the momentum and energy of the particle in its various stationary states, it is of interest to learn in more detail what the particle is doing, for example, what path it is following. At first glance this looks like a simple enough problem—just devise an experiment to determine the position of the particle at various times.

The exact nature of an observation may seem relatively trivial, but it has turned out to be quite important. The quantum theory and the theory of relativity are both based on a close analysis of the act of observation. Let us see how the quantum theory considers the simple process of looking at a particle with a microscope.

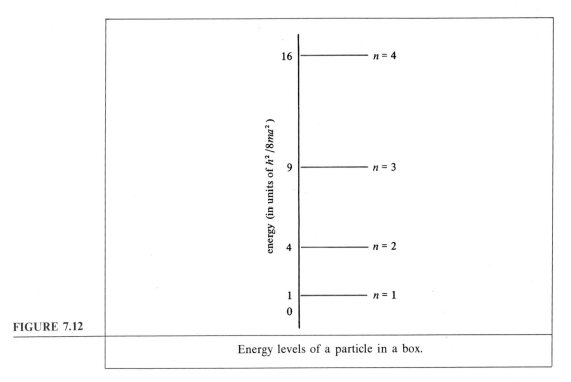

FIGURE 7.12 Energy levels of a particle in a box.

Observation requires light. At least one photon must strike the particle and be scattered into the lens of a microscope and from there into the eye or onto a photographic plate where the event is registered. The photon can follow any one of many possible paths, two of which are indicated in Figure 7.13. Because of the Compton effect some of the momentum of the photon is transferred to the particle, which recoils along some other path (e.g., one of the dotted lines). For objects of ordinary size the effect is unimportant, but for a very small particle, such as an electron, the resulting change is significant.

EXAMPLE 5

Compare the energy of a photon of violet light whose wavelength is 4.0×10^{-7} m with the kinetic energy of an electron whose speed is 1.20×10^6 m s^{-1}.

From Example 4 the kinetic energy of the electron is 6.56×10^{-19} J.
The energy of the photon may be obtained from its frequency since $E = h\nu$ and the frequency ν can be calculated using Eq. 7.2.

$$\nu = \frac{c}{\lambda} = \frac{3.00 \times 10^8 \text{ m s}^{-1}}{4.0 \times 10^{-7} \text{ m}} = 7.5 \times 10^{14} \text{ s}^{-1}$$

$$E = h\nu = (6.626 \times 10^{-34} \text{ J s})(7.5 \times 10^{14} \text{ s}^{-1})$$
$$= 5.0 \times 10^{-19} \text{ J}$$

Thus the energy of the photon is approximately the same as the kinetic energy of the electron.

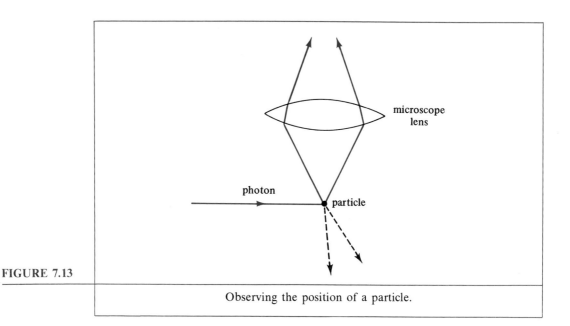

FIGURE 7.13 Observing the position of a particle.

Because momentum is transferred from the photon to the particle, one cannot determine the position of a particle without altering its momentum. Furthermore, this change in momentum may be in any one of many possible directions. Thus the attempt to determine the position of the particle introduces an uncertainty into its momentum. There is, in fact, a connection between the two: *The more closely we determine the position, the greater the uncertainty in momentum.* A quantitative treatment shows that if Δx is the uncertainty in position and Δp_x is the uncertainty in momentum, then their product is *at least* of the order of magnitude of Planck's constant.[4] This result, called the **uncertainty principle,** was discovered in 1927 by Werner Heisenberg.

It is important to realize that the uncertainty here is inherent in the nature of matter and does not depend on the measurement technique. By using poor techniques, one may have large uncertainties in both position and momentum, but no matter how improved experimental techniques become, the accuracy of measurements cannot exceed the limitation described by the uncertainty principle. Many attempts have been made to devise methods that will give $(\Delta x)(\Delta p_x)$ less than h; their failures have only strengthened confidence in the validity of the uncertainty principle.

As another illustration of the uncertainty principle, consider a particle con-

---

[4] Because of diffraction effects, the uncertainty in the position of the particle is of the same order of magnitude as the wavelength of the light ($\Delta x \simeq \lambda$). The Compton effect introduces an uncertainty into the momentum of the particle of the order of magnitude of the momentum of the photon ($\Delta p_x \simeq h/\lambda$). Thus $(\Delta x)(\Delta p_x) \simeq (\lambda)(h/\lambda) = h$.

As used here, Δx and Δp_x are rough estimates or ranges of uncertainty. In the precise formulation of the uncertainty principle, Δx and Δp_x are root-mean-square uncertainties, and $(\Delta x)(\Delta p_x) \geq h/4\pi$.

fined to a box of length a, so that the range of uncertainty in position is $\Delta x \simeq a$. We have seen that the momentum in the nth stationary state is $\pm nh/2a$, so that the range of uncertainty in momentum is from $-nh/2a$ to $+nh/2a$ or $\Delta p_x \simeq nh/a$. This has its smallest value for the ground state, $n = 1$, where $\Delta p_x \simeq h/a$. Thus $(\Delta x)(\Delta p_x) \simeq (a)(h/a) = h$. One can regard the momentum and hence the energy of the ground state as manifestations of the uncertainty principle, and the same is true of zero-point energies of other systems.

7.8 WAVE FUNCTIONS AND PROBABILITY DENSITIES

An important consequence of the uncertainty principle is that it has been necessary to abandon the concept of a path or orbit for very small particles such as electrons. To say that a particle is moving along a certain path implies that one can observe the position of the particle at various times and predict its position at future times. This is true when we are considering the flight of a baseball, the trajectory of a bullet, or the orbit of a satellite, for our observations need have only negligible effects on their motions.[5] But the observation of a particle of small mass, such as an electron, introduces a relatively large uncertainty into its momentum and hence into its future behavior. It is somewhat as if one could observe a baseball only by hitting it with a bat—the concept of the trajectory of a baseball would lose its meaning, and of course spectators would be required to play a more active role.

One can, however, both calculate and measure the **probability** of finding an electron in a particular region of space. This probability interpretation starts from the amplitude of the de Broglie wave, or *wave function,* designated by the symbol ψ (lowercase Greek letter psi). According to the probability interpretation suggested by Max Born in 1926, the *probability density is given by ψ^2, the square of the wave function.*[6]

As an illustration of the probability density, consider the particle in a one-dimensional box. For the three lowest states the wave function, ψ, is shown in the upper part of Figure 7.14. (These curves are the same as the solid curves in Figure 7.11.) The probability density, ψ^2, is shown in the lower part of Figure 7.14. These curves are obtained by squaring the wave function ψ shown directly above. Note that although ψ may have negative regions, corresponding to troughs in the de Broglie wave, ψ^2 is never negative.

---

[5] Even here things are not quite so simple. The momentum of photons from the sun alters satellite orbits to a significant extent.

[6] The reasoning which led to the probability interpretation may be outlined as follows. For light of a given frequency, each photon has energy $h\nu$, and thus the total energy is proportional to the number of photons. But, according to Maxwell's electromagnetic theory of light, the energy carried by a light wave is proportional to the square of the amplitude of the wave. Therefore the number of photons is proportional to the square of the amplitude. For light of low intensity the number of photons in a particular region of space may be less than one, requiring that we interpret the square of the amplitude as giving the probability of finding a photon in that region. Extension of this idea to electrons gives the probability interpretation described above. Experimental evidence supporting the probability interpretation for electrons is presented in the next chapter.

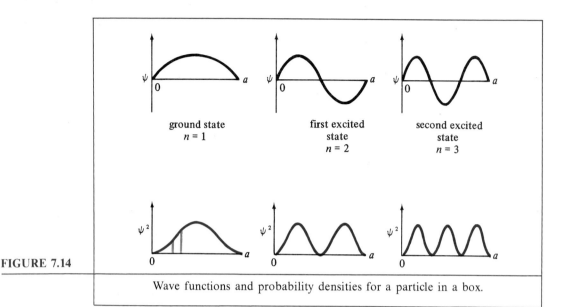

FIGURE 7.14 Wave functions and probability densities for a particle in a box.

The interpretation of the probability density curves is somewhat analogous to the interpretation of the Maxwell-Boltzmann distribution curves encountered in Chapter 4. As indicated by the shaded portions in the lower left diagram of Figure 7.14, the area under the curve between any two points gives the probability of finding the particle between these points. The larger the area, the more likely we are to find the particle in that region. Thus, if the two shaded regions are compared for the ground state in Figure 7.14, the particle is twice as likely to be in the one on the right. For regions of the same size (e.g., $0.01a$), the particle is most likely to be in the middle of the box, where ψ^2 is a maximum, and least likely to be near one end or the other. In the first excited state the particle is most likely to be found in the regions around $\frac{1}{4}a$ and $\frac{3}{4}a$ and least likely to be found near the center or at either end. The curves are drawn so that the total area under each probability density curve is one, corresponding to certainty that the particle is somewhere inside the box. Such functions are said to be normalized.

An important property of the probability density for stationary states is that it is time-independent: The probability of finding the particle in a given region of space does not depend on when the observation is made.

Although the language of wave functions and probability densities may be unfamiliar, these concepts are worth mastering, for they are basic to all modern descriptions of atoms and molecules. In the next chapter we consider how they are used to describe the electronic structure of atoms.

PROBLEMS

7.1 The crests of a series of traveling water waves pass a fixed point at a rate of one crest every 20 s. The distance between the crest and the next trough is

20 m. Each crest is 4 m higher than the succeeding trough. Calculate the frequency (in hertz), wavelength (in meters), speed (in meters per second), and amplitude (in meters).

7.2 Two wave sources in a water tank, 30 cm apart, move up and down in phase. Each generates waves whose wavelength is 20 cm. At a point on the water surface that is 40 cm from one source and 50 cm from the other, are the waves from the two sources in phase or out of phase?

7.3 In the right photograph of Figure 7.2, what points on the water surface are
(a) equidistant from the two sources?
(b) closer to the left source by one-half of a wavelength?
(c) closer to the left source by one wavelength?
At which of the above points is there constructive interference? Destructive interference? How do the regions of constructive and destructive interference differ in appearance in the photograph?

7.4 Calculate the frequency (in hertz) of yellow light of wavelength 5890 Å. Calculate the energy (in joules) of a photon of this frequency.

7.5 The human eye can detect a weak flash of light in which as few as 5 to 15 photons are absorbed. Calculate the corresponding energy, assuming blue light of wavelength 4700 Å.

7.6 The terms "intensity of light" and "frequency of light" are sometimes confused. Explain how they differ.

7.7 The hydrogen chloride molecule absorbs light of wavelength 3.46×10^{-6} m when it changes from the state with vibrational quantum number $v = 0$ to the state with $v = 1$.
(a) In what part of the spectrum is this absorption (see Figure 7.6)?
(b) Calculate the vibrational frequency and zero-point energy of the HCl molecule.

7.8 The photoelectric effect is observed for a certain metal only if the light is of wavelength shorter than 6280 Å. Calculate the threshold frequency (in hertz) and the energy (in joules) required to remove an electron from the metal.

7.9 The threshold frequency for the photoelectric effect with metallic sodium is 4.39×10^{14} Hz. For light of wavelength 3650 Å, a measurement of the kinetic energy of the electrons gave the result 2.56×10^{-19} J. Using these data, calculate Planck's constant, h.

7.10 Calculate the momentum (in kilogram-meters per second) of a photon of blue light of wavelength 4600 Å. Calculate the speed (in meters per second) of an electron having the same wavelength.

7.11 Calculate the de Broglie wavelength of
  (a) a particle whose mass is 1.0 g traveling at a speed of 1.0 m s^{-1}
  (b) a hydrogen molecule traveling at a speed of 1.92×10^3 m s^{-1}
  (c) an electron accelerated by a potential difference of 40 kilovolts (kV) in an electron diffraction apparatus. (The energy of an electron accelerated by a potential difference of one volt (V) is 1.60×10^{-19} J.)

7.12 Calculate the difference in energy between the ground state and first excited state of a particle in a one-dimensional box if
  (a) $m = 1.0$ g and $a = 1.0$ cm
  (b) $m =$ electron mass and $a = 1.0$ Å
  What is the ratio of the (b) answer to the (a) answer?

7.13 Sketch the fourth harmonic of a vibrating string. Sketch ψ and ψ^2 for the analogous excited state of a particle in a one-dimensional box. At what points is ψ^2 a maximum? A minimum?

7.14 When N different states have the same energy, we say that the energy level is N-fold degenerate. For a particle in a cubical box of edge a:
  (a) What is the degeneracy of the energy level having an energy of $6h^2/8ma^2$?
  (b) Specify the values of the quantum numbers n_x, n_y, and n_z for each state in part (a).
  (c) What is the energy of the lowest energy level having sixfold degeneracy?

CHAPTER 8
ATOMIC STRUCTURE

8

Flames, gas-discharge tubes, and electric sparks and arcs emit light that is characteristic of the substances present in contrast to blackbody radiation that is dependent only on the temperature. Familiar examples are the different colors emitted by electric discharge tubes that are used in advertising displays ("neon signs"). The light of various frequencies emitted by a substance is called its **emission spectrum.**

Spectra have important analytical applications. For example, small amounts of sodium chloride (or any other sodium compound) impart a bright yellow color to a flame. This yellow light provides a sensitive test for the presence of sodium. But perhaps the most important application of atomic spectra has been the determination of the allowed electron energy levels for the various atoms. With these experimental data as a basis, it has been possible to reach a detailed understanding of the electronic structures of atoms.

8.1 ATOMIC SPECTRA

Emission spectra can be analyzed by passing the emitted light through a prism, which disperses the light into its various frequencies. From there the light is directed into a light meter (in a spectrophotometer) or onto a photographic plate (in a spectrograph). When a narrow slit is used to define the light source (Figure 8.1), the characteristic frequencies emitted by an atom appear as lines on a photographic plate, and therefore the spectrum of an atom is sometimes called a **line spectrum.** Figure 8.2 shows the emission spectrum of the hydrogen atom in the visible and near-ultraviolet regions.

Under certain conditions an **absorption spectrum** may be observed. When light consisting of a broad range of frequencies, such as blackbody radiation or light from an incandescent lamp, is passed through a gas, certain characteristic frequencies are absorbed. These frequencies are missing when the emerging light is analyzed with a spectrophotometer or spectrograph. Figure 8.3 shows the absorption spectrum of the sodium atom in the ultraviolet region.

Because light darkens a photographic plate, the emission spectrum appears as a series of dark lines, and the absorption spectrum as a series of bright lines. Under direct visual observation (in the visible region) the opposite holds true: The emission spectrum is a series of bright lines, and the absorption spectrum is a series of dark lines against a bright background. The lines in the absorption spectrum have the same frequencies as the lines in the emission spectrum.

8.1 ATOMIC SPECTRA

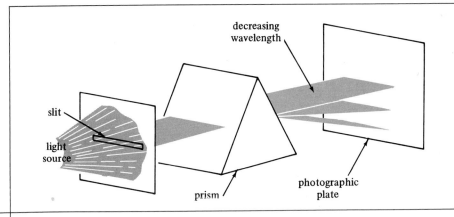

FIGURE 8.1 Formation of a line spectrum by a prism spectrograph.

The modern interpretation of atomic spectra was introduced by Niels Bohr in 1913. In this interpretation, each line represents the energy difference between two energy levels. When an atom excited by high-energy collisions in the electric discharge or flame changes from a state of higher energy, E_h, to a state of lower energy, E_l, it emits a photon whose frequency is related to the energy difference by the equation

$$E_h - E_l = h\nu \tag{8.1}$$

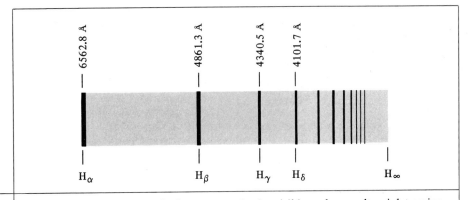

FIGURE 8.2 Emission spectrum of the hydrogen atom in the visible and near ultraviolet region (Balmer series). H_∞ gives the theoretical position of the series limit. (SOURCE: G. Herzberg, *Atomic Spectra and Atomic Structure*, 1st ed., Prentice-Hall, Englewood Cliffs, N.J., 1937, p. 5, and 2nd ed., Dover, New York, 1944, p. 5.)

FIGURE 8.3

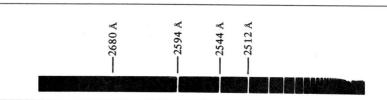

Photograph of the absorption spectrum of the sodium atom in the ultraviolet region, beginning with the fourth line of the principal series. [SOURCE: *H. Kuhn,* Zeitschrift für Physik, **76**, *785 (1932)*.]

A photon of the same frequency is absorbed in the transition from the state of lower energy to the state of higher energy, and thus absorption of light is just the reverse of emission.

From measurements of the frequencies of the various lines, one can calculate the energy differences between the various energy levels. The results of these calculations are usually summarized by an energy level diagram for the atom.

8.2 THE SPECTRUM OF ATOMIC HYDROGEN

Of special interest is the spectrum of the simplest atom, hydrogen, with only one electron. It consists of several series of lines, each series being named after its discoverer. The Lyman series is in the far ultraviolet region, the Balmer series (also shown in Figure 8.2) is in the visible and near ultraviolet, and the Ritz-Paschen series is in the infrared (see Figure 8.4). There are also several other series in the infrared region. The Lyman and Balmer series have also been observed as absorption spectra. Within each series the lines get weaker and closer together as

FIGURE 8.4

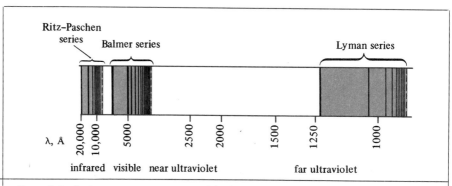

Part of the hydrogen atom spectrum. The intensity is indicated by the thickness of the lines. The dashed lines correspond to the series limits, at which a continuous spectrum joins the series.

they approach a *series limit* at the high-frequency (short wavelength) end; in Figure 8.4 the positions of the series limits are indicated by dashed lines.

Figure 8.5 shows the energy level diagram of the hydrogen atom derived from the spectrum. As noted earlier, the state of lowest energy is called the ground state, and the other states are called excited states. Transitions from the various excited states to the ground state give the Lyman series. The other series arise from transitions from higher excited states to the second, third, and succeeding states. The different energy levels are conveniently designated by a number, n, called the **principal quantum number,** which is assigned the value *1 for the ground state, 2 for the first excited state,* and so on. The energies of the various states are usually expressed in electron volts. [An electron volt is the energy required to raise the potential of an electron by 1 volt (V). One electron volt (eV) is equal to 1.602×10^{-19} J. For conversion from energy per atom to energy per mole of atoms, 1 eV atom^{-1} corresponds to 96.48 kJ mol^{-1}.]

As one goes to higher values of n the energies of the excited states get closer and closer together, and at a certain point (corresponding to the various series limits) the discrete levels give way to a continuous distribution of energy levels. If this dividing point is taken as zero on the energy scale, then energies in the continuous region are positive and energies in the discrete region are negative, as indicated on the left side of Figure 8.5.

The quantitative treatment of the hydrogen atom spectrum (which leads to Figure 8.5) starts from measurements of the wavelengths and frequencies of the various lines in the spectrum. The frequencies of all of the lines in the hydrogen spectrum are found to obey a very simple relationship

$$\nu = cR_H \left(\frac{1}{n_l^2} - \frac{1}{n_h^2} \right) \tag{8.2}$$

where c is the velocity of light, n_l is the principal quantum number of the lower state, n_h is the principal quantum number of the higher state, and R_H is a constant called the Rydberg constant for hydrogen (after Johannes R. Rydberg).[1] From the observed wavelengths this constant is found to be 1.09678×10^7 m^{-1}.

From Eq. 8.1 it follows that

$$E_h - E_l = h\nu = hcR_H \left(\frac{1}{n_l^2} - \frac{1}{n_h^2} \right) \tag{8.3}$$

and

$$E_h + \frac{hcR_H}{n_h^2} = E_l + \frac{hcR_H}{n_l^2} \tag{8.4}$$

As n_h and E_h may vary independently of n_l and E_l, each side of this equation is

---

[1] A relationship of this type was first discovered in 1885 by Johann J. Balmer for the Balmer series ($n_l = 2$). The integers n_l and n_h were later identified as the principal quantum numbers of the two states.

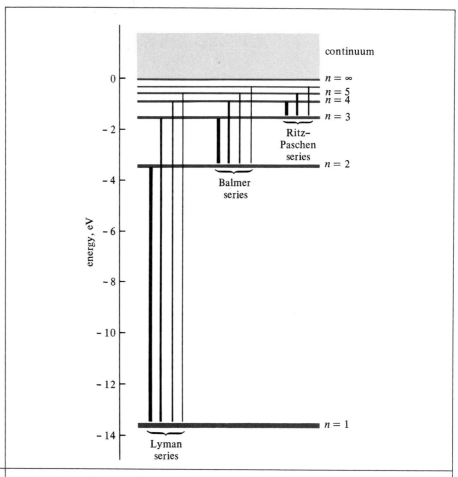

FIGURE 8.5 Energy level diagram of the hydrogen atom. The different series can be expressed mathematically by Eq. 8.2 with different values of n_l: Lyman series ($n_l = 1$), Balmer series ($n_l = 2$), Ritz-Paschen series ($n_l = 3$).

a constant whose value may be chosen arbitrarily. The usual convention sets this constant equal to zero. Then

$$E_h = -\frac{hcR_H}{n_h^2} \quad (8.5)$$

and

$$E_l = -\frac{hcR_H}{n_l^2} \quad (8.6)$$

or in general the energy, E_n, of each state is

$$E_n = -\frac{hcR_H}{n^2} \tag{8.7}$$

where n is the principal quantum number. As n increases to very high values ($n \to \infty$), E_n increases and approaches zero.

The constants in the numerator can be multiplied out:

$$\begin{aligned}hcR_H &= (6.6262 \times 10^{-34} \text{ J s})(2.9979 \times 10^8 \text{ m s}^{-1})(1.09678 \times 10^7 \text{ m}^{-1}) \\ &= 2.1787 \times 10^{-18} \text{ J} \\ &= 13.60 \text{ eV}\end{aligned}$$

Thus one may write Eq. 8.7 in the form

$$E_n = -\frac{13.60}{n^2} \text{ eV} \qquad \text{where } n = 1, 2, 3, \text{ etc.} \tag{8.8}$$

and one may express Eq. 8.8 graphically by Figure 8.5. Note that this simple equation for the energy of the various states of the hydrogen atom was obtained from *experimental* measurements of the hydrogen atom spectrum—nothing has been said up to this point about the electron or proton nor of their attraction for each other. The next step was the development of a theory of the hydrogen atom consistent with Eq. 8.8.

8.3 QUANTUM NUMBERS FOR THE HYDROGEN ATOM

The development of an adequate theory of the hydrogen atom has been both the simplest and the most important problem in atomic and molecular structure—the simplest because there is only one electron and one nucleus (containing one proton) to consider, and the most important because the concepts and language used to describe more complex atoms and molecules are taken from the theory of the hydrogen atom.

In the preceding chapter we saw that a wave function, ψ, and a probability distribution function, ψ^2, are necessary for a modern description of electronic behavior. To determine the allowed energy states for a particle in a box, it was sufficient to use the equation relating the de Broglie wavelength to the momentum of the particle, together with the analogy to a vibrating string. For the hydrogen atom this simple method no longer suffices because the negative electron is attracted to the positive nucleus by the coulombic force, the force of attraction between electric charges of opposite sign. An equation discovered by Erwin Schrödinger in 1926 allows one to take this force into account. Solution of the Schrödinger equation requires rather sophisticated mathematical techniques. It will be sufficient here to indicate that the equation expresses certain conditions that ψ must satisfy, and hence it is used to calculate ψ in a manner somewhat analogous to finding the roots of an algebraic equation. However, the results obtained by Schrödinger for the hydrogen atom are relatively easy to describe.

First of all, only certain states are allowed. Each state is characterized by a wave function that is called an **orbital** of the hydrogen atom. The characteristic energy of each state may be expressed by the formula

$$E_n = -\frac{(\mu e^4/8\epsilon_0^2 h^2)}{n^2} \tag{8.9}$$

where n (the principal quantum number) must be a positive integer (1, 2, 3, etc.), e is the electronic charge, ϵ_0 is the permittivity of vacuum (Appendix D-5), h is Planck's constant, and μ, the reduced mass, is $mM/(m + M)$, where m is the mass of the electron and M is the mass of the proton. (Because M is much larger than m, the reduced mass is approximately equal to the mass of the electron.)

Not only does this equation have the same relation between E_n and n as determined from the spectrum, but insertion of the physical constants e, ϵ_0, h, m, and M leads to the same numerical factor as found experimentally (Eq. 8.8).

[The first successful attack on the problem of the hydrogen atom was made by Niels Bohr in 1913, when he introduced the concept of stationary states and an equation (8.1) for determining their energies. He proposed that the electron moved around the nucleus in a circular path or orbit. From a quantum condition placed on the angular momentum of the electron, he found that only certain orbits were allowed, namely, those with radii given by the expression

$$r = \frac{n^2 h^2 \epsilon_0}{\pi \mu e^2}$$

In the ground state, $n = 1$, the radius of the orbit was $h^2\epsilon_0/\pi\mu e^2 = 0.529$ Å. (This distance is called an atomic unit or bohr.) In the excited states the radius was larger than this by factors of $2^2 = 4$, $3^2 = 9$, and so on. The Bohr theory led to an expression for the energy identical to Eq. 8.9 and hence in agreement with the experimental result, Eq. 8.8. The triumph of the Bohr orbits was short-lived; extension to atoms with more than one electron led to results differing from the experimental results. Even for the hydrogen atom it was later found that the angular momentum was incorrect. Despite various modifications of the original theory, such as the introduction of elliptical orbits, it was superseded in 1926 by Schrödinger's wave theory and a mathematically equivalent form based on matrices developed by Werner Heisenberg. It should be mentioned, however, that Bohr continued to play an important role in this area and was active in developing the new quantum theory.]

The coulombic force enters the Schrödinger equation for the hydrogen atom as the potential energy, which is inversely proportional to the distance between the electron and the proton, r. As r increases, the potential energy approaches zero. Thus the zero-point on the energy scale has a simple physical interpretation. It corresponds to $r = \infty$; that is, the electron has just enough kinetic energy to escape from the nucleus, leading to dissociation or ionization of the atom. The energy required to ionize a hydrogen atom from its ground state, $0 - (-13.60 \text{ eV}) = 13.60 \text{ eV}$, is called the **ionization potential** of the hydrogen atom.

In the positive-energy region the electron has more than enough kinetic energy to escape from the nucleus into free space. This additional kinetic energy is not quantized, and thus any positive energy is allowed, leading to the continuum of positive energies shown in Figure 8.5.

The solution of the Schrödinger equation for the hydrogen atom leads to two quantum numbers in addition to the principal quantum number. First, the **azimuthal quantum number,** or **angular momentum quantum number,** l, is zero or a positive integer. It is usually denoted by a letter equivalent as shown below.

$$l = 0 \quad 1 \quad 2 \quad 3 \quad 4 \quad 5 \cdots$$
$$\text{letter} = s \quad p \quad d \quad f \quad g \quad h \cdots$$

The orbitals are designated by the principal quantum number followed by the letter equivalent of the azimuthal quantum number. For example, a $4f$ orbital has $n = 4, l = 3$. (As its name indicates, the azimuthal or angular momentum quantum number is related to the angular momentum of the electron.)

The quantum rule restricting the possible values of l is $l = 0, 1, 2, \ldots n - 1$. In other words, for any value of n, there are orbitals with values of l running from zero to $n - 1$. Thus, for $n = 1, l = 0$, and only a $1s$ orbital is possible. For $n = 2$, $l = 0$ or 1, and there are both $2s$ and $2p$ orbitals. For $n = 3, l = 0, 1$, or 2, and there are $3s, 3p$, and $3d$ orbitals.

The third quantum number, called the **magnetic quantum number,** m_l, is zero or a positive or negative integer. *The quantum rule restricting the possible values of m_l is $m_l = +l, +l - 1, \ldots -l$.* Thus m_l ranges from a highest allowed value of $+l$ to a lowest allowed value of $-l$. For an s orbital $l = 0$ and hence $m_l = 0$. For a p orbital $l = 1$ and hence m_l may be $+1, 0$, or -1. For a d orbital $l = 2$ and hence m_l may be $+2, +1, 0, -1$, or -2. (When an atom is in a magnetic field, the behavior of the electron depends on its value of m_l, whence the name magnetic quantum number.)

Table 8.1 lists the various orbitals for the lowest three energy levels of the hydrogen atom. Each set of three quantum numbers designates a particular orbital. Note that only one orbital has $n = 1$, four different orbitals have $n = 2$, and nine different orbitals have $n = 3$. In general there are n^2 orbitals having the same value of n and hence the same energy.

TABLE 8.1

ORBITALS FOR $n = 1, 2,$ or 3
$n = 3 \quad 3s(m_l = 0), 3p(m_l = +1, 0, -1), 3d(m_l = +2, +1, 0, -1, -2)$
$n = 2 \quad 2s(m_l = 0), 2p(m_l = +1, 0, -1)$
$n = 1 \quad 1s(m_l = 0)$

8.4 DESCRIPTION OF HYDROGEN ATOM ORBITALS

Like the rules given above for the various quantum numbers, the orbital shapes are determined by solving the Schrödinger equation. All s orbitals have round shapes (spherical symmetry) and thus they can be described by their dependence on a single variable, r, the distance of the electron from the nucleus. Figure 8.6 shows graphs of ψ and ψ^2 as a function of r for the 1s, 2s, and 3s orbitals. Note the appearance of a single node in the 2s orbital, and two nodes in the 3s orbital. (For the 2s orbital the node is a spherical surface; for the 3s orbital the nodes are two concentric spherical surfaces.)

Another useful representation of ψ_{1s}^2 is the cross section or slice shown in Figure 8.7. The intensity of the color is much heavier near the center to indicate that, comparing small volume elements of equal size at different distances, the electron is more likely to be found closer to the nucleus. The highest probability per unit volume is at the nucleus, as is also shown in Figure 8.6.

The representation of the probability density shown in Figure 8.7 resembles a cloud, and in a probability sense the electron is indeed smeared out into an electron cloud. Thus one may speak of the probability density as representing the density of the electron cloud in various regions of space; that is, ψ^2 is the *electron*

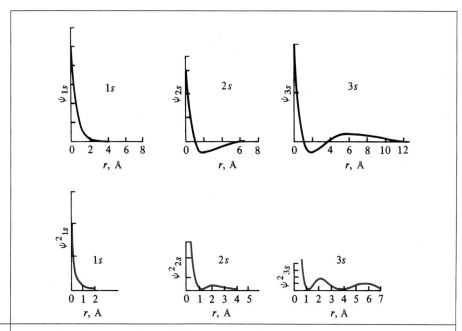

FIGURE 8.6

Wave functions, ψ, and probability densities, ψ^2, for the 1s, 2s, and 3s states of the hydrogen atom.

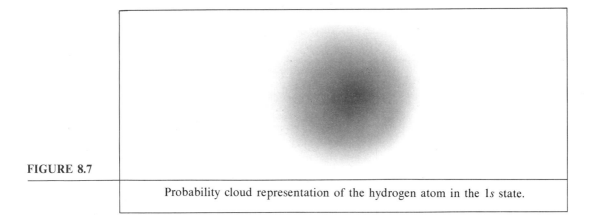

FIGURE 8.7 Probability cloud representation of the hydrogen atom in the 1s state.

density. Thus for the 1s orbital the highest electron density is at the nucleus. Nodes are positions of zero electron density, and hence the spherical nodal surfaces of the 2s and 3s orbitals represent regions of zero electron density between regions of higher electron density. The probability interpretation should be kept in mind in using the terms electron cloud and electron density, for so far as is known a fraction of an electron is never observed in an experiment.

Another property of some interest is the probability of finding the electron at a given distance from the nucleus, r, or, more precisely, within a thin spherical shell whose inner radius is r and whose outer radius is $r + \Delta r$. Because the volume of such a shell is $4\pi r^2 (\Delta r)$, this probability is given by an area of height $4\pi r^2 \psi^2$ and width Δr. The function $4\pi r^2 \psi^2$, called the **radial distribution function,** $D(r)$, is shown in Figure 8.8 for the 1s, 2s, and 3s orbitals. For the 1s orbital the highest probability occurs at 0.529 Å. (This happens to be the same as the radius of the first Bohr orbit.) For the 2s and 3s orbitals the highest probabilities occur at considerably greater distances. Thus as the hydrogen atom is progressively excited to states of higher energy, the electron is most likely to be found farther from the nucleus, leading eventually to ionization.

It should be emphasized that the diagrams for the 1s orbital in Figures 8.6 and 8.8 both refer to a single probability distribution, namely, that shown in Figure 8.7. Different curves are obtained because we have asked different questions. In Figure 8.6 we are comparing volume elements of the same size, that is, electron densities; in Figure 8.8, we are comparing thin spherical shells of the same thickness but different volumes. The situation is somewhat analogous to population densities and populations. Rhode Island has about eight times as many people per square mile as California, but California has a population about 20 times as large as that of Rhode Island.

There are three different 2p orbitals. Figure 8.9 shows the "shapes" of these orbitals—how they depend on direction at a given distance from the nucleus. They are designated as $2p_x$, $2p_y$, and $2p_z$, because each is concentrated along one

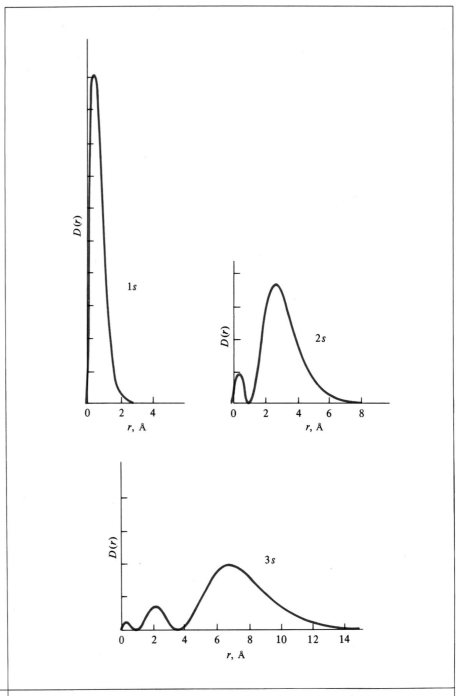

FIGURE 8.8 Radial distribution functions for the 1s, 2s, and 3s orbitals.

8.4 DESCRIPTION OF HYDROGEN ATOM ORBITALS

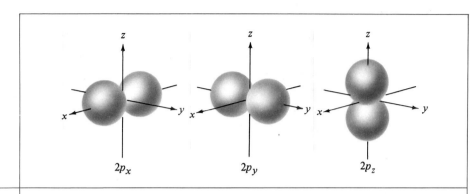

FIGURE 8.9 Variation with direction of the $2p_x$, $2p_y$, and $2p_z$ orbitals.

of a set of Cartesian axes (x, y, or z) with the origin ($x = 0$, $y = 0$, $z = 0$) at the nucleus.

Figure 8.10 is a contour diagram for a cross section of the $2p_z$ orbital, showing contours for seven different values of ψ. It is zero at the nucleus and at all points on the xy plane (the plane passing through the nucleus and perpendicular to the z axis). The $2p_z$ orbital has its largest positive and negative values on the z

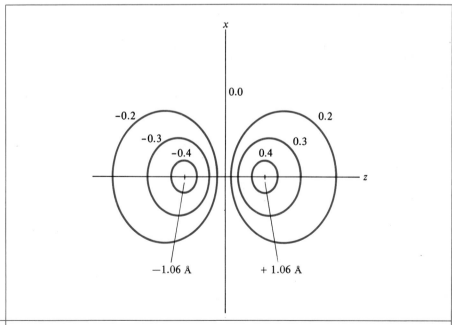

FIGURE 8.10 Contours on the xz plane for the $2p_z$ orbital. The numerical value of ψ is shown next to each contour.

axis (at $z = +1.06$ Å and at $z = -1.06$ Å, respectively). The $2p_x$ and $2p_y$ orbitals are similarly concentrated about the x and y axes, respectively. Their nodal surfaces are the yz and xz planes, respectively.

Figure 8.11 shows the radial distribution function, $D(r)$, for any one of the $2p$ orbitals. The position of the maximum (the most probable distance of the electron from the nucleus) is at $r = 2.12$ Å, which is the same as the radius of the Bohr orbit for $n = 2$.

The $3p$ orbitals have the same general shape as the $2p$ orbitals, except that each has a spherical nodal surface (like the $2s$ orbital) as well as a planar node. The $3d$ orbitals have somewhat more complicated shapes; they are shown in Figure 21.4, and are discussed in Section 21.2.

8.5 ELECTRON SPIN

In addition to the quantum numbers n, l, and m_l, there is a fourth quantum number of particular importance in atoms with more than one electron. This is the **spin quantum number,** m_s, almost the ultimate in simplicity—it can have only two possible values, $+\frac{1}{2}$ or $-\frac{1}{2}$. Each electron behaves as if spinning about its axis (analogous to the rotation of the earth about its axis every 24 hours). The spin quantum number determines the direction of the spin: When m_s is $+\frac{1}{2}$, the spin is oriented one way, usually symbolized by an arrow pointing up (↑); when m_s is $-\frac{1}{2}$ the spin is oriented the other way (↓). One can think of these most simply as clockwise and counterclockwise spinning of the electron.

Associated with the electron's spin is a magnetic dipole. Each electron behaves like a very small compass needle, with a north-seeking pole and a south-seeking pole. Since the spin can have only two possible directions, the same is true of the magnetic dipole, and the same symbols (↑ and ↓) may be used to represent the orientations of the electron's magnetic dipole. The familiar magnetic properties of iron and steel result from the magnetic dipoles of the electrons in the iron atoms.

Electron spin was postulated in 1925 by two students at the University of Leiden, George E. Uhlenbeck and Samuel A. Goudsmit, as a single hypothesis explaining several puzzling features of atomic spectra. The concept of electron spin provided a simple explanation of observations made several years earlier (1922) by Otto Stern and Walter Gerlach. In the Stern-Gerlach experiment, a beam of silver atoms from a high-temperature furnace was passed through a magnetic field (Figure 8.12). The magnets were shaped differently (one was flat and the other had a knife-edge) so as to give an inhomogeneous field, that is, one varying from one point to another.

In a uniform magnetic field the net deflecting force on a magnetic dipole is zero, because the two poles experience equal forces in opposite directions. This is not true in an inhomogeneous field, where the field at one pole differs from the field at the other pole. Depending on the orientation of the magnetic dipole, the net deflecting force will vary in both magnitude and direction.

Stern and Gerlach found that the beam of silver atoms was split: Half of the

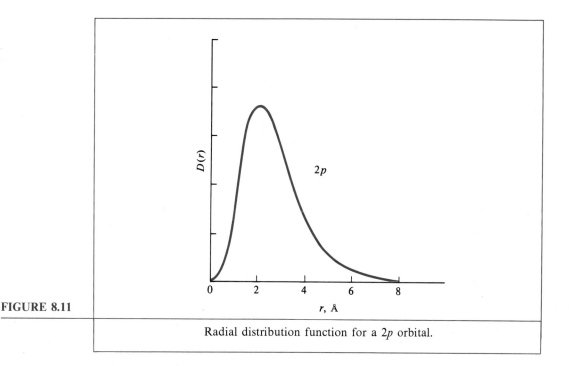

FIGURE 8.11 Radial distribution function for a 2p orbital.

atoms were deflected upward, all by the same amount, and half were deflected downward, all by the same amount. This result showed that the silver atom's magnetic dipole can have only two possible orientations, and these must be in opposite directions. In short, the orientation of the dipoles was *quantized in space*.

This is just the result to be expected on the basis of electron spin. In a silver atom there are 47 electrons. As we shall see in more detail later, all but one electron have their spins paired (↑ ↓), so that their magnetic effects cancel one

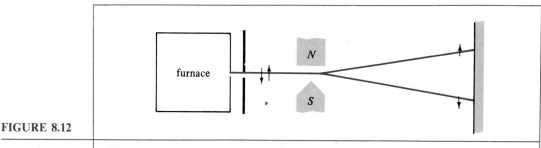

FIGURE 8.12 The Stern-Gerlach experiment. The arrows indicate the directions of the magnetic moments of the atoms. *N* and *S*, magnets.

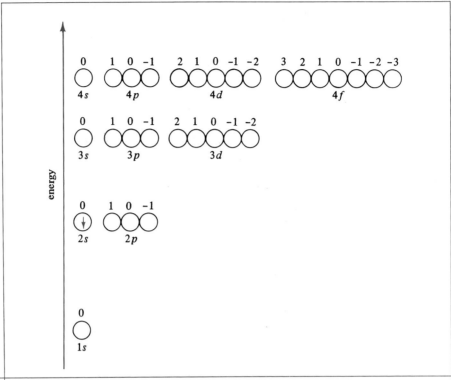

FIGURE 8.13

An orbital diagram for the hydrogen atom ($n = 1, 2, 3,$ or 4). The energy scale has been distorted to provide space for the various symbols. The value of m_l is indicated above each circle, and the value of m_s is indicated by the direction of the arrow within the circle.

another. Thus the magnetism of the silver atom is due to a single electron, and this can have only two orientations (↑ and ↓) corresponding to $m_s = +\frac{1}{2}$ and $m_s = -\frac{1}{2}$. In the beam of silver atoms from the furnace, half of the atoms have $m_s = +\frac{1}{2}$ and are deflected in one direction, all by the same amount. The other atoms, with $m_s = -\frac{1}{2}$, are deflected in the opposite direction, all by the same amount.

We can now summarize the various possible states of the hydrogen atom corresponding to different values of the quantum numbers n, l, m_l, and m_s by an orbital diagram (Figure 8.13). Each orbital is symbolized by a circle, m_l is shown above the circle, and m_s is indicated by an arrow placed within the circle. In the example shown (an excited state of the hydrogen atom), the quantum numbers are $n = 2$, $l = 0$, $m_l = 0$, and $m_s = -\frac{1}{2}$.

8.6 THE PAULI EXCLUSION PRINCIPLE

Thus far we have considered only the electronic structure of the hydrogen atom, with a single electron. What can we say about atoms with more than one electron?

We might expect that in their ground states, many-electron atoms would have all of the electrons in the 1s orbital, which has the lowest energy. Many different experiments show that this does not happen. The 1s orbital can have at most only two electrons, and then only if the spin quantum numbers of the electrons are opposite in sign (↑ ↓). The same is true of every other orbital.

The above restrictions on the number and pairing of electrons in an orbital were first noted by Wolfgang Pauli and summarized in what is known as the **Pauli exclusion principle:** *No two electrons in an atom can have exactly the same values for all four quantum numbers.* Therefore in the lithium atom (atomic number 3) the two electrons in the 1s orbital ($n = 1, l = 0, m_l = 0$) must have opposite spins ($m_s = +\frac{1}{2}$ and $m_s = -\frac{1}{2}$), that is, their spins are paired. The third electron in the lithium atom must have a different set of quantum numbers. It goes into the orbital of next lowest energy, the 2s orbital, and has quantum numbers $n = 2$, $l = 0$, $m_l = 0$, $m_s = +\frac{1}{2}$ (or $m_s = -\frac{1}{2}$).

Why do atoms obey the restriction expressed by the Pauli exclusion principle? We do not know. It is of course easy to construct analogies (e.g., an atom is like an apartment building in which each apartment is restricted to a single person or a couple without children), but these do not really explain why nature behaves this way. A lithium atom with all three electrons in the 1s orbital would not violate any other known physical law. But such an atom has never been observed. At least for the present we must accept the exclusion principle as a fundamental physical law that, like the law of conservation of energy, admits of neither exceptions nor explanations.

As we shall see later, the Pauli exclusion principle applies to molecules as well as to atoms. All aspects of chemistry are profoundly affected by this principle.

8.7 THE SHIELDING EFFECT FOR MANY-ELECTRON ATOMS

The presence of more than one electron in an atom results in a change in the relative energies of the different orbitals. In contrast to the hydrogen atom, where the 2s and 2p orbitals have the same energy, for atoms with two or more electrons the 2s orbital has a *lower energy* than the 2p orbital. For principal quantum number $n = 3$, the energies increase in the order $3s < 3p < 3d$. (The 3d orbital energy is, in fact, so high that it actually lies above the 4s orbital energy.) In general, *for a given principal quantum number n, the energy of the orbital increases with increasing azimuthal quantum number l.* This phenomenon is commonly called the **shielding effect** after its explanation (given below).

Figure 8.14 illustrates both the shielding effect and the Pauli exclusion principle for the ground state of the sodium atom. Note that the 11 electrons of the sodium atom are distributed with two paired electrons in each of the five orbitals of lowest energy, and there is one electron in the 3s orbital. Also note that the energies increase in the order $1s < 2s < 2p < 3s < 3p < 4s < 3d$.

This effect of l on the energy of an orbital can be deduced for sodium atoms by examining the atomic spectrum of sodium. (Part of this spectrum was shown in

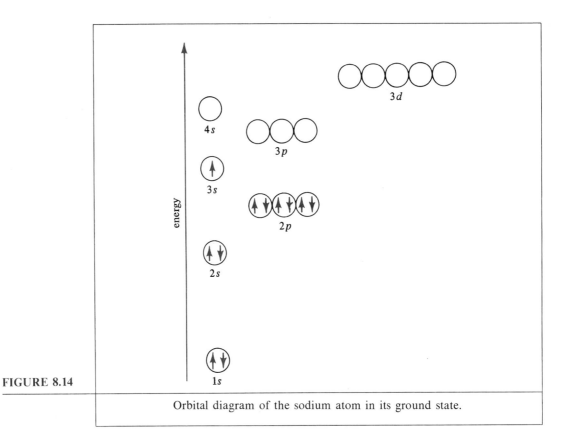

FIGURE 8.14 Orbital diagram of the sodium atom in its ground state.

Figure 8.3.) An energy level diagram resulting from this analysis is shown in Figure 8.15. In this diagram each energy level is labeled with the orbital designation of the outermost electron. (In all of the energy levels shown here, the inner 10 electrons remain in the 1s, 2s, and 2p orbitals.)

The original analysis of the spectrum of the sodium atom gave only a set of energy levels, which were not divided neatly into s, p, d, and f types as shown in Figure 8.15. The type of orbital occupied in each energy level has been determined in part by placing the sodium light source between the pole pieces of a strong magnet. The magnetic field affects the sodium atoms in such a way that each energy level splits into two or more levels of slightly different energy. This causes each spectral line to split into several separate lines, a phenomenon discovered in 1896 by Pieter Zeeman. From observations of the **Zeeman effect,** one can determine which energy levels are of s type, which are of p type, and so on.

Consider, for example, the bright yellow line in the emission spectrum of sodium. On close examination under high resolution, it is found to consist of two separate lines very close together, called a **doublet.** Thus the bright yellow line of sodium is really two lines of wavelengths 5890 and 5896 Å. This is explained by a splitting of the energy level of the excited state (3p) into two levels very close

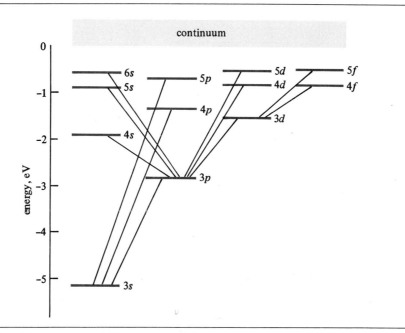

FIGURE 8.15

Energy levels of the sodium atom. Horizontal lines represent the energies of the lowest 12 energy levels. Connecting lines represent observed transitions. It is conventional to place all s states in one column, all p states in a second column, and so on. The energy of each transition is represented by the vertical distance between the upper and lower states. In fixing the energy scale the energy of the ionized atom ($Na^+ + e^-$) has been taken as zero.

together. In the one of higher energy (indicated by ↑ in Figure 8.16) the spin and orbital angular momentum of the outermost electron are in the same direction. In the other (↓) they are in the opposite direction. (This interaction of spin and orbital angular momentum is called spin-orbit coupling.) In the ground state ($3s$) the orbital angular momentum is zero, so the two spin states have the same energy.

In the presence of a magnetic field the $3s$ energy level splits into two levels of slightly different energy. The lower $3p$ energy level splits into two levels also, and the upper $3p$ energy level splits into four levels. The resultant splitting of the two yellow sodium lines is shown in Figure 8.17.

Why does the energy of an orbital in a many-electron atom depend on l, the azimuthal quantum number? The main cause is the shielding effect of the inner electrons. Each electron in an outer orbital not only is attracted by the positive nucleus, but is to a lesser extent repelled by the inner shells of electrons. Thus the inner electrons partially shield or screen the nucleus, decreasing the net attractive force on the outer electrons. Since the net attractive force is smaller, the orbital

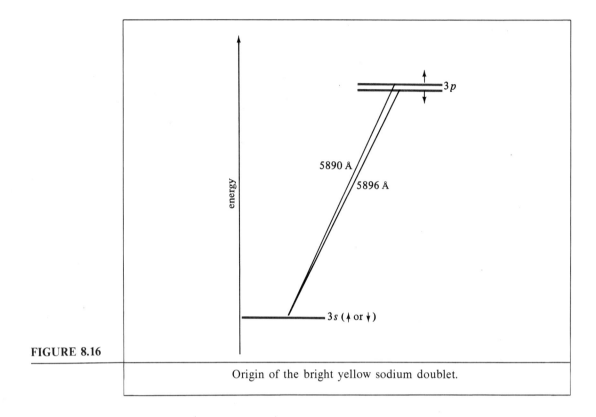

FIGURE 8.16

Origin of the bright yellow sodium doublet.

energy is higher (less stable). Within a shell of given quantum number n, the orbital of highest energy is the one of highest l, because it has a single peak in its radial distribution function at a rather long distance from the nucleus (e.g., the $2p$ orbital shown in Figure 8.11). The shielding effect is smallest for s orbitals, which have small but substantial peaks in their radial distribution functions at short distances from the nucleus (e.g., the $2s$ and $3s$ orbitals shown in Figure 8.8[2]). Thus electrons in s orbitals are best able to penetrate the inner shells into regions close to the nucleus, giving a higher net attractive force on the electrons and causing the s orbitals to be the most stable (lowest energy).

8.8 ELECTRON CONFIGURATIONS

In place of an orbital diagram such as that shown in Figure 8.14, the electronic structure of an atom may be summarized by a brief notation called the **electron configuration.** The number of electrons in each orbital is indicated by a right

---

[2] Although the radial distribution functions for the orbitals of a many-electron atom differ in detail from the hydrogen atom orbitals shown in Figure 8.8 and 8.11, they have the same general shape and the same number of peaks.

8.9 THE BUILDING-UP PRINCIPLE AND THE PERIODIC LAW

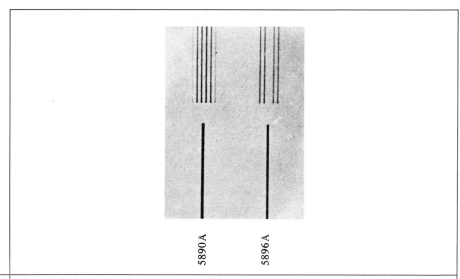

FIGURE 8.17

Photograph of the Zeeman splitting of the yellow sodium doublet at 5890 and 5896 Å. Top, with magnetic field; bottom, without magnetic field. (Source: *E. Back and A. Landé,* Zeemaneffekt and Multiplettstruktur der Spektrallinien, *Verlag von Julius Springer, Berlin, Springer-Verlag, New York, 1925, Tafel I, Bild 2.*)

upper index. Thus the electron configuration of the sodium atom in its ground state is $1s^2 2s^2 2p_x^2 2p_y^2 2p_z^2 3s^1$. An index of 1 is usually omitted.

It is even more convenient to combine orbitals of the same n and l as a *subshell* (for example, the $2p$ subshell consists of the $2p_x$, $2p_y$, and $2p_z$ orbitals). Thus we may write the electron configuration of sodium as $1s^2 2s^2 2p^6 3s$.

We can also simplify the writing of electron configurations by using the symbols of the group 0 elements (called the **noble gases**) to represent the electron configurations of these atoms. (These elements, their symbols, and their atomic numbers are listed in the right column of the periodic table inside the front cover.) For example, the electron configuration of the lithium atom would be [He]$2s$, where [He] represents $1s^2$. Because the neon atom has the electron configuration $1s^2 2s^2 2p^6$, the electron configuration of the sodium atom would be [Ne]$3s$.

8.9 THE BUILDING-UP PRINCIPLE AND THE PERIODIC LAW

We now have enough information to predict correctly the electron configuration of the ground state atoms of many of the chemical elements, and to correlate the electron configurations of the atoms with the positions of the elements in the periodic table. Each atom in its ground state should have all of its electrons in the orbitals of lowest energy, subject to the restriction on occupancy expressed

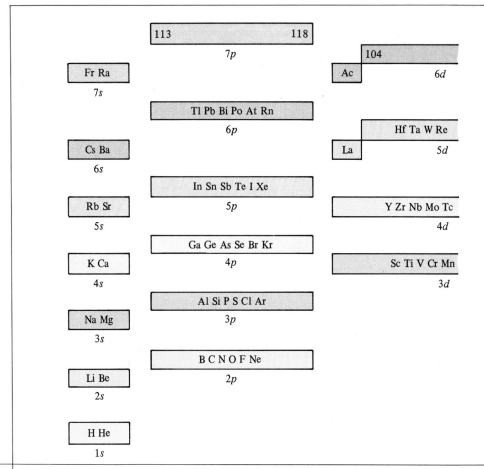

FIGURE 8.18 Periodic table showing the electronic structures of the atoms. [SOURCE: H. C. Longuet-Higgins, *Journal of Chemical Education*, **34**, 31 (1957).]

by the Pauli exclusion principle. Knowing that the orbital energies increase in the sequence $1s < 2s < 2p < 3s < 3p < 4s < 3d$, as explained in Section 8.7, we can *build up* the electron configurations for the atoms of the first 20 elements, as shown in Table 8.2.

For each electron configuration we can draw the corresponding orbital diagram similar to Figure 8.14 for sodium. One way of summarizing quite simply both the electron configurations and the orbital diagrams for all the elements is shown in Figure 8.18. For example, in Figure 8.18 chlorine (Cl) is the fifth element in the $3p$ subshell. This indicates that it has five $3p$ electrons, with all orbitals of lower energy filled, as is also indicated by the electron configuration in Table 8.2.

8.9 THE BUILDING-UP PRINCIPLE AND THE PERIODIC LAW

112		
	Th Pa U Np Pu Am Cm Bk Cf Es Fm Md No Lr	
	5f	
Os Ir Pt Au Hg		
	Ce Pr Nd Pm Sm Eu Gd Tb Ds Ho Er Tm Yb Lu	
	4f	
Ru Rh Pd Ag Cd		
Fe Co Ni Cu Zn		

When an atom has an incomplete p, d, or f subshell, there may be several different electron configurations that do not violate the Pauli exclusion principle. For example, how are the three $2p$ electrons in the nitrogen atom ($Z = 7$) distributed among the $2p$ orbitals? A study of the nitrogen atom spectrum shows that in the ground state there is one electron in each $2p$ orbital, with their spins all aligned in the same direction (e.g., $m_s = +\frac{1}{2}$ for each of the three $2p$ electrons). These electrons are said to be *unpaired*. The orbital diagram is shown in Figure 8.19. Other diagrams not violating the Pauli exclusion principle (such as two paired electrons in the $2p_x$ orbital and one electron in the $2p_y$ orbital) represent excited states of the nitrogen atom.

This kind of behavior, exhibited by the ground states of all atoms with

TABLE 8.2

ELECTRON CONFIGURATIONS OF ATOMS OF THE FIRST 20 ELEMENTS

Name	Symbol	Z	Electron configuration
Hydrogen	H	1	$1s$
Helium	He	2	$1s^2$ or [He]
Lithium	Li	3	[He]$2s$
Beryllium	Be	4	[He]$2s^2$
Boron	B	5	[He]$2s^2 2p$
Carbon	C	6	[He]$2s^2 2p^2$
Nitrogen	N	7	[He]$2s^2 2p^3$
Oxygen	O	8	[He]$2s^2 2p^4$
Fluorine	F	9	[He]$2s^2 2p^5$
Neon	Ne	10	[He]$2s^2 2p^6$ or [Ne]
Sodium	Na	11	[Ne]$3s$
Magnesium	Mg	12	[Ne]$3s^2$
Aluminum	Al	13	[Ne]$3s^2 3p$
Silicon	Si	14	[Ne]$3s^2 3p^2$
Phosphorus	P	15	[Ne]$3s^2 3p^3$
Sulfur	S	16	[Ne]$3s^2 3p^4$
Chlorine	Cl	17	[Ne]$3s^2 3p^5$
Argon	Ar	18	[Ne]$3s^2 3p^6$ or [Ar]
Potassium	K	19	[Ar]$4s$
Calcium	Ca	20	[Ar]$4s^2$

incomplete subshells, is summarized in a statement by Friedrich Hund called **Hund's rule:** *In the ground state, electrons are distributed among the different orbitals of a subshell so as to provide the highest possible number of electrons with the same spin quantum number, m_s.* Electron configurations that violate Hund's rule represent excited states.

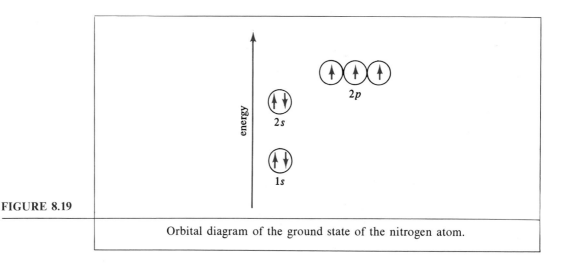

FIGURE 8.19

Orbital diagram of the ground state of the nitrogen atom.

Electrons go into different orbitals because, having like charges, they repel one another. This repulsion is minimized when they are in different orbitals that occupy different regions of space. Considering the nitrogen atom once again: Nothing is to be gained by putting two electrons in, say, the $2p_x$ orbital, because the $2p_x$, $2p_y$, and $2p_z$ orbitals all have the same energy. On the contrary, pairing the electrons in the $2p_x$ orbital puts two electrons in the same region of space. Thus a state of lower energy is obtained when there is only one electron in each $2p$ orbital. In the oxygen atom, which has one more electron ($Z = 8$), the additional electron must go into one of the $2p$ orbitals and have opposite spin ($m_s = -\frac{1}{2}$), so that the oxygen atom has two paired electrons in one $2p$ orbital and one electron in each of the other $2p$ orbitals with the same spin. Thus the oxygen atom has two unpaired electrons.

Considering now the electron configurations of the elements beyond calcium, we would expect that building up the next 10 elements, from scandium (Sc) to zinc (Zn), will fill the five orbitals of the $3d$ subshell ($m_l = +2, +1, 0, -1,$ or -2), and this is indeed the case, as indicated in Figure 8.18. Two of these elements, chromium (Cr) and copper (Cu), are somewhat peculiar in that the electron configurations one would expect, [Ar]$3d^44s^2$ and [Ar]$3d^94s^2$, respectively, represent excited states. The ground state configurations are [Ar]$3d^54s$ and [Ar]$3d^{10}4s$. Several other surprising configurations occur among the elements of still higher atomic number. (Detailed electron configurations for all of the elements are tabulated in Appendix E.) The basic reason for this seemingly unorthodox behavior is that the orbitals involved ($3d$ and $4s$ for chromium and copper) have about the same energy, so that relatively minor factors determine the ground state configuration. Further consideration of these problems will be postponed to Chapter 20.

Beyond zinc, which has the configuration [Ar]$3d^{10}4s^2$, the next six elements (gallium, Ga, to krypton, Kr) fill the $4p$ subshell; then the $5s$ subshell is filled, and so on.

The most striking feature of Figure 8.18 is the periodicity in the electron configurations of the outermost electrons, a periodicity that mirrors almost exactly that of the periodic table developed by Mendeleev and Meyer, as discussed in Chapter 6. For example, the elements along the left-hand side of Figure 8.18 each have one electron in an s orbital, and the elements along the left-hand side of the ordinary chemical periodic table (on the inside front cover) in group IA are the same ones, namely, hydrogen, lithium, sodium, potassium, rubidium, cesium, and francium. The 14 rare earth or lanthanide metals (cerium, Ce, to lutetium, Lu), which were a puzzling footnote to the periodic table, fall naturally into place as the 14 elements required to fill the seven orbitals ($m_l = +3, +2, +1, 0, -1, -2,$ or -3) of the $4f$ subshell. Because Figure 8.18 is drawn like an orbital diagram to indicate increasing orbital energies, it is upside down compared to the conventional periodic table.

It follows that the fundamental explanation of the periodic law, which provides a systematic way of treating so much of chemistry, is to be found in the electronic structure of the atoms.

8.10 ELECTRON CORRELATION

In the preceding section each atom was characterized by an electron configuration based on hydrogenlike orbitals. Although this treatment is sufficient to explain many important features of atomic structure and atomic spectra, for all atoms with more than one electron it is an *approximation* that can be very useful but does not give highly accurate results.

The basic problem with the orbital approximation is that it treats the electrons as independent particles. Each electron is considered as moving in the *average* electric field created by all the other electrons. But, because like charges repel, each electron repels every other electron. The repulsion energy between electrons is least when they are far apart, and therefore electrons tend to *correlate* their positions: If one electron is in a particular region of the atom, the other electrons are much more likely to be found in other regions. The independent-particle or orbital approximation neglects electron correlation.

It is rather as though we attempted to describe a busy intersection by an independent-motorist approximation. Because there is a high probability of finding an automobile traveling along each of the intersecting streets at any time, we would expect to have frequent collisions. But, in fact, drivers usually correlate their positions with the aid of signal lights.

Figure 8.20 shows the effect of electron correlation on the helium atom. Both curves describe the probability of observing the two electrons at various distances, r_{12}, between them. One curve is obtained by using the orbital approximation. The most probable distance (the maximum in the curve) is at 0.53 Å, and the average distance is 0.69 Å. The other curve is obtained by using a much more complicated and accurate wave function, and we can take it as being essentially identical with the probability curve for the real atom. It has a maximum at 0.58 Å, and the average distance is 0.75 Å. These results show that the orbital approximation underestimates the average distance between electrons. Because of this, the orbital approximation overestimates the electron–electron repulsion energy.

For electrons in different orbitals, correlation is greater when the electrons have the same spin than when they have opposite spin, an effect which can be traced to the Pauli exclusion principle. The greater tendency of electrons having the same spin to avoid one another appears to be the reason why electrons in singly occupied orbitals have the same spin in atomic ground states—why, for example, the 2p electrons in the nitrogen atom's ground state have the same spin (↑ ↑ ↑ , not ↑ ↑ ↓).

To estimate the energy of the helium atom, first consider the energy of the helium ion, He$^+$. The one-electron ions (He$^+$, Li^{2+}, etc.) can be treated in the same way as the hydrogen atom, taking into account the higher attractive force between the electron and the nucleus of electric charge Ze. The resulting expression for the energy is

$$E_n = -\frac{13.60 Z^2}{n^2} \text{ eV} \tag{8.10}$$

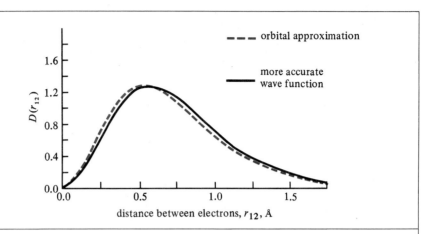

FIGURE 8.20 Calculated electron–electron radial distribution functions, $D(r_{12})$, for the helium atom. [SOURCE: L. S. Bartell and R. M. Gavin, Jr., Journal of the American Chemical Society, **86**, 3496 (1964). Copyright 1964 by the American Chemical Society. Reprinted by permission of the copyright owner.]

For the ground state of He^+ the energy is, therefore, $-(13.6) \times (4) = -54.4$ eV, and the energy required to remove the electron from the nucleus (called the **second ionization potential** of helium) is 54.4 eV.

$$He^+ \longrightarrow He^{2+} + e^- \qquad \Delta E = 54.4 \text{ eV}$$

If both electrons of the helium atom behaved the same as the electron in He^+ (in other words, if there were no electron–electron repulsion), the energy of the ground state of a helium atom would be $2(-54.4) = -108.8$ eV. The energy required to remove one electron from a helium atom (called the **first ionization potential** of helium) would then be 54.4 eV. But experimentally it is found that the first ionization potential is only 24.6 eV.

$$He \longrightarrow He^+ + e^- \qquad \Delta E = 24.6 \text{ eV}$$

On adding the last two equations, we obtain

$$He \longrightarrow He^{2+} + 2e^- \qquad \Delta E = 54.4 + 24.6 = 79.0 \text{ eV}$$

Thus removal of both electrons from a helium atom requires 79.0 eV of energy. If the energy of the separated nucleus and electrons ($He^{2+} + 2e^-$) is taken as zero, then the energy of the helium atom is -79.0 eV.

It is interelectronic repulsion that causes the first ionization potential to be so much smaller than the second ionization potential. An alternative manner of speaking is to say that each electron partially shields the nucleus from the other electron (see Section 8.7). When the interelectronic repulsion is included in the energy calculation with the orbital approximation, a fairly good result (-77.9 eV) is obtained for the energy of the atom, provided that one uses the best possible orbitals. These orbitals are obtained by a method of successive approximations called the **self-consistent field** method. One assumes specific orbitals (for example, hydrogenlike orbitals) for all but one electron. Then one solves the Schrödinger equation for the remaining electron moving in the average electric field created by the nucleus and all the other electrons. The other orbitals are varied in turn until no further improvement is obtained, indicating that the solution to the Schrödinger equation is self-consistent. (The best orbitals are called Hartree-Fock orbitals, named after the developers of this method, Douglas R. Hartree and Vladimir A. Fock.)

The energy difference between the orbital approximation (using Hartree-Fock orbitals) and the experimental value is called the **correlation energy**. For the helium atom it is $-77.9 - (-79.0) = 1.1$ eV. More accurate wave functions, which allow for correlation between the electrons, give better results. The best wave function obtained to date for the helium atom yields an energy essentially identical with the experimental energy.

8.11 ELECTRON DENSITIES AND ATOMIC RADII

In a many-electron atom, each electron contributes to the total electron "cloud." Figure 8.21 shows a comparison of the calculated and experimental radial distribution function, $D(r)$, for the electrons in the argon atom. (Compare with Figure 8.8, which shows $D(r)$ for the hydrogen atom.) The experimental curve was obtained by electron diffraction (Section 7.5), and the calculated curve is based on the orbital approximation.

Three peaks are clearly distinguished. The innermost is mainly due to the two $1s$ electrons, the middle peak represents the eight $2s$ and $2p$ electrons, and the eight $3s$ and $3p$ electrons have a broad outer maximum. Because the radial distribution function falls off slowly at large distances, there is no single distance one can characterize as the radius for the free atom, but most of the electron cloud is within 1 Å of the nucleus and practically all of it is within a radius of 2 Å.

Information on distances between atoms in collisions and distances between atoms in a molecule is used to calculate atomic radii. For example, the noble gases (group 0 elements) are all monatomic gases. The internuclear distance of closest approach of two noble gas atoms during a collision can be calculated from various measurements, such as the deviations from the perfect gas equation (see Section 4.6). Half this distance is taken as the collision radius, or van der Waals radius, of the noble gas atom. Radii for the various noble gas atoms are shown in Table 8.3.

As we go from lower to higher atomic numbers the electrons in each shell are attracted more strongly by the higher positive charge on the nucleus, and each shell is pulled in closer to the nucleus. For example, the peak in the radial distri-

8.11 ELECTRON DENSITIES AND ATOMIC RADII

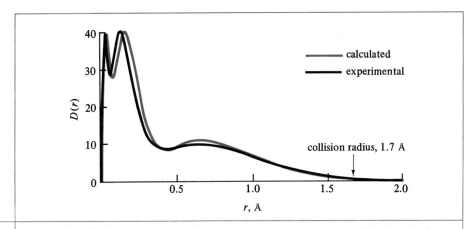

FIGURE 8.21 The radial distribution function for electrons in the argon atom. Calculated values were obtained by using the orbital approximation; experimental values were obtained by using electron diffraction. [SOURCE: L. S. Bartell and L. O. Brockway, Physical Review, **90**, 833 (1953).]

bution function of the $1s$ electron in the hydrogen atom is at 0.53 Å, whereas the $1s^2$ peak in argon ($Z = 18$) is only about one-eighteenth as far from the nucleus, or about 0.03 Å. Thus within each row (period) of the periodic table the size of the atom decreases from left to right.

With the next higher atomic number following each noble gas, a new shell of higher principal quantum number is added. For a higher principal quantum number the probability distribution function is shifted farther from the nucleus and therefore the atomic radius is increased. Thus within each column (group) of the periodic table the size of the atom increases with increasing atomic number. This effect is illustrated by the atomic radii in Table 8.3.

TABLE 8.3 VAN DER WAALS RADII OF NOBLE-GAS ATOMS

Atom	Atomic number	Highest principal quantum number	Radius (Å)
Helium	2	1	1.3
Neon	10	2	1.4
Argon	18	3	1.7
Krypton	36	4	1.8
Xenon	54	5	2.0
Radon	86	6	?

8.12 IONIZATION POTENTIALS AND ELECTRON AFFINITIES

The energy required to remove one electron from a gaseous atom, called the **ionization potential**,[3] is a property that gives considerable insight into the nature of atoms. The most important method of determining ionization potentials is an experimental one based on the atomic spectrum, as described for hydrogen in Sections 8.2 and 8.3. We have already seen that hydrogen and helium have ionization potentials of 13.6 and 24.6 eV, respectively. The ionization processes can be represented by

$$H = H^+ + e^- \quad \Delta E = +13.6 \text{ eV}$$
$$He = He^+ + e^- \quad \Delta E = +24.6 \text{ eV}$$

Following helium the ionization potential drops to the very low value of 5.4 eV for lithium, then gradually rises to a second peak at the second noble gas, neon. (These data represent the energy change per atom. To obtain the energy change per mole of atoms, 1 eV atom^{-1} corresponds to 96.48 kJ mol^{-1}.)

Figure 8.22 shows how the ionization potential changes with atomic number. The main features are a peak at each noble gas, followed by a sharp drop to a very low value for the first atom of the next period, then a gradual rise to the next noble gas. There are small breaks following filled subshells; for example, boron ($Z = 5$) has a lower ionization potential than beryllium ($Z = 4$). There are also small breaks following each half-filled subshell; for example, oxygen ($Z = 8$) has a lower ionization potential than nitrogen ($Z = 7$). For atoms in the same group of the periodic table, there is usually a decrease in ionization potential with increasing atomic number. Thus the peaks of the noble-gas atoms get progressively lower as we go toward higher atomic numbers.

The various features of Figure 8.22 can be explained in terms of the electronic structures of the atoms and the variations in size of the atoms discussed in the preceding section. Within each period the nuclear charge increases as we go from left to right and therefore the outermost electrons are more strongly attracted to the nucleus. Thus it becomes progressively more difficult to remove an electron from an atom as we go across a period from left to right. Following the noble gas at the end of a period the next electron is forced into an orbital of higher energy, and hence an electron is more easily removed. To a lesser extent the same thing happens when a subshell is filled. The breaks that occur at half-filled subshells are caused by the increased electron–electron repulsion when two electrons are forced to occupy the same orbital.

Many atoms attract an additional electron, forming a negative ion. For

---

[3] More precisely, the first ionization potential. Second, third, etc., ionization potentials are known for many atoms. For example, the second ionization potentials of helium, lithium, and beryllium are 54.4, 75.6, and 18.2 eV.

8.12 IONIZATION POTENTIALS AND ELECTRON AFFINITIES

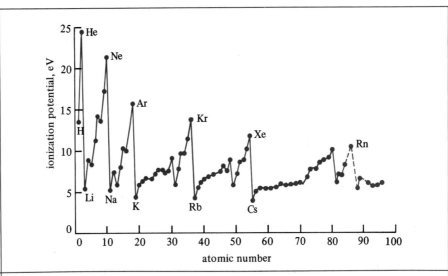

FIGURE 8.22 Dependence of ionization potential on atomic number. As the atomic number is an integer, the lines connecting the experimental points have no physical meaning, and are inserted only to exhibit trends.

example, the chlorine atom and an electron combine to form the very stable chloride ion, Cl^-:

$$Cl + e^- = Cl^- \quad \Delta E = -3.61 \text{ eV}$$

The energy evolved in this reaction ($-\Delta E = 3.61$ eV) is called the **electron affinity** of the chlorine atom. Some calculated and experimental electron affinities are listed in Table 8.4. The atoms with the highest electron affinities are those in group VIIA, immediately preceding the noble gases, called the **halogens**—fluorine, chlorine, bromine, and iodine. (The electron affinity of astatine, another halogen, is not known.)

The electron affinity of the atom can be regarded as simply the ionization potential of the negative ion,

$$Cl^- = Cl + e^- \quad \Delta E = 3.61 \text{ eV}$$

and electron affinities can be determined experimentally by spectroscopic methods. The theoretical value for hydrogen in Table 8.4 is based on a very accurate wave function; the others are based on the orbital approximation plus estimated corrections for electronic correlation and relativistic effects.

The atoms of the third period (sodium to chlorine) have higher electron affinities than the corresponding atoms of the second period (lithium to fluorine). The explanation of this phenomenon is not obvious, and in fact the discovery that

TABLE 8.4

ELECTRON AFFINITIES		
Atom	Theoretical values (eV)[a]	Experimental values (eV)[a]
Hydrogen	0.75415	0.77 ± 0.02
Lithium	0.58	...
Boron	0.30	...
Carbon	1.17	1.25 ± 0.03
Nitrogen	−0.27[b]	...
Oxygen	1.22	1.465 ± 0.005
Fluorine	3.37	3.448 ± 0.005
Sodium	0.78	...
Aluminum	0.49	...
Silicon	1.39	...
Phosphorus	0.78	...
Sulfur	2.12	2.07 ± 0.07
Chlorine	3.56	3.613 ± 0.003
Bromine	...	3.363 ± 0.003
Iodine	...	3.063 ± 0.003

[a] Three dots indicate an unknown value.
[b] The negative value for nitrogen indicates that N$^-$ is unstable, spontaneously decomposing to N + e^- with the evolution of 0.27 eV of energy.

fluorine has a lower electron affinity than chlorine was met with considerable incredulity. One can now say, with the benefit of hindsight, that electron–electron repulsion is greater in the small $n = 2$ shell than in the larger $n = 3$ shell, and this more than offsets the nuclear attraction term, which decreases within each group as the atomic number and size of the atom increase.[4] Of course, qualitative statements such as this are not as valuable as the quantitative predictions that can be made from the theory of the electronic structure of atoms, as shown in Table 8.4.

8.13 ELECTRON-DOT SYMBOLS

It is convenient to have graphic symbols for the electron configurations of the various atoms. In an **electron-dot symbol** the **kernel** (nucleus and all inner shells) is represented by the symbol for the element, and dots are used to represent the outermost electrons (called **valence** electrons, since they are responsible for the combining tendency, or valence, of atoms). These symbols will be especially useful in describing the electronic structure of molecules, to be considered in the next chapter.

---

[4] This is probably an oversimplification. The investigators who calculated the theoretical values for the first and second periods listed in Table 8.4 [E. Clementi, A. D. McLean, D. L. Raimondi, and M. Yoshimine, *Physical Review*, **133**, A1274 (1964)] state: "The physical reason why the electron affinity increases from first to second row presumably involves a balance of factors such as shielding and penetration effects, correlation effects, and the average distance of the valence electrons from the nucleus."

Electron-dot symbols for the first 18 elements are shown below.

H·   He:

Li·   Be: Ḃ: ·Ċ: ·N̈:   ·Ö: :F̈:   :N̈e:

Na·   Mg: Al̇:   ·Ṡi: ·P̈:   ·S̈:   :C̈l: :Är:

When two electrons have their spins paired they are placed together (:); otherwise they are placed at different positions. Thus the symbol ·N̈: conveys the information that the nitrogen atom has five valence electrons, of which three are unpaired. Note that, for the elements in groups IA–VIIA of the periodic table, the group number is the same as the total number of valence electrons.

Just where the electrons are placed is arbitrary (e.g., ·Ö: or :Ö·) because the electron-dot symbols are not intended to depict spatial relationships. What is shown, quickly and graphically, is the number and spin properties of the valence electrons. Periodic relationships are clearly evident, although the position of helium is ambiguous. Because it has two valence electrons, one might place it with the group IIA atoms (beryllium, magnesium, etc.). Because it has a filled electronic shell, one might place it with neon. As the physical and chemical properties of helium are closely similar to neon, the latter choice is made in the periodic table. Why are the group IIA atoms so different from helium? This interesting question will be discussed in the next chapter.

PROBLEMS

8.1 On the basis of Bohr's interpretation of atomic spectra, explain why spectral lines have the same frequencies in absorption and emission.

8.2 Calculate the wavelengths of the first two lines (those of longest λ) in the Balmer series and in the Lyman series.

8.3 (a) Calculate the energy required to ionize an excited hydrogen atom in which the electron occupies the $n = 4$ energy level.
(b) What is the energy change per mole of hydrogen atoms?

8.4 Which subshells in the following list violate the quantum rules?
3d, 4s, 2d, 1s, 1p, 4p

8.5 If $l = 3$, what can you say about n? If $m_l = 3$, what can you say about l?

8.6 If $l = 3$, what values can m_l have?

8.7 How many different orbitals are there in the 4d subshell? In the 4f subshell?

8.8 (a) Sketch two cubical volume elements of the same edge length at different distances from the nucleus. Compare the volumes of the two volume elements.
(b) Sketch two spherical-shell volume elements of the same thickness at different distances from the nucleus. Compare the volumes of the two volume elements.

8.9 (a) For the hydrogen atom in its ground state, what is the most probable distance of the electron from the nucleus?
(b) In an excited state of the hydrogen atom, is the most probable distance of the electron from the nucleus smaller, larger, or about the same as in the ground state?

8.10 The lowest transition shown in Figure 8.15 (from $3p$ to $3s$) gives the yellow sodium line. Will the lines from $4p$ and $5p$ to $3s$ be at shorter or longer wavelengths? Will the first lines of the other series be at shorter or longer wavelengths?

8.11 Write the ground state electron configurations of
(a) boron (B), $Z = 5$
(b) aluminum (Al), $Z = 13$
(c) gallium (Ga), $Z = 31$

8.12 Write the electron configurations of the following and indicate those that are isoelectronic (i.e., those that have the same number of electrons):
Ne, K^+, O^{2-}, Cl^-, F^-, Na^+

8.13 What would be the electron configuration of the sodium atom in its first excited state?

8.14 By referring to Figure 8.18, write the electron configurations of the following atoms in their ground states: vanadium (V), $Z = 23$; iron (Fe), $Z = 26$; arsenic (As), $Z = 33$; strontium (Sr), $Z = 38$; iodine (I), $Z = 53$. Compare your answers with the electron configurations listed in Appendix E.

8.15 What is the total number of unpaired electrons for each of the following atoms?
Ne, Na, S, Cl, Ca ($Z = 10$, 11, 16, 17, and 20, respectively.)

8.16 If an atomic beam of each of the atoms in Problem 8.15 were passed through an inhomogeneous magnetic field (Stern-Gerlach experiment), which ones would not be split?

8.17 The most probable distance between the nucleus and an electron is 0.26 Å in He^+, and 0.30 Å in He. Why is the distance larger in He?

PROBLEMS

8.18 Based on the trends in atomic radii discussed in Section 8.11, which element has the smallest atoms? Which has the largest?

8.19 In the series of elements He, Li, Be, B ($Z = 2$, 3, 4, and 5, respectively), Li has the lowest first ionization potential, but Be has the lowest second ionization potential. Why?

8.20 Draw graphs showing the variation of ψ along the z-axis for the $1s$, $2s$, and $2p_z$ orbitals of the hydrogen atom.

8.21 Compare the curves of ψ obtained in the preceding problem with those of the particle in a box (Figure 7.14). Note both similarities and differences.

CHAPTER 9
MOLECULAR STRUCTURE AND CHEMICAL BONDING

9

Although the description of the electronic structure of molecules proceeds along lines similar to the description of atoms in the preceding chapter, the presence of more than one nucleus complicates the picture. Then, too, there are many more different molecules than atoms to consider. Only 106 different elements are known, but their known combinations number in the millions. To facilitate the description of molecular structure, various levels of approximation are used. The crudest approaches are easy to use, and they emphasize the most important features of molecular structure. More accurate methods require more time and more skill but are necessary for a detailed understanding of molecular structure. We shall consider the simplest methods first and then go on to consider more accurate methods later.

Regardless of one's approach to molecular structure, the fundamental objective is to explain known molecular properties, and to predict and measure the unknown. Thus we shall begin our study with a discussion of some of the more important molecular properties.

9.1 MOLECULAR PROPERTIES

An important property of a molecule is its formula, which indicates the number of each type of atom in the molecule. Experimental methods of determining molecular formulas have been discussed in Chapters 2 and 3. One of the tasks of any theory of molecular structure is to explain why, for example, hydrogen atoms combine to form diatomic molecules, H_2. It is also of interest to know why certain molecules have not been found. Of course one must consider the possibility that unknown molecules may be synthesized by new reactions. The theory of molecular structure is useful in suggesting interesting new compounds and routes to their preparation. Indeed, the discovery of noble gas compounds such as the xenon fluorides a few years ago was a dramatic confirmation of predictions from theory.

A second important property of a molecule is its structure. Adjacent pairs of atoms in a molecule are held together by strong forces called **chemical bonds.** Furthermore, although the atomic nuclei are constantly vibrating, they tend to have a certain definite arrangement in space relative to one another, called the **molecular structure,** or **molecular geometry.** For example, in the water molecule the average distance between the oxygen nucleus and either of the hydrogen nuclei (the **O—H bond length**) is 0.96 Å. The average angle defined by the nuclei, with the oxygen at the apex (called the **HOH bond angle**) is 104.5° (Figure 9.1).

9.1 MOLECULAR PROPERTIES

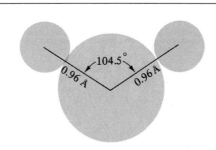

FIGURE 9.1

Structure of the water molecule. The average O—H bond length is 0.96 Å, and the average HOH bond angle is 104.5°. The circles have radii proportional to the respective bond radii (see Section 9.9); they provide an estimate of the relative sizes of the atoms.

Bond lengths and bond angles are determined from diffraction patterns (both electron diffraction and X-ray diffraction are used) and also from molecular spectra (the frequencies at which molecules absorb or emit light) in the microwave and infrared regions. Information about molecular structure can often be obtained from other experimental evidence; examples will be described later.

A third important property of a molecule is the enthalpy required to break each of the bonds in the molecule to form separated atoms. For example, consider the process in which the O—H bonds in water molecules are broken, with the formation of hydrogen atoms and oxygen atoms. If each substance is in the gaseous state at 25°C, the enthalpy required for this process is 926.88 kJ per mole of water:

$$H_2O(g) = 2H(g) + O(g) \qquad \Delta H_1 = 926.88 \text{ kJ} \qquad (9.1)$$

As two identical oxygen-hydrogen bonds are broken in this process, the enthalpy required to break each bond, called the **average O—H bond enthalpy**, is $\frac{1}{2}(926.88) = 463.44$ kJ.

One may just as well work with energies instead of enthalpies, obtaining the **average O—H bond energy.** The $\Delta(PV)$ term that distinguishes ΔH from ΔE (see Eq. 5.19) is usually relatively small, and therefore bond energies are not very different from bond enthalpies. In fact it is fairly common practice to work with bond enthalpies and call them bond energies.

A direct measurement of ΔH_1 for Eq. 9.1 is not possible, but, with the aid of Hess's law (Section 5.7), ΔH_1 can be determined indirectly from the following data (all at 25°C):

$$H_2(g) = 2H(g) \qquad \Delta H_2 = 435.89 \text{ kJ} \qquad (9.2)$$

$$O_2(g) = 2O(g) \qquad \Delta H_3 = 498.31 \text{ kJ} \qquad (9.3)$$

$$2H_2(g) + O_2(g) = 2H_2O(g) \qquad \Delta H_4 = -483.67 \text{ kJ} \qquad (9.4)$$

(The first two quantities are obtained from the absorption spectra of hydrogen and oxygen, respectively, and the last is determined calorimetrically.) Because reaction 1 is equal to reaction $2 + (\frac{1}{2} \times \text{reaction } 3) - (\frac{1}{2} \times \text{reaction } 4)$,

$$\begin{aligned}\Delta H_1 &= \Delta H_2 + \tfrac{1}{2}\Delta H_3 - \tfrac{1}{2}\Delta H_4 \\ &= 435.89 + \tfrac{1}{2}(498.31) - \tfrac{1}{2}(-483.67) \\ &= 926.88 \text{ kJ}\end{aligned}$$

Molecules also have the properties emphasized in the preceding chapter for atoms—absorption and emission spectra, ionization potentials, electron affinities, and electron density distributions. Even the bulk properties of matter in the solid, liquid, and gas states—such as density, viscosity, heat capacity, and melting and boiling points—depend on the properties of the individual molecules and the interactions between them. In addition, an understanding of how and why chemical reactions occur requires a knowledge of the alterations in molecular structure that take place during the reactions as some chemical bonds are broken and other chemical bonds are formed. All aspects of chemistry are profoundly affected by molecular structure and chemical bonding.

One of the most important types of chemical bond (and perhaps the easiest to understand) is the ionic bond, so we shall consider this type first.

9.2 SIMPLE IONIC BONDING THEORY

In the preceding chapter we saw that the electronic structures of the noble gases were similar in that all but helium have eight electrons in their outermost shell. Because the noble gases are extremely unreactive chemically, the presence of eight electrons, or an octet, in the outermost shell must constitute a very stable structure. Additional evidence of the stability of the octet of electrons in the outermost shell is provided by the high ionization potentials of the noble gases as compared to those of the elements of group IA, and the strong tendency of the group VIIA elements to gain an electron. For example, the sodium atom in group IA has a relatively low ionization potential because by losing its valence electron (see Section 8.13) it achieves the stable neon structure with its outer octet ($1s^2 2s^2 2p^6$). The chlorine atom in group VIIA has a high electron affinity because the chloride ion, Cl^-, has the argon structure with its outer octet ($1s^2 2s^2 2p^6 3s^2 3p^6$).

It is thus easy to see in a qualitative way why sodium and chlorine should combine to form sodium chloride, NaCl. The chlorine atom gains an electron at the expense of the sodium atom, so that the molecule might be best represented as Na^+Cl^-. Or, making use of electron-dot symbols, the combination of a sodium atom and a chlorine atom may be represented as

$$\text{Na}\cdot \; + \; \cdot\ddot{\underset{..}{\text{Cl}}}: \; = \; \text{Na}^+ : \ddot{\underset{..}{\text{Cl}}} :^-$$

Because the two ions have unlike charges, they attract each other. This type of bond is called an **ionic bond.**

Each of the elements in group IA below hydrogen (called the **alkali metals**) combines with each of the group VIIA elements (called the **halogens**) in the same way. Both the positive and negative ions thus formed have noble gas electronic structures, and they are held together by ionic bonds. For example,

$$\text{Li}\cdot + \cdot\ddot{\underset{..}{\text{Br}}}: = \text{Li}^+ : \ddot{\underset{..}{\text{Br}}}:^-$$

$$\text{K}\cdot + \cdot\ddot{\underset{..}{\text{F}}}: = \text{K}^+ : \ddot{\underset{..}{\text{F}}}:^-$$

Li^+ has the helium structure, F^- the neon structure, K^+ the argon structure, and Br^- the krypton structure. Except for Li^+, which has only two electrons, each ion has the outer shell of eight electrons characteristic of the noble gases beyond helium.

Similarly, the elements in group IIA (the **alkaline earth metals**), with two loosely held electrons, combine with the elements of group VIA (the **chalcogens**) to form doubly charged ions:

$$\text{Mg}: + \cdot\ddot{\underset{.}{\text{O}}}: = \text{Mg}^{2+} : \ddot{\underset{..}{\text{O}}}:^{2-}$$

It takes two halogen atoms to remove the two valence electrons from a group IIA atom, so the general formula of an alkaline earth metal halide is MX_2 (where M is the metal atom and X is the halogen). For example,

$$\text{Mg}: + 2\cdot\ddot{\underset{..}{\text{F}}}: = \text{Mg}^{2+}[:\ddot{\underset{..}{\text{F}}}:^-]_2$$

One chalcogen atom requires two alkali metal atoms, so the general formula of the compounds formed between these groups is M_2X. For example,

$$2\text{Cs}\cdot + \cdot\ddot{\underset{.}{\text{S}}}: = [\text{Cs}^+]_2 : \ddot{\underset{..}{\text{S}}}:^{2-}$$

Note that to form an electrically neutral compound the total number of electrons lost by one element must equal the total number of electrons gained by the other element.

The electric charge on an ion, which is responsible for its tendency to form ionic bonds with ions of opposite sign, is called its **ionic valence.** (The term *valence* is used in a general way to denote the tendency of an atom to combine with other atoms.) Thus the ionic valence of Cs^+ is $+1$, and of S^{2-} is -2.

It is clear from the above reactions that atoms combine in exactly the ratios required to give each ion a noble gas structure. For example, magnesium fluoride has the formula MgF_2, and not MgF or Mg_2F. We can summarize the tendency of group IA and IIA elements to form ionic bonds with the elements of group VIA and VIIA by the **octet rule:** *Each atom gains or loses electrons to form an ion having an outer octet of electrons.* (Because the first noble gas, helium, has only two electrons, elements near helium in the periodic table are exceptions to the octet rule. For example, lithium and beryllium form Li^+ and Be^{2+}, respectively, having the helium structure, $1s^2$.)

The coulombic attraction between oppositely charged ions causes ionic molecules to form larger aggregates. The alkali metal halides form crystals in which there is one metal ion (M^+) for each halide ion (X^-), so that the simplest formula is MX, but each metal ion is surrounded by several halide ions, and each halide ion is surrounded by several metal ions. The ions are arranged in a geometrical pattern that imparts to the crystal its plane faces and sharp edges.

Most of the alkali metal halides have the same structure as sodium chloride (Figure 9.2). Except for the relatively few ions on the surface of the crystal, each Cl^- ion is surrounded by six Na^+ ions, whose nuclei lie at the corners of an octahedron—four Na^+ ions around the Cl^- at the corners of a square, a fifth Na^+ ion above the Cl^-, and a sixth Na^+ ion below. The arrangement of Cl^- ions around each Na^+ ion is the same—six Cl^- ions at the corners of an octahedron.

It is impossible to pick out individual NaCl molecules in a crystal, because all six neighbors of each ion are equally distant. One might regard the crystal as one giant molecule, with a molecular formula Na_NCl_N, where N is the number of sodium and chloride ions in the crystal. Because this is an awkward notation, it is conventional to use simplest formulas instead of molecular formulas for ionic compounds.

Other ionic molecules, such as MgF_2, form crystals with more complicated structures. They will be discussed in Chapter 17, along with methods for determining the structures of crystals.

9.3 INTERNUCLEAR DISTANCES IN IONIC CRYSTALS

A convenient way to discuss internuclear distances is to treat each ion as a hard sphere with a specific radius. Let us denote the radius of a potassium ion as R_{K^+}, and the radius of a fluoride ion as R_{F^-}. If now the positive and negative ions in an ionic crystal (such as potassium fluoride) are in contact with one another, then the internuclear distance in the crystal should be the sum of the ionic radii. If we denote the internuclear distance in a potassium fluoride crystal as R_{KF}, then $R_{KF} = R_{K^+} + R_{F^-}$.

That this is a fairly good approximation is shown by taking differences in the internuclear distances of pairs of alkali metal halide crystals with a common anion, for example, KX and NaX, where X is one of the halogens. Because

$$R_{KX} - R_{NaX} = (R_{K^+} + R_{X^-}) - (R_{Na^+} + R_{X^-}) = R_{K^+} - R_{Na^+}$$

the differences should be the same for the various anions. The experimental data are

$$R_{KF} - R_{NaF} = 2.66 - 2.31 = 0.35 \text{ Å}$$

$$R_{KCl} - R_{NaCl} = 3.14 - 2.81 = 0.33 \text{ Å}$$

$$R_{KBr} - R_{NaBr} = 3.29 - 2.98 = 0.31 \text{ Å}$$

$$R_{KI} - R_{NaI} = 3.53 - 3.23 = 0.30 \text{ Å}$$

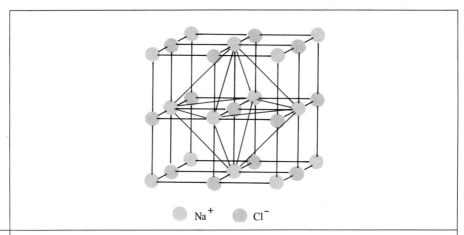

FIGURE 9.2 Structure of the sodium chloride crystal. The octahedral arrangement of the Na$^+$ ions around the central Cl$^-$ ion is indicated in color.

It may be seen that the differences are approximately the same. Except for lithium compounds, other pairs of alkali metal halides give results of about the same consistency.

But just what is the radius of each ion? The structural data on, say, NaCl (obtained by the diffraction of X rays) provide only the internuclear distance, 2.31 Å. What part of this distance is R_{Na^+}, and what part is R_{Cl^-}? One method of answering this question is to consider crystals of isoelectronic ions, that is, ions with the same electron configuration. An example would be NaF in which both Na$^+$ and F$^-$ have 10 electrons. The higher nuclear charge of the Na$^+$ should pull in the electrons closer to the nucleus, so that R_{Na^+} is smaller than R_{F^-}. Estimates of the magnitude of the "effective nuclear charge" experienced by an electron in the outer shell (taking into account both the nuclear charge and the shielding effect of the other electrons) lead to a division of the observed distance of 2.31 Å into ionic radii of 0.95 and 1.36 Å for the sodium and fluoride ions, respectively. Other methods of calculating ionic radii give similar results.

Some radii of ions in crystals based on this method are listed in Table 9.1 and illustrated in Figure 9.3. Within each horizontal row the ions have the same number of electrons, and the nuclear charge increases from left to right. Note how the ionic radius decreases from left to right as the electrons are pulled in toward the nucleus by its higher positive charge. Note also that within each column the ionic radius increases as the total number of electrons increases.

9.4 BOND ENERGIES OF IONIC MOLECULES

In a qualitative sense ionic bonds owe their strength to the coulombic force of attraction between ions of opposite sign. With the aid of the hard-sphere approximation, one may calculate the energy of this coulombic attraction quantitatively

TABLE 9.1 **SOME IONIC CRYSTAL RADII**[a]

		Li⁺	Be²⁺
		0.60	0.31
O²⁻	F⁻	Na⁺	Mg²⁺
1.40	1.36	0.95	0.65
S²⁻	Cl⁻	K⁺	Ca²⁺
1.84	1.81	1.33	0.99
Se²⁻	Br⁻	Rb⁺	Sr²⁺
1.98	1.95	1.48	1.13
Te²⁻	I⁻	Cs⁺	Ba²⁺
2.21	2.16	1.69	1.35

[a] Values expressed in ångströms.

and use it to estimate the bond energies of alkali metal halide gas molecules. As we shall see, the results are in good agreement with bond energies determined experimentally, thus providing further evidence for the validity of the hard-sphere approximation.

For hard spheres the energy required to separate the ions is simply the *coulombic energy of attraction,* and this is proportional to the product of the electric charge on each ion. The energy is also inversely proportional to the distance (R) between their centers (nuclei). (The electric charge on a sphere may be treated as though the entire charge were concentrated at the center of the sphere.) For an alkali metal halide molecule the resulting formula for the attraction energy takes the very simple form

$$\Delta E = \frac{14.40}{R} \text{ eV Å} \tag{9.5}$$

where R is expressed in ångströms. This equation is derived from Coulomb's law (see Appendix G-2). It gives the energy change for the *dissociation of the molecule into positive and negative ions,*

$$\text{MX(g)} = \text{M}^+(\text{g}) + \text{X}^-(\text{g}) \tag{9.6}$$

For the NaCl gaseous molecule, the internuclear distance R is 2.51 Å, so ΔE is $14.40/2.51 = 5.74$ eV for dissociation *to ions.*

At high temperatures, however, sodium chloride dissociates to *atoms* rather than ions as evidenced by the yellow color it imparts to flames. (The yellow light emitted by a sodium atom at high temperatures results from an electronic transition of its valence electron, as discussed in Section 8.7.) Thus it is of more interest to calculate the energy change for the process

$$\text{NaCl(g)} = \text{Na(g)} + \text{Cl(g)} \tag{9.7}$$

9.4 BOND ENERGIES OF IONIC MOLECULES

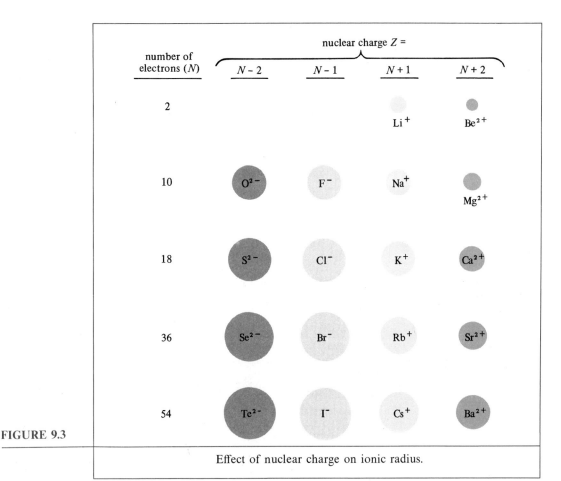

FIGURE 9.3

Effect of nuclear charge on ionic radius.

This can be calculated indirectly, because Eq. 9.7 is the sum of:

$$NaCl(g) = Na^+(g) + Cl^-(g) \quad \Delta E_1 = 5.74 \text{ eV} \quad (9.8)$$

$$Na^+(g) + e^-(g) = Na(g) \quad \Delta E_2 = -5.14 \text{ eV} \quad (9.9)$$

$$Cl^-(g) = Cl(g) + e^-(g) \quad \Delta E_3 = 3.61 \text{ eV} \quad (9.10)$$

$$NaCl(g) = Na(g) + Cl(g) \quad \Delta E_4 = 5.74 - 5.14 + 3.61$$
$$= 4.21 \text{ eV}$$

(In the preceding paragraph ΔE_1 was calculated from Eq. 9.5. Because Eq. 9.9 is just the reverse of the ionization of the sodium atom, the value of ΔE_2 is the negative value of the ionization potential of sodium, tabulated in Appendix E. The value of ΔE_3 is equal to the electron affinity of chlorine—see Table 8.4.)

The value of 4.21 eV calculated above for dissociation to atoms is in excellent agreement with the experimental results, 4.24 ± 0.05 eV. Because the calculated value is based on the hard-sphere approximation, the good agreement with experiment provides evidence that this is indeed a good approximation. Calculations by this method on other alkali metal halides give results of similar accuracy. We may conclude that the hard-sphere approximation provides a useful approach to an understanding of bond energies as well as to an understanding of internuclear distances in ionic crystals. (All the above energies are for single atoms or molecules. One may convert to energies per mole by using the fact that 1 eV per atom or per molecule is equal to 96.48 kJ mol^{-1}.)

A more detailed discussion of the theory of ionic bonding is given in Section 17.6 in connection with a more extensive description of the structures of ionic crystals.

9.5 COVALENT BONDING—LEWIS STRUCTURES

There are many molecules in which atoms of the same element or similar elements are bonded to one another. For example, fluorine atoms combine, producing diatomic molecules:

$$2F = F_2$$

In this kind of molecule there should be little tendency for ions to form. Although each fluorine atom has a substantial electron affinity, if one atom gains an electron the other must lose one—both cannot achieve the neon structure through transfer of electrons. How, then, can one explain the attractive force between like atoms or similar atoms?

A solution to the problem of describing bonds between like atoms is to consider that the atoms cooperate, sharing *two* electrons. An electron-dot representation of this process is

$$:\!\ddot{F}\!\cdot\; +\; \cdot\!\ddot{F}\!: \;=\; :\!\ddot{F}\!:\!\ddot{F}\!:$$

The shared pair of electrons (shown between the kernels) is counted as if it belonged to both atoms, so that in a formal sense each has the neon structure. Cooperative satisfaction of the combining tendency or valence of an atom is called **covalence**, and the bond thus formed is called a shared electron-pair bond, or **covalent bond**. When two atoms share one pair of electrons, as in F_2, the bond is called a **single bond**.

A dash is often used instead of two dots to represent a shared electron pair. Either symbol

$$:\!\ddot{F}\!\!-\!\!\ddot{F}\!: \quad \text{or} \quad :\!\ddot{F}\!:\!\ddot{F}\!:$$

is a satisfactory way of denoting the fluorine molecule. The concept of electron-pair bonds and the above method of representing covalent molecules were first proposed by Gilbert N. Lewis, and these representations are called **Lewis structures.**

Because each fluorine atom in the F_2 molecule has acquired the neon structure, the molecule shows no tendency to combine with a third fluorine atom, and the molecule F_3 is neither expected nor observed. Thus covalent bonding, like ionic bonding, fulfills the combining tendency of an atom.

To describe exactly how atoms share electrons requires consideration of the electronic orbitals in the molecule. Before taking up this aspect of covalent bonding, we shall first consider the general problem of writing Lewis structures for a variety of molecules and show how some molecular properties can be related to covalent bonding in a relatively simple way.

The general procedure for writing a Lewis structure is as follows:

1. Determine or estimate the molecular structure and place the symbols of the kernels to show the relative positions of the nuclei. (Even when the molecular structure is not known, it is often possible to make a reasonable guess or guesses, based on one's knowledge of the known structures of related molecules.) Because there must be at least one bond between each pair of adjacent nuclei to hold them together, place a dash (indicating an electron-pair bond) between adjacent nuclei.
2. Add the number of valence electrons of all atoms in the molecule to obtain the total number of electrons to be assigned. (If a molecule is a positive or negative ion, the total number of electrons must be adjusted for the number of electrons lost or gained when the ion is formed from neutral atoms.)
3. From this total number of electrons subtract the number already shown as single bonds between adjacent nuclei, and distribute the remainder in accordance with the octet rule: Each atom tends to have the noble gas structure of eight outer electrons, counting both shared and unshared electrons. (Of course, the hydrogen atom is an exception to the octet rule, because the neighboring noble gas, helium, has only two electrons. As a hydrogen atom needs only one electron to attain the helium structure, it forms only one covalent bond.)

To illustrate this procedure, consider chlorine monoxide, Cl_2O. It is known to have a bent structure, with each chlorine atom bonded to the oxygen atom. The kernels are accordingly placed as follows with a single bond between each Cl and the O:

$$Cl \diagdown \diagup Cl$$
$$O$$

The total number of valence electrons is $(2 \times 7) + 6 = 20$. The two single bonds require 4 electrons so that there are (20 − 4) or 16 electrons left to be assigned. To satisfy the octet rule we assign 4 unshared electrons to the oxygen atom and 6 to each chlorine. Thus a reasonable structure is

$$\ddot{\underset{..}{Cl}} \diagdown \underset{..}{\overset{..}{O}} \diagup \ddot{\underset{..}{Cl}}$$

As another example consider the nitrogen molecule, N_2. For a diatomic molecule there is no problem concerning kernel locations,

N—N

The total number of valence electrons is $2 \times 5 = 10$. After deducting two electrons for the one single bond, we have eight electrons or four pairs to be assigned. If two pairs are assigned to each atom, we obtain

$$:\!\ddot{N}\!-\!\ddot{N}\!:$$

As each N then has only six electrons, this structure violates the octet rule. One cannot add more electrons, as there are only 10 valence electrons in the molecule. The only solution is for the atoms to share more than one pair of electrons. The situation may be improved by moving an unshared pair into the region between the nuclei, forming a **double bond** (two shared electron pairs):

$$:\!\ddot{N}\!=\!N\!:$$

Then one of the atoms has a valence octet. By moving an unshared pair from the other nitrogen atom into the region between the kernels, one finally attains a satisfactory structure in which both atoms obey the octet rule:

$$:\!N\!\equiv\!N\!: \quad \text{or} \quad :N::\!:N:$$

Thus the nitrogen atoms are held together by three shared electron pairs, called a **triple bond**.

In these examples, the fluorine and chlorine atoms (with seven valence electrons) each form one single bond, the oxygen atom (with six valence electrons) forms two single bonds, and the nitrogen atom (with five valence electrons) forms a triple bond. These results can be generalized as the $8 - N$ *rule* (read "eight minus N"): *If an atom has N valence electrons, it often forms $8 - N$ covalent bonds.* The reason for the $8 - N$ rule is that when atoms contribute equally to shared pairs, an atom gains one electron from each bond, and $8 - N$ bonds are necessary to reach an outer octet. (When the $8 - N$ rule is used, a double bond counts as two covalent bonds and a triple bond as three covalent bonds.)

Those elements most likely to obey the $8 - N$ rule are found in groups IVA–VIIA of the periodic table (see Table 9.2). Even within these groups there are

9.6 VIOLATIONS OF THE OCTET RULE

TABLE 9.2   **THE 8 − N RULE**

Group	N	Number of covalent bonds	Bond types
IVA	4	8 − 4 = 4	Four single Two double Double + two single Triple + single
VA	5	8 − 5 = 3	Three single Double + single Triple
VIA	6	8 − 6 = 2	Two single Double
VIIA	7	8 − 7 = 1	Single

exceptions to the 8 − N rule, but it is obeyed often enough to be of considerable use, particularly in organic chemistry. (Exceptions occur when one atom contributes both electrons to the shared pair, as described in the next section.)

For groups IVA–VIA there are two or more combinations of bond type that satisfy the 8 − N rule. They are listed in Table 9.2. (Groups IA–IIIA are not listed—they rarely obey the 8 − N rule.)

As an illustration of the usefulness of the 8 − N rule, consider formaldehyde, H_2CO. Carbon, in group IVA, usually forms 8 − 4 = 4 covalent bonds. In this molecule the only bond types compatible with the bivalency of oxygen (in group VIA) and the univalency of hydrogen are a double bond to the oxygen atom, and a single bond to each hydrogen atom. Thus the Lewis structure of formaldehyde is

$$\begin{array}{c} H \\ \diagdown \\ C=\ddot{\underset{\ddot{}}{O}} \\ \diagup \\ H \end{array}$$

9.6 VIOLATIONS OF THE OCTET RULE

Besides hydrogen, which has been noted above as an exception to the octet rule, there are two important molecular types having covalent bonding in which the octet rule is violated.

First, the elements in groups IA–IIIA often form molecules in which the number of electrons in one atom's valence shell is smaller than eight. For example, the group IA elements form a series of dimers with a single covalent bond, as in K—K. Second, the elements beyond the first short period sometimes form molecules in which the number of electrons in one atom's valence shell exceeds eight; for example, phosphorus forms two important fluorides, PF_3 and PF_5. Phosphorus trifluoride obeys the octet rule:

$$:\!\overset{..}{\underset{..}{F}}\!-\!\overset{..}{\underset{|}{P}}\!-\!\overset{..}{\underset{..}{F}}\!:$$
$$:\!\overset{..}{\underset{..}{F}}\!:$$

and one might expect that the valence of phosphorus would be saturated in this molecule. But in fact the PF_3 molecule combines with fluorine to form a pentafluoride:

$$PF_3 + F_2 = PF_5$$

In the Lewis structure for the pentafluoride the phosphorus atom has ten electrons:

(Lewis structure of PF_5 with five F atoms around central P)

The fact that some exceptions to the octet rule do exist should not be overemphasized—the octet rule is obeyed so often that it is a very useful generalization.

9.7 DONOR–ACCEPTOR BONDS AND FORMAL CHARGE

Both electrons of a covalent bond may come from one atom, none from the other atom. To illustrate this unequal partnership, we shall work out the Lewis structure for sulfuric acid, H_2SO_4. The atoms are known to be arranged in the following manner:

(Skeletal structure of H_2SO_4: two H—O groups and two O atoms bonded to central S)

As both sulfur and oxygen have six valence electrons per atom, the total number of valence electrons is $(5 \times 6) + (2 \times 1) = 32$. After one electron pair is placed between each pair of adjacent kernels, as shown in the previous structure, the remaining $(32 - 12)$ or 20 electrons are then distributed to obtain an octet around each O and S kernel:

(Lewis structure of H_2SO_4 with lone pairs shown on each O)

9.7 DONOR–ACCEPTOR BONDS AND FORMAL CHARGE

Note that two oxygen atoms have their normal bivalence, but that the other oxygens are univalent and the sulfur is quadrivalent. This is because the sulfur atom contributes *both* electrons to its bond with each univalent oxygen atom. Such bonds are called **donor–acceptor bonds.** The presence of these donor–acceptor bonds in a molecule may be indicated by assigning an electric charge to each atom in the molecule. The simplest way to do this is to assume that each pair of shared electrons is shared equally between the two atoms. This leads to zero charge on each atom unless there is a donor–acceptor bond in the molecule. Charges assigned in this way are called **formal charges.**

To assign formal charges, we begin by noting that an atom having N valence electrons has a charge on its kernel of $+N$ (in electronic charge units). For example, in H_2SO_4 the charge on the sulfur kernel is $+6$. If its four pairs of shared electrons are shared equally between the sulfur and oxygen atoms, each pair contributes -1 to the sulfur atom, or a total of -4. Thus the net charge on the sulfur atom is $+2$. In a similar way, we find that the oxygen atoms to which no hydrogen atoms are bonded have formal charges of -1 (-6 due to the six unshared electrons, -1 due to the shared pair, and $+6$ due to the kernel charge). The formal charges of the other atoms are all zero. The Lewis structure including the formal charges is therefore

$$H-\overset{..}{\underset{..}{O}}\diagdown\overset{2+}{S}\diagup\overset{..}{\underset{..}{O}}:^-$$
$$H-\overset{..}{\underset{..}{O}}\diagup\diagdown\overset{..}{\underset{..}{O}}:^-$$

The reason for the formal charges is that the sulfur atom contributes both electrons to its bonds with the univalent oxygen atoms. One may consider each bond on the right as a combination of a covalent bond and an ionic bond, and it is sometimes called a **semipolar double bond.** Other terms are "coordinate link" and "coordinate covalent bond."

An equation summarizing the method of assigning formal charges is

$$\text{formal charge} = (\text{kernel charge}) - (\text{number of unshared electrons}) - \tfrac{1}{2}(\text{number of shared electrons}) \tag{9.11}$$

Another example of a donor–acceptor bond is found in the ammonium ion, NH_4^+, which is formed by addition of a hydrogen ion, H^+, to the ammonia molecule, NH_3.

$$H-\underset{\underset{H}{|}}{\overset{..}{N}}-H + H^+ = H-\underset{\underset{H}{|}}{\overset{\overset{H}{|}}{\overset{+}{N}}}-H$$

Although the nitrogen atom donates both electrons to bind the H^+, in the resulting complex ion the N—H bonds are identical, and one cannot specify any particular one as a donor–acceptor bond. However, there is a formal charge on the nitrogen atom. Using Eq. 9.11, it is

$$+5 - 0 - \tfrac{1}{2}(8) = +1$$

Formal charge is particularly useful in discussing polyatomic ions because it is a simple way of assigning the net charge on the complex ion to one or more atoms. Examples are BF_4^-, SO_4^{2-}, and ClO^-:

$$\begin{array}{ccc}
:\!\ddot{F}\!: & :\!\ddot{O}\!:^- & \\
\;\;\;| & \;\;\;| & \\
:\!\ddot{F}\!-\!\underset{\underset{\displaystyle:\ddot{F}:}{|}}{\overset{-}{B}}\!-\!\ddot{F}\!: & \;^-\!:\!\ddot{O}\!-\!\underset{\underset{\displaystyle:\ddot{O}:_-}{|}}{\overset{2+}{S}}\!-\!\ddot{O}\!:^- & :\!\ddot{Cl}\!-\!\ddot{O}\!:^-
\end{array}$$

There is a connection between formal charge and violations of the $8 - N$ rule. When the $8 - N$ rule is obeyed for one of the atoms in groups IVA–VIIA, the formal charge on that atom is zero. Conversely, when the $8 - N$ rule is violated, the formal charge usually has a nonzero value.

Assignment of formal charges is a rather arbitrary procedure, whence the adjective "formal." To obtain a more accurate idea of actual net charges, one must take into account the polarity of each bond. This topic will be discussed in Section 9.13.

9.8 COVALENT BOND ANGLES AND THE VSEPR THEORY

A simple approach to molecular geometry is the valence-shell electron-pair repulsion theory (VSEPR theory). This theory is based on the coulombic repulsion between electrons, which causes the various pairs of electrons to repel one another, and therefore to take up positions as far apart as possible. (There is also, of course, a repulsion between the two electrons in each pair. This point will be considered later; it is not necessary to take it into account in the VSEPR theory.)

Because of electron-pair repulsions, in any polyatomic molecule the bond angles about a central atom depend on the *sum* of (1) the number of *atoms* bonded to the central atom, and (2) the number of unshared pairs of electrons ("lone pairs") on the central atom. Let us first compare the predictions of the VSEPR theory with the experimentally determined bond angles for those molecules in which there are no lone pairs on the central atom (so we need consider only the number of atoms bonded to the central atom).

Two Bonded Atoms; No Lone Pairs on the Central Atom

In molecules of this type, the repulsion between the bonding electrons to the two neighboring atoms is minimized when the bonded atoms are on opposite sides of the central atom, and therefore *the bond angles are 180°*. As examples, consider $BeCl_2$, CO_2, and acetylene, C_2H_2.

$$:\!\ddot{\underset{..}{Cl}}\!-\!Be\!-\!\ddot{\underset{..}{Cl}}\!: \qquad :\!\ddot{O}\!=\!C\!=\!\ddot{O}\!: \qquad H\!-\!C\!\equiv\!C\!-\!H$$

The repulsion is between two single bonds in $BeCl_2$, between two double bonds in CO_2, and between a single bond and a triple bond in C_2H_2. (Each carbon atom is treated in turn as the "central atom," and thus both HCC bond angles are 180°.) All of these molecules are linear, that is, the nuclei lie along a straight line.

Three Bonded Atoms; No Lone Pairs on the Central Atom

Here the repulsion between the bonding electrons to the three neighboring atoms is minimized when the four nuclei lie in the same plane, *with bond angles of 120°, or close to 120°*. Examples include BCl_3 and formaldehyde, H_2CO.

Both molecules are planar. In BCl_3 the Cl—B—Cl bond angles are all 120°. In formaldehyde the HCO bond angles are both 121.7°, and the HCH bond angle is 116.6°. (This difference in bond angles indicates that double-bond–single-bond repulsion is somewhat larger than the repulsion between two single bonds, as might be expected from the larger number of electrons involved.)

Four Bonded Atoms; No Lone Pairs on the Central Atom

Four single bonds are farthest apart when they are directed toward the corners of a regular tetrahedron (Figure 9.4). This is the observed molecular structure of many molecules and ions, such as CH_4, NH_4^+, CCl_4, and SiF_4 (Figure 9.5). The bond angle in each molecule, called the **tetrahedral angle**, is 109°28′, or about 109.5°. (Small deviations from the tetrahedral angle may occur when different atoms are bonded to the central atom. For example, in CH_2F_2, the HCH angle is 111.9° ± 0.4°, and the FCF angle is 108.3° ± 0.1°.)

Five Bonded Atoms; No Lone Pairs on the Central Atom

For PCl_5, in which there are five chlorine atoms bonded to the central phosphorus atom, the electron pairs are farthest apart when the molecule has a trigonal bipyramidal structure (Figure 9.6). Three chlorine atoms and the phosphorus atom lie in a plane with Cl—P—Cl angles of 120°. By analogy with the earth, these chlorine atoms are said to occupy **equatorial** positions. The other two chlorine atoms are situated perpendicular to this plane (one above and one below) in what are

FIGURE 9.4

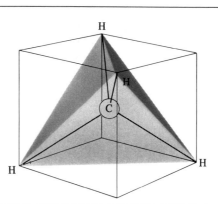

A tetrahedron inscribed in a cube. A tetrahedron may be constructed from a cube as follows: On each face of the cube connect the opposite corners labeled H by straight lines. The resulting figure is a regular tetrahedron with four faces, each of which is an equilateral triangle. The center of the tetrahedron (C) can be obtained from the point of intersection of the body diagonals connecting opposite corners of the cube. The HCH angles of 109°28′ can be calculated using the theorem of Pythagoras and the laws of trigonometry. [If the cube length is l, then the diagonal of a cube face is $2^{1/2}l$, the body diagonal of the cube is $3^{1/2}l$, and the C—H distance is $3^{1/2}l/2$. From the law of cosines the HCH angle is $\cos^{-1}(-\frac{1}{3}) = 109°28′$.]

FIGURE 9.5

Molecular structure of (left to right): methane (CH_4), ammonium ion (NH_4^+), carbon tetrachloride (CCl_4), silicon tetrafluoride (SiF_4).

9.8 COVALENT BOND ANGLES AND THE VSEPR THEORY

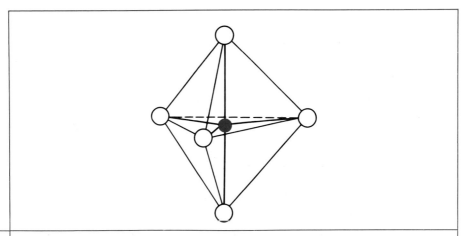

FIGURE 9.6 Structure of PCl$_5$. The five chlorine atoms are at the corners of a trigonal bipyramid with the phosphorus atom at the center. The two chlorine atoms in axial positions are at the top and bottom; the three chlorine atoms in equatorial positions occupy the central belt around the phosphorus atom.

called **axial** positions. This structure minimizes the total repulsion between the five pairs of electrons bonding the chlorine atoms to the phosphorus atom.

Six Bonded Atoms; No Lone Pairs on the Central Atom

When six atoms are bonded to the central atom as in SF$_6$, the six electron pairs are farthest apart when the molecule has an octahedral geometry with 90° bond angles (Figure 9.7). Four fluorine atoms and the sulfur atom lie in a plane, with a fifth F atom situated perpendicular to the plane directly above the S atom, and a sixth F atom directly below the S atom.

The results obtained above are summarized in Table 9.3.

TABLE 9.3 BOND ANGLES AT A CENTRAL ATOM HAVING NO LONE PAIRS

Number of bonded atoms	Predicted geometry	Predicted bond angle	Example
2	Linear	180°	BeCl$_2$
3	Trigonal planar	120°	BCl$_3$
4	Tetrahedral	109.5°	CH$_4$
5	Trigonal bipyramidal	120° and 90°	PCl$_5$
6	Octahedral	90°	SF$_6$

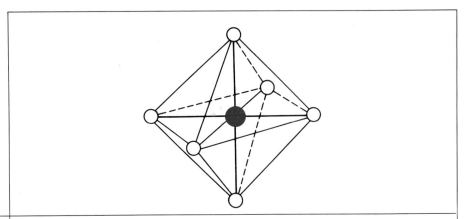

FIGURE 9.7

Structure of SF_6. The six fluorine atoms lie at the corners of an octahedron with the sulfur in the center.

One or More Lone Pairs on the Central Atom

Studies of the structures of many molecules have shown that each lone pair on a central atom appears to occupy a position as if it were a neighboring atom. Thus it is rather easy to extend the VSEPR theory to include such cases. One simply adds the number of lone pairs on the central atom to the number of neighboring atoms. The bond angles are then the same as for this number of neighboring atoms with no lone pairs. For example, if there is *one* lone pair on the central atom and *two* neighboring atoms, the total number is *three*. The bond angles on the central atom would then be close to 120°, that is, similar to the bond angles for *three* bonded atoms with *no* lone pairs. An example is nitrosyl chloride, ClNO.

$$\ddot{\underset{..}{Cl}}\diagdown\\ \ddot{N}=\ddot{\underset{..}{O}}$$

We expect it to be a bent molecule, with a bond angle of about 120°. (The experimental value is 116°.)

As another example, consider phosphorus trichloride, PCl_3.

$$:\!\ddot{Cl}\!-\!\ddot{P}\!-\!\ddot{Cl}\!:\\ |\\ :\!\ddot{Cl}\!:$$

With *one* lone pair on the central atom and *three* neighboring atoms, the total number is *four*. Thus the bond angles on the central atom should be about equal to the tetrahedral angle, 109.5°, that is, similar to bond angles in molecules with *four* bonded atoms and *no* lone pairs.

9.8 COVALENT BOND ANGLES AND THE VSEPR THEORY

When there are lone pairs on the central atom, the general pattern of bond angles suggests that the repulsions between electron pairs are in the order

$$L-L > L-S > S-S$$

where L denotes a lone pair and S a shared pair. That is, the repulsion between two lone pairs is larger than the repulsion between a lone pair and a shared pair, which in turn is larger than the repulsion between two shared pairs. As an example of this effect, consider the isoelectronic series of molecules CH_4, NH_3, and H_2O (Figure 9.8). In each case the sum of the bonded atoms and the number of lone pairs on the central atom is four, yet the known bond angles are 109.5° for H—C—H, 107.3° for H—N—H and 104.5° for H—O—H, as shown in Figure 9.8. Because of the greater repulsion by the unshared pair of electrons in NH_3, the shared pairs are forced closer together, decreasing the bond angle from the tetrahedral value to 107.3°. In H_2O, with two lone pairs, the effect is even larger, and the shared pairs are forced even closer together, reducing the bond angle to 104.5°.

Why does a lone pair exert a greater repulsion than a shared pair? A reasonable explanation is based on the fact that a lone pair is bonded to only one nucleus, whereas a shared pair is bonded to two nuclei. Therefore a lone pair would be expected to occupy a larger volume than a shared pair, and it would be expected to exert a greater repulsive effect on neighboring pairs of electrons.

Other examples of the VSEPR theory will be considered in Chapter 19.

One final point is in order. Before applying the VSEPR theory, it is necessary to draw a Lewis structure to determine the correct numbers of bonded atoms and lone pairs on the central atom. At this point one may guess incorrectly, and draw a structure with the wrong bond angle. However, application of the VSEPR theory will lead to the correct bond angle, at which point the Lewis structure is simply redrawn. For example, if nitrosyl chloride had been drawn as a linear molecule, application of the theory would still show that the molecule has a bent structure.

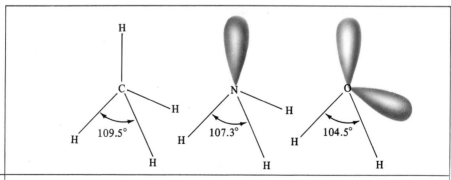

FIGURE 9.8

Molecular structures of methane, ammonia, and water. Unshared pairs of valence electrons around the central atom are indicated by colored lobes.

9.9 COVALENT BOND RADII

Internuclear distances for covalent bonds can be taken as sums of covalent bond radii of the atoms, much as internuclear distances in ionic crystals can be taken as sums of crystal radii of the ions. Here there is no problem involving division of the internuclear distance between the two atoms, since many internuclear distances are known for bonds between like atoms, and the covalent radius is then simply one-half the internuclear distance. For example, the experimental value for the internuclear distance in the Cl_2 molecule,

$$:\ddot{Cl}-\ddot{Cl}:$$

is 1.988 Å, and the covalent single-bond radius of chlorine is therefore $1.988/2 = 0.994$ Å.

Double bonds are shorter than single bonds between the same atoms, and triple bonds are shorter still. In the series of molecules

```
    H H              H       H
    | |               \     /
H—C—C—H              C=C            H—C≡C—H
    | |               /     \
    H H              H       H

  ethane            ethylene          acetylene
```

the carbon–carbon internuclear distances are 1.533, 1.334, and 1.204 Å, respectively. Thus different covalent radii are required for single, double, and triple bonds. The shorter bonds reflect the additional bonding effect of the second and third pairs of shared electrons. For the same reason more energy is required to break a double bond than the corresponding single bond, and breaking a triple bond requires still higher energy.

A graph of covalent radii versus atomic number is shown in Figure 9.9. Within each row of the periodic table the radius decreases with increasing atomic number, showing the effect of the higher nuclear charge. Within each group of the periodic table the radius increases on going to elements of higher atomic number, showing the effect of adding additional electronic shells. These same trends were noted in Section 8.11 for atomic radii.

As an example of the accuracy of covalent radii consider the HCN molecule

$$H-C\equiv N:$$

Triple-bond covalent radii for carbon and nitrogen are 0.602 and 0.547 Å, respectively. The sum of these radii, 1.149 Å, is very close to the experimental carbon–nitrogen internuclear distance in HCN, 1.153 Å.

Covalent radii and ionic crystal radii should not be confused, as they are quite different. For example, the covalent radius of Cl is 0.994 Å, whereas the ionic radius of Cl^- is 1.81 Å.

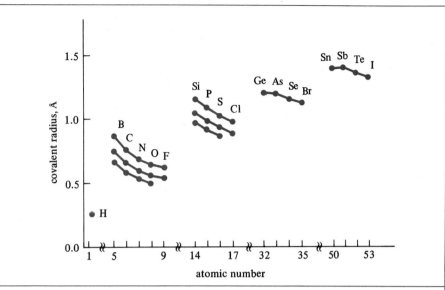

FIGURE 9.9 Covalent single-bond radii of nonmetallic elements. For the elements of the first and second short periods the graph also shows double-bond radii (middle curves) and triple-bond radii (lowest curves). (SOURCE: *From Linus Pauling:* The Nature of the Chemical Bond, *Third Edition.* © *1960 by Cornell University. Used by permission of the publisher, Cornell University Press.*)

9.10 RESONANCE

The sulfur dioxide molecule has a bent structure. Each sulfur–oxygen bond is 1.432 ± 0.001 Å long, and the OSO bond angle is 119.5°. The molecule has 18 valence electrons, and there are two ways of writing a Lewis structure obeying the octet rule:

To account for the fact that the two SO bonds are equivalent in length and strength and to retain the Lewis structures, we adopt the convention that the two structures together represent the electronic distribution in SO_2. The SO_2 molecule is said to be a **resonance hybrid** of the above two structures, and this is indicated by use of the double-headed arrow. The actual electronic distribution is intermediate between the two structures.

It is important to realize that resonance does not imply an oscillation back and forth between the different Lewis structures. The properties of the molecule are time independent; that is, they are the same one moment as the next with the exception that the nuclei are not rigidly fixed but are continually vibrating. An analogy is sometimes made between resonance hybrids and hybrids in genetics: A mule is not a horse one minute and a donkey the next.

Other examples of resonance structures for benzene, C_6H_6, and the nitrate ion, NO_3^-, are as follows:

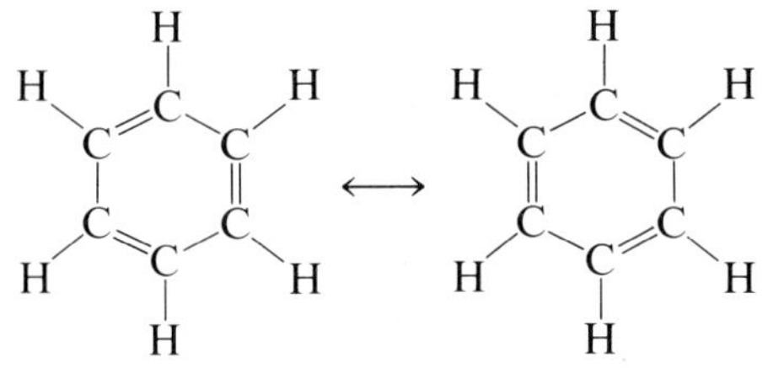

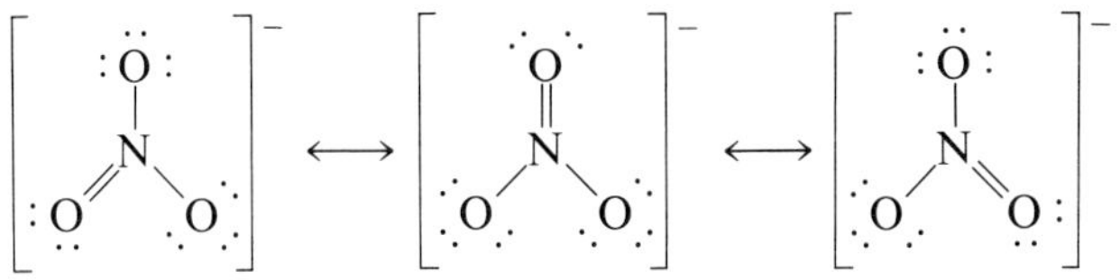

Molecules that can be represented by more than one Lewis structure (resonance hybrids) have different properties than one would predict from any one of the Lewis structures making up the hybrid. Thus the carbon–carbon bonds in benzene are stronger and shorter than they would be if their properties were midway between those of single and double bonds. Twice the single-bond covalent radius of carbon is 1.544 Å, and twice the double-bond radius is 1.334 Å. The average is 1.439 Å. But the experimental value for the carbon–carbon bond distance in benzene is only 1.397 ± 0.001 Å. Similarly, the experimental value of the energy of the benzene molecule, in comparison to the separated atoms, is lower than the value calculated for one of the structures shown above. This extra stability of the molecule is called the **resonance energy**. For benzene the resonance energy is about 155 kJ mol^{-1}.

The explanation of resonance energy is that the electrons are not confined as much as they would be if the molecule had a single Lewis structure; therefore the electron–electron repulsion energy is lower. Thus one pair of electrons in NO_3^- is spread over the entire molecule. In benzene, three pairs of electrons are spread or delocalized over the entire carbon ring.

9.11 POLAR BONDS AND ELECTRONEGATIVITY

Most chemical bonds between unlike atoms are not completely covalent, for the electrons are not shared equally. One atom is somewhat positive, and the other is somewhat negative. On the other hand, the bond may not be completely ionic, since electron transfer from one atom to the other may not be complete. Thus the magnitude of the net charge on each atom is less than in an ionic bond. These bonds, of a character intermediate between covalent and ionic, are called **polar bonds.** The ability of a bonded atom to attract electrons and become the negative partner in a polar bond is called its **electronegativity,** χ (lower-case Greek letter chi). If, in an A—B bond, atom A is more electronegative than atom B, electrons will be partly transferred in the direction of the ionic structure $A:^- B^+$. If B is the more electronegative, the direction will be toward $A^+:B^-$.

The first electronegativity scale was developed by Linus Pauling. He based his scale on bond energies. Because a polar bond can be described as a resonance hybrid of a covalent structure and an ionic structure,

$$A:B \longleftrightarrow A:^- B^+$$

the bond energy should be greater than the energy of a purely covalent bond between the same pair of atoms. The difference is called the **ionic resonance energy,** Δ. The main difficulty here is estimating what the bond energy would be if the bond were purely covalent. As the bonds between like atoms (A—A and B—B) *are* covalent, it is reasonable to take some kind of average of the A—A and B—B bond energies as an estimate of the bond energy of a purely covalent A—B bond. The simplest of these averages is the arithmetic mean.

If we denote the A—A, B—B, and A—B bond energies by $E(A—A)$, $E(B—B)$, and $E(A—B)$, respectively, then the ionic resonance energy of the A—B bond is

$$\Delta_{AB} = E(A—B) - \tfrac{1}{2}[E(A—A) + E(B—B)] \qquad (9.12)$$

Pauling reasoned that ionic resonance energies should be related to the electronegativity differences between the two atoms. He found that, although ionic resonance energies calculated as above are not an additive function, their square roots are approximately additive, that is, $\Delta_{AB}^{1/2} + \Delta_{BC}^{1/2} \approx \Delta_{AC}^{1/2}$, where atom B has an electronegativity intermediate between that of atoms A and C.[1] Thus $\Delta_{AB}^{1/2}$ can be taken as a measure of electronegativity difference:

$$|\chi_A - \chi_B| = 0.102\, \Delta_{AB}^{1/2} \qquad (9.13)$$

where Δ is in kilojoules per mole. (The vertical bars on the left-hand side of this equation denote absolute value.) To obtain numbers in a convenient region, Pauling took the ionic resonance energy in electron volts per molecule, and assigned an electronegativity of 4.0 to fluorine at the upper end of the scale. (The factor of 0.102 in Eq. 9.13 converts the ionic resonance energy from kilojoules per mole to electron volts per molecule. Table 9.4 lists the electronegativities on this scale for the atoms arranged in the same manner as in the periodic table. The following example illustrates the method used to calculate these electronegativities.

---

[1] Electronegativity differences are necessarily additive, since $(x_A - x_B) + (x_B - x_C) = (x_A - x_C)$. Thus any measure of electronegativity difference must also be additive.

TABLE 9.4 THE PAULING ELECTRONEGATIVITIES OF THE ELEMENTS

H 2.1																	
Li 1.0	Be 1.5											B 2.0	C 2.5	N 3.0	O 3.5	F 4.0
Na 0.9	Mg 1.2											Al 1.5	Si 1.8	P 2.1	S 2.5	Cl 3.0
K 0.8	Ca 1.0	Sc 1.3	Ti 1.5	V 1.6	Cr 1.6	Mn 1.5	Fe 1.8	Co 1.8	Ni 1.8	Cu 1.9	Zn 1.6	Ga 1.6	Ge 1.8	As 2.0	Se 2.4	Br 2.8
Rb 0.8	Sr 1.0	Y 1.2	Zr 1.4	Nb 1.6	Mo 1.8	Tc 1.9	Ru 2.2	Rh 2.2	Pd 2.2	Ag 1.9	Cd 1.7	In 1.7	Sn 1.8	Sb 1.9	Te 2.1	I 2.5
Cs 0.7	Ba 0.9	La-Lu 1.1–1.2	Hf 1.3	Ta 1.5	W 1.7	Re 1.9	Os 2.2	Ir 2.2	Pt 2.2	Au 2.4	Hg 1.9	Tl 1.8	Pb 1.8	Bi 1.9	Po 2.0	At 2.2
Fr 0.7	Ra 0.9	Ac-No 1.1–1.4														

Source: Adapted from Linus Pauling: *The Nature of the Chemical Bond*, Third Edition. © 1960 by Cornell University. Used by permission of the publisher, Cornell University Press.

EXAMPLE 1

(a) Calculate the difference in electronegativity between chlorine and fluorine atoms from the following experimental bond energies: $E(\text{Cl}-\text{F})$, 252 kJ mol^{-1}; $E(\text{Cl}-\text{Cl})$, 239 kJ mol^{-1}; $E(\text{F}-\text{F})$, 155 kJ mol^{-1}. (b) Taking 4.0 as the electronegativity of fluorine, and assuming that $\chi_{\text{Cl}} < \chi_{\text{F}}$, calculate the electronegativity of chlorine.

(a) From Eq. 9.12, the ionic resonance energy of the Cl—F bond is

$$\Delta_{\text{ClF}} = E(\text{Cl}-\text{F}) - \tfrac{1}{2}[E(\text{Cl}-\text{Cl}) + E(\text{F}-\text{F})]$$
$$= 252 - \tfrac{1}{2}(239 + 155)$$
$$= 55 \text{ kJ mol}^{-1}$$

Inserting this result in Eq. 9.13 gives

$$|\chi_{\text{Cl}} - \chi_{\text{F}}| = 0.102(55)^{1/2} = 0.76$$

(b) Because $|\chi_{\text{Cl}} - \chi_{\text{F}}| = 0.76$,

$$\chi_{\text{Cl}} - \chi_{\text{F}} = \pm 0.76$$

and

$$\chi_{\text{Cl}} = \chi_{\text{F}} \pm 0.76 = 4.0 \pm 0.76$$

Taking $\chi_{\text{Cl}} < \chi_{\text{F}}$, we obtain

$$\chi_{\text{Cl}} = 4.0 - 0.76 = 3.2$$

9.11 POLAR BONDS AND ELECTRONEGATIVITY

> When calculations of this type are carried through for other molecules containing Cl, slightly different results are obtained, and the best average value for the electronegativity of chlorine is about 3.0.

Because Eq. 9.13 provides only the *absolute value* of the electronegativity difference, one may well wonder, How can the correct sign of $\chi_A - \chi_B$ be obtained without making some arbitrary assumption? This is a question that cannot be answered by the ionic resonance energy method, and one must resort to other lines of evidence, such as the Mulliken scale described below.

The electronegativity scale can be used to estimate bond energies, as shown by the following example.

EXAMPLE 2

Estimate the H—Cl bond energy from the electronegativities listed in Table 9.4 and the following experimental bond energies: $E(\text{H—H})$, 431.8 kJ mol^{-1}; and $E(\text{Cl—Cl})$, 239 kJ mol^{-1}.

To calculate the ionic resonance energy of an H—Cl bond, use Eq. 9.13 in the form

$$\Delta_{AB}^{1/2} = \frac{|\chi_A - \chi_B|}{0.102}$$

or

$$\Delta_{AB} = \frac{(\chi_A - \chi_B)^2}{(0.102)^2}$$

This equation will give Δ_{AB} in units of kilojoules per mole.

From the electronegativities in Table 9.4, $\chi_{Cl} - \chi_H = 0.9$, so that

$$\Delta_{HCl} = \frac{(0.9)^2}{(0.102)^2} = 78 \text{ kJ mol}^{-1}$$

From Eq. 9.12,

$$\begin{aligned} E(\text{H—Cl}) &= \Delta_{HCl} + \tfrac{1}{2}[E(\text{H—H}) + E(\text{Cl—Cl})] \\ &= 78 + \tfrac{1}{2}(431.8 + 239) \\ &= 413 \text{ kJ mol}^{-1} \end{aligned}$$

This result is in fair agreement with the experimental value of 427 kJ mol^{-1}. (The error is about 3%.)

Examination of the electronegativities in Table 9.4 shows certain systematic trends. Within each period there is a general increase in electronegativity from left to right. Within each group there is a general decrease from top to bottom. Fluorine, in the upper right corner, has the highest electronegativity; cesium and francium, in the lower left corner, have the lowest electronegativity.

The trends in electronegativity with position in the periodic table are rather similar to the trends of ionization potential and electron affinity described in Section 8.12. In fact, one electronegativity scale (devised by Robert S. Mulliken and denoted χ_M) defines an atom's electronegativity as the mean of its ionization potential, I, and its electron affinity, E.

$$\chi_M = \tfrac{1}{2}(I + E) \tag{9.14}$$

The reason this definition provides a measure of electronegativity is that I measures the tendency of an atom to retain its most loosely held electron, and E measures the tendency of an atom to gain an additional electron. To become the negative partner in a polar bond, an atom should have a strong tendency to avoid loss of an electron to its partner (large I), and a strong tendency to gain an electron from its partner (large E); the average value of I and E is thus an appropriate measure of an atom's electronegativity. Table 9.5 lists the Mulliken electronegativities for some univalent atoms. (For atoms of higher valence, a correction for the valence state of the atom must be applied to Eq. 9.14.)

It is of course important to distinguish clearly between electron affinity and electronegativity. Electron affinity measures the tendency of an *atom in the gas phase*, far from any other atoms, to become a negative *ion*. Electronegativity measures the tendency of an *atom in a molecule* to become the negative *partner* in a polar bond to a neighboring atom.

For those atoms listed in Table 9.5, the Pauling and Mulliken electronegativities are compared in Figure 9.10. It may be seen that there is a rough proportionality between the two scales.

An advantage to the Mulliken scale is that it offers an answer to the question, Why is fluorine more electronegative than chlorine, even though chlorine has the higher electron affinity? The answer is that fluorine has a much higher ionization potential, which more than compensates for its lower electron affinity (see Table 9.5).

Polar bonds are not only stronger, they are also shorter than if they were

TABLE 9.5	MULLIKEN ELECTRONEGATIVITIES OF SOME UNIVALENT ATOMS			
	Atom	I (eV)	E (eV)	χ_M (eV)
	F	17.42	3.45	10.44
	Cl	13.01	3.61	8.31
	Br	11.84	3.36	7.60
	I	10.45	3.06	6.76
	H	13.60	0.75	7.18
	Li	5.39	0.58	2.99
	Na	5.14	0.78	2.96

9.11 POLAR BONDS AND ELECTRONEGATIVITY

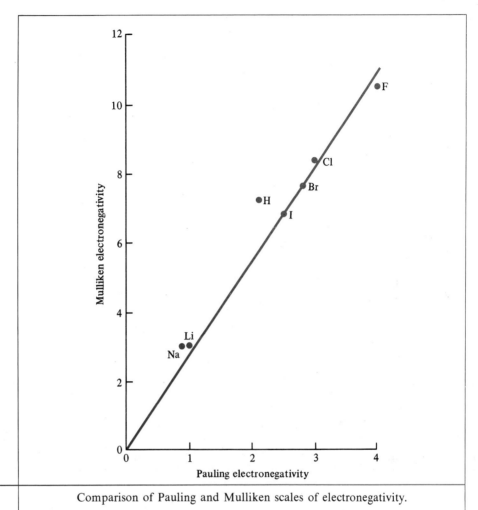

FIGURE 9.10 Comparison of Pauling and Mulliken scales of electronegativity.

purely covalent. For example, carbon and nitrogen have covalent radii of 0.77 and 0.74 Å, respectively. Their sum, 1.51 Å, is 0.04 Å longer than the typical carbon–nitrogen single-bond distance of 1.47 Å. Fairly good estimates of the bond lengths of many polar bonds can be obtained using the equation

$$R_{AB} = R_A + R_B - c|\chi_A - \chi_B| \tag{9.15}$$

where R_{AB} is the calculated bond length and R_A and R_B are the covalent radii of atoms A and B. Coefficient c (selected to give the best agreement with experimental data) is 0.08 for single bonds when one or both atoms are in the first short period. In the example cited

$$R_{CN} = 0.77 + 0.74 - 0.08|2.5 - 3.0|$$
$$= 1.51 - 0.04 = 1.47 \text{ Å}$$

which is identical with the experimental result.

9.12 MOLECULES IN ELECTRIC FIELDS

Experimental evidence supporting the concept of polar bonds can be obtained by investigating the behavior of molecules in electric fields. Figure 9.11 shows an electric condenser, consisting of two parallel metal plates that may be electrically charged by connecting them to a storage battery. When the condenser is charged, an electric field is set up with field strength equal to the applied voltage (V) divided by the distance (d) between the plates. Under the influence of this field an ion will move toward the plate of opposite sign.

Consider now the behavior of an electric dipole—a positive charge and a negative charge of the same magnitude, separated by a distance R. Because the net charge of the dipole is zero, it will not move toward either plate. However, the positive end will be attracted by the negative plate, and the negative end will be attracted by the positive plate, causing the dipole to *rotate,* as shown in Figure 9.11. The dipole will rotate until it is aligned with its negative end nearest the positive plate, and its positive end nearest the negative plate.

Polar molecules are electric dipoles, and they tend to align themselves in this way in an electric field. Thus, if hydrogen chloride gas is injected into a charged condenser, the molecules will align themselves with the positive end (hydrogen) closer to the negative plate, and the negative end (chlorine) closer to the positive plate. This orientation of polar molecules affects the electric field between the two plates because the electric field due to the dipoles is opposed to the electric field due to the charge on the plates.

If the metal plates are charged to a voltage V prior to introduction of the polar substance, and the metal plates are then disconnected from the battery, the voltage will drop to some *lower* value V' when the polar substance is added. Just how much the voltage changes depends on what substance is added, and on its temperature and pressure. The ratio $\epsilon = V/V'$ is a characteristic property of a substance called the **dielectric constant.** Table 9.6 lists the dielectric constants of a few substances. Note that the dielectric constant of a substance is always a positive dimensionless number greater than 1. The experimentally determined dielectric constants are used to obtain information on the polarity of molecules.

The dielectric constants in Table 9.6 show that even nonpolar molecules such as O_2 and N_2 have dielectric constants slightly larger than 1. The reason for this behavior is that all molecules are affected by the presence of an electric field. The electrons in a molecule are attracted toward the positive plate, and the nuclei are attracted toward the negative plate. Thus the electric field *induces* an electric dipole. A molecule that was nonpolar in the absence of a field becomes polar in the presence of a field. (When the electric field is reduced to zero, the molecule becomes nonpolar again.) This effect is called **polarization.** Even molecules that are polar in the absence of a field are subject to polarization.

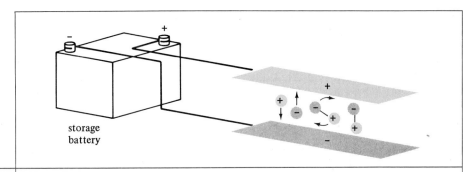

FIGURE 9.11 Ions and dipoles in a parallel-plate condenser. The ions move toward the oppositely charged plates, whereas the dipoles align themselves in the electric field.

Temperature affects the dielectric constant because molecular collisions tend to disrupt the alignment of polar molecules in an electric field. As the temperature is raised, increasing thermal motion of the molecules produces a more random orientation, reducing the dielectric constant of the substance. As shown in Table 9.6, the dielectric constant of liquid water drops from 80.4 at 20°C to 78.5 at 25°C.

9.13 DIPOLE MOMENTS

An important characteristic of any electric dipole is its **dipole moment**, μ (lowercase Greek letter mu). If Q is the magnitude of the electric charge at either end of the dipole (i.e., $+Q$ at one end and $-Q$ at the other end), and R is the distance

TABLE 9.6

DIELECTRIC CONSTANTS AT 1 atm PRESSURE		
Substance	Temperature (°C)	Dielectric constant
Gases		
Oxygen	20	1.000495
Nitrogen	20	1.000537
CO_2	20	1.000922
HCl	0	1.0046
NH_3	0	1.0072
Liquids		
CCl_4	25	2.23
Benzene	25	2.27
Chlorobenzene	25	5.62
Methyl ether	25	5.02
Methyl alcohol	25	32.6
Ethyl alcohol	25	24.3
Water	20	80.4
	25	78.5

between the charges, then the dipole moment is defined as

$$\mu = QR \tag{9.16}$$

When a nonpolar molecule is placed in an electric field, a dipole is induced by the field. The induced dipole moment, μ_I, is proportional to the electric field strength V/d:

$$\mu_I = \alpha \frac{V}{d}$$

where α is a proportionality constant called the **polarizability** of the molecule.

Polar molecules are also polarizable, and therefore the net dipole moment of a polar molecule is the sum of two terms: the dipole moment in the absence of a field, called the **permanent dipole moment,** μ_0, and the induced dipole moment, μ_I. By measuring the dielectric constant of a substance at various temperatures, it is possible to distinguish between the contributions of the permanent and induced dipole moments to the dielectric constant, and hence to measure both the permanent dipole moment and the polarizability of a molecule. (The analysis is rather complicated and is omitted here.)

We shall be interested primarily in permanent dipole moments, commonly called simply **dipole moments.** Table 9.7 lists the dipole moments of some molecules in the gaseous state. The traditional unit of measurement for dipole moments is the debye, D, defined in terms of coulomb meters as $1\text{ D} = 3.336 \times 10^{-30}$ C m. The debye is named after Peter J. W. Debye, who developed the method of determining dipole moments from dielectric constants.

One may reach an understanding of the connection between dipole moment

TABLE 9.7

PERMANENT DIPOLE MOMENTS OF MOLECULES IN THE GASEOUS STATE	
Substance	μ_0 (D)
LiH	5.88
HF	1.74
HCl	1.07
HBr	0.79
HI	0.38
ClF	0.88
BrCl	0.57
BrF	1.29
H_2O	1.82
H_2S	0.98
NH_3	1.47
CO_2	0.00
SO_2	1.61
CH_4	0.00
CH_2Cl_2	1.62
CCl_4	0.00
BF_3	0.00

and bond polarity by assuming a simple model for the electric dipole of a polar molecule. In this model one assumes that each charge of magnitude Q is equal to δe, where e is the magnitude of the electronic charge, and δ is some fraction, called the **fractional ionic character,** to be determined from the dipole moment. (If the bond in a molecule such as HCl were purely covalent, δ would be zero; if the bond were ionic, δ would be 1. One expects to find a δ somewhere between zero and 1.) One further assumes that the charges are centered on the nuclei, and therefore the distance between the charges, R, is equal to the internuclear distance, which can be determined experimentally. Applying Eq. 9.16 to the permanent dipole moment, μ_0, and substituting δe for Q, one obtains

$$\mu_0 = \delta e R \tag{9.17}$$

The following example illustrates the calculation of fractional ionic character.

EXAMPLE 3

Calculate the fractional ionic character of the HCl molecule from its dipole moment of 1.07 D and its internuclear distance of 1.28 Å. (The electronic charge unit, e, is 1.602×10^{-19} C.)

First, we convert the dipole moment from debyes to coulomb meters:

$$\mu_0 = (1.07 \text{ D})(3.336 \times 10^{-30} \text{ C m D}^{-1})$$
$$= 3.57 \times 10^{-30} \text{ C m}$$

Next, we solve Eq. 9.17 for δ:

$$\delta = \frac{\mu_0}{eR}$$

Substituting the various quantities given in the problem, we obtain

$$\delta = \frac{3.57 \times 10^{-30} \text{ C m}}{(1.602 \times 10^{-19} \text{ C})(1.28 \times 10^{-10} \text{ m})}$$
$$= 0.17$$

Thus HCl behaves as though there is 0.17 unit of charge on each atom. Which atom is positive and which negative? Because chlorine's electronegativity is larger than hydrogen's, we infer that the charges are +0.17 on H and −0.17 on Cl.

If the bond in HCl were covalent, these charges would each be zero; if the bond were ionic they would be +1 and −1. Accordingly, one may say that the fractional ionic character of the bond in HCl is 0.17, or 17%. This result may be expressed in terms of Lewis structures by saying that the HCl molecule is a resonance hybrid of a covalent structure and an ionic structure

$$\text{H—}\ddot{\underset{..}{\text{Cl}}}: \longleftrightarrow \text{H}^+ : \ddot{\underset{..}{\text{Cl}}}:^-$$

with contributions of 83% and 17%, respectively.

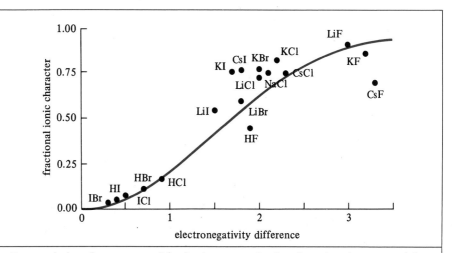

FIGURE 9.12 Curve relating the amount of ionic character of a bond to the electronegativity difference of the two atoms. Experimental points, based upon observed values of the electric dipole moment of diatomic molecules, are shown for 18 bonds. The electronegativity is taken from the Pauling scale (Section 9.11). (SOURCE: *From Linus Pauling:* The Nature of the Chemical Bond, *Third Edition.* © *1960 by Cornell University. Used by permission of the publisher, Cornell University Press.*)

Figure 9.12 shows how the fractional ionic character is related to the magnitude of the difference in electronegativity between the two atoms, $|\chi_A - \chi_B|$. As expected from the foregoing discussion, the fractional ionic character is small when the atoms have nearly the same electronegativity, and large when the electronegativities are substantially different.

What is surprising is that the fractional ionic character calculated as above does not reach 1.00 even for molecules such as LiF. The fault lies with the assumption that the charges $+\delta e$ and $-\delta e$ are symmetrically distributed. Examination of the detailed electron density distribution of LiF shows that the ions polarize each other, so that each ion is to some extent a dipole (e.g., the electrons in F^- are pulled toward the Li^+, creating a small induced dipole in the opposite direction from the overall dipole moment of the molecule). These induced dipoles decrease the net dipole moment of the molecule, and thus decrease the fractional ionic character calculated by the above method. Unshared pairs of electrons may also contribute to the net dipole moment of the molecule, because, in general, they are not symmetrically distributed.

For an accurate comparison of theory and experiment, one must compare the experimental value of μ_0 with the dipole moment calculated from the electron density distribution in the molecule, ψ^2. This has been done for some small molecules such as LiH, and the results are in good agreement.

In polyatomic molecules, only the net dipole moment may be determined experimentally. The dipole moment of each polar bond in a polyatomic molecule

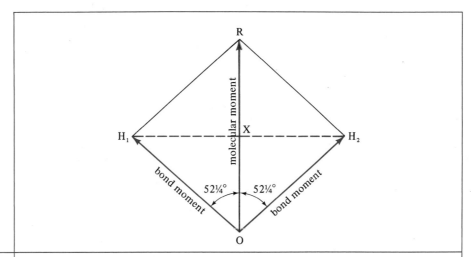

FIGURE 9.13 Bond moment calculation for the water molecule. The bond moments are OH_1 and OH_2, and OR is the resultant molecular moment. In the right triangle OXH_2, $\cos(52\frac{1}{4}°) = OX/OH_2$ or $OH_2 = OX/\cos(52\frac{1}{4}°) = [(1.82\ D)/2]/0.612 = 1.49\ D$.

is known as a **bond moment** and the net dipole moment may be viewed as the vector sum of the bond moments of all the polar bonds in the molecule.[2] For example, the dipole moment of the water molecule is determined experimentally to be 1.82 D. By using the average HOH bond angle of 104.5°, we can construct a vector diagram as shown in Figure 9.13. The OH bond moment is simply the length of one side of the parallelogram; it is found to be 1.49 D.

In a linear symmetrical molecule, such as CO_2, the bond dipoles are of equal moment in the opposite direction. Therefore the net dipole moment of the molecule is exactly zero, even though the individual bonds are polar. Other examples of molecules with zero dipole moments because of symmetry are BF_3 and CCl_4 (Figure 9.14).

In general, molecules of formula MX_n having any of the molecular geometries listed in Table 9.3 will have zero dipole moments because of symmetry.

9.14 MOLECULAR ORBITALS AND CHEMICAL BONDING

Although the treatment of chemical bonding in terms of Lewis structures is enormously useful in chemistry, it fails to answer some important questions. Just how do atoms "share" their electrons in such a way that each atom achieves the electronic structure of a noble gas? How are the dot-and-dash symbols for electrons related to the electron density distribution of the molecule? Why do different covalent bonds have different bond energies? To answer such questions we must

---

[2] That is, ignoring contributions from unshared pairs.

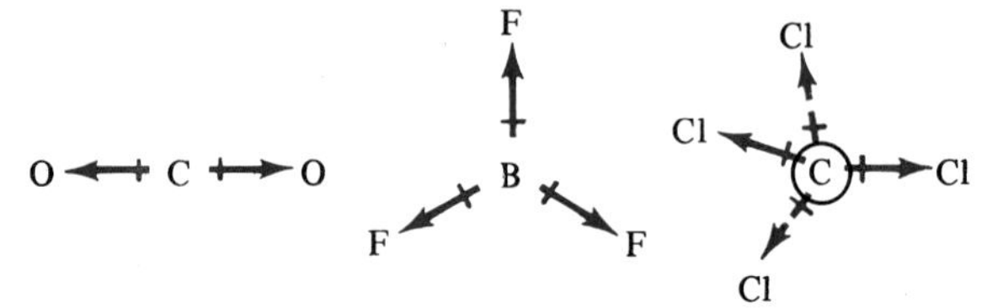

FIGURE 9.14

Molecules with zero dipole moments because of symmetry: carbon dioxide, boron trifluoride, and carbon tetrachloride. Bond dipoles are shown as arrows directed from the positive atom toward the negative atom.

go beyond the simple picture of molecular structure described above, and consider the wave function, ψ, and the probability density, ψ^2, for a molecule. In this section we shall consider the main features of the simplest and most important way of describing molecular wave functions.

Molecular wave functions may be considered at different levels of complexity. The simplest approach is the orbital approximation, analogous to the orbital approximation previously described for atoms (Chapter 8). Each electron occupies one of the orbitals of the molecule. There are two electrons at most per orbital, and electrons in the same orbital have opposite spins ($\uparrow\downarrow$), as required by the Pauli exclusion principle. In the ground state of the molecule, the electrons occupy the orbitals of lowest energy.

Atoms are complicated enough, and it might be thought that the problem of determining the correct orbitals for molecules—called **molecular orbitals** (MO's)—would be intractable. However, good approximations to the orbitals of a molecule are obtained simply by superimposing the orbitals of the separated atoms. The resulting orbitals are called **linear combination of atomic orbitals** molecular orbitals, abbreviated LCAO-MO.

There are two important reasons why LCAO-MO's are good molecular orbitals. First, when an electron is near one nucleus, it is mainly affected by that nucleus, and therefore an atomic orbital centered on that nucleus describes the behavior of the electron quite well. Second, atomic orbitals are simply de Broglie waves, and, as we have seen previously (Chapter 7), waves constructively interfere when they are in the same place with the same phase. Thus atoms combine so that, to the maximum extent possible, their atomic orbitals are in the same region of space with the same sign. This is called the **criterion of maximum overlapping.**

Figure 9.15 shows contour diagrams of the electron density of the hydrogen molecule at various internuclear distances. As the atoms come together, their orbitals overlap, leading to an increase in the amplitude of the wave function (hence an increase in the electron density) near the nuclei and between the nuclei. This increases the electron-nucleus attraction energy, causing the total energy of the molecule to be less than that of the isolated atoms. At very short internuclear distances, however, the nucleus-nucleus repulsion energy becomes so large that the total energy rises steeply. At the point of minimum energy the internuclear distance can be identified with the bond length. The bond energy is the depth of

9.14 MOLECULAR ORBITALS AND CHEMICAL BONDING

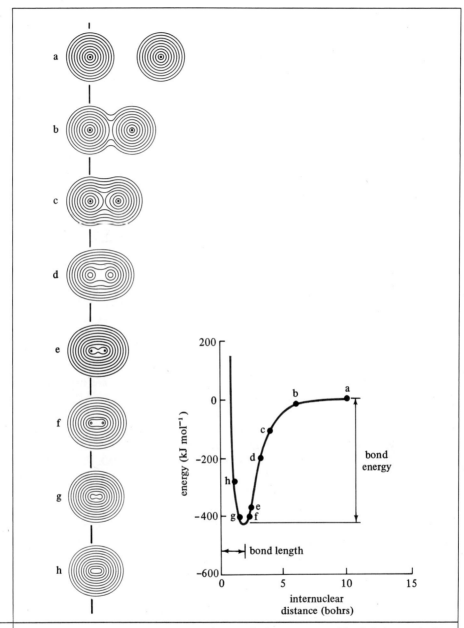

FIGURE 9.15 Formation of a hydrogen molecule by two hydrogen atoms. The left figures (a–h) show electron density contours at various internuclear distances as the two atoms come together. (The outer contours correspond to an electron density of 4.9×10^{-4} electron b^{-3}, where 1 bohr, abbreviated b, is 0.529 Å. The electron density at each succeeding inner contour is larger by a factor of 2.) The right figure shows the energy of the system at the same internuclear distances a–h. [SOURCE: *A. C. Wahl, Scientific American,* **222**, *No. 4, 54 (1970)*.]

the energy curve below the horizontal line to the right (where the atoms are so far apart that they have essentially no effect on each other). Both experimental measurements and calculations with very accurate wave functions show that in H_2 the bond length is 0.741 Å and the bond energy is 458.1 kJ mol^{-1}.

Calculations based on the LCAO-MO described above give a bond length of 0.85 Å and a bond energy of 261 kJ mol^{-1}. The best molecular orbital (self-consistent field MO) gives the correct bond length, and a bond energy of 351 kJ mol^{-1}. These results indicate the accuracy of the molecular orbital approximation and of the approximation of a molecular orbital as a linear combination of atomic orbitals.[3]

9.15 MOLECULAR ORBITALS IN DIATOMIC MOLECULES FORMED BY LIKE ATOMS

To obtain a more comprehensive and detailed understanding of molecular orbitals, it is necessary to consider the various types of molecular orbitals formed from atomic orbitals (AO's). For simplicity, the following discussion will be restricted to bonding between like atoms to form homonuclear diatomic molecules.

In homonuclear diatomic molecules the molecular orbitals are classified by (1) symmetry type, (2) whether bonding or antibonding, and (3) the atomic orbitals from which they are formed. The most important symmetry types are σ (sigma) and π (pi). A σ MO is symmetric around the internuclear axis (i.e., it has the same symmetry as a cylinder, a sausage, or a football). The MO in H_2 considered in the preceding section is an example of a σ orbital. In general, combining two s AO's on different nuclei always forms a σ MO.

A σ MO is also formed by combining two p AO's, provided that the p orbitals are oriented end to end. With the internuclear axis in the z direction, these are the p_z AO's. Figure 9.16 shows how a σ MO is formed in this way. If the positive lobe of each AO is nearest the other AO, then overlap of the two positive lobes forms a large positive lobe between the nuclei. (Recall that the + and − signs refer to the amplitude of the de Broglie wave, not to electric charge.)

A π MO is formed by side-to-side (rather than end-to-end) mixing of atomic p orbitals, as shown in Figure 9.17. With the internuclear axis in the z direction, these are the p_x (or p_y) AO's. The positive lobes of the AO's merge to form a large positive lobe of the MO, and the negative lobes of the AO's merge to form a large

---

[3] Failure to describe electronic correlation (see Section 8.10) is the source of the error of 458.1 − 351 = 107 kJ mol^{-1}, and this difference is known as the correlation energy of the hydrogen molecule. The valence-bond wave function, introduced by Walter Heitler and Fritz London in 1927 in the first successful attack on the H_2 problem, provides that when one electron is near one nucleus, the other electron is near the other nucleus. Correlation of this type is referred to as alternant or **left-right correlation**. The Heitler-London function gives a bond length of 0.80 Å and a bond energy of 303 kJ mol^{-1}. The best wave function of valence bond form gives the correct bond length and a bond energy of 399 kJ mol^{-1}. Further improvements are obtained with more complicated wave functions, which provide that when one electron is on one side of the internuclear axis, the other electron is on the other side (**angular correlation**), and that when one electron is near the center of the molecule, the other electron is in the outer regions (**in-out correlation**).

FIGURE 9.16

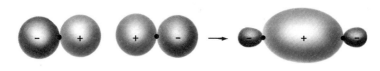

Formation of a σ molecular orbital by end-to-end overlap of two p_z atomic orbitals. (The internuclear axis is taken in the z direction.)

negative lobe of the MO on the opposite side of the internuclear axis. Note that this MO is not symmetrical around the internuclear axis. Note also that there is a nodal plane *containing* the internuclear axis—this is the characteristic feature of a π MO. For the π_x MO (formed from the p_x AO's), the nodal plane is the yz plane. For the π_y MO (formed from the p_y AO's), the nodal plane is the xz plane.

In general there are two ways of combining a pair of atomic orbitals to form a molecular orbital. Each orbital is simply a de Broglie wave, and waves can be superimposed to form new waves (Section 7.1). Superposition of waves leads to different results depending on their relative phase. If two waves are in phase, they reinforce each other to produce a wave of greater amplitude (constructive interference). If out of phase, they tend to cancel each other (destructive interference).

These two ways of combining AO's are illustrated in Figure 9.18 for a pair of 1s AO's. When the orbitals are added in phase, there is constructive interference, and the σ1s **bonding molecular orbital** is formed. It is characterized by a region of high electron density between the nuclei (Figure 9.19), and attraction to this region is what holds the nuclei together.

When the orbitals are combined out of phase, there is destructive interference, and an **antibonding molecular orbital** is formed. The electron density between the nuclei is low (Figure 9.19). In fact, there is a nodal plane midway between the nuclei and *perpendicular to* the internuclear axis. (This nodal plane is the characteristic feature of an antibonding MO.) Repulsion between the relatively bare nuclei tends to dissociate the molecule, hence the term antibonding. To indicate that an MO is antibonding, an asterisk is inserted following the symmetry

FIGURE 9.17

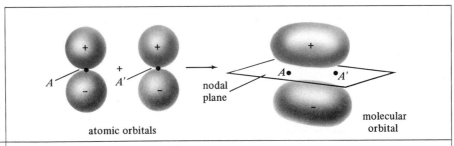

Formation of a π molecular orbital by side-to-side overlap of two p_x (or p_y) atomic orbitals.

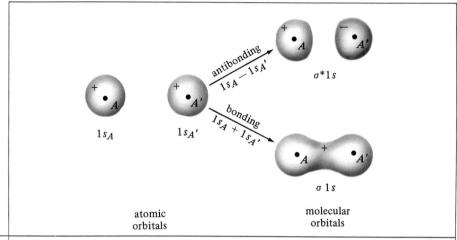

FIGURE 9.18 Formation of bonding and antibonding molecular orbitals from $1s$ atomic orbitals. A and A', nuclei.

type; thus the antibonding MO in Figures 9.18 and 9.19 is designated σ^*1s.

Because there are three $2p$ atomic orbitals on each atom, there are six $2p$ orbitals available in a diatomic molecule, and they can be combined to form six MO's as shown in Figure 9.20. In the upper part of the figure the two $2p_z$ orbitals are directed toward each other, resulting in end-to-end overlap and formation of σ MO's. Addition forms the bonding $\sigma 2p$ MO; subtraction forms the antibonding σ^*2p MO.

In the lower part of Figure 9.20 the two $2p_x$ AO's are directed perpendicular to the internuclear axis, resulting in side-to-side overlap and formation of π MO's. Addition forms a bonding $\pi_x 2p$ MO; subtraction forms an antibonding $\pi_x^* 2p$ MO.

In the same way, the two $2p_y$ AO's (directed perpendicular to the plane of the paper) form a $\pi_y 2p$ MO and a $\pi_y^* 2p$ MO. (Note the use of x and y subscripts to distinguish the different π and π^* MO's.)

Now that we have become familiar with the different types of MO's, we are in a position to consider the MO description of the chemical bonding in some homonuclear diatomic molecules (i.e., diatomic molecules formed by like atoms).

9.16 MOLECULAR ORBITAL DESCRIPTION OF DIATOMIC MOLECULES FORMED BY LIKE ATOMS

To determine which MO's are occupied by electrons in the ground state of a diatomic molecule, it is necessary to know the relative energies of the different MO's. The following generalizations are useful: (1) Energies of MO's increase in the same order as energies of the AO's used to form them: $1s < 2s < 2p$, and so on. (2) For a given symmetry type and AO, the bonding MO has lower energy

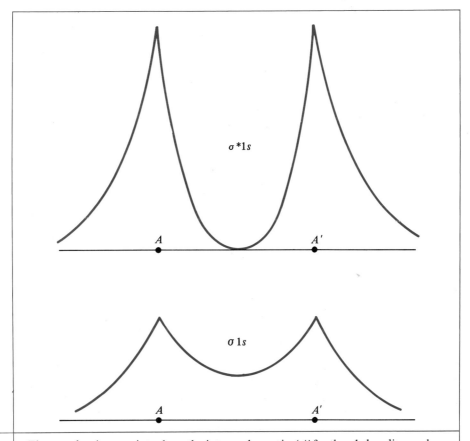

FIGURE 9.19 Electron density at points along the internuclear axis AA' for the $\sigma 1s$ bonding molecular orbital (lower figure) and the σ^*1s antibonding molecular orbital (upper figure).

than the antibonding MO. For example, the $\sigma 1s$ MO is of much lower energy than the σ^*1s MO (Figure 9.21). Thus in H_2^+ and H_2 the antibonding MO is empty, and the bonding MO contains one and two electrons, respectively. For He_2^+ (the helium molecule-ion) the third electron must enter the antibonding MO. And for "He_2" (which is unstable) the antibonding MO is filled by a pair of electrons of opposite spin.

If the *net number of bonding electrons* is defined as the number of electrons in bonding MO's less the number in antibonding MO's, a useful measure of bond strength is obtained. In the cases cited the net numbers are 1, 2, 1, and 0 for H_2^+, H_2, He_2^+, and "He_2," respectively. These numbers are roughly proportional to the bond energies, which are 268, 458, 259, and 0 kJ mol^{-1}, respectively. (The statement that the bond energy is zero in "He_2" is another way of saying that a diatomic helium molecule does not exist.)

A pair of $2s$ AO's can be combined in a manner similar to that described

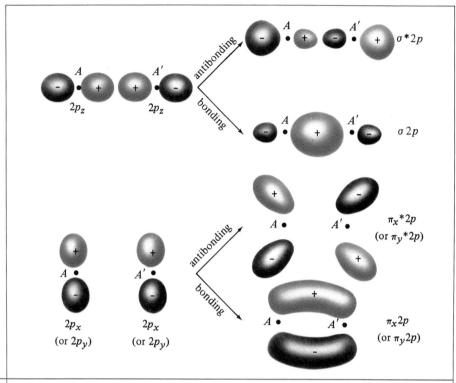

FIGURE 9.20 Mixing 2p atomic orbitals on different nuclei to form bonding and antibonding molecular orbitals. The internuclear axis AA' is taken in the z direction.

above for 1s orbitals. The bonding MO is $\sigma 2s$, and the antibonding MO is σ^*2s. In Li_2 the $\sigma 2s$ MO is filled by a pair of electrons with opposite spin, and the net number of bonding electrons is 2. In the unstable "Be_2 molecule" the σ^*2s MO is also filled by a pair of electrons, and the net number of bonding electrons is zero. Table 9.8 illustrates the electron configurations for the molecules considered pre-

TABLE 9.8 ELECTRON CONFIGURATIONS OF SOME DIATOMIC MOLECULES CONTAINING σ MO's

Molecule	Electron configuration	Net number of bonding electrons
H_2^+	$\sigma 1s$	1
H_2	$(\sigma 1s)^2$	2
He_2^+	$(\sigma 1s)^2 \sigma^*1s$	1
"He_2"	$(\sigma 1s)^2(\sigma^*1s)^2$ or HeHe	0
Li_2	$HeHe(\sigma 2s)^2$	2
"Be_2"	$HeHe(\sigma 2s)^2(\sigma^*2s)^2$ or BeBe	0

9.16 MOLECULAR ORBITAL DESCRIPTION OF DIATOMIC MOLECULES FORMED BY LIKE ATOMS

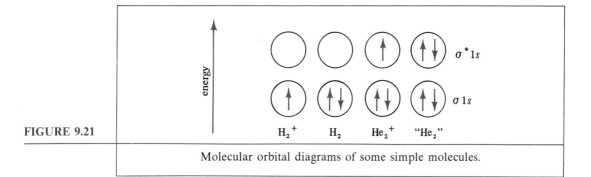

FIGURE 9.21 Molecular orbital diagrams of some simple molecules.

viously, with HeHe and BeBe designating the electron configurations of two helium or two beryllium atoms, respectively.

The MO's of next higher energy are those formed from the $2p$ AO's. Spectroscopic studies of homonuclear diatomic molecules show that these MO's increase in energy in the order $\pi 2p < \sigma 2p < \pi^* 2p < \sigma^* 2p$, as shown in Figure 9.22 for the B_2 molecule. (The explanation of this particular order is complicated and is omitted here.) Note that there are two $\pi 2p$ MO's ($\pi_x 2p$ and $\pi_y 2p$), as well as two $\pi^* 2p$ MO's ($\pi_x^* 2p$ and $\pi_y^* 2p$). In accordance with Hund's rule, the two electrons in the $\pi 2p$ MO's of the B_2 molecule are "unpaired"—they are in different orbitals with parallel spin, as shown in Figure 9.22. The electron configurations of B_2 and some other diatomic molecules are shown in Table 9.9.

Like B_2, the O_2 molecule has two unpaired electrons. In this case they are in the antibonding $\pi_x^* 2p$ and $\pi_y^* 2p$ orbitals. At the time when molecular orbital theory was first developed, it was already known from studies of the magnetic properties of oxygen that the O_2 molecule has two unpaired electrons. The explanation of this experimental fact provided strong support for the theory.

The diatomic molecules of the elements in the next period have similar electronic structures, with principal quantum number one higher. For example, the electron configuration of the Na_2 molecule (analogous to that of Li_2) is NeNe $(\sigma 3s)^2$.

TABLE 9.9

ELECTRON CONFIGURATIONS OF SOME DIATOMIC MOLECULES CONTAINING BOTH σ AND π MO's		
Molecule	Electron configuration	Net number of bonding electrons
B_2	BeBe$(\pi_x 2p)(\pi_y 2p)$	2
C_2	BeBe$(\pi_x 2p)^2(\pi_y 2p)^2$	4
N_2	BeBe$(\pi_x 2p)^2(\pi_y 2p)^2(\sigma 2p)^2$	6
O_2	BeBe$(\pi_x 2p)^2(\pi_y 2p)^2(\sigma 2p)^2(\pi_x^* 2p)(\pi_y^* 2p)$	4
F_2	BeBe$(\pi_x 2p)^2(\pi_y 2p)^2(\sigma 2p)^2(\pi_x^* 2p)^2(\pi_y^* 2p)^2$	2
"Ne_2"	BeBe$(\pi_x 2p)^2(\pi_y 2p)^2(\sigma 2p)^2(\pi_x^* 2p)^2(\pi_y^* 2p)^2(\sigma^* 2p)^2$, or NeNe	0

FIGURE 9.22

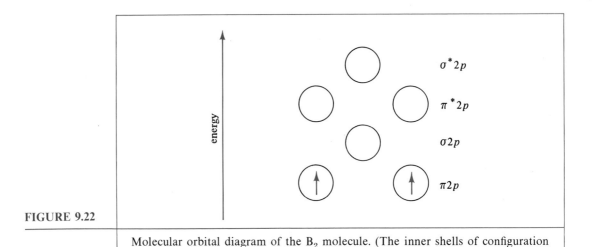

Molecular orbital diagram of the B_2 molecule. (The inner shells of configuration BeBe are not shown.)

9.17 MOLECULAR ORBITALS IN DIATOMIC MOLECULES FORMED BY UNLIKE ATOMS

If the two atoms forming a diatomic molecule have similar (but not identical) atomic numbers, the MO's will closely resemble the homonuclear diatomic MO's described above. For example, the atomic numbers of nitrogen and oxygen are 7 and 8, respectively, and the nitric oxide molecule, NO, has MO's very much like those of N_2 and O_2. However, because of the higher positive charge on the oxygen nucleus, the bonding MO's are somewhat unsymmetrical. They have larger amplitude near the oxygen nucleus, smaller amplitude near the nitrogen nucleus. Thus an electron in a bonding MO is more likely to be found near the oxygen nucleus than near the nitrogen nucleus. (The opposite is true of an antibonding MO. Because more of the oxygen AO has been used to form the bonding MO, less is available to form the antibonding MO.) This unequal sharing of electrons, together with the different electric charges on the nuclei, creates a dipole moment in the molecule, as discussed in Section 9.13.

The pattern of orbital energies in the NO molecule is similar to that of homonuclear diatomic molecules. Because NO has one more electron than N_2, this last electron must enter one of the antibonding π^* MO's. Whereas the net number of bonding electrons in N_2 is six, it drops to five for NO, and of course four for O_2. Thus one would expect the bonding to be weaker in NO than in N_2, and weaker still in O_2. These expectations are borne out by the experimental data on bond lengths and bond energies (Table 9.10).

When the atomic numbers differ substantially, the AO's combining to form MO's are those of similar energy, and these AO's may have different principal quantum numbers. For example, in the hydrogen fluoride molecule, HF, the

TABLE 9.10 **BOND ENERGIES AND BOND LENGTHS OF N_2, NO, AND O_2**

Molecule	Bond energy (kJ mol^{-1})	Bond length (Å)
N_2	942	1.098
NO	627	1.151
O_2	494	1.208

atomic numbers are 1 and 9, respectively. The fluorine $2p_z$ AO has about the same energy as the hydrogen $1s$ AO, and these two AO's also have the correct symmetry to form a bonding σ MO. The other occupied MO's in HF are simply the fluorine $1s$, $2s$, $2p_x$, and $2p_y$ AO's.

In heteronuclear diatomic molecules, the MO's are numbered sequentially within each symmetry type. (Asterisks to indicate antibonding MO's are usually omitted.) Having a total of 10 electrons, the HF molecule has 5 occupied MO's, whose relative energies are shown in Figure 9.23. The 1σ MO is the fluorine $1s$ AO, the 2σ MO is the fluorine $2s$ AO, and the 3σ MO is the bonding MO formed by the hydrogen $1s$ and fluorine $2p_z$ AO's. (These same MO's also form the unoccupied antibonding 4σ MO.) The 1π MO's are the fluorine $2p_x$ and $2p_y$ AO's, each doubly occupied. (The latter orbitals are of π symmetry because they are directed along the x and y axes, respectively, perpendicular to the internuclear axis, and thus their nodal planes contain the internuclear axis.)

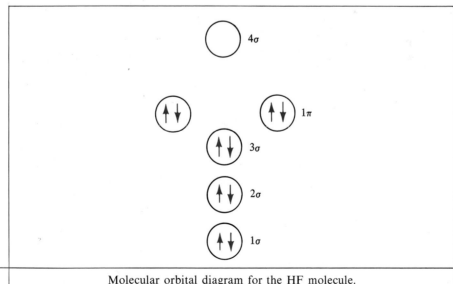

FIGURE 9.23

Molecular orbital diagram for the HF molecule.

9.18 VALENCE BOND THEORY

Although molecular orbital theory has been extended to molecules containing more than two atoms, the language of valence bond theory is more commonly used to describe polyatomic molecules. In valence bond theory, electrons in atomic orbitals on adjacent atoms are paired to form covalent bonds, leading to a theoretical basis for the covalent bonds symbolized by Lewis structures.

Valence bond theory began in 1927 with the work of Walter Heitler and Fritz London on the hydrogen molecule. The Heitler-London wave function may be described as follows. If the two hydrogen nuclei are represented by A and A′, then one electron (with spin ↑) is placed in the $1s_A$ AO, and the other electron (with spin ↓) occupies the $1s_{A'}$ AO. The wave function also includes the possibility that the electron with spin ↑ occupies the $1s_{A'}$ AO, and the electron with spin ↓ occupies the $1s_A$ AO. In either case each electron is in a different AO and thus the electrons are able to avoid each other, minimizing their mutual repulsion. As mentioned in Section 9.14, the valence-bond wave function gives a better energy for the molecule than the MO wave function, where both electrons are in the same MO spread over the entire molecule.

For a polyatomic molecule, the valence-bond wave function must be consistent with the molecular geometry. As a result, the AO's are often not those with which we are already familiar, but are instead *hybrid* AO's formed by mixing different AO's. The most important hybrid AO's are those appropriate to linear, triangular, and tetrahedral molecules.

9.19 LINEAR MOLECULES AND sp HYBRID ORBITALS

The elements of group IIA commonly form molecules having a linear structure, such as beryllium chloride, $BeCl_2$. The Lewis structure is

$$:\!\ddot{C}l\!-\!Be\!-\!\ddot{C}l\!:$$

The bond angle is 180° because of the repulsion between the two covalent bonds, as explained in Section 9.8.

According to valence bond theory, the beryllium atom needs one electron in a singly occupied AO to form a covalent bond with one chlorine atom, and another electron in a different singly occupied AO to form a covalent bond with the other chlorine atom. Yet the electron configuration of the beryllium atom in its ground state is $1s^22s^2$—both the $1s$ and $2s$ orbitals are doubly occupied. To obtain two singly occupied AO's, a $2s$ electron must be *promoted* to a vacant $2p$ orbital, giving the excited state $1s^22s2p$.

Even this configuration is not completely satisfactory, for if one bond were formed using a $2s$ orbital, and the other bond were formed using a $2p$ orbital, then the two bonds would be expected to have different bond lengths. In fact the two bonds have exactly the same length, and are equivalent in all respects.

9.19 LINEAR MOLECULES AND sp HYBRID ORBITALS

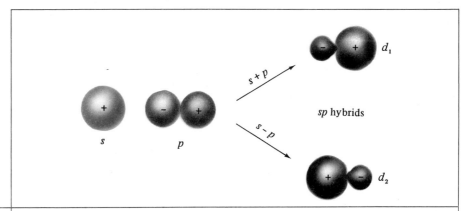

FIGURE 9.24

Mixing an s and a p orbital on the same atom to form digonal or sp hybrid orbitals.

To obtain equivalent atomic orbitals, the beryllium $2s$ and $2p$ AO's are combined or *hybridized*. (Recall that the s and p orbitals are de Broglie waves that interfere constructively when they are in phase and destructively when they are out of phase. Whereas the s orbital has positive amplitude in all directions from the nucleus, the p orbital has positive amplitude in one direction and negative amplitude in the opposite direction.)

The two ways of combining an s and a p orbital on the same atom are shown in Figure 9.24. One combination ($d_1 = s + p$) has a large lobe to the right (where s and p are both positive) and a small lobe to the left (where s is positive and p is negative). The other combination ($d_2 = s - p$) is just the opposite, as the minus sign reverses the sign (changes the phase) of the p orbital. Thus d_1 and d_2 are two new orbitals, concentrated in opposite directions, and each capable of forming a strong bond with a chlorine atom. They are called **digonal hybrids** or **sp hybrids.** (The small negative lobes are relatively unimportant and will be ignored from here on.)

Energy is required to promote the electron from the $2s$ to the $2p$ orbital, and still more energy is required for hybridization. This investment is more than recovered when the beryllium atom forms strong bonds with two chlorine atoms (see Figure 9.25). As d_1 and d_2 are in opposite directions, maximum overlap between these two orbitals and the p orbitals of the two Cl atoms occurs when the molecule is linear. Structural studies show that the $BeCl_2$ molecule is indeed linear and symmetric; the Cl—Be—Cl bond angle is 180°. Most of the other group IIA and IIB metal dihalides are also linear and symmetric. (Some of the heavier members of this series are not linear. The bent molecules include CaF_2, SrF_2, $SrCl_2$, and the barium dihalides. Bonding in these molecules probably involves d orbitals.)

It is not necessary that the beryllium atom actually pass through the states shown in Figure 9.25 to form digonal bonds—the process is probably much more

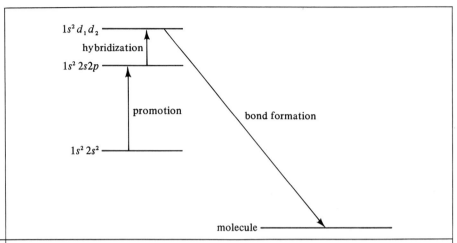

FIGURE 9.25 Energy diagram illustrating the role of bond formation in recovering the energy required for promotion and hybridization of a beryllium atom.

direct. It is digonal hybridization in the *molecule* that is important. The stepwise process provides one way of understanding the role of the valence electrons of the atom in bond formation and the energy changes involved.

Helium has an electronic structure analogous to the group IIA atoms, for it also has two outer electrons in an *s* orbital. But, because there is no $1p$ orbital, promotion can occur in the helium atom only with an increase in principal quantum number ($1s \rightarrow 2s$ or $1s \rightarrow 2p$). The energy required is so large that helium is very unreactive. It is therefore placed with the noble gases instead of with the group IIA elements in the periodic table.

Among the most important molecules with 180° bond angles are certain carbon compounds having either a triple bond adjacent to a single bond (—C≡) or two adjacent double bonds (=C=). These molecules also exhibit *sp* hybridization.

For example, the acetylene molecule, C_2H_2, is linear. Its Lewis structure is H—C≡C—H. To describe the bonding in this molecule, we begin with the fact that a carbon atom in its ground state has electron configuration $1s^2 2s^2 2p^2$. This can be written $1s^2 2s^2 2p_x 2p_y$ to emphasize that the $2p$ electrons are in different $2p$ orbitals, say, $2p_x$ and $2p_y$. According to valence bond theory, a carbon atom can form four bonds only if a $2s$ electron is promoted to the vacant $2p_z$ orbital, giving the excited state $1s^2 2s 2p_x 2p_y 2p_z$.

How can such a carbon atom form a linear molecule? If the internuclear axis is taken in the *z* direction, then the $2s$ and $2p_z$ AO's form two *sp* hybrid AO's, so that the electron configuration becomes $1s^2 d_1 d_2 2p_x 2p_y$. One of these two *sp* hybrid orbitals is concentrated about the internuclear axis in one direction, the other in the opposite direction. Thus each carbon atom can use its *sp* hybrid orbitals to form bonds with its neighbors: one bond to the adjacent hydrogen atom, and one

FIGURE 9.26

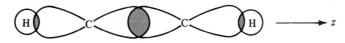

Formation of σ bonds in acetylene by overlap of carbon sp hybrid orbitals and hydrogen $1s$ orbitals.

bond to the other carbon atom. Figure 9.26 illustrates the overlap of the various atomic orbitals. The bonds thus formed are symmetrical about the internuclear axis and are therefore σ bonds.

There remain one p_x and one p_y orbital on each carbon atom (with one electron in each orbital) directed perpendicular to the internuclear axis. The p_x AO's overlap to form one π bond, and the p_y AO's overlap to form a second π bond. (These bonds are π bonds because the p_x and p_y orbitals have nodal planes containing the internuclear axis. Overlap is of the side-to-side type illustrated in Figure 9.17.) Therefore the two carbon atoms in acetylene are bonded by one σ bond and two π bonds.

As another example of sp hybrid orbitals, consider a carbon atom bonded to two neighboring atoms by double bonds as in carbon dioxide.

$$\ddot{\text{O}}\!=\!\text{C}\!=\!\ddot{\text{O}}\!:$$

The molecule is linear, and the carbon atom uses sp hybrid orbitals to form a σ bond to each oxygen atom. There is also a π bond to each oxygen atom: One π bond is formed by overlap of the carbon p_x AO with a p_x AO on one oxygen atom, the other is formed by overlap of the carbon p_y AO with a p_y AO on the other oxygen atom. (In general, for polyatomic molecules a single bond is of the σ type, a double bond consists of one σ and one π bond, and a triple bond consists of one σ and two π bonds.)

9.20 TRIANGULAR MOLECULES AND sp^2 HYBRID ORBITALS

There are a number of molecules having three atoms or groups bonded to a central atom, in which the molecule has a planar triangular shape. If the three atoms or groups are identical, the bond angles are 120°; if not, the bond angles are usually close to 120°. Typical bond arrangements exhibiting this molecular geometry are

(In the last example above, the unshared pair of electrons occupies one of the three positions around the central atom.) As explained in Section 9.8, molecules

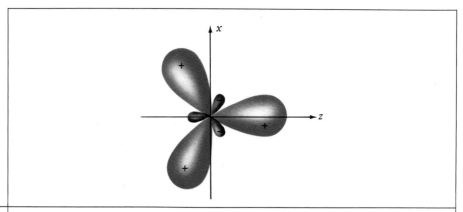

FIGURE 9.27   The three trigonal or sp^2 hybrid atomic orbitals. Each has a large positive lobe and a small negative tail in the opposite direction.

having these bond arrangements adopt a planar triangular shape owing to the repulsion between the different pairs of electrons.

According to valence bond theory, the central atom in a planar triangular molecule uses three hybrid orbitals, called **trigonal hybrids** or **sp^2 hybrids,** to form σ bonds to its neighbors (or to provide an orbital for an unshared pair of electrons). For example, consider the boron trichloride molecule, BCl_3. One Lewis structure (there are three other important resonance structures) is

$$\overset{\displaystyle :\ddot{Cl}:}{\underset{:\ddot{Cl}:\quad :\ddot{Cl}:}{|\;B}}$$

Starting from the boron atom ground state $1s^2 2s^2 2p$, one electron can be promoted to a vacant $2p$ orbital. If the occupied $2p$ orbitals are designated $2p_x$ and $2p_z$, the electron configuration of the excited state is $1s^2 2s 2p_x 2p_z$. Having three singly occupied atomic orbitals, a boron atom in this state should be able to form bonds to three neighboring atoms. But why should these bonds be oriented at 120° to one another?

The answer is that the $2s$, $2p_x$, and $2p_z$ AO's can be mixed, producing the three hybrid orbitals shown in Figure 9.27. They are concentrated in and near the xz plane, and their large positive lobes are oriented at an angle of 120° to one another. They are therefore suitable for forming σ bonds with the chlorine atoms. (The fact that these hybrid AO's are formed from one s and two p AO's gives rise to the notation sp^2.)

As another example of sp^2 hybrid AO's, consider the ethylene molecule, C_2H_4. The Lewis structure is

$$\underset{H\quad\quad H}{\overset{H\quad\quad H}{\diagdown C=C \diagup}}$$

9.21 TETRAHEDRAL MOLECULES AND sp³ HYBRID ORBITALS

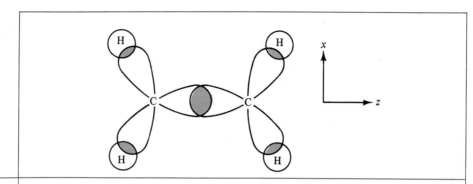

FIGURE 9.28

Formation of σ bonds in ethylene by overlap of carbon sp^2 hybrid atomic orbitals with each other and with the hydrogen $1s$ orbitals.

All six nuclei lie in the same plane, and the bond angles are about 120°. (To be precise, the HCH angle is 116.6°, and the HCC angle is 121.7°.)

Since each carbon atom forms a total of four bonds, it is convenient to start from the excited state having four singly occupied AO's, $1s^2 2s 2p_x 2p_y 2p_z$. When the $2s$, $2p_x$, and $2p_z$ orbitals are mixed to form three trigonal hybrids (designated t_1, t_2, and t_3), the electron configuration becomes $1s^2 t_1 t_2 t_3 2p_y$. The trigonal hybrids of each carbon atom can then form σ bonds to the two neighboring hydrogen atoms and to the other carbon atom, as shown in Figure 9.28. There remain the singly occupied $2p_y$ orbitals, one on each carbon atom, directed perpendicular to the molecular plane. They overlap to form a π bond, as shown in Figure 9.29.

9.21 TETRAHEDRAL MOLECULES AND sp³ HYBRID ORBITALS

Many important molecules have the tetrahedral structure shown in Figures 9.4 and 9.5. Bond arrangements with this structure include molecules having four

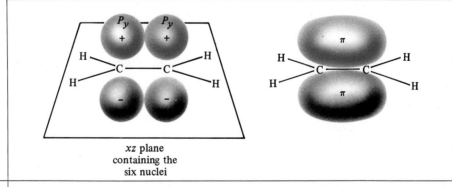

FIGURE 9.29

Ethylene. Left: the σ-bond framework and the p_y orbitals of the two carbon atoms. Right: overlap of the p_y orbitals to form a π bond between the two carbon atoms.

atoms (or groups of atoms) attached to a central atom. Unshared pairs may also occupy tetrahedral positions around the central atoms:

$$-\overset{|}{\underset{|}{X}}- \quad -\overset{..}{\underset{|}{X}}- \quad -\overset{..}{\underset{|}{X}}:$$

If the four neighboring atoms (or groups of atoms) are identical, then all of the bond angles around the central atom have the tetrahedral value, 109.5°; if not, the bond angles are usually about 109.5°. (Some important exceptions for structures with unshared pairs are described below.)

As an example of a tetrahedral molecule, consider methane, CH_4. The central carbon atom forms four bonds to the hydrogen atoms arranged at the corners of a tetrahedron. To do so, it needs four half-filled atomic orbitals having tetrahedral orientation. These can be obtained, starting from the carbon atom excited state $1s^2 2s 2p_x 2p_y 2p_z$, by mixing all four singly occupied AO's. The large lobe of each of the hybrid orbitals formed in this way is indeed directed toward one corner of a regular tetrahedron (Figure 9.30). These hybrid AO's are called **tetrahedral hybrids** or sp^3 **hybrids** (because they are formed from one s and three p AO's). Each of these sp^3 hybrid AO's forms a σ bond to one of the hydrogen atoms.

Because the molecular geometry of carbon compounds is especially important in chemistry, the results of valence bond theory for carbon are summarized in Table 9.11.

Because the atoms of groups VA and VIA have three and two unpaired electrons, respectively, promotion is not necessary for the normal group valences of three and two. Whether or not hybridization occurs is another matter. If hybridization did not occur, the bonds would be formed by p orbitals, concentrated along a set of Cartesian axes at 90° to one another, and thus molecules of these atoms would have 90° bond angles. As shown in Table 9.12, the elements of higher atomic number in these groups have bond angles close to 90°, but the hydrides of oxygen and nitrogen have much larger bond angles. Other oxygen and nitrogen compounds show similarly large bond angles.

Partial hybridization provides a description of these large bond angles. If in

TABLE 9.11 | **BONDING CHARACTERISTICS OF CARBON**

No. of atoms bonded to C	Bonds	Hybrid orbitals[a]	Unhybridized orbitals[a]	Bond angles	Geometry	Bond type
4	4 Single	$sp^3(4)$	None	109.5°	Tetrahedral	4σ
3	2 Single and 1 double	$sp^2(3)$	$p(1)$	120°	Planar	2σ and $1\sigma + 1\pi$
2	1 Single and 1 triple	$sp(2)$	$p(2)$	180°	Linear	1σ and $1\sigma + 2\pi$
2	2 Double	$sp(2)$	$p(2)$	180°	Linear	$1\sigma + 1\pi$ each

[a] Number of orbitals of each type shown in parentheses.

9.21 TETRAHEDRAL MOLECULES AND sp^3 HYBRID ORBITALS

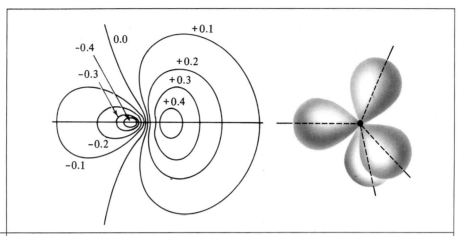

FIGURE 9.30 Contour diagram of a tetrahedral or sp^3 hybrid orbital (left figure). There are four of these orbitals, each with its positive lobe directed toward a different corner of a tetrahedron (right figure).

the process of forming an ammonia molecule the five valence electrons of the nitrogen atom occupied tetrahedral orbitals (one electron in each of three orbitals, and a pair of electrons in the fourth orbital), the nitrogen atom could form strong bonds with three hydrogen atoms, and the bond angles would be tetrahedral. The fact that the bond angle is actually about 2° smaller indicates that sp^3 hybridization is not quite complete. The smaller bond angle in H_2O (104.5°) indicates somewhat less hybridization, with the bonding orbitals intermediate between pure p (bond angle 90°) and sp^3 (bond angle 109.5°).

Whether this description is correct is not yet known. It describes but does not explain the molecular structure. Why should the bonding orbitals be partially hybridized in H_2O and NH_3, but not in H_2S and PH_3? Some of the factors involved are repulsions between the electron pairs, repulsion between the positive ends of the polar OH or NH bonds, and the greater ability of a $2s$ than of a $3s$ orbital to overlap with a hydrogen orbital, as compared to the corresponding p orbitals.

What is needed, of course, are accurate wave functions that yield theoretical bond angles in quantitative agreement with the experimental bond angles. The

TABLE 9.12

BOND ANGLES IN GROUP VA AND VIA HYDRIDES			
Substance	Bond angle	Substance	Bond angle
NH_3	107.3° ± 0.2°	H_2O	104.5° ± 0.1°
PH_3	93.3° ± 0.2°	H_2S	92.2° ± 0.1°
AsH_3	91.8° ± 0.3°	H_2Se	91.0° ± 1°
SbH_3	91.3° ± 0.3°	H_2Te	89.5° ± 1°

results obtained recently with SCF molecular orbitals have been very successful in this regard. Attempts to describe the essential aspects of these wave functions in simple verbal terms have been somewhat less successful.

PROBLEMS

9.1 (a) How many electrons does each of the following atoms gain when an ion with an outer octet of electrons is formed? What is the resulting electric charge on each of the resulting ions?
F, N, S, Br
(b) How many electrons does each of the following atoms lose when an ion with an outer octet is formed? What is the resulting electric charge on each of the resulting ions?
Li, Ca, Al, K

9.2 Assuming ionic bonding, what is the formula of the compound formed by
(a) K and S?
(b) Ca and S?
(c) Ca and Br?

9.3 In ionic compounds, aluminum loses three electrons and nitrogen gains three electrons.
(a) Write the formulas of aluminum fluoride, aluminum oxide, aluminum nitride, and magnesium nitride.
(b) What is the ionic valence of each ion in these compounds?
(c) Which ion would you expect to have the larger ionic radius, Al^{3+} or N^{3-}?

9.4 Use the electric charges on the ions in Problem 9.1 to determine the charges on the following ions, assuming ionic bonding.
(a) Cu ion in $CuBr_2$
(b) Fe ion in FeS
(c) Fe ion in $FeBr_3$
(d) NO_3 ion in $Ca(NO_3)_2$
(e) SO_4 ion in $Al_2(SO_4)_3$

9.5 In the NaCl crystal (Figure 9.2), how many sodium ions are there adjacent to a chloride ion located
(a) in the interior of the crystal?
(b) on an edge?

9.6 (a) From the data in Table 9.1, calculate the distance between adjacent K^+ and F^- ions in a crystal of potassium fluoride, KF.
(b) Consider a cubic crystal of KF and a cubic crystal of NaCl, both containing the same number of ions. Which crystal is larger?

PROBLEMS

9.7 In the following pairs of ions, which ion has the larger ionic radius? K^+ and Cl^-, Na^+ and Mg^{2+}, Na^+ and K^+

9.8 The internuclear distance in the NaBr molecule is 2.64 Å. The ionization potential of sodium is 5.14 eV, and the electron affinity of bromine is 3.36 eV. Using the hard-sphere approximation, calculate ΔE for each of the following reactions:
$$NaBr(g) = Na^+(g) + Br^-(g)$$
$$NaBr(g) = Na(g) + Br(g)$$

9.9 (a) Write a Lewis structure obeying the octet rule and the $8 - N$ rule for each of the following molecules:
HCN, F_2O, OCS, ClNO
(b) Estimate the bond angle in each molecule.

9.10 (a) Write a Lewis structure obeying the octet rule and the $8 - N$ rule for each of the following molecules:
N_2H_4, F_2CO, CH_2Cl_2, HNCO
(b) Estimate the following bond angles:

HNN bond angle in N_2H_4
FCO bond angle in F_2CO
HCCl bond angle in CH_2Cl_2
HNC bond angle in HNCO
NCO bond angle in HNCO

(c) Two of these molecules are planar. Which ones are they?

9.11 The compound dioxirane, H_2CO_2, has recently been discovered to be an important component of smog. Draw a Lewis structure for dioxirane. (The carbon atom and two oxygen atoms form a three-membered ring.)

9.12 (a) What is the formal charge on the sulfur atom in the sulfite ion, SO_3^{2-}?

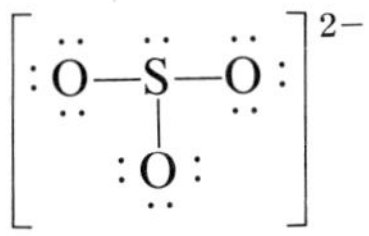

$$\left[\begin{array}{c} \ddot{\mathrm{O}}-\mathrm{S}-\ddot{\mathrm{O}}\!: \\ | \\ :\ddot{\mathrm{O}}\!: \end{array} \right]^{2-}$$

(b) Is the sulfite ion planar or nonplanar?

9.13 Write Lewis structures (including formal charges) for H_3NBH_3, $(CH_3)_3NO$, $(CH_3)_2SO$, and H_3PO_4. Which bonds are of the donor–acceptor type? Which atoms are donors and which are acceptors?

9.14 For the PCl$_5$ molecule (Figure 9.6), designate a chlorine atom in an axial position as Cl^a, and a chlorine atom in an equatorial position as Cl^e. What is the numerical value of each of the following bond angles?
Cl^e—P—Cl^e Cl^e—P—Cl^a Cl^a—P—Cl^a

9.15 The isoelectronic molecules CO and N$_2$ have the highest bond energies of any diatomic molecules (1070 and 942 kJ mol^{-1}, respectively). Suggest an explanation.

9.16 Draw resonance structures for the nitrite ion, NO$_2^-$ (two structures), and the carbonate ion, CO$_3^{2-}$ (three structures). Show all formal charges.

9.17 The bond energies of F$_2$ and Br$_2$ are 155 and 190 kJ mol^{-1}, respectively. Using these data and the electronegativities in Table 9.4, calculate the BrF bond energy.

9.18 On the basis of the electronegativities in Table 9.4, which of the following bonds are nonpolar? In the polar bonds which atom is the negative partner?
H—Na   H—Br   H—P   N—O   N—C   C—S

9.19 Calculate the fractional ionic character of the O—H bond from its bond moment (1.49 D) and internuclear distance (0.96 Å).

9.20 Which of the following molecules have zero dipole moments because of symmetry?
NF$_3$, BF$_3$, CO$_2$, SO$_2$

9.21 On a diagram showing energy versus internuclear distance (see Figure 9.15), draw curves that illustrate a diatomic molecule having a strong short bond and a diatomic molecule having a weak long bond.

9.22 For each of the following pairs of waves, which will interfere constructively, which destructively, and which constructively in one region and destructively in another?

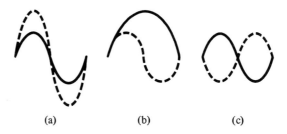

(a)   (b)   (c)

9.23 How does a sigma molecular orbital differ from a pi molecular orbital?

9.24 (a) For a homonuclear diatomic molecule, which of the following molecular orbitals has a nodal plane which is perpendicular to the internuclear axis and midway between the nuclei?

$\sigma 1s \quad \sigma^*1s \quad \sigma 2p \quad \pi 2p$

(b) Which of the above molecular orbitals has a nodal plane passing through the nuclei?

9.25 On diagrams similar to Figure 9.22, draw molecular orbital diagrams of the N_2 and O_2 molecules.

9.26 What are the electron configurations of N_2^+ and O_2^+? What is the net number of bonding electrons in each? Which should have the higher bond energy:

N_2^+ or N_2? $\quad O_2^+$ or O_2?

9.27 Which atomic orbitals combine to form the bonding molecular orbitals in Na_2, K_2, KNa, Cl_2, Br_2, BrCl?

9.28 What is the electron configuration of the lowest excited state of
(a) a magnesium atom having two unpaired electrons?
(b) an aluminum atom having three unpaired electrons?
(c) a silicon atom having four unpaired electrons?

9.29 How do sp, sp^2, and sp^3 hybrids differ? What are the characteristic bond angles for each type of hybridization? Give an example of a molecule that illustrates each type.

9.30 (a) Draw a Lewis structure for cyanogen, C_2N_2. (The molecule is linear, with the nitrogen atoms on the ends, NCCN.)
(b) How many pi bonds are there in the cyanogen molecule?
(c) What kind of hybrid orbitals are used by the carbon atoms in cyanogen to form their sigma bonds: sp, sp^2, or sp^3?

9.31 The most important resonance structures of nitrous oxide, N_2O, are

$:N=N=O: \longleftrightarrow :N\equiv N-\ddot{O}:$

(a) How many pi bonds are in each structure?
(b) What hybrid orbitals are used by the central nitrogen atom in each structure to form sigma bonds to the other atoms?
(c) Estimate the NNO bond angle.
(d) What is the formal charge on each atom in each structure?

9.32 From the following thermodynamic data calculate the average N—H bond enthalpy in NH_3:

$$H_2(g) = 2H(g) \quad \Delta H = 435.9 \text{ kJ}$$
$$N_2(g) = 2N(g) \quad \Delta H = 944.7 \text{ kJ}$$
$$N_2(g) + 3H_2(g) = 2NH_3(g) \quad \Delta H = -92.4 \text{ kJ}$$

9.33 (a) Sketch a cube. Place four Na^+ ions and four Cl^- ions alternately at the eight corners of the cube. If the Na^+—Cl^- distance is a cm, what is the volume of your cube?
(b) Since each of the eight ions in part (a) is a member of eight adjoining cubes, what is the relationship of the volume of your cube to the molecular volume?
(c) The density of solid NaCl is 2.165 g cm^{-3}. Calculate the Na^+—Cl^- internuclear distance a.

9.34 From the ionic radii in Table 9.1 and the atomic masses, determine which ionic crystal has the higher density, NaCl or KCl. Then calculate the ratio of the densities.

9.35 (a) The dissociation energy of the cesium bromide molecule, to form a cesium atom and a bromine atom, is 4.17 eV:
$$CsBr(g) = Cs(g) + Br(g) \quad \Delta E = 4.17 \text{ eV}$$
Given that the ionization potential of cesium is 3.89 eV, and the electron affinity of bromine is 3.36 eV, calculate ΔE in eV for dissociation to form ions:
$$CsBr(g) = Cs^+(g) + Br^-(g)$$
(b) Using this result and Eq. 9.5, calculate the internuclear distance in the CsBr molecule.

9.36 As stated in Section 9.9, the covalent radius of chlorine is smaller than the ionic radius of Cl^-.
(a) Suggest an explanation.
(b) On the basis of your explanation, do you expect the covalent radius of sodium to be larger or smaller than the ionic radius of Na^+?

9.37 Propylene and cyclopropane have the same molecular formula, C_3H_6. Propylene is a bent molecule with a CCC bond angle of about 120°. In cyclopropane the three carbon atoms form a three-membered ring with a CCC bond angle of 60°. Draw a Lewis structure for each molecule.

9.38 Draw two resonance structures obeying the octet rule and the $8-N$ rule for the pyridine molecule, C_5H_5N. (The carbon and nitrogen atoms are joined in a ring, with one hydrogen atom bonded to each carbon atom.)

9.39 (a) Write a Lewis structure for allene, H_2CCCH_2.
(b) How do the two pi bonds in this molecule differ in their orientation?
(c) If the left carbon atom uses its $2p_x$ orbital to form a pi bond with the central carbon atom, which atomic orbital is used by the right carbon atom to form its pi bond?
(d) How, then, are the hydrogen atoms at one end of the molecule arranged relative to the hydrogen atoms at the other end?

CHAPTER 10
LIQUIDS AND LIQUID SOLUTIONS

10

In previous chapters we have noted that substances can exist in the gaseous, liquid, or solid state and that, as the temperature is lowered, real gases condense to form liquids which on further cooling solidify. The solid state is characterized by high cohesion and rigidity; the liquid state by high cohesion, lack of rigidity, and a comparatively low resistance to flow; and the gaseous state by low cohesion, lack of rigidity, and essentially a negligible resistance to flow. Liquids and solids are nearly incompressible, in marked contrast to gases. An increase in pressure from 1 to 2 atm decreases the volume of most liquids or solids by less than 0.01%.

In this chapter we shall examine the conditions under which changes of state occur and the use of liquids as solvents. Solutions provide a convenient means for mixing controlled amounts of reagents and for examining the resulting chemical reactions. The physical properties of solutions are of interest in that they may be used to determine the molecular masses of dissolved substances and to study the effects of different molecules on each other by varying the molecular environment.

10.1 VAPOR PRESSURE

A liquid at a temperature near its boiling point will evaporate entirely if placed in an open container, whereas in a closed container only a portion of the liquid will enter the gas phase. This behavior can be readily investigated using the apparatus shown in Figure 10.1. First let us evacuate the system by opening valve A, thus reducing the pressure to 0.00 atm. After evacuation, A is closed and a small amount of a liquid is introduced by means of valve B. As some of the liquid evaporates, the pressure (as indicated by the pressure gauge, or manometer) will rise from zero to a constant value. (For carbon tetrachloride at 30°C, the pressure will reach 0.19 atm.) If we now add more liquid and keep the temperature constant, we find that the pressure does not change. Thus we conclude that the pressure exerted by the vaporized liquid is *independent of the amount of liquid present* and *independent of the surface area of the liquid*. Different substances behave similarly except that a different pressure is noted for each. The pressure exerted by a gaseous substance in equilibrium with its liquid phase is known as the **equilibrium vapor pressure,** or simply the **vapor pressure** of the substance. (The terms *gas* and *vapor* are essentially synonymous. *Vapor* is often used to denote a substance in the gaseous state when it is in equilibrium with the liquid, or when the gas is at a temperature below the boiling point of the substance.)

Although the vapor pressure of a substance depends on the temperature,

10.1 VAPOR PRESSURE

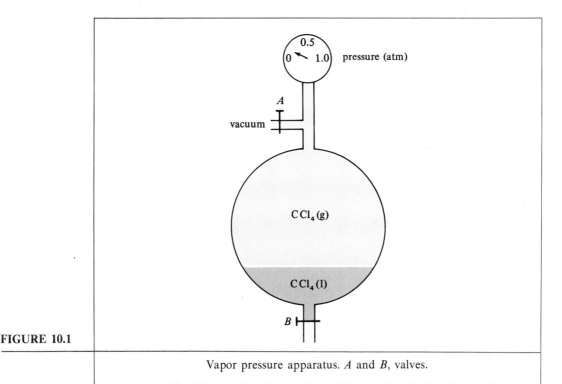

FIGURE 10.1 Vapor pressure apparatus. *A* and *B*, valves.

one obtains the same value for the vapor pressure whether one approaches the temperature of the measurement from a lower temperature or from a higher temperature. The effect of temperature on the vapor pressure of several substances is shown in Figure 10.2 for the pressure range 0–1200 Torr. Note the very rapid increase in vapor pressure as the temperature is increased.

We might next ask: What happens to a system containing a liquid and its vapor when the pressure is not equal to the vapor pressure? In general, if the applied pressure is lower than the vapor pressure, the liquid will completely evaporate, and, so long as the applied pressure remains below the vapor pressure, the substance will exist entirely in the gaseous state. On the other hand, if the applied pressure is higher than the vapor pressure, the gas will completely condense to the liquid state, and, so long as the applied pressure remains above the vapor pressure, the substance will exist entirely in the liquid state. Only when the applied pressure is equal to the vapor pressure can the liquid and gaseous states of a pure substance coexist.

As a specific example of changes of state, consider the following experiment. If we place 1.0 millimole (mmol) of CCl_4 in a 250-mL flask (Figure 10.1) at 30°C, we find that all of the CCl_4 exists as a gas and that the pressure is 0.099 atm.[1] We

---
[1] The pressure may be calculated assuming the ideal gas law:

$$P = \frac{nRT}{V} = \frac{(1.0 \times 10^{-3}\,\text{mol})(0.082\,\text{L atm mol}^{-1}\,\text{K}^{-1})(303\,\text{K})}{0.250\,\text{L}} = 0.099\text{ atm}$$

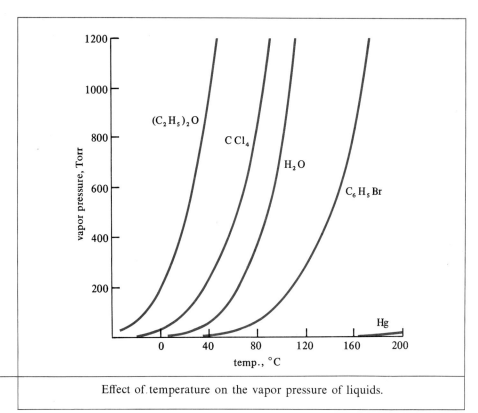

FIGURE 10.2

Effect of temperature on the vapor pressure of liquids.

may now increase the pressure of the CCl_4 gas by introducing mercury through valve B, thus compressing the gas to a smaller volume. The pressure rises, as shown in Figure 10.3 until the volume occupied by the gas is 131 mL and the pressure is 0.19 atm (the vapor pressure at this temperature). Further additions of mercury decrease the volume occupied by the gas but the pressure remains unchanged at 0.19 atm. The explanation for this peculiar behavior is that the decrease in volume of the gas phase is accompanied by formation of liquid CCl_4, which can be observed in the flask when the gas volume is small.[2] When all the CCl_4 has been converted to liquid, compression of the nearly incompressible liquid phase is difficult, and small additions of mercury cause tremendous increases in pressure.

Although we have used carbon tetrachloride as an example, other liquids behave in a manner similar to that shown in Figure 10.3. To summarize, (1) every pure substance has a vapor pressure that depends only on the temperature, and (2) at a given temperature the liquid and gas phases can coexist at this pressure only. Compression or expansion of a substance containing coexisting liquid and gas phases at constant temperature changes the relative amounts of liquid and

[2] The volume of 1.0 mmol of liquid CCl_4 at 30°C is only 0.1 mL, so it is necessary to provide a break in the scale of the plot of Figure 10.3 to show this small volume.

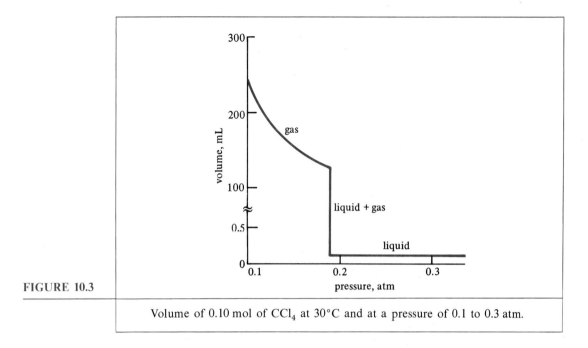

FIGURE 10.3

Volume of 0.10 mol of CCl_4 at 30°C and at a pressure of 0.1 to 0.3 atm.

gas, but no change in pressure occurs until the substance is completely converted to the gaseous or liquid state.

Changes of temperature at a fixed applied pressure may also result in the condensation of a gas or evaporation of a liquid. Condensation of a gas to the liquid phase occurs when a gas is cooled to the point where the vapor pressure is equal to the applied pressure. Conversely, boiling occurs when a liquid is heated until its vapor pressure equals the applied pressure. Thus, for a given pressure, there is a temperature at which boiling of a liquid will occur, the **boiling point** (bp). The *normal* boiling point is the temperature at which the vapor pressure is 1 atm or 760 Torr or 1.01325×10^5 Pa.

Phase transitions between the solid and liquid phases of a pure substance also occur at constant temperature, provided that the pressure is held constant. The **normal melting point** is the temperature at which the solid changes to the liquid phase when the pressure is 1 atm.

The constant temperatures provided by two-phase systems of pure substances are frequently useful in the laboratory, as in calibrating thermometers. The Celsius temperature scale was defined originally in terms of the normal melting point of ice as 0°C and the normal boiling point of water as 100°C.

10.2 PHASE DIAGRAMS

The experimental data on phase changes are reported in various tables or by mathematical equations, but it is much easier to visualize possible phase changes when the data are presented in the form of a **phase diagram.** The phase diagram

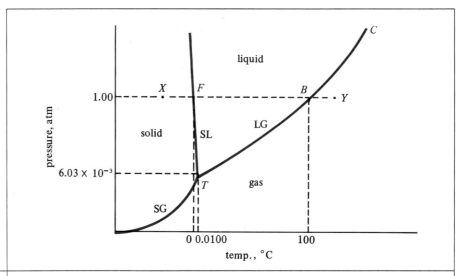

FIGURE 10.4

Phase diagram for water (not drawn to scale). C is the critical point, T is the triple point, F is the normal freezing point, and B is the normal boiling point.

for water is shown in Figure 10.4. The different regions in this diagram indicate the temperatures and pressures where only one phase (gas, liquid, or solid) may exist at equilibrium. Thus it is possible to vary the pressure and temperature independently within the limits of each region and still have only a single phase. We therefore say that there are two degrees of freedom (pressure and temperature).

The solid curves separating the three one-phase regions represent the temperatures and pressures at which the phases adjacent to the curve can exist in equilibrium with each other. For example, curve LG shows the temperatures and pressures at which the liquid and gas are in equilibrium. Note that when two phases are present we have only one degree of freedom (pressure or temperature), because if the temperature is fixed, the pressure is also determined. Curve LG terminates at the **critical point,** C. The temperature and pressure at point C are known as the **critical temperature** and the **critical pressure.** Above the critical temperature it is not possible to distinguish between the liquid and gaseous states. For water, the critical temperature and critical pressure are 347°C and 217.7 atm, respectively.

Curve SL shows the temperatures and pressures at which solid (ice) and liquid water may coexist. (Ice may exist in several different crystalline forms at high pressures that, for simplicity, have not been included in the phase diagram.) Curve SG represents the temperatures and pressures at which solid and gas may coexist. The point of intersection of the three curves (T) is called the **triple point,** and it is the only temperature and pressure at which ice, liquid, and gaseous water may coexist in equilibrium. At this point there are no degrees of freedom. For water, the triple point is at 0.0100°C and 6.03×10^{-3} atm (or 611 Pa or 4.58 Torr).

If we represent the number of phases by p and the number of degrees of freedom by f, we can summarize the behavior of water (or any other pure substance) by the equation

$$p + f = 3 \tag{10.1}$$

The usefulness of a phase diagram in predicting the behavior of a substance may be illustrated by considering what happens to ice at $-10\,°C$ and 1 atm pressure (point X on the diagram) if it is heated at a constant rate while maintaining a pressure of 1 atm. The temperature gradually rises along the dashed line XY. At point F, the temperature remains constant at $0\,°C$ until all the ice is converted to liquid. Then the temperature of the liquid increases until point B is reached ($100\,°C$). The temperature now remains constant until all the liquid is converted to gas, and then the temperature again rises.

Water is an unusual liquid in that an increase in the pressure on a mixture of ice and water will cause the ice to melt. Thus line SL in Figure 10.4, separating the solid and liquid phases, slopes slightly to the left with increasing pressure. (In the phase diagrams of most pure substances, line SL slopes to the right with increasing pressure.) The sport of ice skating and the flow of glaciers are possible because of this atypical property of water.

The phase diagram of carbon dioxide shown in Figure 10.5 shows two important differences from the phase diagram for water. First, the SL line slopes to the right with increasing pressure, which is the case for all solids that expand upon melting. Second, heating solid CO_2 at 1 atm pressure increases the temperature until the SG line is reached (at $-78.5\,°C$), where the solid is converted directly to a gas without going through the liquid phase.

10.3 KINETIC MOLECULAR THEORY OF LIQUIDS

The kinetic theory enabled us to provide a reasonable explanation of the gas laws in terms of a molecular model (Chapter 4). If we extend this theory to liquids, we find that it is quite useful even though it does not provide us with a quantitative law to predict the properties of liquids.

In applying the kinetic theory to gases, we found that the average translational kinetic energy of the molecules is proportional to the absolute temperature. Thus we might reasonably expect that the absolute temperature would also be a measure of the vigor of molecular motion in liquids and solids. The rigidity of solids implies that the molecules are held in fixed positions. Therefore molecular motion in solids involves oscillations of the molecules about their mean positions, as represented schematically in Figure 10.6. As the temperature is increased, the motion increases in vigor until the molecules acquire sufficient energy to overcome the attractive forces holding them in fixed positions. This temperature corresponds to the melting point and the loss of the rigid structure. As the temperature

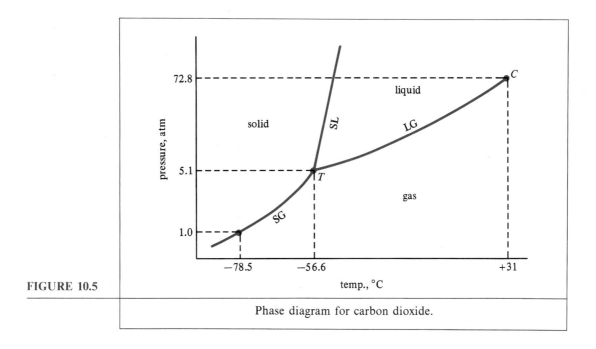

FIGURE 10.5 Phase diagram for carbon dioxide.

is increased further, the molecules move more rapidly and exchange positions more readily (Figure 10.6). When the boiling point of the liquid is reached, the molecules have acquired sufficient energy to overcome the attractive forces holding them together in the liquid phase and they escape into the gas phase.

The available experimental evidence is in agreement with the concept that the average translational kinetic energy of molecules in the liquid state is the same ($\frac{3}{2}k_B T$) as in the gaseous state. Furthermore, there is a wide distribution of molecular speeds, which may be represented by the Maxwell-Boltzmann distribution (Figure 4.3). Thus the essential difference between liquids and gases is the

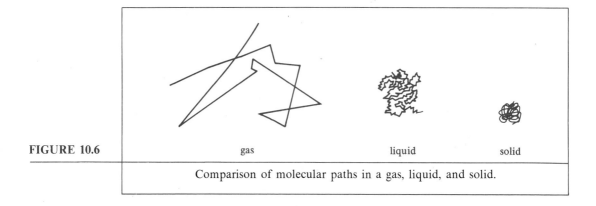

FIGURE 10.6 Comparison of molecular paths in a gas, liquid, and solid.

very restricted translational motion in liquids caused by the close proximity of other molecules.

The kinetic molecular theory can be used to explain why liquids exert an equilibrium vapor pressure. When a liquid is in contact with its vapor, there are numerous collisions of gas molecules with the liquid surface because of the rapid motion of the molecules in the gas phase. Many of these collisions will result in gas molecules entering the liquid and becoming a part of it. Molecules of the liquid also have motion, and a certain number of them will have sufficient energy to overcome the attractive forces of the neighboring molecules and escape into the gaseous state. Thus there is a continuous interchange of molecules between the liquid and the vapor state. When the number of molecules going from liquid to gas per second is equal to the number going from gas to liquid per second, we say that the system is at equilibrium. Under these conditions the number of molecules in the gas phase remains constant and the pressure exerted by these molecules is constant; hence there is an equilibrium vapor pressure.

The kinetic molecular theory can also be used to explain the effect of temperature on equilibrium vapor pressure. An increase in temperature causes an increase in the number of molecules with sufficient energy to escape from the liquid. Because more molecules escape into the vapor phase, the number of molecules in the vapor phase will increase. When the number of molecules in the vapor phase is sufficiently large, the number of molecules colliding with the liquid surface and sticking to it during each second will equal the number of molecules leaving the liquid surface per second. Thus at the higher temperature a new equilibrium is established with a higher equilibrium vapor pressure. Furthermore, because the number of molecules with sufficient energy to escape from the liquid increases rapidly with increasing temperature, the equilibrium vapor pressure of a liquid also increases rapidly with increasing temperature (Figure 10.2).

10.4 INTERMOLECULAR FORCES

One can also use the kinetic molecular theory to relate the boiling point of a liquid to the strength of the intermolecular forces in the liquid. If the attractive force between neighboring molecules is strong, then:

1. Relatively few molecules have sufficient energy to escape from the liquid at a given temperature;
2. The vapor pressure of the liquid is relatively low;
3. A high temperature is required to increase the vapor pressure to 1 atm; and
4. The substance has, therefore, a high normal boiling point.

Conversely, if the attractive force between neighboring molecules is weak, then the molecules more easily escape into the gaseous state, and the substance has a low boiling point.

For many types of substance there is a connection between molecular mass and boiling point; hence, a connection between molecular mass and the strength of the attractive forces between neighboring molecules. In methane (CH_4, molecular mass 16 and bp $-161\,°C$), the intermolecular forces are weak; in *n*-octane (C_8H_{18}, molecular mass 114 and bp $126\,°C$), the intermolecular forces are much stronger. This connection is intuitively reasonable, because one tends to think of forces in terms of gravitational attraction and therefore expects that a higher temperature should be required to vaporize a heavier substance. Calculations of the gravitational forces between molecules show, however, that they are negligible compared to other intermolecular forces.

The most important intermolecular force in nonpolar molecules is called the **dispersion force** or **London force**,[3] caused by the mutual polarization of neighboring molecules. At any instant the electrons in a molecule may be distributed in a nonuniform manner. The region with an overabundance of electrons is negative, and the region with a deficiency of electrons is positive. The resulting electric dipole polarizes a neighboring molecule by attracting electrons toward its positive region and repelling electrons away from its negative region, as shown below (left pair of molecules).

The electric dipole created in the second molecule likewise polarizes the first molecule, and thus there is a mutual polarization of the two molecules.

At some later instant the directions of polarization may differ, as shown above (right pair of molecules), so that over a long period of time the electric dipoles average to zero in symmetrical molecules such as CH_4, which have zero permanent dipole moments. Nevertheless, the synchronization of the directions of the instantaneous dipoles allows the negative region of one molecule to be near the positive region of the other molecule, resulting in a force of attraction between the molecules.

The strength of the London force between two molecules depends on the polarizabilities of the molecules, which in turn depend (at least in an approximate way) on the numbers of electrons in the molecules. Thus *n*-C_8H_{18} with 66 electrons is much more easily polarizable than CH_4 with only 10 electrons, and the attractive force between two octane molecules at a given distance is much greater than the attractive force between two methane molecules at the same distance. And as boiling point varies with the strength of intermolecular forces on the one hand, and molecular mass varies approximately with the number of electrons in a

---

[3] The term *dispersion force* is used because the same basic cause (molecular polarization) is also responsible for the dispersion of light. The term *London force* is used because the theory of this force was first developed by Fritz London (in 1930).

molecule on the other hand, there results the connection between boiling point and molecular mass noted above. The chain of proportionality is:

bp α London force α number of electrons α molecular mass

For molecules having a permanent dipole moment, such as ClF, there is also an attractive force between the permanent electric dipoles of neighboring molecules, which tend to align themselves in opposite directions, but even here the London force is dominant.

In many substances, hydrogen plays a unique role in linking different molecules. This particular type of intermolecular force is called the **hydrogen bond.** For example, in crystalline HF the molecules are arranged in zigzag chains extending from one side of the crystal to the other:

$$\cdots H\diagdown_{F}\diagup^{F}\diagdown H\cdots H\diagdown_{F}\diagup^{F}\diagdown H\cdots H\diagdown_{F}\diagup^{F}\diagdown H\cdots$$

The dotted lines connecting the hydrogen atom of one molecule with the fluorine atom of another molecule represent the attractive force of the hydrogen bond.[4] This force has its origin in the highly polar nature of the HF bond, which makes the hydrogen end positive and the fluorine end negative. Consequently, there is a coulombic force of attraction between the positive hydrogen of one HF molecule and the negative fluorine of another.

The strength of a hydrogen bond increases with increasing charge and with decreasing radius of the negative atom. Thus hydrogen bonding is important primarily when the hydrogen atom is bonded to a small, highly electronegative atom, particularly fluorine, oxygen, or nitrogen.

Like other intermolecular forces and forces between atoms not bonded to one another in the same molecule, the hydrogen bond is usually weak compared to the force of a covalent or ionic bond. For example, the energy of the hydrogen bond in HF is about 29 kJ mol^{-1}, whereas the dissociation energy of the HF molecule is 565 kJ mol^{-1}. Nevertheless, hydrogen bonding is strong enough to determine the structures of many crystalline substances, and it has an important effect on many other properties. Figure 10.7 shows that both the melting points and boiling points of HF, H_2O, and NH_3 are abnormally high in comparison to their homologs, the result of hydrogen bonding in both the solid and liquid states. In addition, as we recall from Section 4.6, the boiling point of water is abnormally high when compared to its van der Waals a constant, another manifestation of hydrogen bonding in liquid water.

---

[4] To emphasize the hydrogen bonds, unshared pairs are omitted from these drawings.

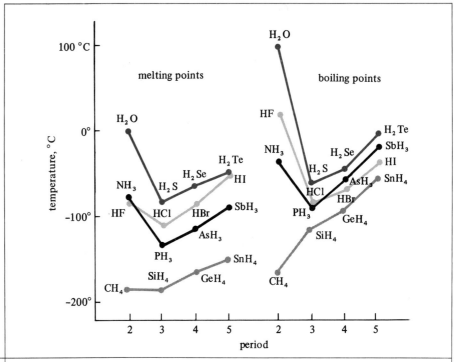

FIGURE 10.7

Melting points and boiling points of hydrides of groups IVA, VA, VIA, and VIIA. (SOURCE: *From Linus Pauling:* The Nature of the Chemical Bond, *Third Edition.* © *1960 by Cornell University. Used by permission of the publisher, Cornell University Press.*)

Hydrogen bonding in ice results in each water molecule being bonded to four other water molecules:

```
            H
             \
              O—H
              ⋮
    H         H         H
     \        |        /
      O—H ⋯ O—H ⋯ O—H
              ⋮
              H
               \
                O—H
```

Each oxygen atom in the interior of the crystal is surrounded by four other oxygen atoms in a nearly regular tetrahedral arrangement. Each hydrogen atom lies between two oxygen atoms, closer to one than the other. (The dashes represent short, strong covalent bonds, while the dots represent longer, weaker hydrogen bonds.)

Liquid water has a somewhat similar but less ordered structure. The structures of liquid water and ice will be discussed in more detail in Chapter 17.

10.5 CONCENTRATION

As previously noted, one of the most important properties of a liquid is its ability to dissolve gases, solids, or other liquids to form solutions. In general the substance present in largest amount is called the **solvent,** whereas the substances present in smaller amounts are called **solutes.** To describe a particular solution we must specify, in addition to the variables needed to describe the state of a pure substance (usually taken as pressure and temperature), the composition of the solution, called its **concentration.** The most commonly used methods of expressing concentration are the mole fraction, the molality, and the molarity.

The **mole fraction** (X) of a component of a solution is merely the fraction of the total molecules that are of one type. For example, if a solution contains n_1 mol of substance 1 and n_2 mol of substance 2, the mole fraction of substance 1 is

$$X_1 = \frac{n_1}{n_1 + n_2} \tag{10.2}$$

Because the mole fraction is a ratio of like quantities, it is dimensionless. Note that in a binary solution (only two components), it is necessary only to specify the mole fraction of one component. The mole fraction of the other component is fixed as

$$X_1 + X_2 = \frac{n_1}{n_1 + n_2} + \frac{n_2}{n_1 + n_2} = 1 \tag{10.3}$$

For solutions containing more than two components, it is sufficient to specify the mole fractions of all but one component of the solution to fix the composition of the solution, because, if X_i is the mole fraction of the ith component, then

$$\sum_i X_i = 1$$

We will use the convention that the subscript 1 designates the solvent and that subscripts 2 or larger designate solutes.

The molal concentration or **molality** (c_m) of a solute is defined as the *number of moles of solute per kilogram of the solvent.* The symbol m is used to represent moles per kilogram of solvent. Thus a $0.20m$ solution of NaCl in water is one that contains 0.20 mol of NaCl dissolved in exactly 1 kg of water. It is important to note that both the mole fraction and the molality of a solution depend on the relative masses of the substances used in preparing the solution and, hence, are independent of changes in temperature.

The most commonly used unit of concentration is the molar concentration or *molarity* (c_M), with the symbol M used to represent moles per liter of solution (see Section 2.8). The molarity unit is usually used because it is convenient to weigh out known masses of solute and to dilute them with solvent to known volumes using calibrated flasks. Note that, because the volume of a solution generally increases with increasing temperature, the molarity of a solution generally decreases with increasing temperature.

EXAMPLE 1

A solution that is 10.0% by mass NaOH in water has a density of 1.109 g mL^{-1}. Calculate the mole fraction of NaOH, the molality of NaOH, and the molarity of NaOH in this solution.

(a) By definition

$$X_{NaOH} = \frac{n_{NaOH}}{n_{NaOH} + n_{H_2O}}$$

In 100 g of solution there would be 10.0 g of NaOH and 90.0 g of water. Therefore

$$n_{NaOH} = \frac{10.0 \text{ g}}{40.0 \text{ g mol}^{-1}} = 0.250 \text{ mol}$$

$$n_{H_2O} = \frac{90.0 \text{ g}}{18.0 \text{ g mol}^{-1}} = 5.00 \text{ mol}$$

$$X_{NaOH} = \frac{0.250 \text{ mol}}{0.250 \text{ mol} + 5.00 \text{ mol}} = 4.76 \times 10^{-2}$$

(b) By definition $c_m = n \text{ (kg H}_2\text{O)}^{-1}$. From part (a) we know there is 0.250 mol of NaOH and 90.0 g of water in 100 g of solution. Therefore the number of moles of NaOH per kilogram of water is

$$c_m = \frac{0.250 \text{ mol}}{90.0 \text{ g of H}_2\text{O}} \times \frac{1000 \text{ g of H}_2\text{O}}{\text{kg of H}_2\text{O}} = 2.78m$$

(c) By definition (Eq. 2.3),

$$c_M = \frac{n}{V}$$

In 100 g of solution there is 0.250 mol of NaOH, as calculated in part (a). The volume of this solution is

$$\frac{100 \text{ g}}{1.109 \text{ g mL}^{-1}} = 90.2 \text{ mL} = 0.0902 \text{ L}$$

Therefore

$$c_M = \frac{0.250 \text{ mol}}{0.0902 \text{ L}} = 2.77M$$

> Note that in this example the numerical value of the molarity differs only slightly from the molality, 2.78m. In general, molarity and molality are about the same for dilute aqueous solutions.

10.6 VAPOR PRESSURE LOWERING

Some properties of solutions, known as **colligative** properties, *depend on the concentration* (hence on the relative number of solute particles present) *and are independent of the kind of solute*. We shall first consider one such property, the vapor pressure lowering. Related colligative properties, such as freezing point depression, boiling point elevation, and osmotic pressure, are considered in subsequent sections.

If the vapor pressure of a liquid is measured before and after the introduction of a nonvolatile solute, we find that the vapor pressure is lowered when the solute is added. A quantitative relationship between the vapor pressure lowering and the relative amount of solute added was proposed by François Marie Raoult in 1888 as a result of his experimental investigations on ethyl ether solutions. To minimize the contribution of the solute to the vapor pressure of the solutions, Raoult used high boiling substances as solutes. If we let P_1° represent the vapor pressure of the pure ether and P_1 represent the vapor pressure above an ether solution at the same temperature, the relative lowering of the vapor pressure due to the addition of the solute may be defined as $(P_1^\circ - P_1)/P_1^\circ$. A few representative values obtained by Raoult are given in Table 10.1.

It is apparent from these results that the relative lowering of the vapor pressure is approximately equal to the mole fraction of solute (X_2) and is *independent of the nature of the solute*. If the relative vapor pressure lowering were precisely equal to X_2, then $(P_1^\circ - P_1)/P_1^\circ X_2$ would be exactly 1. The average value of this ratio obtained by Raoult for 14 solutes in ether solution was 0.98. Therefore Raoult proposed that

$$\frac{P_1^\circ - P_1}{P_1^\circ} = X_2 \tag{10.4}$$

TABLE 10.1

VAPOR PRESSURE LOWERING OF ETHYL ETHER SOLUTIONS AT 16°C[a]

Solute[b]	X_2	$\dfrac{P_1}{P_1^\circ}$	$\dfrac{P_1^\circ - P_1}{P_1^\circ}$
Aniline	0.148	0.846	0.154
Ethyl benzoate	0.096	0.909	0.091
Methyl salicylate	0.092	0.914	0.086
Nitrobenzene	0.060	0.945	0.055

[a] Vapor pressure of pure ether at 16°C (P_1°) is 374 Torr.
[b] Vapor pressure of the pure solutes at 16°C is equal to or less than 4 Torr.

Equation 10.4 is one method of expressing the relationship known as Raoult's law. A simpler form of Raoult's law is obtained by combining Eqs. 10.3 and 10.4.

$$\frac{P_1^\circ - P_1}{P_1^\circ} = X_2 = 1 - X_1$$

or

$$1 - \frac{P_1}{P_1^\circ} = 1 - X_1$$

or

$$P_1 = X_1 P_1^\circ \tag{10.5}$$

In this form, Raoult's law may be interpreted in terms of the kinetic molecular theory. As the mole fraction of solvent is increased, there are relatively more solvent molecules (and fewer solute molecules) present; therefore the rate at which solvent molecules escape from the solution increases, thus increasing the equilibrium vapor pressure.

Let us now consider the application of Raoult's law to binary solutions in which both components are volatile. If both components obey Raoult's law, then $P_1 = X_1 P_1^\circ$ and $P_2 = X_2 P_2^\circ$. The total pressure is

$$P_T = P_1 + P_2 = X_1 P_1^\circ + X_2 P_2^\circ \tag{10.6}$$

The partial pressure of propylene dibromide (P_P) and of ethylene dibromide (P_E) in equilibrium with solutions of different composition of the two components at 85°C is plotted against the mole fraction of the propylene dibromide (X_P) in Figure 10.8. The straight lines labeled P_P and P_E represent the predicted partial pressures if Raoult's law is obeyed by each component and the circles represent the experimentally observed values.[5] The agreement between experiment and prediction is good.

If Raoult's law is obeyed by each component of a solution throughout the entire range of mole fractions, the solution is said to be *ideal*. From a molecular point of view this implies that a molecule of A in the solution has the same tendency to escape into the vapor phase whether it is surrounded by A molecules, B molecules, or by a mixture of A and B molecules. Therefore we conclude that the attractive forces between A and A or A and B must be essentially the same. In agreement with this rationalization we find that the heat of mixing two components to form an ideal solution is zero and that the volume of the resulting solution is equal to the sum of the volumes of the two components.

---

[5] The partial pressures may be determined by measuring the total pressure and by analyzing the vapor above the solution.

10.6 VAPOR PRESSURE LOWERING

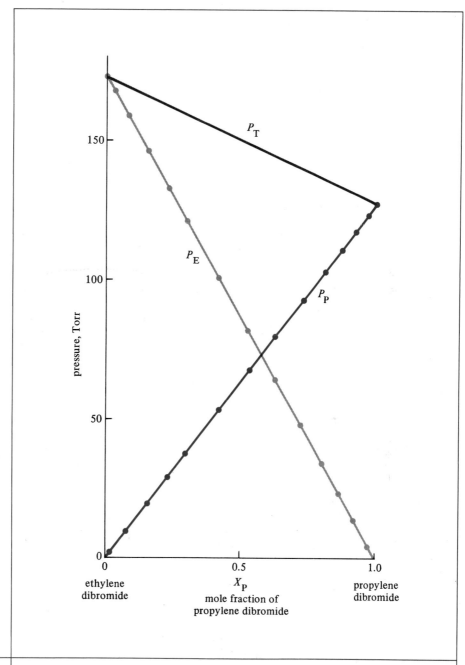

FIGURE 10.8 Vapor pressures of the system ethylene dibromide (E)–propylene dibromide (P) at 85°C. P_T, total vapor pressure.

EXAMPLE 2

A solution is prepared by mixing 1.0 mol of solute and 4.0 mol of solvent. The vapor pressures of pure solute and solvent are 35 and 95 Torr, respectively. Calculate the partial pressures and total pressure of the gas phase in equilibrium with the solution, assuming that the solution is ideal.

$$X_1 = \frac{4.0 \text{ mol}}{4.0 \text{ mol} + 1.0 \text{ mol}} = 0.80$$

$$X_2 = \frac{1.0 \text{ mol}}{4.0 \text{ mol} + 1.0 \text{ mol}} = 0.20$$

$$P_1 = X_1 P_1^\circ = (0.80)(95 \text{ Torr}) = 76 \text{ Torr}$$

$$P_2 = X_2 P_2^\circ = (0.20)(35 \text{ Torr}) = 7.0 \text{ Torr}$$

$$P_T = 76 \text{ Torr} + 7.0 \text{ Torr} = 83 \text{ Torr}$$

Although all gases approach ideal behavior at low pressures, the vast majority of solutions show deviations from Raoult's law for the solute even at low concentrations of solute. As shown in Figures 10.9 and 10.10, vapor pressures may be greater or less than the vapor pressures predicted by Raoult's law. The acetone–carbon disulfide system shown in Figure 10.9 is said to exhibit a positive deviation from Raoult's law because the vapor pressure of each component is higher than that calculated from Raoult's law. Solutions that exhibit positive deviations from Raoult's law may show other deviations from ideal behavior as well. In particular the volume of the solution is usually greater than the sum of the volumes of its components, and there is an absorption of heat on mixing. The acetone–chloroform system shown in Figure 10.10 exhibits negative deviations from Raoult's law in that the vapor pressure of each component is less than that calculated from Raoult's law. The volume of the solution is less than the sum of the volumes of its components, and heat is evolved when the components are mixed.

One of the factors contributing to nonideal behavior is the difference in attractive forces between solute–solute molecules and solute–solvent molecules. The absorption of heat on mixing for solutions exhibiting positive deviations leads us to the conclusion that the attractive forces between like molecules is greater than the attractive forces between unlike molecules. In cases where this difference is quite large the two liquids may be immiscible (i.e., they form separate liquid phases). In contrast, the evolution of heat noted when preparing solutions that exhibit negative deviations from Raoult's law may be explained as due to greater attractive forces between unlike molecules than between like molecules. In extreme cases the greater attractive forces between the unlike molecules may lead to formation of a solute–solvent compound.

It is important to note that, even when marked deviations from Raoult's law exist over a portion of the mole fraction range, the vapor pressure of the *solvent* approaches the value predicted by Raoult's law as its mole fraction approaches

10.6 VAPOR PRESSURE LOWERING

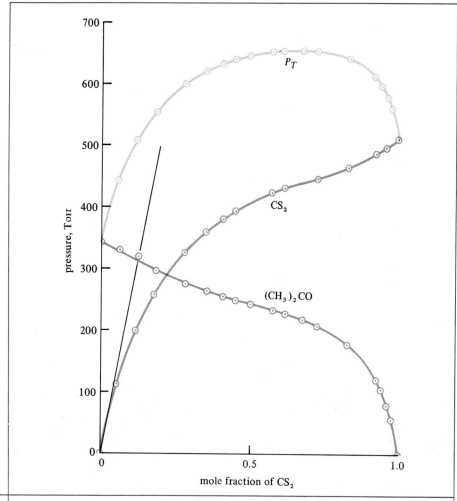

FIGURE 10.9 Partial pressure of carbon disulfide (CS_2) and of acetone [$(CH_3)_2CO$] and total vapor pressure (P_T) of acetone–carbon disulfide solutions at 35.2°C.

unity. Thus Raoult's law may be used to predict the vapor pressure of the solvent component in nonideal solutions that are very *dilute* (i.e., contain only small amounts of a solute).

The vapor pressure of the *solute* can be estimated using Henry's law,[6]

$$P_2 = kX_2 \tag{10.7}$$

---

[6] This equation was proposed by William Henry in 1803 to explain his experimental data on the increase in solubility of various gases in water with increasing pressure of the gas above the solution.

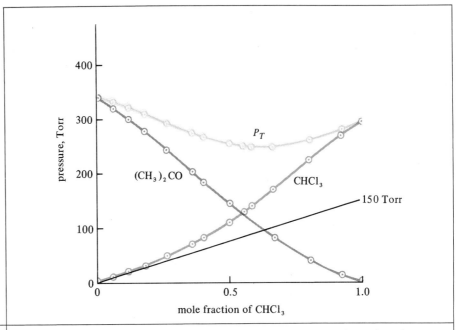

FIGURE 10.10 Partial pressure of acetone [$(CH_3)_2CO$] and of chloroform ($CHCl_3$) and total vapor pressure (P_T) of acetone–chloroform solutions at 35.2°C.

For ideal solutions, Henry's law is equivalent to Raoult's law and $k = P_2°$. Even for solutions that show positive or negative deviations from Raoult's law, the solute obeys Henry's law in very dilute solutions. To evaluate k from the experimental data, a straight line is drawn tangent to the P_2 versus X_2 curve in the vicinity of $X_2 = 0$ (see solid lines in Figures 10.9 and 10.10). The value of k is the value of P_2 of the tangent line at unit mole fraction (150 Torr in Figure 10.10).

For ideal solutions and many nonideal solutions, the mole fraction of the more volatile component is larger in the vapor phase than in the liquid with which it is in equilibrium. As a result it is frequently possible to separate a binary solution into its components by a process known as **fractional distillation.** As an example of this method of separation, consider Figure 10.11, which shows a plot of the boiling point (lower curve) of solutions of benzene and toluene against the mole fraction of benzene. The upper curve represents the mole fraction of benzene in the vapor phase that is in equilibrium with the boiling liquid. Thus a solution containing 20 mol % benzene will boil when the temperature reaches 102°C (point A). Reading from the upper curve at the same temperature, the vapor above the solution at 102°C would contain 38 mol % benzene (point B). If vapor of this composition were condensed, the resulting liquid would boil at a temperature of 96°C (point C) and the vapor produced would have a composition of 60 mol % benzene (point D). Further evaporations and condensations eventu-

10.7 BOILING POINT ELEVATION AND FREEZING POINT DEPRESSION

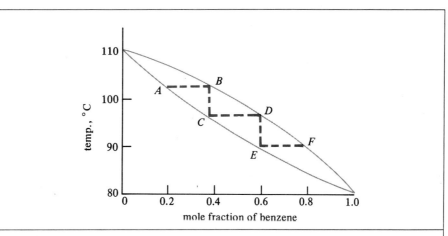

FIGURE 10.11 Composition of the vapor phase (upper curve) and liquid phase (lower curve) at the boiling point of benzene–toluene solutions.

ally would lead to pure benzene. In practice the process of fractional distillation usually involves a column containing some coarse packing, such as glass beads, above the boiling liquid to allow for repeated evaporation and condensation, as indicated by the dashed lines. With efficient columns, it is possible to effect a high degree of separation of a benzene–toluene solution.

10.7 BOILING POINT ELEVATION AND FREEZING POINT DEPRESSION

The effect of a nonvolatile solute upon the vapor pressure of a liquid may be expressed in a graphical form similar to the phase diagram for the pure solvent. In Figure 10.12 the phase diagram for water has been modified by adding the vapor pressure of an aqueous solution of a nonvolatile solute, curve $T'B'$. (The vapor pressures for curve $T'B'$ can be measured or they can be calculated from the vapor pressure of water at each temperature and the mole fraction of water, using Raoult's law if the solution is ideal.)

We can predict (using Figure 10.12) that the addition of a nonvolatile solute to boiling water will result in a reduction of the vapor pressure of water in equilibrium with the solution and that boiling will cease. However, if the solution is then heated, boiling will begin again at a higher temperature when the vapor pressure again reaches 1 atm. The magnitude of the increase in temperature (ΔT_b in Figure 10.12) required to produce boiling is found to be proportional to the mole fraction (X_2) or to the molality[7] (c_m) of the solute for dilute solutions. This

---

[7] One can show that c_m is proportional to X_2 in dilute solutions as follows. By definition $X_2 = n_2/(n_1 + n_2)$. In dilute solution, $n_1 \gg n_2$ and $n_1 + n_2 \approx n_1$. Therefore $X_2 \approx n_2/n_1$. If the molecular mass of the solvent is designated M_1, then kilograms of solvent = $n_1 M_1/1000$, and $c_m = n_2/(n_1 M_1/1000) = (n_2/n_1)(1000/M_1) \approx X_2(1000/M_1)$. For a given solvent, M_1 is a constant and therefore c_m is proportional to X_2 in dilute solutions.

relationship may be expressed by the following equation, applicable to dilute solutions:

$$\Delta T_b = K_b c_m \tag{10.8}$$

where the proportionality constant K_b is called the **boiling point constant** of the solvent. The results of many experimental investigations indicate that K_b is a constant for each solvent and that it is independent of the type of solute. Values of K_b in good agreement with experimentally observed values may be calculated from theoretical considerations involving only properties of the solvent. Table 10.2 lists values of K_b and the boiling points of several liquids. (Note that temperature differences are the same in either °C or K units.)

If we add a nonvolatile solute to water (or any other liquid) at its freezing point, we find that it is necessary to lower the temperature of the solution before freezing will occur. Thus introduction of a solute causes a depression of the freezing point of the liquid. For dilute solutions, the freezing point depression, ΔT_f, is related to the concentration of solute, c_m, as follows:

$$\Delta T_f = K_f c_m \tag{10.9}$$

where the proportionality constant K_f is called the **freezing point constant** of the solvent. The value of K_f is constant for any given solvent and is independent of the solute used. Table 10.2 lists values of K_f and the freezing point (fp) of several liquids. The following example illustrates how measurements of the freezing point of a solution may be used to evaluate the molecular mass of the solute.

EXAMPLE 3

The freezing point of a solution of 6.34 g of naphthalene dissolved in 250 g of benzene is 4.467°C. Calculate the molecular mass of naphthalene, using the constants in Table 10.2.

The molality of the solution can be calculated from Eq. 10.9 and from the measured lowering of the freezing point:

$$c_m = \frac{\Delta T_f}{K_f} = \frac{(5.482 - 4.467)\,K}{5.12\,K\,m^{-1}}$$
$$= 0.198 m$$

In 250 g of benzene there is $(0.198 \text{ mol}/1000 \text{ g}) \times 250 \text{ g} = 0.0495$ mol of naphthalene. From the data given in the problem, this amount of naphthalene has a mass of 6.34 g. Therefore 1 mol has a mass of

$$\frac{6.34 \text{ g}}{0.0495 \text{ mol}} = 128 \text{ g mol}^{-1}$$

and the molecular mass of naphthalene is 128.

10.7 BOILING POINT ELEVATION AND FREEZING POINT DEPRESSION

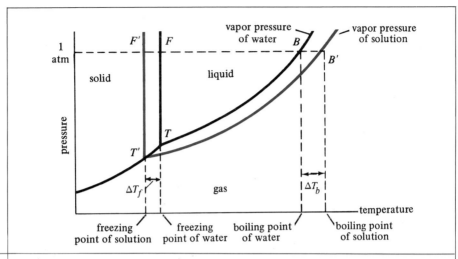

FIGURE 10.12 Elevation of boiling point and lowering of freezing point caused by addition of a nonvolatile solute to water (not drawn to scale). At 1 atm pressure, the boiling point increases by ΔT_b (from B to B') and the freezing point decreases by ΔT_f (from F to F'). (The triple point also changes from T to T'.)

The connection between the three colligative properties of freezing point lowering, boiling point elevation, and vapor pressure lowering is illustrated in Figure 10.12. Just as the lower vapor pressure curve of the solution (curve $T'B'$) raises the normal boiling point (to B'), it necessarily lowers the triple point of the solution (to T'). And because the properties of the solid and liquid states are only slightly affected by pressure changes of the order of 1 atm, the ice-solution equilibrium curve (curve $T'F'$) is nearly vertical, paralleling the solid-liquid curve of pure water (curve TF). Thus the effect of solute in lowering the triple point to T' is practically the same as its effect in lowering the normal freezing point to F'.

It should be emphasized that Eqs. 10.8 and 10.9 for the boiling point elevation and the freezing point depression are valid only in dilute solutions and only when the solvent is in equilibrium with its pure gaseous or pure solid phase. To the extent that a solute is nonvolatile, the gaseous phase in equilibrium with the

TABLE 10.2

BOILING-POINT AND FREEZING-POINT CONSTANTS

Solvent	bp (°C)	K_b (K m^{-1})	fp (°C)	K_f (K m^{-1})
Acetic acid	118.5	3.07	16.6	3.90
Benzene	80.1	2.67	5.482	5.12
Camphor	—	—	176.0	40.0
Carbon tetrachloride	76.8	5.03	−23.0	29.8
Water	100.0	0.512	0.0	1.86

liquid at the boiling point will contain primarily solvent molecules. However, at the freezing point there is a possibility that a solid solution, rather than the solid phase of the solvent, could separate from the liquid solution. Equation 10.9 is valid only when the solid phase in equilibrium with the solution is pure solvent. Thus it is necessary to test the solid phase that separates to assure that it is not a solid solution. This can readily be done by determining the melting point of a portion of the separated solid.

An interesting practical application of freezing point lowering is the use of "antifreeze" in automobiles having liquid cooling systems. When water freezes it expands, and the expansion is sufficient to cause engine blocks to crack. Therefore, in climates where temperatures fall below 0°C, it is necessary to use additives to lower the freezing point. Any solute would lower the freezing point, but other requirements for a good antifreeze include low volatility and low corrosion of the metal in the cooling system. The most popular antifreeze is ethylene glycol, $C_2H_6O_2$.

10.8 OSMOTIC PRESSURE

The molecular masses of most substances may be determined from vapor density measurements or from the freezing point lowering or boiling point elevation of solutions of the substance in an appropriate solvent. However, the extremely large molecules that occur in living organisms cannot be vaporized without decomposition, and only extremely dilute aqueous solutions (on a molal basis) can be prepared. The boiling point elevation or freezing point lowering of these solutions is much too small to be determined with any precision. But another colligative property of solutions, the osmotic pressure, may be used to estimate the molecular masses of large molecules.

As shown in Figure 10.13, when a solution is separated from pure solvent by a membrane that is permeable to the solvent but impermeable to the solute, solvent will diffuse through the membrane into the solution until the vapor pressures of the two liquids become identical. This phenomenon is known as **osmosis** (from the Greek *osmos*, "push").

To prevent this transfer of solvent into the solution, one can increase the external pressure on the solution, tending to drive the solvent in the other direction. If the external pressure is large enough, an equilibrium is established where there is no net flow of solvent in either direction. The pressure difference between the two sides of the apparatus at equilibrium is called the **osmotic pressure** of the solution, symbol π.

A very narrow tube, called a capillary tube, is used to determine the direction of flow of solvent through the membrane. When solvent diffuses through the membrane into the solution, the level of solvent in the capillary drops. If the pressure on the solution is too high, the level of solvent in the capillary tube will rise because of transfer of solvent from the solution to the solvent compartment. At equilibrium the height of solvent remains stationary.

The osmotic pressures of aqueous solutions of sucrose (0.10–1.0m) at 25°C

10.8 OSMOTIC PRESSURE

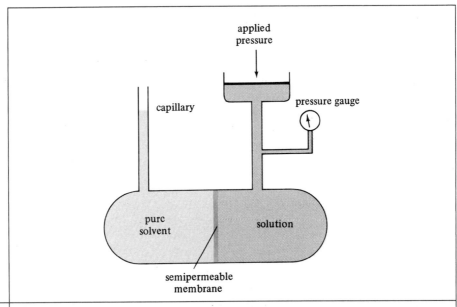

FIGURE 10.13

Osmotic pressure apparatus. The semipermeable membrane is permeable to solvent but not to the solute. Flow of solvent into the solution may be detected by a drop of the level in the capillary tube. The pressure is increased on the solution until there is no net transfer of solvent through the semipermeable membrane. The resulting pressure on the solution recorded by the gauge is equal to the osmotic pressure of the solution.

are recorded in Table 10.3. The osmotic pressure is directly proportional to the molal concentration of sucrose, as shown by the constancy of the values π/m given in column 3 of the table. Other solutes may be used, and the value of π/m is found to be independent of the nature of the solute in dilute solutions. In the lower section of Table 10.3, the osmotic pressure of a $0.200m$ sucrose solution is recorded for several temperatures between 0° and 60°C. The third column indicates that values of π/T are nearly constant, so that the osmotic pressure is approximately proportional to the absolute temperature. If we combine these two observations, we obtain $\pi = Rc_m T$, where R is a proportionality constant. For very dilute aqueous solutions, the molality is approximately equal to the molarity, so that we may write

$$\pi = \frac{n}{V} RT \quad \text{or} \quad \pi V = nRT \tag{10.10}$$

The values of R recorded in Table 10.3 have been calculated from the experimental data, and it is surprising to note that, in addition to being relatively constant, as the temperature is increased they approach the value of 0.082 L atm K^{-1} mol^{-1} obtained from the ideal gas equation. This agreement was originally noted in

TABLE 10.3 — OSMOTIC PRESSURE OF AQUEOUS SUCROSE SOLUTIONS

Measurements at 25 °C

Molality (c_m)	π (atm)	π/c_m (atm m^{-1})
0.10	2.63	26.3
0.20	5.20	26.0
0.40	10.30	25.8
0.60	15.62	26.0
0.80	21.25	26.6
1.00	27.05	27.0

Measurements on a 0.200m solution

T (K)	π (atm)	$100 \times \pi/T$ (atm K^{-1})	R (L atm K^{-1} mol^{-1})
273.2	4.72	1.73	0.0865
298.2	5.20	1.74	0.0870
313.2	5.16	1.65	0.0825
333.2	5.44	1.63	0.0815

1885 by Jacobus H. van't Hoff. It was subsequently shown that the laws of thermodynamics require that an ideal dilute solution obey Eq. 10.10, with a value of R identical to that of the ideal gas equation.

The following example indicates the usefulness of osmotic pressure measurements to obtain molecular masses of large molecules.

EXAMPLE 4

An aqueous solution containing 5.0 g of horse hemoglobin in exactly 1 L of solution exhibited an osmotic pressure of 1.80 cm of water at 25 °C. Calculate the molecular mass of the hemoglobin.

Since π, V, R, and T are known, n may be calculated from Eq. 10.10. The molecular mass would then be $5.0/n$. It is first necessary to convert the measured osmotic pressure from centimeters of water to the same unit of pressure used for R (atmospheres). This can be done by using Eq. 3.6. For columns of two different liquids exerting the same pressure,

$$P_L = h_1 d_1 g = h_2 d_2 g$$

(The acceleration of gravity, g, is the same for the two liquids.) Thus

$$h_2 = h_1 \frac{d_1}{d_2}$$

Using the densities of water and mercury (1.00 and 13.6 g cm^{-3}, respectively),

$$h_{Hg} = h_{H_2O} \frac{d_{H_2O}}{d_{Hg}} = 1.80 \text{ cm} \times \frac{1.00 \text{ g cm}^{-3}}{13.6 \text{ g cm}^{-3}} = 0.132 \text{ cm}$$

> Converting to atmospheres gives
>
> $$\pi = \frac{0.132 \text{ cm Hg}}{76.0 \text{ cm Hg atm}^{-1}} = 1.74 \times 10^{-3} \text{ atm}$$
>
> From Eq. 10.10,
>
> $$n = \frac{\pi V}{RT} = \frac{(1.74 \times 10^{-3} \text{ atm})(1.000 \text{ L})}{(0.08206 \text{ L atm K}^{-1} \text{mol}^{-1})(298 \text{ K})}$$
> $$= 7.12 \times 10^{-5} \text{ mol}$$
>
> The mass of 1 mol is therefore
>
> $$\frac{5.0 \text{ g}}{7.12 \times 10^{-5} \text{ mol}} = 7.0 \times 10^{4} \text{ g mol}^{-1}$$
>
> and the molecular mass of horse hemoglobin is 7.0×10^4.
>
> Note that the freezing point lowering of this solution would be $\Delta T_f = (1.86 \text{ K } m^{-1})(7.12 \times 10^{-5} m) = 1.32 \times 10^{-4}$ K. Such a small temperature difference would be extremely difficult to detect, let alone to measure with any precision.

Osmotic pressure measurements yield information on the number of moles of solute that cannot pass through the membrane; so if there is a small amount of a low molecular mass impurity that is permeable to the membrane, it will not interfere with the determination of the molecular mass of large solute molecules that are impermeable to the membrane.

10.9 SOLUBILITY OF NONELECTROLYTES

Pairs of liquids, such as benzene and toluene, that can be combined in any amounts to give homogeneous solutions are said to be *miscible in all proportions,* or simply **miscible.** If the concentration of dissolved solute increases until it reaches a constant value, even though some solute remains as a separate phase, we say that the solution is **saturated.** (Solutions that contain smaller concentrations than a saturated solution are called **unsaturated.**) For example, ordinary sugar (sucrose) is quite soluble in water; but if we increase the amount of sugar used we find that eventually solid sugar remains in contact with the solution. No matter how long we wait, the concentration of sugar in the solution remains constant. The concentration of solute in the saturated solution is known as the **solubility.** The solubility of a substance depends on the nature of the solvent and solute as well as on the temperature and pressure.

Pairs of liquids that form ideal solutions are always miscible, and, as we have previously noted, there is no heat of mixing. Thus the attractive forces between like molecules and between unlike molecules must be equal, or nearly so. Most examples of ideal solutions involve nonpolar compounds in which the bonds

10.10 ELECTROLYTES

As more information became available on the colligative properties of solutions of a variety of solutes in different solvents, it became apparent that aqueous solutions of some solutes behaved in an anomalous manner. This abnormal behavior is illustrated by the wide range of freezing points of $0.100m$ aqueous solutions shown in Table 10.4. The first four and the last two compounds all give solutions that freeze in the vicinity of $-0.186°C$ in agreement with the value[8] calculated by means of Eq. 10.9. The remaining compounds listed in Table 10.4 give aqueous solutions whose freezing points are abnormally low. If the freezing point depression is a measure of the concentration of solute particles, then larger than normal freezing point depressions signify that the concentration of solute particles is larger than that calculated from the amount of solute used. In other words, a solute molecule may dissociate to two or more particles, each of which can act to lower the freezing point. It should be noted that solutions of these same solutes in some other solvents (in which dissociation does not occur) exhibit normal behavior.

TABLE 10.4 FREEZING POINT OF $0.100m$ AQUEOUS SOLUTIONS

Solute	fp (°C)	$i = \dfrac{\Delta T_f}{1.86 \times 0.100}$	Class[a]
Acetone, $(CH_3)_2CO$	-0.185	1.00	NE
Ethyl alcohol, C_2H_5OH	-0.184	0.99	NE
Glycerol, $C_3H_8O_3$	-0.186	1.00	NE
Sucrose, $C_{12}H_{22}O_{11}$	-0.188	1.01	NE
Hydrogen chloride, HCl	-0.352	1.89	SE (H^+, Cl^-)
Sodium chloride, NaCl	-0.348	1.87	SE (Na^+, Cl^-)
Sodium hydroxide, NaOH	-0.342	1.84	SE (Na^+, OH^-)
Silver nitrate, $AgNO_3$	-0.332	1.78	SE (Ag^+, NO_3^-)
Sodium sulfate, Na_2SO_4	-0.459	2.46	SE ($2Na^+$, SO_4^{2-})
Barium hydroxide, $Ba(OH)_2$	-0.469	2.52	SE (Ba^{2+}, $2OH^-$)
Lanthanum chloride, $LaCl_3$	-0.65	3.5	SE (La^{3+}, $3Cl^-$)
Ammonia, NH_3	-0.189	1.02	WE (NH_4^+, OH^-)
Acetic acid, CH_3CO_2H	-0.189	1.02	WE (H^+, $CH_3CO_2^-$)

[a] Nonelectrolyte (NE), weak electrolyte (WE), or strong electrolyte (SE).

[8] As indicated previously, theoretical considerations allow us to evaluate K_f from properties of the solvent (water) so that we may safely assume that the normal freezing point of $0.100m$ aqueous solutions is $-0.186°C$.

10.10 ELECTROLYTES

We may use the freezing point depression to obtain information on the dissociation of solutes by rewriting Eq. 10.9 as follows:

$$\Delta T_f = K_f i c_m \qquad (10.11)$$

where i is the average number of particles formed by the dissociation of each solute molecule, and c_m is the solute molality calculated from its molecular formula. Thus ic_m is the concentration of solute particles in the solution. If all the solute molecules dissociated into fragments, i would be a whole number equal to the number of fragments per molecule. Values of i calculated from Eq. 10.11 and from the freezing points in Table 10.4 are recorded for each compound in the table. The first four and the last two compounds have i values of approximately 1 (normal behavior), whereas the other compounds have much larger i values, indicating dissociation.

What effect do changes in concentration have on the dissociation of a solute? The freezing points of aqueous solutions of NaCl from 0.00100 to $1.00m$ are recorded in Table 10.5, together with calculated values of i. It is apparent that the i values approach a value of 2 as the solution becomes more dilute. By similar experiments, we find that HCl, NaOH, and $AgNO_3$ also give i values of approximately 2, Na_2SO_4 and $Ba(OH)_2$ give i values approaching 3, and $LaCl_3$ gives i values approaching 4 at low concentrations.

The dissociation of dissolved solutes was first proposed by Svante Arrhenius in 1887 in his theory of electrolytic dissociation. Arrhenius, while a doctoral candidate at the University of Uppsala, proposed that the electric conductivity of aqueous solutions could be accounted for by assuming that solutes dissociate into positive and negative ions that carry the electric current. This ability of some aqueous solutions to conduct electricity can be demonstrated by using the apparatus shown in Figure 10.14. Liquids such as benzene, carbon tetrachloride, and kerosene do not conduct electricity, and pure water is a very poor conductor. Some aqueous solutions are no better conductors than water, some are only slightly better, and some are relatively good conductors. Thus compounds may be classified into two main groups: **Electrolytes** are compounds whose aqueous solutions are better conductors than water, while **nonelectrolytes** are compounds whose aqueous solutions do not conduct electricity better than water. Electrolytes are subdivided into **strong electrolytes**—those that produce aqueous solutions

TABLE 10.5	FREEZING POINT OF AQUEOUS NaCl SOLUTIONS		
	Molality (c_m)	fp (°C)	$i = \dfrac{\Delta T_f}{1.86 \times c_m}$
	1.00	-3.37	1.81
	0.100	-0.348	1.87
	0.0100	-0.0360	1.94
	0.00100	-0.00366	1.97

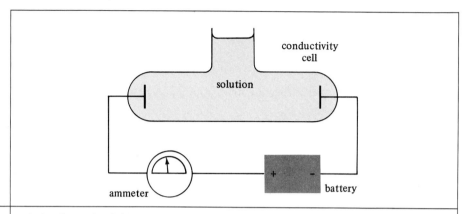

FIGURE 10.14 A simple conductivity apparatus, used to measure the electric conductivity of liquids or solutions. The conductivity of the solution in the cell is proportional to the electric current registered by the ammeter.

that are relatively good conductors—and **weak electrolytes**—those whose aqueous solutions are only slightly better conductors than water.

If we test the conductivity of the aqueous solutions listed in Table 10.4, we find the results listed in the last column. The first four compounds are nonelectrolytes (NE), the next seven are strong electrolytes (SE), and the last two are weak electrolytes (WE). For each electrolyte the formulas of its ions are shown in parentheses. For the strong electrolytes the total number of ions formed by dissociation of each molecule of solute is equal to the i value approached at low concentration. Thus each molecule of sodium sulfate, Na_2SO_4, dissociates into two Na^+ ions and one SO_4^{2-} ion, and its i value approaches 3 at low concentration.

The solutions of the weak electrolytes contain only a few ions, as evidenced by the slight electrical conductivity and the almost normal freezing point ($i = 1.02$). Thus only a small fraction of the CH_3CO_2H molecules in $0.100m$ CH_3CO_2H dissociate to form H^+ and $CH_3CO_2^-$. The fraction of the molecules that dissociate is known as the **degree of dissociation** (α). It can be determined from freezing point data as shown in the following example.

EXAMPLE 5

From the freezing point data in Table 10.4 calculate the concentration of H^+ and $CH_3CO_2^-$ in $0.100m$ CH_3CO_2H. What is the degree of dissociation of CH_3CO_2H in $0.100m$ CH_3CO_2H?

The dissociation of 1 mol of CH_3CO_2H would give 1 mol of H^+ and 1 mol of $CH_3CO_2^-$, that is,

$$CH_3CO_2H = CH_3CO_2^- + H^+$$

In this example, we have 0.100 mol of CH_3CO_2H dissolved in 1 kg of water. Therefore after x mol of CH_3CO_2H has dissociated we will have

$$[CH_3CO_2H] = 0.100 - x \quad [H^+] = x \quad \text{and} \quad [CH_2CO_2^-] = x$$

The total concentration of solute particles is

$$[CH_3CO_2H] + [H^+] + [CH_3CO_2^-] = 0.100 - x + x + x = 0.100 + x$$

From Eq. 10.11 we can calculate the concentration of solute species, that is,

$$ic_m = \frac{\Delta T_f}{K_f} = \frac{0.189 \text{ K}}{1.86 \text{ K } m^{-1}} = 0.102 m$$

Therefore $0.100 + x = 0.102$, and $x = 0.002$.

$$\alpha = \frac{\text{moles } CH_3CO_2H \text{ dissociated}}{\text{moles } CH_3CO_2H \text{ added}} = \frac{0.002}{0.100} = 0.02$$

Thus 2% of the acetic acid molecules dissociate into ions, and 98% remain as undissociated CH_3CO_2H molecules.

Although it is possible to detect the presence of undissociated molecules in aqueous solutions of weak electrolytes, efforts to detect undissociated molecules in dilute aqueous solutions of strong electrolytes such as HCl and NaCl have been unsuccessful. Because strong electrolytes appear to be completely dissociated in dilute solution, the Arrhenius concept that less than integral values of i correspond to a combination of undissociated molecules and ions requires some modification. In looking for alternative explanations, we might note that the equations used to predict the colligative properties of solutions are in reasonably good agreement with experimental values up to concentrations of $1.0m$ or greater for nonelectrolytes where the attractive or repulsive forces between solute molecules would be small. However, in electrolyte solutions the attractive and repulsive forces between ions would be much larger. A theoretical investigation of these interionic forces was carried out by Peter J. W. Debye and Erich Hückel in 1923, and their calculations indicated that the ions would not be randomly distributed throughout the solvent as nonelectrolytes would be. They found that positive ions tend to cluster around a negative ion, and vice versa, with the overall result being a reduction in the effective concentration of ions in solution. Furthermore, as the solutions become more dilute the ions will be farther apart and the clustering tendency will be reduced, so that in very dilute solutions the clustering effect will be negligible and the i values will approach integral values. Thus our present view of strong electrolytes is that they are completely dissociated and that deviations from ideal behavior in dilute solutions are due to the clustering effect predicted by Debye and Hückel.

10.11 SOLUBILITY OF ELECTROLYTES IN WATER

Ionic compounds are not readily soluble in nonpolar solvents but usually dissolve in water to an appreciable extent, owing to their tendency to dissociate into ions in aqueous solution. To provide an explanation for the solubility of ionic compounds and their dissociation into ions in some solvents but not in others, we must consider the energy requirements involved in ion separation and in ion–solvent interaction.

In Section 9.4 we calculated the energy required to dissociate a gaseous NaCl molecule into the gaseous ions as 5.74 eV. In view of this large energy requirement, why does NaCl dissociate into ions in aqueous solution? First we note that the ions would be separating in a water medium (dielectric constant $\epsilon = 78.5$) instead of in a vacuum ($\epsilon = 1$). The attraction energy between a positive electric charge and a negative electric charge is inversely proportional to the dielectric constant (see Appendix D-5), and therefore the energy required to separate the ions in water is only

$$\frac{5.74 \text{ eV}}{78.5} = 0.073 \text{ eV}$$

This small energy requirement is more than provided for by the energy released when the ions attract water molecules.

The strong attractive forces between an ion and the water dipole result in one or more water molecules being strongly bound to each ion—the ion is said to be **solvated.** Since the exact number of water molecules per ion is usually not known, it is customary to use the simple formula for the ion (Na^+) in aqueous solution with the understanding that there are several molecules of water tightly bound to the ion. A better notation (not often used) would be to write $Na^+(aq)$, and thus we could express the dissolving of NaCl in water by

$$NaCl(s) = Na^+(aq) + Cl^-(aq)$$

Although the attractive forces between solute and solvent are somewhat weaker for nonionic solutes, solvation strongly influences the solubility of many solutes.

10.12 CHEMICAL REACTIONS AND NET REACTIONS

We say that a *chemical reaction occurs whenever a new chemical species is formed or when the amounts of the chemical species in a reaction vessel are altered.* Usually the results of such a reaction are summarized by writing a **net reaction,** which shows only the chemical species that react and the products that are formed as a result of the reaction. The following examples will illustrate some of the common types of net reactions.

Solubility of Substances in Water

All gases, liquids, and solids are soluble to some extent in water—some more so than others. Table sugar (sucrose, $C_{12}H_{22}O_{11}$) is quite soluble in water and is a nonelectrolyte. The net reaction

$$C_{12}H_{22}O_{11}(s) = C_{12}H_{22}O_{11}(aq)$$

summarizes this result although the (aq) is often omitted when it is clear that the solvent is water.

Hydrogen chloride gas (HCl) is readily soluble in water.

$$HCl(g) = H^+(aq) + Cl^-(aq)$$

Since the dissolved HCl is a strong electrolyte, it is shown as H^+ and Cl^- in the net reaction. Aqueous solutions of HCl are called hydrochloric acid.

Precipitation Reactions

If a sodium sulfate (Na_2SO_4) solution is added to a solution of barium chloride ($BaCl_2$), a white solid ($BaSO_4$) separates from the solution. (This type of reaction, in which a solid separates from a solution, is called **precipitation.**) Since Na_2SO_4 and $BaCl_2$ are strong electrolytes, the ionic reaction would be

$$2Na^+ + SO_4^{2-} + Ba^{2+} + 2Cl^- = BaSO_4(s) + 2Na^+ + 2Cl^-$$

Since NaCl is a strong electrolyte, we write Na^+ and Cl^- on the right-hand side of the equation. However, we note that the Na^+ and Cl^- appear on both sides of the equation and are not involved in the chemical reaction. Therefore the net reaction is

$$Ba^{2+} + SO_4^{2-} = BaSO_4(s)$$

The same net reaction would apply whenever a solution containing SO_4^{2-} ions is added to a solution containing Ba^{2+} ions, regardless of the substances used to furnish the SO_4^{2-} and Ba^{2+} ions.

Dissociation of Weak Electrolytes

Acetic acid, CH_3CO_2H, is a liquid at room temperature. When it is dissolved in water, we find that the resulting solution contains acetic acid molecules and very small concentrations of hydrogen ion and acetate ion. (CH_3CO_2H is a weak electrolyte.) We can represent these results by the following two net reactions.

$$CH_3CO_2H(l) = CH_3CO_2H(aq)$$
$$CH_3CO_2H = H^+ + CH_3CO_2^-$$

Note that a net reaction merely indicates what new species are formed. It does not indicate the extent of the reaction, that is, it does not imply that all the acetic acid molecules are dissociated into hydrogen ions and acetate ions.

Formation of a Weak Electrolyte

When solutions of hydrochloric acid (HCl) and sodium hydroxide (NaOH) are mixed, the resulting solution will contain H^+, Cl^-, Na^+ and OH^- before any reaction occurs. Since H^+ and OH^- can combine to form water (a weak electrolyte), the net reaction would be

$$H^+ + OH^- = H_2O$$

Note that Na^+ and Cl^- are not involved in the reaction since they are present as ions in the reactants and in the final solution.

It is essential to distinguish clearly between the concepts of solubility and electrolyte strength. The solubility of a substance indicates how much of the substance can be dissolved in a given amount of solvent. It gives no indication whether the substance in solution is dissociated or undissociated.

The terms *weak electrolyte* and *strong electrolyte* are used to indicate what happens to the substance once it is in solution. A strong electrolyte is completely dissociated to ions; a solution of a weak electrolyte contains both undissociated molecules and ions.

The following two examples illustrate problems involving the stoichiometry of solutions of strong electrolytes. It may be helpful to review Section 2.8 on solutions and molarity before studying these examples.

EXAMPLE 6

A solution is prepared by mixing 50 mL of $0.10M$ HCl and 50 mL of $0.20M$ NaCl. Both HCl and NaCl are strong electrolytes. List all ions and molecules present, together with their molar concentrations, in the above solution.

Because these substances are strong electrolytes, the only solutes present in this solution are H^+, Na^+, and Cl^-. The number of moles of H^+ in the HCl solution is obtained by multiplying the initial molarity by the initial volume, $[H^+]_i V_i$. It is the same as the number of moles of H^+ in the final solution, $[H^+]_f V_f$. Thus

$$[H^+]_i V_i = [H^+]_f V_f$$

or

$$[H^+]_f = [H^+]_i \frac{V_i}{V_f}$$

When mixing dilute solutions, the final volume is approximately equal to the sum of the volumes of the solutions that are mixed. In this case the final volume is about 100 mL. Therefore

$$[H^+] = 0.10M \times \frac{50 \text{ mL}}{100 \text{ mL}} = 0.050M$$

10.12 CHEMICAL REACTIONS AND NET REACTIONS

$$[Na^+] = 0.20M \times \frac{50 \text{ mL}}{100 \text{ mL}} = 0.10M$$

In calculating the concentration of chloride ion, it is necessary to include Cl^- from both the HCl and the NaCl solutions.

$$[Cl^-] = \left[0.10M \times \frac{50 \text{ mL}}{100 \text{ mL}}\right] + \left[0.20M \times \frac{50 \text{ mL}}{100 \text{ mL}}\right] = 0.15M$$

EXAMPLE 7

If 20 mL of $0.20M$ Na_2SO_4 and 30 mL of $0.10M$ $Ba(NO_3)_2$ are mixed, how many grams of barium sulfate will precipitate? What are the concentrations of Na^+, NO_3^-, and SO_4^{2-} in the final solution?

The solutions that are mixed contain $(0.20M)(0.020 \text{ L}) = 0.0040$ mol of SO_4^{2-} ion, and $(0.10M)(0.030 \text{ L}) = 0.0030$ mol of Ba^{2+}. Therefore sulfate ion is in excess, and Ba^{2+} ion is the limiting reagent in the reaction

$$Ba^{2+} + SO_4^{2-} = BaSO_4(s)$$

The number of moles of solid barium sulfate that precipitate is 0.0030 mol. From the atomic masses on the inside back cover, the molecular mass of $BaSO_4$ is calculated to be 233.4. Therefore $(0.0030 \text{ mol})(233.4 \text{ g mol}^{-1}) = 0.70$ g of $BaSO_4$ will precipitate.

The amount of SO_4^{2-} remaining in solution is $0.0040 - 0.0030 = 0.0010$ mol, and the concentration of sulfate ion is

$$[SO_4^{2-}] = \frac{0.0010 \text{ mol}}{0.050 \text{ L}} = 0.020M$$

The concentration of Na^+ in the $0.20M$ Na_2SO_4 solution is $2 \times 0.20 = 0.40M$; its concentration in the final solution is

$$[Na^+] = 0.40M \times \frac{20 \text{ mL}}{50 \text{ mL}} = 0.16M$$

Similarly,

$$[NO_3^-] = 0.20M \times \frac{30 \text{ mL}}{50 \text{ mL}} = 0.12M$$

There will also be a *very* small amount of barium ion in solution, because no substance is completely insoluble, and the precipitation of Ba^{2+} is almost but not quite complete. Calculating the concentration of Ba^{2+} in this system requires an understanding of chemical equilibria, which will be considered in Chapter 12.

PROBLEMS

10.1 Estimate the normal boiling points of all the liquids (except mercury) from the vapor pressure curves shown in Figure 10.2.

10.2 On the basis of the data in Table 3.1, what is the boiling point of water if the applied pressure is 9.2 Torr?

10.3 A flask containing liquid water and air at 25°C is attached to a vacuum pump capable of reducing the pressure to less than 1 Torr. The vapor pressure of water at 25°C is 23.8 Torr. What will happen to the water?

10.4 At 35°C the vapor pressure of ethanol is 100 Torr. If 0.020 mol of ethanol were introduced into a 500-mL evacuated container at 35°C, what fraction of the ethanol would remain in the liquid state? (Neglect the volume occupied by the liquid phase.)

10.5 Calculate the molar volume of water in each of its three phases at the triple point. The densities of liquid water and ice at 0.01°C are 1.000 and 0.917 g mL^{-1}, respectively. The vapor pressure of water at 0.01°C is 6.03 × 10^{-3} atm.

10.6 Does ice melt at a temperature above or below 0°C
(a) at 0.5 atm pressure?
(b) at 2 atm pressure?

10.7 Explain how the movement of glaciers could be affected by the slope of the solid–liquid line in the phase diagram of water.

10.8 Illustrate the following path on a sketch of the phase diagram of water. At what points would a change of state occur?
(a) Water at 25°C and 1 atm pressure is cooled to −10°C (P constant).
(b) The pressure is then decreased to 1 × 10^{-3} atm (T constant).
(c) The temperature is increased to 25°C (P constant).
(d) The pressure is increased to 1 atm (T constant).

10.9 Consider a container in which ice, liquid water, and steam coexist in equilibrium. For each of the following changes in conditions, state which phase (or phases) will remain.
(a) The temperature is decreased, with the pressure held constant.
(b) The pressure is decreased, with the temperature held constant.

10.10 By using the kinetic molecular theory, explain why the vapor pressure of a pure liquid substance
(a) depends on the temperature
(b) is independent of the surface area of the liquid
(c) is independent of the volume of the gas phase

10.11 In which of the following liquids is hydrogen bonding important?
H_2O, CH_4, PH_3, NH_3

10.12 (a) Calculate the mole fraction, the molality, and the molarity of sucrose ($C_{12}H_{22}O_{11}$) in a solution at 20°C that was prepared by dissolving 5.00 g of sucrose in 95.0 g of water. (The density of the solution at 20°C is 1.02 g mL^{-1}.)
(b) If the solution were warmed to 30°C, which of the above concentrations would *not* change in value?

10.13 How many grams of sodium hydroxide (NaOH) are required to prepare 200 mL of 0.50M NaOH?

10.14 What volume of 0.800M NaOH is required to prepare 400 mL of 0.250M NaOH?

10.15 Calculate $(P_1^\circ - P_1)/P_1^\circ X_2$ for each solution listed in Table 10.1.

10.16 Benzene and toluene form a solution that is nearly ideal. At 25°C the vapor pressure of benzene is 94 Torr, whereas that of toluene is 29 Torr. What would be the total vapor pressure at 25°C of a liquid solution containing equimolar amounts of benzene and toluene? Calculate the mole fraction of benzene in the vapor phase.

10.17 A solution of a nonvolatile solute dissolved in benzene has a vapor pressure of 79 Torr at 25°C. The vapor pressure of pure benzene at the same temperature is 94 Torr.
(a) What is the mole fraction of benzene in the solution?
(b) If the solution was prepared by dissolving 6.49 g of solute in 0.27 mol of benzene, what is the molecular mass of the solute?

10.18 To obtain a constant temperature bath at 0°C (as for chilling champagne), ice and liquid water are mixed. To obtain a constant temperature bath at a temperature below 0°C (as for making ice cream in a hand freezer), salt is added to the mixture of ice and liquid water. Explain these phenomena with reference to Figure 10.12.

10.19 Solutions containing 0.10 mol of a nonvolatile solute in exactly 1 kg of each of the solvents in Table 10.2 were prepared. Arrange the solutions in decreasing order of freezing point and then in increasing order of boiling point. What reasons can you suggest for the use of different solvents for molecular mass determinations?

10.20 A solution prepared by dissolving 2.0 g of compound A in 100 g of water has a freezing point of −0.93°C. Assuming that the solution is ideal, what is the molecular mass of A?

10.21 Ethylene glycol, $C_2H_6O_2$, is a nonelectrolyte commonly added to water in motor vehicles having liquid cooling systems to lower the freezing point and thereby prevent damage, such as cracked engine blocks, in cold weather. How many moles of ethylene glycol must be added per kilogram of water to lower the freezing point to $-15.0°C$?

10.22 Suggest a design for a process that uses the phase transformation $H_2O(l) \rightarrow H_2O(s)$ to convert seawater to fresh water.

10.23 Proteins are macromolecules of high molecular mass. Assuming a molecular mass of 100,000, calculate the freezing point depression and the boiling point elevation of a 1.0% (by mass) aqueous solution of a protein. What would be the osmotic pressure of this solution at 25°C (in centimeters of H_2O)?

10.24 A $0.100m$ solution of HF in water has a freezing point $-0.201°C$.
(a) Calculate the concentration of H^+ and F^- in $0.100m$ HF.
(b) What is the degree of dissociation of HF in this solution?

10.25 (a) Which of the following pairs of substances can coexist in solution at high concentration at room temperature?
(b) For those which cannot, write a balanced equation to represent the net reaction that occurs.
Na^+ and Cl^-
H^+ and $CH_3CO_2^-$
H^+ and OH^-
Ba^{2+} and Cl^-
Ba^{2+} and SO_4^{2-}

10.26 If 0.040 mol of solid sodium sulfate (Na_2SO_4) is dissolved in 250 mL of water, what will be the concentrations of Na^+ and SO_4^{2-} in the resulting solution? Write an equation for the reaction.

10.27 (a) When 40 mL of $0.50M$ $BaCl_2$ is added to 60 mL of $0.20M$ K_2SO_4, what is the equation for the net reaction that occurs?
(b) How many moles of solid $BaSO_4$ will precipitate?
(c) What are the concentrations of K^+, Ba^{2+}, and Cl^- in solution?

10.28 When 50 mL of $0.10M$ $BaCl_2$ is added to 50 mL of a sulfuric acid solution, practically all of the sulfate ion is precipitated as $BaSO_4$. If the solution now contains $0.020M$ Ba^{2+}, what was the molarity of the H_2SO_4? (Assume that the final volume is 100 mL.)

10.29 (a) At 25°C the vapor pressure of benzene is 94 Torr, whereas that of toluene is 29 Torr. For a particular benzene–toluene solution, the partial pressures of benzene and toluene in the gas phase are the same. Assuming that Raoult's law is obeyed, calculate the mole fraction of benzene in the liquid phase.

(b) What is the total vapor pressure of this solution?

10.30 Vapor density measurements of acetic acid show that the acetic acid molecules in the gas phase consist mainly of dimers, with each dimer containing two CH_3CO_2H units. Draw Lewis structures for acetic acid and its dimer, showing two hydrogen bonds in the dimer.

10.31 (a) The vapor pressure of carbon disulfide (CS_2) is 433 Torr at 30°C, whereas that of a solution of 19.40 g of iodine in 38.0 g of CS_2 is 378 Torr. Assuming that Raoult's law is obeyed and that iodine is nonvolatile, estimate the molecular mass of iodine in the solution.

(b) Suggest reasons for any difference between your calculated value and the molecular mass based on the molecular formula I_2.

10.32 (a) A solution of 5.65 g of benzoic acid in 50.0 g of benzene has a boiling point 1.25°C higher than that of pure benzene. Using the value of K_b for benzene in Table 10.2, calculate the molecular mass of benzoic acid in this solution.

(b) The molecular formula of benzoic acid is usually written $C_6H_5CO_2H$. How many $C_6H_5CO_2H$ units are there in each molecule of benzoic acid in benzene solution?

CHAPTER 11
THE CHEMICAL ELEMENTS AND THEIR COMPOUNDS

11

As an interlude following the development of the basic principles of atomic and molecular structure, we present here a brief overview of the chemistry of the elements. The intent is to illustrate principles already developed and to provide a basis sufficiently broad for development of additional principles in later chapters. Chemical principles do not operate in a vacuum—they relate to the properties of real substances, and a full understanding requires some knowledge of the more important substances.

The wide variety of properties exhibited by chemical substances makes this a complicated subject. Depending to some extent on one's approach, it may seem a bewildering assemblage of dull factual material or a grand profusion of subtle interrelations and fascinating anomalies.

It is helpful not to attempt memorizing the facts of chemistry, which, in view of the complexity of the subject, is a hopeless task anyway. To the extent that a given point is important, it will be encountered again in other contexts—you will learn it without trying.

A better approach is to look for explanations in terms of basic principles, for interconnections having some theoretical significance, for similarities and dissimilarities, and for applications of special interest. The most useful tools are the periodic law and the principles of atomic and molecular structure that provide the theoretical foundation of the periodic law.

11.1 METALS, METALLOIDS, AND NONMETALS

A **metal** is an element possessing certain characteristic properties: metallic luster or sheen, high electrical and thermal conductivity, and the mechanical properties of malleability (capable of being pressed or beaten into various shapes) and ductility (capable of being drawn into wires). In the periodic table, metals are found along the left side, extending to the right as far as beryllium and aluminum in the short periods and gallium in the first long period (see Figure 11.1). There is also a group of elements called **metalloids** having metallic properties to a slight degree. They include boron and silicon in the short periods and germanium, arsenic, and selenium in the first long period. Substances not having metallic properties to an appreciable degree are called **nonmetals.** They are found along the right side of the periodic table, extending to the left as far as carbon and phosphorus in the short periods and bromine in the first long period. These classifications are obviously somewhat arbitrary. For example, antimony is a fairly good conductor of electricity, but it is also brittle; it is sometimes classed as a metal, sometimes as a metalloid.

Metals owe their properties of high electrical and thermal conductivity to loosely held electrons, called **conduction electrons,** which move through the solid or liquid carrying electric charge and kinetic energy. Consequently metallic properties are associated with low ionization potentials (see Figure 8.22). The conductivity of a metal decreases with increasing temperature because the increased vibrational motion of the atoms interferes with the motion of the electrons.

Metalloids have much lower conductivity than metals and are often called **semiconductors.** Their conductivity increases very rapidly with increasing temperature. Only a very small fraction of the electrons are conduction electrons at ordinary temperatures, but as the temperature rises many more electrons acquire sufficient energy to move through the solid. The electrical properties of semiconductors have found important applications in recent years in electronic circuits. Silicon and germanium, modified by addition of very small amounts of other elements, are used as transistors, the functional equivalent of vacuum tubes.

In a nonmetallic solid, practically none of the electrons have sufficient energy to move through the solid, the electrical conductivity is extremely low, and the substance is a good insulator. (The theory of electrical conduction in the solid state is considered in more detail in Section 17.11.)

Many elements exist in different physical forms, called **allotropes,** under different conditions. At ordinary temperatures, tin is a silvery metal called white tin but, below 13.2°C, it slowly changes to a brittle gray metalloid. Two unstable forms of selenium are known, both of which are red and nonmetallic. The stable form is a gray metalloid. Gray selenium has the important property of photoconductivity—its electrical conductivity is much higher when illuminated than in the dark.

11.2 HYDROGEN

Hydrogen occupies a unique position at the beginning of the periodic table. It resembles the other group IA elements in some respects and the group VIIA elements (halogens) in other respects. Like the group IA metals, it has one valence electron in an s orbital, and it forms compounds having formulas analogous to those of the group IA metals (HCl and LiCl, H_2SO_4 and Li_2SO_4, etc.). It differs from the group IA metals in having a much higher ionization potential, and it is a nonmetal. Moreover, as we have seen in Chapter 9, the bond in HCl is mainly covalent, whereas the LiCl bond is mainly ionic. Like the halogens (F, Cl, etc.), hydrogen has one less electron than a noble gas. Although there are many other points of resemblance to a halogen (nonmetal, forms diatomic molecules, and forms ionic compounds such as lithium hydride, LiH, with the group IA metals), hydrogen has a lower electronegativity and reacts much less readily than a halogen.

Liquid hydrogen has a very low boiling point, $-252.8°C$ or 20.4 K. Hydro-

PERIODIC TABLE OF THE ELEMENTS

IA										
H 1	IIA									
Li 3	Be 4							VIII		
Na 11	Mg 12	IIIB	IVB	VB	VIB	VIIB				
K 19	Ca 20	Sc 21	Ti 22	V 23	Cr 24	Mn 25	Fe 26	Co 27	Ni 28	
Rb 37	Sr 38	Y 39	Zr 40	Nb 41	Mo 42	Tc 43	Ru 44	Rh 45	Pd 46	
Cs 55	Ba 56	* 57-71	Hf 72	Ta 73	W 74	Re 75	Os 76	Ir 77	Pt 78	
Fr 87	Ra 88	† 89-103	Rf 104	Ha 105	106					

*	La 57	Ce 58	Pr 59	Nd 60	Pm 61	Sm 62	Eu 63	Gd 64
†	Ac 89	Th 90	Pa 91	U 92	Np 93	Pu 94	Am 95	Cm 96

FIGURE 11.1 Periodic table of the elements, showing the classification of the elements as metals, metalloids, and nonmetals.

11.2 HYDROGEN

		IIIA	IVA	VA	VIA	VIIA	0
							He 2
		B 5	C 6	N 7	O 8	F 9	Ne 10
IB	IIB	Al 13	Si 14	P 15	S 16	Cl 17	Ar 18
Cu 29	Zn 30	Ga 31	Ge 32	As 33	Se 34	Br 35	Kr 36
Ag 47	Cd 48	In 49	Sn 50	Sb 51	Te 52	I 53	Xe 54
Au 79	Hg 80	Tl 81	Pb 82	Bi 83	Po 84	At 85	Rn 86

Tb 65	Dy 66	Ho 67	Er 68	Tm 69	Yb 70	Lu 71
Bk 97	Cf 98	Es 99	Fm 100	Md 101	No 102	Lr 103

☐ metals   ◩ metalloids   ■ nonmetals

gen has the lowest molecular mass and hence the lowest density of any gas. This property has found some application in balloons and dirigible airships, but hydrogen burns so readily that large amounts present a serious fire hazard.[1]

A convenient laboratory method for obtaining small amounts of hydrogen is the reaction of metals such as zinc with hydrochloric acid (see Figure 11.2):

$$Zn(s) + 2H^+ = Zn^{2+} + H_2(g)$$

The products of the reaction are gaseous hydrogen (which is only slightly soluble in water or aqueous solutions) and a solution of zinc chloride.

Hydrogen is also formed by the reaction of a group IA metal (e.g., sodium) with water:

$$2Na(s) + 2H_2O(l) = 2Na^+ + 2OH^- + H_2(g)$$

Besides hydrogen an aqueous solution of sodium hydroxide is produced in this reaction.

Electrolysis of water is a source of both hydrogen and oxygen (see Figure 1.6):

$$2H_2O(l) = 2H_2(g) + O_2(g)$$

A strong electrolyte such as sodium sulfate, Na_2SO_4, is added to increase the electrical conductivity.

Like most gases of any importance, hydrogen of high purity is available in steel cylinders under high pressure, and this is the source most commonly used in laboratory work.

Hydrocarbons (compounds containing carbon and hydrogen only) from petroleum and natural gas wells are the major source of industrial hydrogen. Methods used to produce hydrogen from hydrocarbons include a process called catalytic steam—hydrocarbon reforming. For example, when hydrogen is formed from methane (CH_4) by this process, the reaction is

$$CH_4(g) + 2H_2O(g) = CO_2(g) + 4H_2(g)$$

To accelerate an otherwise slow reaction, elevated temperatures and a catalyst are used. (A **catalyst** is a substance that causes a reaction to proceed more rapidly and that can be recovered unchanged at the end of the reaction.) For this reaction metallic nickel has been found to be a good catalyst.

Although hydrogen is not ordinarily very reactive, it forms compounds with all of the elements except the noble gases. It forms more compounds than any other element, with carbon being a close second. Its compounds include many of

---

[1] The airship *Hindenburg*, a hydrogen-filled dirigible, caught fire and burned while landing at Lakehurst, N.J., on May 6, 1937, following a transatlantic crossing. Thirty-six lives were lost.

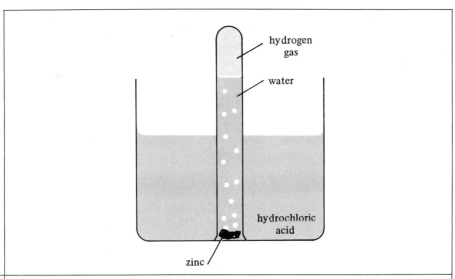

FIGURE 11.2

The reaction of zinc with hydrochloric acid. The acid seeps under the edge of the test tube and reacts with the zinc, forming bubbles of hydrogen gas.

paramount importance in industry, for example, water, ammonia (NH_3), sulfuric acid (H_2SO_4), and many hydrocarbons such as methane and the octanes (C_8H_{18}). Almost all compounds of biological importance contain hydrogen.

The combustion of hydrogen

$$H_2(g) + \tfrac{1}{2}O_2(g) = H_2O(g) \qquad \Delta H° = -241.84 \text{ kJ}$$

is highly exothermic, and the hydrogen–oxygen torch produces temperatures as high as 2800°C. The combustion of hydrogen with oxygen or fluorine is used for rocket propulsion. At elevated temperatures in the presence of a suitable catalyst, hydrogen reacts with nitrogen to form ammonia:

$$N_2(g) + 3H_2(g) = 2NH_3(g)$$

Ammonia is widely used as a fertilizer, and this reaction is the major use of industrial hydrogen. (Because the prices of petroleum and natural gas have increased dramatically in recent years, industrial hydrogen has also become much more expensive, and so have substances, such as ammonia, that require hydrogen for their synthesis.)

Compounds containing carbon–carbon double bonds, such as unsaturated oils, react with hydrogen:

$$\mathrm{\underset{}{>}C=C\underset{}{<} + H_2 = -\underset{H}{\overset{|}{C}}-\underset{H}{\overset{|}{C}}-}$$

and the hydrogenation of vegetable oils is important in the preparation of foods such as margarine and peanut butter.

Because hydrogen is of intermediate electronegativity (2.1 on the Pauling scale), it is the negative partner in bonds with most metals (e.g., Li^+H^-) and the positive partner in bonds with many nonmetals (e.g., H^+F^-). Bonds with elements of the same electronegativity (e.g., phosphorus) should be purely covalent. Phosphine, PH_3, has a small dipole moment, 0.58 D, which may be mainly due to the unshared pair of electrons on phosphorus. The polarity of the carbon–hydrogen bond has been a matter of considerable controversy. Judging from the electronegativity of carbon (2.5) the bond should be only slightly polar and in the direction H^+C^-. There is considerable evidence that the bond moment is small, about 0.4 D, but the direction of the polarity is not clear, and it may not be the same in all molecules. The dipole moment of methane is zero because of symmetry.

11.3 ACIDS, BASES, AND SALTS

There is an important class of substances, called **acids,** whose aqueous solutions have the following properties: (a_1) taste sour; (a_2) react with certain metals such as zinc to produce gaseous hydrogen; (a_3) give characteristic colors to certain substances called **indicators,** for example, turn litmus[2] pink; and (a_4) lose these properties on reaction with a base. Important acids include hydrochloric acid (HCl), nitric acid (HNO_3), sulfuric acid (H_2SO_4), phosphoric acid (H_3PO_4), and acetic acid (CH_3CO_2H). Acetic acid gives vinegar its sour taste.

A **base** is a member of another class of substances whose aqueous solutions have the following characteristics: (b_1) taste bitter; (b_2) feel slippery to the touch; (b_3) give certain colors to indicators, for example, turn litmus blue; and (b_4) lose these properties on reaction with an acid. Important bases include sodium hydroxide (NaOH), potassium hydroxide (KOH), calcium hydroxide [$Ca(OH)_2$], and ammonia (NH_3).

The reaction between an acid and a base is called **neutralization.** The substances formed in this reaction (besides water) are called **salts.** Thus the neutralization reaction can be written in general as

acid + base = water + salt

Examples of neutralization are

$$HCl(aq) + KOH(aq) = H_2O + KCl(aq) \tag{11.1}$$

$$H_2SO_4(aq) + 2NaOH(aq) = 2H_2O + Na_2SO_4(aq) \tag{11.2}$$

$$2HNO_3(aq) + Ca(OH)_2(aq) = 2H_2O + Ca(NO_3)_2(aq) \tag{11.3}$$

Evaporation of the water from the solutions formed in these reactions yields the

---

[2] Litmus is a mixture of dyes originally obtained from certain lichens. It is usually used as an acid–base indicator in the form of strips of absorbent paper colored with litmus.

solid salts: potassium chloride (KCl), sodium sulfate (Na_2SO_4), and calcium dinitrate [$Ca(NO_3)_2$], usually called simply calcium nitrate.

As part of his theory of electrolytic dissociation (described in the preceding chapter), Arrhenius showed that acids have still another property in common, namely, in aqueous solutions they all dissociate to some extent, forming hydrogen ion, H^+. It is to H^+ that acids owe their distinctive properties (a_1 to a_4). Arrhenius proposed that *an acid* could be defined as *any substance that dissociates to produce H^+ in aqueous solution*. (Although the hydrogen ion in water is often written as H^+, it is undoubtedly hydrated and more correctly should be referred to as hydronium ion, H_3O^+.

$$H^+ + H_2O = H_3O^+$$

As a matter of convenience we shall usually write H^+ for hydrogen ion in aqueous solutions and assume it to be understood that the ion is hydrated.)

Similarly, bases dissociate to form hydroxide ion, OH^-, which gives them their distinctive properties (b_1 to b_4). Arrhenius proposed that *a base* can be defined as *any substance that dissociates to produce OH^- in aqueous solution*. Upon neutralization the H^+ from an acid combines with the OH^- from a base to form water. Thus each of the neutralization reactions in Eqs. 11.1–11.3 may be represented by the single reaction

$$H^+ + OH^- = H_2O \tag{11.4}$$

The resulting salt solution contains the anion of the acid (e.g., Cl^-) and the cation of the base (e.g., K^+). These ions need not be shown in the equation because they undergo no change in amount or form.

Ammonia in aqueous solution produces OH^- through reaction with the water, which can be written

$$NH_3 + H_2O = NH_4^+ + OH^-$$

At one time an aqueous solution of ammonia was called ammonium hydroxide, with formula NH_4OH. This served to emphasize the basic properties of the solution, but it required the supposition of a molecule, NH_4OH, whose existence was highly doubtful. Such a solution is now called aqueous ammonia, formula $NH_3(aq)$.

Not all substances containing hydrogen are acidic; for example, CH_4 does not have acidic properties. In general hydrogen atoms attached to carbon atoms are not acidic. Thus, in acetic acid, only the hydrogen of the —CO$_2$H group (the carboxyl group) is acidic, and 1 mol of acetic acid can neutralize only 1 mol of hydroxide ion to form 1 mol of acetate ion, CH$_3$CO$_2^-$:

$$CH_3CO_2H + OH^- = H_2O + CH_3CO_2^-$$

Most bases contain an OH group associated with a metallic ion, whereas most acids consist of hydrogen and a nonmetal, or hydrogen and oxygen and another nonmetal. The simple halogen acids (HF, HCl, HBr, and HI) do not contain oxygen, but most other acids are oxyacids, with the acidic hydrogens bonded to oxygen. Examples are acetic acid (discussed before) and sulfuric acid,

$$\ddot{\underset{\ddot{\underset{H}{O}}}{\overset{\ddot{O}:}{\underset{|}{\overset{|}{S}}}}}\underset{}{\overset{}{-}}\overset{H}{\underset{}{\overset{\nearrow}{O}}}$$

Given an OH group bonded to another atom, it is usually acidic if the atom is a nonmetal,

$$XOH = XO^- + H^+$$

and basic if the atom is a metal,

$$MOH = M^+ + OH^-$$

There are also some elements of intermediate electronegativity (usually a metalloid or an element near the metalloids in the periodic table, such as beryllium, aluminum, and tin) whose hydroxides show both acidic and basic properties. They are called **amphoteric** hydroxides. For example, beryllium hydroxide is only slightly soluble in water, but it dissolves in either hydrochloric acid or aqueous sodium hydroxide:

$$Be(OH)_2(s) + 2H^+ = 2H_2O + Be^{2+}$$
$$Be(OH)_2(s) + 2OH^- = 2H_2O + BeO_2^{2-}$$

In the first reaction beryllium hydroxide acts as a base by reacting with H$^+$; in the second reaction it acts as an acid by reacting with OH$^-$. The salts formed by evaporation of the solutions to dryness are beryllium chloride (BeCl$_2$) and sodium beryllate (Na$_2$BeO$_2$), respectively.

The following examples illustrate the solution stoichiometry of acid–base reactions.

EXAMPLE 1

What is the concentration of a hydrochloric acid solution if 40.0 mL of the solution is required to neutralize the base in 50.0 mL of 0.200M NaOH?

We first note that the process of neutralization requires 1 mol of NaOH for each mole of HCl as shown in the equation for the net reaction:

$$H^+ + OH^- = H_2O$$

The 50.0 mL, or 0.0500 L, of 0.200M NaOH contains (0.200 mol L^{-1})(0.0500 L) = 0.0100 mol of NaOH. Therefore we must have 0.0100 mol of HCl in 40.0 mL, or 0.0400 L, of solution:

$$\text{molarity of HCl} = \frac{0.0100 \text{ mol}}{0.0400 \text{ L}} = 0.250M$$

EXAMPLE 2

If 50.0 mL of 0.400M NaOH is required to neutralize the acid in 20.0 mL of an H_2SO_4 solution, what is the molarity of the H_2SO_4 solution?

We first calculate the number of moles of NaOH, then the number of moles of H_2SO_4 required for neutralization, and finally the molar concentration of the H_2SO_4 solution.

$$\text{moles of NaOH} = (0.400 \text{ mol L}^{-1})(0.0500 \text{ L}) = 0.0200 \text{ mol}$$

The net reaction is the same as in the preceding example, but because each mole of H_2SO_4 produces 2 mol of H^+, only $\frac{1}{2}$ mol of H_2SO_4 is required for each mole of NaOH:

$$\text{moles of } H_2SO_4 = (\tfrac{1}{2})(0.0200 \text{ mol}) = 0.0100 \text{ mol}$$

$$\text{molarity of } H_2SO_4 = \frac{0.0100 \text{ mol}}{0.0200 \text{ L}} = 0.500M$$

11.4 THE NOBLE GASES

The group of elements appearing along the right side of the periodic table are familiar in the sense that we are constantly breathing small amounts of all of them (see Table 11.1). Helium is a major constituent of the sun and other stars; next to hydrogen, it is the most abundant element in the universe. Its main source here on earth is certain natural gas fields, where it occurs at concentrations as high as 8%.

In spite of their occurrence in the atmosphere these elements were not discovered and isolated until the late nineteenth century. The spectrum of helium was first observed in 1868 in the sun's chromosphere. Nothing else was known of

TABLE 11.1 | **COMPOSITION OF DRY AIR**[a]

Substance	Vol. (%)	Substance	Vol. (%)
Helium	0.00052	Nitrogen	78.09
Neon	0.0018	Oxygen	20.95
Argon	0.934	Carbon dioxide	0.03[b]
Krypton	0.00011	Methane	0.0002
Xenon	0.0000086	Hydrogen	0.00005
Radon	6.0×10^{-18}[b]	Nitrous oxide	0.00005
		Ozone	0.000001[b]

[a] At sea level in open country.
[b] Average value—varies from one locality to another.

its properties, and the name assigned (from the Greek *helios*, "sun") used the suffix *-ium*, indicating that it was thought to be a metal. No other nonmetal has this ending, and the name "helion" would be more appropriate.

The most important characteristic of the noble gases is their relative inertness toward chemical combination. The gases are all monatomic, and low temperatures are required to condense them to the liquid state. Helium has the lowest boiling point of any liquid, $-268.9\,°C$ or 4.2 K. These gases were known as the inert gases until 1962, when Neil Bartlett discovered that xenon combines with platinum hexafluoride, PtF_6. Since then several compounds of krypton, xenon, and radon have been synthesized. These elements are now usually referred to as the rare gases or the noble gases. (The term *noble* is used in the anthropomorphic sense indicating that these gases remain aloof from other elements.)

The relative inertness of the noble gases provides one of the best examples of the periodic law. This lack of reactivity of the noble gases has been shown to follow in a direct way from the electronic structures of the atoms.

Although helium has twice the density of hydrogen, it is much preferred for lighter-than-air craft because it does not burn. Helium and argon are used to provide a chemically inert atmosphere in metallurgy. Electric discharge tubes used in advertising displays contain helium, neon, argon, mercury vapor, or mixtures of these gases for various colors.

11.5 THE ALKALI METALS

Group IA elements beyond hydrogen, called the **alkali metals,** have similar electronic structures, with one valence electron in an *s* orbital outside a kernel having the electronic structure of a noble gas. Their first ionization potentials and electronegativities are the lowest of any group.

The solid elements are all soft metals with low melting points and low densities, indicative of weak interatomic bonds. Three of them (lithium, sodium, and potassium) are less dense than water (0.53, 0.97, and 0.86 g cm^{-3}, respectively).

The metals are so highly reactive that they are often stored under kerosene to minimize reactions with air. A freshly cut surface of sodium metal is silvery in appearance but, in contact with air, a dull oxide coating quickly forms.

$$4Na(s) + O_2(g) = 2Na_2O(s)$$

The metals react vigorously with water to form hydrogen gas and aqueous hydroxide solutions called **alkalies.** For example,

$$2K(s) + 2H_2O(l) = 2K^+ + 2OH^- + H_2(g)$$

The single valence electron of an alkali metal atom is easily lost to form the univalent cation. The common compounds of these elements (halides, nitrates, sulfates, carbonates, etc.) are all colorless crystals or white powders that are readily soluble in water. They are all strong electrolytes. Their compounds are colored only when the anion is colored. For example, $KMnO_4$ owes its dark purple color to the permanganate ion, MnO_4^-.

In general the salts of the alkali metals have high melting and boiling points, and the liquids are good conductors of electricity. Electrolysis of the halides at temperatures above the melting points yields the alkali metal and halogen, and this is the chief method of producing the alkali metals in the elemental state. For example, electrolysis of liquid sodium chloride forms liquid sodium metal and gaseous chlorine:

$$2NaCl(l) = 2Na(l) + Cl_2(g)$$

11.6 THE ALKALINE EARTH METALS

Group IIA elements, often called the **alkaline earth metals,** are harder and have higher densities and higher melting points than the alkali metals. Beryllium has the highest melting point, 1283°C, and magnesium the lowest, 650°C. Even though the Group IIA elements are much less reactive than the Group IA metals, they are still so reactive that they are never found in the metallic state in nature.

The metals may be prepared by electrolysis of their molten chlorides. They are all good conductors of electricity, and magnesium metal is used in lightweight alloys[3] for structural purposes. It is also used in flashbulbs for photographic purposes because it emits a brilliant flash of light when it reacts with oxygen at elevated temperatures to form the oxide.

$$2Mg(s) + O_2(g) = 2MgO(s)$$

The properties of beryllium compounds differ markedly from those of the other group IIA elements. Beryllium compounds are extremely poisonous, whereas Mg^{2+} and Ca^{2+} are essential to plant and animal growth. The bones and

---

[3] An alloy is a metallic material containing two or more elements.

teeth of animals are mainly $Ca_5(PO_4)_3OH$ (hydroxyapatite). Magnesium is the central atom of the chlorophyll molecule, which plays a key role in photosynthesis and plant life.

The elements of Group IIA readily lose two electrons to form the $+2$ ions and, with the exception of beryllium, the aqueous chemistry of these elements is that of the $+2$ ion. In aqueous solution their compounds are strong electrolytes. The chlorides and nitrates are extremely soluble in water, whereas the hydroxides, carbonates, sulfates, and phosphates are much less soluble. Indeed the carbonate and phosphate compounds of calcium and barium are very slightly soluble, as is barium sulfate, $BaSO_4$.

Calcium carbonate occurs in nature as the mineral limestone. On heating in a lime kiln, limestone decomposes:

$$CaCO_3(s) = CaO(s) + CO_2(g)$$

The calcium oxide, called *lime*, is used mainly as a constituent of mortar and Portland cement. Eggshells and the shells of mollusks (oysters, clams, etc.) are mainly calcium carbonate.

11.7 THE GROUP IIIA ELEMENTS

Group IIIA elements (boron, aluminum, gallium, indium, and thallium) have similar electronic structures in the valence shell, namely, two *s* electrons and one *p* electron. Boron is a metalloid, and the others are metals. Boron is quite hard and has a high melting point, $\sim 2250\,°C$. The melting point of aluminum, $659\,°C$, is near that of magnesium. The other metals have relatively low melting points.

Each atom in this group has three valence electrons, and the characteristic ionic valence is $+3$. They form nitrides of formula MN, oxides of formula M_2O_3, and fluorides of formula MF_3. Of the hydroxides [formula $M(OH)_3$], that of boron is acidic, those of aluminum and gallium are amphoteric, and those of indium and thallium are basic.

Boric oxide, B_2O_3, is a constituent of borosilicate glass, commonly used for laboratory glassware, baking dishes, and the like, under trade names such as Pyrex and Kimax. Boron forms oxyacids such as H_3BO_3 (orthoboric acid) and HBO_2 (metaboric acid). The sodium salt of tetraboric acid, $Na_2B_4O_7 \cdot 10H_2O$ (sodium tetraborate decahydrate) is found in substantial deposits as the mineral borax.

Aluminum is a metal of low density (2.70 g cm^{-3}), widely used in the aircraft industry and elsewhere as a lightweight structural metal. It is a familiar household item in the form of cooking utensils, beverage cans, toothpaste tubes, and aluminum foil. A good conductor of electricity, it is often used for electrical transmission lines. Aluminum occurs widely in various minerals and rocks, usually as aluminum silicates. The oxide, Al_2O_3, occurs as the mineral corundum and, in hydrated forms, as bauxite. Aluminum metal is produced by electrolysis of bauxite dissolved in molten cryolite, Na_3AlF_6:

$$2Al_2O_3(l) = 4Al(l) + 3O_2(g)$$

Important aluminum compounds are the chloride, $AlCl_3$; sulfate, $Al_2(SO_4)_3$; hydroxide, $Al(OH)_3$; and potassium aluminum sulfate, $KAl(SO_4)_2 \cdot 12H_2O$, called *alum*.

Gallium, indium, and thallium are relatively rare and are not widely used. Their chemistry is generally similar to that of aluminum.

11.8 CARBON—ORGANIC CHEMISTRY

Most substances in biological systems contain carbon, and the term **organic chemistry** is used for the chemistry of carbon compounds. At one time it was thought that these substances were fundamentally different from the substances found in the nonliving world of rocks and minerals—some vital force in living organisms was supposedly necessary for their creation. During the nineteenth century it was discovered that this was not true. Carbon from limestone, hydrogen from water, oxygen from air, and other elements from nonliving sources are sufficient to create any carbon compound.

Elemental carbon has two allotropic forms: diamond and graphite. Diamond is the hardest naturally occurring substance,[4] whereas graphite (a constituent of "lead" of pencils) is relatively soft and is used as a lubricant. The differences in physical properties result from structural differences. In graphite (Figure 11.3) the atoms are arranged in layers, with strong chemical bonds between the atoms within each layer, but only weak forces between different layers.[5] In diamond (Figure 11.4) each carbon atom is connected by single bonds to four other atoms at the corners of a regular tetrahedron. The resulting three-dimensional framework is extremely rigid. Diamonds are used industrially as an abrasive because of their hardness and socially as gems because of their brilliant optical properties and high cost. At very high temperatures and pressures, graphite may be converted to diamond.

Although carbon is a nonmetal, graphite is a fairly good conductor of electricity and is used for that purpose in some storage batteries and electrolysis cells. Carbon is extremely nonvolatile. At 1 atm pressure and 4000 K, graphite vaporizes directly from solid to gas, a process called **sublimation**. (Liquid carbon forms only at higher pressures.)

---

[4] Boron nitride, BN, is isoelectronic with carbon. It has been prepared in a form having the diamond structure, and in this form it is even harder than diamond.

[5] For many years it was thought that graphite owed its good lubricating properties to the ease with which the different layers slip past one another. When high-altitude airplanes were developed, it was found that graphite is a poor lubricant at low pressures. Further investigation showed that adsorbed water vapor and other gases from the atmosphere are of importance to its lubricating qualities.

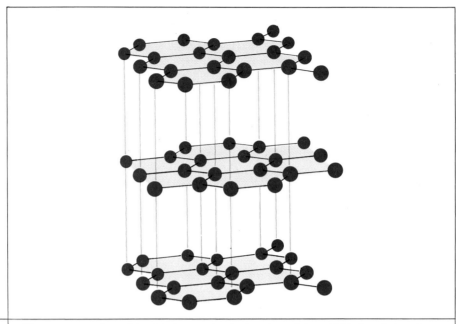

FIGURE 11.3 The structure of graphite. Each atom in the interior of the solid is bonded to three other atoms by one double bond and two single bonds. Bond angles are 120°.

Because it has four valence electrons, carbon is usually quadrivalent, as discussed in Chapter 9. In the normal saturated hydrocarbons the carbon atoms are linked by single bonds in a chain. For example, the structure of normal butane (*n*-butane) is

$$\begin{array}{c} \text{H} \quad \text{H} \quad \text{H} \quad \text{H} \\ | \quad\; | \quad\; | \quad\; | \\ \text{H—C—C—C—C—H} \\ | \quad\; | \quad\; | \quad\; | \\ \text{H} \quad \text{H} \quad \text{H} \quad \text{H} \end{array}$$

A brief notation for this structure, called a condensed structural formula, is $CH_3CH_2CH_2CH_3$.

Names and boiling points of the C_1–C_8 normal saturated hydrocarbons are listed in Table 11.2. Although these compounds are sometimes called **straight-chain hydrocarbons** and are usually so drawn in a Lewis structure, the carbon chains actually form zigzag patterns. The bond angles around each carbon atom are close to the tetrahedral value, 109.5°.

Hydrocarbons with double and triple carbon–carbon bonds are called **unsaturated.** Important examples are

11.8 CARBON—ORGANIC CHEMISTRY

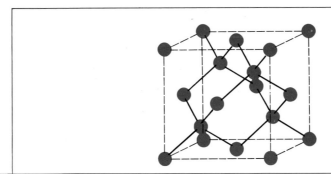

FIGURE 11.4

The diamond structure. Each atom in the interior of the solid is bonded to four other atoms by single bonds. Bond angles are 109.5°.

$$\begin{array}{c}H\\ \end{array}\!\!C\!=\!C\!\begin{array}{c}H\\ \end{array} \quad \text{or} \quad CH_2\!=\!CH_2 \qquad H\!-\!C\!\equiv\!C\!-\!H \quad \text{or} \quad HC\!\equiv\!CH$$

ethylene   acetylene

Unsaturated hydrocarbons undergo a variety of **addition reactions;** for example, ethylene adds hydrogen to form ethane, water to form ethyl alcohol, and chlorine to form ethylene dichloride:

$$\begin{array}{c}H\\H\end{array}\!\!C\!=\!C\!\begin{array}{c}H\\H\end{array} + H_2 = H\!-\!\underset{\underset{H}{|}}{\overset{\overset{H}{|}}{C}}\!-\!\underset{\underset{H}{|}}{\overset{\overset{H}{|}}{C}}\!-\!H \quad \text{or} \quad CH_3CH_3$$

ethane

$$\begin{array}{c}H\\H\end{array}\!\!C\!=\!C\!\begin{array}{c}H\\H\end{array} + H_2O = H\!-\!\underset{\underset{H}{|}}{\overset{\overset{H}{|}}{C}}\!-\!\underset{\underset{H}{|}}{\overset{\overset{H}{|}}{C}}\!-\!\ddot{\underset{..}{O}}\!-\!H \quad \text{or} \quad CH_3CH_2OH$$

ethyl alcohol

$$\begin{array}{c}H\\H\end{array}\!\!C\!=\!C\!\begin{array}{c}H\\H\end{array} + Cl_2 = H\!-\!\underset{\underset{:\ddot{Cl}:}{|}}{\overset{\overset{H}{|}}{C}}\!-\!\underset{\underset{:\ddot{Cl}:}{|}}{\overset{\overset{H}{|}}{C}}\!-\!H \quad \text{or} \quad CH_2ClCH_2Cl$$

ethylene dichloride

TABLE 11.2

NORMAL SATURATED HYDROCARBONS		
Name	Formula	bp (°C)
Methane	CH_4	−161
Ethane	C_2H_6	−89
Propane	C_3H_8	−42
n-Butane	C_4H_{10}	−1
n-Pentane	C_5H_{12}	36
n-Hexane	C_6H_{14}	69
n-Heptane	C_7H_{16}	98
n-Octane	C_8H_{18}	126

By comparison saturated hydrocarbons are usually not very reactive, and elevated temperatures or ultraviolet light are necessary to obtain reasonably rapid rates for **substitution reactions,** such as

$$CH_3CH_3 + Cl_2 = CH_3CH_2Cl + HCl$$

As a consequence of the reluctance of C—H and C—C single bonds to react, the saturated hydrocarbon part of a molecule is usually unaltered in a chemical reaction. For example, ethyl chloride readily undergoes a substitution reaction with an aqueous base:

$$CH_3CH_2Cl + OH^- = CH_3CH_2OH + Cl^-$$

For this reason it is convenient to describe most organic molecules as combinations of an inert part and one or more chemically reactive parts called **functional groups.** Substances having the same functional group or groups form a **class** of organic compounds. Examples are shown in Table 11.3. Examples of saturated hydrocarbon groups are methyl (CH_3), ethyl (C_2H_5), and propyl (C_3H_7). Collectively they are called **alkyl groups.**

TABLE 11.3

SOME ORGANIC FUNCTIONAL GROUPS	
Functional group	Class of compound
—Cl	Chlorides
—OH	Alcohols
—NH_2	Amines
—CO_2H	Carboxylic acids
—CO_2H and —NH_2	Amino acids
$\diagdown$C=C$\diagup$	Alkenes
—C≡C—	Alkynes

11.9 ISOMERISM

In some cases a particular combination of atoms can exist as molecules having different structures, called **isomers.** For example, there are two isomers having the formula C_4H_{10}. In one isomer (called *n*-butane) the carbon atoms form a single chain. In the other isomer (called isobutane) the carbon atoms form a branched chain. The two structures are

n-butane   isobutane

These isomers are different chemical substances with different physical and chemical properties.

All of the saturated hydrocarbons with four or more carbon atoms have straight-chain and branched-chain isomers. As the number of carbon atoms increases, the number of ways of forming branched chains increases, and thus the number of isomers increases. There are two butanes, three pentanes, five hexanes, and nine heptanes.

Branched and normal saturated hydrocarbons form a class of compound called **alkanes.** The alkanes are extensively used as fuels and solvents. Gasoline consists mainly of the C_6–C_9 alkanes.

When isomerism occurs because of different ways of linking the atoms in a molecule, as in the previous example, it is called **structural isomerism.** In another type of isomerism called **geometrical isomerism,** the atoms are linked in the same sequence but in a different spatial orientation. For example, there are two butenes in which the carbon atoms are linked in a single chain with the double bond in the center of the molecule:

cis-2-butene   *trans*-2-butene

As a consequence of the planarity of the atoms linked to a double bond (Section 9.8), the four carbon atoms in these molecules are coplanar. If the end carbon atoms are on the same side of the double bond, the compound is called the *cis*-isomer; if the end carbon atoms are on the opposite side of the double bond, the compound is called the *trans*-isomer.

Another type of isomerism, called **optical isomerism,** occurs when four different atoms or groups are bonded to the same central atom in an approximately tetrahedral configuration.[6] Two distinct arrangements in space are possible, each a mirror image of the other. Pairs of molecules of this type are called **enantiomers.** They have the important property of rotating the plane of polarized light, a property referred to as **optical activity.** One enantiomer, designated **dextrorotatory** or (+), rotates the plane of polarization to the right; the other, **levorotatory** or (−), rotates it to the left. An equimolal mixture, called the **racemate,** has no effect on the plane of polarization.

Figure 11.5 shows the configurations of (+)-lactic acid and (−)-lactic acid, for which the Lewis structure is

$$\begin{array}{c} \text{H} \quad \text{O} \quad \quad \text{O} \\ | \quad \;\; | \quad \;\; \| \\ \text{H}-\text{C}-\text{C}-\text{C} \\ | \quad \;\; | \quad \;\; \backslash \\ \text{H} \quad \text{H} \quad \;\; \text{O}-\text{H} \end{array}$$

The four different groups attached to the central carbon atom are —H, —CH$_3$, —OH, and —CO$_2$H. Although the enantiomers look similar (and in fact the two compounds have the same melting point, density, etc.), it is impossible to position mirror-image models so that all of the attached groups have the same orientation. If one of the models in Figure 11.5 is rotated 180° to bring the —OH and —H groups into the same orientation, then the —CH$_3$ and —CO$_2$H groups have opposite orientation. Thus the enantiomers are not superimposable.

These molecules are related in the same way as the right and left hands, which are also mirror images of each other. We are sometimes reminded that they are not superimposable when attempting to put on a pair of gloves. Enantiomers are occasionally referred to as right-handed and left-handed molecules.

The fact that lactic acid exists as enantiomers results from its lack of symmetry. All the molecules shown in Figure 9.5 have planes of symmetry passing through the central atom, such that the half of the molecule on one side of the plane of symmetry is a mirror image of the other half. Lactic acid has no plane of symmetry. Because of this lack of symmetry, a carbon atom to which four different atoms or groups are attached is called **asymmetric.**

It was to explain the phenomenon of optical activity that the tetrahedral carbon atom was first postulated, in 1874, by Jacobus H. van't Hoff and Joseph A. Le Bel independently. Its importance in chemistry was soon recognized, and the truth of Le Bel's and van't Hoff's postulate has since been established by direct structural determinations.

---

[6] The bond angles need not be (and usually are not) those of a regular tetrahedron.

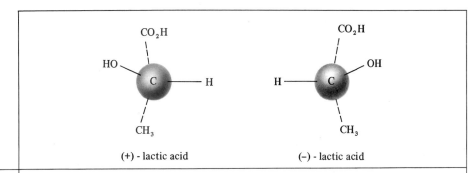

FIGURE 11.5 (+) - lactic acid   (−) - lactic acid

Enantiomers of lactic acid. Solid lines are bonds directed toward us above the plane of the paper; broken lines are bonds directed away from us behind the plane of the paper.

11.10 SILICON, GERMANIUM, TIN, AND LEAD

The group IVA elements following carbon are silicon, germanium, tin, and lead. Silicon and germanium are metalloids; tin and lead are metals. Like carbon, these elements have two s electrons and two p electrons in the valence shell and therefore tend to be quadrivalent. This tendency decreases toward the end of the group, where bivalent compounds of tin and lead are quite stable.

Because the carbon–carbon single bond is very strong, carbon has a strong tendency to form chains of like atoms. As one goes down the group the single bonds between like atoms become much weaker. (The average single-bond enthalpies are C—C, 356; Si—Si, 238; Ge—Ge, 188; and Sn—Sn, 151 kJ mol^{-1}.) As a consequence, the tendency to form chains of like atoms decreases markedly from carbon to tin.

Whereas carbon dioxide is a gas that dissolves in water to produce an acidic solution, silica (SiO_2) is a high-melting solid that is practically insoluble in water. However, SiO_2 is soluble in aqueous solutions of strong bases such as NaOH, so it does have acidic properties, as does GeO_2. The other dioxides, SnO_2 and PbO_2, are amphoteric.

Quartz is one of the naturally occurring forms of silica. Most of the outer crust, or lithosphere, of the earth is composed of silicate minerals (see Table 11.4). Rocks, clays, sands, gravels, and soils are mainly silica and silicates, as are important products of the building industry such as cement, mortar, bricks, ceramics, and various kinds of glass. One of the most common rocks is granite, a heterogeneous mixture consisting mainly of quartz (SiO_2) and various feldspars such as $NaAlSi_3O_8$ and $CaAl_2Si_2O_8$. Minor constituents include micas such as $KAl_3Si_3O_{10}(OH)_2$ and pyroxenes such as $CaMgSi_2O_6$.

Tin and lead are important metals with a variety of uses. A "tin can" is actually made of steel, with a very thin coating of tin to prevent corrosion. Copper–tin alloys are called **bronzes**. Lead is a rather soft metal and is often alloyed

TABLE 11.4	AVERAGE COMPOSITION OF LITHOSPHERE	
	Element	% by mass
	Oxygen	46.6
	Silicon	27.7
	Aluminum	8.1
	Iron	5.0
	Calcium	3.6
	Sodium	2.9
	Potassium	2.6
	Magnesium	2.1
	Titanium	0.6
	Others	0.8

with small amounts of antimony to harden it. Its density, 11.3 g cm^{-3}, is notoriously high. One of the main uses of metallic lead is in storage batteries. Solder is a low-melting alloy of tin and lead used to join copper wires in electronic circuits and the like. Lead compounds are highly toxic. The ancient Romans used lead water pipes and lead wine vessels; in fact, the fall of Rome is attributed by some historians to lead poisoning. For many years tetraethyllead, $Pb(C_2H_5)_4$, has been added to gasoline to provide an even rate of combustion in automobile engines, thus preventing detonation, or "knock." In unleaded fuels other substances not containing lead are now used to accomplish the same purpose.

11.11 THE GROUP VA ELEMENTS

Nitrogen and phosphorus are nonmetals, arsenic and antimony are metalloids, and bismuth is usually considered a metal, although it is brittle and has rather low electrical conductivity. The oxides of nitrogen and phosphorus are acidic, of arsenic and antimony amphoteric, and of bismuth basic.

With its high electronegativity (3.0), nitrogen can form ionic compounds, called **nitrides,** with elements of low electronegativity. Examples are lithium nitride, Li_3N, and magnesium nitride, Mg_3N_2. To some extent phosphorus forms similar compounds, called **phosphides.** Compounds with covalent or slightly polar bonds are more common, and, in accordance with the $8 - N$ rule, the group VA elements are often tervalent.

Elemental nitrogen, the major constituent of air, consists of diatomic molecules, N_2. Nitrogen has a very low boiling point, $-195.8\,°C$ or 77.4 K, and liquid nitrogen is often used to produce low temperatures in the chemical laboratory.

The other elements are solids at ordinary temperatures. Phosphorus has been prepared in several allotropic forms, including white phosphorus, black phosphorus, and several red forms. One of the red forms has a normal sublimation point of $431\,°C$. At this temperature the vapor is tetratomic, P_4; above about

800°C the P_4 molecules dissociate to P_2 molecules. The diatomic molecules have an electronic structure analogous to N_2,

$$:N\equiv N: \qquad :P\equiv P:$$

The striking difference between the physical properties of nitrogen and the other elements of this group reflects the extraordinary strength of the nitrogen–nitrogen triple bond and the relative weakness of the nitrogen–nitrogen single bond. Consequently, liquid and solid nitrogen contain diatomic molecules, with strong forces holding the two atoms together, but with only weak forces between adjacent molecules. In the P_4 molecule (found in white phosphorus and in phosphorus vapor) the atoms lie at the corners of a regular tetrahedron, with each atom joined to the three others by single bonds. In the other forms of phosphorus and in arsenic and bismuth, each atom is linked to three other atoms by single bonds to form solids.

It is a fairly good generalization to say that multiple-bond formation is especially pronounced among the elements of the first short period, whereas the other elements tend to form structures with single bonds. Another example is the contrast between the physical properties of CO_2, with carbon–oxygen double bonds in each molecule, and SiO_2, with silicon–oxygen single bonds in an extended framework structure.

Ammonia is one of nitrogen's most important compounds. Nitrogen is essential to biological processes, and free nitrogen in the air cannot be utilized by most plants. Atmospheric nitrogen is converted to biologically usable nitrogen compounds (the "fixation" of nitrogen) by a process, developed by Fritz Haber, in which ammonia is formed from the elements at high pressures and temperatures in the presence of a metal catalyst:

$$N_2(g) + 3H_2(g) = 2NH_3(g)$$

The ammonia is then used directly as a fertilizer or converted to other compounds such as ammonium sulfate, $(NH_4)_2SO_4$. The Haber process has enormously increased world food production.

Ammonia is a gas under ordinary conditions with a fairly low liquefaction point, $-33°C$. Liquid ammonia has a high dielectric constant (22.4) and as a solvent it is rather similar to water. The alkali metals dissolve in liquid ammonia to give electrically conducting solutions. These solutions slowly decompose to form hydrogen gas and a metal amide, so that the overall reaction is analogous to the reaction of the alkali metals with water, for example,

$$2Na(s) + 2NH_3(l) = H_2(g) + 2Na^+ + 2NH_2^-$$

The amide ion, NH_2^-, plays the same role in liquid ammonia solutions as the hydroxide ion does in aqueous solutions.

Important oxides of nitrogen include nitrous oxide, N_2O; nitric oxide, NO; nitrogen dioxide, NO_2; and nitrogen pentoxide, N_2O_5. The NO and NO_2 molecules are unusual in that each has an odd number of electrons. The chemical substance "NO_2" is actually a mixture of a red-brown monomer, NO_2, and a colorless dimer, N_2O_4, called *nitrogen tetroxide:*

$$2NO_2 = N_2O_4$$

Nitrous acid, HNO_2, and nitric acid, HNO_3, form salts called *nitrites* and *nitrates*, respectively. The resonance structures of the nitrate ion, NO_3^-, are shown in Section 9.10.

The chemistry of phosphorus is rather different from that of nitrogen. One reason for this is the tendency of nitrogen to form multiple bonds, whereas phosphorus tends to form single bonds. Thus N_2O_5 dissolves in water to form the monoprotic nitric acid, a planar molecule whose resonance structures have a nitrogen-oxygen double bond:

$$N_2O_5(s) + H_2O(l) = 2HNO_3(l)$$

Phosphorus pentoxide, P_2O_5, dissolves in excess water to form a solution of orthophosphoric acid, H_3PO_4, usually called simply phosphoric acid. It is a triprotic acid.

Arsenic, antimony, and bismuth resemble one another and phosphorus rather more than phosphorus resembles nitrogen. An important distinction is their toxicity, that of arsenic being notoriously high.

11.12 THE GROUP VIA ELEMENTS

Oxygen and sulfur are nonmetals, selenium and tellurium are metalloids, and polonium is a metal. Sulfur, selenium, and tellurium form a number of oxyacids; polonium forms an amphoteric hydroxide.

The more electronegative elements react with metals to form ionic solids containing the oxide, O^{2-}, sulfide, S^{2-}, selenide, Se^{2-}, and telluride, Te^{2-}, ions.

Some of these compounds, such as MgO, CaO, MgS, and CaS, have the NaCl structure (Figure 9.2). In molecules that are mainly covalent the atoms are often bivalent, in accordance with the $8 - N$ rule.

Elemental oxygen is a major constituent of the atmosphere (see Table 11.1) and is essential to certain important biological functions. It consists of diatomic molecules, O_2. The electronic structure of this molecule is unusual in that two electrons have unpaired spins, a phenomenon that is readily explained by molecular orbital theory (Section 9.16).

Ozone, O_3, is an unstable, highly reactive form of oxygen prepared by the action of light or an electric discharge on diatomic oxygen molecules. The upper atmosphere contains significant quantities of ozone formed by sunlight, the maximum concentration being at an altitude of about 20 to 25 km. Its presence is extremely important to life on earth, for it absorbs practically all of the damaging ultraviolet radiation of short wavelengths. The ozone molecule is nonlinear, the bond angle being 117°. Its resonance structures are analogous to those of SO_2 (see Section 9.10).

Whereas oxygen is a gas with a low boiling point, the other group VIA elements are all solids. Sulfur exists in a variety of allotropic forms. At room temperature the stable form is orthorhombic (usually called simply **rhombic**). At 95°C the rhombic crystals change to monoclinic crystals that melt at 119°C. (Orthorhombic and monoclinic refer to the shapes of the crystals, a topic discussed in Chapter 17.) Both rhombic and monoclinic sulfur contain S_8 molecules, in which the sulfur atoms are joined together in eight-membered rings.

Water, H_2O, is no doubt the single most important and familiar chemical compound. Enormous quantities of water of high purity are available from natural sources, resulting from distillation of seawater by the action of sunlight. The need for water is so great that increasing attention is being paid to reducing the pollution of natural fresh water and to finding economical methods for obtaining water of high purity from the oceans. In the chemical laboratory, water is used mainly as a solvent.

Hydrogen peroxide, H_2O_2, and its salts such as sodium peroxide, Na_2O_2, are highly reactive substances. The electronic structure of H_2O_2 is

$$\begin{array}{c} H \\ \diagdown \\ \ddot{O}-\ddot{O}\colon \\ \diagdown \\ H \end{array}$$

The molecule is nonplanar, with an average angle between the two OOH planes of 94° (see Figure 2.3).

Most minerals are oxygen compounds of various metals and metalloids, and oxygen is the most abundant element in the lithosphere (see Table 11.4). A number of important minerals are sulfides and, to a lesser extent, selenides and tellurides. Because their compounds are important sources of metals, the group VIA elements are sometimes called **chalcogens** (from the Greek *chalkos*, "copper," and *genos*, "birth").

Sulfur, selenium, and tellurium form hydrides analogous to water: H_2S, H_2Se, and H_2Te. They are all extremely toxic gases with foul odors. Sulfur also forms a series of sulfanes of formula H_2S_n ($n = 2$ to 6) containing sulfur–sulfur chains.

Sulfur burns in air to form sulfur dioxide, SO_2, which can be further oxidized with the aid of a catalyst to sulfur trioxide, SO_3. Analogous compounds are formed by selenium and tellurium. The structure of SO_2 was considered in Section 9.10. Gaseous SO_3 contains planar symmetrical SO_3 molecules that have resonance structures analogous to the nitrate ion, NO_3^- (see Section 9.10). Sulfur trioxide reacts with water to form the diprotic acid H_2SO_4 (sulfuric acid). Sulfuric acid is produced in large quantities for a variety of uses.

Violations of the octet rule, previously noted for phosphorus (Section 9.6), occur for sulfur, selenium, and tellurium, which form tetrafluorides and hexafluorides:

sulfur tetrafluoride   sulfur hexafluoride

Two factors involved are the size of the central atom, large enough to accommodate more than four neighboring atoms, and the use of d orbitals by the central atom for bond formation (Chapter 19).

Selenium trioxide, SeO_3, forms selenic acid, H_2SeO_4, analogous to sulfuric acid, but for tellurium trioxide the factors noted in the preceding paragraph are great enough that telluric acid is H_6TeO_6. A similar trend occurs in the neighboring elements of group VA. Phosphoric and arsenic acids are H_3PO_4 and H_3AsO_4, respectively, whereas antimonic acid is H_7SbO_6.

11.13 THE HALOGENS

The group VIIA elements have highly distinctive properties, and their names have the distinctive ending *-ine*. They are all very reactive nonmetals with the possible exception of astatine. All of them have high electronegativities, fluorine being the most electronegative of all the elements. The halogens form gaseous diatomic hydrides that dissolve in water to give acidic solutions. Except for fluorine, they also form a number of oxyacids.

The halogens are often univalent ($8 - N$ rule) and as such the elements have diatomic molecules. Under ordinary conditions, F_2 and Cl_2 are gases, Br_2 is a liquid, and I_2 is a solid. Iodine melts at 114°C, but its vapor pressure at the melting point is rather high (0.12 atm), and iodine sublimes readily at lower temperatures, for example, on a steam plate. The diatomic molecules retain their identity in the liquid and solid phases, so that, for example, an iodine crystal is composed of I_2 molecules packed closely together.

With metals the halogens form a series of salts, such as NaCl, MgCl$_2$, and AlCl$_3$, and the term **halogen** denotes this characteristic (from the Greek *halos,* "salt"). The alkali metal halides are ionic solids under ordinary conditions, most of them having the NaCl structure (Figure 9.2).

The halide ions are important constituents of the oceans (see Table 11.5). Fluorine minerals include fluorspar, CaF$_2$; cryolite, Na$_3$AlF$_6$; and fluorapatite, Ca$_5$(PO$_4$)$_3$F. Bones and teeth contain small amounts of fluoride. Although at high concentrations fluorides are highly toxic, natural water supplies sometimes contain F$^-$ at concentrations as high as 0.0012% or 12 parts per million. It has been found that a concentration of about 1 part per million is optimum[7] for the development of teeth resistant to decay (dental caries), and fluorides are sometimes added to municipal water supplies for this purpose. Seawater contains 1.4 parts per million of F$^-$.

Chloride ion occurs in seawater at a concentration of 1.9%, and in blood at a concentration of 0.27%. Sodium chloride, found in large underground deposits as the mineral halite or rock salt, is used in large amounts primarily for the production of other sodium and chlorine compounds.

Bromides and iodides are mainly obtained from underground salt solutions (brines). The main use of bromine is in gasoline as ethylene dibromide, BrCH$_2$CH$_2$Br, which acts as a lead scavenger. The thyroid gland produces an important iodine compound, the hormone thyroxine, C$_{15}$H$_{11}$O$_4$NI$_4$. Small amounts of sodium iodide are sometimes added to table salt to supplement other sources of iodine in the diet.

Fluorine forms compounds with all the elements except helium, neon, and argon. It often causes violations of the octet rule by other atoms (e.g., PF$_5$ and SF$_6$). With carbon it forms an important series of fluorocarbons analogous to the hydrocarbons. Molecules of tetrafluoroethylene, C$_2$F$_4$, combine to form long chains:

$$n\text{C}_2\text{F}_4 \longrightarrow \left[\begin{array}{cc} \ddot{\text{F}}: & :\ddot{\text{F}}: \\ | & | \\ -\text{C}-\text{C}- \\ | & | \\ :\ddot{\text{F}}: & :\ddot{\text{F}}: \end{array} \right]_n$$

TABLE 11.5	AVERAGE COMPOSITION OF SEAWATER			
	Substance	% by mass	Substance	% by mass
	F$^-$	0.00014	H$_3$BO$_3$	0.0026
	Cl$^-$	1.90	Na$^+$	1.06
	Br$^-$	0.0065	K$^+$	0.038
	I$^-$	0.000004	Mg^{2+}	0.127
	SO$_4^{2-}$	0.265	Ca^{2+}	0.040
	HCO$_3^-$	0.014	Sr^{2+}	0.0008

[7] Higher concentrations cause mottling of the dental enamel.

where n is a very large number. Large molecules of this kind, formed from small like molecules, are called **polymers,** and this type of reaction is called **polymerization.** The polymer of C_2F_4 is marketed under trade names such as Teflon and Fluon.

Chlorine forms a series of oxyacids. Their names and the names of their ions are

HClO hypochlorous acid   ClO$^-$ hypochlorite ion
HClO$_2$ chlorous acid   ClO$_2^-$ chlorite ion
HClO$_3$ chloric acid   ClO$_3^-$ chlorate ion
HClO$_4$ perchloric acid ClO$_4^-$ perchlorate ion

Bromine and iodine form similar compounds, except for HIO_2, which is unknown.

11.14 THE TRANSITION METALS

The remaining 62 elements all have low ionization potentials and all are metals. Their distinguishing characteristic is the filling of d and f subshells (see Figure 8.18). Thus, in the first long period, calcium has two electrons in the $4s$ subshell. Prior to the filling of the $4p$ subshell, there is a transition region of 10 elements (scandium to zinc) in which the $3d$ subshell of 10 electrons is filled.

The second long period has 10 similar elements (yttrium to cadmium) in which the $4d$ subshell is filled. In the third long period, there are 24 transition metals, 14 of them with 1 to 14 electrons in the $4f$ subshell. The latter elements are called the **rare earth metals** (from their oxides which are called **rare earths**) or the **lanthanides** (from their resemblance to the first transition metal in this period, lanthanum). Following the lanthanides the $5d$ subshell is completed with the nine elements hafnium to mercury. A similar sequence occurs in the fourth long period, the first transition metal (actinium) being followed by 14 actinides. The remaining transition metals of this period are unknown, except for the recently discovered elements of atomic numbers 104 (rutherfordium), 105 (hahnium), and 106 (as yet not named).

The d and f subshells do not always fill in the sequential manner encountered with the main group elements (see Appendix E). For example, the copper atom has a complete $3d$ subshell at the expense of the $4s$ subshell, which thus has only one electron. Silver and gold behave similarly. With only one electron in an s orbital, these elements resemble the alkali metals in some respects, and they are designated group IB. Filling of the s orbitals occurs at the next elements, zinc, cadmium, and mercury, which are designated group IIB as they resemble the alkaline earth metals in many ways.

For the transition metals toward the left side of the periodic table (groups IIIB to VIIB) the group number indicates the total number of valence electrons in the outer d and s orbitals. For example, chromium in group VIB has five $3d$ electrons and one $4s$ electron. The nine central elements are combined in one group, partly for historical reasons and partly because there are, indeed, points of similarity in their properties. Designation of these elements as group VIII is traditional—only the metals in the first column (iron, ruthenium, and osmium) have eight electrons in their outer d and s orbitals.

In general, most or all of these valence electrons take part in bonding within the metal, so that the atoms are connected by short strong bonds, and the metals are characterized by high densities, high melting points, high boiling points, and high structural strength. The metals in the center of the third long period have the highest densities of any elements: rhenium, 21.04; osmium, 22.57; iridium, 22.5; and platinum, 21.45 g cm^{-3}. They also have the highest boiling points of any elements: tantalum, 5700; tungsten, 5800; and rhenium, 5900 K.

Several transition elements, particularly iron, are used as structural metals. Iron is usually used in the form of alloys called **steels.** Carbon steels are especially important. Stainless steels are alloys of iron, chromium, and nickel. For over a century a well-developed iron and steel industry has been the hallmark of an industrial society.

The transition elements generally possess the characteristic metallic properties to a high degree. At room temperature silver has the highest electrical conductivity of any metal and that of copper is almost as high. Chromium is used to coat automobile parts because of its luster and resistance to corrosion. The lure of gold and silver is legendary, and the group IB elements are sometimes called the **coinage metals** because of their formerly abundant use in coins.

The transition metals show an unusually wide range of compound formation, as in some cases anywhere from one electron to the total number of valence electrons may take part in bonding. For example, titanium forms four chlorides: TiCl, TiCl$_2$, TiCl$_3$, and TiCl$_4$. The behavior of the oxides and hydroxides toward acids and bases depends not only on the element but on the extent to which the valence electrons have entered into compound formation. Thus chromium forms basic chromous hydroxide, Cr(OH)$_2$; amphoteric chromic hydroxide, Cr(OH)$_3$; and acidic chromic acid, H$_2$CrO$_4$. Because the d and f electrons often absorb visible light in electronic transitions to low-lying excited states, transition metal compounds are often colored. In emeralds the otherwise colorless mineral beryl, Be$_3$Al$_2$Si$_6$O$_{18}$, is colored green by small amounts of chromic oxide. Rubies are corundum, Al$_2$O$_3$, colored red by chromic oxide, while sapphires are corundum colored blue by iron and titanium oxides.

In this chapter we have been able to take only a brief look at the chemistry of the elements, but it may serve to illustrate the extraordinary richness of phenomena displayed by even relatively obscure elements. Before considering descriptive chemistry in more detail, we shall first study the principles of chemical equilibria (Chapters 12–15) and chemical kinetics (Chapter 16), as well as the nature of the solid state (Chapter 17). This will make possible a deeper examination of the chemistry of the main group elements (Chapters 18 and 19), the transition metals (Chapters 20 and 21), and organic chemistry (Chapter 22).

PROBLEMS

11.1 With the aid of Figures 8.22 and 11.1, compare metallic character with ionization potential for the first 20 elements.

11.2 Suggest an explanation for the photoconductivity of gray selenium. How can light increase the electrical conductivity of a metalloid?

11.3 For each of the following properties of hydrogen, state whether hydrogen more closely resembles the first alkali metal (lithium), or the first halogen (fluorine).
(a) Each atom has one valence electron.
(b) It is a gas at ordinary temperatures and pressures.
(c) It is a nonmetal.
(d) It forms an ionic compound with sodium in which hydrogen is the negative partner, Na^+H^-.
(e) It forms a diatomic molecule with Cl that is mainly covalent.
(f) It forms a positive ion, H^+, in aqueous solution.

11.4 How many grams of zinc are required to produce 250 mL of gaseous hydrogen (measured at STP) by reaction with excess hydrochloric acid?

11.5 In the electrolysis of water, if 128 mL of oxygen gas is formed at one electrode, how many milliliters of hydrogen gas are formed at the other electrode? (The two gases are measured at the same temperature and pressure.)
$$2H_2O(l) \rightarrow 2H_2(g) + O_2(g)$$

11.6 Which of the following substances are acids, which are bases, and which are salts?

NaCl	NaOH	NH_3
H_2SO_4	$Ca(NO_3)_2$	HNO_3
$HClO_4$	Na_2SO_4	$KClO_3$
$Ca(OH)_2$	KOH	H_3PO_4
$NaNO_3$	CH_3CO_2H	$(NH_4)_2SO_4$

11.7 A solution was prepared by mixing 20 mL of 0.400M HCl and 30 mL of 0.200M NaOH. What ions were present and at what concentrations in each solution before mixing and in the final solution?

11.8 If 20.0 mL of 0.200M NaOH is required to neutralize an acid solution, how many moles of NaOH were used to neutralize the acid? If the acid was H_2SO_4, how many moles of acid were there?

11.9 What is the concentration of a sodium hydroxide solution if 16.9 mL of the solution is required to neutralize the acid in 45.0 mL of 0.120M HNO_3?

11.10 If 23.6 mL of a potassium hydroxide solution is required to neutralize the acid in 15.0 mL of a 0.200M H_2SO_4 solution, what is the molar concentration of the KOH solution?

11.11 In the latter part of the nineteenth century it was thought that pure nitrogen could be produced by two methods: (1) removal of oxygen, water vapor, and carbon dioxide from air, and (2) thermal decomposition of ammonium nitrite
$$NH_4NO_2(s) = N_2(g) + 2H_2O(g)$$
followed by removal of the water vapor produced in the reaction. A careful investigation showed that samples from these sources had slightly different densities. Suggest an explanation. Which sample had the higher density?

11.12 Compare the weights that can be lifted at STP by two balloons, each of volume 22.4 L, one containing hydrogen and the other helium. Assume that each balloon weighs 10 g and use 29 for the average molecular mass of air.

11.13 Compare the alkali metals and alkaline earth metals with respect to each of the following properties:
(a) melting points
(b) numbers of valence electrons per atom
(c) electric charges of ions in aqueous solution
(d) formulas of chlorides
(e) solubilities of carbonates in water

11.14 Draw three resonance structures for one layer of the graphite structure shown in Figure 11.3. (Each atom forms a double bond to one neighbor and single bonds to the other two neighbors.)

11.15 Write balanced equations for the formation of ethane by addition of hydrogen to
(a) ethylene
(b) acetylene

11.16 Write a sequence of two reactions by which C_2H_5OH may be synthesized from C_2H_6.

11.17 Write Lewis structures for isomers of C_4H_8 with
(a) the carbon atoms in a chain, and a double bond between the first and second
(b) the carbon atoms in a chain, and a double bond between the second and third
(c) the carbon atoms at the corners of a square
Which of these molecules has cis- and trans-isomers?

11.18 Do the following Lewis structures represent the same molecule or different isomers? How many isomers of C_2H_5Cl are there?

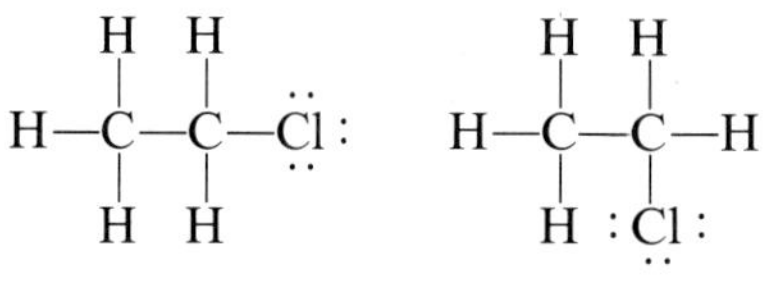

```
    H H              H H
    | |              | |
H—C—C—Cl:       H—C—C—H
    | |              |
    H H              H :Cl:
```

11.19 Draw Lewis structures for
  (a) the two isomers of $C_2H_4Cl_2$
  (b) the four isomers of $C_3H_6Cl_2$

11.20 Which of the following structures describe molecules having enantiomers? A carbon atom is bonded to
  (a) H, F, Cl, and Br atoms
  (b) two H and two Cl atoms
  (c) one H atom and CH_3, NH_2, and CO_2H groups
  (d) one H atom, one Cl atom, and a $=CH_2$ group

11.21 (a) In the synthesis of ammonia from nitrogen and hydrogen, how many tons of ammonia can be made from 1.00 ton of hydrogen?
  (b) Write a balanced equation for the reaction of ammonia and sulfuric acid to form ammonium sulfate. How many tons of ammonium sulfate can be produced from the ammonia from part (a)?

11.22 Draw resonance structures for ozone and for sulfur trioxide.

11.23 Write balanced equations for the reaction of water and
  (a) phosphorus pentoxide to form orthophosphoric acid
  (b) selenium trioxide to form selenic acid
  (c) tellurium trioxide to form telluric acid

11.24 Draw Lewis structures for P_4, S_8, H_2S, H_2S_2, and H_2S_3.

11.25 Draw Lewis structures for the isomers FSSF and SSF_2. Which has a donor-acceptor bond?

11.26 Draw Lewis structures for HClO, $HClO_2$, $HClO_3$, and $HClO_4$. (In each molecule all oxygen atoms are bonded to a central chlorine atom, with the hydrogen atom bonded to one of the oxygens.)

11.27 How do the outermost electronic shells of Cs^+ and Au^+ (both in the third long period) differ? Which do you expect to have the smaller ionic radius? Why?

11.28 From the enthalpy change of the hydrogen–oxygen reaction and the temperature of the hydrogen–oxygen torch, estimate the average heat capacity of water vapor between room temperature and 2800°C.

11.29 If 25 mL of a NaOH solution is required to neutralize 15 mL of $0.20 M$ HCl, and 50 mL of the same NaOH solution is required to neutralize 20 mL of a sulfuric acid solution, what is the molarity of the H_2SO_4?

11.30 Dichloroethylene, $C_2H_2Cl_2$, has two isomers. One isomer has a dipole moment of 2.95 D; the other isomer's dipole moment is exactly zero. Which is the *cis*-isomer, and which is the *trans*?

CHAPTER 12
CHEMICAL EQUILIBRIA

12

Although we have previously treated chemical reactions by assuming that the reaction continued until at least one of the reactants had been used up, there are many reactions that arrive at an equilibrium position in which the amount of product no longer increases even though appreciable amounts of all of the reactants are still available. In the following sections, we shall examine such systems and the principles of chemical equilibrium, together with methods of increasing the yield of a desired product.

12.1 GAS PHASE EQUILIBRIA

At 425°C, hydrogen gas reacts fairly rapidly with gaseous iodine to form hydrogen iodide:

$$H_2(g) + I_2(g) = 2HI(g) \tag{12.1}$$

This reaction may be investigated by introducing, for example, 5.345 mmol of H_2 and 5.345 mmol of I_2 into a 1-L container at 425°C. The progress of the reaction may be followed by observing the decrease in intensity of the violet color (due to I_2) in the reaction mixture. When the reaction has ceased (no further loss of color), the contents of the container may be analyzed, and we find that there are 8.410 mmol of HI, 1.140 mmol of H_2, and 1.140 mmol of I_2. (As required by the stoichiometry of Eq. 12.1, the amounts of H_2 and I_2 that have reacted, 5.345 − 1.140 = 4.205 mmol, are equal to one-half the amount of HI produced.)

If we now place an equivalent amount of HI (2 × 5.345 = 10.690 mmol) in a 1-L container at 425°C, we find that the gas gradually acquires a violet color due to formation of I_2 as the reaction of Eq. 12.1 occurs in the reverse direction. When reaction has ceased (no further increase in color), the contents of the container may be analyzed and we find that again there are 8.410 mmol of HI and 1.140 mmol each of H_2 and I_2. Thus we may prepare the same mixture of HI, I_2, and H_2 starting either with H_2 and I_2 or with HI. This behavior is shown in graphical form in Figure 12.1.

It is evident that this chemical reaction is *reversible*—we can form HI from H_2 and I_2 and we can form H_2 and I_2 from HI. Furthermore, the reactants and products coexist at certain definite concentrations. The situation is very similar to the vaporization of a liquid (or the condensation of a gas) in which, by using

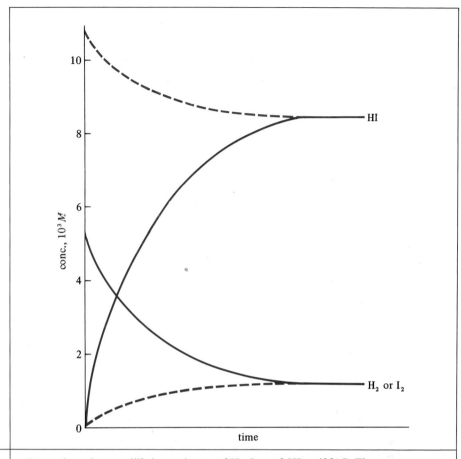

FIGURE 12.1 Formation of an equilibrium mixture of H_2, I_2, and HI at 425°C. The curves represent the concentrations of H_2, I_2, and HI at various times. The solid curves are for an initial mixture of 5.345 mmol of H_2 and 5.345 mmol of I_2 in a 1-L container; the dashed curves are for an initial sample of 10.69 mmol of HI in 1 L.

suitable conditions, we can have liquid and gas coexisting in an equilibrium situation (Chapter 10). We describe the H_2, I_2, HI mixture as an **equilibrium system** because the composition of the system remains unchanged with time and because we can obtain the same composition of the system starting with either the reactants or the products.

When we wish to emphasize the equilibrium aspects of a reaction, it is customary to use opposing arrows:

$$H_2(g) + I_2(g) \rightleftharpoons 2HI(g) \tag{12.2}$$

Often such an equilibrium is referred to as a **dynamic equilibrium,** as it may be shown that the equilibrium position is the result of opposing reactions of equal rates rather than a cessation of reaction.

The dynamic nature of the equilibrium in the reaction of hydrogen and iodine may be demonstrated by using a radioactive isotope of hydrogen, designated H*. A trace amount of H*I is introduced into the equilibrium mixture. After a short time, the HI and I_2 can be frozen out by using a liquid air bath and the gaseous hydrogen can be removed. If the reaction had ceased, all of the radioactive hydrogen would still be in the form of H*I. However, in addition to ordinary hydrogen molecules, the gaseous hydrogen is found to contain H*H and H*H* molecules, showing that reaction has continued to occur.

The equilibrium concentrations (denoted [ ]$_e$) at 425°C for two different reaction mixtures of H_2 and I_2 and for two different initial concentrations of HI are shown in Table 12.1. It is evident that the equilibrium concentrations of H_2, I_2, and HI are dependent on the initial concentrations of the reaction mixture. However there is an extremely simple relationship between the equilibrium concentrations. The quantity $[HI]_e^2/[H_2]_e[I_2]_e$ is a constant at 425°C. This constancy is shown in Table 12.1 where values of

$$K = \frac{[HI]_e^2}{[H_2]_e[I_2]_e} \tag{12.3}$$

are tabulated for each of the experiments listed. Although the values of K vary somewhat owing to experimental error, the values are reasonably constant. K is known as the **equilibrium constant** for the reaction, and its value has been found to be dependent only on the temperature.

(Note that only equilibrium concentrations may be used to calculate K. Ordinarily these concentrations are expressed in moles per liter.)

Experimental investigations of many chemical reactions together with theoretical considerations (Chapter 15) have shown that all reactions tend to approach an equilibrium position governed by an equilibrium constant. To generalize the concept of equilibrium, let us consider the following general equation for any chemical reaction:

$$aA + bB + \cdots \rightleftharpoons cC + dD + \cdots \tag{12.4}$$

TABLE 12.1

H_2, I_2, HI SYSTEM AT 425°C						
$10^3 \times$ Initial concentrations (mol L^{-1})			$10^3 \times$ Equilibrium concentrations (mol L^{-1})			
$[H_2]_i$	$[I_2]_i$	$[HI]_i$	$[H_2]_e$	$[I_2]_e$	$[HI]_e$	K
5.345	5.345	0	1.140	1.140	8.410	54.4
0	0	10.690	1.140	1.140	8.410	54.4
0	0	4.650	0.495	0.495	3.660	54.7
11.33	7.51	0	4.56	0.738	13.54	54.5

12.1 GAS PHASE EQUILIBRIA

(Capital letters represent chemical substances, small letters their numerical coefficients. The dots indicate that as many symbols are included as are required to represent all of the reactants and products.) The **law of chemical equilibrium** states that *the mathematical product of the equilibrium concentrations of the products of the reaction, with each concentration raised to a power equal to its coefficient in the equation, divided by the mathematical product of the equilibrium concentrations of the reactants, with each concentration raised to a power equal to its coefficient in the equation, is equal to a constant.* Symbolically, the law of chemical equilibrium[1] is

$$\frac{[C]_e^c[D]_e^d \cdots}{[A]_e^a[B]_e^b \cdots} = K \tag{12.5}$$

EXAMPLE 1

Express the law of chemical equilibrium for each of the following reactions:
(a) $2H_2(g) + O_2(g) = 2H_2O(g)$
(b) $C_2H_5OH(g) + 3O_2(g) = 2CO_2(g) + 3H_2O(g)$

For (a),

$$K = \frac{[H_2O]^2}{[H_2]^2[O_2]}$$

For (b),

$$K = \frac{[CO_2]^2[H_2O]^3}{[C_2H_5OH][O_2]^3}$$

(Each chemical reaction has a specific numerical value for its equilibrium constant at every temperature. Thus the K's for these two reactions differ from each other and differ from K for the hydrogen–iodine reaction. However, it is convenient to use K as a general symbol for any equilibrium constant. When it is necessary to distinguish between different K's, subscripts or prime marks [K's] can be used.)

Many reactions occur very slowly, so it is not permissible to assume that a system is at equilibrium simply because there is no change in the concentrations of the species making up the system in a reasonable time period. Either one of the following tests may be applied to ascertain whether or not a system is at equilibrium.

1. A system is at equilibrium if one can prepare the same system (each species at the same concentration) starting with either the reactants or the products of the reaction.

---

[1] Also known as the law of mass action or simply the mass law, because, at the time of its discovery (in 1863 by Cato M. Guldberg and Peter Waage), the expression "active mass" was used for concentration.

2. If the same value of the equilibrium constant is obtained for a series of reaction mixtures in which the initial concentration of each reactant and product is varied, then each mixture is at equilibrium at that temperature.

The magnitude of the equilibrium constant is a measure of the extent of reaction at equilibrium. A large value of K signifies that the equilibrium concentration of the products is much greater than that of the reactants, that is, the reaction goes nearly to completion. A value of K near unity implies that at equilibrium appreciable concentrations of both reactants and products are present. A small value of K signifies that the reaction progresses to only a slight extent; that is, at equilibrium the reaction forms only small amounts of products.

It is necessary to associate the equilibrium expression with the direction in which the reaction is written. For example, the formation of HI at 425°C is written

$$H_2 + I_2 \rightleftharpoons 2HI \qquad K = \frac{[HI]_e^2}{[H_2]_e[I_2]_e} = 54.5$$

whereas for the reverse reaction (dissociation),

$$2HI \rightleftharpoons H_2 + I_2 \qquad K' = \frac{[H_2]_e[I_2]_e}{[HI]_e^2} = \frac{1}{K} = 0.0183$$

Thus the equilibrium constant for the reverse reaction (K') is simply the reciprocal of the equilibrium constant for the forward direction (K).

Often the partial pressures of gases are used instead of their concentrations in equilibrium expressions. If we assume that the ideal gas law is valid for the gases, then the concentration of each gas in moles per liter (n/V) is proportional to its partial pressure at constant temperature because $n/V = P/RT$. Note that in the HI equilibrium (Eq. 12.2) the numerical value of the equilibrium constant is independent of the choice of concentration units or partial pressures since

$$K = \frac{[HI]_e^2}{[H_2]_e[I_2]_e} = \frac{\left[\frac{P_{HI}}{RT}\right]^2}{\left[\frac{P_{H_2}}{RT}\right]\left[\frac{P_{I_2}}{RT}\right]} = \frac{P_{HI}^2}{P_{H_2}P_{I_2}}$$

This situation obtains only when the number of molecules of gaseous reactants is equal to the number of molecules of gaseous products. Otherwise the RT terms will not cancel and the numerical value of K is dependent on the choice of units. For clarity, the symbol K_c is sometimes used to indicate the use of concentrations; the symbol K_p indicates the use of partial pressures.

12.2 CALCULATION OF EQUILIBRIUM CONCENTRATIONS

All chemical reactions proceed spontaneously toward an equilibrium condition. Once the equilibrium position is attained, the system remains unchanged unless disturbed by some outside influence, in which case the system will move toward a new equilibrium position. We may make use of the law of chemical equilibrium to determine the direction in which reaction will occur in any given mixture of H_2, I_2, and HI. Let us define a concentration quotient (Q) as having the same form as the equilibrium expression except that equilibrium concentrations are not required to be used. Thus

$$H_2 + I_2 \rightleftharpoons 2HI \qquad Q = \frac{[HI]_i^2}{[H_2]_i[I_2]_i}$$

where we can now employ the initial concentration [ ]$_i$ of the substances used to prepare the reaction mixture. To determine what happens under these conditions, we compare the numerical value of Q with the numerical value of K at the same temperature. If Q is less than K, Q must increase with time until $Q = K$. Thus the reaction will shift toward the right and more HI will be formed. For $Q > K$, the value of Q will decrease with time until $Q = K$. Thus the reaction will shift toward the left and more H_2 and I_2 will be formed. If initially $Q = K$, an equilibrium mixture has been prepared, and no changes in the concentrations will occur.

EXAMPLE 2

If 2.0×10^{-3} mol of hydrogen, 5.0×10^{-2} mol of iodine, and 4.0×10^{-3} mol of HI are placed in a 1.0-L flask at 425°C, will more HI be formed or will some HI dissociate?

To solve the problem, we first calculate Q.

$$Q = \frac{[HI]_i^2}{[H_2]_i[I_2]_i} = \frac{(4.0 \times 10^{-3})^2}{(2.0 \times 10^{-3})(5.0 \times 10^{-2})} = 0.16$$

Since Q is less than $K = 54.5$, more HI will form.

If the numerical value of the equilibrium constant is known, it is possible to calculate the equilibrium concentrations that will result from any mixture of the reactants and products at the same temperature.

EXAMPLE 3

Calculate the equilibrium concentrations of H_2, I_2, and HI if 0.200 mol of H_2 and 0.200 mol of I_2 are placed in a 1.00-L flask at 425°C.

The reaction $H_2 + I_2 = 2HI$ would occur, and the concentration of HI would increase until it attained its equilibrium value. For each mole of H_2 that reacts, 1 mol of I_2 also reacts and 2 mol of HI is formed. Thus, if at equilibrium x mol of H_2 have reacted, there will be $(0.200 - x)$ mol of H_2, $(0.200 - x)$ mol of I_2, and $2x$ mol of HI.

Initially	At equilibrium
$[H_2]_i = 0.200$ mol L^{-1}	$[H_2]_e = (0.200 - x)$ mol L^{-1}
$[I_2]_i = 0.200$ mol L^{-1}	$[I_2]_e = (0.200 - x)$ mol L^{-1}
$[HI]_i = 0$	$[HI]_e = 2x$ mol L^{-1}

At equilibrium,

$$\frac{[HI]_e^2}{[H_2]_e[I_2]_e} = 54.5$$

Therefore

$$\frac{(2x)^2}{(0.200 - x)^2} = 54.5$$

To solve for x we take the square root of both sides of the equation and obtain

$$\frac{2x}{0.200 - x} = \pm 7.38$$

We choose the plus value on the right since the minus value leads to negative concentrations of H_2 and I_2, which are physically impossible.

$$2x = 1.476 - 7.38x$$

$$x = \frac{1.476}{9.38} = 0.157$$

$[HI]_e = 2x = 0.314$ mol L^{-1}

$[H_2]_e = [I_2]_e = 0.200 - 0.157 = 0.043$ mol L^{-1}

In this section we have used the subscript e to emphasize that equilibrium concentrations must be used with the law of chemical equilibrium. However, in the future we shall drop the subscript e in accord with common usage in chemistry.

12.3 CONVENTIONS USED FOR EQUILIBRIUM CONSTANTS

The law of chemical equilibrium as expressed in Eq. 12.5 is valid for ideal gases and ideal solutions. For real gases and real solutions, the value of K shows some variation with the concentrations although K approaches a constant value at low concentrations. Over the years, conventions have been developed for expressing the law of chemical equilibrium more precisely and for using it to calculate the equilibrium concentrations accurately even in cases of large deviations from ideality. In its rigorous form the law of chemical equilibrium is expressed in the same form as Eq. 12.5 but with activities in place of concentrations. The **activity**, *a*, of a chemical substance is defined as the ratio of its effective concentration in the system of interest to its effective concentration in a reference state, or **standard state.** The activity so defined takes into account any deviations from ideality. (We shall not consider just how the "effective concentration" takes into account deviations from ideality, as this subject is best left to a course in physical chemistry.) Because activities are dimensionless ratios, it follows that the equilibrium constant is also dimensionless. However, the numerical value of the equilibrium constant depends on the choice of standard states. The following conventions are commonly used.

1. The standard state for a *solid* or a *liquid substance* is the pure solid or liquid at 1 atm pressure, and thus the activity of a pure solid or liquid is exactly 1.
2. For an *ideal gas* the standard state is the gas at 1 atm pressure. Thus the activity of an ideal gas at 1 atm pressure is 1, and the activity of the ideal gas at other pressures is *numerically* equal to its pressure in atmospheres. Because *real gases* at pressures up to a few atmospheres usually deviate from ideality to only a small extent, we shall ordinarily neglect deviations from ideality and consider the activity of a real gas to be numerically equal to its pressure in atmospheres. (For a gaseous solution, the activity of each gaseous substance is numerically equal to its *partial* pressure in atmospheres.)
3. For solutions we shall choose the *pure solvent* as the standard state for the solvent. In an ideal solution the activity of the solvent is equal to its mole fraction. For the solvent in a nonideal solution we shall neglect deviations from ideality and assume that the activity of the solvent is equal to its mole fraction.
4. The choice of a standard state for the *solute* is complicated by the dissociation of many solutes and by the practice of reporting solute concentrations in terms of molalities (or molarities) rather than as mole fractions.[2] To simplify matters, we shall assume that the activity of each solute is numerically equal to its molality, whether the solution is ideal or nonideal. (In dilute *aqueous* solutions, the molality and molarity are approximately equal. For such solutions, we shall often assume that the activity of each solute is numerically equal to its molarity.)

The conventions outlined above are summarized in Table 12.2. Given these conventions, we shall express pressures as atmospheres, solution concentrations as

---

[2] For a precise treatment, one must extrapolate the behavior of the solute at ordinary concentrations to extremely dilute solutions.

TABLE 12.2	CONVENTIONS FOR EXPRESSING ACTIVITIES
System	Activity
Pure solid	Exactly 1
Pure liquid	Exactly 1
Gas	Numerical value of pressure in atmospheres
Solution:	
Solvent	Mole fraction of solvent
Solute	Numerical value of molality of solute

molalities (or molarities for dilute aqueous solutions), and we shall not include units with the numerical value of K.

Many real solutions exhibit quite large departures from ideal behavior at concentrations even much less than $1m$. However, all solutions approach ideal behavior as they become more dilute, just as real gases more closely approach the behavior of ideal gases at low pressure. Thus, although we are neglecting deviations from ideality of real gases and real solutions, the results of our calculations will usually be reasonably accurate, particularly so for gases at pressures up to a few atmospheres and for dilute solutions, and any errors will not often affect our conclusions. (To obtain more accurate results, one must, with considerable additional effort, include activity corrections in the calculations.)

The following example illustrates the use of the above conventions in writing the law of chemical equilibrium.

EXAMPLE 4

Write the law of chemical equilibrium for the reaction of solid sodium with an excess of water,

$$2Na(s) + 2H_2O(l) = 2Na^+ + 2OH^- + H_2(g)$$

Because sodium is a pure solid its activity is 1. For a dilute aqueous solution the mole fraction of water is approximately 1, and thus the activity of water is approximately 1. Therefore the law of chemical equilibrium can be written:

$$K = \frac{(a_{Na^+})^2(a_{OH^-})^2(a_{H_2})}{(a_{Na})^2(a_{H_2O})^2} \approx [Na^+]^2[OH^-]^2 P_{H_2}$$

To evaluate K we would use the numerical value of the equilibrium molal concentrations (or ordinarily the molar concentrations) of Na^+ and OH^-, and the numerical value of the equilibrium partial pressure of H_2 in atmospheres.

12.4 LE CHÂTELIER'S PRINCIPLE

Let us now investigate methods of controlling the equilibrium position of chemical reactions to produce a desired effect such as increasing the amount of product formed. This may be done qualitatively by using the equilibrium expression or by applying the following principle,[3] enunciated by Henri Louis Le Châtelier in 1884: *If a change is made in any of the factors influencing a system at equilibrium, reaction will occur in the direction that tends to counteract the change made.* The factors influencing an equilibrium are the concentrations or partial pressures of the reactants and products and the temperature. First let us examine the effect of changes in the partial pressures.

At 500°C, ammonia may be synthesized from nitrogen and hydrogen by the reaction

$$N_2(g) + 3H_2(g) \rightleftharpoons 2NH_3(g) \qquad K = \frac{P_{NH_3}^2}{P_{N_2} P_{H_2}^3} = 3.8 \times 10^{-3}$$

Owing to the small value of K, only a small partial pressure of ammonia is produced before the equilibrium position is reached. How, then, can we get more ammonia to form? We note that, when $Q < K$, the reaction goes to the right until $Q = K$. To decrease the value of Q, we can decrease P_{NH_3} by passing the gases through water (which absorbs NH_3 but not appreciable amounts of N_2 or H_2) or by increasing P_{H_2} by adding more H_2. We can arrive at the same conclusion using Le Châtelier's principle. If we decrease P_{NH_3} without affecting P_{H_2} or P_{N_2}, a reaction occurs tending to increase P_{NH_3}; that is, the equilibrium shifts to the right. If P_{H_2} is increased, a reaction occurs tending to decrease P_{H_2} and again the equilibrium shifts to the right.

We now consider the effect of an increase in pressure on the position of the NH_3 equilibrium. If the pressure is increased by compressing the gases to a smaller volume, all the partial pressures will increase by the same factor. As this factor is raised to the fourth power in the denominator, and only to the second power in the numerator, Q will be less than K. Therefore more NH_3 will form until $Q = K$. Or we could note that the pressure was increased so that reaction would occur in the direction that tended to decrease the pressure, that is, to decrease the total number of molecules, and the equilibrium shifts to the right.

However, one must be careful, in applying Le Châtelier's principle, not to generalize the above reasoning to say that *any* increase in pressure causes the equilibrium to shift in the direction that minimizes the number of molecules in the system. For example, let us increase the pressure in the system by adding an unreactive gas such as helium. The partial pressures of reactants and products do not change and hence the equilibrium position is not altered. In this case the higher pressure is caused by a substance (helium) which is not involved in the chemical reaction, and therefore the higher pressure does not affect the equilibrium of that reaction.

---

[3]This principle was proposed independently in 1887 by Ferdinand Braun.

EXAMPLE 5

Assuming that the temperature is held constant, describe how the equilibrium position of the reaction

$$2NO_2(g) \rightleftharpoons 2NO(g) + O_2(g)$$

would be altered by

(a) increasing the partial pressure of NO_2
(b) decreasing the partial pressure of O_2
(c) compressing the gases to a smaller volume by increasing the pressure on the system

If the partial pressure of NO_2 is increased, reaction will occur tending to decrease P_{NO_2}. Therefore, the equilibrium will shift to the right, forming more NO and O_2.

If the partial pressure of O_2 is decreased, reaction will occur tending to increase P_{O_2}. Therefore the equilibrium will shift to the right, forming more NO and O_2.

If the system is compressed to a smaller volume, resulting in an increase in the total pressure, reaction will occur tending to decrease the total pressure, and therefore tending to decrease the total number of molecules in the system. As there are two molecules on the left side of the equation and three molecules on the right side, a shift of the equilibrium to the left reduces the total number of molecules. Thus the equilibrium shifts to the left, forming more NO_2.

Now let us examine the effect of temperature on an equilibrium. (Recall that some reactions, called exothermic, occur with the evolution of heat, and have negative ΔH values. Other reactions, called endothermic, occur with the absorption of heat, and have positive ΔH values.) The formation of hydrogen iodide from its elements (Eq. 12.1) is an exothermic reaction ($\Delta H = -12.6$ kJ):

$$H_2(g) + I_2(g) = 2HI(g) + 12.6 \text{ kJ}$$

From Le Châtelier's principle, we predict that increasing the temperature (thus increasing the energy of the system) will cause the equilibrium to shift in the direction that absorbs energy, that is, to the left. Therefore at higher temperature the equilibrium partial pressure of HI will be smaller, and the equilibrium constant will also be smaller. The experimental values for K are in accord with this prediction: K decreases from 54.5 at 425°C to 45.9 at 490°C.

An example of an endothermic reaction is the formation of nitric oxide (NO) from its elements ($\Delta H = 180$ kJ):

$$N_2(g) + O_2(g) = 2NO(g) - 180 \text{ kJ}$$

As before, increasing the temperature causes the equilibrium to shift in the direction that absorbs energy. Therefore we predict that an increase in temperature will cause a greater amount of NO to be formed and that the value of the equilibrium constant will increase. Again our prediction is in accord with the experimental values: K increases from 0.86×10^{-4} at 1811 K to 64×10^{-4} at 2675 K.

To generalize these results, K *decreases* with increasing temperature for an exothermic reaction, and K *increases* with increasing temperature for an endothermic reaction. For the special case of a reaction for which $\Delta H = 0$, an increase in temperature should not cause any change in the amount of product formed nor in the equilibrium constant. This has also been confirmed by experiment.

Le Châtelier's principle also explains the different slopes of solid–liquid lines in phase diagrams (Section 10.2). In general, an increase in pressure shifts the equilibrium toward the phase occupying the smaller volume, thus tending to counteract the increase in pressure. In the case of the water–ice equilibrium, liquid water has the higher density and occupies the smaller volume. Thus an increase in pressure shifts the equilibrium toward liquid water. To form ice, the temperature must be reduced. Thus the solid–liquid line shifts toward lower temperature with increasing pressure (see Figure 10.4).

For most other substances, including carbon dioxide, the solid phase has higher density than the liquid, and the opposite type of behavior is shown. The solid–liquid line shifts toward higher temperature with increasing pressure (see Figure 10.5).

12.5 DISSOCIATION OF WEAK ELECTROLYTES

Although chemists are interested in chemical reactions in all solvents, the chemical reactions that occur in aqueous systems are of great importance because nearly all biochemical processes, such as the formation and growth of living cells, require the presence of water. These biochemical processes depend on many chemical reactions that, in common with other chemical reactions, may be markedly influenced by the concentration of hydrogen ion in the medium. In fact plants and animals, including man, cannot survive if the hydrogen ion concentration of the medium in which the biological processes occur varies markedly from the normal value. Thus the study of aqueous solutions of acids and bases is of extreme importance to all of us and, in addition, may be used to illustrate the methods applicable to equilibria in general.

As an example of a weak electrolyte, let us consider acetic acid (CH_3CO_2H), which contains the acidic —CO_2H group. As we noted in Chapter 10, aqueous solutions of acetic acid do contain a small concentration of ions owing to dissociation of the acetic acid. Thus

$$CH_3CO_2H \rightleftharpoons CH_3CO_2^- + H^+ \qquad K = \frac{[CH_3CO_2^-][H^+]}{[CH_3CO_2H]} \qquad (12.6)$$

The equilibrium constant for this reaction is called the **dissociation constant** of acetic acid. Table 12.3 presents some experimental evidence that K is a constant for solutions containing different equilibrium concentrations of the ions and the

TABLE 12.3	EQUILIBRIUM CONCENTRATIONS IN ACETIC ACID SOLUTIONS AT 25°C			
Total concentration $10^3 \times c_i$ (mol L^{-1})	$10^3 \times$ [H$^+$] (mol L^{-1})	α	$10^5 \times K$	
0.0280	0.0151	0.539	1.77	
0.1114	0.0365	0.328	1.78	
1.0283	0.1273	0.124	1.80	
5.9115	0.3193	0.0540	1.82	
20.000	0.5975	0.0299	1.84	
100.00	1.3496	0.0135	1.85	
200.00	1.899	0.0095	1.82	

undissociated acid. Column 1 shows the molar concentration of acetic acid (c_i) before dissociation occurs. (This is based on the amount of acetic acid used to prepare the solution.) Column 2 gives the equilibrium concentration of hydrogen ion determined by conductivity measurements as well as by other experimental methods. The degree of dissociation (α) is the fraction of the acetic acid that has dissociated to ions at equilibrium:

$$\alpha = \frac{\text{concentration of molecules that dissociate}}{\text{concentration of molecules before dissociation}} = \frac{[\text{H}^+]}{c_i} \qquad (12.7)$$

The values of α shown in column 3 of Table 12.3 indicate that the fraction of acetic acid in the form of ions decreases with increasing concentration of acetic acid. It is clear that one cannot give any fixed α as characteristic of all acetic acid solutions. The fraction of acetic acid molecules that dissociate varies from over 50% for the most dilute solution to less than 1% for the most concentrated solution listed in Table 12.3. As we shall see, the property that does *not* vary is the dissociation constant, K.

The values of the dissociation constant in column 4 may readily be obtained from the values of c_i and [H$^+$] by noting that the dissociation of one molecule of acetic acid produces one H$^+$ ion and one CH$_3$CO$_2^-$ ion so that, if there is no other source[4] of H$^+$ or CH$_3$CO$_2^-$, then

$$[\text{H}^+] = [\text{CH}_3\text{CO}_2^-]$$

Since the concentration of H$^+$ is equal to the concentration of acetic acid molecules that dissociated, and the sum of the dissociated and undissociated acetic acid concentrations equals the concentration of acetic acid before dissociation,

$$[\text{H}^+] + [\text{CH}_3\text{CO}_2\text{H}] = c_i$$

[4] As we shall see in Chapter 13, the amount of H$^+$ produced by dissociation of water is negligible compared to that resulting from dissociation of the acetic acid.

12.5 DISSOCIATION OF WEAK ELECTROLYTES

or

$$[CH_3CO_2H] = c_i - [H^+]$$

Substituting these expressions into Eq. 12.6 yields an equation for calculating K from the experimental data,

$$K = \frac{[H^+]^2}{c_i - [H^+]} \tag{12.8}$$

As shown in Table 12.3, values of K calculated from Eq. 12.8 are reasonably constant over the concentration range investigated ($2.8 \times 10^{-5} M$ to $0.200 M$ acetic acid). The best average value of K is 1.8×10^{-5}.

Let us now see how to calculate the equilibrium concentrations of all the species involved in the dissociation of a weak electrolyte, providing we know the initial composition of the solution and the numerical value of K.

EXAMPLE 6

Calculate the equilibrium concentration of H^+, $CH_3CO_2^-$, and CH_3CO_2H in aqueous $0.200 M$ CH_3CO_2H. Use $K = 1.8 \times 10^{-5}$ for the dissociation constant of acetic acid.

We know that some acetic acid molecules will dissociate to form ions and that the equilibrium shown in Eq. 12.6 will be established. Let us assume that the only source of H^+ and $CH_3CO_2^-$ is the dissociation of acetic acid. If in 1 L of solution y mol of acetic acid dissociate, y mol of H^+ and y mol of $CH_3CO_2^-$ will be formed, leaving $0.200 - y$ mol of CH_3CO_2H.

Initial concentrations	Equilibrium concentrations
$[CH_3CO_2H]_i = 0.200$	$[CH_3CO_2H] = 0.200 - y$
$[H^+]_i = 0$	$[H^+] = y$
$[CH_3CO_2^-]_i = 0$	$[CH_3CO_2^-] = y$

We now place these equilibrium concentrations in Eq. 12.6.

$$\frac{y^2}{0.200 - y} = 1.8 \times 10^{-5}$$

or

$$y^2 + 1.8 \times 10^{-5} y - 3.6 \times 10^{-5} = 0$$

Equations of the type $ay^2 + by + c = 0$ may be solved by using the quadratic formula:

$$y = \frac{-b \pm \sqrt{b^2 - 4ac}}{2a}$$

In our example, $a = 1$, $b = 1.8 \times 10^{-5}$, and $c = -3.6 \times 10^{-6}$. Therefore

$$y = \frac{-1.8 \times 10^{-5} \pm \sqrt{(3.24 \times 10^{-10}) + (14.4 \times 10^{-6})}}{2}$$

$$= \frac{-1.8 \times 10^{-5} \pm \sqrt{14.4 \times 10^{-6}}}{2}$$

$$= \frac{-1.8 \times 10^{-5} \pm 3.8 \times 10^{-3}}{2} = 1.9 \times 10^{-3}$$

To obtain a positive value for [H$^+$], we must choose the plus sign in the term on the right. Thus $[H^+] = [CH_3CO_2^-] = 1.9 \times 10^{-3} M$.

We note that in the above solution the b and b^2 terms have essentially no effect on the result. This suggests a simpler method of obtaining an appropriate solution. Because K is small, CH_3CO_2H is only slightly ionized, and [H$^+$] or y must be small. As a first approximation, let us assume that y is negligible compared to 0.200, so that $0.200 - y \approx 0.200$. Under these conditions,

$$\frac{y^2}{0.200 - y} \approx \frac{y^2}{0.200} \approx 1.8 \times 10^{-5}$$

and

$$y^2 \approx 0.200 \times 1.8 \times 10^{-5} = 3.6 \times 10^{-6}$$
$$y \approx (3.6 \times 10^{-6})^{1/2} = 1.9 \times 10^{-3}$$

We now check our approximation to see if it is valid: $0.200 - y = 0.200 - (1.9 \times 10^{-3}) = 0.198$. As we assumed 0.200, our error was only 1%. Because K is given only to two significant figures and deviations from ideal behavior are ignored, we shall accept approximations that do not lead to more than a 10% error. Thus the equilibrium concentrations are

$$[H^+] = [CH_3CO_2^-] = 1.9 \times 10^{-3} M$$

and

$$[CH_3CO_2H] = 0.200 - 1.9 \times 10^{-3} \approx 0.20 M$$

Additional examples of the dissociation of weak electrolytes will be discussed in the next chapter.

12.6 HETEROGENEOUS EQUILIBRIA

An equilibrium that involves substances in more than one phase is classified as heterogeneous. A simple example, which we have previously considered, is the vapor pressure of a liquid:

$$H_2O(l) \rightleftharpoons H_2O(g) \qquad K = \frac{a_{H_2O(g)}}{a_{H_2O(l)}} \approx P_{H_2O(g)}$$

The activity of gaseous water is equal to its pressure, and the activity of liquid water is unity (Table 12.2). Thus the pressure of water in the gas phase (or vapor pressure) in equilibrium with liquid water is a constant at a given temperature, as noted in Chapter 10.

Let us now consider the solubility of a solid, such as iodine, in liquid carbon tetrachloride:

$$I_2(s) \rightleftharpoons I_2(\text{in } CCl_4) \qquad K = \frac{a_{I_2(\text{in } CCl_4)}}{a_{I_2(s)}} \approx m_{I_2}$$

Because the activity of solid iodine is 1, the molal concentration of iodine in a solution in equilibrium with solid iodine is a constant, which is in agreement with the experimental observation that the equilibrium concentration of the solute (iodine) in a saturated solution is independent of the amount of excess solute in contact with the saturated solution.

The situation is a little more complex if the dissolved solid dissociates into ions. As most salts in dilute aqueous solution are completely dissociated, we can write an equation expressing the equilibrium between a solid compound and its aqueous solution:

$$AgCl(s) \rightleftharpoons Ag^+ + Cl^- \qquad K = \frac{(a_{Ag^+})(a_{Cl^-})}{a_{AgCl(s)}} \approx [Ag^+][Cl^-]$$

$$Ca(IO_3)_2(s) \rightleftharpoons Ca^{2+} + 2IO_3^- \qquad K = \frac{(a_{Ca^{2+}})(a_{IO_3^-})^2}{a_{Ca(IO_3)_2(s)}} \approx [Ca^{2+}][IO_3^-]^2$$

$$La(IO_3)_3(s) \rightleftharpoons La^{3+} + 3IO_3^- \qquad K = \frac{(a_{La^{3+}})(a_{IO_3^-})^3}{a_{La(IO_3)_3(s)}} \approx [La^{3+}][IO_3^-]^3$$

The equilibrium constant for each of the above reactions is equal to the product of the numerical values of the concentrations of the ions (each with an appropriate exponent). For this reason the equilibrium constant is commonly called the **solubility product,** K_{sp}. Therefore we may write

$$K_{sp} = [Ag^+][Cl^-]$$
$$K_{sp} = [Ca^{2+}][IO_3^-]^2$$
$$K_{sp} = [La^{3+}][IO_3^-]^3$$

in place of the equilibrium expressions given above.

The magnitude of K_{sp} for each substance must be determined by experiment. For example, the solubility of silver chloride in water at 25°C is 1.3×10^{-5} mol L^{-1} of solution; therefore $[Ag^+] = [Cl^-] = 1.3 \times 10^{-5} M$ and

$$K_{sp} \text{ for AgCl} = (1.3 \times 10^{-5})(1.3 \times 10^{-5}) = 1.7 \times 10^{-10}$$

This value of K_{sp} may then be used for any aqueous solution in equilibrium with solid silver chloride at 25°C.

Values of the solubility product of a few compounds are listed in Table 12.4. From the value for K_{sp}, one can calculate the solubility of the compound as in the following example.

EXAMPLE 7

Using the solubility product of calcium iodate listed in Table 12.4, calculate the solubility of calcium iodate in water at 25°C.

$$Ca(IO_3)_2(s) \rightleftharpoons Ca^{2+} + 2IO_3^- \quad K_{sp} = [Ca^{2+}][IO_3^-]^2 = 1.7 \times 10^{-6}$$

Let the solubility of $Ca(IO_3)_2$ be s mol L^{-1}. Since all the s mol of dissolved $Ca(IO_3)_2$ will be dissociated, $[Ca^{2+}] = s$ and $[IO_3^-] = 2s$. Therefore

$$[Ca^{2+}][IO_3^-]^2 = s(2s)^2 = 4s^3 = 1.7 \times 10^{-6}$$

and

$$s = \sqrt[3]{1.7 \times 10^{-6}/4} = 7.5 \times 10^{-3}$$

The solubility of $Ca(IO_3)_2$ is therefore $7.5 \times 10^{-3} M$.

When applying the concept of the solubility product, one must distinguish between three different cases. For solutions containing Ag^+ and Cl^-, these cases are:

1. If $[Ag^+][Cl^-] > K_{sp}$, the system is not at equilibrium. Precipitation of AgCl will occur, reducing the concentrations of Ag^+ and Cl^- until $[Ag^+][Cl^-] = K_{sp}$.
2. If $[Ag^+][Cl^-] < K_{sp}$, the solution is not saturated with AgCl. If solid AgCl is present or is added to the solution, some will dissolve until $[Ag^+][Cl^-] = K_{sp}$.
3. If $[Ag^+][Cl^-] = K_{sp}$, the solution is saturated with AgCl, that is, equilibrium conditions are established.

These three cases are illustrated in Figure 12.2. Any saturated solution of silver chloride is represented by a point on the curve. (A saturated solution prepared by dissolving silver chloride in water has equal concentrations of Ag^+ and Cl^-; it is represented by point E.) Points in the region to the upper right of the curve represent solutions containing too much Ag^+ and Cl^-—precipitation will occur. Points in the region to the lower left of the curve represent unsaturated solutions.

It is often desired to reduce the concentration of one ion to a certain level, and, in accordance with Le Châtelier's principle, this may be accomplished by

TABLE 12.4 SOLUBILITY PRODUCTS AT 25°C

Compound	K_{sp}
Carbonates	
Ag_2CO_3	8.1×10^{-12}
$BaCO_3$	5.0×10^{-9}
$CaCO_3$	4.5×10^{-9}
$CuCO_3$	2.3×10^{-10}
$FeCO_3$	2.1×10^{-11}
Hg_2CO_3 [a]	8.9×10^{-17}
$PbCO_3$	9.8×10^{-12}
$ZnCO_3$	1.0×10^{-10}
Halides	
$AgBr$	5.0×10^{-13}
$AgCl$	1.7×10^{-10}
AgI	4.5×10^{-17}
Hg_2Cl_2 [b]	1.2×10^{-18}
$PbCl_2$	1.7×10^{-5}
Hydroxides or oxides	
Ag_2O [c]	2.0×10^{-8}
$Al(OH)_3$	3×10^{-34}
$Ba(OH)_2$	2.5×10^{-4}
$Ca(OH)_2$	6×10^{-6}
$Cu(OH)_2$	4.8×10^{-20}
$Fe(OH)_2$	7.9×10^{-16}
$Fe(OH)_3$	2×10^{-39}
HgO [d]	4×10^{-26}
$Pb(OH)_2$	1.3×10^{-15}
$Zn(OH)_2$	4×10^{-17}
Iodates	
$AgIO_3$	3.1×10^{-8}
$Ca(IO_3)_2$	1.7×10^{-6}
$La(IO_3)_3$	6.2×10^{-12}
Sulfates	
Ag_2SO_4	1.5×10^{-5}
$BaSO_4$	1.1×10^{-10}
$CaSO_4$	2.4×10^{-5}
$PbSO_4$	1.6×10^{-8}
Sulfides	
Ag_2S	8×10^{-51}
CdS	1×10^{-27}
CoS	9×10^{-22}
CuS	8×10^{-37}
FeS	8×10^{-19}
HgS	2×10^{-53}
NiS	4×10^{-20}
PbS	3×10^{-28}
SnS	1×10^{-26}
ZnS	3×10^{-23}

[a] $Hg_2CO_3(s) = Hg_2^{2+} + CO_3^{2-}$ $K_{sp} = [Hg_2^{2+}][CO_3^{2-}]$
[b] $Hg_2Cl_2(s) = Hg_2^{2+} + 2Cl^-$ $K_{sp} = [Hg_2^{2+}][Cl^-]^2$
[c] $\frac{1}{2}Ag_2O(s) + \frac{1}{2}H_2O = Ag^+ + OH^-$ $K_{sp} = [Ag^+][OH^-]$
[d] $HgO(s) + H_2O = Hg^{2+} + 2OH^-$ $K_{sp} = [Hg^{2+}][OH^-]^2$

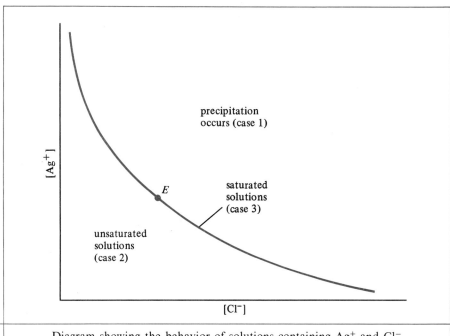

FIGURE 12.2 Diagram showing the behavior of solutions containing Ag^+ and Cl^-.

increasing the concentration of the other ion. For example, if solid silver chloride is in equilibrium with a saturated solution,

$$AgCl(s) \rightleftharpoons Ag^+ + Cl^-$$

and if the concentration of Ag^+ is increased (by addition of a soluble silver salt such as $AgNO_3$), reaction will occur in the direction that tends to decrease the concentration of Ag^+, and the equilibrium will shift to the left. This example of Le Châtelier's principle is called the **common-ion effect.**

EXAMPLE 8

Calculate the solubility of AgCl in $0.010M$ $AgNO_3$ at 25°C, using the solubility product of silver chloride listed in Table 12.4.

Let the solubility of AgCl be s mol L^{-1}. Since the s mol of dissolved AgCl is completely dissociated, and the only source of Cl^- is from dissolved AgCl, we note that $[Cl^-] = s$. However, the silver ion in solution comes from two sources: There is $0.010M$ Ag^+ from dissociation of $AgNO_3$, and s M Ag^+ from dissociation of dissolved AgCl.

$[Ag^+] = 0.010 + s$

> From the small value of K_{sp} we know that s will be small. Let us assume that $0.010 + s \approx 0.010$.
>
> If these values of [Ag$^+$] and [Cl$^-$] are substituted into the expression for K_{sp}, we obtain
>
> $$(0.010)(s) = 1.7 \times 10^{-10}$$
>
> and
>
> $$s = 1.7 \times 10^{-8}$$
>
> Note that our assumption is justified since $(0.010) + (1.7 \times 10^{-8}) \approx 0.010$. The solubility of AgCl in $0.010M$ AgNO$_3$ is therefore $1.7 \times 10^{-8} M$.

The same principles may be applied to the solubility of other electrolytes in water. As in the case of the dissociation of weak acids, we have assumed ideal behavior so that concentrations of ions can be used in place of activities. For this reason the above treatment applies only to slightly soluble substances. If ionic concentrations are high, there may be considerable deviations from ideal behavior, and it is necessary to use more complicated methods.

It is interesting to note that the solubility of many electrolytes may be strongly influenced by the presence of other ions in the solution. For example, many hydroxides, such as Cu(OH)$_2$, are only slightly soluble in water, but are very soluble in HCl solutions because the H$^+$ reacts with the OH$^-$ from the Cu(OH)$_2$, forming water. The reactions are

$$Cu(OH)_2(s) \rightleftharpoons Cu^{2+} + 2OH^-$$

$$OH^- + H^+ \rightleftharpoons H_2O$$

By applying Le Châtelier's principle, we see that as H$^+$ removes a product of the first reaction, it will shift the equilibrium of the first reaction to the right. As OH$^-$ is removed, more Cu(OH)$_2$ dissolves. The net reaction (the sum of the first reaction and twice the second) is

$$Cu(OH)_2(s) + 2H^+ \rightleftharpoons Cu^{2+} + 2H_2O$$

In general, the solubility of any electrolyte is greater in acid solution than in water if the negative ion of the electrolyte is capable of forming a weak electrolyte (water or a weak acid) with H$^+$. A similar increase in solubility is shown when other weak electrolytes are formed. Examples will be treated in more detail in the next chapter.

PROBLEMS

12.1 For each of the following mixtures of H$_2$, I$_2$, and HI at 425°C, is the system at equilibrium? If not, will the concentration of HI increase or decrease? (All concentrations are in moles per liter.)

(a) $[H_2] = [I_2] = [HI] = 1.00 \times 10^{-3}$
(b) $[H_2] = [HI] = 2.00 \times 10^{-2}$; $[I_2] = 1.00 \times 10^{-4}$
(c) $[I_2] = [HI] = 1.14 \times 10^{-3}$; $[H_2] = 8.41 \times 10^{-3}$

12.2 A flask contains hydrogen and iodine gases at 425°C, each at a concentration of 1.00×10^{-3} mol L^{-1}. If one desires to maintain these hydrogen and iodine concentrations, how much HI (in moles per liter) must be added?

12.3 Express the law of chemical equilibrium for each of the following reactions, using the conventions given in Table 12.2:
(a) $2CO(g) + O_2(g) \rightleftharpoons 2CO_2(g)$
(b) $4NH_3(g) + 7O_2(g) \rightleftharpoons 4NO_2(g) + 6H_2O(g)$
(c) $2NO_2(g) \rightleftharpoons N_2O_4(g)$
(d) $H_2(g) + \frac{1}{2}O_2(g) \rightleftharpoons H_2O(g)$

12.4 For each of the following values of K, what can you say about the relative concentrations of reactants and products at equilibrium?
(a) 1
(b) 1×10^{-10}
(c) 1×10^{10}

12.5 Calculate the equilibrium concentrations of H_2, I_2, and HI if 0.300 mol of HI is placed in a 1.00-L flask at 425°C.

12.6 At 2000 K the equilibrium constant is 4.40 for the reaction
$H_2(g) + CO_2(g) \rightleftharpoons H_2O(g) + CO(g)$
If a 1.00-L container initially has only $H_2(g)$ and $CO_2(g)$ present, each at a partial pressure of 0.500 atm, what are the final equilibrium partial pressures (in atmospheres) of each gas?

12.7 Express the law of chemical equilibrium for each of the following reactions, using the conventions given in Table 12.2:
(a) $CaCO_3(s) \rightleftharpoons CaO(s) + CO_2(g)$
(b) $CH_4(g) + 2O_2(g) \rightleftharpoons CO_2(g) + 2H_2O(l)$
(c) $CO_2(s) \rightleftharpoons CO_2(g)$
(d) $FeCl_2(s) + H_2(g) \rightleftharpoons Fe(s) + 2HCl(g)$

12.8 From the equilibrium constant for the synthesis of ammonia at 500°C (Section 12.4), calculate the equilibrium constant for the reverse reaction.

12.9 For each of the following reactions, how would the equilibrium be affected by
(a) increasing the partial pressure of oxygen?
(b) increasing the total pressure (by compressing the gases to a smaller volume, the temperature being maintained constant)?

PROBLEMS

$$N_2(g) + O_2(g) \rightleftharpoons 2NO(g)$$
$$2SO_2(g) + O_2(g) \rightleftharpoons 2SO_3(g)$$
$$2CO_2(g) \rightleftharpoons 2CO(g) + O_2(g)$$

12.10 For each of the following mixtures of N_2, H_2, and NH_3 at 500°C, is the system at equilibrium? If not, will the total pressure of the system, after equilibrium has been reached, be larger or smaller?
(a) $P_{H_2} = P_{N_2} = P_{NH_3} = 0.10$ atm
(b) $P_{N_2} = P_{NH_3} = 2.0 \times 10^{-3}$ atm; $P_{H_2} = 1.0$ atm
(c) $P_{N_2} = P_{H_2} = 0.20$ atm; $P_{NH_3} = 0.0040$ atm

12.11 In which direction will the following equilibria shift
(a) if the volume of the system is decreased?
(b) if the temperature of the system is increased?

$$2SO_2(g) + O_2(g) \rightleftharpoons 2SO_3(g) \qquad \Delta H = -197 \text{ kJ}$$
$$H_2O(l) \rightleftharpoons H_2O(g) \qquad \Delta H = 40.66 \text{ kJ}$$

12.12 Consider the following high-temperature equilibrium:
$$H_2O(g) + C(s) \rightleftharpoons CO(g) + H_2(g)$$
For each of the following changes, state whether the equilibrium shifts to the left, to the right, or is unaffected. (The temperature and total volume are held constant.)
(a) More $H_2O(g)$ is added.
(b) Some $H_2O(g)$ is removed.
(c) More solid carbon is added.
(d) More $CO(g)$ is added.

12.13 (a) From the data in Table 12.3, estimate the concentration (C_i) of the acetic acid solution that is 10% dissociated ($\alpha = 0.100$).
(b) From the value 1.8×10^{-5} for the acetic acid dissociation constant, calculate the concentration of the acetic acid solution that is 10% dissociated.

12.14 The dissociation constant of acetic acid is 1.8×10^{-5}. If a solution is prepared in which $[H^+] = 1.0 \times 10^{-5}M$, $[CH_3CO_2^-] = 0.18M$, and $[CH_3CO_2H] = 0.10M$, is the system at equilibrium? If not, will the following reaction proceed to the left or to the right?
$$CH_3CO_2H = CH_3CO_2^- + H^+$$

12.15 In an aqueous solution of the weak acid HX, measurements of the equilibrium concentrations show that $[H^+] = [X^-] = 2.0 \times 10^{-4}M$ and $[HX] = 0.20M$. Calculate the dissociation constant of HX. Is HX a stronger or weaker acid than acetic acid?

12.16 HX is a weak acid. The hydrogen ion concentration of $0.040M$ HX is $2.0 \times 10^{-5}M$. Calculate the dissociation constant of HX.

12.17 Calculate the numerical value of the equilibrium constant for the equilibrium
$$H^+ + CH_3CO_2^- \rightleftharpoons CH_3CO_2H$$

12.18 Calculate the equilibrium concentrations of H^+, $CH_3CO_2^-$, and CH_3CO_2H in aqueous $0.0200 M\ CH_3CO_2H$.

12.19 Excess solid iodine (I_2) is in equilibrium with its saturated solution in carbon tetrachloride (CCl_4). Will the equilibrium concentration of iodine increase, decrease, or remain the same if more solid iodine is added?

12.20 Excess solid silver acetate is in equilibrium with its saturated solution.
$$CH_3CO_2Ag(s) \rightleftharpoons CH_3CO_2^- + Ag^+$$
A small amount of each of the following substances is added to a separate portion of the mixture. After equilibrium is reestablished, is the concentration of silver ion in the solution larger, smaller, or the same as before?
(a) solid CH_3CO_2Na
(b) solid $AgNO_3$
(c) solid $NaNO_3$
(d) $6M\ HNO_3$

12.21 Excess solid calcium hydroxide is in equilibrium with its saturated solution.
$$Ca(OH)_2(s) \rightleftharpoons Ca^{2+} + 2OH^-$$
A small amount of each of the following substances is added to a separate portion of the mixture. After equilibrium is reestablished, is the concentration of calcium ion in the solution larger, smaller, or the same as before?
(a) solid NaOH
(b) solid $Ca(NO_3)_2$
(c) solid $Ca(OH)_2$
(d) $6M\ HCl$

12.22 From the solubility products in Table 12.4, calculate the solubility in water of
(a) $BaSO_4$
(b) $Ca(OH)_2$

12.23 The solubility of magnesium hydroxide, $Mg(OH)_2$, in water is 9.0×10^{-4} g per 100 mL of water. Calculate the solubility product of $Mg(OH)_2$.

12.24 The salt MX is slightly soluble in water. When 100 mL of a solution containing $0.010M\ M^+$ is mixed with 100 mL of a solution containing $0.010M\ X^-$, 50% of the M^+ ions precipitate as MX(s). Calculate the solubility product of MX. (Assume that the final volume of the solution is 200 mL.)

PROBLEMS

12.25 Calculate the solubility (in moles per liter) of $BaSO_4$ in $0.010M$ $BaCl_2$. ($BaCl_2$ is a strong electrolyte.)

12.26 Calculate the solubility of silver sulfate, Ag_2SO_4, in
(a) water
(b) $0.10M$ $AgNO_3$
Does the presence of the silver nitrate increase or decrease the solubility of silver sulfate?

12.27 In which of the following will precipitation occur?
(a) A solution contains $1 \times 10^{-6}M$ Ag^+ and $2 \times 10^{-3}M$ Cl^-.
(b) A solution contains $2 \times 10^{-2}M$ Ca^{2+} and $2 \times 10^{-2}M$ IO_3^-.
(c) A solution contains $2 \times 10^{-2}M$ Pb^{2+} and $2 \times 10^{-2}M$ Cl^-.

12.28 When two slightly soluble salts are compared, it is usually true that the salt with the smaller solubility product is the less soluble in water. Exceptions to this rule sometimes occur when the salts have different ratios of ions, for example, MX and MX_2, or MX and MX_3. By calculating the solubility of each in water, show that lanthanum iodate, $La(IO_3)_3$, is more soluble than silver chloride, AgCl, even though silver chloride has the larger value of K_{sp}.

12.29 Silver bromide (AgBr) is slightly soluble in water.
$$AgBr(s) \rightleftharpoons Ag^+ + Br^-$$
Its solubility is not affected by the addition of a small volume of $6M$ HNO_3. What conclusion can you draw about the strength of HBr as an acid?

12.30 In Example 3, it was stated that the solution
$$\frac{2x}{0.200 - x} = -7.38$$
leads to negative concentrations of H_2 and I_2. Verify this statement.

CHAPTER 13
SIMULTANEOUS EQUILIBRIA

13

The general subject of chemical equilibrium was introduced in Chapter 12. In this chapter we shall consider several types of equilibria, including systems in which more than one equilibrium is occurring. We shall also develop a systematic approach to evaluating the equilibrium concentrations of substances participating in the equilibria. Although the major emphasis will be on equilibria involving acids and bases, the methods employed can be applied to other types of equilibria.

13.1 GENERALIZED SYSTEMS OF ACIDS AND BASES

The Arrhenius definition (see Section 11.3) of an acid as a compound that dissociates to yield H^+, and of a base as a compound that dissociates to yield OH^-, was quite useful for aqueous solutions. However, it is not useful for solvents other than water.

In 1923, Johannes N. Brønsted and Thomas M. Lowry independently proposed a more general system of acids and bases that may be used in other solvents as well as for water solutions. They defined an acid as any substance that can donate a proton (H^+) and a base as any substance that can accept a proton. Thus when an acid (HA) donates its proton, a base (A^-) is formed. The HA,A^- pair is called a **conjugate acid–base pair.** A^- is the conjugate base of HA, and HA is the conjugate acid of A^-. Brønsted and Lowry further proposed that acid–base reactions take place by the transfer of a proton from an acid (HA) to a base (B) forming the new base (A^-) and the new acid (HB^+).

$$HA + B \rightleftharpoons HB^+ + A^- \qquad (13.1)$$
$$\text{acid(1)} \quad \text{base(2)} \quad \text{acid(2)} \quad \text{base(1)}$$

In the Brønsted-Lowry theory, the Arrhenius concept of dissociation of an acid in water is replaced by the concept of transfer of a proton from the acid to water, acting as a base, as shown in the following examples:

$$HCl + H_2O \rightleftharpoons H_3O^+ + Cl^-$$
$$NH_4^+ + H_2O \rightleftharpoons H_3O^+ + NH_3$$
$$CH_3CO_2H + H_2O \rightleftharpoons H_3O^+ + CH_3CO_2^-$$

13.1 GENERALIZED SYSTEMS OF ACIDS AND BASES

In each of these reactions, the substance H_3O^+ (called **hydronium ion**) is formed. This ion is simply a hydrated H^+ ion ($H^+ + H_2O \rightarrow H_3O^+$). Up till now we have used the symbol H^+ with the understanding that the proton is hydrated in aqueous solution, but in the Brønsted-Lowry theory it is essential to indicate this hydration explicitly by the symbol H_3O^+.

Other examples of acid–base reactions are

$$NH_3 + H_2O \rightleftharpoons NH_4^+ + OH^-$$

$$HCl + NH_3 \rightleftharpoons NH_4^+ + Cl^-$$

$$H_3O^+ + OH^- \rightleftharpoons H_2O + H_2O$$

Note that H_2O can function either as an acid (forming the conjugate base OH^-) or as a base (forming the conjugate acid H_3O^+).

The equilibrium position of these acid–base reactions will favor the formation of the weaker acid and weaker base. For example,

$$\underset{\substack{\text{stronger} \\ \text{acid}}}{HNO_3} + \underset{\substack{\text{stronger} \\ \text{base}}}{H_2O} \rightleftharpoons \underset{\substack{\text{weaker} \\ \text{acid}}}{H_3O^+} + \underset{\substack{\text{weaker} \\ \text{base}}}{NO_3^-}$$

The stronger the acid, the more readily it donates a proton, and the lower the ability of its conjugate base to attract a proton. Thus there is a complementary relation between the strengths of an acid and its conjugate base: the stronger the acid, the weaker its conjugate base, and vice versa.

EXAMPLE 1

For each of the following reactions select the stronger acid and the stronger base.

$$NH_3 + H_2O \rightleftharpoons NH_4^+ + OH^- \qquad K = 1.8 \times 10^{-5}$$

$$CN^- + H_3O^+ \rightleftharpoons HCN + H_2O \qquad K = 2.5 \times 10^9$$

In the first reaction the equilibrium position favors the reactants, so NH_4^+ is a stronger acid than H_2O, and OH^- is a stronger base than NH_3.
In the second reaction the equilibrium position favors the products, so H_3O^+ is a stronger acid than HCN, and CN^- is a stronger base than H_2O.

The relative strengths of Brønsted-Lowry acids and bases are conveniently shown in a table such as Table 13.1. Conjugate acid–base pairs are shown on the same horizontal line. The strongest acid is listed at the top, and the other acids are arranged in order of decreasing acid strength from top to bottom. Because of the complementary relation between acid strength and base strength in a conjugate pair, the bases automatically fall in order of increasing base strength from top to

TABLE 13.1 **RELATIVE STRENGTHS OF BRØNSTED-LOWRY ACIDS AND BASES**

Conjugate acid	Conjugate base
HCl	Cl$^-$
H$_3$O$^+$	H$_2$O
CH$_3$CO$_2$H	CH$_3$CO$_2^-$
NH$_4^+$	NH$_3$
H$_2$O	OH$^-$

(↑ increasing acid strength; ↓ increasing base strength)

bottom. An acid will react with a base at a lower level in the table, but not with a base at a higher level. For example, the following equilibrium lies to the right,

$$H_3O^+ + NH_3 \rightleftharpoons NH_4^+ + H_2O \qquad K > 1$$

while the following equilibrium lies to the left,

$$CH_3CO_2H + Cl^- \rightleftharpoons HCl + CH_3CO_2^- \qquad K < 1$$

In 1923, Gilbert N. Lewis proposed an even more general theory of acids and bases. Lewis defined an acid as a substance that can accept a pair of electrons from a donor substance, which he designated as a base, to form a compound in which an electron-pair bond is shared between the acid and base:

$$\text{Base} + \text{Acid} \rightleftharpoons \text{Electron-pair-bonded compound} \qquad (13.2)$$

$$:\!\ddot{\text{O}}\!-\!\text{H}^- + \text{H}^+ \rightleftharpoons :\!\ddot{\text{O}}\!-\!\text{H}\;|\;\text{H}$$

$$\text{H}\!-\!\ddot{\text{N}}\!:\,\text{H} + \text{H}^+ \rightleftharpoons \left[\text{H}\!-\!\text{N}(\text{H})(\text{H})\!-\!\text{H}\right]^+$$

$$2\,\text{H}\!-\!\ddot{\text{N}}\!:\,\text{H} + \text{Ag}^+ \rightleftharpoons \left[\text{H}\!-\!\text{N}(\text{H})(\text{H})\!-\!\text{Ag}\!-\!\text{N}(\text{H})(\text{H})\!-\!\text{H}\right]^+$$

$$\text{H}\!-\!\ddot{\text{N}}\!:\,\text{H} + \text{B}(\!:\!\ddot{\text{Cl}}\!:)(\!:\!\ddot{\text{Cl}}\!:)\!-\!\ddot{\text{Cl}}\!: \rightleftharpoons \text{H}\!-\!\text{N}(\text{H})(\text{H})\!-\!\text{B}(\!:\!\ddot{\text{Cl}}\!:)(\!:\!\ddot{\text{Cl}}\!:)\!-\!\ddot{\text{Cl}}\!:$$

The different acid–base systems are compared in Table 13.2. By comparing the first two, we see that the definitions of an acid are essentially equivalent, but that the Brønsted-Lowry concept broadens the class of bases to include proton acceptors such as Cl$^-$ and CH$_3$CO$_2^-$. Comparing the second two, we find that the

TABLE 13.2 | **ACID-BASE SYSTEMS**

System	Acid	Base
Arrhenius	Produces H⁺	Produces OH⁻
Brønsted-Lowry	Proton donor	Proton acceptor
Lewis	Electron-pair acceptor	Electron-pair donor

definitions of a base are essentially equivalent, but that the Lewis concept broadens the class of acids to include electron-pair acceptors such as BCl_3 and Ag^+.

The most common acid–base system for aqueous solutions is that of Arrhenius and we shall use it ordinarily, but the other systems will be used when appropriate.

13.2 THE DISSOCIATION OF WATER; SOLUTIONS OF STRONG ACIDS AND STRONG BASES

Although pure water is a poor conductor of electricity, the fact that it has some conductivity indicates that there are ions present. These ions (H^+ and OH^-) arise from the dissociation of water:

$$H_2O \rightleftharpoons H^+ + OH^- \qquad K_w = \frac{(a_{H^+})(a_{OH^-})}{a_{H_2O}} \qquad (13.3)$$

where the subscript w indicates that this is the dissociation constant of water. In accordance with the conventions in Section 12.2, the activity of pure water is unity, and we assume that the activity of the ions is numerically equal to their concentrations. Thus the equilibrium expression for the dissociation of water is usually written

$$K_w = [H^+][OH^-] \qquad (13.4)$$

It is apparent that the dissociation of water must lead to equal amounts of H^+ and OH^-. From a variety of experimental methods, we find that in pure water at 25°C the concentrations of H^+ and OH^- are each $1.00 \times 10^{-7} M$. Therefore $K_w = 1.00 \times 10^{-14}$ at 25°C.

Pure water (or any solution in which $[H^+] = [OH^-]$) is said to be neutral. Solutions in which $[H^+] > [OH^-]$ are defined as acidic, whereas those in which $[OH^-] > [H^+]$ are said to be basic. Because the dissociation of water always leads to equal amounts of H^+ and OH^- ions, acidic or basic solutions arise only when a solute produces or consumes H^+ or OH^-.

The dissociation of water always makes a contribution to the H^+ and OH^- concentrations of a solution. But in most solutions this contribution is so small that it may be neglected (as noted in Section 12.5). This can be explained qualitatively using Le Châtelier's principle. Addition of an acidic substance to water would cause an increase in the H^+ concentration; consequently the equilibrium

(Eq. 13.3) would be shifted to the left, and the concentration of H$^+$ arising from the dissociation of water would be less than $1 \times 10^{-7} M$. Addition of a basic substance to water would contribute OH$^-$ and hence would also cause the equilibrium of Eq. 13.3 to shift to the left, decreasing the dissociation of water. Thus we find that the *maximum* contribution to [H$^+$] or [OH$^-$] from the dissociation of water is $1 \times 10^{-7} M$. Addition of H$^+$ or OH$^-$ from other sources always decreases the amount of water that dissociates to less than 1×10^{-7} mol in each liter.

To find out when one may safely neglect the amount of H$^+$ or OH$^-$ from the dissociation of water, we shall consider the dissociation of water in very dilute solutions of strong acids and strong bases. First, let us calculate the contribution of H$^+$ from the dissociation of water to the total amount of H$^+$ in an aqueous solution of a strong acid, $1.00 \times 10^{-6} M$ HCl. The concentration of H$^+$ is equal to the concentration of H$^+$ produced by the dissociation of water (which we set equal to y), plus the concentration of H$^+$ from the HCl, $1.00 \times 10^{-6} M$. Thus

$$[H^+] = y + (1.00 \times 10^{-6})$$

The only source of OH$^-$ is the dissociation of water. Since we have assumed that y mol L^{-1} of water dissociates, [OH$^-$] = y. By substituting these expressions for [H$^+$] and [OH$^-$] in Eq. 13.4, together with the known value of K_w at 25°C, we obtain

$$K_w = 1.00 \times 10^{-14} = [y + (1.00 \times 10^{-6})]y \tag{13.5}$$

In the preceding paragraph we found that y must be less than $1 \times 10^{-7} M$ when an acid is present. Therefore it is reasonable to assume that, as a result of the solution's acidity, y is so small that it is negligible compared to 1.00×10^{-6}:

$$(1.00 \times 10^{-6}) + y \approx 1.00 \times 10^{-6}$$

If we make this substitution in Eq. 13.5, then

$$y \approx \frac{1.00 \times 10^{-14}}{1.00 \times 10^{-6}} = 1.00 \times 10^{-8}$$

and

$$[OH^-] = 1.00 \times 10^{-8} M$$

We now check our approximation and see that

$$(1.00 \times 10^{-6}) + y = (1.00 \times 10^{-6}) + (1.00 \times 10^{-8})$$
$$= 1.01 \times 10^{-6}$$

So our approximation is good to 1%.

For solutions in which the concentration of H$^+$ is larger than $1.00 \times 10^{-6} M$, the extent of dissociation of water (y) will be still smaller. Thus for acidic solutions in which $[H^+] \geqslant 1 \times 10^{-6}M$, we may neglect the contribution of $[H^+]$ from the dissociation of water. Similarly, for basic solutions in which $[OH^-] \geqslant 1 \times 10^{-6}M$, we may neglect the contribution of $[OH^-]$ from the dissociation of water.

EXAMPLE 2

What are the concentrations of hydrogen ion, hydroxide ion, and sodium ion in a $0.040M$ NaOH solution? (NaOH is a strong base.)

Since the NaOH is completely dissociated in water,

$[Na^+] = 0.040$

$[OH^-] = 0.040 + y$

where y = number of moles of H_2O that dissociate in each liter. Since $y < 1 \times 10^{-7}M$ owing to added OH^-, y is negligible compared to 0.040, and

$[OH^-] \approx 0.040M$

To determine the concentration of H^+, we divide K_w by the concentration of OH^-:

$$[H^+] = \frac{1.0 \times 10^{-14}}{4.0 \times 10^{-2}} = 2.5 \times 10^{-13}M$$

13.3 THE CONCEPT OF pH

In many aqueous sytems the concentration of H^+ is extremely small and it may be expressed in terms of negative powers of 10. An alternative method of expressing these values was suggested by Søren P. L. Sørensen in 1909. He defined a new term, pH, as the negative of the logarithm[1] of the hydrogen ion concentration. Or,

$$\text{pH} = -\log[H^+] \tag{13.6}$$

The interconversion of $[H^+]$ and pH may be illustrated by the following examples:

EXAMPLE 3

What is the pH of a solution if the $[H^+] = 2 \times 10^{-5}M$?

By definition $\text{pH} = -\log[H^+] = -\log(2 \times 10^{-5})$
$= -\log 2 - \log 10^{-5} = -0.30 - (-5.00)$
$= -0.30 + 5.00 = 4.70$

[1] See Appendix C-5 for a discussion of logarithms.

EXAMPLE 4

What is the [H$^+$] of a solution whose pH = 7.60?

$$7.60 = -\log [H^+]$$
$$\log [H^+] = -7.60 = +0.40 - 8.00$$

The antilogarithm of +0.40 is 2.5, and the antilogarithm of −8.00 is 10^{-8}. Therefore

$$[H^+] = 2.5 \times 10^{-8} M$$

Note that the pH of pure water at 25°C is $-\log (10^{-7}) = 7$. Acidic solutions have pH's less than 7, whereas basic solutions have pH's greater than 7.

The symbol p is now used to designate (−log), so that pK = −log K, pAg$^+$ = −log [Ag$^+$], and so on.

13.4 RELATIVE STRENGTHS OF ACIDS AND BASES

The strength of an acid, HA, increases with the ability of the acid to donate a proton and, in aqueous solution, with increasing magnitude of its dissociation constant.

$$HA \rightleftharpoons H^+ + A^- \qquad K_a = \frac{[H^+][A^-]}{[HA]} \qquad (13.7)$$

Values of K_a for some common acids, along with the conjugate bases of the acids, are listed in Table 13.3. We say that the strength of a base increases with increasing ability of the base to attract a proton. Thus we can summarize the relative strengths of the acids and bases as follows: As K_a decreases, the strength of the acid decreases and the strength of its conjugate base increases. Thus any base in Table 13.3 will react with any acid whose K_a value is larger than the K_a value of its conjugate acid, producing the weaker acid. For example, acetate ion will react with nitrous acid to give NO$_2^-$ and CH$_3$CO$_2$H:

$$CH_3CO_2^- + HNO_2 = CH_3CO_2H + NO_2^-$$

It is important to note that we need not devise a new set of dissociation constants for aqueous solutions to use with the Brønsted-Lowry concept of acids and bases. For acetic acid, we write

$$CH_3CO_2H + H_2O \rightleftharpoons H_3O^+ + CH_3CO_2^- \qquad K_a = \frac{(a_{H_3O^+})(a_{CH_3CO_2^-})}{(a_{CH_3CO_2H})(a_{H_2O})}$$

But in dilute aqueous solutions the activity of water is nearly unity, and the

13.4 RELATIVE STRENGTHS OF ACIDS AND BASES

TABLE 13.3 **DISSOCIATION CONSTANTS OF ACIDS AND BASES IN WATER AT 25°C**

Conjugate acid		Conjugate base	K_a
Acetic	CH_3CO_2H	$CH_3CO_2^-$	1.8×10^{-5}
Ammonium ion	NH_4^+	NH_3	5.6×10^{-10}
Carbonic	$H_2CO_3[CO_2(aq) + H_2O]$	HCO_3^-	$K_1 = 4.3 \times 10^{-7}$
	HCO_3^-	CO_3^{2-}	$K_2 = 4.7 \times 10^{-11}$
Hydrochloric	HCl	Cl^-	Very large
Hydrocyanic	HCN	CN^-	6.1×10^{-10}
Hydrofluoric	HF	F^-	6.8×10^{-4}
Hydrogen sulfide	H_2S	HS^-	$K_1 = 1.0 \times 10^{-7}$
	HS^-	S^{2-}	$K_2 = 1.3 \times 10^{-14}$
Nitric	HNO_3	NO_3^-	Very large
Nitrous	HNO_2	NO_2^-	7.1×10^{-4}
Perchloric	$HClO_4$	ClO_4^-	Very large
Phosphoric	H_3PO_4	$H_2PO_4^-$	$K_1 = 7.1 \times 10^{-3}$
	$H_2PO_4^-$	HPO_4^{2-}	$K_2 = 6.3 \times 10^{-8}$
	HPO_4^{2-}	PO_4^{3-}	$K_3 = 4.5 \times 10^{-13}$
Sulfuric	H_2SO_4	HSO_4^-	K_1 = large
	HSO_4^-	SO_4^{2-}	$K_2 = 1.0 \times 10^{-2}$
Water	H_2O	OH^-	$K_w = 1.0 \times 10^{-14}$

Conjugate base		Conjugate acid	K_b
Ammonia	NH_3	NH_4^+	1.8×10^{-5}

activities of the other substances are approximately equal to their molar concentrations, so we have

$$K_a = \frac{[H_3O^+][CH_3CO_2^-]}{[CH_3CO_2H]}$$

where K_a is the dissociation constant (Table 13.3) and $[H_3O^+]$ is equivalent to $[H^+]$.

The relative strengths in aqueous solution of the strong acids (those whose K_a value is large), such as perchloric acid, hydrochloric acid, and nitric acid, cannot be determined. Water, being a good base, causes the equilibrium position for reactions of the type

$$HClO_4 + H_2O \rightleftharpoons H_3O^+ + ClO_4^-$$

to lie so far to the right that we cannot measure the concentrations of the reactants and hence cannot evaluate K_a. Therefore water is said to have a **leveling effect** on the strength of acids such as $HClO_4$, HCl, and HNO_3. We can, however, measure the relative strengths of these acids by using a solvent that is a weaker base than water. Such a solvent is acetic acid. The acetic acid functions as a base for these acids:

$$HX + CH_3CO_2H \rightleftharpoons CH_3CO_2H_2^+ + X^-$$

but the equilibrium positions are not so far to the right as with water, and the equilibrium constants may be determined experimentally. In acetic acid solutions the order of acid strengths is $HClO_4 > HCl > HNO_3$.

13.5 WEAK ACIDS AND BASES; EXACT SOLUTIONS OF EQUILIBRIUM PROBLEMS

A wide variety of equilibrium problems are encountered in chemistry, and it will be useful to review the general kinds of information one should have available to solve them. The amount of each reagent used to prepare the solution is often known. In addition one should have a knowledge of the possible reactions or equilibria in which the solvent and solutes might participate. For each equilibrium that is established, the law of chemical equilibrium provides a needed relationship. Additional relationships may be obtained from the stoichiometry of the reactions or by using equations for material balance and charge balance as described next.

It is, of course, extremely important in solving any problem to keep in mind just what is desired. Often one wishes to calculate the equilibrium concentration of one or more species, given the appropriate equilibrium constants. In other cases one may wish to calculate an equilibrium constant from measurements of the equilibrium concentrations. Or else one may wish to calculate the amounts of various reagents required to achieve some objective.

In general, from the known information a set of simultaneous equations is established that relate the various unknown quantities. Provided there are as many equations as there are unknowns, the equations can be solved by standard algebraic methods to determine all the unknown quantities.

To illustrate this approach, let us consider a specific example, namely, a solution prepared by mixing 0.020 mol of CH_3CO_2Na (the sodium salt of acetic acid) and 0.020 mol of HCl in enough water to make exactly 1 L of solution. Our objective will be to calculate the concentration of each species in solution, given that sodium acetate and hydrochloric acid are soluble and are strong electrolytes and that acetic acid is a weak electrolyte with a dissociation constant K_a of 1.8×10^{-5}.

Before any chemical reaction occurs other than the dissolving of the reagents in the water, the substances in solution will be $CH_3CO_2^-$, Na^+, H^+, and Cl^-, each at a concentration of $0.020 M$. We now consider what additional substances might be present from the solvent or from reactions of the species from the reagents with one another or with the solvent. In the present case we note that all aqueous solutions contain H^+ and OH^- because of the dissociation of water:

13.5 WEAK ACIDS AND BASES; EXACT SOLUTIONS OF EQUILIBRIUM PROBLEMS

$$H_2O \rightleftharpoons H^+ + OH^- \qquad K_w = [H^+][OH^-] \qquad (13.8)$$

Furthermore, acetic acid is a weak electrolyte, so its ions will combine to a large extent:

$$H^+ + CH_3CO_2^- \rightleftharpoons CH_3CO_2H \qquad (13.9)$$

How far will this reaction go? It certainly does not go to completion, as acetic acid is a weak electrolyte, not a nonelectrolyte. The essence of the problem is, therefore, to determine the extent of this reaction. This is not difficult, because we know that the reaction will proceed until equilibrium is established:

$$CH_3CO_2H \rightleftharpoons H^+ + CH_3CO_2^- \qquad K_a = \frac{[H^+][CH_3CO_2^-]}{[CH_3CO_2H]} \qquad (13.10)$$

There are six species in the solution in addition to water molecules, namely, Na^+, $CH_3CO_2^-$, H^+, Cl^-, OH^-, and CH_3CO_2H. The equilibrium concentrations of Na^+ and Cl^- are equal to $0.020M$, for they do not react with any other species. Thus there are four unknown equilibrium concentrations, and we require four equations.

Equations 13.8 and 13.10 provide two of these equations.

The third and fourth equations may be obtained from a knowledge of the composition of the solution. For example, all the 0.020 mol of acetate added to the solution is present as $CH_3CO_2^-$ or CH_3CO_2H. Thus we can write

$$0.020 = [CH_3CO_2^-] + [CH_3CO_2H] \qquad (13.11)$$

This third equation is known as a **material balance equation.**

Our fourth equation, called the **charge balance equation,** may be obtained from the condition that the solution must be electrically neutral. That is, there must be as many positive charges in the solution as there are negative charges. The number of positive charges would be equal to the number of each type of positive ion *times the charge on that ion*. A similar relationship holds for the negative ions. Thus the number of moles of positive charge in 1 L of solution due to Na^+ would be $[Na^+] \times 1 = [Na^+]$. (For a solution containing a dipositive ion, it is necessary to multiply the concentration of that ion by a factor of 2, e.g., $[Ca^{2+}] \times 2$.) In this example the charge balance equation would be:

$$[Na^+] + [H^+] = [Cl^-] + [CH_3CO_2^-] + [OH^-]$$

We have already noted that $[Na^+] = [Cl^-] = 0.020M$ so our charge balance equation reduces to

$$[H^+] = [CH_3CO_2^-] + [OH^-] \qquad (13.12)$$

The four simultaneous equations (13.8, 13.10, 13.11 and 13.12) may be solved by the methods of algebra to obtain an equation with only one unknown quantity:

$$\frac{[H^+]\{[H^+] - (K_w/[H^+])\}}{0.020 - [H^+] + (K_w/[H^+])} = K_a \tag{13.13}$$

Because $K_a = 1.8 \times 10^{-5}$ and $K_w = 1.00 \times 10^{-14}$, Eq. 13.13 may now be solved exactly to obtain $[H^+]$. Substitution of this value in the other equations will yield the other desired concentrations.

13.6 APPROXIMATIONS

Although solution of simultaneous equations by rigorous algebraic methods always leads to correct answers, the exact solution of Eq. 13.13 is difficult, and it is more convenient and enlightening to employ methods of approximation. These are based on estimates of what is essential to the problem and what can be neglected. First of all, one is faced with the problem of estimating the equilibrium position of the various possible reactions. Once this is done the approximations to use are usually fairly obvious, especially after one acquires a certain familiarity with typical problems.

Generally, one first simplifies the charge balance equation because usually $[H^+]$ or $[OH^-]$ will be negligible compared to the concentration of other ions. For example, as the solution is acidic, the $[OH^-]$ from the dissociation of water must be less than $10^{-7} M$. Therefore we assume in Eq. 13.12 that $[OH^-]$ is negligible compared to $[CH_3CO_2^-]$, and our first approximation is, accordingly,

$$[H^+] \approx [CH_3CO_2^-]$$

If we now substitute this result in Eq. 13.11, we obtain $[CH_3CO_2H] = 0.020 - [H^+]$. These values for $[CH_3CO_2^-]$ and $[CH_3CO_2H]$ may then be substituted in Eq. 13.10 to obtain

$$\frac{[H^+]^2}{0.020 - [H^+]} \approx 1.8 \times 10^{-5} \tag{13.14}$$

We can now simplify the solution of Eq. 13.14 by recognizing that the dissociation constant of acetic acid is very small. Therefore, when its ions are mixed at moderate concentrations (such as $0.020 M$), they will largely combine. Consequently $[H^+]$ should be negligible compared to 0.020, and our second approximation is

$$0.020 - [H^+] \approx 0.020$$

13.6 APPROXIMATIONS

We now have

$$\frac{[H^+]^2}{0.020} \approx 1.8 \times 10^{-5}$$

or

$$[H^+]^2 \approx 36 \times 10^{-8}$$

and

$$[H^+] \approx 6.0 \times 10^{-4} M$$

By the first approximation this is also the concentration of acetate ion,

$$[CH_3CO_2^-] \approx 6.0 \times 10^{-4} M$$

The concentration of undissociated acetic acid is calculated from Eq. 13.11:

$$[CH_3CO_2H] = 0.020 - [CH_3CO_2^-] = 0.020 - (6.0 \times 10^{-4})$$
$$= 0.019 M$$

The last unknown is $[OH^-]$; it is obtained from Eq. 13.8:

$$[OH^-] = \frac{K_w}{[H^+]} = \frac{1.0 \times 10^{-14}}{6.0 \times 10^{-4}} = 1.7 \times 10^{-11} M$$

We therefore end up with the following set of equilibrium concentrations:

$$[Na^+] = 0.020 M$$
$$[Cl^-] = 0.020 M$$
$$[CH_3CO_2H] = 0.019 M$$
$$[H^+] = 6.0 \times 10^{-4} M$$
$$[CH_3CO_2^-] = 6.0 \times 10^{-4} M$$
$$[OH^-] = 1.7 \times 10^{-11} M$$

Finally, we show that these results are consistent with the two approximations used to obtain them:

1. The first approximation was that $[OH^-]$ is negligible compared to $[CH_3CO_2^-]$. And indeed, $[OH^-] = 1.7 \times 10^{-11} M$, very much smaller than $[CH_3CO_2^-] = 6.0 \times 10^{-4} M$.
2. The second approximation was that $[H^+]$ is negligible compared to $0.020 M$. And indeed, $[H^+] = 6.0 \times 10^{-4} M$, very much smaller than $0.020 M$.

Whenever the results of a calculation are inconsistent with an approximation used to solve the problem, one must look for a better approximation, or solve the equations by means of successive approximations, or solve the equations exactly.

The previous problem may be solved much more rapidly and easily if one recognizes that the dissociation of water is negligible. One can then use the methods described in Chapter 12, as shown in the following example.

EXAMPLE 5

Calculate the $[H^+]$ of a solution prepared by dissolving 0.020 mol of CH_3CO_2Na and 0.020 mol of HCl in enough water to prepare a liter of solution.

First we calculate the initial concentrations of all the dissolved species before any reactions occur.

$$[Na^+]_i = [CH_3CO_2^-]_i = 0.020 M$$

and

$$[H^+]_i = [Cl^-]_i = 0.020 M$$

Then we write any reactions that might occur, for example,

$$H^+ + CH_3CO_2^- \rightleftharpoons CH_3CO_2H$$

This reaction will not go to completion, so the final solution will be slightly acidic. We therefore assume that $[H^+] > 1 \times 10^{-6} M$ and neglect the dissociation of water. Let $[H^+]$ at equilibrium $= y$. Then as a result of the reaction we find that

$$[CH_3CO_2H] = [H^+]_i - [H^+] = 0.020 - y$$
$$[CH_3CO_2^-] = [CH_3CO_2^-]_i - [CH_3CO_2H] = 0.020 - (0.020 - y)$$

or

$$[CH_3CO_2^-] = y$$

These values may be substituted into the equilibrium expression for K_a to obtain

$$\frac{y^2}{0.020 - y} = 1.8 \times 10^{-5}$$

We then assume that y is negligible compared to 0.020. Therefore

$$y = (0.020 \times 1.8 \times 10^{-5})^{1/2} = 6.0 \times 10^{-4} M = [H^+]$$

Finally, we check our assumptions to see if they are valid. Since $[H^+] > 1 \times 10^{-6} M$, we can neglect the dissociation of water. And $0.020 - (6.0 \times 10^{-4}) = 0.0194 \approx 0.020$, so both assumptions are valid.

13.6 APPROXIMATIONS

The equilibrium concentrations of ions and molecules in an aqueous solution of a weak base may be calculated using the same principles. We shall illustrate the methods with a different type of problem.

EXAMPLE 6

Ammonia, NH_3, is a weak base with a dissociation constant K_b of 1.8×10^{-5}. What are the equilibrium concentrations of ions and of NH_3 in $0.50\ M\ NH_3$? What is the pH of the solution?

Before any reactions occur,

$[NH_3]_i = 0.50 M$

The following equilibrium would be established so that the final solution would be basic:

$$NH_3 + H_2O \rightleftharpoons NH_4^+ + OH^- \qquad K_b = 1.8 \times 10^{-5} = \frac{[NH_4^+][OH^-]}{[NH_3]}$$

Since the solution will be basic, we will assume that $[OH^-] > 1 \times 10^{-6}$ and that the dissociation of water can be neglected.

Let y = number of moles of NH_3 that react to form ions in each liter of solution.

Then

$[NH_4^+] = [OH^-] = y$

and

$[NH_3] = 0.50 - y$

If these values are substituted in the equilibrium expression, we obtain

$$\frac{y^2}{0.50 - y} = 1.8 \times 10^{-5}$$

Since K_b is small, y will be small so we assume $0.50 - y \approx 0.50$. We therefore obtain

$y \approx (0.50 \times 1.8 \times 10^{-5})^{1/2} \approx 3.0 \times 10^{-3} M$.

Thus

$[NH_4^+] = [OH^-] = 3.0 \times 10^{-3} M$

and

$[NH_3] = 0.50 - 0.003 \approx 0.50\ M$

Our two approximations are valid. Since $[OH^-] > 1 \times 10^{-6} M$, the dissociation of water may be neglected. Also, $0.50 - y \approx 0.50$.

To calculate the pH of the solution we note that

$$[H^+] = \frac{1.0 \times 10^{-14}}{[OH^-]} = \frac{1.0 \times 10^{-14}}{3.0 \times 10^{-3}} = 3.3 \times 10^{-12}$$

$$pH = -\log(3.3 \times 10^{-12}) = 11.48$$

13.7 SALT SOLUTIONS

Aqueous solutions of salts are found to be acidic, neutral, or basic depending on the type of salt used. This type of behavior is easily predicted if one considers the effect that added substances might have on the equilibrium between water and its ions. As an example, let us consider a solution of sodium acetate. The acetate ion could react with the H^+ from water to produce some acetic acid. Thus we have two equilibria,

$$H_2O \rightleftharpoons H^+ + OH^-$$

and

$$H^+ + CH_3CO_2^- \rightleftharpoons CH_3CO_2H$$

In accordance with Le Châtelier's principle, removal of H^+ by formation of CH_3CO_2H causes more water to dissociate; therefore $[OH^-] > [H^+]$, and a basic solution is obtained. The net result of the interaction of these two equilibria may be expressed by their sum:

$$H_2O + CH_3CO_2^- \rightleftharpoons CH_3CO_2H + OH^-$$

$$K_h = \frac{[CH_3CO_2H][OH^-]}{[CH_3CO_2^-]} \tag{13.15}$$

A reaction of this type, in which a water molecule splits, is known as **hydrolysis**. (In the Brønsted-Lowry system this is simply another acid–base reaction between the acid H_2O and the base $CH_3CO_2^-$.) The value of the hydrolysis constant (K_h) is not tabulated because it can easily be obtained from the dissociation constants of water and acetic acid. As illustrated next, *the equilibrium constant for the sum of two equilibria is the product of the individual equilibrium constants.*

$$H_2O \rightleftharpoons H^+ + OH^- \qquad K_w = [H^+][OH^-]$$

$$CH_3CO_2^- + H^+ \rightleftharpoons CH_3CO_2H \qquad \frac{1}{K_a} = \frac{[CH_3CO_2H]}{[CH_3CO_2^-][H^+]}$$

$$H_2O + CH_3CO_2^- + \cancel{H^+} \rightleftharpoons CH_3CO_2H + \cancel{H^+} + OH^- \qquad K_h = \frac{[CH_3CO_2H][\cancel{H^+}][OH^-]}{[CH_3CO_2^-][\cancel{H^+}]}$$

$$= \frac{K_w}{K_a} \tag{13.16}$$

Thus

$$K_h = \frac{1.0 \times 10^{-14}}{1.8 \times 10^{-5}} = 5.6 \times 10^{-10}$$

Let us now consider a salt solution that is acidic, such as that of ammonium chloride, NH_4Cl. Like most salts, NH_4Cl is a strong electrolyte. The NH_4^+ tends to remove some OH^- by forming NH_3 and thereby causing an increase in the dissociation of water, with the net result that $[H^+] > [OH^-]$. The two equilibria are

$$H_2O \rightleftharpoons H^+ + OH^-$$

and

$$OH^- + NH_4^+ \rightleftharpoons NH_3 + H_2O$$

The hydrolysis equation for this system would be the sum of the equilibria shown above, or

$$NH_4^+ \rightleftharpoons NH_3 + H^+ \qquad K_h = \frac{[NH_3][H^+]}{[NH_4^+]} \qquad (13.17)$$

Thus NH_4^+ is behaving as an acid. The hydrolysis constant, K_h, is equal to K_w/K_b, where K_b is the equilibrium constant for the reaction

$$NH_3 + H_2O \rightleftharpoons NH_4^+ + OH^- \qquad K_b = \frac{[NH_4^+][OH^-]}{[NH_3]} \qquad (13.18)$$

(The reader may easily verify that $K_h = K_w/K_b$ using the same methods as used for the sodium acetate system.) Thus

$$K_h = \frac{1.0 \times 10^{-14}}{1.8 \times 10^{-5}} = 5.6 \times 10^{-10}$$

We are now in a position to explain the acidity or basicity of salt solutions. The salt of a *strong acid* and a *strong base* will produce a *neutral* solution because the ions of the salt will not react with H^+ or OH^-, and therefore the water equilibrium will remain undisturbed and $[H^+] = [OH^-]$. The salt of a *weak acid* and a *strong base* (such as sodium acetate) will produce a *basic* solution because the negative ion of the salt will increase the dissociation of water by combining with H^+ to form some of the weak acid, leaving an excess of OH^-. The salt of a *strong acid* and a *weak base* (such as ammonium chloride) will produce an *acidic* solution because the positive ion of the salt will increase the dissociation of water by combining with OH^- to form the weak base, leaving an excess of H^+. The salt of a *weak acid* and a *weak base* will increase the dissociation of water since both the weak acid and the weak base will be formed, and the resulting solution will be *acidic, neutral,* or *basic* depending on the *relative weakness* of the acid and base formed. For example, with ammonium acetate the resulting solution is neutral because K_a and K_b are both 1.8×10^{-5}, that is, acetic acid is as weak an acid as aqueous ammonia is a weak base. The three equilibria are

$$H_2O \rightleftharpoons H^+ + OH^-$$

$$H^+ + CH_3CO_2^- \rightleftharpoons CH_3CO_2H$$

$$OH^- + NH_4^+ \rightleftharpoons NH_3 + H_2O$$

The hydrolysis equation is the sum of the above three equilibria:

$$NH_4^+ + CH_3CO_2^- \rightleftharpoons NH_3 + CH_3CO_2H \tag{13.19}$$

The quantitative treatment of the acidity of salt solutions is similar to that of a solution of a weak acid. To illustrate the method, let us calculate the equilibrium concentrations of the substances present in a $1.00 M$ CH_3CO_2Na solution. The following equilibria are established:

$$H_2O \rightleftharpoons H^+ + OH^- \qquad K_w = [H^+][OH^-]$$

$$CH_3CO_2^- + H_2O \rightleftharpoons CH_3CO_2H + OH^- \qquad K_h = \frac{[CH_3CO_2H][OH^-]}{[CH_3CO_2^-]} \tag{13.20}$$

Since $K_h = 5.6 \times 10^{-10}$ is very small, $[OH^-]$ will be small, but we will assume that it is greater than $1 \times 10^{-6} M$ and we will neglect the dissociation of water.

If we let $[OH^-] = y$, then $[CH_3CO_2H] = y$ and $[CH_3CO_2^-] = 1.00 - y \approx 1.00$. We may substitute these values into Eq. 13.20 to obtain

$$\frac{y^2}{1.00} \approx 5.6 \times 10^{-10}$$

or

$$y = (5.6 \times 10^{-10})^{1/2} = 2.4 \times 10^{-5} M.$$

Since $1.00 - y = 1.00 - (2.4 \times 10^{-5}) \approx 1.00$, our second assumption is valid and we note that $[OH^-] = y$ is greater than $1 \times 10^{-6} M$, se we can neglect the amount of OH^- from the first equilibrium above. Thus we have

$$[OH^-] = [CH_3CO_2H] = 2.4 \times 10^{-5} M$$

and

$$[CH_3CO_2^-] = 1.00 M$$

Therefore

$$[H^+] = \frac{1.0 \times 10^{-14}}{[OH^-]} = \frac{1.0 \times 10^{-14}}{2.4 \times 10^{-5}} = 4.2 \times 10^{-10} M$$

Finally, the pH of the solution is $-\log [4.2 \times 10^{-10}]$ or 9.38.

In summary, the equilibrium concentrations in a $1.00M$ CH_3CO_2Na solution are

$$[Na^+] = 1.00M$$
$$[CH_3CO_2^-] = 1.00M$$
$$[OH^-] = 2.4 \times 10^{-5} M$$
$$[CH_3CO_2H] = 2.4 \times 10^{-5} M$$
$$[H^+] = 4.2 \times 10^{-10} M$$
$$pH = 9.38$$

13.8 BUFFERS

In previous sections we have considered solutions of a weak acid, a weak base, and salts of weak acids or weak bases. In this section we consider equilibria in solutions containing both a weak acid and its salt or both a weak base and its salt. Such solutions have the remarkable property that they undergo relatively little change in pH when an acid or a base is added, and it is this property that gives them their name of **buffer solutions.**

To see why they have this property, consider a solution containing a weak acid, HX, and its negative ion X^-. If a base is added, the OH^- is removed by reaction with the acid,

$$OH^- + HX \rightleftharpoons H_2O + X^-$$

If an acid is added, the H^+ is removed by combination with the negative ion,

$$H^+ + X^- \rightleftharpoons HX$$

Thus in neither case does the concentration of H^+ or OH^- change very much, and the pH remains almost constant. (The small changes that do occur result from shifts in the equilibria of these reactions, as the relative amounts of reactants and products are altered.)

To illustrate the quantitative treatment of buffer solutions, consider a solution containing 0.20 mol of CH_3CO_2H and 0.10 mol of CH_3CO_2Na in 1.00 L of solution. The initial concentrations of dissolved species before any reaction occurs or equilibrium is established are

$$[CH_3CO_2H]_i = \frac{0.20 \text{ mol}}{1.00 \text{ L}} = 0.20 M$$

$$[Na^+]_i = [CH_3CO_2^-]_i = \frac{0.10 \text{ mol}}{1.00 \text{ L}} = 0.10 M$$

The following equilibria will be established:

$$CH_3CO_2H \rightleftharpoons CH_3CO_2^- + H^+ \qquad K_a = 1.8 \times 10^{-5}$$

$$H_2O \rightleftharpoons H^+ + OH^- \qquad K_w = 1.0 \times 10^{-14}$$

$$CH_3CO_2^- + H_2O \rightleftharpoons CH_3CO_2H + OH^- \qquad K_h = 5.6 \times 10^{-10}$$

Since K_a is much greater than K_w or K_h, we shall neglect the last two equilibria. We note that K_a is small and that the added $CH_3CO_2^-$ will repress the dissociation of CH_3CO_2H to a very small value.

Let y = number of moles per liter of acetic acid that dissociate. Then

$$[H^+] = y$$

$$[CH_3CO_2H] = [CH_3CO_2H]_i - y = 0.20 - y$$

and

$$[CH_3CO_2^-] = [CH_3CO_2^-]_i + y = 0.10 + y$$

Since y will be very small, we will neglect y compared to 0.20 or 0.10 and write

$$[CH_3CO_2H] \approx 0.20 \quad \text{and} \quad [CH_3CO_2^-] \approx 0.10$$

Thus

$$K_a = \frac{[H^+][CH_3CO_2^-]}{[CH_3CO_2H]} \approx \frac{y \times 0.10}{0.20} = 1.8 \times 10^{-5}$$

and

$$y = \frac{0.20}{0.10} \times 1.8 \times 10^{-5} = 3.6 \times 10^{-5}$$

Our approximations were justified, since this result is negligible compared to 0.20 or 0.10. Thus $[H^+] = 3.6 \times 10^{-5} M$. The pH of this buffer solution is

$$\text{pH} = -\log [H^+] = -\log [3.6 \times 10^{-5}] = 4.44$$

13.8 BUFFERS

To illustrate the tendency of buffer solutions to maintain constant pH, let us consider two cases. First let us dilute the solution used in the preceding calculation to 10 times its original volume by adding water. The only change in our calculations as a result of this dilution is to reduce the concentrations of $[CH_3CO_2^-]$ to $0.010 + [H^+]$, and that of $[CH_3CO_2H]$ to $0.020 - [H^+]$. If $[H^+]$ is negligible compared to 0.010 or 0.020, we find that

$$[H^+] = 1.8 \times 10^{-5} \times \frac{[CH_3CO_2H]}{[CH_3CO_2^-]}$$

$$= 1.8 \times 10^{-5} \times \frac{0.020}{0.010} = 3.6 \times 10^{-5} M$$

Thus dilution by a factor of 10 did not alter the $[H^+]$ of the buffer solution, whereas dilution of a solution of a strong acid by a factor of 10 would reduce the $[H^+]$ by a factor of 10.

Second let us calculate the effect on the $[H^+]$ of introducing 0.010 mol of HCl into the original buffer solution. The added H^+ will react with $CH_3CO_2^-$ to form CH_3CO_2H. Let us assume that the 0.010 mol reacts with 0.010 mol of $CH_3CO_2^-$ to form 0.010 mol of CH_3CO_2H. Then $[CH_3CO_2H] = 0.20 + 0.010 - [H^+] \approx 0.21$, assuming $[H^+]$ is negligible. By similar reasoning

$$[CH_3CO_2^-] = 0.10 - 0.010 + [H^+] \approx 0.09$$

Therefore

$$[H^+] = 1.8 \times 10^{-5} \times \frac{0.21}{0.09} = 4.2 \times 10^{-5} M \quad (pH = 4.38)$$

Thus, addition of 0.010 mol of HCl to 1 L of the acetic acid–sodium acetate buffer solution caused only a slight change in $[H^+]$, that is, from 3.6×10^{-5} to $4.2 \times 10^{-5} M$. This corresponds to a pH change from 4.44 to 4.38. In contrast, addition of 0.010 mol of HCl to 1 L of water causes a change in $[H^+]$ from 1×10^{-7} to $1 \times 10^{-2} M$, that is, a 100,000-fold change. This corresponds to a pH change from 7 to 2.

The ability of blood to maintain a constant pH in the vicinity of 7.4 is due to the presence of buffer systems similar to the acetic acid–sodium acetate buffer discussed above. In fact the pH of venous blood differs by less than 0.02 pH unit from that of arterial blood even though the venous blood has been exposed to the numerous acid- and base-producing reactions in the body cells.

If one is interested in calculating the pH of buffer solutions rather than the $[H^+]$, it is convenient to express K_a and $[H^+]$ in terms of pK_a and pH as follows:

$$HX \rightleftharpoons H^+ + X^- \qquad K_a = \frac{[H^+][X^-]}{[HX]} \qquad (13.21)$$

or

$$[H^+] = K_a \frac{[HX]}{[X^-]}$$

If we take the negative logarithm of both sides of the equation, we obtain

$$-\log[H^+] = -\log K_a - \log \frac{[HX]}{[X^-]}$$

or

$$pH = pK_a + \log \frac{[X^-]}{[HX]} \tag{13.22}$$

13.9 ACID–BASE INDICATORS

The hydrogen ion concentration of a solution may be determined by adding an appropriate acid–base indicator or by use of a pH meter that measures [H$^+$] by an electrochemical process (Chapter 14). Indicators are organic acids or bases that exhibit different colors in their neutral and ionic forms. The colors of the indicators are so intense that only extremely small concentrations of indicators are used, and therefore they do not appreciably alter the H$^+$ concentration or other equilibria in solution.

For example, in acid solutions, bromothymol blue imparts a yellow color to the solution, whereas it turns basic solutions blue. Because it is a weak acid, let us represent this indicator by the formula HIn, and its ionic form as In$^-$. In aqueous solutions containing bromothymol blue, the following equilibrium is established:

$$\underset{\text{yellow}}{\text{HIn}} \rightleftharpoons H^+ + \underset{\text{blue}}{\text{In}^-} \qquad K_{\text{HIn}} = \frac{[H^+][\text{In}^-]}{[\text{HIn}]} \tag{13.23}$$

In strongly acid solutions the equilibrium is shifted to the left and the indicator will be predominately in the undissociated or yellow form, whereas in basic solutions the equilibrium is shifted to the right, the ionic form predominates, and a blue color is imparted to the solution.

The relative amounts of the two colored forms in a solution are directly dependent on the [H$^+$] as

$$\frac{[H^+]}{K_{\text{HIn}}} = \frac{[\text{HIn}]}{[\text{In}^-]} \tag{13.24}$$

If [H$^+$]/K_{HIn} is large, then [HIn]/[In$^-$] is large and the color of the solution will be that due to the acid form (yellow if bromothymol blue is used). If [H$^+$]/K_{HIn} is unity, the two forms are present in equivalent amounts and the mixture of yellow and blue forms turns the solution green.

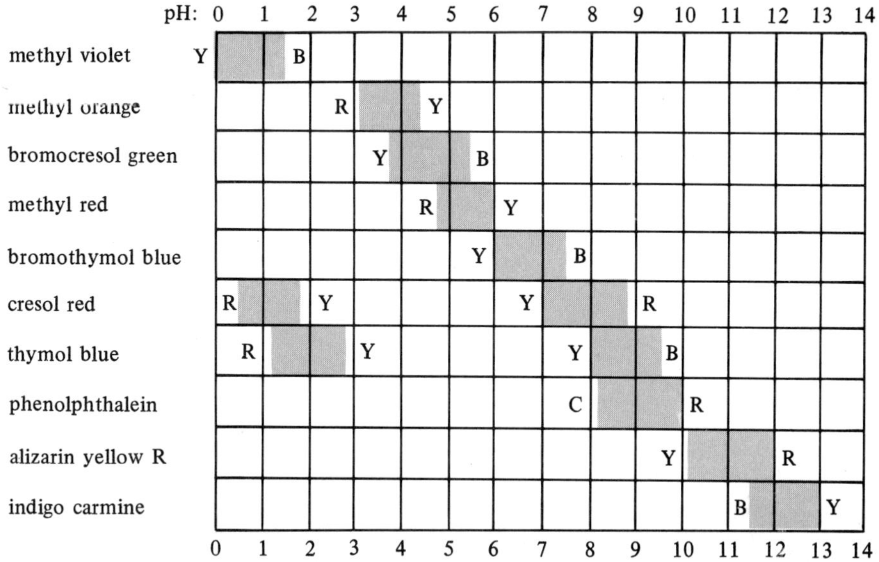

13.10 ACID–BASE TITRATIONS

FIGURE 13.1

	pH:	0	1	2	3	4	5	6	7	8	9	10	11	12	13	14
methyl violet		Y	B													
methyl orange					R	Y										
bromocresol green						Y	B									
methyl red						R	Y									
bromothymol blue								Y	B							
cresol red		R	Y					Y		R						
thymol blue		R		Y					Y	B						
phenolphthalein									C		R					
alizarin yellow R											Y	R				
indigo carmine												B	Y			
		0	1	2	3	4	5	6	7	8	9	10	11	12	13	14

Color changes of indicators. Abbreviations: B = blue, C = colorless, R = red, Y = yellow.

The range of pH in which an indicator may be useful is dependent on the method of determining the ratio of [HIn]/[In⁻]. In general the human eye is capable of detecting a color change between values of [HIn]/[In⁻] from 0.10 to 10. Therefore, when an indicator is observed to change color, the [H⁺] is in the range from $10 K_{HIn}$ to $0.10 K_{HIn}$ (or the pH = pK ± 1). Within this range the pH of a solution may be estimated to an accuracy of about 0.1 by visual observation using appropriate color standards. The colors of some common indicators at different pH's are shown in Figure 13.1.

13.10 ACID–BASE TITRATIONS

A fairly common problem in laboratory work is the determination of the amount or the concentration of an acid or base in a particular solution, usually accomplished by a procedure called an **acid–base titration.** A portion of the solution to be analyzed is placed in a flask, an indicator is added, and then a solution of known concentration is added from a buret (Figure 13.2). For example, we may have 50.0 mL of HCl of unknown concentration in the flask and add 0.100M NaOH from the buret. As the titration proceeds the solution in the flask becomes progressively less acidic because of the neutralization reaction

$$H^+ + OH^- \longrightarrow H_2O$$

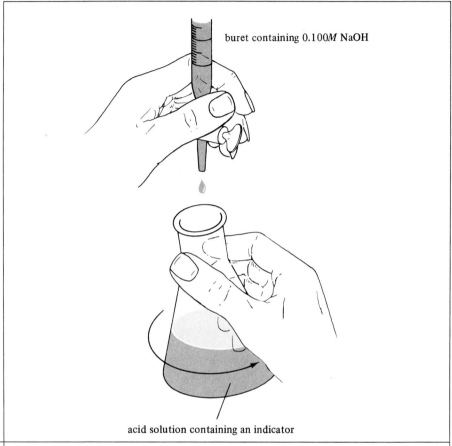

FIGURE 13.2

acid solution containing an indicator

An acid–base titration. The buret is read before any NaOH is added to the acid solution, and it is read again at the end point, when the indicator changes color. The difference between the two readings gives the volume of NaOH solution used.

When the number of moles of base added is equal to the number of moles of acid originally present (a point called the **equivalence point**), the solution is neutral (pH = 7). As more base is added the solution becomes progressively more basic. The change in pH during the titration is shown in Figure 13.3. Note that near the equivalence point the pH changes very rapidly on addition of a small amount of base. This point is therefore easy to detect using indicators (Section 13.9). An indicator that changes color at or near the equivalence point is chosen. The point at which the indicator changes color is called the **end point.**

From the volume of base required to reach the end point, it is an easy matter to calculate the concentration of the HCl solution. For example, if 35.0 mL of 0.100M NaOH are required, we know that at the end point we have added

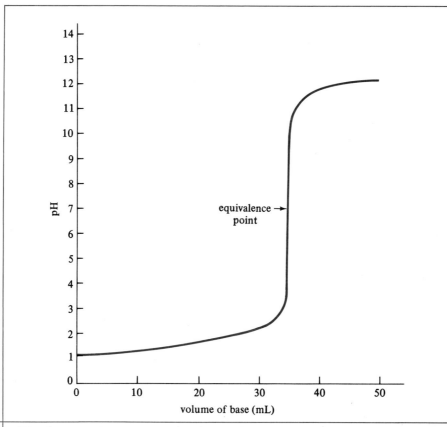

FIGURE 13.3 Titration curve for a strong acid and strong base (0.100 M NaOH added to 50.0 mL of 0.0700 M HCl).

(0.0350 L)(0.100 mol L^{-1}) = 0.00350 mol of NaOH. Thus there must have been 0.00350 mol of HCl present in the original 50.0-mL sample, and therefore the concentration of HCl must have been (0.00350 mol)/(0.0500 L) = 0.0700M.

To see why the pH of the solution changes as shown in Figure 13.3, we may calculate the pH at any point in the titration. For example, after adding 30.0 mL of base, we have neutralized (0.0300 L)(0.100 mol L^{-1}) = 0.00300 mol of H$^+$, leaving 0.00350 − 0.00300 = 0.00050 mol of H$^+$ in solution. As the total volume is now 50.0 + 30.0 = 80.0 mL, the concentration of H$^+$ is (0.00050 mol)/(0.0800 L) = 0.00625M, and the pH is −log(0.00625) = 2.20. This is not much higher than the pH of the original sample, −log(0.0700) = 1.16.

At the equivalence point, all the HCl has been neutralized. There are, of course, small but equal amounts of H$^+$ and OH$^-$ from the dissociation of water. Although both Na$^+$ and Cl$^-$ are present, neither has any effect on this equilibrium, and the pH is the same as that of pure water, namely 7.00.

Beyond the equivalence point the excess OH⁻ makes the solution basic. For example, if 50.0 mL of 0.100M NaOH was added, the number of moles of OH⁻ present would be $0.00500 - 0.00350 = 0.00150$. As the volume would now be $50.0 + 50.0 = 100.0$ mL, the concentration of OH⁻ would be (0.00150 mol)/(0.1000 L) = 0.0150M. To calculate the extremely small concentration of H⁺ present, we resort to the equilibrium condition for the dissociation of water (Eq. 13.4):

$$[H^+] = \frac{1.0 \times 10^{-14}}{[OH^-]} = \frac{1.0 \times 10^{-14}}{0.0150} = 6.7 \times 10^{-13}$$

The pH is $-\log(6.7 \times 10^{-13}) = 12.17$.

Titration of a *weak* acid by a strong base follows a curve of the type shown in Figure 13.4. In comparison to the strong acid–strong base curve, it is displaced toward higher pH levels in the early stages of the titration, and the equivalence point occurs at a pH higher than 7. Beyond the equivalence point the two curves are nearly identical.

The reason for higher pH's in the early stages of the titration is simply that because the acid is weak, it does not dissociate very much, and the concentration of H⁺ is therefore relatively low. To calculate the pH at any point in this region we note that the solution is essentially a buffer solution (Section 13.7), because it contains both the weak acid and its ion at appreciable concentrations. The net reaction during the titration is

$$HX + OH^- \longrightarrow X^- + H_2O \tag{13.25}$$

where HX and X⁻ symbolize the weak acid and its ion, respectively.

The exact curve depends on the dissociation constant of the weak acid. Thus at the halfway point in the titration, half of the acid has been converted to its ion by Eq. 13.25, $[HX] = [X^-]$, $[H^+] = K_a$, and $pH = pK_a$. Figure 13.4 is drawn for the titration of 50.0 mL of 0.0700M CH₃CO₂H, whose dissociation constant is 1.8×10^{-5}. At the halfway point, 17.5 mL of 0.100M NaOH has been added, $[H^+] = 1.8 \times 10^{-5}$, and $pH = -\log(1.8 \times 10^{-5}) = 4.75$. (The theory of buffer solutions explains why the curve is almost horizontal in this region.)

Let us calculate the pH after addition of 30.0 mL of 0.100M NaOH. At this point we have added 0.00300 mol of OH⁻, producing 0.00300 mol of CH₃CO₂⁻, and leaving $0.00350 - 0.00300 = 0.00050$ mol of CH₃CO₂H. As the total volume is now about $30.0 + 50.0 = 80.0$ mL, the concentrations of acetic acid and acetate ion are therefore 0.00625M and 0.0375M, respectively. Substituting these values into the equilibrium condition (Eq. 13.10) gives

$$[H^+] = \frac{1.8 \times 10^{-5}[CH_3CO_2H]}{[CH_3CO_2^-]} = \frac{(1.8 \times 10^{-5})(0.00625)}{0.0375}$$
$$= 3.0 \times 10^{-6}$$

and the pH is $-\log(3.0 \times 10^{-6}) = 5.52$.

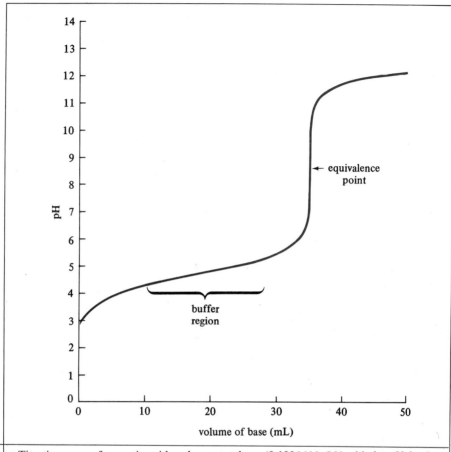

FIGURE 13.4 Titration curve for acetic acid and a strong base (0.100M NaOH added to 50.0 mL of 0.0700M CH$_3$CO$_2$H).

At the equivalence point just enough base has been added to neutralize all the acetic acid, so the solution is identical with a solution of sodium acetate. But a solution of sodium acetate is basic because of hydrolysis (Section 13.7). This is another way of saying that reaction 13.25 does not go to completion when an equal number of moles of OH$^-$ and CH$_3$CO$_2$H are mixed but, instead, reaches an equilibrium position where small but significant amounts of OH$^-$ and CH$_3$CO$_2$H are present.

To calculate the pH at the equivalence point, we note that there is 0.00350 mol of sodium acetate in 0.085 L or 0.041M CH$_3$CO$_2$Na. By using the method outlined in Section 13.7, we find that

$$[OH^-]^2 = (0.041)(5.6 \times 10^{-10})$$

$$[OH^-] = 4.8 \times 10^{-6} M$$

$$[H^+] = \frac{1.0 \times 10^{-14}}{4.8 \times 10^{-6}} = 2.1 \times 10^{-9} M$$

and the pH is $-\log (2.1 \times 10^{-9}) = 8.68$.

A suitable indicator for this titration is phenolphthalein, which changes from colorless to red as the pH changes from 8 to 10. The same indicator is often used for strong acid–strong base titrations. Their titration curves are so steep in this region (see Figure 13.3) that the difference between the end point and equivalence point introduces only a slight error.

13.11 POLYPROTIC ACIDS

Many acids have more than one hydrogen per molecule that may dissociate to produce hydrogen ion. These acids are called **polyprotic acids.** If there are two dissociable hydrogens per molecule of acid, the acid is called a **diprotic acid.** Sulfuric acid is a diprotic acid and it is a strong acid with regard to loss of the first proton in aqueous solution to produce bisulfate ion:

$$H_2SO_4 \longrightarrow H^+ + HSO_4^-$$

It is a slightly weak acid with regard to loss of the second proton:

$$HSO_4^- \rightleftharpoons H^+ + SO_4^{2-} \qquad K = 1.0 \times 10^{-2} \tag{13.26}$$

In general the dissociation of protons from polyprotic acids occurs stepwise and the acids are often weak acids.

As an example of a diprotic acid, let us choose carbonic acid (H_2CO_3). Carbonic acid solutions may be prepared by dissolving carbon dioxide in water. These aqueous solutions of carbon dioxide are extremely important to understand because they provide an explanation of the method of transport of carbon dioxide by the blood from the cells, in which the carbon dioxide is produced, to the lungs, from which it is exhaled to the atmosphere. The concentration of dissolved carbon dioxide in an aqueous solution at 25 °C in equilibrium with gaseous carbon dioxide at 1 atm pressure is $0.0337 M$. This solution is acidic, and its acidic properties may be explained assuming that the dissolved carbon dioxide (which would be solvated) is present as the diprotic acid, H_2CO_3. (As we shall see in Chapter 18, part of the dissolved carbon dioxide is present as H_2CO_3 and part is present as $CO_2(aq)$, but the symbol "H_2CO_3" may be used to represent total dissolved carbon dioxide in both forms to simplify the discussion.)

Just as in the case of sulfuric acid, the acidic hydrogens of carbonic acid dissociate stepwise. The first dissociation may be expressed as

13.11 POLYPROTIC ACIDS

$$H_2CO_3 \rightleftharpoons H^+ + HCO_3^- \quad K_1 = 4.3 \times 10^{-7} \quad (13.27)$$

whereas the second dissociation is given by

$$HCO_3^- \rightleftharpoons H^+ + CO_3^{2-} \quad K_2 = 4.7 \times 10^{-11} \quad (13.28)$$

Thus a solution of carbonic acid contains H^+, H_2CO_3, HCO_3^-, CO_3^{2-} and OH^- from the dissociation of water. Let us utilize Eqs. 13.27 and 13.28 to calculate the fraction of each CO_3-containing species in solution as a function of the pH of the solution.

From Eqs. 13.27 and 13.28 we have

$$[H_2CO_3] = \frac{[H^+][HCO_3^-]}{4.3 \times 10^{-7}} \quad (13.29)$$

$$[CO_3^{2-}] = \frac{4.7 \times 10^{-11}[HCO_3^-]}{[H^+]} \quad (13.30)$$

The material balance equation would be

$$c_i = [H_2CO_3] + [HCO_3^-] + [CO_3^{2-}]$$

where c_i is the concentration of carbonate species used to prepare the solution. The first two equations may be substituted in the material balance equation to obtain

$$c_i = \frac{[H^+][HCO_3^-]}{4.3 \times 10^{-7}} + [HCO_3^-] + \frac{4.7 \times 10^{-11}[HCO_3^-]}{[H^+]}$$

Solving this equation for $[HCO_3^-]$, we obtain

$$[HCO_3^-] = \frac{c_i}{1 + \dfrac{[H^+]}{4.3 \times 10^{-7}} + \dfrac{4.7 \times 10^{-11}}{[H^+]}}$$

This last equation provides the desired basis for determining the relative proportions of the different CO_3-containing species at various pH's. For example, at a pH of 7.0, we have $[H^+] = 1.0 \times 10^{-7}$, and

$$[HCO_3^-] = \frac{c_i}{1 + \dfrac{1.0 \times 10^{-7}}{4.3 \times 10^{-7}} + \dfrac{4.7 \times 10^{-11}}{1.0 \times 10^{-7}}} = \frac{c_i}{1 + 0.23 + (4.7 \times 10^{-4})}$$

$$= 0.81\, c_i$$

That is, 81% of the total of the CO_3-containing species is present as HCO_3^-. Now we may use Eqs. 13.29 and 13.30 to estimate the percentages of H_2CO_3 and CO_3^{2-}.

$$[H_2CO_3] = \frac{1.0 \times 10^{-7}}{4.3 \times 10^{-7}} \times 0.81 c_i = 0.19 c_i$$

$$[CO_3^{2-}] = \frac{4.7 \times 10^{-11}}{1.0 \times 10^{-7}} \times 0.81 c_i = 3.8 \times 10^{-4} c_i$$

Thus at pH = 7.0, we have

81% as HCO_3^-
19% as H_2CO_3
0.038% as CO_3^{2-}

The calculations may be repeated at other pH's. The results are presented graphically in Figure 13.5.

The vertical line at pH = 8.35 corresponds to a solution of $NaHCO_3$, as we shall show in the following paragraph. Below a pH of 8.35, $[CO_3^{2-}]$ will be negligible and we can disregard the second dissociation of H_2CO_3, treating H_2CO_3 solutions in the same manner as we did for a monoprotic acid. Above a pH of 8.35 we need not consider $[H_2CO_3]$ and we can treat the solution the same as we did for a solution containing a monoprotic acid and its salt. The hydrolysis of CO_3^{2-},

$$CO_3^{2-} + H_2O = HCO_3^- + OH^- \qquad K_{h_2} = \frac{K_w}{K_2}$$

may be treated the same as the hydrolysis of the salt of a monoprotic acid.

Thus the only new situation encountered is for a $NaHCO_3$ solution. The chemistry of HCO_3^- is complicated because it can dissociate to give H^+ and CO_3^{2-} (Eq. 13.28) or it can hydrolyze to produce H_2CO_3 and OH^-,

$$HCO_3^- + H_2O \rightleftharpoons H_2CO_3 + OH^- \qquad (13.31)$$

In addition, the H^+ produced by dissociation can combine with the OH^- produced by hydrolysis to form water. The sum of these three equations gives Eq. 13.32.

$$HCO_3^- \rightleftharpoons H^+ + CO_3^{2-} \qquad K_2 = \frac{[H^+][CO_3^{2-}]}{[HCO_3^-]}$$

$$HCO_3^- + H_2O \rightleftharpoons H_2CO_3 + OH^- \qquad K_{h_1} = \frac{K_w}{K_1} = \frac{[H_2CO_3][OH^-]}{[HCO_3^-]}$$

$$H^+ + OH^- \rightleftharpoons H_2O \qquad \frac{1}{K_w} = \frac{1}{[H^+][OH^-]}$$

$$2HCO_3^- \rightleftharpoons H_2CO_3 + CO_3^{2-} \qquad K_d = \frac{[H_2CO_3][CO_3^{2-}]}{[HCO_3^-]^2} \qquad (13.32)$$

13.11 POLYPROTIC ACIDS

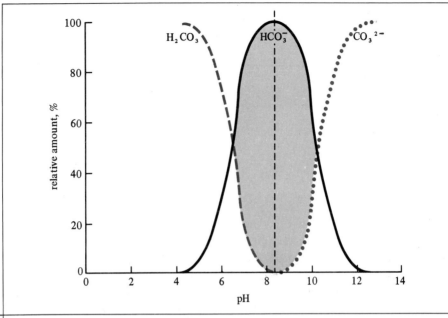

FIGURE 13.5

Effect of pH on the relative amounts of H_2CO_3, HCO_3^-, and CO_3^{2-} in an aqueous solution.

Because we obtained Eq. 13.32 by adding three equilibria, the equilibrium constant K_d is the product of the three equilibrium constants:

$$K_d = \frac{K_2 K_{h_1}}{K_w} = \frac{K_2 K_w}{K_1 K_w} = \frac{K_2}{K_1} = 1.1 \times 10^{-4}$$

The reaction shown in Eq. 13.32 is an example of a type of reaction called **disproportionation,** in which a single substance changes into two different substances. In this case HCO_3^- acts as both an acid and a base in the Brønsted-Lowry sense, with one HCO_3^- losing its proton to another HCO_3^- ion. As this reaction produces H_2CO_3 molecules and CO_3^{2-} ions in equal numbers, we have $[H_2CO_3] = [CO_3^{2-}]$.

Although the disproportionation reaction is the principal equilibrium in the solution, we note that the equilibrium constant for hydrolysis ($K_{h_1} = 2.3 \times 10^{-8}$) is somewhat greater than the equilibrium constant for dissociation ($K_2 = 4.7 \times 10^{-11}$). Therefore the hydrolysis proceeds farther to the right than the dissociation, producing a small excess of OH^-, which makes the solution slightly basic. (For the same reason the concentration of H_2CO_3 is slightly larger than that of CO_3^{2-}, but the difference is so small that $[H_2CO_3] \approx [CO_3^{2-}]$.)

The quickest way to calculate the pH of the solution is to multiply the dissociation constants:

$$K_1 K_2 = \frac{[H^+][HCO_3^-]}{[H_2CO_3]} \times \frac{[H^+][CO_3^{2-}]}{[HCO_3^-]} = \frac{[H^+]^2[CO_3^{2-}]}{[H_2CO_3]}$$

Because $[CO_3^{2-}] \approx [H_2CO_3]$, these terms cancel in the last expression, and

$$[H^+]^2 = K_1 K_2 = (4.3 \times 10^{-7})(4.7 \times 10^{-11})$$
$$[H^+] = \sqrt{K_1 K_2} = 4.5 \times 10^{-9} M$$

or

$$pH = 8.35$$

Thus a solution of $NaHCO_3$ is slightly basic, and its pH is independent of the concentration of $NaHCO_3$.

The following example illustrates a buffer problem for a diprotic acid—in this case a carbonate-bicarbonate buffer.

EXAMPLE 7

Calculate the pH of a buffer solution prepared by dissolving 0.010 mol of $NaHCO_3$ and 0.020 mol of Na_2CO_3 in water and diluting the solution to 400 mL.

The initial concentrations of ions before any reaction occurs other than the dissolution of the $NaHCO_3$ and Na_2CO_3 are

$$[HCO_3^-]_i = \frac{0.010 \text{ mol}}{0.400 \text{ L}} = 0.025 M$$

$$[CO_3^{2-}]_i = \frac{0.020 \text{ mol}}{0.400 \text{ L}} = 0.050 M$$

$$[Na^+]_i = \frac{0.010 \text{ mol} + (2 \times 0.020 \text{ mol})}{0.400 \text{ L}} = 0.125 M$$

The following equilibria would be present:

$HCO_3^- = H^+ + CO_3^{2-}$ $K_2 = 4.7 \times 10^{-11}$

$H_2O = H^+ + OH^-$ $K_w = 1.0 \times 10^{-14}$

$CO_3^{2-} + H_2O = HCO_3^- + OH^-$ $K_{h_2} = \dfrac{K_w}{K_2} = 2.1 \times 10^{-4}$

Since K_{h_2} is much larger than K_2 or K_w, we shall neglect the first two equilibria and select the hydrolysis equilibrium as the most important reaction.
Let y = number of moles of CO_3^{2-} in 1 L that hydrolyze.
Then $[OH^-] = y$, and we obtain

$$[HCO_3^-] = [HCO_3^-]_i + y = 0.025 + y \approx 0.025$$
$$[CO_3^{2-}] = [CO_3^{2-}]_i - y = 0.050 - y \approx 0.050$$

That is, we shall neglect y compared to 0.025 or 0.050.
Using the above values of [OH⁻], [HCO$_3^-$] and [CO$_3^{2-}$],

$$K_{h_2} = \frac{[HCO_3^-][OH^-]}{[CO_3^{2-}]} = \frac{0.025 y}{0.050} = 2.1 \times 10^{-4}$$

and

$$y = 4.2 \times 10^{-4} = [OH^-]$$

We note that y is negligible compared to 0.025 or 0.050, so the approximations were justified.

Finally, we calculate the concentration of H⁺ and the pH.

$$[H^+] = \frac{1.0 \times 10^{-14}}{4.2 \times 10^{-4}} = 2.4 \times 10^{-11} M$$

and

$$pH = -\log(2.4 \times 10^{-11}) = 10.62$$

13.12 SULFIDE PRECIPITATIONS

Most metals form binary compounds with sulfur, called **sulfides.** In qualitative analysis the metallic ions are often separated into groups by means of the differing solubility of their sulfides. This selective precipitation is a very useful procedure, and we shall examine its general features.

To precipitate the sulfides one uses hydrogen sulfide, H$_2$S, which is a gas—a highly toxic gas with a vile odor—at ordinary temperatures and pressures.[2] It is readily soluble in water, and, at 1 atm pressure and 25°C, its solubility is 0.10 M. It is a diprotic acid, considerably weaker than carbonic acid:

$$H_2S \rightleftharpoons H^+ + HS^- \quad K_1 = 1.0 \times 10^{-7} \quad (13.33)$$

$$HS^- \rightleftharpoons H^+ + S^{2-} \quad K_2 = 1.3 \times 10^{-14} \quad (13.34)$$

The metallic sulfides differ markedly in their solubility in water. Whereas the sulfides of the group IA and IIA metals are readily soluble in water, the sulfides of many of the transition metals are only slightly soluble, and the sulfides of Cu^{2+}, Ag$^+$, Hg^{2+}, and Pb^{2+} are practically insoluble. These large differences in solubility may be used to effect a separation of the metals from a solution containing a mixture of different metallic ions. The solution is first acidified to a hydrogen ion concentration of 0.3 M by addition of the appropriate amount of a strong acid such as HCl, and then saturated with H$_2$S. At this point those ions whose sulfides are least soluble will precipitate. For example,

[2] Hydrogen sulfide is often generated within a solution by hydrolysis of an organic compound called thioacetamide, CH$_3$CSNH$_2$:

$$CH_3CSNH_2 + 2H_2O = CH_3CO_2^- + NH_4^+ + H_2S$$

$$Cu^{2+} + H_2S \rightleftharpoons CuS(s) + 2H^+$$

After separation of the sulfide precipitate from the solution, the pH of the solution is increased by addition of a base, such as ammonia, and the solution is again saturated with H_2S. The more soluble metal sulfides (such as ZnS and FeS) now precipitate, leaving the group IA and IIA ions in solution.

A more detailed understanding of the basis of this separation requires consideration of the equilibria involved. When a solid sulfide of formula MS is formed, the following equilibrium is established:

$$MS(s) \rightleftharpoons M^{2+} + S^{2-} \qquad K_{sp} = [M^{2+}][S^{2-}]$$

Table 12.4 lists the solubility products of several metal sulfides. To determine whether precipitation or dissolution will occur under given conditions, we compare the ion product $[M^{2+}][S^{2-}]$ with K_{sp}. If the product of the ion concentrations is larger than K_{sp}, precipitation occurs until the concentrations fall to the levels at which their product equals K_{sp}. If smaller, nothing happens. Thus our problem is to control the S^{2-} concentration. This we can do by adjusting the $[H^+]$, for S^{2-} is the ion of the weak diprotic acid H_2S.

To obtain an expression relating $[S^{2-}]$ to $[H^+]$ and $[H_2S]$, we eliminate the $[HS^-]$ in Eqs. 13.32 and 13.33 by multiplying the equilibrium constants:

$$K_1 K_2 = \frac{[H^+][HS^-]}{[H_2S]} \times \frac{[H^+][S^{2-}]}{[HS^-]} = \frac{[H^+]^2[S^{2-}]}{[H_2S]}$$

Rearranging this equation to obtain an explicit expression for $[S^{2-}]$ produces

$$[S^{2-}] = \frac{K_1 K_2 [H_2S]}{[H^+]^2} = \frac{1.3 \times 10^{-21}[H_2S]}{[H^+]^2} \qquad (13.35)$$

If we have a solution saturated with H_2S, then $[H_2S] = 0.10 M$. If the solution also contains $0.3 M\ H^+$, then Eq. 13.35 reduces to

$$[S^{2-}] = \frac{(1.3 \times 10^{-21})(0.10)}{(0.3)^2} \approx 1 \times 10^{-21} M \qquad (13.36)$$

We may now ask: Which metal ions will precipitate from this solution if, say, $[M^{2+}] = 0.001 M$? To answer this question, we compare the solubility products with the product of the ion concentrations:

$$[M^{2+}][S^{2-}] = (0.001)(1 \times 10^{-21}) = 1 \times 10^{-24}$$

The sulfides whose solubility products are smaller than this number (such as CuS) will precipitate from a solution containing $0.3 M\ H^+$ saturated with H_2S; the others (such as ZnS) will not.

13.13 COMPLEX IONS

Any species in solution that can be formed from two or more simpler species that can exist in solution is known as a complex. If the complex has a positive or negative charge, it is known as a **complex ion.** Many of the metallic ions form complexes with neutral molecules or anions of the nonmetals. The neutral molecules or anions attached to the central metallic ion are known as **ligands.** The silver–ammonia complex is typical,

$$Ag^+ + 2NH_3 \rightleftharpoons Ag(NH_3)_2^+ \qquad K = 1.7 \times 10^7 \text{ at } 25\,°C$$

Because the equilibrium is written to show the formation of the complex ion, the corresponding equilibrium constant is called an association or **formation constant.** From the formation constant one may calculate the ratio of complexed to uncomplexed silver at any given concentration of uncomplexed ammonia. Thus, if the concentration of ammonia $[NH_3]$ is $0.0100M$, then from

$$\frac{[Ag(NH_3)_2^+]}{[Ag^+][NH_3]^2} = 1.7 \times 10^7$$

we find that

$$\frac{[Ag(NH_3)_2^+]}{[Ag^+]} = (1.7 \times 10^7)[NH_3]^2 = (1.7 \times 10^7)(0.0100)^2$$
$$= 1.7 \times 10^3$$

Thus even at this very low concentration of ammonia, practically all the silver is in the complexed form. At higher concentrations of NH_3 the equilibrium is shifted to the right, in accordance with Le Châtelier's principle, and even more of the silver is complexed. It should be noted that the ratio of silver in the two forms does not depend on the total amount of silver present.

Although all cations are hydrated in aqueous solution and therefore could be considered to be complex ions, the term "complex ion" is usually used when the ligand is other than a water molecule. Some of the more common ligands are chloride ion, hydroxide ion, cyanide ion, thiosulfate ion, and ammonia. The Lewis structures of these ligands,

all have one feature in common, namely, the presence of one or more lone pairs of electrons. The reaction of any one of these ligands with a cation to produce a complex ion is simply a Lewis acid–base reaction. In the case of the silver-ammonia complex ion, the Ag^+ is the Lewis acid and the ammonia is the Lewis base.

In many cases complex ion formation proceeds stepwise.[3] The complexes of Hg^{2+} and Cl^- exemplify this phenomenon:

$$Hg^{2+} + Cl^- \rightleftharpoons HgCl^+ \qquad K_1 = 5.5 \times 10^6 \qquad (13.37a)$$

$$HgCl^+ + Cl^- \rightleftharpoons HgCl_2 \qquad K_2 = 3.0 \times 10^6 \qquad (13.37b)$$

$$HgCl_2 + Cl^- \rightleftharpoons HgCl_3^- \qquad K_3 = 7.1 \qquad (13.37c)$$

$$HgCl_3^- + Cl^- \rightleftharpoons HgCl_4^{2-} \qquad K_4 = 10 \qquad (13.37d)$$

The calculation of the equilibrium concentrations of the several complexes is quite similar to that of the calculation of $[H^+]$ in a solution of a polyprotic acid, as shown in the following example.

EXAMPLE 8

Calculate the equilibrium concentrations of all the mercury-containing species in a solution prepared by dissolving 0.0010 mol of $HgCl_2$ in 1.00 L of $0.100M$ HCl.

We have six unknown concentrations, namely: $[Hg^{2+}]$, $[HgCl^+]$, $[HgCl_2]$, $[HgCl_3^-]$, $[HgCl_4^{2-}]$, and $[Cl^-]$. Thus we need two equations in addition to the four equilibrium expressions. These may be obtained by using material balances for mercury and for chlorine. (Alternatively we could use a charge balance for one of these equations.)

All the mercury species are produced from the 0.0010 mol of $HgCl_2$, so we may write

$$0.0010 = [Hg^{2+}] + [HgCl^+] + [HgCl_2] + [HgCl_3^-] + [HgCl_4^{2-}] \qquad (13.38)$$

The total chlorine in the system comes from the chlorine in both $HgCl_2$ and HCl. There is $2 \times 0.0010 = 0.0020$ mol of Cl atoms from the $HgCl_2$, and 0.100 mol of Cl atoms from the HCl, or a total of 0.102 mol of Cl atoms. Thus

$$0.102 = [Cl^-] + [HgCl^+] + 2[HgCl_2] + 3[HgCl_3^-] + 4[HgCl_4^{2-}] \qquad (13.39)$$

(The coefficients of 2, 3, and 4 in this equation are necessary because each mole of $HgCl_2$ contains 2 mol of Cl atoms, etc.)

Although the six equations may be solved exactly, often (as in this example) it is much simpler and faster to solve the equations by using approximations. For instance, from the material balance equations, we note that the smallest possible value of $[Cl^-]$ would be found if all the mercury were fully complexed as $HgCl_4^{2-}$. In this case $[Cl^-] = 0.102 - (4 \times 0.0010) = 0.098M$. The largest possible value of $[Cl^-]$ would be found if none of the mercury were complexed by chloride ion. In this case $[Cl^-] = 0.102M$. Because these extreme values are so close together, we may assume that $[Cl^-] \approx 0.10M$. The extent of formation of the mercury complexes can now be ascertained by using their formation constants.

---

[3] $Ag(NH_3)_2^+$ is also formed stepwise, but $Ag(NH_3)^+$ is unimportant except at very low ammonia concentrations.

13.13 COMPLEX IONS

Because K_1 and K_2 are very large, the first two equilibria (Eqs. 13.37a and 13.37b) must be very far to the right, and therefore $[Hg^{2+}]$ and $[HgCl^+]$ will be negligible compared to $[HgCl_2]$. Thus we may neglect the first two terms on the right-hand-side of Eq. 13.38, and

$$0.0010 \approx [HgCl_2] + [HgCl_3^-] + [HgCl_4^{2-}] \tag{13.40}$$

We now solve this equation by obtaining expressions for the last two unknown concentrations in terms of the first unknown concentration, $[HgCl_2]$, as follows. For the third equilibrium, Eq. 13.37c, the law of chemical equilibrium states that

$$\frac{[HgCl_3^-]}{[HgCl_2][Cl^-]} = K_3$$

or

$$[HgCl_3^-] = K_3[Cl^-][HgCl_2] \tag{13.41}$$

For the fourth equilibrium, Eq. 13.37d, a similar relation holds.

$$\frac{[HgCl_4^{2-}]}{[HgCl_3^-][Cl^-]} = K_4$$

and

$$[HgCl_4^{2-}] = K_4[Cl^-][HgCl_3^-]$$

On substituting the expression previously obtained for $[HgCl_3^-]$, we find that

$$[HgCl_4^{2-}] = K_4 K_3 [Cl^-]^2 [HgCl_2] \tag{13.42}$$

Eqs. 13.41 and 13.42 are now used to solve Eq. 13.40.

$$\begin{aligned} 0.0010 &= [HgCl_2] + K_3[Cl^-][HgCl_2] + K_4 K_3 [Cl^-]^2 [HgCl_2] \\ &= [1 + (7.1 \times 0.10) + (10 \times 7.1 \times 0.10^2)][HgCl_2] \\ &= 2.42[HgCl_2] \end{aligned}$$

$$[HgCl_2] = \frac{1.0 \times 10^{-3}}{2.42} = 4.1 \times 10^{-4} M$$

$$\begin{aligned} [HgCl_3^-] &= K_3[Cl^-][HgCl_2] = (7.1)(0.10)(4.1 \times 10^{-4}) \\ &= 2.9 \times 10^{-4} M \end{aligned}$$

$$\begin{aligned} [HgCl_4^{2-}] &= K_4 K_3 [Cl^-]^2 [HgCl_2] = (10)(7.1)(0.10)^2(4.1 \times 10^{-4}) \\ &= 2.9 \times 10^{-4} M \end{aligned}$$

$$[HgCl^+] = \frac{[HgCl_2]}{K_2[Cl^-]} = \frac{4.1 \times 10^{-4}}{(3.0 \times 10^6)(0.10)} = 1.4 \times 10^{-9} M$$

$$[Hg^{2+}] = \frac{[HgCl^+]}{K_1[Cl^-]} = \frac{1.4 \times 10^{-9}}{(5.5 \times 10^6)(0.10)} = 2.5 \times 10^{-15} M$$

These last two results show that our approximation in neglecting $[HgCl^+]$ and $[Hg^{2+}]$ was justified.

13.14 USE OF ACIDS OR LIGANDS TO DISSOLVE SLIGHTLY SOLUBLE ELECTROLYTES

In Section 12.6 we used Le Châtelier's principle to explain the much greater solubility of $Cu(OH)_2$ in dilute HCl solutions compared to its solubility in water. The two equilibria involved are

$$Cu(OH)_2(s) \rightleftharpoons Cu^{2+} + 2OH^- \qquad K_{sp} = 4.8 \times 10^{-20}$$

and

$$2H^+ + 2OH^- \rightleftharpoons 2H_2O \qquad \frac{1}{K_w^2} = \left[\frac{1}{1.0 \times 10^{-14}}\right]^2 = 1.0 \times 10^{28}$$

The sum of the two equilibria is

$$Cu(OH)_2(s) + 2H^+ \rightleftharpoons Cu^{2+} + 2H_2O \qquad K = \frac{K_{sp}}{K_w^2} = 4.8 \times 10^8$$

The extremely large value of the equilibrium constant for the reaction of $Cu(OH)_2$ with H^+ indicates that $Cu(OH)_2$ is very soluble in dilute acids. Similar methods may be used to predict or calculate the solubility of any electrolyte containing the anion of a weak acid in acidic solutions.

In a similar manner we can use solutions of a ligand to increase the solubility of electrolytes that contain a cation capable of forming a complex ion with the ligand. As an example, let us consider the solubility of silver chloride in water and in $3.0M$ NH_3. The two equilibria involved in $3.0M$ NH_3 are

$$AgCl(s) \rightleftharpoons Ag^+ + Cl^- \qquad K_{sp} = 1.7 \times 10^{-10}$$

and

$$Ag^+ + 2NH_3 \rightleftharpoons Ag(NH_3)_2^+ \qquad K_2 = 1.7 \times 10^7$$

The sum of the two equilibria gives

$$AgCl(s) + 2NH_3 \rightleftharpoons Ag(NH_3)_2^+ + Cl^- \qquad K = K_{sp}K_2 = 2.9 \times 10^{-3}$$

Thus we expect silver chloride to be much more soluble in $3.0M$ NH_3 than in water. In the following example we shall see how much more soluble it is.

EXAMPLE 9

The solubility of silver chloride in water is $1.3 \times 10^{-5}M$. Calculate the solubility of AgCl in $3.0M$ NH_3 given that

$$AgCl(s) + 2NH_3 \rightleftharpoons Ag(NH_3)_2^+ + Cl^- \qquad K = 2.9 \times 10^{-3}$$

Let the solubility of AgCl in $3.0 M$ $NH_3 = s$. Then

$$[Ag(NH_3)_2^+] = [Cl^-] = s$$

The material balance equation for NH_3 species is

$$[NH_3] = [NH_3]_i - 2[Ag(NH_3)_2^+] = 3.0 - 2s$$

Let us assume that $2s$ is negligible compared to 3.0 so that

$$[NH_3] \approx 3.0$$

On substituting these values into the law of chemical equilibrium, we obtain

$$\frac{[Ag(NH_3)_2^+][Cl^-]}{[NH_3]^2} \approx \frac{s^2}{(3.0)^2} = 2.9 \times 10^{-3}$$

$$s = 3.0 \times (2.9 \times 10^{-3})^{1/2} = 0.16 M$$

We now check our approximation for $[NH_3]$.

$$[NH_3] = 3.0 - 2s = 3.0 - (2 \times 0.16) = 2.7 M$$

which is within 10% of our assumed value. If we need a more precise value for s, we can substitute this new value for $[NH_3]$ in our equations and obtain

$$s = 2.7 \times (2.9 \times 10^{-3})^{1/2} = 0.15 M$$

Now $[NH_3] = 3.0 - 2s = 3.0 - 0.3 = 2.7 M$, in agreement with the value used in the last calculations. So the calculated solubility of AgCl in $3.0 M$ NH_3 is $0.15 M$.

PROBLEMS

13.1 What is the conjugate acid of H_2O? What is the conjugate base of H_2O?

13.2 From the following list select all the pairs of conjugate acids and bases: H_3O^+, H_2O, OH^-, O^{2-}, NH_4^+, NH_3, NH_2^-, $H_3SO_4^+$, H_2SO_4, HSO_4^-, SO_4^{2-}, H_2S, HS^-, S^{2-}

13.3 In the following reactions, which substances are acids (according to the Brønsted-Lowry concept)? Which substances are bases?
$H_3O^+ + CH_3CO_2^- = CH_3CO_2H + H_2O$
$HCO_3^- + OH^- = H_2O + CO_3^{2-}$

13.4 (a) In the Brønsted-Lowry system, is a proton acceptor an acid or a base?
(b) In the Lewis system, is an electron-pair acceptor an acid or a base?

13.5 In the Brønsted-Lowry system, which of the following substances does *not* have a conjugate base?
$H_3SO_4^+$, H_2SO_4, HSO_4^-, SO_4^{2-}

13.6 In the Brønsted-Lowry system, which one of the following substances is most likely to function as an acid in some reactions and as a base in other reactions?
$CH_3CO_2^-$, H_2O, H_3O^+, HCl

13.7 In the following reactions, which substances are acids (according to the Lewis concept)? Which substances are bases?
$$BF_3 + F^- \longrightarrow BF_4^-$$
$$SO_3^{2-} + S \longrightarrow S_2O_3^{2-}$$

13.8 Using the list of conjugate acids and bases in Table 13.1, predict the equilibrium position ($\rightarrow$ or $\leftarrow$) of each of the following reactions.
$$CH_3CO_2H + OH^- = H_2O + CH_3CO_2^-$$
$$NH_4^+ + H_2O = H_3O^+ + NH_3$$
$$HCl + NH_3 = NH_4^+ + Cl^-$$
$$2H_2O = H_3O^+ + OH^-$$

13.9 The following reactions proceed in the direction indicated:
$$HF + CH_3CO_2^- \longrightarrow CH_3CO_2H + F^-$$
$$HF + NH_3 \longrightarrow NH_4^+ + F^-$$
$$H_3O^+ + F^- \longrightarrow HF + H_2O$$
Where would you place the HF/F^- pair in the list of conjugate acids and bases in Table 13.1?

13.10 Calculate the hydrogen ion concentration, the hydroxide ion concentration, and the pH of each of the following:
(a) $1.0\,M$ HCl
(b) $2.0 \times 10^{-4}\,M$ HCl
(c) $1.0 \times 10^{-10}\,M$ HCl
(d) $2.0\,M$ NaOH
(e) $6.0\,M$ HCl
(f) $1.0 \times 10^{-4}\,M$ KOH

13.11 Calculate $[H^+]$ and $[OH^-]$ for each of the following aqueous solutions:
(a) pH = 7.00
(b) pH = 3.69
(c) pH = 8.40
(d) pH = 4.25

13.12 Of two hydrochloric acid solutions, one has a pH of 3.00, the other a pH of 4.00. If equal volumes of the two solutions are mixed, what is the pH of the mixture? (Warning: One cannot simply average the two pH's.)

13.13 How many moles of H^+ are excreted per day if a patient eliminates 1200 mL of urine having a pH of 5.7?

13.14 Using the data in Table 13.3, predict the equilibrium position ($\rightarrow$ or $\leftarrow$) of each of the following reactions.
$$HNO_2 + CN^- = HCN + NO_2^-$$
$$HSO_4^- + NO_2^- = HNO_2 + SO_4^{2-}$$
$$SO_4^{2-} + H_2O = HSO_4^- + OH^-$$
$$CH_3CO_2H + NO_3^- = HNO_3 + CH_3CO_2^-$$

13.15 Write the charge balance equation for a solution prepared by mixing
(a) solutions of HCl and NaOH
(b) solutions of NH_3 and NH_4Cl
(c) solutions of NH_3 and H_2SO_4

13.16 Write a material balance equation in nitrogen-containing species for a solution prepared by mixing
(a) 250 mL of $0.200M$ NH_3 and 250 mL of $0.12M$ HCl
(b) 300 mL of $0.150M$ NH_4Cl and 200 mL of $0.10M$ NaOH
(In each case assume that the final volume is 500 mL.)

13.17 A solution is prepared by dissolving 0.010 mol of CH_3CO_2Na and 0.0020 mol of HCl in enough water to form 1.0 L of solution.
(a) What substances are present in the resulting solution?
(b) Write an equation for each equilibrium that is present.
(c) Calculate the concentrations of all substances not involved in the equilibria listed in part (b).
(d) How many concentrations of substances are unknown? How many equations relating these concentrations are needed for an exact calculation of all concentrations? List the equations you would use.
(e) Solve for all the concentrations stating any assumptions that you use.

13.18 What is the equilibrium concentration of hydrogen ion in $0.20M$ HCN?

13.19 Calculate the equilibrium concentration of hydrogen ion in $0.50M$ NH_3.

13.20 Will a $0.1M$ solution of each of the following salts be acidic, neutral, or basic?
$$CH_3CO_2Na, NH_4Cl, NaCl, CH_3CO_2NH_4, NH_4CN$$

13.21 Sodium cyanide is readily soluble in water and is a strong electrolyte. (It is also extremely poisonous.) Write equations to show the equilibria present in a sodium cyanide solution and calculate the value of the hydrolysis constant. (Use $K = 6.1 \times 10^{-10}$ for the dissociation constant of HCN.) Calculate the concentrations of all species present in a $0.10M$ NaCN solution.

13.22 Verify the equation $K_h = K_w/K_b$ for the hydrolysis of NH_4^+ (Section 13.7).

13.23 Which of the following solutions has the highest pH? Which has the lowest pH? Which has the pH closest to 7?
$1M\ CH_3CO_2Na$, $1M\ NH_3$, $1M\ NaCl$, $1M\ NH_4Cl$, $1M\ CH_3CO_2H$

13.24 HX is a strong acid. HY and HZ are both weak acids, but HZ is weaker than HY. Which of the following solutions has the highest pH?
$1M\ NaX$, $1M\ NaY$, $1M\ NaZ$

13.25 A solution is prepared by mixing equal volumes of $0.10M\ CH_3CO_2H$ and $0.10M\ CH_3CO_2Na$. Which of the following approximations are valid? (There may be more than one correct answer.)
$[Na^+] \approx [CH_3CO_2^-]$ $[H^+] \approx [CH_3CO_2^-]$
$[OH^-] \approx [CH_3CO_2H]$ $[CH_3CO_2H] \approx [CH_3CO_2^-]$

13.26 Calculate the pH of the buffer solution described in the preceding problem.

13.27 What are the equilibrium concentrations of ions and of NH_3 in a solution prepared by dissolving 0.010 mol of NH_3 and 0.020 mol of NH_4Cl in 1.0 L of solution?

13.28 A buffer solution is prepared by dissolving 0.0100 mol of NH_4Cl in 100 mL of $0.100M\ NH_3$.
(a) What is the pH of the resulting solution?
(b) What is the pH if 0.0050 mol of HCl is added?
(c) What is the pH if 0.0050 mol of NaOH is added to the original buffer solution?

13.29 To prepare a buffer solution having a pH of 5.00, a student mixes $1.0M\ CH_3CO_2H$ and $1.0M\ CH_3CO_2Na$. If 100 mL of the acetic acid solution is used, how many mL of the sodium acetate solution should be used?

13.30 A solution is prepared by mixing 0.060 mol of NH_4Cl and 0.020 mol of NaOH in sufficient water to give a final volume of 250 mL. Calculate the equilibrium concentrations of NH_4^+ and OH^- in the resulting solution.

13.31 To determine the dissociation constant of an acid, a solution of the acid was titrated with NaOH, and 26.4 mL of $0.20M\ NaOH$ was required to reach the equivalence point. Then a buffer solution was prepared by adding an equal volume of $0.10M\ HCl$. The pH of the resulting solution was 5.86. Calculate the dissociation constant of the acid.

13.32 The dissociation constant of a weak acid is 1.0×10^{-6}. The acid imparts a red color to a solution at $pH = 1.0$ and a yellow color at $pH = 13.0$. What color would you expect this indicator to show in a solution of pH 5? Of pH 6? Of pH 7?

13.33 Indicator HIn is a weak acid, HIn = H$^+$ + In$^-$. At pH = 4, 91% of the indicator is present as the anion In$^-$, and only 9% is present in the weak acid form HIn. At what pH will there be 50% in each form?

13.34 A certain solution turns yellow when tested with methyl red. A separate portion of the solution turns yellow when tested with thymol blue. Using Figure 13.1, what can you say about the pH of the solution?

13.35 (a) When phenolphthalein is used as an indicator in the titration of 50.00 mL of 0.1000M HCl with 0.1000M NaOH, what volumes of NaOH are required to reach the equivalence point (pH = 7) and the end point (pH = 9)?
(b) How many mL of NaOH are required to go from the equivalence point to the end point? (This represents the error in the titration.)
(c) What error (in percentage) is caused by use of phenolphthalein as the indicator?

13.36 (a) Why are different indicators needed for weak acid–strong base and weak base–strong acid titrations?
(b) Of the acid–base indicators shown in Figure 13.1, which is most suitable for titration of 0.10M NH$_3$ with 0.10M HCl?

13.37 A solution is prepared by mixing equal volumes of 0.100M NaOH and 0.100M CH$_3$CO$_2$H. Which of the following approximations are valid? (There may be more than one correct answer.)
  [H$^+$] ≈ [OH$^-$]   [OH$^-$] ≈ [CH$_3$CO$_2$H]
  [Na$^+$] ≈ [CH$_3$CO$_2^-$]   [H$^+$] ≈ [CH$_3$CO$_2^-$]

13.38 If 50.0 mL of 0.200M formic acid (HCO$_2$H) is placed in a flask and titrated with 0.200M NaOH, what is the pH after addition of 50.0 mL of the sodium hydroxide solution? (Formic acid is a monoprotic acid. Its dissociation constant is 1.8×10^{-4}.)

13.39 In some of the following cases the two substances cannot exist in the same solution at high concentrations at room temperature. For each such case write a balanced equation for the net reaction.
(a) HCO$_3^-$ and OH$^-$
(b) CO$_3^{2-}$ and OH$^-$
(c) CO$_3^{2-}$ and H$_2$CO$_3$
(d) HCO$_3^-$ and H$^+$
(e) H$_2$CO$_3$ and H$^+$
(f) H$_2$CO$_3$ and OH$^-$

13.40 Consider the following three solutions: $0.034M\ H_2CO_3$, $1.0M\ NaHCO_3$, and $1.0M\ Na_2CO_3$. In which of these solutions is the following approximation valid?
(a) $[H_2CO_3] \approx [CO_3^{2-}]$
(b) $[H^+] \approx [HCO_3^-]$
(c) $[OH^-] \approx [HCO_3^-]$

13.41 Excess solid calcium carbonate is in equilibrium with its saturated solution.
$$CaCO_3(s) \rightleftharpoons Ca^{2+} + CO_3^{2-}$$
State whether the equilibrium concentration of Ca^{2+} will increase, decrease, or remain the same when $CO_2(g)$ is bubbled into the solution.

13.42 A sample of human blood has a pH of 7.40. What is the concentration of hydrogen ion? Calculate the percentage of the carbonate in the blood sample that is in each of the following forms: H_2CO_3, HCO_3^-, and CO_3^{2-}.

13.43 Phosphoric acid (H_3PO_4) has three acidic hydrogens. Its dissociation constants are $K_1 = 7.1 \times 10^{-3}$, $K_2 = 6.3 \times 10^{-8}$, and $K_3 = 4.5 \times 10^{-13}$. What form of phosphate (H_3PO_4, $H_2PO_4^-$, HPO_4^{2-}, or PO_4^{3-}) is present in largest concentration in each of the following solutions:
(a) pH = 3
(b) pH = 5
(c) pH = 7
(d) pH = 9
(e) pH = 11
(f) pH = 14

13.44 Complete the calculations in Section 13.11 for a $0.100M\ NaHCO_3$ solution by calculating the concentrations of Na^+, H_2CO_3, HCO_3^-, CO_3^{2-}, and OH^-.

13.45 In the following list, which solution has the highest pH? The lowest pH? The pH closest to 8?
$1.0M\ Na_2CO_3$, $0.5M\ NaHCO_3$, $1.0M\ NH_4Cl$, $0.034M\ H_2CO_3$

13.46 Seawater, which contains H_2CO_3, HCO_3^-, and CO_3^{2-}, has a pH of 8.2. Which of the following species has the highest concentration in seawater?
H_2CO_3, HCO_3^-, CO_3^{2-}
Which of these species has the next highest concentration? Calculate the ratio $[HCO_3^-]/[CO_3^{2-}]$ in seawater.

13.47 Calculate the solubility of the following sulfides in a saturated solution of H_2S (1 atm pressure) when $[H^+] = 0.3M$.
(a) CdS
(b) HgS

(c) ZnS
(d) SnS

13.48 Calculate the solubility of ZnS and FeS in a saturated solution of H_2S (1 atm pressure) when $[H^+] = 1 \times 10^{-7}M$.

13.49 If a solution is saturated with H_2S, at what pH will FeS have a solubility of $1.0 \times 10^{-5}M$?

13.50 The equilibrium constant for the formation of the silver–ammonia complex ion is 1.7×10^7 at 25°C.
$$Ag^+ + 2NH_3 \rightleftharpoons Ag(NH_3)_2^+ \qquad K = 1.7 \times 10^7$$
If the concentration of $Ag(NH_3)_2^+$ is $0.10M$, and the concentration of ammonia is $[NH_3] = 1.0M$, what is the concentration of uncomplexed silver ion, $[Ag^+]$?

13.51 Using the equilibrium constant in Section 13.13, calculate the concentration of free chloride ion, $[Cl^-]$, at which $[Hg^{2+}] = [HgCl^+]$.

13.52 Write a balanced equation for the net reaction that occurs when dilute nitric acid is used to dissolve each of the following solid substances:
$$Mg(OH)_2, \ CaCO_3, \ CH_3CO_2Ag$$

13.53 (a) The dissociation constant of weak acid HX is 1.0×10^{-6}. At a certain point in the titration of 50.0 mL of $0.100M$ HX with $0.100M$ NaOH, the pH of the solution is 7.00. At this point, what is the ratio of X^- concentration to HX concentration?
(b) How many mL of the NaOH solution are required to reach this point?

13.54 (a) Show that for a solution of KH_2PO_4,
$$[H^+] = \sqrt{K_1 K_2}$$
(b) Show that for a solution of K_2HPO_4,
$$[H^+] = \sqrt{K_2 K_3}$$

13.55 Using the dissociation constant of HSO_4^-, calculate the pH of a $0.010M$ H_2SO_4 solution.

CHAPTER 14
ELECTRON-TRANSFER REACTIONS

14

There are many different types of chemical reactions. One large group of reactions that involve a proton transfer (acid–base reactions) was discussed in the preceding chapters. In this chapter we consider an important type of reaction involving a transfer of electrons from one substance to another. These electron-transfer reactions were originally called oxidation–reduction reactions, and the terms *oxidation* and *reduction* are still commonly used to identify different aspects of these reactions.

14.1 OXIDATION AND REDUCTION

Oxidation originally referred only to the reaction of oxygen with elements to form oxides. Thus the burning of hydrogen and charcoal in air were classified as oxidation reactions:

$$2H_2(g) + O_2(g) = 2H_2O(g)$$
$$C(s) + O_2(g) = CO_2(g)$$

The reverse reaction, removal of oxygen from an oxide, was known as reduction. For example, red mercuric oxide is reduced to mercury by heating:

$$2HgO(s) = 2Hg(l) + O_2(g)$$

As more chemical reactions were studied, it became apparent that elements other than oxygen could react with metallic elements or hydrogen in a manner similar to that shown by oxygen. Thus zinc reacts with oxygen to form zinc oxide:

$$2Zn(s) + O_2(g) = 2ZnO(s)$$

Zinc also reacts with chlorine gas, dilute aqueous acids, and with cupric ion, Cu^{2+}, as shown in the following equations:

$$Zn(s) + Cl_2(g) = ZnCl_2(s)$$
$$Zn(s) + 2H^+ = Zn^{2+} + H_2(g)$$
$$Zn(s) + Cu^{2+} = Zn^{2+} + Cu(s)$$

The ZnO may be dissolved in dilute acid, and the ZnCl$_2$ in water, to produce solutions that contain Zn^{2+}. Thus the above reactions of Zn metal all have one feature in common, namely, the formation of Zn^{2+} either in an ionic crystal [Zn^{2+}O^{2-} or Zn^{2+}(Cl$^-$)$_2$] or as the solvated ion Zn^{2+}. The formation of Zn^{2+} from a neutral Zn atom requires a loss of two electrons. Therefore **oxidation** may be defined as *loss of electrons*, and zinc is said to be oxidized in each of the above reactions.

The electrons that are lost by zinc during the oxidation step are accepted by the other reactant, which is said to be reduced. Thus, in the reactions shown above, O$_2$, Cl$_2$, H$^+$, and Cu^{2+} accept electrons and are reduced. **Reduction** is therefore defined as *gain of electrons*. The substance accepting the electrons and hence causing oxidation to occur is known as the **oxidizing agent,** or oxidant. The substance losing electrons and hence causing reduction to occur is the **reducing agent,** or reductant. In the preceding reactions the oxidizing agents are O$_2$, Cl$_2$, H$^+$, and Cu^{2+}; in each case the reducing agent is Zn. It should be noted that the oxidizing agent is itself reduced, and the reducing agent is oxidized, in an oxidation–reduction reaction.

The above examples of oxidation–reduction reactions may all be explained in terms of the transfer of electrons, justifying the more modern description of such reactions as electron-transfer reactions. This concept of oxidation–reduction reactions as electron-transfer reactions is analogous to the Brønsted-Lowry concept of acid–base reactions as proton-transfer reactions.

14.2 OXIDATION STATES

The concept of electron loss or electron gain is obvious for the conversion of atoms to monatomic ions. Thus an Na atom can lose an electron to become Na$^+$, whereas an O atom can accept two electrons to form O^{2-}. Over the years the term **oxidation state** has been used to designate the *net electric charge* on an atom or monatomic ion. Thus the oxidation state of atoms of an elemental substance such as Na is zero, whereas the oxidation states of Na$^+$, S^{2-}, Cl$^-$, and Al^{3+} are $+1$, -2, -1, and $+3$, respectively.

This concept has proved so useful in treating electron-transfer reactions and in correlating the chemical properties of ions that it has been extended to molecules and ions containing electron-pair bonds. Here *oxidation state* refers to the net electric charge on an atom when the valence electrons are assigned to bonded atoms according to the following rules:

1. Electrons shared between unlike atoms are assigned to the *more electronegative atom*.
2. Electrons shared by like atoms are divided equally between the atoms.

Because electrons are negative, the net electric charge is then obtained by subtracting the total number of valence electrons on the atom from its kernel charge.[1] For example, the oxygen atom in group VIA has a kernel charge of +6. If the total number of valence electrons (assigned according to the above rules) is 8, the oxidation state of the oxygen atom is $+6 - 8 = -2$.

To assign electrons according to these rules, one must first write the Lewis structure for the molecule or ion, and then use the table of electronegativities (Table 9.4) to determine which atom in each bonded pair is the more electronegative. For example, the Lewis structure of hydrazine, N_2H_4, is shown below. Because nitrogen is more electronegative than hydrogen, electrons shared between N and H are assigned to N. The electron pair shared by the two nitrogen atoms is divided equally, leading to the following assignments:

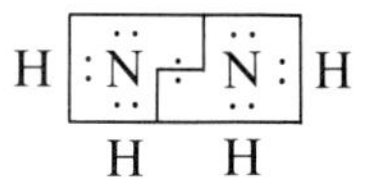

H :N⎯N: H
   H H

The oxidation state of nitrogen is

$$\text{kernel charge }(+5) - \text{valence electrons }(7) = -2$$

The oxidation state of hydrogen is

$$\text{kernel charge }(+1) - \text{valence electrons }(0) = +1$$

As a check on our results, we note that for a neutral molecule the sum of the oxidation states for all atoms is zero:

$$2(-2) + 4(+1) = 0$$

(For an ion, the sum of the oxidation states for all atoms is equal to the net charge on the ion.)

EXAMPLE 1

What are the oxidation states of all atoms in water, carbon dioxide, and the nitrate ion, NO_3^-?

The Lewis structures are shown below. Lines indicate assignment of shared electrons to the more electronegative atom.

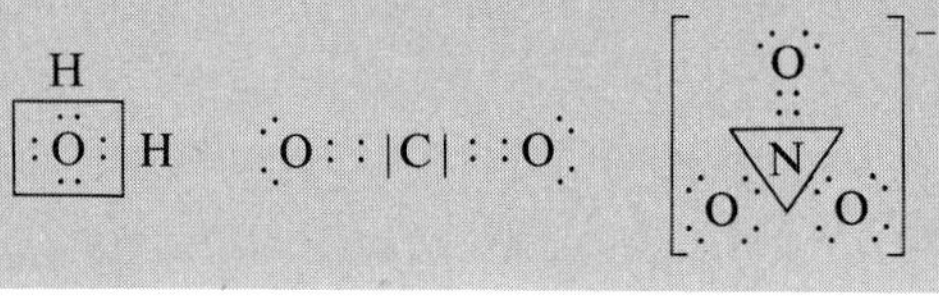

---

[1] As discussed in Section 8.13, the kernel of an atom consists of the nucleus plus its inner shells—all electrons except the electrons in the outermost, or valence, shell. The kernel charge is the electric charge on the kernel in electronic charge units. For an atom having N valence electrons, the kernel charge is $+N$. Thus for the main group elements the kernel charge is the same as the group number in the periodic table. For example, phosphorus in group VA has a kernel charge of +5.

In each molecule, each oxygen atom is assigned eight valence electrons, so its oxidation state is $+6 - 8 = -2$. The oxidation states of H, C, and N are equal to their kernel charges, $+1$, $+4$, and $+5$, respectively. (Only one of the three resonance structures is shown for the nitrate ion; the other structures give the same results.)

If one wishes to show oxidation states with molecular formulas, the oxidation state of each element is shown directly below the symbol for the element.

$$H_2O \qquad CO_2 \qquad NO_3^-$$
$$+1, -2 \qquad +4, -2 \qquad +5, -2$$

Based on these general principles, some specific rules can be given for certain oxidation states. Application of these rules usually gives oxidation states more quickly. (In case of doubt, the general principles should be used.)

1. The oxidation states of atoms in elemental substances are always equal to zero. Thus H_2, Na, C, O_2, and Cl_2 all contain the element in an oxidation state of zero.
2. For a monatomic ion, the oxidation state is equal to the charge on the ion. Thus Ca^{2+} and I^- have oxidation states of $+2$ and -1, respectively.
3. Group IA, IIA, and IIIA metals usually have oxidation states in their compounds of $+1$, $+2$, and $+3$, respectively. For example, in compounds of Na, Mg, and Al the typical oxidation states are $+1$, $+2$, and $+3$, respectively.
4. Hydrogen usually has an oxidation state of $+1$ in its compounds. (Exceptions are the metal hydrides such as lithium hydride, LiH, in which the oxidation state of hydrogen is -1.)
5. Oxygen usually has an oxidation state of -2 in its compounds. There are two important exceptions. (a) In the fluorine oxides, oxygen has a positive oxidation state. An example is F_2O, in which the oxidation state of oxygen is $+2$. (b) In compounds containing an oxygen-oxygen bond, called **peroxides**, oxygen usually has an oxidation state of -1. Examples are hydrogen peroxide, H_2O_2, and sodium peroxide, Na_2O_2.
6. The most typical oxidation state for the halogens in their compounds is -1, though there are many exceptions for all halogens but fluorine. Fluorine always has an oxidation state of -1 in its compounds.
7. For the other elements, a quick procedure is to assign oxidation states to all but one element using rules 1-6. The unknown oxidation state is then determined by difference.

For example, in potassium permangate, $KMnO_4$, one can assign $+1$ to K (rule 3) and -2 to O (rule 5). The total electric charge of the potassium and four oxygen atoms is then $+1 + 4(-2) = -7$. To balance this charge, the oxidation state of the manganese atom must be $+7$.

It should be emphasized that oxidation states arise from a strictly formal assignment of electric charge according to rather arbitrary rules, and they are not intended to represent the actual distribution of charge in a molecule or ion. Nevertheless, oxidation states facilitate both the systematic treatment of the chemistry of different elements and the interpretation of electron-transfer reactions. Thus the +6 oxidation state of S in SO_4^{2-} may be correlated with the six electrons (e^-) given up in the oxidation of elemental sulfur to sulfate in aqueous solution:

$$S + 4H_2O = 8H^+ + SO_4^{2-} + 6e^-$$

The separation of electron-transfer reactions into half-reactions of this type involving only oxidation or reduction will be the subject of the following sections.

14.3 THE CONCEPT OF HALF-REACTIONS

One of the most amazing features of electron-transfer reactions is that many of the reactions may be carried out with the reactants in separate containers so that they do not come into contact with each other. For example, the reaction of zinc metal with cupric ion,

$$Zn(s) + Cu^{2+} = Zn^{2+} + Cu(s) \tag{14.1}$$

may be carried out using the apparatus shown in Figure 14.1. The beaker on the left contains a strip of metallic zinc dipping into a $1M$ $ZnSO_4$ solution, whereas the right-hand beaker contains a strip of copper in a $1M$ $CuSO_4$ solution. The two containers are connected by a tube containing a solution of an electrolyte, such as KCl, called a **salt bridge**. When the zinc strip and copper strip are connected (by means of a copper wire) to the terminals of an ammeter (which measures electric current in amperes), an electric current flows corresponding to a transfer of electrons from the zinc electrode to the copper electrode. If the current is allowed to flow for some time, one finds that the amount of zinc metal and the concentration of Cu^{2+} have decreased, whereas the amount of metallic copper and the concentration of Zn^{2+} have increased. Furthermore, the changes in the number of moles of reactants and products are in accord with Eq. 14.1. The electric current will continue to flow until one of the reactants is used up, the electric connection is broken, or the salt bridge is removed.

The above observations may be explained by considering the chemical changes that occur in each beaker. The flow of electrons from the zinc rod into the external circuit (through the ammeter) and the corresponding change of zinc atoms to zinc ions can be expressed by

$$Zn(s) = Zn^{2+} + 2e^- \tag{14.2}$$

The flow of electrons to the copper rod and the conversion of cupric ions to metallic copper can be represented by

$$Cu^{2+} + 2e^- = Cu(s) \tag{14.3}$$

14.3 THE CONCEPT OF HALF-REACTIONS

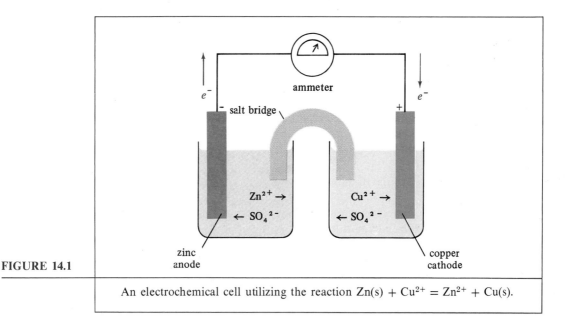

FIGURE 14.1 An electrochemical cell utilizing the reaction $Zn(s) + Cu^{2+} = Zn^{2+} + Cu(s)$.

If the external circuit is broken, the transfer of electrons is stopped and the chemical reaction ceases. The reaction that occurs in each beaker (Eq. 14.2 or 14.3) is known as a **half-reaction.** What, then, is the purpose of the salt bridge? The formation of zinc ions would cause an excess of positive charges in the left container, whereas the removal of cupric ions would result in an excess of negative charges in the right container. This separation of charge would cause the chemical reaction to cease were it not for the salt bridge which provides a path for migration of charged particles (ions) between the two containers so that each container remains electrically neutral. Thus the electric current consists of a flow of electrons in the external circuit (**electron conduction** in metals) and a flow of ions in the solutions (**electrolytic conduction**). Note that this change in the method of conducting electricity from electron flow to ion flow requires chemical reactions (half-reactions) to occur at the metal rods where electrons are removed or introduced. These metal rods are called **electrodes.** The electrode at which *oxidation* occurs (Eq. 14.2) is defined as the **anode;** the electrode at which *reduction* occurs (Eq. 14.3) is defined as the **cathode.** The positive ions migrate toward the copper cathode (to balance the removal of Cu^{2+}) and hence are called **cations,** whereas the negative ions migrate toward the zinc anode (to balance the production of Zn^{2+}) and are called **anions.**

To convert 1 mol of zinc atoms to zinc ions would require the removal of $2 \times 6.0 \times 10^{23}$ electrons or 2 mol of electrons. The quantity of electric charge in coulombs (C) associated with 1 mol of electrons is known as the faraday (F).

$$F = N_0 e$$
$$= (6.0220 \times 10^{23} \text{ electrons mol}^{-1})(1.6022 \times 10^{-19} \text{ C electron}^{-1})$$
$$= 9.648 \times 10^4 \text{ C mol}^{-1}$$

A coulomb is the electric charge carried by an electric current of 1 ampere (A) flowing for 1 s, and therefore 1 C = 1 A s. (For a discussion of the concepts of electric charge and electric current, see Appendix D-5.)

In the above example the electron-transfer reaction occurred spontaneously, producing an electric current. In **electrolysis** the opposite occurs: An electric current from some external source is used to force an electron-transfer reaction to proceed in the direction opposite to its spontaneous direction. Thus by connecting the zinc electrode to the negative terminal, and the copper electrode to the positive terminal, of some external source of sufficiently high voltage, one can force each half-reaction to proceed in the opposite direction. The net reaction becomes the reverse of Eq. 14.1, or

$$Cu(s) + Zn^{2+} = Cu^{2+} + Zn(s)$$

Because oxidation now occurs at the copper electrode, it becomes the anode; the zinc electrode (where reduction occurs) is the cathode.

The positive cations migrate toward the cathode, and the negative anions migrate toward the anode, but these directions are opposite to those shown in Figure 14.1, where the cell operates spontaneously. The Zn^{2+} and Cu^{2+} cations move toward the zinc cathode, and the SO_4^{2-} anions move toward the copper anode.

As another example of electrolysis, consider what happens when an electric current is forced to flow through liquid sodium chloride at a temperature above its melting point of 800°C (Figure 14.2). At one electrode (the anode) electrons are given up by the chloride ions, and gaseous chlorine is formed. The electrons released by the Cl^- ions at the anode move through the electric circuit to the cathode, where they combine with Na^+ ions, forming liquid sodium metal. The half-reactions are

at the anode $\qquad 2Cl^- = Cl_2(g) + 2e^-$

at the cathode $\quad 2Na^+ + 2e^- = 2Na(l)$

The net reaction (the sum of the electrode reactions) is

$$2Na^+ + 2Cl^- = 2Na(l) + Cl_2(g)$$

Note that the Cl^- anions move toward the anode, and the Na^+ cations move toward the cathode. [To conduct electrons to and from the liquid NaCl, the cathode is made of iron, and the anode is made of carbon (graphite). The only function of the iron and carbon is to conduct electricity; they are not involved in the half-reactions of the electrolysis cell.]

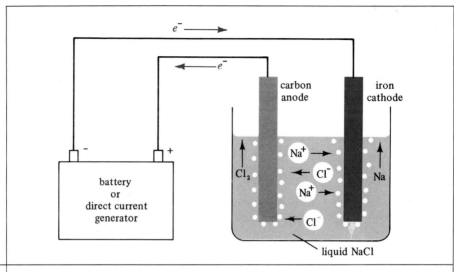

FIGURE 14.2 Electrolysis of liquid sodium chloride. Gaseous chlorine and liquid sodium rise to the surface and are removed separately.

EXAMPLE 2

Aluminum metal is prepared commercially by electrolysis of molten Al_2O_3. How many coulombs are required to produce 100 g of aluminum? How many amperes are required to produce this much aluminum in 1 h?

Since the atomic mass of aluminum is 26.98, in 100 g of aluminum there are

$$\frac{100 \text{ g}}{26.98 \text{ g (mol of atoms)}^{-1}} = 3.71 \text{ mol of atoms}$$

The oxidation state of Al in Al_2O_3 is +3; therefore three electrons are required to form one atom of aluminum, or 3 F are required to form 1 mol of aluminum atoms. Thus 3.71 mol of aluminum atoms requires $(3.71 \text{ mol})(3\ F\ \text{mol}^{-1}) = 11.1\ F$, or

$$(11.1\ F)(9.648 \times 10^4\ C\ F^{-1}) = 1.07 \times 10^6\ C = 1.07 \times 10^6\ A\ s$$

There are 3600 s in 1 h; therefore the current in amperes will be

$$\frac{1.07 \times 10^6\ A\ s}{3600\ s} = 297\ A$$

14.4 BALANCING ELECTRON-TRANSFER REACTIONS

The first step in balancing an equation for an electron-transfer reaction is the same as for any other chemical reaction, namely, the reactants and products must be determined. This may be done by experiment or by reference to previous results reported in books or journals.

Any electron-transfer reaction may be treated conceptually as consisting of two half-reactions, one an oxidation and the other a reduction. This is the basis of the following method of balancing electron-transfer reactions that occur in acidic or neutral aqueous solutions:

1. Separate the reactants and products into half-reactions.
2. Balance each half-reaction with respect to atoms, using only known reactants, products, or other substances present. This is most easily accomplished by first balancing with respect to all atoms except H and O. Then balance O atoms using H_2O as required to provide any extra oxygens. Finally, balance H atoms using H^+.
3. Balance each half-reaction with respect to electric charge by using electrons. Check to see that one half-reaction involves oxidation and that the other involves reduction.
4. Multiply the half-reactions by appropriate factors so that the number of electrons produced is equal to the number of electrons used up.
5. Add the half-reactions to obtain the overall equation, cancel electrons, and cancel equivalent amounts of any other substances appearing on both sides of the equation. If all coefficients are multiples of some integer, divide through by that integer to simplify the result.
6. Check for balance with respect to atoms and charge. If an imbalance is discovered, an error was made in a preceding step.

The following examples illustrate the use of this method.

EXAMPLE 3

Balance $Fe^{3+} + Sn^{2+} \rightarrow Fe^{2+} + Sn^{4+}$.

Step 1.

$$Fe^{3+} \longrightarrow Fe^{2+}$$
$$Sn^{2+} \longrightarrow Sn^{4+}$$

Step 2. No change necessary.

Step 3.

$$Fe^{3+} + e^- \longrightarrow Fe^{2+} \quad \text{(reduction)}$$
$$Sn^{2+} \longrightarrow Sn^{4+} + 2e^- \quad \text{(oxidation)}$$

Step 4. The first half-reaction must be multiplied by 2 to provide for an equal number of electrons in the two half-reactions.

14.4 BALANCING ELECTRON-TRANSFER REACTIONS

$$2Fe^{3+} + 2e^- \longrightarrow 2Fe^{2+}$$
$$Sn^{2+} \longrightarrow Sn^{4+} + 2e^-$$

Step 5.

$$2Fe^{3+} + \cancel{2e^-} + Sn^{2+} \longrightarrow 2Fe^{2+} + Sn^{4+} + \cancel{2e^-}$$
$$2Fe^{3+} + Sn^{2+} = 2Fe^{2+} + Sn^{4+}$$

Step 6. A final check shows one tin and two iron atoms on each side of the equation, and a net charge of +8 on each side.

EXAMPLE 4

In acidic solutions, dichromate ion ($Cr_2O_7^{2-}$), which imparts an orange-red color to aqueous solutions, may be reduced to Cr^{3+}. This reduction can be caused by I^-, which is oxidized to iodine (I_2). Balance

$$Cr_2O_7^{2-} + I^- \longrightarrow Cr^{3+} + I_2 \quad \text{(acidic solution)}$$

Step 1.

$$Cr_2O_7^{2-} \longrightarrow Cr^{3+}$$
$$I^- \longrightarrow I_2$$

Step 2. The dichromate ion must produce $2Cr^{3+}$ to balance chromium atoms. The seven oxygen atoms must form seven H_2O molecules, which, in turn, require $14H^+$ on the left. There must also be $2I^-$ ions to form I_2.

$$Cr_2O_7^{2-} + 14H^+ \longrightarrow 2Cr^{3+} + 7H_2O$$
$$2I^- \longrightarrow I_2$$

Step 3.

$$Cr_2O_7^{2-} + 14H^+ + 6e^- \longrightarrow 2Cr^{3+} + 7H_2O \quad \text{(reduction)}$$
$$2I^- \longrightarrow I_2 + 2e^- \quad \text{(oxidation)}$$

Step 4. As six electrons are required to reduce $Cr_2O_7^{2-}$, the second half-reaction must be multiplied by 3 to balance the electrons:

$$Cr_2O_7^{2-} + 14H^+ + 6e^- \longrightarrow 2Cr^{3+} + 7H_2O$$
$$6I^- \longrightarrow 3I_2 + 6e^-$$

Step 5.

$$Cr_2O_7^{2-} + 14H^+ + \cancel{6e^-} + 6I^- \longrightarrow 2Cr^{3+} + 7H_2O + 3I_2 + \cancel{6e^-}$$
$$Cr_2O_7^{2-} + 14H^+ + 6I^- = 2Cr^{3+} + 7H_2O + 3I_2$$

Step 6. A final check shows 2Cr, 7O, 14H, 6I, and a net charge of +6 on each side of the equation.

A slight modification of the above method is required to balance electron-transfer reactions in basic solutions. After balancing H atoms with H^+ (step 2), add the same number of OH^- ions to each side of the equation, and then combine H^+ and OH^- to form H_2O. This procedure removes H^+ from the equation for the reaction—a desirable modification since H^+ is present in extremely low concentrations in basic solutions.

EXAMPLE 5

In basic solutions, permanganate ion (MnO_4^-), which imparts a purple color to aqueous solutions, is reduced by tin (Sn) to manganese dioxide, MnO_2, a dark solid precipitate, and the tin is oxidized to $Sn(OH)_3^-$. Balance

$$MnO_4^- + Sn \longrightarrow MnO_2 + Sn(OH)_3^- \quad \text{(basic solution)}$$

Step 1.

$$MnO_4^- \longrightarrow MnO_2$$
$$Sn \longrightarrow Sn(OH)_3^-$$

Step 2. In the first half-reaction, O atoms are balanced by adding $2H_2O$ to the right, and then H atoms are balanced by adding $4H^+$ to the left:

$$MnO_4^- + 4H^+ \longrightarrow MnO_2 + 2H_2O$$

Now we add $4OH^-$ to each side and replace $4H^+ + 4OH^-$ by $4H_2O$:

$$MnO_4^- + 4H_2O \longrightarrow MnO_2 + 2H_2O + 4OH^-$$

After canceling $2H_2O$ from each side, we obtain

$$MnO_4^- + 2H_2O \longrightarrow MnO_2 + 4OH^-$$

In the second half-reaction, we first obtain

$$Sn + 3H_2O \longrightarrow Sn(OH)_3^- + 3H^+$$

We then add $3OH^-$ to each side, replace $3H^+ + 3OH^-$ by $3H_2O$, and cancel $3H_2O$ from each side, which leads to

$$Sn + 3OH^- \longrightarrow Sn(OH)_3^-$$

Step 3.

$$MnO_4^- + 2H_2O + 3e^- \longrightarrow MnO_2 + 4OH^- \quad \text{(reduction)}$$
$$Sn + 3OH^- \longrightarrow Sn(OH)_3^- + 2e^- \quad \text{(oxidation)}$$

Step 4. Multiplying the first half-reaction by 2 and the second by 3 gives

14.4 BALANCING ELECTRON-TRANSFER REACTIONS

$$2MnO_4^- + 4H_2O + 6e^- \longrightarrow 2MnO_2 + 8OH^-$$
$$3Sn + 9OH^- \longrightarrow 3Sn(OH)_3^- + 6e^-$$

Step 5.

$$2MnO_4^- + 4H_2O + \cancel{6e^-} + 3Sn + 9OH^- \longrightarrow$$
$$2MnO_2 + 8OH^- + 3Sn(OH)_3^- + \cancel{6e^-}$$

Canceling $8OH^-$ as well as $6e^-$ from each side gives

$$2MnO_4^- + 4H_2O + 3Sn + OH^- = 2MnO_2 + 3Sn(OH)_3^-$$

Step 6. A final check shows 2Mn, 3Sn, 13O, 9H, and a net charge of -3 on each side.

A final example illustrates the method when the same substance is both oxidized and reduced.

EXAMPLE 6

Hydrogen peroxide (H_2O_2) in aqueous solution decomposes slowly to yield H_2O and O_2. Balance

$$H_2O_2 \longrightarrow H_2O + O_2 \quad \text{(neutral solution)}$$

Step 1.

$$H_2O_2 \longrightarrow O_2$$
$$H_2O_2 \longrightarrow H_2O$$

Step 2.

$$H_2O_2 \longrightarrow O_2 + 2H^+$$
$$2H^+ + H_2O_2 \longrightarrow 2H_2O$$

Step 3.

$$H_2O_2 \longrightarrow O_2 + 2H^+ + 2e^- \quad \text{(oxidation)}$$
$$2H^+ + H_2O_2 + 2e^- \longrightarrow 2H_2O \quad \text{(reduction)}$$

Step 4. No change necessary.

Step 5.

$$H_2O_2 + \cancel{2H^+} + H_2O_2 + \cancel{2e^-} \longrightarrow O_2 + \cancel{2H^+} + \cancel{2e^-} + 2H_2O$$
$$2H_2O_2 = O_2 + 2H_2O$$

Step 6. Final check: 4H, 4O, and zero net charge on each side.

14.5 THE STANDARD POTENTIAL OF AN ELECTROCHEMICAL CELL

The term **electrochemical cell** is applied to any apparatus using an electron-transfer reaction to obtain a flow of electrons in an external circuit (as in Figures 14.1 and 14.2). Electrochemical cells are extremely important from both a practical and a theoretical viewpoint because they provide (1) a mechanism for interconverting chemical energy and electrical energy, (2) information on the equilibrium position of many reversible reactions, and (3) a means of determining the relative strengths of oxidizing and reducing agents.

The electric current produced in a cell as a result of a chemical reaction may be used to do work. Electrical work (w_{el}) is defined as the product of charge (Q) and potential difference (V) through which the charge is transferred (Appendix D-6).

$$w_{el} = QV \tag{14.4}$$

[This relationship is analogous to the more familiar one of an object falling under the influence of gravity, where the amount of work that the object can do is the product of its weight (analogous to charge) and the height or elevation difference through which it falls (analogous to potential difference).]

If n mol of electrons is transferred through the external circuit of a cell, the charge Q transferred would be nF. If the difference in potential between the two electrodes is $\Delta\mathcal{E}$, then the electrical work that the cell can do is given by

$$w_{el} = nF\Delta\mathcal{E} \tag{14.5}$$

The potential difference between the electrodes depends on the conditions under which the cell is operated, particularly on the electric current drawn from the cell. When the potential difference is measured with a potentiometer, the electric current can be reduced to an extremely small value. (This is accomplished by opposing the cell's potential difference with a variable potential from another cell. When the magnitude of the opposing potential is too small, reaction occurs as written; when it is too large, reaction occurs in the reverse direction—see Figure 14.3. When the two potential differences are identical, no reaction occurs, and no current flows.) As the current approaches zero, the potential difference approaches a constant value called the **electromotive force** of the cell, abbreviated emf. Since the emf is measured in volts, it is commonly called the cell voltage. From now on, we shall assume that $\Delta\mathcal{E}$ has been measured with a potentiometer, and that it is therefore the emf of the cell. We shall also use the terms *emf, cell potential,* and *cell voltage* as synonymous.

The cell voltage depends on what half-reactions are occurring at the electrodes, and it also depends on the concentrations (for solutes) and partial pressures (for gases) of the reactants and products of the cell reaction. Let us examine some experimental observations on a typical electrochemical cell (such as the

14.5 THE STANDARD POTENTIAL OF AN ELECTROCHEMICAL CELL

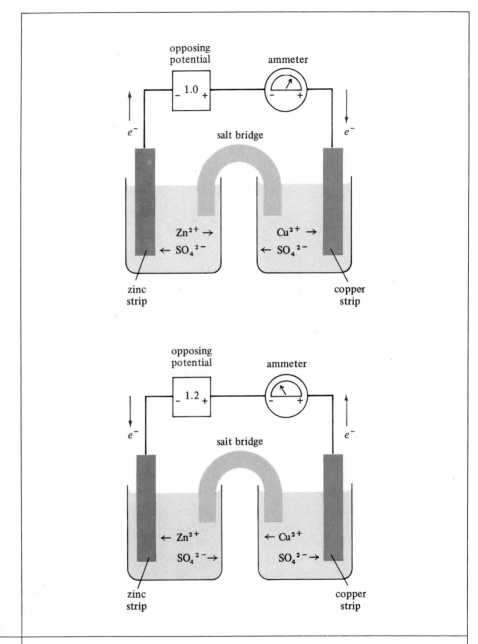

FIGURE 14.3 Effect of an opposing external potential on electron flow in a cell. In the top figure, the magnitude of the opposing potential (1.0 V) is smaller than that of the cell (1.1 V), and electrons flow from zinc to copper. In the bottom figure, just the opposite occurs when the opposing potential has a magnitude of 1.2 V, larger than that of the cell.

zinc–copper cell in Figure 14.1) so that we may understand its operation and how experimental data on such cells may be utilized to provide useful information.

As noted earlier (Section 14.3), a current of electrons flows from the anode (Zn electrode) to the cathode (Cu electrode) when Zn reacts with Cu^{2+} (Eq. 14.1). Because electrons are negatively charged particles, they will flow toward the electrode with the higher positive potential (Cu electrode). This we can easily confirm experimentally by connecting a potentiometer to the two electrodes. If the concentrations of Cu^{2+} and Zn^{2+} are each $1M$, the potentiometer will read 1.10 V, that is, the Cu electrode will be $+1.10$ V with respect to the Zn electrode. This result may be shown by writing

$$Zn(s) + Cu^{2+}(1M) = Cu(s) + Zn^{2+}(1M) \qquad \Delta\mathcal{E} = 1.10 \text{ V} \qquad (14.6)$$

The fact that the copper electrode is positive with respect to the zinc electrode is indicated by the $+$ and $-$ signs in Figure 14.1. Note that electrons flow away from the negative zinc anode, and toward the positive copper cathode. (This may be confusing if you have learned that the anode is the positive electrode and the cathode is the negative electrode—definitions that work well for electronic circuits but not for electrochemical cells. The correct definitions of anode and cathode are those given in Section 14.3.)

We adopt the following convention to determine the sign of the cell voltage: If the reaction *as written* will occur spontaneously in the electrochemical cell when its electrodes are connected through an external circuit, we assign $\Delta\mathcal{E}$ a *positive* sign. Thus, for reaction 14.6, which is spontaneous as written, $\Delta\mathcal{E} = +1.10$ V. (Note that the positive electrode is on the right-hand side of the equation when the voltage is positive.) If we had written the equation in the reverse direction, it would *not* be spontaneous *as written*. Thus according to our convention we would report $\Delta\mathcal{E} = -1.10$ V for the reaction

$$Cu(s) + Zn^{2+}(1M) = Zn(s) + Cu^{2+}(1M) \qquad \Delta\mathcal{E} = -1.10 \text{ V} \qquad (14.7)$$

If we now extend our experimental observations, we find that the voltage of the zinc–copper cell increases if the concentration of cupric ion is increased or if the concentration of zinc ion is decreased. Changes in the mass or surface area of the electrodes do not alter the cell voltage.

Because the voltage produced by an electrochemical cell depends on the concentrations of the substances involved in the electron-transfer reaction, it is customary to define the **standard voltage** ($\Delta\mathcal{E}°$) of a cell as the measured voltage of the cell when all the reactants and products are present at unit activity (see Table 12.2). We then find that[2]

---

[2] The reaction could be written

$$Zn(s) + Cu^{2+}(a = 1) = Cu(s) + Zn^{2+}(a = 1)$$

However, it is customary to omit the value of the activity when it is unity.

14.6 STANDARD ELECTRODE POTENTIALS

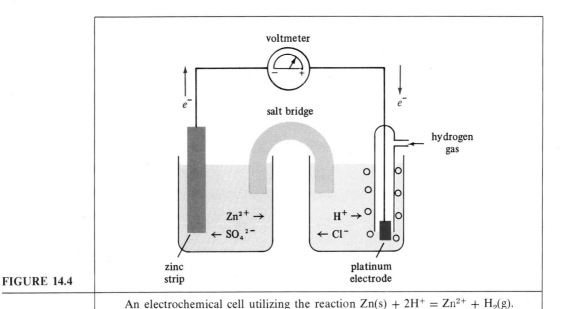

FIGURE 14.4 An electrochemical cell utilizing the reaction $Zn(s) + 2H^+ = Zn^{2+} + H_2(g)$.

$$Zn(s) + Cu^{2+} = Cu(s) + Zn^{2+} \qquad \Delta\mathcal{E}° = +1.10 \text{ V} \qquad (14.8)$$

If the beaker containing the copper electrode (Figure 14.1) is replaced with one containing HCl and a platinum electrode over which hydrogen gas is bubbled (Figure 14.4), electrons flow from the zinc electrode to the platinum electrode, where H^+ is reduced to H_2. (The platinum is not involved in the net reaction, and it is therefore known as an *inert* electrode. Its function is to conduct electrons to H^+ and to catalyze the half-reaction.) The platinum electrode is 0.76 V more positive than the zinc electrode when all substances present are at unit activity, that is,

$$Zn(s) + 2H^+ = Zn^{2+} + H_2(g) \qquad \Delta\mathcal{E}° = +0.76 \text{ V} \qquad (14.9)$$

It is apparent that the standard voltages of all electron-transfer reactions could be determined, providing suitable cells could be constructed to measure them. Indeed, such values are of considerable interest in that they are a measure of the tendency for the reaction to occur.

14.6 STANDARD ELECTRODE POTENTIALS

The problem of tabulating the standard voltages of electron-transfer reactions is considerably simplified if we assume that the voltage of a cell is the sum of two voltages, each associated with one of the half-reactions in the cell. To obtain numerical values of the various half-cell voltages, it is necessary to assign an arbitrary half-cell voltage to one half-reaction as a standard reference point from which all other half-cell voltages are measured. By convention *the hydrogen half-cell has been chosen as the standard*. Its potential is assigned a value of *exactly zero* when all substances are present at unit activity, that is,

$$2H^+ + 2e^- = H_2(g) \qquad \mathcal{E}° = 0 \text{ V, exactly}$$

(To distinguish between the standard potential of a cell and the standard potential of a half-cell, we shall use $\Delta\mathcal{E}°$ for the former, and $\mathcal{E}°$ for the latter.)

The standard voltage of the zinc–zinc ion half-cell can then be evaluated from Eq. 14.9. We note that Eq. 14.9 may be obtained by adding the following two half-reactions:

$$\begin{array}{ll} Zn(s) = Zn^{2+} + 2e^- & \mathcal{E}°_{Zn} = ? \\ 2H^+ + 2e^- = H_2(g) & \mathcal{E}° = 0 \text{ V} \\ \hline Zn(s) + 2H^+ = Zn^{2+} + H_2(g) & \Delta\mathcal{E}° = \mathcal{E}°_{Zn} + 0 = +0.76 \text{ V} \end{array}$$

or $\mathcal{E}°_{Zn}$ is $+0.76$ V.

By international convention it is customary to report standard potentials for half-cells as reduction reactions.[3] Therefore the standard potential for the zinc ion–zinc metal half-cell (which is the *reverse* of the half-reaction as written above) would be -0.76 V:

$$Zn^{2+} + 2e^- = Zn(s) \qquad \mathcal{E}° = -0.76 \text{ V}$$

Note that when a cell reaction or half-cell reaction is reversed, the sign of the voltage is changed. We may now evaluate the $\mathcal{E}°$ value for the copper half-cell from Eq. 14.8:

$$\begin{array}{ll} Zn(s) = Zn^{2+} + 2e^- & \mathcal{E}° = +0.76 \text{ V} \\ Cu^{2+} + 2e^- = Cu(s) & \mathcal{E}°_{Cu} = ? \\ \hline Zn(s) + Cu^{2+} = Zn^{2+} + Cu(s) & \Delta\mathcal{E}° = +0.76 + \mathcal{E}°_{Cu} = +1.10 \text{ V} \end{array}$$

Thus $\mathcal{E}°_{Cu} = +1.10 - 0.76 = +0.34$ V.

There is a connection between reduction potentials and the potentials of the electrodes. For example, the copper half-cell reduction potential is 1.10 V positive with respect to the zinc half-cell reduction potential, just as the copper electrode is 1.10 V positive with respect to the zinc electrode. Therefore standard reduction potentials are often called **standard electrode potentials.**

---

[3] This practice is by no means uniform. When the opposite convention is used, the half-reactions are oxidations, and the potentials are called oxidation potentials.

14.6 STANDARD ELECTRODE POTENTIALS

TABLE 14.1   **STANDARD ELECTRODE POTENTIALS IN ACIDIC SOLUTION AT 25°C**

Half-reaction	$\mathcal{E}°$ (V)
$K^+ + e^- = K(s)$	−2.925
$Na^+ + e^- = Na(s)$	−2.714
$Mg^{2+} + 2e^- = Mg(s)$	−2.36
$Al^{3+} + 3e^- = Al(s)$	−1.66
$Zn^{2+} + 2e^- = Zn(s)$	−0.763
$Fe^{2+} + 2e^- = Fe(s)$	−0.44
$Co^{2+} + 2e^- = Co(s)$	−0.277
$Sn^{2+} + 2e^- = Sn(s)$	−0.136
$Pb^{2+} + 2e^- = Pb(s)$	−0.126
$2H^+ + 2e^- = H_2(g)$	0
$Sn^{4+} + 2e^- = Sn^{2+}$	+0.15
$Cu^{2+} + e^- = Cu^+$	+0.153
$Cu^{2+} + 2e^- = Cu(s)$	+0.34
$I_2(s) + 2e^- = 2I^-$	+0.536
$O_2(g) + 2H^+ + 2e^- = H_2O_2$	+0.682
$Fe^{3+} + e^- = Fe^{2+}$	+0.771
$Hg_2^{2+} + 2e^- = 2Hg(l)$	+0.788
$Ag^+ + e^- = Ag(s)$	+0.799
$2Hg^{2+} + 2e^- = Hg_2^{2+}$	+0.92
$NO_3^- + 4H^+ + 3e^- = NO(g) + 2H_2O$	+0.96
$Br_2(l) + 2e^- = 2Br^-$	+1.065
$O_2(g) + 4H^+ + 4e^- = 2H_2O$	+1.229
$Cr_2O_7^{2-} + 14H^+ + 6e^- = 2Cr^{3+} + 7H_2O$	+1.33
$Cl_2(g) + 2e^- = 2Cl^-$	+1.36
$MnO_4^- + 8H^+ + 5e^- = Mn^{2+} + 4H_2O$	+1.51
$F_2(g) + 2e^- = 2F^-$	+2.87

When two half-reactions are combined, it is often necessary to double or triple the coefficients of one half-reaction to cancel electrons. In such a case the standard electrode potential does *not* change, because the $\mathcal{E}°$ value measures the tendency for reaction to occur, and it is not dependent on the number of electrons transferred. For example, the following half-reactions have the same $\mathcal{E}°$ value.

$$\left. \begin{array}{l} Ag^+ + e^- = Ag(s) \\ 2Ag^+ + 2e^- = 2Ag(s) \end{array} \right\} \mathcal{E}° = +0.799 \text{ V}$$

The standard electrode potentials of a number of half-cells are tabulated in Table 14.1. Because each half-reaction in the table may be combined with any other half-reaction to obtain an electron transfer reaction, and there are 26 × 25/2, or 325, such combinations, it is possible to predict the standard potentials of 325 reactions from the 26 potentials reported in the table. These standard potentials are useful in that they enable one to predict whether or not the reactions will occur, as shown in the following examples.

EXAMPLE 6

Predict whether or not the following reactions will occur as written if all substances are present at unit activity:
(a) $Cu^{2+} + Fe(s) = Fe^{2+} + Cu(s)$
(b) $Cu^{2+} + 2Ag(s) = 2Ag^+ + Cu(s)$

(a) To get Fe(s) on the left and Cu(s) on the right, we must add the Cu^{2+}–Cu(s) half-reaction listed in Table 14.1 to the reverse of the Fe^{2+}–Fe(s) half-reaction. On reversing the latter half-reaction, its $\mathcal{E}°$ changes from -0.44 V to $+0.44$ V.

$$Cu^{2+} + 2e^- = Cu(s) \qquad \mathcal{E}° = +0.34 \text{ V}$$
$$Fe(s) = Fe^{2+} + 2e^- \qquad \mathcal{E}° = +0.44 \text{ V}$$
$$\overline{Cu^{2+} + Fe(s) = Fe^{2+} + Cu(s)} \qquad \Delta\mathcal{E}° = +0.78 \text{ V}$$

Because $\Delta\mathcal{E}°$ is positive, the reaction will go as written when all substances are present at unit activity.

(b) From Table 14.1, we note that

$$Cu^{2+} + 2e^- = Cu(s) \qquad \mathcal{E}° = +0.34 \text{ V}$$
$$Ag^+ + e^- = Ag(s) \qquad \mathcal{E}° = +0.799 \text{ V}$$

To obtain the desired reaction, it is necessary to write the second half-reaction in the opposite direction and multiply its coefficients by 2.

$$2Ag(s) = 2Ag^+ + 2e^- \qquad \mathcal{E}° = -0.799 \text{ V}$$
$$Cu^{2+} + 2e^- = Cu(s) \qquad \mathcal{E}° = +0.34 \text{ V}$$
$$\overline{Cu^{2+} + 2Ag(s) = 2Ag^+ + Cu(s)} \qquad \Delta\mathcal{E}° = -0.46 \text{ V}$$

Because $\Delta\mathcal{E}°$ is negative, the reaction will not go as written when all substances are present at unit activity. However, the reverse reaction will occur spontaneously:

$$2Ag^+ + Cu(s) = Cu^{2+} + 2Ag(s) \qquad \Delta\mathcal{E}° = +0.46 \text{ V}$$

The conventions regarding the use of standard electrode potentials are summarized as follows:

1. The standard electrode potential refers to the half-reaction written as a reduction, that is,

    oxidized form + ne^- = reduced form

2. The half-cell potential of the standard hydrogen electrode is by convention zero, and the half-cell potentials of other electrodes are measured relative to this standard.
3. The sign of the half-cell potential is reversed when the half-reaction is written in the opposite direction.

4. The potential of the half-reaction is not dependent on the number of electrons involved, and therefore multiplying a half-reaction by a positive number does not alter the potential.
5. As a cell reaction is the sum of two half-reactions, the cell potential is the sum of the two half-cell potentials.

It should be emphasized that the $\Delta \mathcal{E}°$ value for an electron-transfer reaction measures the tendency for reactants *in their standard states* to go to products *in their standard states*. In Section 14.8 we shall investigate reactions in which the reactants and products are not in their standard states.

14.7 THE RELATIVE STRENGTH OF OXIDIZING AGENTS AND REDUCING AGENTS

When a set of half-reactions is arranged in standard form as in Table 14.1, all oxidizing agents are on the left side of the equations and all reducing agents are on the right side. Because the half-reactions are arranged in order of increasing standard electrode potential, each oxidizing agent will oxidize all reducing agents appearing *above and to the right of it* in the table. For example, Cu^{2+} will oxidize Fe(s) to Fe^{2+}, as shown in detail in Example 6. The Cu^{2+} ion is therefore said to be a *stronger* oxidizing agent than Fe^{2+}.

Conversely, each reducing agent will reduce all oxidizing agents appearing *below and to the left of it* in the table. For example, Fe(s) will reduce Cu^{2+} to Cu(s). Metallic iron is therefore said to be a *stronger* reducing agent than metallic copper. The general pattern is:

stronger oxidizing agent + stronger reducing agent ⟶
weaker oxidizing agent + weaker reducing agent

In Table 14.1 the strongest oxidizing agent is gaseous fluorine. It will oxidize any of 25 reducing agents listed, from K(s) to Mn^{2+}. The strongest reducing agent in this table is metallic potassium. It will reduce any of 25 oxidizing agents listed, from Na^+ to $F_2(g)$.

All metals appearing above hydrogen (those with negative $\mathcal{E}°$) will dissolve in a strong acid, if the H^+ concentration is $1M$, to produce the corresponding metallic ion and hydrogen gas. Such metals are called **base metals.** (Some metals, such as lead, react very slowly unless a catalyst is present. The $\Delta \mathcal{E}°$ values show the direction in which reaction occurs, but give no information on how fast the reaction is.) Metals appearing below hydrogen (those with positive $\mathcal{E}°$) are called **noble metals.** They do not dissolve in solutions containing H^+ unless some stronger oxidizing agent is present. For example, metallic copper does not dissolve in hydrochloric acid, but it does dissolve in nitric acid which contains the stronger oxidizing agent NO_3^-. (The $\mathcal{E}°$ value for the nitric acid half-reaction is listed at +0.96 V in Table 14.1.)

Instead of beginning with a table of standard electrode potentials, and predicting the direction in which a reaction will occur, one can begin with information on the directions of several reactions, and determine the relative strengths of the oxidizing and reducing agents involved. This type of problem is illustrated in the following example.

EXAMPLE 7

> Given that the following reactions occur spontaneously as written when all reactants and products are initially present at unit activity, arrange the oxidizing agents in order of decreasing strength.
>
> $$2Fe^{3+} + 2I^- = 2Fe^{2+} + I_2(s)$$
>
> $$O_2(g) + 4H^+ + 4Fe^{2+} = 4Fe^{3+} + 2H_2O$$
>
> From the first reaction we conclude that Fe^{3+} is a stronger oxidizing agent than $I_2(s)$. From the second equation we find that $O_2(g)$ is a stronger oxidizing agent than Fe^{3+}.
>
> Combining these two results we obtain the relative strengths of the oxidizing agents involved in these reactions as
>
> $$O_2(g) > Fe^{3+} > I_2(s)$$

14.8 EFFECT OF CONCENTRATION AND PRESSURE ON CELL VOLTAGE; DETERMINATION OF EQUILIBRIUM CONSTANTS

The standard voltage of a cell ($\Delta \mathcal{E}°$) is a measure of the tendency for reactants in their standard states to form products in their standard states. For example, the equation

$$Fe^{2+} + Ag^+ = Fe^{3+} + Ag(s) \qquad \Delta \mathcal{E}° = +0.028 \text{ V} \qquad (14.10)$$

tells us that silver metal will form if a solution is prepared that contains each of the ions (Ag^+, Fe^{2+}, and Fe^{3+}) at unit activity, which we can assume to be about $1.0M$. However, if all of the concentrations are $0.10M$, we find that the reaction is reversed: Silver metal will dissolve to produce silver ion. To predict whether or not reactions will occur when reactants and products are not in their standard states, we must examine the effect of the concentrations of these substances on the voltage of the cell.

Experimental measurements of the cell voltage of many cells indicate that the voltage is a function of the concentrations of the reactants and products involved in the cell reaction. This dependence of voltage on concentration may be expressed for the general reaction

$$aA + bB + \cdots = cC + dD + \cdots$$

by the equation

14.8 EFFECT OF CONCENTRATION AND PRESSURE ON CELL VOLTAGE

$$\Delta \mathcal{E} = \Delta \mathcal{E}° - \frac{2.3026RT}{nF} \log Q \tag{14.11a}$$

or, at 25°C,

$$\Delta \mathcal{E} = \Delta \mathcal{E}° - \frac{0.059}{n} \log Q \tag{14.11b}$$

In these equations $\Delta \mathcal{E}$ is the cell voltage under nonstandard conditions, $\Delta \mathcal{E}°$ is the standard potential of the cell in volts, n is the number of moles of electrons transferred in the reaction as written, and

$$Q = \frac{(a_C)^c(a_D)^d \cdots}{(a_A)^a(a_B)^b \cdots} \approx \frac{[C]^c[D]^d \cdots}{[A]^a[B]^b \cdots} \tag{14.12}$$

The first form of the equation for Q uses the activity, a, of each substance; in the second form of the equation, activities have been replaced by concentrations. (If a reactant or product is a gas, one uses the partial pressure of the gas in atmospheres. If a reactant or product is a pure liquid or pure solid, the activity is 1.) Note that Q has the same form as the equilibrium constant expression, but one uses the actual activities (or concentrations) in place of the equilibrium activities (or concentrations).

The constant 0.059 (more precisely, 0.05916) in Eq. 14.11b is calculated from $2.3026RT/F$ for a temperature T of 298.15 K, or 25°C. R is the gas constant and F is the Faraday constant. The factor 2.3026 arises from the conversion of the natural logarithm of Q to its common logarithm (to the base 10). Equation 14.11 was discovered by Walther Nernst in 1889. It has been confirmed by extensive experimental data on many cells, in addition to being derivable from the laws of thermodynamics.

Let us now apply the Nernst equation to the reaction represented by Eq. 14.10 when the ions are all present at $0.10M$ concentrations instead of $1.0M$.

$$Fe^{2+}(0.10M) + Ag^+(0.10M) = Fe^{3+}(0.10M) + Ag(s) \tag{14.13a}$$

$$\Delta \mathcal{E} = 0.028 - \frac{0.059}{n} \log \frac{[Fe^{3+}]}{[Fe^{2+}][Ag^+]} \tag{14.13b}$$

To convert 1 mol of Fe^{2+} to Fe^{3+}, 1 mol of electrons must be transferred to Ag^+, so that n is 1.

$$\Delta \mathcal{E} = 0.028 - \frac{0.059}{1} \log \frac{0.10}{(0.10)(0.10)} = 0.028 - 0.059 \log 10$$

$$= 0.028 - 0.059 = -0.031 \text{ V}$$

Thus we find that the voltage is negative, and we predict that the reaction will proceed in the reverse direction, in agreement with the experimental results noted above.

Let us examine Eq. 14.13b in a little more detail. When the ratio $[Fe^{3+}]/[Fe^{2+}][Ag^+]$ is large, $\Delta\mathcal{E}$ will be negative, so the reaction will proceed spontaneously from right to left. If this ratio is small, $\Delta\mathcal{E}$ will be positive, and the reaction will proceed from left to right.

Evidently there must be a concentration ratio that will make the last term of Eq. 14.13b just equal to -0.028 V so that the voltage of the cell will be zero. If such is the case, electrons will not be transferred and a chemical reaction will not occur. [That this is a true chemical equilibrium can be verified experimentally. Because a small decrease in the concentration ratio (from the equilibrium value) will result in a positive value of $\Delta\mathcal{E}$, the reaction will proceed from left to right, causing an increase in the concentration ratio, and $\Delta\mathcal{E}$ will approach zero. When $\Delta\mathcal{E}$ reaches zero, the reaction will again cease, and the system will be at chemical equilibrium. If, on the other hand, the value of the concentration ratio is such that $\Delta\mathcal{E}$ is negative, the reaction will proceed from right to left, the concentration ratio will decrease, and the voltage $\Delta\mathcal{E}$ will again approach zero.]

We may therefore conclude that the condition for chemical equilibrium to exist is that $\Delta\mathcal{E} = 0$ and that under such conditions the equilibrium concentrations will be given by the equation

$$\Delta\mathcal{E} = 0 = 0.028 - \frac{0.059}{1} \log \frac{[Fe^{3+}]_e}{[Fe^{2+}]_e[Ag^+]_e}$$

However, the concentration ratio in the previous equation is merely the equilibrium constant, K, for the reaction of Eq. 14.10. Therefore

$$0 = 0.028 - 0.059 \log K$$

and

$$\log K = \frac{0.028}{0.059} = 0.47$$

$$K = 3.0$$

This reasoning may be extended to other reactions by utilizing Eq. 14.11 and noting that when the system is at equilibrium, $\Delta\mathcal{E} = 0$, and Q is equal to the equilibrium constant:

$$0 = \Delta\mathcal{E}° - \frac{0.059}{n} \log K$$

or

$$\Delta\mathcal{E}° = \frac{0.059}{n} \log K \text{ at } 25°C \tag{14.14}$$

14.8 EFFECT OF CONCENTRATION AND PRESSURE ON CELL VOLTAGE

Thus a positive $\Delta\mathcal{E}°$ value corresponds to a value of K greater than unity, whereas a negative $\Delta\mathcal{E}°$ value indicates that the value of K is less than unity.

The following examples indicate the use of $\Delta\mathcal{E}°$ values obtained from measurements on appropriate cells in evaluating equilibrium constants for the corresponding chemical reactions.

EXAMPLE 8

Calculate $\Delta\mathcal{E}°$ and K for the reaction $Zn(s) + 2H^+ \rightleftharpoons Zn^{2+} + H_2(g)$.

From Table 14.1, we obtain

$$Zn(s) = Zn^{2+} + 2e^- \qquad \mathcal{E}° = +0.763 \text{ V}$$
$$2H^+ + 2e^- = H_2(g) \qquad \mathcal{E}° = 0 \text{ V}$$

We add these two half-reactions to obtain

$$Zn(s) + 2H^+ = Zn^{2+} + H_2(g) \qquad \Delta\mathcal{E}° = +0.763 \text{ V}$$

Because

$$\Delta\mathcal{E}° = \frac{0.059}{n} \log K$$

we write

$$0.763 = \frac{0.059}{2} \log K$$

and

$$\log K = \frac{2 \times 0.763}{0.059} = 26$$

$$K = 1 \times 10^{26}$$

EXAMPLE 9

Calculate the pressure of hydrogen gas that would be required to maintain equilibrium between solid zinc and a solution that is $0.10M$ in zinc ion and $0.20M$ in hydrogen ion.

From Example 8, we find that

$$K = 1 \times 10^{26} = \frac{[Zn^{2+}]P_{H_2}}{[H^+]^2}$$

$$P_{H_2} = \frac{1 \times 10^{26}[H^+]^2}{[Zn^{2+}]} = \frac{(1 \times 10^{26})(0.20)^2}{0.10} = 4 \times 10^{25} \text{ atm}$$

Such a large result illustrates the strong tendency for zinc metal to dissolve in a solution containing $0.20M$ H^+ and $0.10M$ Zn^{2+}.

The use of electrochemical cell potentials to evaluate equilibrium constants is particularly important when K is very large or very small. In such cases the evaluation of K from the measured equilibrium concentrations would be extremely difficult owing to the very large or very small values of the concentrations or pressures (as in the preceding example). In an electrochemical cell, one may use convenient concentrations (0.10 to 1.0M) and measure the voltage produced.

The solubility product for silver chloride in aqueous solution may be calculated from the voltage of an electrochemical cell using the following half-reactions:

$$Ag(s) = Ag^+ + e^-$$

$$AgCl(s) + e^- = Ag(s) + Cl^-$$

The sum of the two half-reactions gives

$$AgCl(s) = Ag^+ + Cl^- \qquad K_{sp} = [Ag^+][Cl^-]$$

The first half-cell may be constructed by placing a silver electrode in 0.0010M AgNO$_3$. The second half-cell may be constructed by placing a silver electrode in a 0.0010M HCl solution that has been saturated with solid silver chloride. The measured voltage of this cell is found to be $\Delta\mathcal{E} = -0.222$ V.

The standard cell potential, $\Delta\mathcal{E}°$, may be calculated using Eq. 14.11, with $Q = [Ag^+][Cl^-] = (0.0010)(0.0010)$:

$$-0.222 = \Delta\mathcal{E}° - \frac{0.059}{1}\log(0.0010)^2$$

$$\Delta\mathcal{E}° = -0.222 - 0.354 = -0.576 \text{ V}$$

K_{sp} may now be calculated using Eq. 14.14:

$$\log K_{sp} = \frac{n\Delta\mathcal{E}°}{0.059} = \frac{(1)(-0.576)}{0.059} = -9.8$$

$$K_{sp} = 1.6 \times 10^{-10}$$

14.9 PRIMARY CELLS, STORAGE CELLS, AND FUEL CELLS

The production of an electric current by means of a chemical reaction is accomplished by using a primary cell, a storage cell, or a fuel cell. A **primary cell** is an electrochemical cell (as in Figure 14.1) in which a chemical reaction proceeds with a flow of electrons (discharge) in an external circuit. A disadvantage of most primary cells is that the supply of reactants is limited and that, once the reactants are used up, the cell is no longer useful. A **storage cell** is a primary cell of large capacity that can be returned after use to its original state (charged) by an opposing external voltage. A **fuel cell** is similar to a primary cell except that there is a provision for replacement of the oxidizing and reducing agents as they are depleted by the cell reaction. The term **battery** is usually used to indicate a group of cells connected together, although the term is sometimes used for a single cell.

The most common primary cell is the "dry cell" used in flashlights. It consists of a zinc cylinder that contains a graphite rod in the center surrounded by a moist paste of ammonium chloride (NH_4Cl), manganese dioxide (MnO_2), and zinc chloride ($ZnCl_2$) (Figure 14.5). The zinc container is the anode and the graphite rod serves as the cathode. For small electrical currents the electrode reactions are

anode

$$Zn(s) = Zn^{2+} + 2e^-$$

cathode

$$2NH_4^+ + 2MnO_2(s) + 2e^- = 2MnO(OH)(s) + 2NH_3$$

The cell reaction is

$$Zn(s) + 2MnO_2(s) + 2NH_4^+ = Zn^{2+} + 2MnO(OH)(s) + 2NH_3$$

The dry cell generates about 1.5 V.

Alkaline cells have several advantages over ordinary dry cells: They have a longer shelf life, are usable at lower temperatures, provide a larger operating voltage with large current drains, and can be recharged. The cell consists of a zinc anode and a manganese dioxide cathode with a solution of potassium hydroxide as the electrolyte. The cell reaction is

$$Zn(s) + 2MnO_2(s) + 2H_2O = 2MnO(OH)(s) + Zn(OH)_2(s)$$

The voltage of the cell is 1.5 V.

The small batteries used in hearing aids, digital watches, cameras, and the like contain zinc anodes and mercuric oxide or silver oxide cathodes. The electrolyte solution contains sodium hydroxide or potassium hydroxide. With a mercuric oxide cathode the cell voltage is 1.35 V, and the cell reaction is

$$Zn(s) + HgO(s) = ZnO(s) + Hg(l)$$

The most common storage cell is the lead storage cell, which consists of a lead (Pb) electrode and a lead dioxide (PbO_2) electrode partially immersed in an aqueous sulfuric acid solution, as shown in Figure 14.6. On discharge, its reactions are

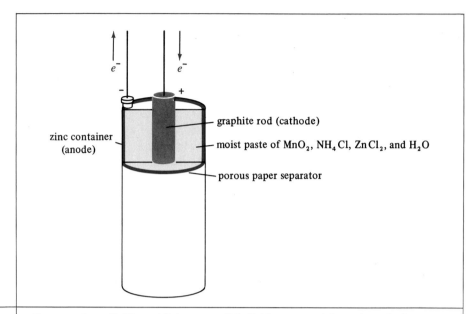

FIGURE 14.5

Common dry cell. The oxidizing agent (MnO_2) is separated from the reducing agent (Zn) by porous paper wet with a solution of NH_4Cl and $ZnCl_2$.

anode

$$Pb(s) + HSO_4^- = PbSO_4(s) + H^+ + 2e^-$$

cathode

$$PbO_2(s) + 3H^+ + HSO_4^- + 2e^- = PbSO_4(s) + 2H_2O$$

The cell reaction is

$$Pb(s) + PbO_2(s) + 2H^+ + 2HSO_4^- = 2PbSO_4(s) + 2H_2O$$

The voltage of the cell is 2 V. A typical automobile storage battery consists of six cells, connected together to deliver 12 V.

As the cell discharges, slightly soluble lead sulfate is formed which adheres to the surface of the electrodes, and the concentration of H_2SO_4 decreases. This decrease in concentration causes a decrease in the density of the solution. Thus the density may be used to determine the extent of charge or discharge of the cell. After use the cell may be charged by connecting it to an external voltage and forcing the reverse reaction to occur.

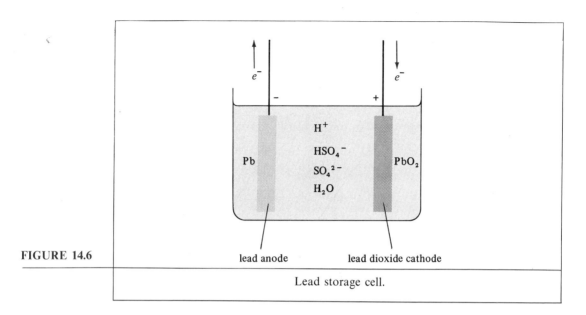

FIGURE 14.6 Lead storage cell.

Except in an electrochemical cell, the conversion of chemical energy to electric energy usually involves a three-stage process: combustion of a fuel to form thermal energy; conversion of thermal energy to mechanical work in a heat engine; and, finally, conversion of mechanical work to electrical energy by means of a generator. The overall efficiency of such a three-stage process is about 35% for a steam plant using coal as a fuel. The direct conversion of chemical energy to electrical energy in fuel cells is capable of efficiencies on the order of 70 to 80%. Thus, there has been considerable interest in developing fuel cells. In general fuel cells produce about 1 V per cell.

One successful fuel cell utilizes the electron-transfer reaction of hydrogen with oxygen to form water (Figure 14.7). The electrodes are made of porous carbon containing suitable catalysts. The anode contains finely divided platinum or palladium, whereas the cathode contains oxides of cobalt, silver, or gold. These electrodes dip into an aqueous solution of potassium hydroxide. When hydrogen gas is introduced into the anode chamber and air (or oxygen) flows into the cathode chamber, an electrical current is obtained. The half-reactions are:

anode

$$2H_2(g) + 4OH^- = 4H_2O + 4e^-$$

cathode

$$O_2(g) + 2H_2O + 4e^- = 4OH^-$$

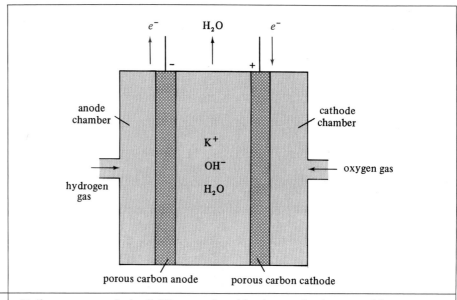

FIGURE 14.7 Hydrogen-oxygen fuel cell. Water produced by the reaction is removed by evaporation.

The cell reaction is

$$2H_2(g) + O_2(g) = 2H_2O$$

PROBLEMS

14.1 (a) Draw a Lewis structure for each of the following substances, then assign an oxidation state to each atom.
 H_2S SO_4^{2-} CH_3CO_2H
(b) What is the *average* oxidation state of the two carbon atoms in an acetic acid molecule?

14.2 By using the rules for common oxidation states in Section 14.2, assign an oxidation state to each element in the following substances:
(a) Al_2O_3
(b) H_3PO_4
(c) C_2F_4
(d) CCl_4
(e) $CaSO_4$
(f) SO_3^{2-}
(g) KNO_3
(h) ClO_4^-

14.3 What type of reagent is required for each of the following changes: an oxidizing agent, a reducing agent, or some other type of reagent?
(a) $CrO_4^{2-} \rightarrow Cr^{3+}$
(b) $H_2S \rightarrow S_8$
(c) $NH_3 \rightarrow N_2H_4$
(d) $SO_2 \rightarrow SO_3^{2-}$

14.4 In an electrolysis experiment, metallic copper is deposited at the cathode from a solution of cupric sulfate, $CuSO_4$. The half-reaction is $Cu^{2+} + 2e^- = Cu(s)$. If a current of 0.20 A flows for 24 h, how many grams of copper will be deposited?

14.5 When a solution of sodium sulfate is electrolyzed with inert electrodes, the half-reactions that occur at the electrodes are
$$2H_2O + 2e^- = 2OH^- + H_2(g)$$
$$2H_2O = 4H^+ + O_2(g) + 4e^-$$
(a) At which electrode is gaseous oxygen produced, the anode or cathode?
(b) Toward which electrode do Na^+ ions migrate, the electrode at which gaseous hydrogen is produced, or the electrode at which gaseous oxygen is produced?
(c) Write a balanced equation for the net reaction.
(d) How many faradays would be required to produce 0.25 g of hydrogen? How many coulombs?

14.6 An electrochemical cell makes use of the spontaneous reaction $Zn(s) + Cl_2(g) \rightarrow Zn^{2+} + 2Cl^-$. The $Cl_2(g)$ reacts at a carbon (graphite) electrode.
(a) Which electrode is the anode, zinc or carbon?
(b) Toward which electrode do Zn^{2+} ions migrate in the solution?
(c) Toward which electrode do electrons move?
(d) With respect to the external circuit, which electrode should be marked positive?
(e) In this reaction, what substance is the oxidizing agent? What substance is the reducing agent?
(f) How many moles of $Cl_2(g)$ are required to produce an electric current of 0.10 A for 125 h?

14.7 Diagram an electrochemical cell that makes use of the spontaneous reaction $Fe(s) + 2Ag^+ \rightarrow Fe^{2+} + 2Ag(s)$. Solutions of $AgNO_3$ and $FeSO_4$ (both strong electrolytes) may be used as sources of Ag^+ and Fe^{2+} ions. Label the anode and the cathode. Show clearly the migration of electrons, cations, and anions. Write the electrode reactions that occur at the anode and at the cathode.

14.8 Write a balanced half-reaction for each of the following:
   (a) $IO_3^- = I^-$ (acid solution)
   (b) $NO_3^- = NO(g)$ (acid solution)
   (c) $NO_3^- = NH_4^+$ (acid solution)
   (d) $H_2S = HSO_4^-$ (acid solution)
   (e) $BrO^- = Br^-$ (basic solution)
   (f) $HO_2^- = O_2(g)$ (basic solution)
   (g) $Zn(s) = Zn(OH)_3^-$ (basic solution)

14.9 Write a complete balanced equation for each of the following reactions. In each case indicate the oxidizing agent, the reducing agent, the substance oxidized, and the substance reduced.
   (a) $MnO_4^- + Fe^{2+} = Mn^{2+} + Fe^{3+}$ (acid solution)
   (b) $I_2(s) = I^- + IO^-$ (basic solution)

14.10 Complete and balance the following electron-transfer reactions that occur in acidic aqueous solutions:
   $Al(s) + H^+ = H_2(g) + Al^{3+}$
   $I_2(s) + H_2S = I^- + S(s)$
   $Cu(s) + NO_3^- = Cu^{2+} + NO(g)$
   $Br^- + H_2O_2 = Br_2(l) + H_2O$
   $Cr_2O_7^{2-} + Fe^{2+} = Fe^{3+} + Cr^{3+}$
   $S_2O_3^{2-} + I_3^- = I^- + S_4O_6^{2-}$
   $NH_4^+ + NO_3^- = N_2O(g)$
   $Fe(s) + Fe^{3+} = Fe^{2+}$
   $HNO_2 = NO(g) + NO_3^-$

14.11 Complete and balance the following electron-transfer reactions that occur in basic aqueous solutions:
   $Al(s) + H_2O = Al(OH)_4^- + H_2(g)$
   $SO_3^{2-} + ClO_3^- = SO_4^{2-} + Cl^-$
   $Cl_2(g) = Cl^- + ClO_3^-$
   $N_2H_4 + Cu(OH)_2(s) = N_2(g) + Cu(s)$
   $P_4(s) + OH^- = PH_3(g) + H_2PO_2^-$

14.12 Predict whether each of the following reactions will occur spontaneously as written or in the reverse direction if all substances and ions are present at unit activity.
   (a) $Fe(s) + 2H^+ = Fe^{2+} + H_2(g)$
   (b) $2Fe^{2+} + 2H^+ = 2Fe^{3+} + H_2(g)$
   (c) $2Ag(s) + 2H^+ = 2Ag^+ + H_2(g)$
   (d) $Hg^{2+} + Hg(l) = Hg_2^{2+}$
   (e) $3Cl_2(g) + 2NO(g) + 4H_2O = 6Cl^- + 2NO_3^- + 8H^+$
   (f) $Br_2(l) + 2Cl^- = Cl_2(g) + 2Br^-$

PROBLEMS

14.13 From the following data, calculate $\Delta\mathcal{E}°$ for the reaction
$$Ni(s) + Cu^{2+} = Cu(s) + Ni^{2+}$$
(a) $Cu(s) + 2Ag^+ = 2Ag(s) + Cu^{2+}$ $\Delta\mathcal{E}° = 0.46$ V
(b) $2Ag^+ + Ni(s) = Ni^{2+} + 2Ag(s)$ $\Delta\mathcal{E}° = 1.05$ V

14.14 Cadmium (Cd) and mercury (Hg) are both metals, mercury being a liquid. Cadmium dissolves in $1M$ HCl to form Cd^{2+} and gaseous hydrogen. Mercury is not soluble in $1M$ HCl, requiring a more powerful oxidizing agent to oxidize it to Hg^{2+}. Arrange the following half-reactions in the order in which they should appear in a standard table (strongest oxidizing agent at the bottom of the list).
$$2H^+ + 2e^- = H_2(g); \quad Hg^{2+} + 2e^- = Hg(l); \quad Cd^{2+} + 2e^- = Cd(s)$$

14.15 Given that the following reactions occur spontaneously as written when all reactants and products are initially present at unit activity, arrange the half-reactions in the order in which they should appear in a standard table.
$$H_2O_2 + Br_2(l) = 2H^+ + O_2(g) + 2Br^-$$
$$Cl_2(g) + 2Br^- = 2Cl^- + Br_2(l)$$

14.16 The following reactions occur spontaneously as written when all substances are initially present at unit activity
$$2Fe^{3+} + Fe(s) = 3Fe^{2+}$$
$$2Cu^+ = Cu(s) + Cu^{2+}$$
(a) Which substance is the better oxidizing agent, Fe^{2+} or Fe^{3+}?
(b) Which substance is the better oxidizing agent, Cu^+ or Cu^{2+}?

14.17 An electrochemical cell uses the reaction
$$Zn(s) + 2Ag^+ = 2Ag(s) + Zn^{2+}$$
(a) Calculate $\Delta\mathcal{E}°$ from the data in Table 14.1.
(b) Calculate $\Delta\mathcal{E}$ if $[Zn^{2+}] = 0.20M$ and $[Ag^+] = 3.0 \times 10^{-4}M$.
(c) Calculate the equilibrium constant for the cell reaction at 25°C.

14.18 (a) From the data in Table 14.1, calculate $\Delta\mathcal{E}°$ for the zinc–chlorine cell described in Problem 14.6.
(b) Calculate $\Delta\mathcal{E}$ if the partial pressure of chlorine is 0.25 atm, $[Zn^{2+}] = 0.050M$, and $[Cl^-] = 6.0 \times 10^{-3}M$.
(c) Calculate the equilibrium constant for the cell reaction at 25°C.

14.19 A solution contains Ag^+ at low concentration. A silver wire is dipped into the solution forming a half-cell, which is connected to a standard hydrogen half-cell. The potential of the silver wire is $+0.56$ V with respect to the hydrogen electrode. Calculate the concentration of Ag^+ in the solution.

14.20 Calculate $\Delta\mathcal{E}°$ and the equilibrium constant at 25°C for each of the following reactions:

(a) $Hg^{2+} + Hg(l) = Hg_2^{2+}$
(b) $3Zn(s) + 2NO_3^- + 8H^+ = 3Zn^{2+} + 2NO(g) + 4H_2O$

14.21 What is the oxidizing agent and what is the reducing agent in
(a) the dry cell?
(b) the lead storage cell?
In each case, what changes in oxidation state occur in the anode half-reaction and in the cathode half-reaction?

14.22 A six-cell lead storage battery produces a current of 1.00 A for 100 h at 12 V. Calculate
(a) the number of coulombs transferred
(b) the mass of metallic lead oxidized to lead sulfate in each cell
(c) the electric work done by the battery on its surroundings

14.23 An aluminum-air fuel cell has been proposed to power an electric automobile.
(a) Write the cell reaction if $O_2(g)$ is the oxidizing agent and the aluminum metal (in contact with a basic solution) is oxidized to $Al(OH)_3$.
(b) Calculate $\Delta\mathcal{E}°$ for the cell reaction in part (a) from the following data:
$Al(OH)_3(s) + 3e^- = Al(s) + 3OH^-$ $\mathcal{E}° = -2.30$ V
$O_2(g) + 2H_2O + 4e^- = 4OH^-$ $\mathcal{E}° = +0.40$ V

14.24 Complete and balance the following electron-transfer reactions:
$CuS(s) + NO_3^- = Cu^{2+} + S(s) + NO(g)$ (acid solution)
$Fe(OH)_2(s) + O_2(g) = Fe(OH)_3(s)$ (basic solution)

14.25 An electrochemical cell is constructed by placing a silver wire in an aqueous solution containing Ag^+ at unit activity and another silver wire in a solution saturated with AgBr that contains Br^- at unit activity. A salt bridge connects the two solutions. The cell generates 0.77 V, with the silver wire in the bromide solution being negative compared to the silver wire in the silver ion solution.
(a) Write the half-reactions that occur at the anode and at the cathode. (Note that AgBr is a slightly soluble salt.)
(b) Calculate the standard electrode potential for the half-reaction that occurs in the compartment containing bromide ion.
(c) Calculate the solubility product of silver bromide.

14.26 (a) A cell has a zinc electrode immersed in $1.0M$ $ZnSO_4$, and a silver electrode immersed in $1.0M$ $AgNO_3$. Using the data in Table 14.1, calculate $\Delta\mathcal{E}°$. Which electrode is the anode?
(b) Calculate $\Delta\mathcal{E}$ if sufficient potassium iodide, KI, were added to the silver nitrate solution to precipitate almost all of the Ag^+ as $AgI(s)$, increasing $[I^-]$ to $1.0M$. (The solubility product of silver iodide is 4.5×10^{-17}.) Which electrode is now the anode?

CHAPTER 15
THERMODYNAMICS AND CHEMICAL EQUILIBRIA

15

The methods developed in preceding chapters allow us to treat chemical equilibria in a precise way. From measurements of a set of equilibrium concentrations, one may calculate the equilibrium constant, K. Then, for a given set of initial conditions at the same temperature, this constant may be used to predict in which direction the reaction will proceed to reach equilibrium and what the final concentrations will be.

But more insight is needed to answer questions such as: Why does a particular chemical reaction proceed spontaneously in a certain direction? Why is a chemical equilibrium at some specified temperature characterized by a particular value of K? We can obtain clues to the answers by examining quantitatively how the vapor pressure of a substance and the equilibrium constant of a reaction depend on the temperature.

15.1 TEMPERATURE DEPENDENCE OF VAPOR PRESSURES AND EQUILIBRIUM CONSTANTS

The vapor pressure of a substance is always larger at higher temperatures. Figure 15.1(a) shows some typical vapor pressure data. Although the vapor pressure does not vary linearly with temperature, and the lines have marked curvature, *the logarithm of the vapor pressure does vary almost linearly with the reciprocal of the absolute temperature.* Figure 15.1(b) shows a graph of $R \ln P$ versus $1/T$, where R is the gas constant (8.314 J K^{-1} mol^{-1}), T is the absolute temperature, and P is numerically equal to the vapor pressure in atmospheres.[1] Over these limited ranges of temperature the lines are essentially straight. As we shall see, the slope of each line is $-\Delta H°$, the negative of the standard enthalpy (or heat) of vaporization at 1 atm pressure. (We recall that the degree sign, °, means that all substances are in their standard states, which were defined in Sections 5.8 and 12.2.)

These relationships can be expressed by a simple equation:

$$R \ln P = -\frac{\Delta H°}{T} + C \tag{15.1}$$

---

[1] The term $\ln P$ is the logarithm of P to the base e and is called the **natural logarithm** (see Appendix C-5).

15.1 TEMPERATURE DEPENDENCE OF VAPOR PRESSURES AND EQUILIBRIUM CONSTANTS

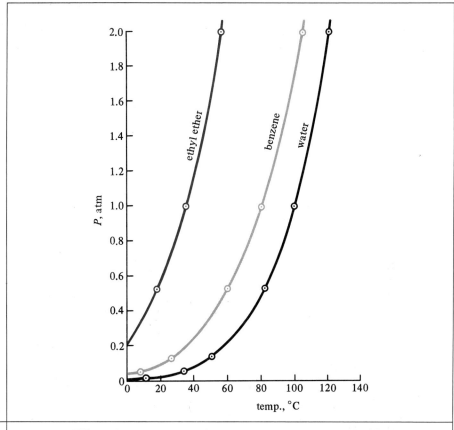

FIGURE 15.1(a)

Effect of temperature on the vapor pressure of three liquids.

where C is a constant. For example, the equation that fits the benzene line shown in Figure 15.1(b) is

$$R \ln P = -\frac{32.6 \times 10^3 \text{ J mol}^{-1}}{T} + 92 \text{ J K}^{-1} \text{ mol}^{-1} \tag{15.2}$$

The standard enthalpy of vaporization of benzene is therefore 32.6 kJ mol^{-1}.

Just what is this constant C? From the laws of thermodynamics it may be shown that C is the change in an important property of the substance called the **entropy** (S), which will be defined in the next section. Just as $\Delta H°$ is the increase in enthalpy on vaporization ($H°_{\text{gas}} - H°_{\text{liquid}}$), $\Delta S°$ is the increase in entropy on vaporization ($S°_{\text{gas}} - S°_{\text{liquid}}$). Thus Eq. 15.1 becomes

$$R \ln P = -\frac{\Delta H°}{T} + \Delta S° \tag{15.3}$$

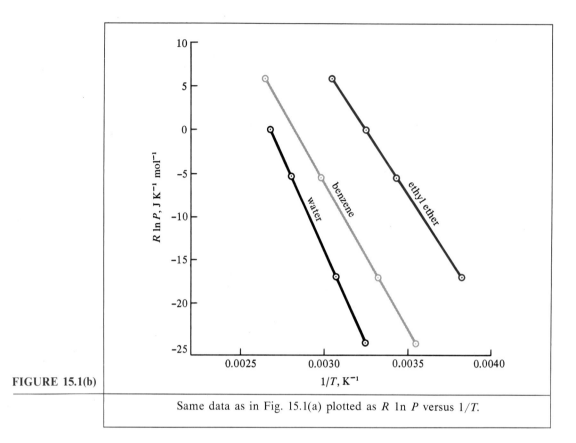

FIGURE 15.1(b) Same data as in Fig. 15.1(a) plotted as $R \ln P$ versus $1/T$.

By comparing Eq. 15.2 with Eq. 15.3, we find that the **standard entropy of vaporization** of benzene is 92 J K^{-1} mol^{-1}. (Note that entropy, like volume and enthalpy, is an extensive property. Thus the standard entropy of vaporization of 2 mol of benzene is 184 J K^{-1}.)

We see, therefore, that the vapor pressure depends on two factors: a constant factor, $\Delta S°$, and a factor, $\Delta H°/T$, in which the effect of the constant $\Delta H°$ term is modified by the absolute temperature. At low temperatures T is small, $\Delta H°/T$ is large, and the more important factor is $\Delta H°/T$. At high temperatures T is large, $\Delta H°/T$ is small, and the more important factor is $\Delta S°$. At all temperatures the vapor pressure depends on a balance between the two.

To gain some insight into the way these factors operate, consider a 1-L flask into which various substances can be introduced. If some oxygen gas is placed in the flask, the thermal energy of the molecules carries them into all parts of the flask, so that the distribution of molecules quickly becomes uniform, and the chance of finding a particular O_2 molecule in any 1-mL region of volume is 1 in 1000.

If some liquid water (say, 1 mL) is placed in the flask, a small fraction of it will evaporate until the equilibrium vapor pressure is reached. Like the oxygen molecules, the water molecules tend to distribute themselves uniformly throughout the flask. Opposing this tendency are the attractive forces between the molecules, which keep practically all the water molecules confined in the liquid, the volume of which is now slightly less than 1 mL.

If the temperature is increased, the molecules have more thermal energy to overcome the attractive forces. The rate of evaporation increases, the vapor pressure rises, and the distribution of molecules becomes more nearly uniform. At sufficiently high temperature the vapor pressure becomes so high that all of the water vaporizes, and the distribution of water molecules in the flask is completely uniform.

In Eq. 15.3 the entropy term, $\Delta S°$, represents the tendency of the molecules to be distributed throughout the flask, and the enthalpy term, $-\Delta H°$, represents the attractive forces between the molecules. Division of the $-\Delta H°$ term by T represents the effect of higher thermal energies in overcoming these attractive forces.

A similar equation applies to the equilibrium constant. Figure 15.2 shows that $R \ln K$ is a linear function of $1/T$ for two chemical reactions

$$2HI(g) = H_2(g) + I_2(g)$$

and

$$N_2(g) + O_2(g) = 2NO(g)$$

The appropriate equation expressing this relation, derived from the laws of thermodynamics, is

$$R \ln K = -\frac{\Delta H°}{T} + \Delta S° \tag{15.4}$$

Here $\Delta H°$ is the standard enthalpy of reaction ($H°_{products} - H°_{reactants}$), and $\Delta S°$ is the standard entropy change ($S°_{products} - S°_{reactants}$). Therefore the standard enthalpy of reaction can be measured indirectly, using only the temperature dependence of the equilibrium constant. The values of $\Delta H°$ calculated from the slopes of the lines in Fig. 15.2 are 12.6 kJ mol^{-1} (HI reaction) and 120 kJ mol^{-1} (for formation of NO). (We are assuming that $\Delta H°$ and $\Delta S°$ are constant over the temperature range studied. This is usually a good assumption if the temperature range is not large. If $\Delta H°$ and $\Delta S°$ do vary with temperature, one must use a more complicated version of Eq. 15.4 that takes into account the different heat capacities of the reactants and products.)

The standard enthalpy of reaction may be calculated from the values of K at two different temperatures. Let K_A be the equilibrium constant at temperature T_A, and let K_B be the equilibrium constant at temperature T_B. If each set of values of

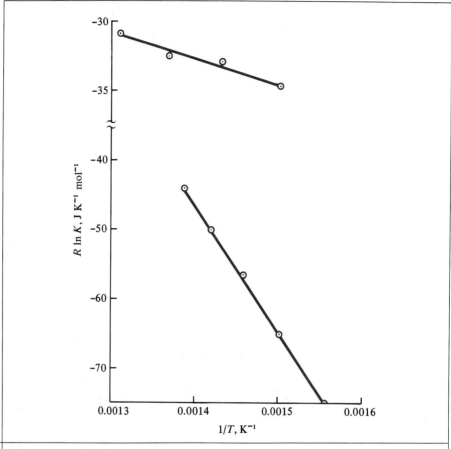

FIGURE 15.2 Plot of $R \ln K$ versus $1/T$ for (a) $2HI(g) = H_2(g) + I_2(g)$ (upper line); (b) $N_2(g) + O_2(g) = 2\,NO(g)$ (lower line). For (b) the values of $1/T$ should be reduced by 0.0010.

K and T is substituted in Eq. 15.4, then

$$\Delta S° = R \ln K_A + \frac{\Delta H°}{T_A}$$

and

$$\Delta S° = R \ln K_B + \frac{\Delta H°}{T_B}$$

Therefore

$$R \ln K_A + \frac{\Delta H°}{T_A} = R \ln K_B + \frac{\Delta H°}{T_B}$$

15.1 TEMPERATURE DEPENDENCE OF VAPOR PRESSURES AND EQUILIBRIUM CONSTANTS

or

$$\Delta H° \left[\frac{1}{T_A} - \frac{1}{T_B} \right] = R(\ln K_B - \ln K_A)$$

Because $\ln K_B - \ln K_A = \ln (K_B/K_A)$, we obtain

$$\Delta H° = \frac{R \ln (K_B/K_A)}{\left[\frac{1}{T_A} - \frac{1}{T_B} \right]}$$

The term in the denominator may be expanded to give

$$\Delta H° = \frac{T_A T_B R \ln (K_B/K_A)}{T_B - T_A} \tag{15.5a}$$

As the equilibrium constant for the vaporization of a liquid is equal to the liquid's vapor pressure, we may use a similar equation to determine the standard enthalpy of vaporization from the vapor pressures at two temperatures.

$$\Delta H° = \frac{T_A T_B R \ln (P_B/P_A)}{T_B - T_A} \tag{15.5b}$$

An application of Eq. 15.5a is shown in the following example.

EXAMPLE 1

Calculate the standard enthalpy of reaction ($\Delta H°$) for

$$H_2(g) + \tfrac{1}{2}O_2(g) = H_2O(g)$$

if $K = 2.19 \times 10^6$ at 1400 K and 5.21×10^5 at 1500 K.

We may substitute these values into Eq. 15.5a to calculate $\Delta H°$. Let T_A be 1400 K, and T_B be 1500 K. Then K_A is 2.19×10^6, and K_B is 5.21×10^5.

$$\Delta H° = \frac{(1400 \text{ K})(1500 \text{ K})(8.314 \text{ J K}^{-1} \text{ mol}^{-1}) \ln [(5.21 \times 10^5)/(2.19 \times 10^6)]}{(1500 - 1400) \text{ K}}$$

$$= -251 \text{ kJ mol}^{-1}$$

Thus not just the equilibrium of liquid with vapor but chemical equilibria in general depend only on the temperature and the changes in enthalpy and entropy. Enthalpy was discussed in detail in Chapter 5. In the rest of this chapter we consider how entropy is measured, its connection with molecular behavior, the second and third laws of thermodynamics, and finally the derivation of Eqs. 15.3 and 15.4 from the laws of thermodynamics.

15.2 ENTROPY

In previous chapters we have seen that at low temperatures most substances exist as solid crystals in which the atoms and molecules are arranged in an orderly structure. When heated to sufficiently high temperatures, solids melt, and in the liquid state, although the atoms and molecules are still in contact, they have a much less orderly arrangement. On heating to yet higher temperatures the liquid vaporizes to the gaseous state, which is characterized by a high degree of disorder. The molecules of a gas are on the average quite far apart, but they are in rapid motion and occasionally collide, so that a diagram showing the locations of the molecules at any instant looks completely random. Thus there is an intimate connection among disorder, thermal energy, and temperature.

It is useful for several purposes to have a quantitative measure of disorder, in particular the change in disorder of a substance on going from one state (A) to another (B). At first sight the heat absorbed, q, would appear to be a useful measure. However, the temperature must also be taken into account. A substance at a low temperature is in a highly ordered state with little thermal energy, and an additional amount of energy increases the disorder quite a bit, whereas one might expect that addition of the same amount of energy at a higher temperature would have a smaller effect. Thus it is plausible that the measure of change in disorder should be of the form q/T, the ratio of the heat absorbed to the absolute temperature at which the conversion occurs. Finally, one must consider the process by which the substance is changed from state A to state B, as in general q varies from one path to another, as shown in Section 5.2. The kind of process appropriate here is called **reversible**.

A reversible process is one in which a very slight change in conditions will reverse the direction of the process. For example, a gas sealed in a cylinder by a piston (see Figure 5.3) will expand if the external pressure, P_E, is maintained always slightly less than the pressure of the gas, P. That is, $P_E = P - \delta$, where δ is an arbitrarily small quantity. If the external pressure is now increased by 2δ to $P'_E = P + \delta$, the pressure inside the cylinder is smaller than the external pressure by δ, and the direction in which the piston is moving will be reversed. In principle, the quantity δ can be made as small as desired, and the limiting path, as δ approaches zero, is called a reversible process. In going from state A to state B by a reversible process, the system passes through a succession of intermediate states each of which is in an equilibrium condition. All real processes are to some degree irreversible, but it is possible to calculate the heat absorbed in a reversible process, q^{rev}.

As another example of reversibility, consider heat transfer between a system and its surroundings. To heat a system the temperature of the surroundings, T_{surr}, need be only slightly higher than the temperature of the system, T, or

$$T_{\text{surr}} = T + \delta$$

Then, if the temperature of the surroundings is decreased by the very small quantity 2δ to $T'_{surr} = T - \delta$, the direction of heat transfer is reversed. As δ approaches zero, the process approaches a reversible process.

We are now in a position to define the entropy, which will be used as a quantitative measure of disorder. *When a system changes from state A to state B by a reversible process, then the change in entropy (S) is*

$$\Delta S = \frac{q^{rev}}{T} \tag{15.6}$$

Nothing is said here about the path taken except that it must be reversible. What do we find if we compare different reversible processes leading from the same initial state A to the same final state B? The results for ΔS are always the same! Thus entropy (like volume, energy, and enthalpy) is a *state function*: Entropy is a *property* of a system, and the difference in entropy can be taken as the difference between the entropies of the initial and final states, $\Delta S = S_B - S_A$.

Two aspects of the definition of entropy require special comment.

1. The emphasis on reversible processes in the definition of entropy should not lead to the conclusion that irreversible processes are of no interest here—quite the contrary. It is in dealing with irreversible processes that the concept of entropy is most useful.
2. The statement that entropy, as defined by Eq. 15.6, is a measure of the disorder of a system requires some justification. Arguments advanced above to make this seem plausible are of course no substitute for proof. In Section 15.6 we shall examine in detail the connection between entropy and disorder and show that it is capable of experimental verification.

15.3 ENTROPIES OF CHEMICAL SUBSTANCES

At the normal boiling point of a liquid, the vapor pressure is exactly 1 atm. If 1 mol of the liquid is evaporated reversibly at its normal boiling point, the external pressure being maintained at 1 atm, then the heat absorbed, q^{rev}, is the same as the standard enthalpy of vaporization, ΔH°_{vap} (see Eq. 5.16). Thus the standard entropy of vaporization is[2]

$$\Delta S^\circ_{vap} = \frac{q^{rev}}{T_{bp}} = \frac{\Delta H^\circ_{vap}}{T_{bp}} \qquad [n, T, P] \tag{15.7a}$$

As shown by the last column of Table 5.3, several liquids with boiling points ranging from 158.9 K (HCl) to 1073 K (NaCl) have about the same molar entropy of vaporization as measured at the normal boiling points. Other liquids give simi-

[2]Symbols in brackets indicate that the equation is valid only if these properties are constant.

lar results. This regularity is called **Trouton's rule:**[3] The molar entropy of vaporization of a liquid at its boiling point is about 90 J K^{-1} mol^{-1}. It is interpreted to mean that the increase in molecular disorder on vaporization is about the same for all liquids. A few liquids, including water, have abnormally high entropies of vaporization. This suggests either that these liquids have especially well-ordered structures and low entropies, or that the vapors are unusually disordered with high entropies, or both. There is a considerable body of evidence showing that the first interpretation is correct. As mentioned in Section 10.4, the structure of liquid water (and also of ice) is based on the attractive forces (called hydrogen bonds) between the hydrogen atoms of each water molecule and the oxygen atoms of neighboring water molecules. This topic is discussed further in Section 17.10.

A relation similar to Eq. 15.7a applies to the melting or fusion process, so that

$$\Delta S°_{\text{fus}} = \frac{q^{\text{rev}}}{T_{\text{mp}}} = \frac{\Delta H°_{\text{fus}}}{T_{\text{mp}}} \qquad [n, T, P] \qquad (15.7b)$$

The data in Table 5.3 for the standard entropies of fusion range from 9.6 to 38 J K^{-1} mol^{-1} (for zinc and methyl ether, respectively). In general, standard entropies of fusion are much smaller than standard entropies of vaporization.

If n mol of a substance is heated reversibly from T_A to T_B at constant pressure without undergoing a phase change, the heat absorbed may be calculated from Eq. 5.13:

$$q^{\text{rev}} = nC_P \Delta T \qquad [n, C_P, P] \qquad (15.8)$$

However, we cannot apply Eq. 15.6 to calculate ΔS because the temperature changes during the process. What we must do is to divide each small increment of energy absorbed by the temperature at which it is absorbed, and sum the resultant values over the temperature range from T_A to T_B. Mathematically this may be stated as

$$\Delta S = \int_{T_A}^{T_B} \frac{nC_P}{T} dT \qquad [n, P] \qquad (15.9)$$

The symbol $\int$ is called the integral sign, and the expression on the right-hand side of this equation is called the definite integral over T of nC_P/T between T_A and T_B. When C_P is constant over the temperature range from T_A to T_B, Eq. 15.9 can be changed by mathematical techniques to the simpler form

$$\Delta S = nC_P \ln \frac{T_B}{T_A} \qquad [n, C_P, P] \qquad (15.10)$$

---

[3]Unlike the laws of thermodynamics, which hold rigorously under all conditions, Trouton's rule is an approximation, useful though imprecise.

15.3 ENTROPIES OF CHEMICAL SUBSTANCES

Although Eq. 15.9 may also be solved when C_P varies with T, the solution is more complicated and will not be taken up here. When C_P does not change markedly with temperature, an approximate value of ΔS may be obtained by using an average value for C_P in Eq. 15.10, as in the following example.

EXAMPLE 2

What is the entropy change when exactly 1 mol of liquid chlorine is heated from its normal freezing point to its normal boiling point? For liquid chlorine at its freezing point of 172.2 K, C_P is 67.0 J K^{-1} mol^{-1}. At its boiling point of 239.1 K, C_P is 65.8 J K^{-1} mol^{-1}.

We use Eq. 15.10 and the average value of C_P to obtain ΔS:

$$\Delta S = (1.000 \text{ mol}) \left(\frac{67.0 + 65.8}{2} \text{ J K}^{-1} \text{ mol}^{-1} \right) \left(\ln \frac{239.1 \text{ K}}{172.2 \text{ K}} \right)$$
$$= 21.8 \text{ J K}^{-1}$$

This result is very close to the accurate value of 21.89 J K^{-1} obtained from the exact solution, considering the variation of C_P with T.

The entropy of any chemical substance may be obtained from calorimetric data and the **third law of thermodynamics:** *The entropy of any chemical substance in the perfect crystalline state at 0 K is zero.* Thus one may calculate the entropy of chlorine gas at 298.15 K ($S°_{298.15}$) as the sum of five entropy changes, namely: (1) the entropy change due to heating the crystalline solid from 0 K to its normal melting point ($\Delta S°_s$), (2) the entropy of melting ($\Delta S°_{s \to l}$), (3) the entropy change due to heating the liquid from the melting point to the normal boiling point ($\Delta S°_l$), (4) the entropy of vaporization ($\Delta S°_{l \to g}$), and (5) the entropy change due to heating the gas from the boiling point to 298.15 K ($\Delta S°_g$).

$$S°_{298.15} = \Delta S°_s + \Delta S°_{s \to l} + \Delta S°_l + \Delta S°_{l \to g} + \Delta S°_g \tag{15.11}$$

$\Delta S°_s$, $\Delta S°_l$, and $\Delta S°_g$ may be calculated using Eq. 15.9, and $\Delta S°_{s \to l}$ and $\Delta S°_{l \to g}$ may be calculated using Eqs. 15.7b and 15.7a, respectively. Representative values for the several entropy changes are given in Table 15.1 to indicate the relative magnitudes of the entropy changes.

The entropy of a reactant or product may be calculated from the entropy change in a chemical reaction, provided that the entropies of the other substances are known. As an example, let us calculate the entropy of hydrogen iodide using the entropy change for the reaction $H_2 + I_2 = 2HI$ and the entropies of H_2 and I_2 gases. For the reaction

$$H_2(g) + I_2(g) = 2HI(g)$$

the entropy change in the neighborhood of 700 K is

$$\Delta S° = 15.65 \text{ J K}^{-1}$$

TABLE 15.1 — ENTROPY OF CHLORINE (J K^{-1} mol^{-1})

ΔS°_s	70.71
$\Delta S^\circ_{s \to l}$	37.20
ΔS°_l	21.89
$\Delta S^\circ_{l \to g}$	85.40
ΔS°_g	7.36
Entropy of Cl$_2$(g) at 25°C and 1 atm	222.56

The entropies of H$_2$ and I$_2$ gases at 700 K are 155.5 and 292.5 J K^{-1} mol^{-1}, respectively. Since

$$\Delta S^\circ = 2 S^\circ_{HI} - S^\circ_{H_2} - S^\circ_{I_2}$$

the entropy of HI is found to be

$$S^\circ_{HI} = \tfrac{1}{2}(15.65 + 155.5 + 292.5) = 231.8 \text{ J K}^{-1} \text{ mol}^{-1} \text{ at 700 K}$$

Entropies of a few chemical substances at 25°C and 1 atm are listed in Table 15.2 and a more extensive listing is given in Appendix F.

In contrast to the entropies of liquids and solids, the entropy of a gas depends markedly on its pressure or volume as well as on the temperature. For the isothermal expansion of n mol of an ideal gas at a temperature T from P_A, V_A, to P_B, V_B, the entropy change may be calculated starting with the first law of thermodynamics,

$$\Delta E = q + w$$

The work (w) done by the system in going from state A to state B may be calculated by the methods of integral calculus. The result is

$$w = -nRT \ln \frac{V_B}{V_A} \qquad [n, T] \tag{15.12}$$

Because the energy of an ideal gas is independent of pressure or volume at constant temperature (Section 5.9), $\Delta E = 0$. Therefore

$$q^{rev} = \Delta E - w = 0 + nRT \ln \frac{V_B}{V_A} \qquad [n, T]$$

and

$$\Delta S = \frac{q^{rev}}{T} = nR \ln \frac{V_B}{V_A} \qquad [n, T] \tag{15.13a}$$

TABLE 15.2 ENTROPY OF SOME CHEMICAL SUBSTANCES AT 25°C AND 1 ATM[a]

Substance	Entropy (J K^{-1} mol^{-1})	Substance	Entropy (J K^{-1} mol^{-1})
Al(s)	28.3	Cl$_2$(g)	223.0
Br$_2$(l)	152.2	H$_2$(g)	130.6
CaCl$_2$(s)	104.6	H$_2$O(l)	69.9
C(graphite)	5.70	HCl(g)	186.8
C(diamond)	2.38	I$_2$(s)	116.1
CO$_2$(g)	213.7	MgCO$_3$(s)	65.9
CH$_4$(g)	186.2	MgO(s)	26.9
C$_2$H$_4$(g)	219.2	O$_2$(g)	205.0
C$_2$H$_5$OH(l)	161	SO$_2$(g)	248.1
CCl$_4$(l)	214.4	ZnO(s)	44

[a] Gases are in the ideal gas state.

Alternatively, Eq. 15.13a may be expressed in terms of pressure by using the ideal gas law:

$$\Delta S = nR \ln \frac{(nRT/P_B)}{(nRT/P_A)} = nR \ln \frac{P_A}{P_B} \qquad [n, T] \qquad (15.13b)$$

Like the ideal gas equation, these equations can be used with only slight error for real gases under ordinary conditions.

As an example of the use of this last equation, consider the isothermal expansion of exactly 2 mol of hydrogen from a pressure of exactly 1 atm to a pressure of exactly 0.5 atm:

$$\Delta S = (2 \text{ mol})(8.314 \text{ J K}^{-1} \text{ mol}^{-1})[\ln (1 \text{ atm} /0.5 \text{ atm})]$$
$$= 11.53 \text{ J K}^{-1}$$

Note that ΔS for this process is the same at any temperature.

15.4 ENTROPY CHANGES IN SPONTANEOUS IRREVERSIBLE PROCESSES

In the Joule experiment (see Figure 5.9), a flask containing a gas is connected by means of a stopcock to an evacuated flask. When the stopcock is opened the gas rushes into the evacuated flask, equalizing the pressure in the two flasks. This is a good example of a spontaneous irreversible process. A small change in the conditions of the experiment could not reverse the direction of this process. What, then, is the change in entropy?

To have a specific set of conditions, let there be 2 mol of hydrogen at a pressure of 1 atm in one flask at the beginning of the experiment, and let the two flasks have the same volume, so that the final pressure is 0.5 atm. The boundary of the system is most conveniently defined to include both flasks and the stopcock, and, as we saw in Section 5.9, $q = 0$, $w = 0$, and $\Delta E = 0$. Because entropy is a state function, and because the initial and final states are the same as in the example at the end of the preceding section, the change of entropy of the system must be 11.53 J K^{-1}. Furthermore, the surroundings are completely unaffected by this process, and therefore the change in entropy of the surroundings is zero. Thus, for the irreversible process of the Joule experiment, the total entropy change is 11.53 J K^{-1}.

$$\Delta S_{\text{system}} + \Delta S_{\text{surr}} = 11.53 + 0 = 11.53 \text{ J K}^{-1}$$

Here is a striking difference between a reversible and an irreversible process. In a reversible process $q_{\text{system}}^{\text{rev}} = -q_{\text{surr}}^{\text{rev}}$, $T_{\text{system}} = T_{\text{surr}}$, and therefore $\Delta S_{\text{system}} = -\Delta S_{\text{surr}}$ or

$$\Delta S_{\text{system}} + \Delta S_{\text{surr}} = 0 \quad [\text{rev}] \tag{15.14}$$

As a second example of a spontaneous irreversible process, consider a system consisting of liquid water and steam at 100°C and 1 atm pressure. Let it be connected by a copper bar to a mixture of liquid water and ice at 0°C and 1 atm pressure that constitute part of the surroundings (see Figure 15.3). The copper bar will transmit thermal energy from the system to the surroundings, so that, as the steam condenses to liquid water, the ice will melt. The enthalpies of fusion and vaporization of water are 6.01 and 40.66 kJ mol^{-1}, respectively. Therefore, when 1 mol of steam condenses, 40.66 kJ are transmitted to the ice-water mixture, causing (40.66/6.01) = 6.77 mol of ice to melt.

The change of entropy of the steam and water system depends only on the initial and final states, and therefore it is the negative of the entropy of vaporization of 1 mol of water, -109.0 J K^{-1}. As the change of entropy of the ice-water mixture also depends only on the initial and final states, it is simply the entropy of fusion of 6.77 mol of ice or (6.77 mol)(6.01 kJ mol^{-1})/(273.15 K) = 149 J K^{-1}. The copper bar is in the same state at the end as it is at the beginning (i.e., 100°C at one end, 0°C at the other, and various intermediate temperatures in between) so that its entropy change is zero.

Thus

$$\begin{aligned}\Delta S_{\text{system}} + \Delta S_{\text{surr}} &= (-109.0 + 149) \text{ J K}^{-1} \\ &= 40 \text{ J K}^{-1}\end{aligned}$$

In both of these experiments involving irreversible processes, there is an increase in total entropy (system plus surroundings). Examination of many irreversible processes has shown that this is always true, and these experimental results are summarized by the **second law of thermodynamics:** *In a spontaneous irreversible process, the total entropy of the system and its surroundings increases.*

$$\Delta S_{\text{system}} + \Delta S_{\text{surr}} > 0 \quad [\text{irrev}] \tag{15.15}$$

15.4 ENTROPY CHANGES IN SPONTANEOUS IRREVERSIBLE PROCESSES

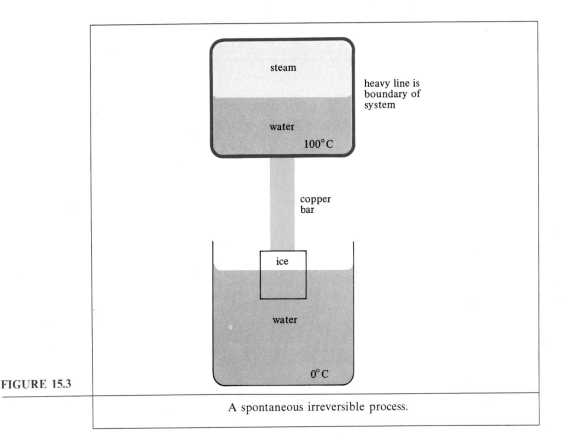

FIGURE 15.3

A spontaneous irreversible process.

We have already seen that in a reversible or equilibrium process, the total entropy of the system and its surroundings remains unchanged. The change in total entropy thus provides a quantitative measure of the spontaneity, or irreversibility, of any process—the closer the change in total entropy is to zero, the closer the process is to being reversible or the closer the system is to equilibrium.

When does the total entropy of the system and its surroundings decrease? Never! If one calculates that for some macroscopic process of interest the *total* entropy will decrease, one can be quite sure that this process will not occur. After the Joule experiment, the hydrogen will not move spontaneously into one flask, leaving the other flask evacuated. For the system in Figure 15.3, thermal energy will not flow from the 0°C end of the bar to the 100°C end, causing ice to form in the lower reservoir and steam in the upper. The following is an alternative statement of the second law: *Heat cannot of itself pass from a colder to a hotter body.* (Nor can it be made to do so without a change in the surroundings that counterbalances the entropy change of the system.)

Thus the change in total entropy provides a valuable criterion for determining whether or not some process of interest will occur spontaneously. It also provides a connection between the properties of matter and the concept of time. The future not only lies ahead, it lies in the direction of an increase in total entropy.

The direction of spontaneous change is, of course, a familiar one, and it is usually taken for granted unless one views the ludicrous antics of a film run backwards. But, in dealing with the unfamiliar, entropy is a useful tool for determining whether or not a particular process will be spontaneous. And the fact that the total entropy change is zero in any reversible process provides a key to the study of chemical equilibria.

One final point should be emphasized regarding the laws of thermodynamics. They were discovered through experiment, and by reflection on experimental results. Because they are fundamental laws that hold rigorously under all conditions, their empirical basis may seem dissatisfying in a philosophical sense. It might be nice to think that they were formulated through pure reason, but such is not the history of science.

15.5 HEAT ENGINES

Heat engines have been of special importance in connection with the second law of thermodynamics. It was an examination of the efficiency of a certain kind of heat engine that led N. L. Sadi Carnot, in 1824, to the discoveries from which the concept of entropy (introduced by Rudolf Clausius) and the second law later emerged. Without attempting to trace the historical development of the subject, let us use the second law to calculate the efficiency of a heat engine.

The essentials of a heat engine are an apparatus (such as a cylinder and piston enclosing a working fluid, e.g., steam or air) that undergoes a cyclical process during the course of which it absorbs heat from and does work on its surroundings. A particularly important cycle is the **Carnot cycle** shown in Figure 15.4. It consists of four steps. At first consider that each step is performed in a reversible manner.

1. An isothermal expansion at a temperature T_2 (A to B).
2. An expansion, called **adiabatic,** in which the system is thermally insulated from its surroundings, so that $q = 0$. As work is being done on the surroundings, the gas cools from T_2 to T_1 (B to C).
3. An isothermal compression at the lower temperature T_1 (C to D).
4. An adiabatic compression in which the temperature rises from T_1 to T_2 (D to A).

Let q_2 and q_1 be the amounts of heat absorbed at T_2 and T_1, respectively (q_1 is, of course, a negative quantity). Then the entropy changes in the four steps are q_2/T_2, 0, q_1/T_1, and 0. Because in a cyclical process the change in any state function, including the entropy change, is zero, we have

$$\frac{q_2}{T_2} + \frac{q_1}{T_1} = 0 \tag{15.16}$$

The energy change for the entire process must also be zero. Applying the first law

15.5 HEAT ENGINES

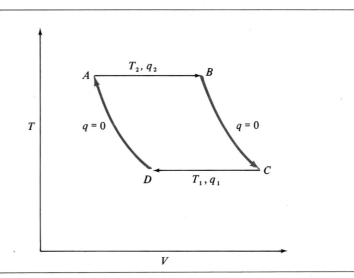

FIGURE 15.4

The Carnot cycle. AB is an isothermal expansion at temperature T_2, BC is an adiabatic expansion to a lower temperature T_1, CD is an isothermal compression at temperature T_1, and DA is an adiabatic compression in which the system returns to its original state A. The quantities of heat absorbed by the system in these steps are q_2, 0, q_1, and 0, respectively.

of thermodynamics, we obtain

$$\Delta E = 0 = q_2 + q_1 + w$$

or

$$-w = q_2 + q_1 \tag{15.17}$$

where w is the *net* amount of work done on the system by the surroundings (w is a negative quantity).

From Eq. 15.16, we find that

$$q_1 = -\frac{q_2 T_1}{T_2} \tag{15.18}$$

and, by substituting this value in Eq. 15.17, we obtain

$$-w = q_2 - \frac{q_2 T_1}{T_2} = \frac{q_2 T_2 - q_2 T_1}{T_2} \tag{15.19}$$

or

$$\frac{-w}{q_2} = \frac{T_2 - T_1}{T_2} \quad \text{[rev]} \tag{15.20}$$

This last expression, the ratio of the net work done by the system ($-w$) to the heat absorbed at the higher temperature (q_2), is called the **efficiency** of the engine. It depends only on the ratio of the difference in absolute temperatures between which the engine is working ($T_2 - T_1$) to the higher temperature (T_2).

If $T_2 = 400$ K and $T_1 = 300$ K, $-w/q_2 = 100/400 = 0.25$ or 25% efficiency. If $q_2 = 100$ J, then $-w = 25$ J and $q_1 = -75$ J. Thus, for every 100 J absorbed at 400 K, 75 J must be discharged at 300 K, and only 25 J of work is done by the system.

What about a real engine, which necessarily operates in a somewhat irreversible manner? Because entropy is a state property, the sum of the entropy changes of the system is still zero. The increase in total entropy of the system and surroundings must therefore be manifested as an increase in the entropy of the surroundings. For the same q_2, less work is done and, at the lower temperature, more heat is discharged than in a reversible process. In the example given above we might have, say, $q_1 = -76$ J and $-w = 24$ J. Thus in a real engine the efficiency must be less than that given by Eq. 15.20, that is,

$$\frac{-w}{q_2} < \frac{T_2 - T_1}{T_2} \quad \text{[irrev]} \tag{15.21}$$

The theory of a refrigerator or heat pump is based on a Carnot cycle operating in the reverse direction. For the reversible process, q_1, q_2, and w simply have opposite signs, and Eq. 15.20 is applicable. Thus, if the surroundings do 25 J of work *on* the system, it can absorb 75 J at 300 K and discharge 100 J at 400 K ($w = 25$ J, $q_1 = 75$ J, $q_2 = -100$ J).

15.6 ENTROPY AND PROBABILITY

Why does nature always exhibit the behavior summarized by the second law of thermodynamics? From one point of view this is a manifestation of the tendency toward disorder. When a hammer strikes an anvil, the coordinated motion of the atoms in the hammer is converted to the disordered motion of the atoms in a now hotter hammer and anvil. The mechanical energy of the hammer is dissipated as thermal energy, only part of which can ever be recovered as mechanical energy.

From a closely related point of view, this behavior is a manifestation of the tendency toward the state of highest probability. Consider the Joule experiment once more. The probability of the final state, in which the molecules are divided equally between the two flasks, is very close to certainty, which in the theory of probability is assigned the number 1. Now, what is the probability that a particular molecule will be in the left flask? As there is equal probability of its being in either flask, the answer is $\frac{1}{2}$. What is the probability that *two* particular molecules are in the left flask? From the theory of probability, the probability that two independent events will both occur is the *product* of the probabilities of the two events, so the answer is $\frac{1}{2} \times \frac{1}{2} = \frac{1}{4}$. It is exactly the same as the probability that, on flipping two coins, both coins will land heads up; or that, on flipping one coin twice, heads will occur both times.

Thus the probability, W, that all of the molecules in 2 mol of hydrogen will be simultaneously in the left flask is

$$W = (\tfrac{1}{2})^{2N_0}$$

where N_0 is Avogadro's number, 6.02×10^{23}. It is not impossible that such an event might occur, it is just highly improbable. In fact, the probability is the same as the probability that 12.04×10^{23} coins will all land heads up (or that one coin flipped 12.04×10^{23} times will land heads up each time). That this is indeed very close to an impossibility can be seen by experimenting with a coin and trying to obtain a run of, say, 10 successive heads.

Changes of entropy can be connected to changes of probability through the relation

$$S = k_B \ln W \qquad (15.22)$$

where $k_B = R/N_0$ is the Boltzmann constant. In the Joule experiment described above,

$$\begin{aligned}
\Delta S &= k_B \ln W^{\text{final}} - k_B \ln W^{\text{initial}} \\
&= k_B \ln 1 - k_B \ln (\tfrac{1}{2})^{2N_0} \\
&= 0 - 2N_0 k_B \ln (\tfrac{1}{2}) \\
&= 2R \ln 2 = 11.53 \text{ J K}^{-1}
\end{aligned}$$

This is identical with the result obtained earlier (Section 15.3).

Intervention of an intelligent being can result in apparent violations of the second law. James Clerk Maxwell proposed a hypothetical experiment in which a demon operates a stopcock or gate between the two flasks (Figure 15.5). As the operation of the gate can, in principle, be made frictionless, no work is required. And yet "Maxwell's demon" can perform miracles: By closing the gate at the proper times, he can prevent molecules from leaving the left flask; by opening the gate at the proper times he can allow all of the molecules in the right flask to move into the left flask. Thus, Maxwell's demon is able to restore the apparatus to its original state with an apparent decrease in the total entropy of the system and its surroundings. Alternatively, the demon can sort the faster molecules into one flask, and the slower molecules into the other, so that one side gets hot and the other cold, again in apparent violation of the second law.

A detailed examination of this problem has shown that the information required by the demon carries with it an entropy, and that when this is taken into account the second law is still obeyed. So far as is known the second law of thermodynamics is applicable to the operations of intelligent creatures and to human affairs, and, in fact, the concept of entropy and the second law have assisted in developing the theory of information and its transmission.

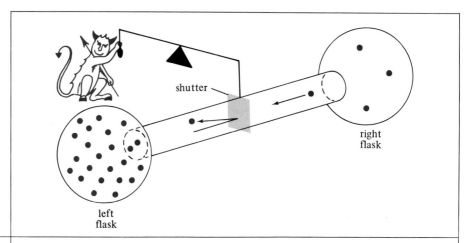

FIGURE 15.5 Maxwell's demon. He has just prevented a molecule from leaving the left flask, and he is about to raise the shutter and allow the molecule approaching from the right to pass on through. Soon all the molecules will be in the left flask.

15.7 THE GIBBS FREE ENERGY

Will a specified chemical reaction be spontaneous? If so, under what conditions? Under what conditions will the reactants and products be at equilibrium? These questions can always be answered correctly by considering the change in total entropy, as shown in the preceding section. But in one respect the entropy is an awkward criterion, for one must consider not only the entropy change of the system of interest but also the entropy change of the surroundings.

It is much more convenient if one need consider only some property of the system itself. Under the conditions most frequently of interest, namely, constant temperature and constant pressure, the property having the desired characteristics is the **Gibbs free energy** (G). Named after its discoverer, J. Willard Gibbs, it is defined by the equation

$$G = H - TS \tag{15.23}$$

Because G is composed of state functions, it is itself a state function.

There are two important applications of the Gibbs free energy to any chemical reaction or physical process occurring at constant temperature and pressure.

1. The condition for equilibrium is that ΔG is exactly zero.
2. For any spontaneous reaction or process, ΔG is negative.

Furthermore, the magnitude of ΔG provides a quantitative measure of the maximum useful work that can be obtained from the reaction or process.

15.7 THE GIBBS FREE ENERGY

We shall now proceed to derive these statements from the first and second laws of thermodynamics.

For any process or reaction at constant temperature T, the system changes from state A (of enthalpy H_A and entropy S_A) to state B (of enthalpy H_B and entropy S_B). Therefore $G_A = H_A - TS_A$, $G_B = H_B - TS_B$, and

$$\Delta G = G_B - G_A = (H_B - TS_B) - (H_A - TS_A)$$
$$= (H_B - H_A) - T(S_B - S_A)$$

or

$$\Delta G = \Delta H - T\Delta S \quad [T] \quad (15.24)$$

If the process also occurs at constant pressure, $\Delta H = \Delta E + P\Delta V$ (Eq. 5.10) and thus

$$\Delta G = \Delta E + P\Delta V - T\Delta S \quad [P, T] \quad (15.25)$$

From the first law of thermodynamics,

$$\Delta G = q + w + P\Delta V - T\Delta S \quad [P, T] \quad (15.26)$$

If, as often happens, the only kind of work done in this process is expansion or contraction against an applied pressure (no electrical work, etc.), then $w = -P\Delta V$ (Eq. 5.6), and

$$\Delta G = q - T\Delta S \quad [P, T, \text{only } PV \text{ work}] \quad (15.27)$$

This equation is simplified still further by noting that if the path is reversible, $q^{\text{rev}} = T\Delta S$. Thus, under the several restrictions noted,

$$\Delta G = 0 \quad [\text{rev}, P, T, \text{only } PV \text{ work}] \quad (15.28)$$

When a certain quantity of ice, at $0\,°C$ and 1 atm pressure, is heated reversibly so that it melts, forming liquid water at the same pressure and temperature, what is ΔG? Under these conditions the last equation applies, and $\Delta G = 0$. Likewise $\Delta G = 0$ for the reversible vaporization of liquid water at $100\,°C$ and 1 atm pressure.

What is the free-energy change in a spontaneous chemical reaction, such as the reaction of zinc with copper ions, $Zn(s) + Cu^{2+} \longrightarrow Zn^{2+} + Cu(s)$? Let us first suppose that the reaction is carried out reversibly by separating the reactants in an electrical cell (see Figure 14.1), with the voltage of the cell, $\Delta \mathcal{E}$, balanced by an opposing voltage. For many cell reactions it is possible to approach reversibility very closely—a change in the opposing voltage as small as $1\,\mu V$ (0.000001 V) will cause a detectable change in the direction in which the electric current is flowing, thus reversing the direction of the reaction.

When the reaction occurs under the usual conditions of constant pressure and temperature, Eq. 15.26 applies. But now the total work done by the surroundings on the system, w, consists of two terms—the pressure-volume work as before, $-P\Delta V$, and the electrical work. Designating the latter as w' (often called the **net useful work**),

$$w = -P\Delta V + w' \tag{15.29}$$

Substituting this expression and Eq. 5.6 into Eq. 15.26 gives

$$\Delta G = q + w' - T\Delta S \quad [P, T] \tag{15.30}$$

For reversible operation of the cell, $q^{\text{rev}} = T\Delta S$, and

$$\Delta G = w' \quad [\text{rev}, P, T] \tag{15.31}$$

How is w' related to $\Delta \mathcal{E}$? When one zinc atom reacts, two electrons flow through the circuit. When 1 mol of zinc atoms reacts, $2N_0$ electrons flow through the circuit. They carry an electric charge of $2N_0 e$, where e is the electronic charge, moving through a potential difference of $\Delta \mathcal{E}$, and thus the electrical work done *by* the cell is $(2N_0 e)(\Delta \mathcal{E})$ or $2F\Delta\mathcal{E}$, where $F = N_0 e$ is the Faraday constant. Thus the electrical work done *on* the cell is

$$w' = -2F\Delta\mathcal{E} \quad [\text{rev}, P, T]$$

In general, when n electrons are transferred in the cell reaction,

$$w' = -nF\Delta\mathcal{E} \quad [\text{rev}, P, T] \tag{15.32}$$

Therefore

$$\Delta G = w' = -nF\Delta\mathcal{E} \quad [\text{rev}, P, T] \tag{15.33}$$

If the reaction occurs spontaneously in the direction indicated, $\Delta\mathcal{E}$ is positive, w' is negative, and ΔG is negative. This can be visualized as in Figure 15.6: As $\Delta G = G_B - G_A$ is negative, the final state has a lower free energy than the initial state. The direction of a spontaneous reaction, from higher to lower free energy, is analogous to water running downhill. Of course, if the reactants and products are at equilibrium, as in a dead battery, $\Delta\mathcal{E} = 0$ and $\Delta G = 0$. And, if ΔG is positive, $\Delta\mathcal{E}$ is negative and the reaction is spontaneous in the opposite direction.

As free energy is a state function, changes in free energy depend only on the initial and final states. In a process or reaction that is spontaneous under the conditions of constant temperature and pressure, ΔG is negative whether the cell is operated reversibly or irreversibly—in fact, whether the reaction does or does

15.7 THE GIBBS FREE ENERGY

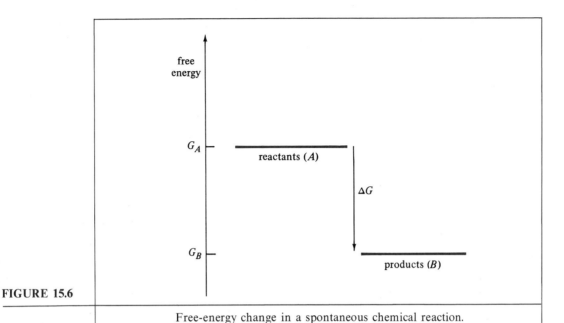

FIGURE 15.6 Free-energy change in a spontaneous chemical reaction.

not occur in a cell. Thus free energy is the desired property of the *system* which determines what kind of process occurs at constant temperature and pressure. If $\Delta G = 0$, the process or reaction is at equilibrium; if ΔG is negative, the process or reaction is spontaneous. When carried out reversibly in an electrical cell, the electrical work done by the cell is $-w' = -\Delta G = nF\Delta\mathcal{E}$. When carried out irreversibly, the work done by the cell can be shown to be somewhat less ($-w' < -\Delta G$). Thus *the free-energy change is a measure of the maximum useful work that can be obtained from the chemical reaction*. Exactly how much less work is done by a cell depends on how much the actual conditions differ from reversibility. In the extreme case where the reactants are simply mixed together, as when a strip of zinc is placed in a solution containing cupric ions, w' is, of course, zero.

Finally, it will be of interest to calculate the free-energy change for a process at constant temperature but not at constant pressure, namely, the isothermal expansion of n mol of a perfect gas. On applying Eq. 15.24, first it is necessary to determine ΔH:

$$\Delta H = H_B - H_A = (E_B + P_B V_B) - (E_A + P_A V_A)$$

From Boyle's law, $P_B V_B = P_A V_A$ at constant T and amount. As shown in Section 5.9, $E_B = E_A$. Therefore $\Delta H = 0$. Second, $\Delta S = nR \ln (P_A/P_B)$ from Eq. 15.13b. Therefore

$$\Delta G = \Delta H - T\Delta S = -nRT \ln (P_A/P_B) \qquad [n, T]$$

or

$$\Delta G = nRT \ln(P_B/P_A) \quad [n, T] \tag{15.34}$$

15.8 STANDARD FREE-ENERGY CHANGES

For many purposes it is convenient to consider free-energy changes for chemical reactions based on a single set of reference points, and these points are, of course, the standard states of the chemical substances mentioned earlier (Sections 5.8 and 12.2). The initial state is the reactants in their standard states, and the final state is the products in their standard states. Then Eq. 15.24 becomes

$$\Delta G° = \Delta H° - T\Delta S° \quad [T] \tag{15.35}$$

Thus, from standard enthalpy and entropy changes at the temperature of interest, one may compute the standard free-energy change.

To illustrate this calculation, consider the hydration of ethylene at 25°C:

$$C_2H_4(g) + H_2O(l) \longrightarrow C_2H_5OH(l)$$

Using Eq. 5.25 and the standard enthalpies of formation listed in Table 5.4 for these substances, we may calculate $\Delta H°$ for this reaction:

$$\begin{aligned}
\Delta H° &= 1 \text{ mol} \times \Delta H_f°(C_2H_5OH, l) - 1 \text{ mol} \times \Delta H_f°(C_2H_4, g) \\
&\quad - 1 \text{ mol} \times \Delta H_f°(H_2O, l) \\
&= (1 \text{ mol})(-277.63 \text{ kJ mol}^{-1}) - (1 \text{ mol})(52.283 \text{ kJ mol}^{-1}) \\
&\quad - (1 \text{ mol})(-285.84 \text{ kJ mol}^{-1}) \\
&= -44.07 \text{ kJ}
\end{aligned}$$

The entropies, $S°$, of ethylene, water, and ethyl alcohol in their standard states at 25°C, listed in Table 15.2, give an entropy change of

$$\begin{aligned}
\Delta S° &= 1 \text{ mol} \times S°(C_2H_5OH, l) - 1 \text{ mol} \times S°(C_2H_4, g) - 1 \text{ mol} \\
&\quad \times S°(H_2O, l) \\
&= (1 \text{ mol})(161 \text{ J K}^{-1} \text{ mol}^{-1}) - (1 \text{ mol})(219.2 \text{ J K}^{-1} \text{ mol}^{-1}) \\
&\quad - (1 \text{ mol})(69.9 \text{ J K}^{-1} \text{ mol}^{-1}) \\
&= -128 \text{ J K}^{-1}
\end{aligned}$$

$$\begin{aligned}
\Delta G° &= \Delta H° - T\Delta S° \\
&= -44.07 \times 10^3 \text{ J} - (298.15 \text{ K})(-128 \text{ J K}^{-1}) \\
&= -5.9 \times 10^3 \text{ J} = -5.9 \text{ kJ}
\end{aligned}$$

As this is a negative quantity, the hydration of ethylene is a spontaneous reaction at 25°C and 1 atm pressure, and the reaction in the opposite direction (the dehydration of ethyl alcohol) is not spontaneous at this temperature. Experimentally it is found that these statements are correct. The acid-catalyzed hydration of ethylene is the main source of ethyl alcohol for industrial purposes.

For a reaction occurring in an electrochemical cell, the standard free-energy change is directly related to the standard voltage. This follows from Eq. 15.33. If all substances are in their standard states, this equation becomes

$$\Delta G° = w' = -nF\Delta\mathcal{E}° \qquad [\text{rev}, P, T] \tag{15.36}$$

For example, in Section 14.5 we found that $\Delta\mathcal{E}° = 1.10$ V for the reaction

$$Zn(s) + Cu^{2+} = Cu(s) + Zn^{2+}$$

In this reaction, $n = 2$, and therefore

$$\Delta G° = -(2 \text{ mol})(9.648 \times 10^4 \text{ C mol}^{-1})(1.10 \text{ V})$$
$$= -212 \times 10^3 \text{ J} = -212 \text{ kJ}$$

Because $\Delta\mathcal{E}°$ is related to the equlibrium constant for the reaction (Eq. 14.14), so is $\Delta G°$. Combining Eqs. 14.14 and 15.36 (with the numerical coefficient in Eq. 14.14 expressed to four significant figures), we obtain

$$\Delta G° = -nF\Delta\mathcal{E}° = -nF\left(\frac{0.05916 \log K}{n}\right) \text{ at } 25°C$$
$$= -(9.648 \times 10^4)(0.05916)(\log K) \text{ J}$$
$$= -5.708 \times 10^3 \log K \text{ J at } 25°C \tag{15.37}$$

As an illustration of the use of this equation, consider the dissociation of acetic acid:

$$CH_3CO_2H = H^+ + CH_3CO_2^-$$

At 25°C, $K = 1.8 \times 10^{-5}$, $\log K = -4.745$, and $\Delta G° = (-5.708 \times 10^3)(-4.745)$ J $= 27.08$ kJ. Therefore when each substance is in its standard state (unit activity, or a concentration of approximately 1 m), the reaction is spontaneous in the direction opposite to that written above.

To obtain Eq. 15.37 in a more general form, we note that the coefficient of the logarithmic term in the Nernst equation (Eq. 14.11a), applicable to any temperature, is $-2.3026RT/nF$, where R is the gas constant. Thus the general forms of Eqs. 14.14 and 15.37 are

$$\Delta\mathcal{E}° = \frac{2.3026RT}{nF}\log K = \frac{RT}{nF}\ln K \tag{15.38}$$

$$\Delta G° = -2.3026RT\log K = -RT\ln K \tag{15.39}$$

Combining Eq. 15.39 with Eq. 15.35 yields the equation connecting the equilibrium constant to the standard changes in enthalpy and entropy, introduced at the beginning of this chapter,

$$\Delta G° = -RT \ln K = \Delta H° - T\Delta S°$$

or

$$R \ln K = -\frac{\Delta H°}{T} + \Delta S° \tag{15.40}$$

Equation 15.40 is identical with Eq. 15.4.

For the equilibrium between a liquid or solid and its vapor, the equilibrium constant is equal to the vapor pressure, and Eq. 15.40 becomes

$$R \ln P = -\frac{\Delta H°}{T} + \Delta S° \tag{15.41}$$

which is the same as Eq. 15.3.

When a compound is formed from its elements, the standard change in free energy is called the **standard free energy of formation,** and the subscript f is used. Thus

$$\Delta G_f° = \Delta H_f° - T\Delta S_f° \tag{15.42}$$

(These are usually tabulated for a temperature of 25°C or $T = 298.15$ K.) A table of standard enthalpies of formation, standard free energies of formation, and entropies of various compounds and elements at 25°C is given in Appendix F. Since free energy is a state function, one may determine the standard free-energy change for any chemical reaction by an equation analogous to Eq. 5.32:

$$\Delta G° = \Sigma \Delta G_f° \text{(products)} - \Sigma \Delta G_f° \text{(reactants)} \tag{15.43}$$

In this equation, the standard free energy of formation per mole of each substance is multiplied by the coefficient of that substance in the balanced equation for the reaction. Then the sum of these terms for the reactants is subtracted from the sum of terms for the products.

A similar equation may be used to obtain the change in entropy for a chemical reaction

$$\Delta S° = \Sigma S° \text{(products)} - \Sigma S° \text{(reactants)} \tag{15.44}$$

EXAMPLE 3

What are the changes in free energy, enthalpy, and entropy for the reaction

$2SO_2(g) + O_2(g) = 2SO_3(g)$

if all substances are in their standard states (1 atm pressure) at 25°C? (This reaction is an important step in the manufacture of sulfuric acid.)

From Appendix F we obtain the standard free energies of formation, standard enthalpies of formation, and entropies of these substances. When the ΔG_f° values are substituted in Eq. 15.43, we find that

$$\Delta G^\circ = (2 \text{ mol})(-3.71 \text{ kJ mol}^{-1}) - (2 \text{ mol})(-300.2 \text{ kJ mol}^{-1})$$
$$- (1 \text{ mol})(0 \text{ kJ mol}^{-1})$$
$$= -141.8 \text{ kJ}$$

In a similar way, by using Eq. 5.25 and the ΔH_f° values in Appendix F, we obtain the enthalpy change for the reaction.

$$\Delta H^\circ = (2 \text{ mol})(-395.8 \text{ kJ mol}^{-1}) - (2 \text{ mol})(-296.8 \text{ kJ mol}^{-1})$$
$$- (1 \text{ mol})(0 \text{ kJ mol}^{-1})$$
$$= -198.0 \text{ kJ}$$

The entropy change can be obtained from the S° values listed in Appendix F and Eq. 15.44.

$$\Delta S^\circ = (2 \text{ mol})(256.7 \text{ J K}^{-1} \text{ mol}^{-1}) - (2 \text{ mol})(248.1 \text{ J K}^{-1} \text{ mol}^{-1})$$
$$- (1 \text{ mol})(205.0 \text{ J K}^{-1} \text{ mol}^{-1})$$
$$= -187.8 \text{ J K}^{-1}$$

Note that $\Delta G^\circ = \Delta H^\circ - T\Delta S^\circ$, except for a very small difference due to round-off error.

$$-141.8 \text{ kJ} = -198.0 \text{ kJ} - (298.15 \text{ K})(-187.8 \text{ J K}^{-1})$$
$$= -198.0 \text{ kJ} + 56.0 \text{ kJ}$$

Thus although 198.0 kJ is released by the reaction, 56.0 kJ is required to form the more ordered product, leaving 141.8 kJ as the driving force for the reaction.

PROBLEMS

For all problems, assume that the temperature is 25°C unless otherwise specified.

15.1 The equation that fits the ethyl ether line shown in Figure 15.1(b) is

$$R \ln P = -\frac{29.1 \times 10^3 \text{ J mol}^{-1}}{T} + 94.6 \text{ J K}^{-1} \text{ mol}^{-1}$$

What are ΔH° and ΔS° for the vaporization of 1 mol of ethyl ether?

15.2 (a) Liquid water having a mass of 0.500 g is placed in a 1.0-L flask. Calculate the number of moles of water vapor in the flask and the percentage of water in vapor form at 25°C and at 100°C. (The vapor pressure of water is 24 Torr at 25°C.)
(b) Do these data show that the distribution of water molecules becomes more uniform as the temperature increases?

15.3 The dissociation constant of ammonia is 1.37×10^{-5} at $0°C$ and 1.89×10^{-5} at $50°C$. Calculate $\Delta H°$ for the reaction
$$NH_3 + H_2O = NH_4^+ + OH^-$$
(Assume that $\Delta H°$ is constant in this temperature range.)

15.4 Ethyl alcohol has a normal boiling point of $78°C$ and a heat of vaporization of 39.4 kJ mol^{-1}. Calculate the vapor pressure of ethyl alcohol (in atmospheres and Torr) at $25°C$. (Assume that $\Delta H°$ is constant in this temperature range.)

15.5 Consider ice, liquid water, and water vapor in equilibrium at the triple point. Which phase has the highest entropy per mole?

15.6 Which of the following thermodynamic properties is a measure of the disorder of a system?
   enthalpy, entropy, heat capacity, Gibbs free energy

15.7 Calculate the change in entropy when 1.00 mol of liquid water is heated from $0°C$ to $100°C$ at 1 atm pressure. (Use 75.3 J K^{-1} mol^{-1} for the heat capacity, C_p, of liquid water.)

15.8 What is the entropy of any chemical substance in the perfect crystalline state at 0 K?

15.9 The heat capacity, C_p, of liquid carbon tetrachloride at $25°C$ is 132 J K^{-1} mol^{-1}. If 2.50 mol of $CCl_4(l)$ is heated from $25.00°C$ to $29.00°C$, what is ΔS?

15.10 In an irreversible process, what happens to the total entropy of the system and its surroundings?

15.11 For any cyclic process, the change in any thermodynamic state function of a system (such as V, H, S, or G) is zero. Why?

15.12 Calculate the efficiency of an engine using a Carnot cycle and operating between
   (a) $100°C$ and $25°C$
   (b) $357°C$ and $0°C$

15.13 (a) A real engine operating between $80°C$ and $20°C$ absorbs 280 J of heat at the higher temperature and does 35 J of work during each cycle. How much heat is discharged at the lower temperature during each cycle?
   (b) Calculate the efficiency of an engine using a Carnot cycle and operating

between the same temperatures as the engine in part (a). Does the real engine in part (a) violate the second law of thermodynamics?

15.14 Exactly 2 mol of oxygen gas is compressed from 1.00 to 2.00 atm at a constant temperature of 27°C. Calculate ΔE, ΔH, ΔS, and ΔG. (Assume ideal gas behavior.)

15.15 The heat of fusion of ice is 6.01 kJ mol^{-1}. Calculate $\Delta H°$, $\Delta S°$, and $\Delta G°$ when 2.00 mol of ice melts at 0°C.

15.16 (a) When 1 mol of liquid water vaporizes at 100°C,
$$H_2O(l) = H_2O(g)$$
What is the numerical value of ΔG if the process is carried out reversibly?
(b) If this process is carried out irreversibly, is ΔG larger, smaller, or the same as the answer in part (a)?

15.17 For each of the following processes, is ΔG positive, zero, or negative? (In each case define the system carefully.)
(a) Liquid water evaporates at 25°C against an applied pressure equal to its equilibrium vapor pressure of 24 Torr.
(b) Hydrogen and oxygen burn to form water in an oxyhydrogen torch.
(c) Hydrogen and oxygen are generated by electrolysis of water.
(d) Glucose ($C_6H_{12}O_6$) burns in oxygen to form carbon dioxide and water.
(e) Under the action of sunlight, carbon dioxide and water combine in a plant leaf to form glucose and oxygen.

15.18 At low temperature a chemical reaction is spontaneous as written. At high temperature the reaction is spontaneous in the reverse direction. Assuming that $\Delta H°$ and $\Delta S°$ are independent of temperature, what can you say about the sign (+ or −) of each?

15.19 Reactions A, B, C, and D differ in the signs of their $\Delta H°$ and $\Delta S°$ values as shown below:

Reaction	$\Delta H°$	$\Delta S°$
A	+	+
B	+	−
C	−	+
D	−	−

At constant temperature and pressure, which of these reactions are
(a) definitely spontaneous?
(b) possibly spontaneous?
(c) definitely not spontaneous?

15.20 Using the data in Tables 5.4 and 15.2, calculate $\Delta H°$, $\Delta S°$, and $\Delta G°$ for each of the following reactions:
(a) $H_2(g) + Cl_2(g) = 2HCl(g)$
(b) $CH_4(g) + 2O_2(g) = CO_2(g) + 2H_2O(l)$

15.21 Calculate $\Delta G°$ for the following electrochemical cell reactions:
$2Ag^+ + Cu(s) = 2Ag(s) + Cu^{2+}$ $\quad\quad \Delta\mathcal{E}° = +0.46$ V
$O_2(g) + 4H^+ + 4Fe^{2+} = 2H_2O + 4Fe^{3+}$ $\quad\quad \Delta\mathcal{E}° = +0.46$ V

15.22 Calculate $\Delta G°$ for each of the following reactions:
$Ba^{2+} + SO_4^{2-} = BaSO_4(s)$ $\quad\quad K = 9.1 \times 10^9$
$H_2O = H^+ + OH^-$ $\quad\quad K = 1.0 \times 10^{-14}$

15.23 From the data in Table 14.1, calculate $\Delta G°$ for the reaction
$2Sn^{2+} = Sn(s) + Sn^{4+}$

15.24 When a certain metal, M, dissolves in an acidic solution, the equation for the reaction is
$M(s) + 2H^+ = M^{2+} + H_2(g)$
The standard free-energy change, $\Delta G°$, is -147 kJ.
Calculate the standard electrode potential, $\mathcal{E}°$, for the half-reaction
$M^{2+} + 2e^- = M(s)$

15.25 Using data in Appendix F, calculate $\Delta G°$ and K at 25°C for each of the following reactions
$N_2(g) + O_2(g) = 2NO(g)$
$CaCO_3(s) = CaO(s) + CO_2(g)$
$2Al(s) + Fe_2O_3(s) = 2Fe(s) + Al_2O_3(s)$
$3C_2H_2(g) = C_6H_6(l)$
$CH_4(g) + 4Cl_2(g) = CCl_4(l) + 4HCl(g)$

15.26 (a) Using data in Appendix F, calculate $\Delta H°$, $\Delta S°$, and $\Delta G°$ for the reaction $C_2H_4(g) + H_2(g) = C_2H_6(g)$.
(b) In the expression $\Delta G° = \Delta H° - T\Delta S°$, which term on the right-hand side favors the reaction as written? Which term favors the reverse direction?

15.27 The entropy of crystalline iodine, I_2, is 131.6 J K^{-1} mol^{-1} at its melting point of 114°C. The enthalpy of fusion of iodine is 15.8 kJ mol^{-1}. At the boiling point of 183°C, the enthalpy of vaporization is 41.7 kJ mol^{-1}. From these data and the heat capacity of liquid I_2, 80.3 J K^{-1} mol^{-1}, calculate the entropy of $I_2(g)$ at 183°C.

15.28 (a) Calculate the entropy change for the irreversible process that occurs when exactly 2 mol of water at $-10\,°C$ freezes to form ice at $-10\,°C$. A reversible path to accomplish this process is:

$$H_2O(l)(-10\,°C) = H_2O(l)(0\,°C)$$
$$H_2O(l)(0\,°C) = H_2O(s)(0\,°C)$$
$$H_2O(s)(0\,°C) = H_2O(s)(-10\,°C)$$

The enthalpy of fusion of ice is 6.01 kJ mol^{-1}. The heat capacity of liquid water is 75 J K^{-1} mol^{-1}, whereas that for ice is 38 J K^{-1} mol^{-1}.

(b) Explain why the calculated entropy change is negative when the process occurs spontaneously. What is the net overall entropy change for the system and its surroundings?

CHAPTER 16
CHEMICAL KINETICS

16

In Chapter 15 we examined the criteria for predicting whether a chemical reaction will occur spontaneously and, if it does, the extent to which it will occur. We found that, if we know the equilibrium constant or the change in free energy for a possible reaction, we can predict the maximum yield of products that can be obtained.

However, many reactions that are feasible on thermodynamic considerations ($\Delta G < 0$) occur so slowly that no appreciable amounts of products are obtained in reasonable time periods. For example, common fuels (coal, oil, and wood) and foods (sugar, carbohydrates, and proteins) are thermodynamically unstable in the presence of oxygen with respect to formation of carbon dioxide and water, as shown below for sucrose (common table sugar):

$$C_{12}H_{22}O_{11}(s) + 12O_2(g) = 12CO_2(g) + 11H_2O(l)$$
$$\Delta G° = -5796.9 \text{ kJ at } 25°C$$

Yet, as we all know by visual observation, at ordinary temperatures these substances do not react rapidly with the oxygen in air.

In this chapter we shall consider the factors that determine the rate of a reaction, the use of rate information to elucidate possible pathways whereby reactants are converted to products, and some theories that are useful in explaining rates of reactions. This field of study is known as chemical kinetics.

16.1 RATE OF REACTION

The amount of reaction (product formed or reactant used up) occurring in unit time is a measure of the rate of the reaction. Before formulating a more precise definition of the rate of a reaction, let us consider some of the problems that arise in a simple case. Dinitrogen pentoxide (N_2O_5) decomposes to produce nitrogen dioxide and oxygen,

$$2N_2O_5 \longrightarrow 4NO_2 + O_2 \tag{16.1}$$

To follow the course of this reaction we could measure the amount of the reactant or a product as it changes with time. It is, of course, necessary to consider the stoichiometry of the reaction:

16.1 RATE OF REACTION

TABLE 16.1 DECOMPOSITION OF N_2O_5 IN CCl_4 SOLUTION AT 45°C

t (s)	$[N_2O_5]$ (M)	Rate $= -\dfrac{1}{2}\dfrac{d[N_2O_5]}{dt}$ ($M\,s^{-1}$)	$k = \dfrac{\text{rate}}{[N_2O_5]}$ (s^{-1})
0	2.33	7.3×10^{-4}	3.1×10^{-4}
184	2.08	6.7×10^{-4}	3.2×10^{-4}
319	1.91	6.0×10^{-4}	3.1×10^{-4}
526	1.67	5.1×10^{-4}	3.1×10^{-4}
867	1.36	4.1×10^{-4}	3.0×10^{-4}
1198	1.11	3.5×10^{-4}	3.2×10^{-4}
1877	0.72	2.2×10^{-4}	3.1×10^{-4}

$$\text{moles } O_2 \text{ produced} = \tfrac{1}{4} \times \text{moles } NO_2 \text{ produced}$$
$$= \tfrac{1}{2} \times \text{moles } N_2O_5 \text{ reacted}$$

Usually reactions are studied in the liquid phase where the volume is approximately constant, and it is customary to measure the amounts of substances in terms of their molar concentrations. We shall restrict our treatment to such constant volume systems.

In Table 16.1 values of the concentration of N_2O_5 at various times are recorded for a solution of N_2O_5 in CCl_4 at 45°C. These data may be used to calculate the average rate of change in N_2O_5 concentration. If we take the first two times as an example, we obtain

$$\frac{\Delta[N_2O_5]}{\Delta t} = \frac{(2.08 - 2.33)M}{(184 - 0)\,s} = -1.36 \times 10^{-3}\,M\,s^{-1} \qquad (16.2a)$$

(This result is negative because $[N_2O_5]$ is decreasing.) Since 2 mol of NO_2 is produced for every mole of N_2O_5 used up, the average rate of formation of NO_2 would be

$$\frac{\Delta[NO_2]}{\Delta t} = 2 \times 1.36 \times 10^{-3}\,M\,s^{-1} = 2.72 \times 10^{-3}\,M\,s^{-1} \qquad (16.2b)$$

(This result is positive because $[NO_2]$ is increasing.) Furthermore, 1 mol of O_2 is produced for every 2 mol of N_2O_5 that reacts, and therefore

$$\frac{\Delta[O_2]}{\Delta t} = \frac{1}{2} \times 1.36 \times 10^{-3}\,M\,s^{-1} = 6.8 \times 10^{-4}\,M\,s^{-1} \qquad (16.2c)$$

To obtain a *single* average rate of reaction, we may write

$$\text{average rate} = -\frac{1}{2}\frac{\Delta[N_2O_5]}{\Delta t} = \frac{1}{4}\frac{\Delta[NO_2]}{\Delta t} = \frac{\Delta[O_2]}{\Delta t} \qquad (16.3)$$

$$= 6.8 \times 10^{-4}\,M\,s^{-1}$$

In Eq. 16.3, the average rate of change of each substance's concentration is divided by the coefficient of that substance in the balanced equation for the reaction, Eq. 16.1. Further, because it is customary to consider the rate of reaction as a positive quantity, it is necessary to change the sign of a reactant's rate of change in concentration, and thus a minus sign appears in the second term of Eq. 16.3.

If we proceed to calculate the average rate of decomposition of N_2O_5 for each time interval in Table 16.1, we find that the average rate decreases as the reaction proceeds. Because our objective is to relate the rate of reaction to the concentrations of the substances in solution, it is necessary to measure the rate at a known concentration (or a known time). To do this we must use a very short time interval; that is, we determine the value of the average rate as Δt approaches zero. This can be done by plotting the concentration of N_2O_5 versus time as shown in Figure 16.1, and then constructing straight lines that are tangent to the curve obtained. For example, the slope of the tangent to the curve at 526 s represents the instantaneous rate of change of N_2O_5 concentration at 526 s. It may be expressed in mathematical terms as $d[N_2O_5]/dt$, called the **derivative** of $[N_2O_5]$ with respect to t. (The symbol Δ has been replaced by the symbol d to indicate the limiting value of the ratio $\Delta[N_2O_5]/\Delta t$ as Δt approaches zero.)

To evaluate the slope at 526 s, we pick any two points on the tangent, for example, $2.20 M$ at 0 s and $0.00 M$ at 2150 s. (These two points are at the ends of the tangent shown in Figure 16.1.) The formula given in Appendix C-3 for the slope of a straight line is slope $= (y_2 - y_1)/(x_2 - x_1)$. On applying this formula to the points selected, we obtain

$$\text{slope} = \frac{d[N_2O_5]}{dt} = \frac{(0.00 - 2.20)M}{(2150 - 0)\text{ s}} = -1.02 \times 10^{-3} M\,\text{s}^{-1}$$

Of course, the instantaneous rates of change of the NO_2 and O_2 concentrations will be different because of the different stoichiometric coefficients in Eq. 16.1. To obtain a single instantaneous rate of reaction, one resorts to the same method used earlier for the average rate of reaction. The result is

$$\text{rate} = -\frac{1}{2}\frac{d[N_2O_5]}{dt} = \frac{1}{4}\frac{d[NO_2]}{dt} = \frac{d[O_2]}{dt} \tag{16.4}$$

where the term *instantaneous* has been dropped. (From now on it will be assumed that "rate" means "instantaneous rate.") Thus at 526 s, the rate of reaction is

$$\text{rate} = -\frac{1}{2}\frac{d[N_2O_5]}{dt} = -\tfrac{1}{2}(-1.02 \times 10^{-3}) = 5.1 \times 10^{-4} M\,\text{s}^{-1}$$

Values of the rate of reaction calculated as above are given in column 3 of Table 16.1.

It is now possible to generalize our definition of rate. The **rate of reaction** is

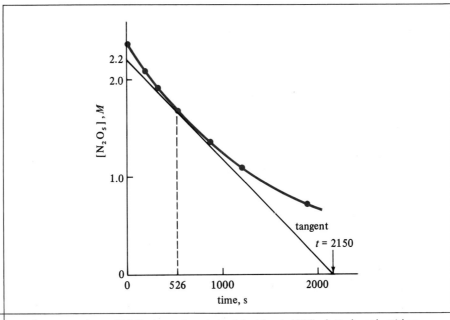

FIGURE 16.1 Concentration of N_2O_5 in carbon tetrachloride at 45°C plotted against time.

defined as the *instantaneous rate of change in concentration of a product or reactant after correction for the stoichiometry of the reaction.* For the general reaction

$$aA + bB + \cdots \longrightarrow cC + dD + \cdots \quad (16.5a)$$

the rate of reaction is defined by

$$\text{rate} = -\frac{1}{a}\frac{d[A]}{dt} = -\frac{1}{b}\frac{d[B]}{dt} = \frac{1}{c}\frac{d[C]}{dt} = \frac{1}{d}\frac{d[D]}{dt} = \cdots \quad (16.5b)$$

The rate of reaction must be evaluated from experimental data as we cannot evaluate it from theoretical considerations at the present time. (Although it is now conventional to define a single rate of reaction for any given reaction [Eq. 16.5b], investigators sometimes report rates in terms of the rate of appearance of a product or the rate of disappearance of a reactant, without correcting for the stoichiometry of the reaction. Thus one must be careful to determine how a rate has been defined in each case.)

16.2 RATE LAWS

Often there is a simple mathematical relationship between the rate of a reaction and the concentration of one or more species in the reaction mixture. This relationship must be determined experimentally and it is known as the **rate law** for the reaction.

For example, we find that, for the decomposition of N_2O_5 at constant temperature, the rate of reaction divided by the concentration of unreacted dinitrogen pentoxide is a constant as shown in column 4 of Table 16.1. Thus we may write

$$\text{rate} = -\frac{1}{2}\frac{d[N_2O_5]}{dt} = k[N_2O_5] \tag{16.6}$$

with $k = 3.1 \times 10^{-4}\,\text{s}^{-1}$ at 45°C. The proportionality constant (k) is called the **rate constant** and has a definite numerical value that is dependent on the temperature.

As another example, consider the reaction of bromate ion (BrO_3^-) with bromide ion (Br^-) in acidic solutions.

$$BrO_3^- + 5Br^- + 6H^+ \longrightarrow 3Br_2 + 3H_2O \tag{16.7a}$$

The rate law for this reaction is

$$\text{rate} = -\frac{d[BrO_3^-]}{dt} = k[BrO_3^-][Br^-][H^+]^2 \tag{16.7b}$$

The exponent to which the concentration of each species is raised in the rate law is known as the **reaction order** *of that species.* In the above example the reaction is first order in BrO_3^-, first order in Br^-, and second order in H^+. The decomposition of N_2O_5 (Eq. 16.6) is first order in N_2O_5. Note that the reaction order may or may not be the same as the corresponding stoichiometric coefficient in the equation for the reaction. For example, there are two molecules on the left side of Eq. 16.1, but the rate law is first order in N_2O_5.

In general, for any reaction

$$aA + bB + \cdots \longrightarrow cC + dD + \cdots \tag{16.8a}$$

the rate law may be of the form

$$\text{rate} = k[A]^x[B]^y[C]^z \cdots \tag{16.8b}$$

where the numerical exponents x, y, z, and so on, must be determined experimentally. Because the exponent of each concentration term is known as the reaction order for that species, the reaction is xth order with respect to A, yth order with respect to B, and so on. The "overall" order of the reaction is equal to the sum of the exponents, that is, $x + y + z + \cdots$. For example, the overall order of the bromate–bromide reaction, Eq. 16.7b, is fourth order.

Reaction orders determined experimentally are usually so close to integers that, if we take into account possible experimental errors, they may be assigned

16.3 METHOD OF INITIAL RATES

integral values as in Eqs. 16.6 and 16.7b. However, a few rate laws are known with fractional reaction orders such as $\frac{1}{2}$, $\frac{3}{2}$, and $\frac{2}{3}$. Of course, if the concentration of a substance has no effect on the rate, its reaction order is zero, and that substance does not appear in the rate law. For example, the bromate-bromide rate law, Eq. 16.7b, is zero order in Br_2, a reaction product. The rate law is usually zero order in all reaction products.

16.3 METHOD OF INITIAL RATES

The order of a reaction with respect to any molecule or ion present in the reaction mixture may be determined by investigating the effect of a change in concentration of the molecule or ion on the initial rate of reaction. The analysis of the data is most convenient if all but one concentration is held constant. Thus one usually conducts a series of experiments, in which each concentration is varied in turn. Then the effect on the initial rate of a change in a particular substance's concentration gives the order of that substance in the rate law.

For example, if doubling the concentration of a substance doubles the initial rate, the rate is directly proportional to the concentration of the substance, and the rate law is first order in that substance. On the other hand, if doubling the concentration increases the rate by a factor of four, the rate is proportional to the square of the concentration, and the rate law is second order in the substance whose concentration was changed.

It is generally not possible to vary the concentration of solvent over more than a narrow range, and therefore the solvent's reaction order is usually not determined.

The determination of the experimental rate law by the method of initial rates is illustrated in the following example.

EXAMPLE 1

The initial rate of disappearance of bromate ion (Eq. 16.7a) was determined for each of four reaction mixtures as shown in the following table.

Experiment number	$[H^+]$ (M)	$[BrO_3^-]$ (M)	$[Br^-]$ (M)	Rate ($M\,s^{-1}$)
1	0.0080	0.0010	0.10	2.5×10^{-8}
2	0.0080	0.0010	0.20	4.7×10^{-8}
3	0.0040	0.0010	0.20	1.2×10^{-8}
4	0.0080	0.0020	0.10	5.4×10^{-8}

What is the rate law for the reaction and what is the numerical value of the rate constant?

From experiments 1 and 2 we note that increasing the Br^- concentration by a factor of 2 (with $[H^+]$ and $[BrO_3^-]$ constant) increases the rate by a factor of $(4.7 \times 10^{-8})/(2.5 \times 10^{-8}) = 1.9$, or approximately 2, so that the reaction is first order in Br⁻. From experiments 2 and 3 (with [BrO₃⁻] and [Br⁻] constant) a decrease in H⁺ concentration by a factor of 2 causes a reduction in rate of $(4.7 \times 10^{-8})/(1.2 \times 10^{-8}) = 3.9$, or approximately 4, so that the reaction is second order in H⁺.

From experiments 1 and 4 (with [H⁺] and [Br⁻] constant) we note that doubling the concentration of BrO_3^- causes an increase in rate of $(5.4 \times 10^{-8})/(2.5 \times 10^{-8}) = 2.2$, or about 2, so that the reaction is first order in BrO_3^-.

We may summarize the above results as follows:

$$\text{rate} = -\frac{d[BrO_3^-]}{dt} = k[BrO_3^-][Br^-][H^+]^2$$

The numerical value of k may now be obtained by using any one of the four sets of data. For example, using experiment 1 we obtain

$$k = \frac{\text{rate}}{[BrO_3^-][Br^-][H^+]^2} = \frac{2.5 \times 10^{-8} M\,s^{-1}}{(0.0010M)(0.10M)(0.0080M)^2}$$
$$= 3.9 M^{-3}\,s^{-1}$$

16.4 INTEGRATED FORM OF THE FIRST-ORDER RATE LAW

In many cases it is possible to determine the order of a reaction and to evaluate the rate constant by using linear graphs. This procedure is particularly important for a first-order reaction.

The general form of the first-order rate law for a reactant species A is

$$-\frac{1}{a}\frac{d[A]}{dt} = k[A] \tag{16.9a}$$

where a is the stoichiometric coefficient of A in the balanced equation for the reaction. The methods of integral calculus may be used to convert Eq. 16.9a to the following equation:

$$\ln\frac{[A]_0}{[A]} = akt \tag{16.9b}$$

where $[A]_0$ and $[A]$ are the concentrations of A at time 0 and any subsequent time t, respectively, and ln represents the natural logarithm to base e. Since $\ln x = 2.303 \log x$, where $\log x$ is the logarithm to base 10, this equation may be written in the alternative form

$$\log\frac{[A]_0}{[A]} = \frac{akt}{2.303} \tag{16.9c}$$

To determine if a particular rate law is first order, the experimental data are plotted as the logarithmic term (either ln or log) versus time. If the rate law is

16.4 INTEGRATED FORM OF THE FIRST-ORDER RATE LAW

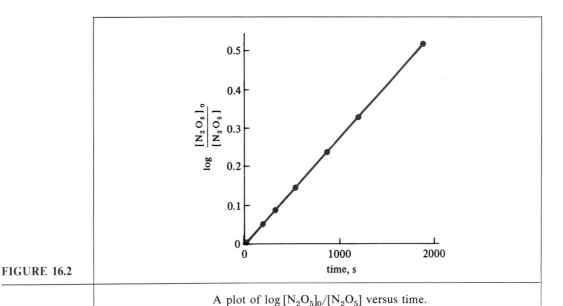

FIGURE 16.2

A plot of log [N$_2$O$_5$]$_0$/[N$_2$O$_5$] versus time.

indeed first order, the experimental points will fall on or near a straight line. (If a curve is obtained, the rate law is not first order.) This method is more accurate and easier to use than the method of drawing tangents to curves as in Figure 16.1.

As an example of the use of Eq. 16.9b, in Figure 16.2 the experimental data of Table 16.1 are plotted on a graph of the log term versus time. A linear relationship is obtained, so the reaction is first order in N$_2$O$_5$. The slope of the straight line is equal to $2k/2.303$, since a is 2 in this reaction. On evaluating the slope we obtain 2.70×10^{-4} s^{-1}. Therefore

$$k = \frac{2.303}{2} \times \text{slope} = \frac{2.303}{2} \times 2.70 \times 10^{-4} \text{ s}^{-1}$$
$$= 3.11 \times 10^{-4} \text{ s}^{-1}$$

Another test of a first-order reaction is to measure the time for one-half of the limiting reagent to be used up, called the **half-life** (symbol $t_{1/2}$). *For a first-order reaction the half-life is independent of the initial concentration of the reactant,* as shown in the following equations. If [A]$_0$ is the initial concentration of N$_2$O$_5$ at time 0, then [A]$_0$/2 will be the concentration of N$_2$O$_5$ at time $t_{1/2}$. When these values are substituted in Eq. 16.9b, we obtain

$$\ln \frac{[A]_0}{[A]_0/2} = ak\, t_{1/2}$$

and

$$t_{1/2} = \frac{\ln 2}{ak} = \frac{0.693}{ak} \tag{16.10}$$

This expression is valid for all first-order reactions. Thus the time required to reduce the concentration of N_2O_5 from $12.0 \times 10^{-4} M$ to $6.0 \times 10^{-4} M$ would be the same as that to reduce the concentration from $2.0 \times 10^{-4} M$ to $1.0 \times 10^{-4} M$.

There are many first-order chemical reactions. All decay reactions of radioactive nuclei (see Chapter 24) are first order. Many chemical reactions of higher order can be reduced to **pseudo-first-order** reactions by a suitable choice of concentrations of the reactants. In a pseudo-first-order reaction, the initial concentration of one reactant (say, reactant A) is made much smaller than the initial concentrations of the other substances that affect the rate. Under these conditions the other concentrations are practically constant. If now the rate law is first order in A, the reaction will behave as if it were a first-order reaction, regardless of the reaction orders with respect to the other substances.

For example, the rate law for the reduction of bromate ion (Eq. 16.7b) may be reduced to a pseudo-first-order reaction by using $0.0010 M$ $KBrO_3$, $0.10 M$ $HClO_4$, and $0.10 M$ KBr. Since the concentrations of H^+ and Br^- will remain essentially constant while the BrO_3^- is used up, the reaction will be pseudo-first order in BrO_3^-. Thus

$$-\frac{d[BrO_3^-]}{dt} = k[BrO_3^-][Br^-][H^+]^2 = k_0[BrO_3^-]$$

where

$$k_0 = k[Br^-][H^+]^2 = k \times 0.10 \times 0.10^2 = 1 \times 10^{-3} k$$

Note that the pseudo-first-order rate constant k_0 depends on the concentrations of H^+ and Br^-. For example, if the concentration of Br^- were doubled, the reaction would still behave as if it were first order, but it would go faster, with a rate constant (k_0) twice as large.

Again the time required for the bromate ion concentration to change from $0.0010 M$ to $0.0005 M$ would be the same as that required to change from $0.0005 M$ to $0.00025 M$. Thus the half-life would be a constant during an experiment. Of course, in another experiment with different concentrations of H^+ and Br^-, the half-life would be different.

16.5 INTEGRATED FORM OF THE SECOND-ORDER RATE LAW

As an example of a second-order reaction, consider the decomposition of gaseous hydrogen iodide:

$$2HI \longrightarrow H_2 + I_2$$

16.5 INTEGRATED FORM OF THE SECOND-ORDER RATE LAW

The experimental rate law is

$$\text{rate} = -\frac{1}{2}\frac{d[\text{HI}]}{dt} = k[\text{HI}]^2$$

The reaction is therefore second order in hydrogen iodide. If the rate is expressed in mol L^{-1} s^{-1}, and the concentration of HI is expressed in mol L^{-1}, the units of k are mol^{-1} L s^{-1}, or M^{-1} s^{-1}.

For the general case of a reaction of the type,

$$a\text{A} \longrightarrow \text{products} \tag{16.11a}$$

with a rate law of the form

$$\text{rate} = -\frac{1}{a}\frac{d[\text{A}]}{dt} = k[\text{A}]^2 \tag{16.11b}$$

the methods of integral calculus may be applied to obtain the following equation:

$$\frac{1}{[\text{A}]} = akt + \frac{1}{[\text{A}]_0} \tag{16.11c}$$

In this equation $[\text{A}]_0$ is the initial concentration of A, and $[\text{A}]$ is the concentration of A at time t. Thus a plot of $1/[\text{A}]$ versus time would yield a straight line whose slope would be ak.

Figure 16.3 shows such a plot for the dimerization of cyclopentadiene (C_5H_6) in benzene solution at 25.1°C.

$$2C_5H_6 \longrightarrow C_{10}H_{12}$$

The plot of $1/[C_5H_6]$ against time is linear, indicating a second-order reaction. Thus the rate law would be

$$\text{rate} = -\frac{1}{2}\frac{d[C_5H_6]}{dt} = k[C_5H_6]^2$$

The rate constant k may be evaluated from the slope of the straight line, which is $1.02 \times 10^{-4} M^{-1}$ s^{-1}. The slope is equal to $2k$, so k is $5.10 \times 10^{-5} M^{-1}$ s^{-1}.

The half-life method may also be used with reactions whose rate law is given by Eq. 16.11b, as shown below. Initially the concentration of reactant A is $[\text{A}]_0$. At time $t_{1/2}$, 50% of A has reacted and the concentration of A will be $[\text{A}]_0/2$. When these values are substituted in Eq. 16.11c, we obtain

$$\frac{1}{[\text{A}]_0/2} = akt_{1/2} + \frac{1}{[\text{A}]_0}$$

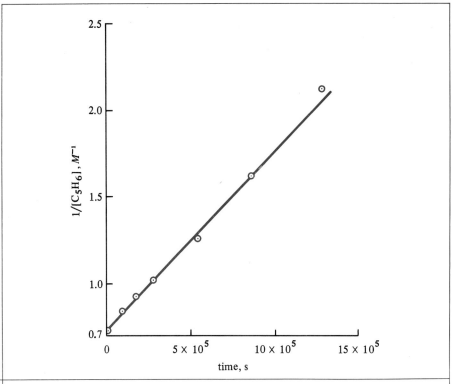

FIGURE 16.3 A linear graph for a second-order reaction: the dimerization of cyclopentadiene in benzene solution.

or

$$t_{1/2} = \frac{1}{ak[A]_0} \tag{16.12}$$

Thus for a reaction that is second order with respect to A, the half-life is *inversely proportional to the initial concentration of A*.

For example, if the concentration of A decreases from an initial value of $0.40M$ to $0.20M$ in 20 min, how long will it take for its concentration to drop to $0.10M$? According to Eq. 16.12, the second half-life ($0.20M \longrightarrow 0.10M$) will be twice the first half-life ($0.40M \longrightarrow 0.20M$), and thus it will take 40 min for the concentration of A to change from $0.20M$ to $0.10M$. (The total elapsed time required to change from $0.40M$ to $0.10M$ is therefore $20 + 40$, or 60 min.)

Equation 16.12 is valid for reactions of the type shown in Eq. 16.11a or for reactions that can be reduced to that type by a suitable choice of concentrations. For example, if the reaction is

$$A + B \longrightarrow \text{products}$$

we could start with equal concentrations of A and B, in which case they would remain equal throughout the reaction. If the rate law is first order in A and first order in B, it reduces to Eq. 16.11b,

$$\text{rate} = -\frac{d[A]}{dt} = k[A][B] = k[A]^2$$

Equations 16.11c and 16.12 are also valid under these conditions.

16.6 REACTION RATES AND EQUILIBRIUM

Many chemical reactions approach an equilibrium position in which appreciable amounts of reactants as well as products are present, as noted in Chapter 12 for the reaction of hydrogen with iodine at 425°C:

$$H_2(g) + I_2(g) \rightleftharpoons 2HI(g) \qquad K = \frac{[HI]_e^2}{[H_2]_e[I_2]_e} = 54.5 \qquad (16.13)$$

The rate of the forward reaction, R_f, in systems with no HI present has been shown to be

$$R_f = \frac{1}{2}\frac{d[HI]}{dt} = k_f[H_2][I_2] \qquad (16.14)$$

where k_f is $3.35 \times 10^{-2} M^{-1} s^{-1}$ at 425°C. At the same temperature the rate of the reverse reaction, R_r, in the absence of H_2 or I_2 is

$$R_r = -\frac{1}{2}\frac{d[HI]}{dt} = k_r[HI]^2 \qquad (16.15)$$

where $k_r = 6.2 \times 10^{-4} M^{-1} s^{-1}$.

Under conditions where all three components (H_2, I_2, and HI) are present, the *net* rate of the forward reaction will be equal to the rate of the forward reaction less the rate of the reverse reaction, or

$$\frac{1}{2}\frac{d[HI]}{dt} = R_f - R_r = k_f[H_2][I_2] - k_r[HI]^2 \qquad (16.16)$$

Note that the first term can be evaluated in the early stages of the reaction between H_2 and I_2 when the concentration of HI would be extremely low. The second term can be evaluated from the early stages of the decomposition of HI when the H_2 and I_2 concentrations would be extremely low. At all times the rate of reaction conforms to Eq. 16.16.

At chemical equilibrium the net rate of reaction is zero (i.e., there is no change in the concentration of HI with time). Thus

$$\frac{1}{2}\frac{d[HI]}{dt} = 0 = k_f[H_2]_e[I_2]_e - k_r[HI]_e^2 \tag{16.17}$$

where the subscript e refers to the equilibrium concentration. Equation 16.17 may be rearranged to give

$$\frac{k_f}{k_r} = \frac{[HI]_e^2}{[H_2]_e[I_2]_e} = K \tag{16.18}$$

Thus the equilibrium constant for the reaction is equal to the rate constant for the forward reaction divided by the rate constant for the reverse reaction. At 425°C, K for formation of HI would be $(3.35 \times 10^{-2} M^{-1} s^{-1})/(6.2 \times 10^{-4} M^{-1} s^{-1}) = 54.0$, which compares favorably with the value of 54.5 calculated from the equilibrium concentrations. For all other cases investigated the principal conclusion obtained above is also valid: *The law of chemical equilibrium is identical to the equation obtained by equating the rate for the forward reaction to the rate in the reverse direction. Furthermore, the magnitude of the equilibrium constant is equal to the rate constant in the forward direction divided by the rate constant in the reverse direction.*

16.7 EFFECT OF TEMPERATURE ON THE RATE CONSTANT

The rates of most chemical reactions increase rapidly with rising temperature. This increase may be accompanied by a change in the rate law, but usually it is due only to the effect of temperature on the magnitude of the rate constant. As we shall see, the effect of temperature on the rate constant has been useful in developing theories of rate processes.

A typical example is shown in Figure 16.4, where the rate constant for the formation of HI (k_f of Eq. 16.14) has been plotted against the temperature. The shape of the curve is very similar to that obtained for the vapor pressure of a liquid as a function of temperature (Figure 15.1).

An even more impressive test of this similarity may be obtained by plotting the data so that a linear relationship results. We have noted that a plot of $R \ln K$ versus $1/T$ is linear (Figure 15.2) in accordance with Eq. 15.4:

$$R \ln K = -\frac{\Delta H°}{T} + \Delta S°$$

or

$$\ln K = -\frac{\Delta H°}{RT} + \frac{\Delta S°}{R}$$

16.7 EFFECT OF TEMPERATURE ON THE RATE CONSTANT

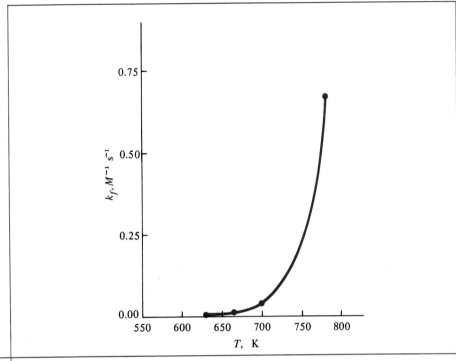

FIGURE 16.4 The rate constant for the formation of hydrogen iodide as a function of temperature.

In 1889 Arrhenius reasoned that, because $K = k_f/k_r$, a reasonable equation relating the rate constant to the absolute temperature would be

$$\ln k = -\frac{E_a}{RT} + \ln A \qquad (16.19)$$

In this equation E_a and A are constants, independent of temperature. (E_a plays the same role as $\Delta H°$ in Eq. 15.4, and $\ln A$ plays the same role as $\Delta S°/R$.) This equation is applicable to any rate constant, including k_f and k_r. Of course, each rate constant has its own particular set of values of A and E_a.

In Figure 16.5 values of $\ln k_f$ for the formation of hydrogen iodide are plotted against the reciprocal of the absolute temperature. The validity of the equation is confirmed by the linear relationship obtained. An alternative form of Eq. 16.19 is the famous Arrhenius equation for the effect of temperature on the rate constant:

$$k = Ae^{-E_a/RT} \qquad (16.20)$$

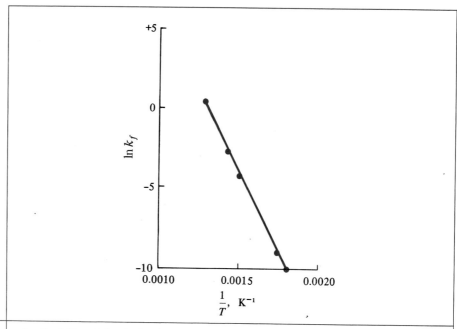

FIGURE 16.5 Applicability of the Arrhenius equation to the temperature dependence of the rate constant for formation of hydrogen iodide.

Because of the position of the constant A in this equation, A is called the **preexponential factor.** (Another name for A is the frequency factor.) The constant E_a is called the **activation energy.**

The activation energy may be evaluated by substituting experimental values of the rate constant at two different temperatures into Equation 16.21.

$$E_a = \frac{T_A T_B R \ln(k_B/k_A)}{T_B - T_A} \tag{16.21}$$

where k_A is the rate constant at temperature T_A, and k_B is the rate constant at temperature T_B. (The derivation of Eq. 16.21 from Eq. 16.19 is similar to the derivation of Eq. 15.5 from Eq. 15.4.)

To determine the preexponential factor A from the same data, we rearrange Eq. 16.19 as follows:

$$\ln A - \ln k = \ln\left(\frac{A}{k}\right) = \frac{E_a}{RT}$$

Because $\ln x = 2.303 \log x$, we may write

16.7 EFFECT OF TEMPERATURE ON THE RATE CONSTANT

$$\log\left(\frac{A}{k}\right) = \frac{E_a}{2.303RT} \tag{16.22}$$

This equation is applicable to any temperature. To calculate A, one simply substitutes a value for k at a particular temperature in Eq. 16.22, along with the activation energy computed from Eq. 16.21.

EXAMPLE 2

The rate constant for the formation of hydrogen iodide is $6.6 \times 10^{-5} M^{-1} s^{-1}$ at 575 K and $3.21 \times 10^{-2} M^{-1} s^{-1}$ at 700 K. Evaluate E_a and A in the Arrhenius equation.

Equation 16.21 may be used to evaluate E_a. If T_A is the lower temperature and T_B is the higher temperature, this equation becomes

$$E_a = \frac{(575\ K)(700\ K)(8.314\ J\ K^{-1}\ mol^{-1}) \ln\left(\frac{3.21 \times 10^{-2}\ M^{-1}\ s^{-1}}{6.6 \times 10^{-5}\ M^{-1}\ s^{-1}}\right)}{700\ K - 575\ K}$$

$$= 166\ kJ\ mol^{-1}$$

To obtain A, we substitute k at either temperature, along with E_a, in Eq. 16.22. Using k_B and T_B, we obtain

$$\log \frac{A}{3.21 \times 10^{-2}\ M^{-1}\ s^{-1}} = \frac{166 \times 10^3\ J\ mol^{-1}}{(2.303)(8.314\ J\ K^{-1}\ mol^{-1})(700\ K)}$$

$$= 12.4$$

$$\frac{A}{3.21 \times 10^{-2}\ M^{-1}\ s^{-1}} = 2.5 \times 10^{12}$$

$$A = (2.5 \times 10^{12})(3.21 \times 10^{-2}\ M^{-1}\ s^{-1})$$

$$= 8.0 \times 10^{10}\ M^{-1}\ s^{-1}$$

Thus we may express the temperature dependence of the rate constant for formation of HI by

$$k = 8.0 \times 10^{10} e^{-166 \times 10^3/RT} M^{-1}\ s^{-1}$$

Note that the units of A correspond to those of the rate constant, and the units of E_a correspond to those of RT. Thus E_a is an energy term, as the name activation energy implies.

In the preceding section we found that the equilibrium constant was equal to the ratio of the rate constants in the forward and reverse directions. If we take the natural logarithm of both sides of Eq. 16.18, we obtain

$$\ln K = \ln k_f - \ln k_r$$

Then multiplying by R yields

$$R \ln K = R \ln k_f - R \ln k_r$$

If we now utilize Eqs. 15.40 and 16.19, we obtain

$$-\frac{\Delta H^\circ}{T} + \Delta S^\circ = R \ln A_f - \frac{E_{a(f)}}{T} - R \ln A_r + \frac{E_{a(r)}}{T} \tag{16.23}$$

where subscripts f and r denote the forward and reverse directions, respectively. If we equate the terms in $1/T$, we find that

$$\Delta H^\circ = E_{a(f)} - E_{a(r)} \tag{16.24}$$

Equating the terms that do not involve T gives

$$\Delta S^\circ = R \ln \frac{A_f}{A_r} \tag{16.25}$$

Thus the enthalpy change for the reaction is equal to the difference in the activation energies for the forward and reverse reactions. The enthalpy change for the formation of HI (Eq. 16.13), calculated from Figure 15.2, is -12 kJ mol^{-1}. Therefore $E_{a(r)}$ is $166 - (-12) = 178$ kJ mol^{-1}. It is necessary to provide 166 kJ of energy to produce the reaction, but, when it does occur, then 178 kJ of energy is released, giving a net release of 12 kJ for the reaction. This is shown graphically in Figure 16.6. Theoretical interpretations of the highest energy state in Figure 16.6 are given in the next two sections. As we shall see, this activation energy requirement is the principal reason that reactions do not occur almost instantaneously.

16.8 THE COLLISION THEORY OF REACTION RATES

In its simplest form, the **collision theory** assumes that *the rate of an elementary reaction is equal to the number of collisions between the reactant molecules, multiplied by the fraction of collisions that result in formation of products.* An **elementary reaction** is a reaction that occurs in a single step. For example, if the formation of C and D from A and B is an elementary reaction,

$$A + B \longrightarrow C + D \tag{16.26}$$

then the rate of reaction will be

$$\text{rate} = Z_{AB} f_e \tag{16.27}$$

16.8 THE COLLISION THEORY OF REACTION RATES

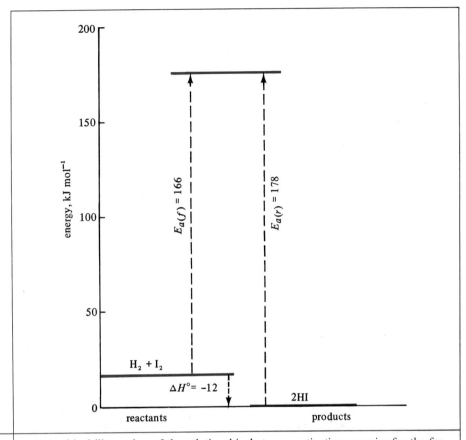

FIGURE 16.6 A graphical illustration of the relationship between activation energies for the forward and reverse reactions and the overall enthalpy change for $H_2 + I_2 \rightleftharpoons 2\,HI$.

where Z_{AB} is the total number of collisions between A and B molecules per second per cubic meter, and f_e is the fraction of collisions that are effective in forming the products C and D. (In most collisions, the reactant molecules do not undergo any chemical change: $A + B \longrightarrow A + B$.)

We can now use the results of the kinetic molecular theory to calculate the number of collisions between A and B. As a first approximation, let us consider the A and B molecules to be rigid spheres of radius r_A and r_B. (The radius of a gaseous molecule can be calculated from the gas's van der Waals b constant, as discussed in Section 4.6.) A collision will result whenever the center of an A molecule is within a distance of $(r_A + r_B)$ of the center of a B molecule. The mathematical treatment is somewhat simplified if we consider the B molecules as stationary points and first calculate the number of collisions that would result owing to the motion of a single A molecule of *effective* radius $r_A + r_B$. If the speed of the A molecule is u_A, then in 1 s the A molecule sweeps out a volume $V = u_A \pi (r_A + r_B)^2$. It collides with all the B molecules whose centers are within this volume (Figure 16.7). Therefore each A molecule experiences VN_B collisions per second, where N_B is the number of B molecules per cubic meter.

To obtain the *total* number of collisions per second between A and B molecules in 1 m³, we must multiply the number of collisions per second for one A molecule by N_A, the number of A molecules per cubic meter:

$$Z_{AB} = VN_A N_B = u_A \pi (r_A + r_B)^2 N_A N_B \tag{16.28}$$

A more exact treatment (which considers the relative movement of both the A and B molecules) leads to replacing the speed of the A molecule, u_A, by the *mean relative speed*, $\bar{u}_r$, of the two molecules. [The average speed of a molecule may be obtained from the kinetic theory and is given by the equation $\bar{u} = (8RT/\pi M)^{1/2}$, where M is the molecular mass. The mean relative speed of two molecules is $\bar{u}_r = (\bar{u}_A^2 + \bar{u}_B^2)^{1/2}$.] With this change, Z_{AB} becomes

$$Z_{AB} = \bar{u}_r \pi (r_A + r_B)^2 N_A N_B \tag{16.29}$$

If we substitute the value of Z_{AB} from Eq. 16.29 in Eq. 16.27, we find that the collision theory predicts that the rate of a bimolecular reaction is

$$\text{rate} = Z_{AB} f_e = \bar{u}_r \pi (r_A + r_B)^2 f_e N_A N_B \tag{16.30a}$$

If r_A and r_B are in meters, and $\bar{u}_r$ is in meters per second, this formula gives the rate in units of molecules per cubic meter. To obtain a formula in terms of the more common units of moles per liter, we must multiply by a factor of $10^3 N_0$, where N_0 is Avogadro's number, and replace N_A and N_B by the molar concentrations [A] and [B], respectively.[1] Thus we find that

$$\text{rate} = 10^3 N_0 \bar{u}_r \pi (r_A + r_B)^2 f_e [A][B] \tag{16.30b}$$

This equation shows that collision theory leads to a second-order rate law (first order in A and first order in B) for a bimolecular reaction.

To go beyond this point, we must analyze f_e, the fraction of collisions that are effective in forming products. It is usually considered to be a product of two factors, a *steric factor p* and a *temperature factor* $e^{-E_a/RT}$:

$$f_e = pe^{-E_a/RT} \tag{16.31}$$

---

[1] First, as there are 10^3 L in 1 m³, and N_0 molecules in 1 mol, we have $N_A = 10^3 N_0 [A]$ and $N_B = 10^3 N_0 [B]$. On substituting these expressions into Eq. 16.30a, the result is

$$\text{rate} = (10^3 N_0)^2 \bar{u}_r \pi (r_A + r_B)^2 f_e [A][B]$$

Next, to convert this rate from molecules per cubic meter to moles per liter, we must divide it by $10^3 N_0$. The factor of $10^3 N_0$ in the denominator cancels one of the factors of $10^3 N_0$ in the numerator, giving Eq. 16.30b.

16.8 THE COLLISION THEORY OF REACTION RATES

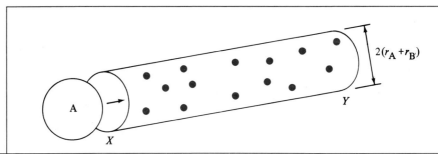

FIGURE 16.7

Collisions of a single A molecule of effective cross-sectional area $\pi(r_A + r_B)^2$ with stationary point B molecules. In the volume swept out, there are 14 point B molecules. Thus there would be 14 collisions during the time required for A to go from X to Y.

1. The steric factor p is typically much less than 1 because in only a small fraction of the collisions are the molecules properly oriented for reaction to occur. For example, in the reaction $A + BC \longrightarrow AB + C$, one would expect that atom A must strike the B end of the BC molecule to obtain the AB product. (Collisions in which atom A strikes the C end of the BC molecule will be ineffective because C is in the way, preventing formation of the AB bond.) Thus of the many possible orientations of reactant molecules, only some will lead to products. The steric factor p is *the fraction of collisions having favorable orientation for reaction to occur.*
2. The temperature factor $e^{-E_a/RT}$ represents *the fraction of collisions having energy of relative motion greater than or equal to the activation energy E_a.* (This result can be derived from the kinetic molecular theory, but the proof is rather complicated and it will not be given here.) Thus only high-energy collisions of molecules lead to the formation of products. If the energy of relative motion is smaller than E_a, no reaction can occur. One may therefore regard the activation energy as an *energy barrier*—the collisional energy must be at least as large as the activation energy for reaction to occur.

When Eqs. 16.30b and 16.31 are combined, we obtain the following expression for the rate of reaction:

$$\text{rate} = 10^3 N_0 \bar{u}_r \pi (r_A + r_B)^2 p e^{-E_a/RT}[A][B] \qquad (16.32)$$

The terms multiplying [A][B] are equal to the second-order rate constant, k.

$$k = 10^3 N_0 \bar{u}_r \pi (r_A + r_B)^2 p e^{-E_a/RT} \qquad (16.33)$$

This result has the same form as the Arrhenius equation. The preexponential factor A is equal to the product of all terms multiplying the exponential:

$$A = 10^3 N_0 \bar{u}_r \pi (r_A + r_B)^2 p \qquad (16.34)$$

One can now use this equation to calculate the steric factor p, as shown in the following example.

EXAMPLE 3

The gas phase reaction of nitric oxide and ozone,

$$NO + O_3 \longrightarrow NO_2 + O_2$$

has been investigated experimentally at $-75°C$ and $-43°C$. The rate law is first order in NO and first order in O_3.

$$\text{rate} = k[NO][O_3]$$

This reaction is believed to consist of a single elementary reaction identical with the stoichiometric equation given above. The preexponential factor A, calculated by means of Eq. 16.22, is $8.0 \times 10^8 M^{-1} s^{-1}$, and the effective radius $(r_{NO} + r_{O_3})$ is 1.7×10^{-10} m. Calculate the steric factor p.

The average temperature of the rate experiments is $-59°C$ or 214 K, and the molecular masses of NO and O_3 are 30 and 48, respectively. Using the formula for average speed given above, we obtain

$$\bar{u}_{NO} = \sqrt{\frac{(8)(8.314 \text{ kg m}^2 \text{ s}^{-2} \text{ K}^{-1} \text{ mol}^{-1})(214 \text{ K})}{(3.14)(30 \times 10^{-3} \text{ kg mol}^{-1})}}$$
$$= 389 \text{ m s}^{-1}$$

A similar calculation for ozone gives $\bar{u}_{O_3} = 307$ m s^{-1}. The mean relative speed of the two molecules is then obtained from the formula given above as

$$\bar{u}_r = \sqrt{(389 \text{ m s}^{-1})^2 + (307 \text{ m s}^{-1})^2}$$
$$= 496 \text{ m s}^{-1}$$

[As shown in the derivation of Eq. 16.34, the numerical factors require values of $\bar{u}_r$ in meters per second, and $r_A + r_B$ in meters, to obtain A in the correct units of liters per mole second (M^{-1} s^{-1}).]

Substituting the various factors in Eq. 16.34, we obtain

$$8.0 \times 10^8 = (10^3)(6.02 \times 10^{23})(496)(3.14)(1.7 \times 10^{-10})^2 p$$
$$= 2.7 \times 10^{10} p$$

Thus the steric factor p is $(8.0 \times 10^8)/(2.7 \times 10^{10}) = 0.030$, indicating that in only 3.0% of the collisions between NO and O_3 molecules are the molecules oriented properly for reaction to occur.

The analysis of reaction rates in terms of molecular collisions is an illuminating one, for it enables us to see how the reaction rate is influenced by such factors as molecular radius, relative orientation of the reactant molecules, temperature, and activation energy. However, it is no doubt an oversimplification to state that there is a sharp change in reactivity at the activation energy threshold, such that molecules colliding with proper orientation and with energy of relative motion at or above E_a will always react, while molecules colliding with energy of relative motion below E_a will never react. In more sophisticated forms of collision theory, the probability of reaction varies in a smooth manner with increasing energy of relative motion above the activation energy threshold.

An unsatisfactory feature of collision theory is the lack of any method of predicting the steric factor p for a reaction whose rate has not been measured. The transition state theory described in the next section may be used to predict the preexponential factor A without introduction of an unknown steric factor.

16.9 THE TRANSITION-STATE THEORY

The **transition-state** or **activated-complex** theory of reaction rates was developed into its present form chiefly by Henry Eyring and co-workers. In the transition-state theory, it is assumed that the reactant molecules are in equilibrium with an activated complex that decomposes to give the reaction products. [The activated complex is identified by using the double-dagger superscript ($\ddagger$).]

$$A + BC \rightleftharpoons ABC^\ddagger \qquad K^\ddagger = \frac{[ABC^\ddagger]}{[A][BC]} \qquad (16.35a)$$

$$ABC^\ddagger \longrightarrow AB + C \qquad (16.35b)$$

The activated complex has unique characteristics that distinguish it from a normal molecule, and, hence, the above equilibrium from a normal equilibrium. The atoms of the reactant molecules have become partially bonded to each other in such a manner that the complex may decompose to form the products, and there is sufficient energy concentrated in the bond (or bonds) to be broken so that the bond will break during one vibration (i.e., about 10^{-13} s).

The activation energy (E_a) that is required to form the activated complex is obtained by a collisional process. This activation energy is required because (1) at least one bond in the reactant molecules must be broken; and (2) the electron clouds of the reactant molecules will tend to repel each other as the molecules approach, and, at closer distances, the nuclear charges will repel each other. Therefore only highly energetic molecules can approach closely enough for the reaction to occur.

Estimates of the activation energy for $H + H_2 \rightarrow H_2 + H$ have been made by calculating the potential energy of the system of three hydrogen atoms for various internuclear distances to find the lowest energy barrier that exists between reactants and products. These calculations show that the linear configuration of atoms H---H---H gives the path of minimum energy. To distinguish between the H atoms, we shall designate them as A, B, and C. As the atom A approaches the molecule BC the energy of the system gradually increases, and, when A is close to

B, an AB bond begins to form. Concurrently the BC bond is weakened and the distance between B and C increases. Eventually a point is reached where the A---B---C activated complex (ABC‡) may either decompose to form products or return to form reactants. This progression from reactants to activated complex to products is shown in the form of a graph (Figure 16.8) of potential energy versus the **reaction coordinate,** where the reaction coordinate is essentially a measure of the progress of the reaction. The progress of the reaction is measured by the decrease in the A-to-BC distance in the initial phase of the reaction and by the increase in the AB-to-C distance in the latter phase of the reaction. The reverse reaction (if it occurs to any extent) must go through the same activated complex. The reaction coordinate follows the path of lowest energy between reactants and products. All other paths necessitate higher energies so that the situation resembles that of a pass between two valleys separated by a mountain range.

In summary, *the transition state theory assumes that the products are formed by decomposition of the activated complex that is in equilibrium with the reactants. Furthermore, it predicts that all activated complexes decompose at the same rate,* that is,

$$\text{rate} = \frac{k_B T}{h}[\text{ABC}^‡] \tag{16.36}$$

where k_B is the Boltzmann constant and h is Planck's constant. (The numerical value of $k_B T/h$ at room temperature, 300 K, is 6.25×10^{12} s^{-1}.)

Because ABC‡ is in equilibrium with A and BC, we may replace [ABC‡] by $K^‡$[A][BC] and obtain

$$\text{rate} = \frac{k_B T}{h} K^‡[\text{A}][\text{BC}] \tag{16.37}$$

Thus we predict that the rate law for the bimolecular reaction will be first order in A and first order in BC, as it must be to agree with experimental results. $K^‡$ can be evaluated by theoretical methods, thus eliminating the necessity of assuming an arbitrary steric factor as in the collision theory. However, the methods of evaluating $K^‡$ are beyond the scope of this text.

16.10 REACTION MECHANISMS

In previous chapters we have written equations for chemical reactions to show how many moles of each reactant were required to produce a specified number of moles of product. Now we want to investigate how these reactions take place: what bonds are broken, what new bonds are formed, and in what order. *A* **reaction mechanism** *is a postulated sequence of elementary reactions that results in the conversion of reactants to products.* An elementary reaction is a single molecular

16.10 REACTION MECHANISMS

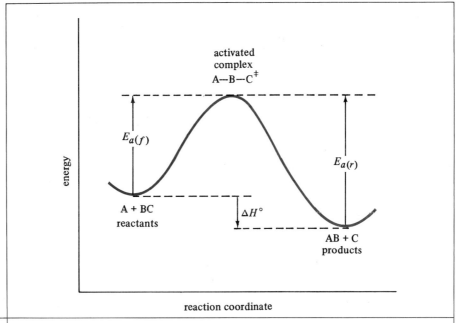

FIGURE 16.8 Energy changes accompanying a chemical reaction. The top of the curve corresponds to the activated complex (ABC‡) for the reaction.

event or "step" in the reaction. It is characterized as unimolecular, bimolecular, or trimolecular, depending on the number of molecules that react.

Unimolecular reactions are reactions involving a single molecule. They are typically decomposition reactions, and they have first-order rate laws.

$$A \longrightarrow B + C \qquad -\frac{d[A]}{dt} = k[A]$$

Most elementary reactions are bimolecular in that they occur by the collision of two molecules and hence obey a second-order rate law. The reaction of two different molecules can be expressed symbolically as

$$A + B \longrightarrow products \qquad -\frac{d[A]}{dt} = k[A][B]$$

Trimolecular reactions involving the collision of three molecules are known and they obey a third-order rate law. One reaction of this type is

$$2A + B \longrightarrow products \qquad -\frac{1}{2}\frac{d[A]}{dt} = k[A]^2[B]$$

Elementary reactions of molecularity greater than three have not been discovered, and even trimolecular reactions are rather rare, no doubt because the chance of three or more molecules colliding simultaneously is much smaller than the chance of two molecules colliding.

Note that the concept of molecularity is restricted to the individual elementary reactions in a reaction mechanism; it is not used to describe the overall stoichiometry of a reaction.

One of the principal methods of testing proposed mechanisms is to compare the rate law predicted by the mechanism with the experimentally determined rate law. If they are in agreement, the mechanism is possible. If they are not, the mechanism must be rejected.

Simple Reactions

In some cases the reaction mechanism consists of a single elementary reaction that is identical with the stoichiometric equation for the reaction. An example is the reaction of nitric oxide (NO) with ozone (O_3),

$$NO + O_3 \longrightarrow NO_2 + O_2 \tag{16.38}$$

The experimental rate law for this reaction is

$$\text{rate} = k[NO][O_3] \tag{16.39}$$

The proposed mechanism for this reaction is the capture of an oxygen atom by an NO molecule during the collision of NO with O_3. Thus the proposed mechanism is the single step given by Eq. 16.38. Because this is a bimolecular elementary reaction, it should have a rate law that is first order in NO and first order in O_3, in agreement with the experimental rate law given by Eq. 16.39.

Initial Slow Step

Most mechanisms consist of two or more steps. Although the variety of mechanisms is so great that it is difficult to classify them in any simple way, there are a few fairly common types of mechanism. One important type consists of an initial slow step, followed by one or more fast steps, or *follow-reactions*. Because the chemical reaction can proceed only as fast as the first slow step, this step acts as a reaction "bottleneck" that determines the rate of the overall reaction. Furthermore, the molecularity of the slow step determines the rate law of the reaction. (The terms *slow* and *fast* are used in a relative sense. The slow step may actually be quite fast, but it is significantly slower than the subsequent steps in the reaction mechanism.)

As an example of this type of mechanism, consider the reaction of acetone with hypobromite ion in basic aqueous solution:

$$CH_3COCH_3 + BrO^- \longrightarrow CH_3COCH_2Br + OH^- \tag{16.40}$$

16.10 REACTION MECHANISMS

The experimentally obtained rate law is

$$\text{rate} = k[CH_3COCH_3][OH^-] \qquad (16.41)$$

The reaction cannot proceed in a single step because the experimental rate is independent of the concentration of one of the reactants, BrO^-. Furthermore, the rate depends on the concentration of a reaction product, OH^-. Substances, such as OH^-, that increase the rate of a reaction without being used up in the reaction (although they may be consumed and then regenerated) are known as **catalysts**. (In this case, the reaction is said to be **autocatalytic**, as it produces its own catalyst and therefore catalyzes itself.)

The following mechanism has been proposed for the reaction:

$$CH_3COCH_3 + OH^- \xrightarrow{k_2} CH_3COCH_2^- + H_2O \quad \text{(slow)}$$

$$H_2O + CH_3COCH_2^- + BrO^- \longrightarrow CH_3COCH_2Br + 2OH^- \quad \text{(fast)}$$

The sum of the two proposed steps gives the correct overall reaction (Eq. 16.40), and the rate law predicted from the mechanism is

$$\text{rate} = k_2[CH_3COCH_3][OH^-]$$

which has the same form as the experimental rate law. The rate constant for the slow step, k_2, must then be identical with the experimental rate constant k. Note that, although BrO^- is a reactant, the rate is independent of the concentration of BrO^- as it does not enter the reaction pathway until after the slow, rate-determining step.

Additional support for the proposed mechanism has been obtained by using hypochlorite ion (ClO^-) or hypoiodite ion (IO^-) in place of BrO^-. All three reactions have the same form of rate law and the same value of the experimental rate constant k.

Substances like $CH_3COCH_2^-$ are called **intermediates**; they form during one step of the reaction mechanism, then react in another step. In devising acceptable reaction mechanisms, one looks for chemically reasonable intermediates.

Pre-equilibrium Followed by a Slow Step

In this type of mechanism, the first step is fast in both directions, and thus an equilibrium is established for this step. (Typically the first step creates one or more intermediates.) The next step is relatively slow, and it determines the reaction rate.

As an example of this type of mechanism, consider the hypochlorite ion–iodide ion reaction.

Hypochlorite ion reacts rapidly with I^- in aqueous solutions to form hypoiodite ion and Cl^-.

$$ClO^- + I^- \longrightarrow IO^- + Cl^- \qquad (16.42)$$

The experimentally observed rate law is

$$\text{rate} = k \frac{[I^-][ClO^-]}{[OH^-]} \tag{16.43}$$

It is evident that the reaction cannot occur in the single bimolecular step,

$$ClO^- + I^- \longrightarrow IO^- + Cl^-$$

for, if it did, the rate law would be

$$\text{rate} = k'[ClO^-][I^-]$$

Therefore we must rule out a single-step process and investigate mechanisms consisting of more than one elementary reaction.

The following three-step mechanism is in agreement with the experimental rate law, as we shall show:

(1) $ClO^- + H_2O \rightleftharpoons HOCl + OH^-$ (rapid, reversible)

$$K_1 = \frac{[HOCl][OH^-]}{[ClO^-]}$$

(2) $I^- + HOCl \xrightarrow{k_2} HOI + Cl^-$ (slow)

(3) $OH^- + HOI \rightleftharpoons H_2O + IO^-$ (rapid)

The sum of the three reactions gives the overall reaction (Eq. 16.42). Because the second reaction is much slower than the other two, it will determine the overall rate of reaction. It is therefore the rate-determining step.

The rate of the second reaction is

$$\text{rate}(2) = k_2[I^-][HOCl]$$

for it is assumed to be an elementary reaction. The concentration of the intermediate HOCl may be calculated from the equilibrium constant for the first reaction:

$$[HOCl] = K_1 \frac{[ClO^-]}{[OH^-]}$$

If we substitute this value in the rate law for reaction 2 and note that the overall rate is equal to the rate of reaction 2, we obtain

$$\text{rate} = k_2 K_1 [I^-] \frac{[ClO^-]}{[OH^-]}$$

This rate law predicted from the mechanism has the same form as the experimental rate law, Eq. 16.43. Thus the mechanism is a possible way in which the reaction could occur. If this is the correct mechanism, then the experimental rate constant is the product of the rate constant for step 2 and the equilibrium constant for step 1: $k = k_2 K_1$.

Chain Reactions

Reactions whose mechanisms consist of repetitive steps involving highly reactive intermediates are known as **chain reactions.** A chain reaction is initiated by a relatively slow step that produces a highly reactive intermediate. In a subsequent step (or steps) this intermediate attacks a reactant molecule to form a product molecule and also to regenerate itself or another reactive intermediate. Because the intermediate is regenerated, it can attack reactant molecules over and over again. Eventually the intermediate is removed by some chain-terminating step.

A good example of a chain reaction is the hydrogen–chlorine reaction, $H_2 + Cl_2 \rightarrow 2HCl$, which occurs rapidly at high temperatures or under the influence of light. The mechanism of the reaction is

$$Cl_2 \longrightarrow 2Cl \qquad \text{chain initiation}$$

$$\left.\begin{array}{l} Cl + H_2 \longrightarrow HCl + H \\ H + Cl_2 \longrightarrow HCl + Cl \end{array}\right\} \text{chain propagation}$$

$$2Cl \longrightarrow Cl_2 \qquad \text{chain termination}$$

At high temperatures, or on absorbing ultraviolet light, chlorine molecules dissociate in the chain initiation step, forming chlorine atoms. These chlorine atoms are then involved in a pair of chain-propagating steps, the sum of which is the overall reaction. As the second chain-propagating step regenerates chlorine atoms, we have a chain which can repeat itself many times. (A chlorine atom from step 3 attacks another H_2 molecule in step 2, etc.) Eventually two chlorine atoms combine, terminating the chain.

An important application of chain reactions is **polymerization,** which is the basis of the plastics industry. Polymerization is usually initiated by dissociation of a substance such as an organic peroxide to form free radicals, which are highly reactive molecules having odd numbers of electrons. Each free radical then reacts with a monomer such as ethylene to form a larger free radical. As this free radical attacks additional monomer molecules it may grow to great length, and is then called a high polymer. Eventually a chain-terminating step stops the growth of the free radical. For the polymerization of ethylene, this sequence of reactions may be represented symbolically as

$$R_2 \longrightarrow 2R\cdot \qquad \text{chain initiation}$$

$$\left.\begin{array}{l} R\cdot + CH_2{=}CH_2 \longrightarrow RCH_2{-}CH_2\cdot \\ RCH_2{-}CH_2\cdot + CH_2{=}CH_2 \longrightarrow R(CH_2)_3{-}CH_2\cdot \\ R(CH_2)_3{-}CH_2\cdot + CH_2{=}CH_2 \longrightarrow R(CH_2)_5{-}CH_2\cdot \\ \qquad\qquad\qquad\qquad \text{etc.} \end{array}\right\} \text{chain propagation}$$

$$2R(CH_2)_n\text{—}CH_2\cdot \longrightarrow R(CH_2)_{2n+2}R$$
$$2R(CH_2)_n\text{—}CH_2\cdot \longrightarrow R(CH_2)_nCH_3 + R(CH_2)_{n-1}CH\text{=}CH_2$$

chain termination

In these equations a dot represents the odd electron of a free radical.

Supporting Evidence for Mechanisms

It should be emphasized that a mechanism is not proved to be the correct mechanism when the rate law derived from the mechanism is found to be consistent with the experimental rate law (though this is an important first step in establishing a mechanism). In some cases there may be two or more different mechanisms consistent with the experimental rate law. The following list includes some of the additional lines of evidence used to determine if a mechanism is correct.

1. Identification of intermediates. In some cases it is possible to obtain direct experimental evidence for the presence of a reaction intermediate, and even to measure the concentration of the intermediate during various stages of the reaction.
2. One can vary the chemical nature of the reactant molecules, and see how this affects the rate law and the reaction products. In some cases the same intermediate, or similar intermediates, may appear in many different reactions. For example, the $CH_3COCH_2^-$ intermediate mentioned earlier is a member of a general class of substance called "carbanions," which have wide importance in organic chemistry.
3. Isotope studies are frequently useful. For example, at high temperatures acetaldehyde decomposes, forming methane and carbon monoxide:

    $$CH_3CHO \longrightarrow CH_4 + CO$$

    Decomposition of mixtures of acetaldehyde and deuterated acetaldehyde, CD_3CDO (where D is deuterium, or *heavy hydrogen*), produces not only CH_4 and CD_4 but also the partially deuterated methanes CH_3D and CD_3H. If the mechanism were a simple one-step unimolecular decomposition, the only methanes formed would be CH_4 and CD_4. Therefore a more complex mechanism is indicated.

    A chain mechanism involving methyl radical, $CH_3\cdot$ or $CD_3\cdot$, is in agreement with these isotope studies and with the experimental rate law (half order in acetaldehyde). The steps that form the partially deuterated methanes are

    $$CH_3\cdot + CD_3CDO \longrightarrow CH_3D + CD_3CO\cdot$$
    $$CD_3\cdot + CH_3CHO \longrightarrow CD_3H + CH_3CO\cdot$$

4. Reactions involving enantiomers (*optical isomers*) often provide helpful mechanistic information. For example, if the reactant is optically active,

is the product also optically active? If so, how does its configuration compare with that of the reactant molecule?

5. In favorable cases one can study the individual steps of a reaction mechanism. This has become important with the development of new experimental techniques such as flash photolysis and molecular beams.

16.11 CATALYSTS AND ENZYMES

The feasibility of producing reasonable yields of products in a desired time period involves both the equilibrium and the rate behavior of the proposed reaction. As an example, let us consider the conversion of elemental nitrogen to nitrogen compounds, called **nitrogen fixation.** These compounds are widely used as fertilizers. The most important process for nitrogen fixation is the reaction of nitrogen and hydrogen to form ammonia:

$$N_2(g) + 3H_2(g) = 2NH_3(g)$$

The standard free energy of formation of NH_3 is -16.7 kJ mol^{-1} at 25°C. Thus the equilibrium favors the formation of ammonia, but under ordinary conditions the reaction is so slow that mixtures of N_2 and H_2 can be kept for years with no perceptible formation of ammonia. This is a typical case where one should look for a catalyst to increase the rate of attaining the equilibrium position. Indeed, the **Haber process,** the industrial process for producing ammonia, uses an iron catalyst, high temperatures (550–600°C), and high pressures (100–200 atm).

In contrast to these extreme conditions of temperature and pressure, certain bacteria in the root nodules of legumes produce a catalyst that causes the conversion of N_2 to NH_3 to occur at room temperature and atmospheric pressure. Such naturally occurring catalysts in biological systems are known as **enzymes,** and they are the most specific and efficient catalysts known for many reactions. Enzymes are proteins (see Section 23.1). They often contain transition metals such as iron, cobalt, molybdenum, and titanium. The enzymes that fix nitrogen are called nitrogenases, and they contain molybdenum and iron.

One of the methods used to investigate the behavior of enzymes as catalysts is to synthesize known chemical compounds of structure similar to the enzyme and to examine how structural changes in the model compound alter the rate of reaction. Recently, it has been shown that some titanium, molybdenum, and tungsten complexes catalyze the conversion of molecular nitrogen to ammonia at atmospheric pressure and room temperature. It is hoped that further investigations of such inorganic complexes, as well as further study of the structures and functions of the nitrogenases, will lead to a better understanding of nitrogen fixation, and perhaps to the development of new catalysts for this important process.

The primary function of a catalyst or enzyme in enhancing the rate of a chemical reaction is to provide a path of lower activation energy between reactants and products. For example, the uncatalyzed decomposition of hydrogen

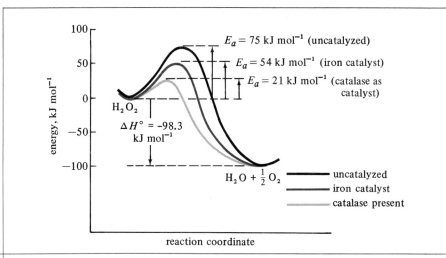

FIGURE 16.9 Energy changes accompanying the decomposition of H_2O_2 in the presence and absence of catalysts.

peroxide to water and oxygen,

$$H_2O_2 \longrightarrow H_2O + \tfrac{1}{2}O_2$$

proceeds by a pathway involving an activation energy of 75 kJ mol^{-1}. In the presence of an iron catalyst a second pathway is used with an activation energy of 54 kJ mol^{-1}, and in the presence of the liver enzyme, catalase, a third pathway is used with an activation energy less than 21 kJ mol^{-1}, as shown in Figure 16.9.

As noted previously, the activation energy has a marked effect on the rate constant because it appears in the exponential term of the Arrhenius equation, $k = Ae^{-E_a/RT}$. We can calculate the effect of changes in E_a on the rate constant if we assume that the preexponential factor A is a constant. Under such conditions at 25°C, we find that a decrease in the activation energy of 1 kJ would increase the rate constant by a factor of 1.50. For a decrease in activation energy of x kJ the rate constant would increase by a factor of 1.50^x.

It is important to note that the equilibrium position of a reaction is not altered by the presence of a catalyst. The catalyst merely provides an alternative pathway between reactants and products, and accelerates the rates of both the forward and the reverse reactions so that the equilibrium position is reached in a shorter time. It may also be shown that if a system is at equilibrium, then for each path, whether catalyzed or uncatalyzed, the rate of formation of products is equal to the rate at which products return to reactants.

Catalysis by enzymes is discussed in more detail in Section 23.5.

PROBLEMS

16.1 The concentration of a reactant A changes from 0.0400 to 0.0350 M in 20 min. What is the average rate of change of A in unit time? Give your answer using both the minute and the second as the unit of time. What additional information would be required before the average rate of the reaction could be determined?

16.2 From the data in Table 16.1, calculate the average rate of change in the concentration of N_2O_5 between 526 and 867 s. What would be the average rate of change in concentration of NO_2 during this same period?

16.3 For the following reaction in aqueous solution,
$$Cr_2O_7^{2-} + 14H^+ + 6Fe^{2+} = 2Cr^{3+} + 7H_2O + 6Fe^{3+}$$
express the rate of reaction in terms of the rate of change in concentration of $Cr_2O_7^{2-}$, H^+, Fe^{2+}, Cr^{3+}, and Fe^{3+}.

16.4 On the basis of the rate law given by Eq. 16.43 for the reaction $ClO^- + I^- \rightarrow IO^- + Cl^-$, what effect would each of the following changes have on the rate of reaction?
(a) The concentration of I^- is doubled.
(b) The concentration of OH^- is doubled.
(c) A solution containing I^-, ClO^-, and OH^- is diluted with an equal volume of water, decreasing each concentration by a factor of 2.

16.5 Acetone reacts with iodine in acidic solutions to give an iodinated product
$$CH_3COCH_3 + I_2 \longrightarrow CH_3COCH_2I + H^+ + I^-$$
The experimentally determined rate law for the reaction at 25°C is
rate = $k[CH_3COCH_3][H^+]$
where $k = 2.73 \times 10^{-5} M^{-1} s^{-1}$. Calculate the initial rate of the reaction in each of the following solutions at 25°C:

$[I_2]$ (M)	$[CH_3COCH_3]$ (M)	$[H^+]$ (M)	Initial rate ($M s^{-1}$)
(a) 0.010	0.100	0.010	?
(b) 0.010	0.100	0.100	?
(c) 0.010	0.010	0.010	?
(d) 0.100	0.100	0.010	?

(e) Calculate the concentrations of reactants and products and the rate of the reaction when half the iodine in solution (a) has reacted.

16.6 For the gas-phase reaction $2NO + 2H_2 = N_2 + 2H_2O$ the rate law is
$$\frac{d[N_2]}{dt} = k[NO]^2[H_2]$$
(a) How is the initial rate of this reaction changed when the initial [NO] is increased by a factor of 3 (keeping initial [H_2] constant)?

(b) How is the initial rate of this reaction changed when the initial $[H_2]$ is increased by a factor of 3 (keeping initial [NO] constant)?

(c) If the initial rate of change of NO concentration is
$$\frac{d[NO]}{dt} = -5.8 \times 10^{-6} M\,s^{-1}$$
what is the rate of change of N_2 concentration, $d[N_2]/dt$?

16.7 If concentrations are measured in molarities and time is measured in seconds, what are the units of the rate constant for
(a) a first-order reaction?
(b) a second-order reaction?
(c) a third-order reaction?

16.8 The following experimental data were obtained for the reaction
$$2NO + 2H_2 \longrightarrow N_2 + 2H_2O$$

$10^2 \times$ Initial [NO] (M)	$10^2 \times$ Initial $[H_2]$ (M)	$10^5 \times$ Initial rate ($M\,s^{-1}$)
0.60	0.40	0.58
1.20	0.40	2.3
1.20	0.80	4.6

(a) What is the order of the reaction with respect to NO? With respect to H_2?
(b) Write a rate law in agreement with the data and evaluate the rate constant.

16.9 The following experimental data were obtained for the reaction $A + 2B \rightarrow$ products.

Initial [A]	Initial [B]	Initial rate
0.20 M	0.30 M	0.006 M min^{-1}
0.40 M	0.30 M	0.012 M min^{-1}
0.40 M	0.60 M	0.024 M min^{-1}

(a) What is the order of the reaction with respect to A?
(b) What is the order of the reaction with respect to B?

16.10 The following reaction occurs slowly in aqueous solution:
$$S_2O_8^{2-} + 2I^- \longrightarrow 2SO_4^{2-} + I_2$$
(a) From the data given below, what is the rate law for this reaction?

Initial $[S_2O_8^{2-}]$ (M)	Initial $[I^-]$ (M)	Initial rate ($M\,s^{-1}$)
0.010	0.016	4.4×10^{-7}
0.010	0.0080	2.2×10^{-7}
0.0050	0.016	2.2×10^{-7}

(b) Calculate the rate constant.

PROBLEMS

16.11 Using the rate constant in Section 16.2 for the decomposition of N_2O_5 at 45°C, calculate
(a) the time required for the decomposition of 50% of a sample of N_2O_5 at this temperature
(b) the percentage decomposition of a sample of N_2O_5 at this temperature in exactly 1 h

16.12 A certain substance A decomposes by a first-order rate law, A → products. After 750 s has elapsed, 22% of a sample of A has decomposed. Calculate the first-order rate constant.

16.13 A certain substance A decomposes by the reaction 2A → B + C.
(a) If a graph of $\log[A]_0/[A]$ versus time is linear, what is the order of the reaction with respect to A?
(b) If a graph of $1/[A]$ versus time is linear, what is the order of the reaction with respect to A?

16.14 For the reaction $H^+ + OH^- \rightarrow H_2O$, the rate law is
rate = $k[H^+][OH^-]$
with $k = 1.3 \times 10^{11} M^{-1} s^{-1}$ at 25°C. What time would be required for 50% of the H^+ to react
(a) if at time zero a solution contains $1.0 \times 10^{-3} M$ HCl and $1.0 \times 10^{-3} M$ NaOH? (Note that $[H^+]$ remains equal to $[OH^-]$ as the reaction proceeds.)
(b) if at time zero a solution contains $1.0 \times 10^{-3} M$ HCl and $1.0 M$ NaOH? (Note that $[OH^-]$ does not change appreciably during the reaction.)

16.15 The rate law for the reaction 2A → B + C is

$$-\frac{1}{2}\frac{d[A]}{dt} = k[A]^2$$

When the initial concentration is $[A]_0 = 1.0 \times 10^{-4} M$, the time required for the disappearance of 50% of A is 26 s (that is, $[A] = 0.50 \times 10^{-4} M$ after 26 s).
(a) What is the length of time required for $[A]$ to change from $0.50 \times 10^{-4} M$ to $0.25 \times 10^{-4} M$?
(b) Calculate the rate constant k.

16.16 The rate law for the reaction A + B → C + D is rate = $k[A][B]$. When the initial concentrations are $[A]_0 = 1.0 \times 10^{-4} M$ and $[B]_0 = 0.10 M$, the time required for the disappearance of 50% of A is 38 s (that is, $[A] = 0.50 \times 10^{-4} M$ after 38 s).
(a) What is the length of time required for $[A]$ to change from $0.50 \times 10^{-4} M$ to $0.25 \times 10^{-4} M$?

(b) If the initial concentration of B is changed to $[B]_0 = 0.20M$, calculate the time required for the disappearance of 50% of A.
(c) Calculate the rate constant k.

16.17 Assume that the rate law in the forward direction in Problem 16.5 is also valid at equilibrium. Write a rate law for the reverse reaction that is consistent with the law of chemical equilibrium.

16.18 Aqueous solutions of nitrous acid (HNO_2) decompose to form nitric oxide (NO) and nitric acid:
$$3HNO_2 = 2NO + H^+ + NO_3^- + H_2O$$
The rate law for the forward reaction is
$$\text{forward rate} = \frac{k_f[HNO_2]^4}{[NO]^2}$$
The rate law for the reverse reaction is
reverse rate = $k_r[H^+][NO_3^-][HNO_2]$
Show that the equation obtained by equating the forward and reverse rates is identical to the law of chemical equilibrium for this reaction.

16.19 Calculate the activation energy of a reaction whose rate constant at 37°C is exactly double its rate constant at 27°C.

16.20 What is the activation energy for the decomposition of N_2O_5 if the rate constant is 3.1×10^{-4} s^{-1} at 318 K, and 3.0×10^{-3} s^{-1} at 338 K?

16.21 If the activation energy of a reaction is zero, what effect does temperature have on the rate constant for the reaction?

16.22 The reaction A + B → C + D is exothermic ($\Delta H° = -20.0$ kJ mol^{-1}), and its activation energy in the forward direction is 30 kJ mol^{-1}. What is the activation energy for the reverse reaction?

16.23 What is the minimum value of the activation energy in the forward direction if a reaction is endothermic by 15 kJ mol^{-1}?

16.24 The rate constant of a particular chemical reaction is $7.2 \times 10^{-3} M^{-1}$ s^{-1} at 25°C, and the activation energy is 45 kJ mol^{-1}.
(a) Calculate the rate constant at 50°C.
(b) Calculate the preexponential factor A.

16.25 When the temperature changes from 25°C to 35°C, the rate constant of a particular chemical reaction increases by a factor of 3. Which of the following also increases by a factor of 3? (There may be more than one correct answer.)

(a) the factor $e^{-E_a/RT}$
(b) the preexponential factor A
(c) the steric factor
(d) the average molecular speed
(e) the average translational kinetic energy of the molecules
(f) the number of molecular collisions per cubic meter per second
(g) the proportion of molecular collisions having at least the minimum energy, E_a, required for reaction

16.26 Sketch the molecules NO and O_3 in a collision that has
(a) favorable orientation for the reaction $NO + O_3 \rightarrow NO_2 + O_2$
(b) unfavorable orientation for this same reaction
In (a) indicate where a bond is broken and where a bond is formed. (NO_2 and O_3 are symmetrical bent molecules.)

16.27 How does an increase in temperature affect the various terms in the collision-theory rate constant (Eq. 16.33)?

16.28 The reaction
$$2HCrO_4^- + 3HSO_3^- + 5H^+ = 2Cr^{3+} + 3SO_4^{2-} + 5H_2O$$
follows the rate law
$$\text{rate} = k[HCrO_4^-][HSO_3^-]^2[H^+]$$
Does the mechanism for the reaction consist of one step or more than one step?

16.29 When substances A and B are mixed, a reaction occurs with the following mechanism:
$$A + B \longrightarrow C \quad \text{(fast)}$$
$$B + C \longrightarrow D + E + A \quad \text{(slow)}$$
(a) What is the stoichiometric equation for the net reaction?
(b) Which substance is a catalyst for this reaction?
(c) Which substance is an intermediate in this reaction?

16.30 For the gas-phase reaction $2NO_2 + O_3 \rightarrow N_2O_5 + O_2$
$$\text{rate} = k[NO_2][O_3]$$
Devise a two-step mechanism consistent with this rate law. For an intermediate, use NO_3. State which step is fast, and which step is slow.

16.31 The reaction $A + B \rightarrow C + D$ is catalyzed by a substance X. If the mechanism consists of the following two steps,
$$A + X \longrightarrow C + Y \quad \text{(slow)}$$
$$B + Y \longrightarrow D + X \quad \text{(rapid)}$$
what is the rate law for the catalyzed reaction?

16.32 The gas-phase decomposition of phosgene, $COCl_2 \rightarrow CO + Cl_2$, is catalyzed by chlorine. Assuming that the mechanism is
$$Cl_2 \rightleftharpoons 2Cl \quad \text{(rapid, reversible equilibrium)}$$

$$COCl_2 + Cl \xrightarrow{k'} COCl + Cl_2 \quad \text{(slow, rate-determining step)}$$
$$COCl \longrightarrow CO + Cl \quad \text{(fast)}$$

we can write
$$K = \frac{[Cl]^2}{[Cl_2]}$$
and
$$\text{rate} = -\frac{d[COCl_2]}{dt} = k'[COCl_2][Cl]$$

(a) If the experimental rate law is
$$\text{rate} = -\frac{d[COCl_2]}{dt} = k_{obs}[COCl_2]^a[Cl_2]^b$$
what are the orders a and b?

(b) Write an equation for the observed rate constant, k_{obs}, in terms of k' and K.

16.33 In aqueous solutions methyl acetate, MA, hydrolyzes to give methyl alcohol and acetic acid. The reaction is catalyzed by hydrogen ion.
$$CH_3CO_2CH_3 + H_2O = CH_3CO_2H + CH_3OH$$
A $0.701 M$ solution of methyl acetate containing $0.100 M$ HCl was prepared and the following data were obtained:

Time (min)	[MA]	Time (min)	[MA]
0	$0.701 M$	620	$0.470 M$
200	$0.617 M$	1515	$0.271 M$
280	$0.584 M$	1705	$0.243 M$
445	$0.529 M$		

The reverse reaction can be neglected.
(a) Plot the data assuming that the reaction is first order in methyl acetate (see Figure 16.2).
(b) Plot the data assuming that the reaction is second order in methyl acetate (see Figure 16.3).
(c) Which plot gives the best linear relationship? Is the reaction first order or second order in methyl acetate?
(d) Evaluate the rate constant from the slope of the linear plot.

16.34 (a) For the gas-phase reaction
$$NO_2 + CO \longrightarrow NO + CO_2$$
the experimental value of the activation energy is 132 kJ mol^{-1}, and the experimental preexponential factor is $1.2 \times 10^{10} M^{-1} s^{-1}$. This reaction is believed to occur in a single step. Using a temperature of 600 K and an effective radius, $r_A + r_B$, of 1.7×10^{-10} m, calculate the steric factor p.
(b) For this temperature, calculate the fraction of collisions, f_e, that are effective in forming products.

CHAPTER 17
THE SOLID STATE

17

The solid state is distinguished by rigidity, or high resistance to flow, which finds use in various structural materials. Solids have other interesting properties that will be considered below in some detail. Because of these properties, solids have many practical applications, but we shall be concerned here mainly with studies of the arrangements of atoms, ions, and molecules in crystalline solids, which have led to greatly increased understanding of chemical bonding and intermolecular forces.

17.1 MACROSCOPIC PROPERTIES OF SOLIDS

Crystals

Most pure substances in the solid state have characteristic crystalline shapes bounded by plane faces and sharp edges (Figure 17.1). For each crystalline substance the various faces are formed at certain characteristic angles. By measuring these angles we can classify the crystal in one of seven crystal systems, described in the next section.

Many substances occur naturally in crystalline form. Crystals are usually prepared in the laboratory from a liquid by cooling or from a solution by cooling or by evaporating the solvent. In general, larger crystals are formed when the crystals are grown slowly. Gems, such as diamonds, rubies, and amethysts, are large crystals with appealing optical properties. Solids in a microcrystalline or powder form reveal their crystalline nature only under a microscope.

The volume of a crystal varies with pressure and temperature, but to a very much smaller extent than the volume of a gas. In general, crystal dimensions change more in one direction than another with a change in temperature or pressure. The variation in properties with direction is called **anisotropy.**

Prior to the discovery of methods for determining the internal structure of crystals, it was inferred from the shapes and anisotropy of crystals that the atoms, ions, or molecules must be arranged in ordered patterns. Some remarkably accurate guesses of the internal structures of certain crystals were based on their external form. In recent years, electron microscope pictures of crystals formd by very large molecules have provided direct evidence that the development of crystal faces results from an ordered array of the molecules within the crystal (Figure 17.2). X-ray diffraction (Section 17.3) is the standard technique for determining the internal structure of crystals.

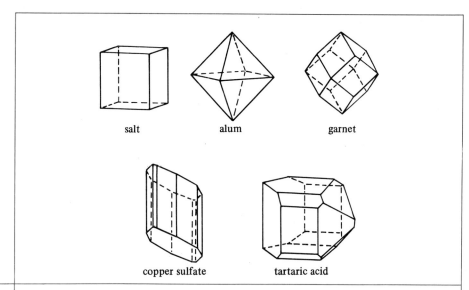

FIGURE 17.1 Crystals of some common substances. (SOURCE: *A. F. Wells,* The Third Dimension in Chemistry, *The Clarendon Press, Oxford, 1956, p. 63.*)

The heat capacities of the solid elements obey the law of Petit and Dulong, and for solid compounds Kopp's law is obeyed (Section 5.4). Both laws are consequences of the effect of temperature in increasing the vibrational motion of individual atoms or ions.

At sufficiently high temperatures the vibrational motion of the atoms, ions, or molecules overcomes the forces holding them in an ordered array, and the crystal melts. For a pure crystalline substance the change to the liquid state occurs at a definite temperature (the melting point). A solid can also vaporize at temperatures below the melting point, a process called **sublimation.** For most solids the vapor pressure is quite low, even at the melting point, but in some cases, as with solid carbon dioxide (dry ice), the equilibrium vapor pressure exceeds 1 atm before melting occurs, and the liquid is not formed except at higher pressures. Sublimation is sometimes used to separate a substance from impurities.

Amorphous Solids

Amorphous solids, or **glasses,** are characterized by disordered atomic or molecular arrangements, akin to those of liquids. In fact, they may be regarded as supercooled liquids, at temperatures so low that their resistance to flow is like that of a solid. On heating, an amorphous solid changes to a liquid gradually over a wide range of temperature. This behavior differs markedly from the abrupt melting of true crystals. It is this characteristic that makes possible the art of glassblowing. Many important solid materials, such as plastics, fibers, rubber, and various kinds of glass, are amorphous.

FIGURE 17.2

Photograph of a very small crystal of tobacco necrosis virus, taken by the electron microscope. Individual virus molecules, which are about 250 Å in diameter, are clearly visible in regular arrays. Magnification 84,000. [SOURCE: *L. W. Labaw and R. W. G. Wyckoff,* Journal of Ultrastructure Research, **2,** *9 (1958)*.]

External features are an uncertain guide to determining the amorphous or crystalline nature of a solid. Glass may be cut to a crystalline shape, and the edges of a crystal may be removed by grinding. X-ray diffraction (Section 17.3) is the best method for determining the degree of order in a solid.

Amorphous solids slowly change to crystalline form, a process called **devitrification.** For example, when elemental sulfur is melted and heated to above 160°C, it changes rather abruptly from a light yellow, mobile liquid to a dark brown, viscous liquid, as the S_8 rings break open and link up to form long chains. When this liquid is quickly chilled, a rubbery solid, called plastic sulfur, is formed. On standing for a few days the amorphous solid becomes brittle as the sulfur reverts to its more stable crystalline form.

By comparison, devitrification of glass is extremely slow. Glass utensils several thousand years old are known, dating from the early Egyptian and Babylonian civilizations. Obsidian, a naturally occurring glass from volcanoes, usually black and shiny, was widely used by the early inhabitants of North America for arrowheads, tools, and ornaments. Because of devitrification, glass older than the Tertiary period (ca. 7×10^7 years ago) is rare.

17.2 THE SEVEN SYSTEMS OF CRYSTALS

Both the external features of a crystal (the crystalline faces and the angles between them) and its internal structure (the arrangement of the atoms, ions, or molecules) are most easily expressed in terms of the **unit cell.** This is the fundamental building block of the crystal: The crystal is composed of a succession of identical unit cells extending in each direction. Each unit cell is composed of atoms, ions, or molecules arranged in a particular geometrical pattern. The basic pattern of the unit cell is repeated throughout the crystal. Thus the shape of the unit cell determines the shape of the crystal. Conversely, by measuring the angles between crystalline faces, one can determine the form of the unit cell.

Only seven distinct crystalline forms are possible, and they are shown in Figure 17.3. The most symmetrical form is the cubic system, based on a unit cell that is a cube. At the corner of any unit cell three edges meet, and if their lengths are designated a, b, and c, respectively, then in the cubic system $a = b = c$. The angles between the edges of a unit cell are designated α, β, and γ, such that α is the angle formed by sides b and c, β is the angle formed by a and c, and γ is the angle formed by a and b. In the cubic system the angles are all right angles ($\alpha = \beta = \gamma = 90°$).

The other systems are less symmetrical. For example, in the tetragonal system the angles are all right angles, but one edge (designated c) is either shorter or longer than the others ($a = b \neq c$).

Interfacial angles suffice to determine the shape of the unit cell and hence the crystal system, but in general the determination of the unit cell dimensions a, b, and c requires study of the X-ray diffraction pattern of the crystal.

17.3 X-RAY DIFFRACTION

One of the most direct lines of evidence for the wave nature of light is the interference or diffraction pattern formed when light passes through a set of holes or slits (see Section 7.1). A condition for obtaining a diffraction pattern is that the wavelength of the light must be of the same order of magnitude as the distance between slits. X rays are light rays of very short wavelength (ca. 1×10^{-8} to 1×10^{-9} cm), about the same as the distances between layers of atoms in crystals. Therefore the scattering of X rays by a crystal produces a diffraction pattern. This was first demonstrated in 1912 by Walter Friedrich and Paul Knipping, following a suggestion by Max von Laue. During the next few years this phenomenon was used to determine the structures of many crystals by the Braggs, William Lawrence Bragg and his father, William Henry Bragg.

To see how a crystal forms a diffraction pattern, consider the scattering of a beam of X rays striking a crystal at an angle θ (Figure 17.4). If some of the X rays are reflected in the same way as light striking a mirror, they will leave the crystal at an angle θ. But the X rays reflected by different layers of atoms travel different distances. The extra distance for rays reflected from a lower layer is shown by the colored line in Figure 17.4.

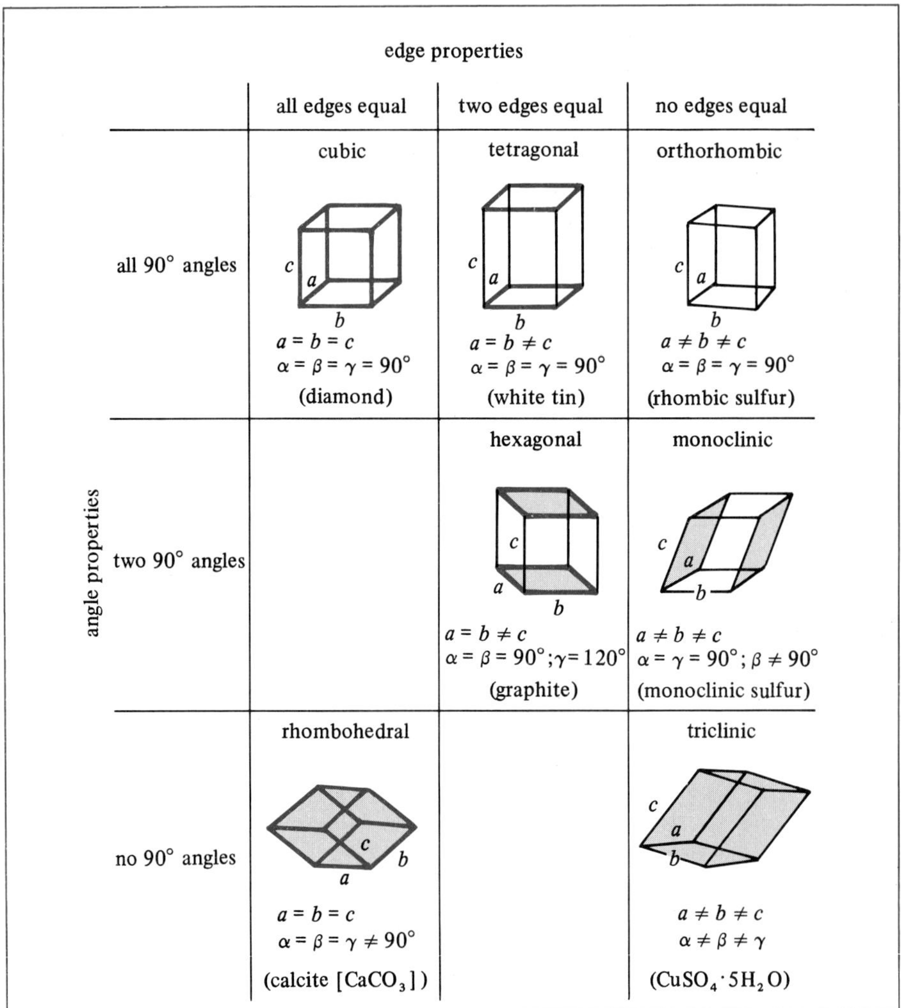

	edge properties		
angle properties	all edges equal	two edges equal	no edges equal
all 90° angles	cubic $a = b = c$ $\alpha = \beta = \gamma = 90°$ (diamond)	tetragonal $a = b \neq c$ $\alpha = \beta = \gamma = 90°$ (white tin)	orthorhombic $a \neq b \neq c$ $\alpha = \beta = \gamma = 90°$ (rhombic sulfur)
two 90° angles		hexagonal $a = b \neq c$ $\alpha = \beta = 90°; \gamma = 120°$ (graphite)	monoclinic $a \neq b \neq c$ $\alpha = \gamma = 90°; \beta \neq 90°$ (monoclinic sulfur)
no 90° angles	rhombohedral $a = b = c$ $\alpha = \beta = \gamma \neq 90°$ (calcite [$CaCO_3$])		triclinic $a \neq b \neq c$ $\alpha \neq \beta \neq \gamma$ ($CuSO_4 \cdot 5H_2O$)

FIGURE 17.3 Unit cells of the seven crystal systems. Colored edges indicate equal lengths. Colored planes indicate nonrectangular faces.

If θ is such that this extra distance is exactly one wavelength, then constructive interference of the different reflected rays occurs, and an intense beam of X rays is reflected at these angles. Constructive interference also occurs when the extra distance is 2λ, or 3λ, and so on. At other angles, destructive interference occurs. Thus the condition for constructive interference is $l = n\lambda$, where l is the extra distance traveled by the lower ray, n is an integer (1, 2, 3, . . .) called the **order** of the reflection, and λ is the wavelength of the X rays. By trigonometry $l = 2d \sin \theta$, where d is the distance between layers, and $\sin \theta$ is the sine of angle θ.

17.3 X-RAY DIFFRACTION

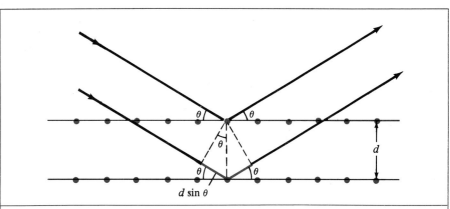

FIGURE 17.4

Reflection of X rays by a crystal. Dots represent atoms. The X rays strike each layer of atoms at an angle θ. The distance between layers is d. Extra distance traveled by lower X ray (color segment) is $l = 2d \sin \theta$.

Thus the condition for reflection is

$$n\lambda = 2d \sin \theta$$

This is known as the Bragg equation. The apparatus used by the Braggs to determine the reflection angles is shown in Figure 17.5.

As an example of the use of this equation, consider the reflection by a sodium chloride crystal of X rays of wavelength $\lambda = 0.586$ Å from a palladium target (Figure 17.6). The smallest angle that gives an intense reflection is $\theta = 5°58'$, for which $\sin \theta = 0.104$. This must be the first-order reflection ($n = 1$). Therefore $d = \lambda/(2 \sin \theta) = (0.586 \text{ Å})/2(0.104) = 2.82$ Å. The second-order reflection occurs at the angle whose sine is twice as great ($\sin \theta = 0.208$, for which $\theta = 12°00'$), and so on.

By studying the diffraction patterns of X rays striking the crystal from various directions, we can determine the detailed structure of the crystal. An apparatus that does this rather simply and directly is the two-wavelength microscope invented by Martin J. Buerger. An X-ray diffraction pattern is first obtained. Then holes are drilled in a metal plate at the same positions as the intense reflections of the X-ray diffraction pattern. Finally, a second diffraction pattern is obtained by passing a beam of visible light through the holes in the metal plate, which acts like the pinhole arrangement described in Section 7.1. This second diffraction pattern represents the original crystal, magnified by a factor of about 10^6. Thus one can "photograph" the individual atoms in a crystal. (The first and second diffraction patterns are somewhat analogous to the "negative" and "positive" in ordinary photography.) An example of this method using a crystal of marcasite (one of the forms of FeS_2) is shown in Figure 17.7. Note that it shows not only the positions of

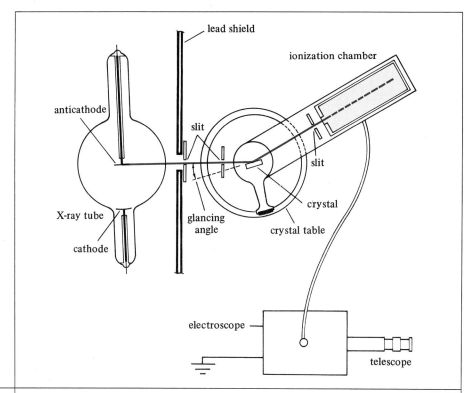

FIGURE 17.5 The original X-ray spectrometer designed by W. H. Bragg. "A collimated beam from the X-ray tube fell on the face of the crystal and was reflected through slits into the recording ionization chamber, which was filled with a heavy gas (methyl bromide) to increase ionization. The outer case was at a potential of several hundred volts, and the ionization was measured by driving the charge onto a coaxial wire connected to a tilted gold-leaf electroscope. It was with this instrument that my father made his pioneer investigations on the X-ray spectra from anticathodes of a number of different metals, a project that formed the basis for H. G. J. Moseley's subsequent work on atomic number; the early determinations of crystal structure, for which I was mainly responsible, were also made with this instrument." (*From "X-Ray Crystallography" by Sir Lawrence Bragg. Copyright © 1968 by Scientific American, Inc. All rights reserved.*)

the various atoms but also the electron density distribution because it is the electrons in the crystal that scatter the X rays.

Digital computer methods are now commonly used to determine crystal structures from the observed X-ray diffraction patterns. Internuclear distances can be determined to an accuracy of the order of ± 0.001 Å. Electron density maps are becoming increasingly useful in understanding the electronic structures of molecules and crystals.

17.3 X-RAY DIFFRACTION

FIGURE 17.6

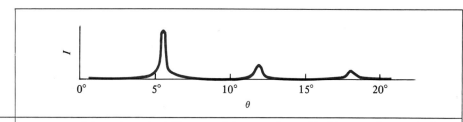

Reflected intensity (I) of X rays striking a sodium chloride crystal at angle θ.

FIGURE 17.7

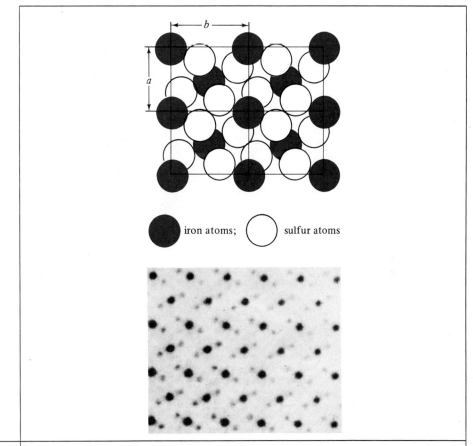

Bottom: Photograph of the atoms in marcasite (FeS$_2$), an orthorhombic crystal, taken by the two-wavelength microscope looking along the crystallographic c axis (magnification 2.8 million). The larger dark circular areas are iron atoms (26 electrons each); the fainter darkened areas are sulfur atoms (16 electrons each). Top: Drawing of the crystal structure of marcasite showing four unit cells. [SOURCE: M. J. Buerger, Journal of Applied Physics, **21**, *916* (*1950*).]

17.4 METALLIC CRYSTAL STRUCTURES

The metals all have crystal structures characterized by high coordination numbers. In two of the three most common structures the coordination number is 12; that is, each atom is surrounded by 12 other atoms, all the same distance away. The atoms are arranged in the most efficient ways of packing spheres into the

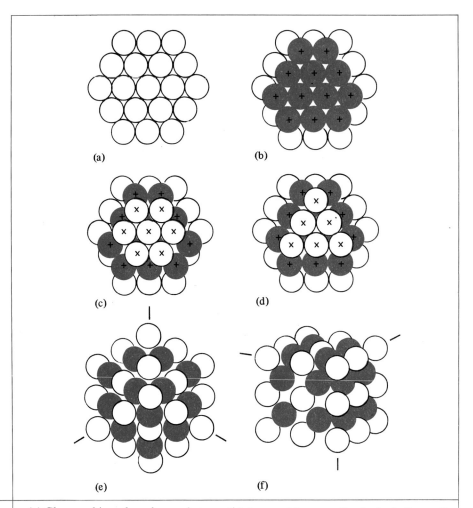

FIGURE 17.8 (a) Close packing of one layer of atoms. (b) A second layer, resting in the hollows of the first. (c) Three layers, forming a hexagonal close-packed structure (third layer directly above the first). (d) Third layer shifted to an alternate set of hollows in the second layer, giving a different structure (cubic close-packed). (e) The same as structure (d), with more atoms added. (f) Another view of structure (e), to bring out the cubic character of the structure. (SOURCE: C. Bunn, *Crystals: Their Role in Nature and in Science*, Academic Press, New York, 1964, p. 52.)

17.4 METALLIC CRYSTAL STRUCTURES

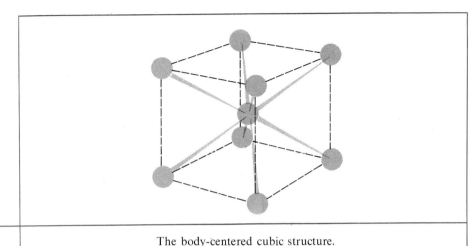

FIGURE 17.9 The body-centered cubic structure.

smallest possible space. One crystal structure of coordination number 12 is called **hexagonal close-packed,** the other **cubic close-packed.**

In either case the spheres are tightly packed into layers, as shown in Figure 17.8(a), an arrangement familiar in the game of pocket billiards. Note that, except for the spheres on the outside, each sphere is surrounded by six others in a hexagonal arrangement. When a second layer is packed on top of the first, as shown in Figure 17.8(b), each sphere nestles in a depression formed by three spheres in the bottom layer. If a third layer is added directly over the first, as in Figure 17.8(c), the structure is hexagonal close-packed. Note that a sphere within the second layer [marked with a plus (+) in these figures] is touching 12 others: 3 in the bottom layer, 3 in the top layer, and 6 in the same layer. Additional layers can be added in the same pattern, with the fourth layer directly over the second, the fifth over the first and third, and so on. Beryllium, magnesium, and many other metals have this arrangement of atoms. The crystal belongs to the hexagonal system, hence the designation hexagonal close-packed.

A slightly different arrangement is obtained if the third layer is placed in a different set of depressions, as shown in Figure 17.8(d), so that it is *not* directly over the first layer. (Each sphere in the interior of the second layer is still touching 12 others, but the 3 above are not directly over the 3 below.) Additional layers can then be added to repeat this sequence, with the fourth directly over the first, the fifth over the second, the sixth over the third, and so on. This arrangement of atoms is found in aluminum, copper, silver, gold, and many other metals. The crystal belongs to the cubic system, hence the designation cubic close-packed. The unit cell is a cube with one atom at each corner, and also one atom in the middle of each face of the cube. The terms "cubic close-packed" and "face-centered cubic" are synonymous.

The third common metal structure (which is *not* a close-packed structure) is body-centered cubic (Figure 17.9). The coordination number is 8, each atom being surrounded by eight others at the corners of a cube. In addition, there are six other atoms only slightly (15%) farther away, one at the center of the cube adjoining each face. Among the many metals with this arrangement of atoms are the group IA metals.

Under different conditions of temperature and pressure, some metals are found to change from one of these structures to another. Calcium and strontium are known in all three forms. For a given metal the internuclear distance between neighboring atoms is found to be about 3% shorter in the body-centered cubic structure than in a close-packed structure.

17.5 THE METALLIC BOND

The elements of groups IA and IB form both diatomic and monatomic molecules in the gas phase. As each atom of these elements has only one valence electron, the diatomic molecules have covalent single bonds. One may, therefore, compare bonding in metallic crystals with covalent bonding by comparing the properties of these elements in the solid state and in the corresponding diatomic molecules.

Table 17.1 compares the energy required to convert diatomic molecules and crystals to atoms. For the solid state the appropriate quantity is the heat of sublimation to gaseous atoms; for the diatomic molecules the appropriate quantity is the dissociation energy. The data show that there is a general correspondence between the two properties: the higher the dissociation energy, the higher the heat of sublimation. Thus there must be a close connection between the covalent bond and the metallic bond.

The similarity between covalent and metallic bonding is emphasized in the resonating-valence-bond theory of metals, proposed by Linus Pauling. (This theory is an extension of the valence bond theory described in Section 9.18.) Each atom in a crystal of potassium, for example, is considered to use its valence elec-

TABLE 17.1 **PROPERTIES OF GROUP IA AND IB ELEMENTS**

Element[a]	Dissociation energy of diatomic molecule (kJ mol^{-1})	Heat of sublimation of metal to gaseous atoms [kJ (mol of atoms)$^{-1}$]
Cesium	38.0	78.2
Rubidium	47	82.0
Potassium	49.6	89.5
Sodium	69.5	108
Lithium	100.9	161
Silver	160	286
Copper	196	339
Gold	222	354

[a] Elements are arranged in order of increasing dissociation energy.

tron to form a covalent bond with one of its eight neighbors. A schematic representation in two dimensions is

```
      |
—K    K    K
      
K—K        K
      |
—K    K—K
```

One must also consider resonance structures in which the central atom is bonded to one of its other neighbors, such as

```
      |
K          K—K
      
K     K    K—
      |
—K    K    K
                |
```

The net result is that each atom is bonded to a particular one of its eight neighboring atoms in only one-eighth of the resonance structures, so that in a formal sense each bond is only one-eighth of a single bond. (Of course, resonance stabilization causes each bond to be somewhat stronger than if the wave function corresponded to only a single valence bond structure.) This conclusion is in keeping with the rather long internuclear distance in the metal: 4.54 Å for potassium metal as compared to only 3.93 Å in the diatomic molecule, K_2.

Somewhat different views of the behavior of electrons in metals are discussed in Sections 17.11 and 17.12.

17.6 IONIC CRYSTAL STRUCTURES

The properties of ions and ionic crystals were discussed briefly in Chapter 9. It was shown there that it is a fairly good approximation to treat ions as hard spheres, and the same point of view will be taken here. The radii of some ions are shown in Table 9.1 and Figure 9.3.

The coulombic force of attraction between opposite electric charges causes ionic compounds to form crystals in which each positive ion is surrounded by negative ions, and each negative ion is surrounded by positive ions. Examples are the NaCl structure (Figure 9.2) and the CsCl structure (Figure 17.10), with coordination numbers 6 and 8, respectively. In NaCl, each ion is surrounded by six ions of opposite charge, located at the corners of an octahedron; in CsCl each ion is surrounded by eight ions of opposite charge, located at the corners of a cube.

Why do these two alkali metal halides crystallize in different structures? The energy of attraction is larger when each ion is surrounded by more ions of opposite charge, so that this factor should favor the structure with coordination number 8. And yet the only group IA halides having this structure at ordinary temperature and pressure are CsCl, CsBr, and CsI. (The rubidium salts ordinarily have

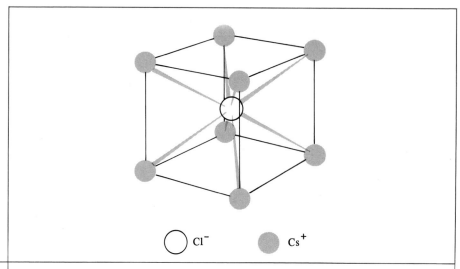

FIGURE 17.10

The structure of cesium chloride. The Cl^- ion at the center is surrounded by eight Cs^+ ions at the corners of the cube, and vice versa. (The eight Cl^- ions surrounding a Cs^+ ion are at the centers of the eight cubes adjoining each corner, only one of which is shown.)

the NaCl structure, but, under very high pressures, RbCl, RbBr, and RbI change to the CsCl structure.)

Clues to solving this puzzle are (1) cations are generally smaller than anions (see Figure 9.3), and (2) only the relatively large cations, Rb^+ and Cs^+, form crystals with coordination number 8. When the cation is very small, then anion–anion contact results, as shown in Figure 17.11. The anion–anion repulsion then outweighs the extra attraction energy of the higher coordination number. By trigonometric calculation it may be shown that, in the CsCl structure, anion–anion contact occurs whenever the ratio of the cation radius to the anion radius is less than 0.732, that is,

$$R_+/R_- < 0.732$$

In the NaCl structure, anion–anion contact occurs only at a much smaller ratio, namely,

$$R_+/R_- < 0.414$$

Thus anion–anion contact explains qualitatively why only the larger cations crystallize with coordination number 8. To obtain a quantitative theory of the alkali metal halides, one must use a soft-sphere model, including the van der Waals repulsion between adjacent ions.

17.6 IONIC CRYSTAL STRUCTURES

FIGURE 17.11 Anion–anion contact.

The radius ratio is also important in the crystal structures of compounds of the group IIA metals with the group VIA nonmetals (the alkaline earth chalcogenides). The chalcogenide ions have about the same radii as the neighboring halide ions (e.g., O^{2-}, 1.40 Å; F^-, 1.36 Å). But the alkaline earth ions are much smaller than the neighboring alkali metal ions (e.g., Mg^{2+}, 0.65 Å; Na^+, 0.95 Å). Consequently, the radius ratio in the alkaline earth chalcogenides is so small that the CsCl structure does not occur. Most of these compounds (e.g., MgO and MgS) have the NaCl structure. And in those cases where the radius ratio is smallest, the substances crystallize with a coordination number of only 4. (Anion–anion contact does not occur for coordination number 4 unless the radius ratio is less than 0.225.)

There are two slightly different ionic crystal structures with coordination number 4, and they are named after the two zinc sulfide minerals, sphalerite and wurtzite. In sphalerite (also known as zinc blende) the larger S^{2-} ions are arranged in cubic close-packing, with the smaller Zn^{2+} ions in the interstices (cavities) between the sulfide ions. The resulting structure resembles that of diamond, with alternating Zn^{2+} and S^{2-} ions (Figure 17.12). Wurtzite has a similar structure; the only difference is that the S^{2-} ions are arranged in hexagonal close-packing (Figure 17.13). In both structures, each ion is surrounded by four ions of opposite charge, located at the corners of a tetrahedron. Other examples include BeO and MgTe with the wurtzite structure, and BeS, BeSe, and BeTe with the sphalerite structure.

All of the examples cited above are simple binary compounds, with one anion per cation. In ionic crystals of formula MX_2, such as CaF_2, there are two anions per cation, with the result that the coordination number of the cation must be twice the coordination number of the anion. The sets of coordination numbers

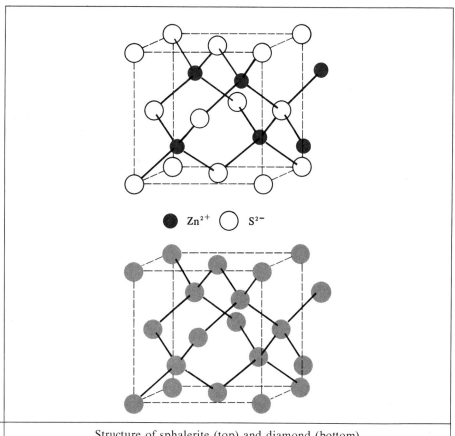

FIGURE 17.12 Structure of sphalerite (top) and diamond (bottom).

commonly found are 2:1, 4:2, 6:3, and 8:4. The first of these necessarily describes a crystal consisting of discrete molecules, such as the CO_2 crystal. (These molecular crystals are described in Section 17.9.) Coordination numbers 4 and 2 are found in the crystalline forms of SiO_2. In one of these forms (cristobalite) the silicon ions have the diamond arrangement, with an oxide ion inserted halfway between each pair of adjacent silicons (Figure 17.14). Thus, each silicon is surrounded by four oxygens, located at the corners of a tetrahedron, and each oxygen is intermediate between two silicons; hence the coordination numbers of silicon and oxygen are 4 and 2, respectively. The resulting structure may be viewed as a set of SiO_4 tetrahedra, with each tetrahedron connected to another at each corner. (Instead of viewing SiO_2 as an ionic crystal, made up of Si^{4+} ions and O^{2-} ions, one can view it as a gigantic covalent molecule. Each silicon atom forms four bonds, and each oxygen atom forms two bonds, conforming to the $8 - N$ rule. Of course, the bonds are actually polar bonds, intermediate between purely ionic and purely covalent bonds, as discussed in Section 9.11.)

FIGURE 17.13

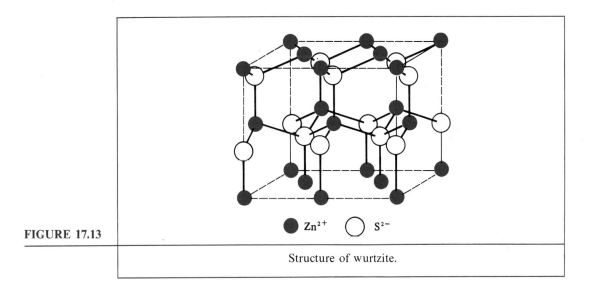

Zn²⁺ S²⁻

Structure of wurtzite.

The other crystalline forms of SiO_2, quartz and tridymite, have similar structures, but the tetrahedra are interconnected in different patterns. In quartz, they are connected in helices, which may be either left-handed or right-handed, so that there are two kinds of quartz crystals, each a mirror image of the other (Figure 17.15). Many minerals contain SiO_4 and AlO_4 tetrahedra or AlO_6 octahedra linked together; an example is sodalite, $Na_4Al_3Si_3O_{12}Cl$ (Figure 17.16).

FIGURE 17.14

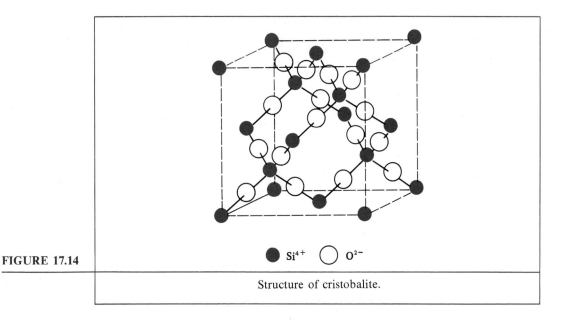

Si^{4+} O^{2-}

Structure of cristobalite.

FIGURE 17.15

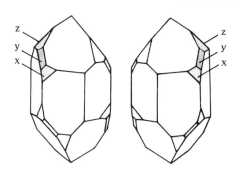

Left- and right-handed quartz crystals. Some of the small faces (facets), such as x, y, and z, are inclined in opposite directions so that each crystal is a mirror image of the other. (SOURCE: *C. Bunn,* Crystals: Their Role in Nature and in Science, *Academic Press, New York, 1964, p. 188.*)

FIGURE 17.16

A model of sodalite, $Na_4Al_3Si_3O_{12}Cl$. Each SiO_4 tetrahedron is attached at its corners through shared oxygens to four adjacent AlO_4 tetrahedra. Large spheres represent chloride ions; sodium ions are not shown. (SOURCE: *From Linus Pauling:* The Nature of the Chemical Bond, *Third Edition.* © *1960 by Cornell University. Used by permission of the publisher, Cornell University Press.*)

17.6 IONIC CRYSTAL STRUCTURES

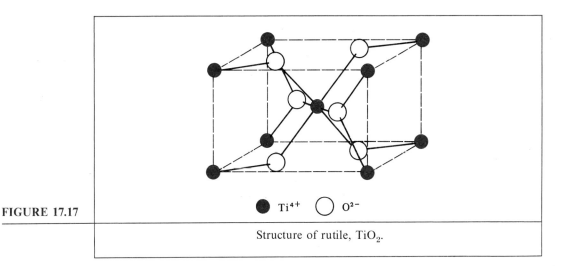

FIGURE 17.17

Structure of rutile, TiO_2.

● Ti^{4+} ○ O^{2-}

Coordination numbers of 6 and 3 are found in crystals of TiO_2, rutile (Figure 17.17). (Note the progressive increase in coordination number for the group IVA dioxides, CO_2, SiO_2, and TiO_2, as the radius ratio increases.) Each titanium is surrounded by six oxygens, located at the corners of an octahedron, and each octahedron is linked at each corner to two other octahedra, forming a three-dimensional framework structure, so that the coordination numbers of titanium and oxygen are 6 and 3, respectively.

Coordination numbers of 8 and 4 are found in crystals of fluorite, CaF_2 (Figure 17.18). Each Ca^{2+} ion is surrounded by eight F^- ions, located at the corners of a cube, and each F^- ion is surrounded by four Ca^{2+} ions, located at the corners of a tetrahedron.

Many metal dioxides and difluorides have the structures described above, with the radius ratio determining which structure is formed. For example, in the alkaline earth difluorides, BeF_2 has the cristobalite structure (4:2), MgF_2 the rutile structure (6:3), and CaF_2, SrF_2, and BaF_2 the fluorite structure (8:4). All of the alkali metal chalcogenides (e.g., Na_2O) crystallize with the antifluorite structure, in which the cations occupy the F^- positions and the anions the Ca^{2+} positions of fluorite.

Crystals of salts containing complex ions necessarily have somewhat more complicated structures, the internal structure of the complex ion being an additional matter for concern. In calcite, one of the crystalline forms of $CaCO_3$, the Ca^{2+} and CO_3^{2-} ions are arranged somewhat like the ions in NaCl, but the triangular CO_3^{2-} distorts the structure to produce a rhombohedral crystal (Figure 17.19); $NaNO_3$ has the same structure.

In some cases the complex ion is approximately spherical in shape, or it attains effectively spherical symmetry by rotation or random orientation. For example, the ammonium ion, NH_4^+, usually behaves like a spherical ion of about the same radius as Rb^+. The ammonium halides, NH_4Cl, NH_4Br, and NH_4I, all

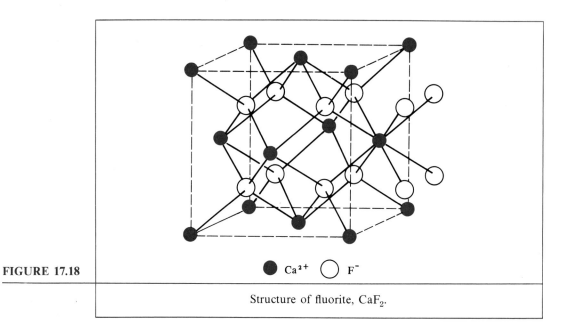

FIGURE 17.18 Structure of fluorite, CaF$_2$.

● Ca^{2+} ○ F$^-$

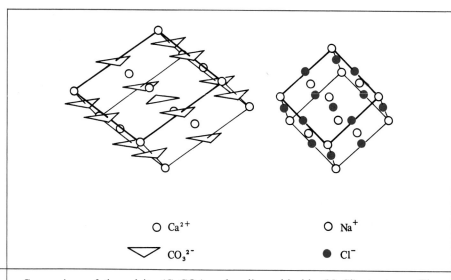

FIGURE 17.19 Comparison of the calcite (CaCO$_3$) and sodium chloride (NaCl) structures. The triangular carbonate ions are arranged parallel to one another. (SOURCE: *A. G. Ward*, The Nature of Crystals, *Blackie & Son Limited, Glasgow, 1945, p. 76.*)

○ Ca^{2+} ▽ CO$_3^{2-}$ ○ Na$^+$ ● Cl$^-$

have the NaCl structure; CsCN and CsSH have the CsCl structure, with CN⁻ or SH⁻ ions occupying the anion positions; $Co(NH_3)_6I_2$ has the fluorite structure (with $[Co(NH_3)_6]^{2+}$ ions in the cation positions); and K_2PtCl_6 has the antifluorite structure (with $[PtCl_6]^{2-}$ ions in the anion positions).

17.7 THE BORN-HABER CYCLE

An important property of an ionic crystal is the energy required to break the crystal apart into individual ions, called the **crystal lattice energy** or **cohesive energy**. It can be measured indirectly using a thermochemical cycle, the Born-Haber cycle, named after Max Born and Fritz Haber, who proposed it independently in 1919. For sodium chloride the Born-Haber cycle is

$$\begin{array}{ccc} NaCl(c) & \xrightarrow{A} & Na^+(g) + Cl^-(g) \\ D \uparrow & & \downarrow B \\ Na(c) + \tfrac{1}{2}Cl_2(g) & \xleftarrow{C} & Na(g) + Cl(g) \end{array}$$

The quantity of interest is the energy change for process A.[1] But when a system passes through any set of changes and is returned to its initial state, the sum of the energy changes must be zero, according to the first law of thermodynamics:

$$\Delta E_A + \Delta E_B + \Delta E_C + \Delta E_D = 0$$

The energy changes in steps B, C, and D can be readily calculated from the known ionization potential of sodium and the electron affinity of chlorine (step B), the sublimation energy of sodium metal and the dissociation energy of diatomic chlorine molecules (step C), and the standard energy of formation of crystalline sodium chloride (step D). These results can then be inserted in the equation above to obtain ΔE_A. The value obtained in this way for the crystal lattice energy of NaCl is 776.6 kJ mol⁻¹ at 25°C. Figure 17.20 shows how the crystal lattice energy varies with the distance between adjacent ions for alkali metal halides having the NaCl structure.

The hard-sphere approximation leads to the following equation for the energy required to dissociate a diatomic ionic molecule to its ions (Section 9.4):

$$\Delta E = \frac{14.40}{R} \text{ eV Å}$$

where R is the internuclear distance in ångströms. When the hard-sphere approximation is applied to an ionic crystal, the same equation must be used for the attraction of each cation to each anion throughout the crystal. In addition, the

---

[1] The earliest use of the Born-Haber cycle was the calculation of the electron affinity, at that time an unknown quantity, with the aid of estimates of the crystal lattice energy.

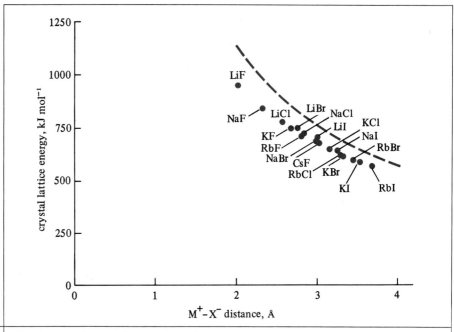

FIGURE 17.20 Lattice energies and internuclear distances of alkali metal halide crystals with the NaCl structure. The dashed line is based on the hard-sphere model.

repulsion between each pair of cations and the repulsion between each pair of anions must be included. The crystal lattice energy is then the sum of all such terms.

Although this is a formidable problem, it is considerably simplified by the ordered structure of the crystal, which allows one to calculate the distance between each pair of ions from the distance between two adjacent ions. When this is done, the crystal lattice energy, per mole of NaCl, is found to be given by an equation of the same form, but with an additional factor, M:

$$\Delta E = \frac{M(14.40)}{R} \text{ eV Å} \tag{17.1}$$

where R is the shortest anion–cation internuclear distance in ångströms.

The factor M is called the **Madelung constant** after E. Madelung, who solved the problem outlined above in 1918, clearing the way for the work of Born and Haber in the following year. For the NaCl crystal structure, M is 1.748; for the CsCl structure, M is 1.763.

By applying the preceding equation to crystalline NaCl, for which R is 2.82 Å, one obtains $\Delta E = (1.748)(14.40)/(2.82) = 8.93$ eV. As 1 eV = 96.48 kJ mol^{-1}, the calculated crystal lattice energy is thus (8.93) ×

(96.48) = 862 kJ mol^{-1}. This result is higher than the experimental value by 11%. More sophisticated calculations that take into account the van der Waals repulsive force between ions (called the **soft-sphere model**) generally agree with the experimental values for the alkali metal halides within about ±2%.

We may conclude that the hard-sphere model provides a satisfactory qualitative theory of ionic crystals, but a soft-sphere model is necessary to obtain precise results. The very good agreement between the soft-sphere results and the experimental values for the crystal lattice energy supports the interpretation of the solid alkali metal halides as ionic crystals.

17.8 COVALENT NETWORK CRYSTALS

Some crystals are held together by a network of covalent bonds, which extends from one side of the crystal to the other. The most common types of network are the two-dimensional layer structure (with the layers loosely stacked on top of one another) and the three-dimensional framework extending throughout the crystal. Graphite (Figure 11.3) and diamond (Figure 11.4), the two allotropic forms of carbon, are good examples of these two types. Other group IVA elements with the diamond structure are silicon, germanium, and gray tin.

The group VA elements arsenic, antimony, and bismuth form crystals having a hexagonal double layer or "puckered sheet" structure somewhat similar to graphite (Figure 17.21). Each atom is singly bonded to three atoms in the other half of the double layer, giving bond angles of 97° (As), 96° (Sb), and 94° (Bi), as compared to 120° in graphite.

The binary compounds formed by the elements of groups IIIA and VA have structures similar to those of the group IVA elements, although the IIIA–VA bonds are, of course, somewhat polar rather than purely covalent. Like the group IVA elements, several of these compounds are useful semiconductors. Boron nitride, BN, crystallizes ordinarily in a hexagonal graphitelike structure, with alternating boron and nitrogen atoms. It is white and slippery, and is called **white graphite**. A cubic form with the diamondlike sphalerite structure has been synthesized under conditions of very high pressure and temperature. It is even harder than diamond, which is the hardest naturally occurring substance. Many other group IIIA–VA compounds (e.g., AlP and GaAs) have the sphalerite structure. The nitrides of aluminum, gallium, and indium have the closely rated wurtzite structure. In contrast, the group IIIB–VA compounds ScN, YN, and LaN have the NaCl structure.

17.9 MOLECULAR CRYSTALS

A molecular crystal consists of individual molecules packed together in an orderly array. The forces holding the molecules in this array are the relatively weak forces of the van der Waals type (see Section 4.6). Because these forces are weak, molecular crystals generally have low melting points.

The crystal lattice reflects the shape of the individual molecules. If they are

FIGURE 17.21

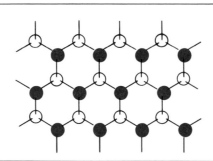

The structure of a double layer in crystalline arsenic, antimony, or bismuth. The dark spheres form an upper layer, and the white ones form a lower layer.

spherical, or nearly so, the crystal has a close-packed structure. The simplest molecular crystals are those of the noble gases (neon, argon, krypton, and xenon), which are cubic close-packed. Methane (CH_4) and hydrogen sulfide (H_2S) have the same structure. Some plant viruses are approximately spherical in shape and form cubic close-packed crystals. Solid hydrogen, H_2, is hexagonal close-packed. Iodine, I_2, has a structure very similar to cubic close-packed, but the cylindrical shape of the iodine molecule introduces a distortion, so that the crystals have orthorhombic symmetry (Figure 17.22). Each iodine atom has one neighbor (its partner in the I_2 molecule) at a distance of 2.68 Å; the distance to the nearest iodine atom in another I_2 molecule is 3.56 Å.

In general the molecular structure is practically the same in a molecular crystal as in the gas phase. However, there are some exceptions. Biphenyl, for

FIGURE 17.22

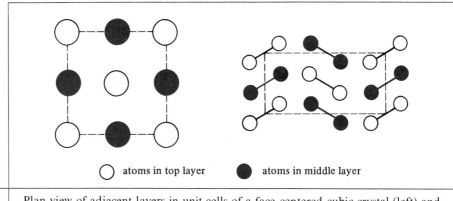

Plan view of adjacent layers in unit cells of a face-centered cubic crystal (left) and an iodine crystal (right). (SOURCE: *A. F. Wells*, Structural Inorganic Chemistry, *3rd ed., The Clarendon Press, Oxford, 1962, p. 149.*)

which one resonance structure is

$$\text{H-C} \begin{array}{c} \text{H} \\ \diagup \\ \text{C=C} \\ \diagdown \\ \text{C-C} \\ \diagup \\ \text{H} \end{array} \begin{array}{c} \text{H} \\ \diagdown \\ \text{C-C} \\ \diagup \\ \text{H} \end{array} \begin{array}{c} \text{H} \\ \diagup \\ \text{C-C} \\ \diagdown \\ \text{C=C} \\ \diagup \\ \text{H} \end{array} \text{C-H}$$

is planar in the crystal. In the gas phase, although each phenyl ring is still planar, the two rings are no longer coplanar. They are twisted about the central carbon–carbon bond at an angle of 42° with respect to each other.

17.10 HYDROGEN-BONDED CRYSTALS

General aspects of hydrogen bonding were discussed in Section 10.4. Many crystalline substances have structures that show the influence of hydrogen bonding. For example, in crystalline formic acid, HCOOH, the molecules are linked together in long chains by hydrogen bonds:

(In this drawing unshared pairs are omitted in order to emphasize the hydrogen bonds, shown as dotted lines.)

Ammonium fluoride, NH_4F, has the wurtzite structure, with alternate NH_4^+ and F^- ions. The ammonium ion is oriented with its hydrogens pointed toward the four neighboring fluoride ions, so that the ions are all interconnected by hydrogen bonds. A two-dimensional representation of the tetrahedral structure is

Note the contrast between NH$_4$F and the other ammonium halides with the NaCl structure.

Ice also has a structure similar to wurtzite. A two-dimensional representation is

$$
\begin{array}{c}
\text{H} \\
\backslash \\
\text{O---H} \\
\vdots \\
\text{H} \quad \text{H} \quad \text{H} \\
\backslash \quad \backslash \quad \backslash \\
\text{O---H}\cdots\text{O---H}\cdots\text{O---H} \\
\vdots \\
\text{H} \\
\backslash \\
\text{O---H}
\end{array}
$$

A three-dimensional representation of the ice structure is shown in Figure 17.23. Each proton lies between two oxygens, closer to one than the other. The shorter distance is 1.01 Å, the longer is 1.75 Å. Because of the hydrogen bonds, ice has a very open structure and low density (0.917 g cm^{-3}).

When ice melts, some of the hydrogen bonds are broken, enabling the water molecules to approach more closely to one another, with a consequent increase in density to 1.000 g cm^{-3}. The structure of liquid water is not nearly so well established as that of ice, but it is thought to have small regions resembling the ice structure. Each oxygen atom is typically hydrogen-bonded to four other oxygen atoms in a roughly tetrahedral configuration, but the long-range order apparent in Figure 17.23 is practically gone.

Hydrogen bonding is very important in substances of biological interest, such as proteins and nucleic acids, which are discussed in Chapter 23.

17.11 THE FREE-ELECTRON THEORY OF METALS

Metals are characterized by properties such as high electric and thermal conductivities. These properties are due to the ease with which the relatively free electrons travel through the metal, carrying electric charge and kinetic energy, as has been directly demonstrated by the Tolman effect (discovered by Richard C. Tolman). A metal wire, wound around a cylinder, is rotated at very high speed. The cylinder is then quickly stopped by applying a brake. Because of their inertial mass the electrons tend to keep on moving, so that a surge of electric current passes through the wire. By measuring the current, one may calculate the charge-to-mass ratio of the current-carrying particles in the metal. The results agree with the charge-to-mass ratio of the electron as determined by the Thomson and other experiments (see Section 6.2).

This leads to a picture of a metal as consisting of a set of positive ions immersed in an "electron gas" or "sea of electrons." For example, sodium metal contains Na$^+$ ions arranged in a body-centered cubic lattice, surrounded by the

17.11 THE FREE-ELECTRON THEORY OF METALS

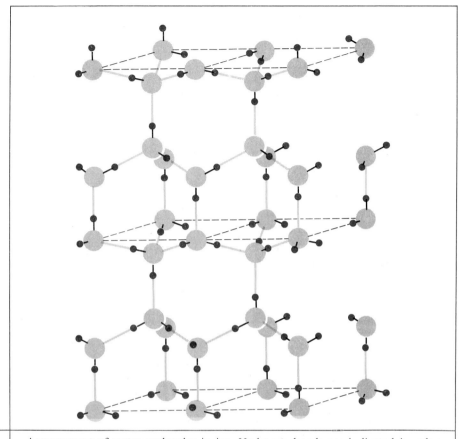

FIGURE 17.23 Arrangement of water molecules in ice. Hydrogen bonds are indicated in color. (SOURCE: *From Linus Pauling:* The Nature of the Chemical Bond, *Third Edition.* © *1960 by Cornell University. Used by permission of the publisher, Cornell University Press.*)

free valence electrons, one per sodium atom. The free electrons must occupy most of the space in the metal, because the ionic radius of Na$^+$ is 0.95 Å, whereas one-half the internuclear distance in sodium metal is 1.86 Å (Figure 17.24).

One difficulty with this picture is that it seems inconsistent with the heat capacities of metals, which are only slightly larger than the value of $3R = 25$ J K^{-1} (mol of atoms)$^{-1}$ to be expected from the effect of temperature in creasing the vibrational energy of the positive ions (see Section 5.5). If a free-electron gas is also present, why should it not contribute an additional $\frac{3}{2}R = 12.5$ J K^{-1} (mol of atoms)$^{-1}$, the same as the heat capacity of the other monatomic gases at constant volume?

The simplest theory providing a good description of metallic properties, including heat capacities, was proposed by Arnold Sommerfeld. It is an extension

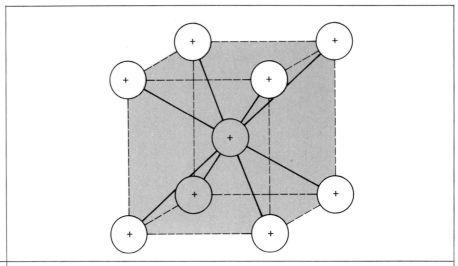

FIGURE 17.24 Sodium metal viewed as containing positive sodium ions immersed in a free-electron gas.

of molecular orbital theory (see Section 9.14) to a metal crystal, and it utilizes the energy levels of a particle in a box (Section 7.6) and the Pauli exclusion principle (Section 8.6). Each electron in the free-electron gas occupies one of the energy levels of a particle in a three-dimensional box. The general expression for these energy levels, given by Eq. 7.14, depends on three quantum numbers n_x, n_y, and n_z, each a positive integer, and the dimensions of the box a, b, and c. To simplify the treatment, consider that the metal is in the shape of a cube, so that $a = b = c$. Also, let $n^2 = n_x^2 + n_y^2 + n_z^2$. As the electrons are free, the potential energy can be taken as zero, and the total energy E is equal to the kinetic energy E_k. Thus Eq. 7.14 becomes

$$E = \frac{h^2 n^2}{8ma^2} \tag{17.2}$$

where m is the mass of the electron.

Because of the Pauli exclusion principle, only two electrons can occupy an energy level, and they must have opposite spins. Thus the state of the free-electron gas of lowest energy will be as shown schematically in Figure 17.25. If there are N electrons, then the lowest $N/2$ energy levels will be filled. The energy of the highest occupied level is called the **Fermi energy,** E_F. It has been shown that the Fermi energy is

$$E_F = \frac{h^2}{8m}\left(\frac{3N}{\pi a^3}\right)^{2/3} \tag{17.3}$$

17.11 THE FREE-ELECTRON THEORY OF METALS

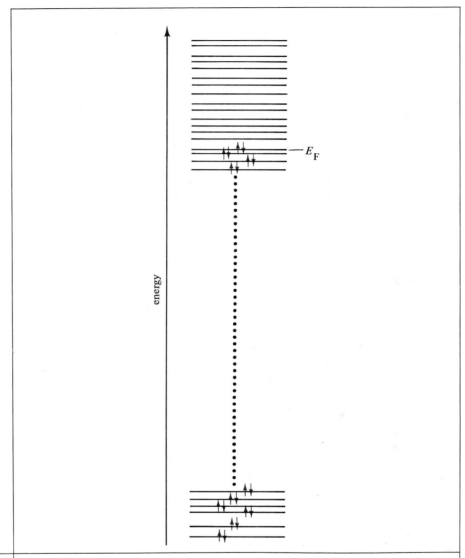

FIGURE 17.25

Lowest state of a free-electron Fermi gas. Only the highest and lowest energy levels are shown, as the number of occupied levels is of the order of 10^{23}.

Thus the Fermi energy is determined by the number of free electrons per unit volume, N/a^3, which can be calculated for each metal from its density and the number of valence electrons per atom. For sodium, the Fermi energy calculated by this equation is 5.0×10^{-19} J or 3.1 eV, and for other metals it is of the same order of magnitude.

These energies greatly exceed the energies available at ordinary tempera-

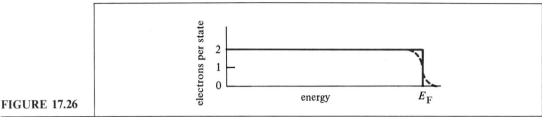

FIGURE 17.26 Distribution of electrons among different energy levels at 0 K (solid line) and room temperature (broken line).

tures due to thermal vibrations. The average vibrational energy of one of the positive ions is $3k_BT$, which, at 25°C, is $3(1.38 \times 10^{-23} \text{ J K}^{-1})(298 \text{ K}) = 1.23 \times 10^{-20}$ J or 0.077 eV. Therefore most of the electrons remain in the lowest energy levels except at extremely high temperatures. When a metal is heated from, say, 25° to 26°C, very few electrons are excited into higher energy levels, and thus the free-electron gas has little effect on the heat capacity at 25°C.

Figure 17.26 shows the distribution of electrons at 0 K (solid line) in a different way. An energy level having $E \leq E_F$ has two electrons; the other levels have none. At room temperature a few electrons occasionally acquire sufficient thermal energy to move from the energy levels just below E_F to the levels just above E_F, producing the average distribution of electrons shown by the broken line.

The statistical distribution of identical particles obeying the Pauli exclusion principle was developed by Enrico Fermi and Paul A. M. Dirac. The Fermi-Dirac distribution shown in Figure 17.26 is clearly much different from the Maxwell-Boltzmann distribution for an ordinary gas (Figure 4.3). Because the free-electron gas obeys Fermi-Dirac statistics, it is often called the **free-electron Fermi gas.**

As kinetic energy is related to velocity, the average distribution of energies, shown in Figure 17.26, implies an average distribution of velocities as shown in Figure 17.27 (black line). The velocity distribution is useful in explaining the electrical conductivity of metals. When a metal wire is connected to the positive and negative electrodes of a battery, the electrons are subject to an electric force. They are attracted toward the positive electrode and repelled by the negative

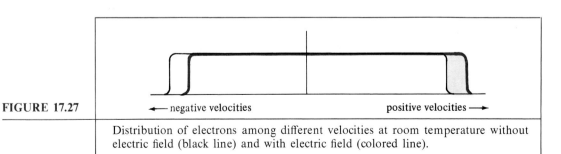

FIGURE 17.27 Distribution of electrons among different velocities at room temperature without electric field (black line) and with electric field (colored line).

electrode. This increases the number of electrons moving toward the positive electrode, and decreases the number moving in the opposite direction. This effect of the electric field on the velocity distribution is shown in Figure 17.27 (colored line). Because more electrons are moving in one direction than the other, an electric current flows toward the positive electrode. (As the electrons move out of the wire they are replaced by an equal number of electrons moving into the wire from the negative electrode.)

The large number of closely spaced, unoccupied levels just above the Fermi energy is essential to the phenomenon of electrical conduction in metals. As shown in Figure 17.27, there is a net shift of electrons with high negative velocities to still higher positive velocities. They can acquire these higher positive velocities only by moving into the various energy levels above the Fermi energy. This situation is somewhat analogous to ocean waves. The water molecules at the surface can move up and down under the influence of wind and tide, generating waves that carry energy. The molecules in the interior cannot do so, as the neighboring sites are already occupied by other molecules. This analogy has led to the term "Fermi sea of electrons."

The main sources of resistance to the flow of an electric current in a metal are interference by the positive ions (which, as they vibrate back and forth, move into the way of the conduction electrons), crystal defects, and impurities. Defects and impurities depend on the particular sample, whereas the vibrational motion of the positive ions depends on the temperature. Figure 17.28 shows the effect of these two factors on the electrical resistance of sodium. Note that the resistance at very low temperatures is less than 0.1% of the resistance at room temperature.[2]

17.12 THE ELECTRONIC BAND THEORY

To obtain a more accurate theory of electrons in solids, it is necessary to take into account the coulombic attraction between the negative electrons and the positive ions. The main effect of this attraction is to divide the energy levels into groups or bands, separated by gaps. The resulting theory is called the **electronic band theory**. It is particularly useful in explaining the properties of insulators and semiconductors.

Figure 17.29 compares the energy bands in a typical metal, an insulator, and a semiconductor. In a metal, one band of closely spaced energy levels (called the **conduction band**) is only partly filled, so that electrical conduction is possible by the mechanism outlined above. In an insulator the bands of lower energy are completely filled, the bands of higher energy are completely empty, and in between there is a large energy gap. With no partly filled bands, electrical conduction cannot occur. It is as though the ocean were covered by a tight-fitting lid, eliminating surface waves.

A semiconductor is similar to an insulator, but the energy gap between the

[2] For some metals and alloys, the electrical resistance completely disappears below a very low temperature that is characteristic for the metal (e.g., below 4.15 K for mercury), a phenomenon called **superconductivity**. An electric current in a loop of metal persists indefinitely, provided the temperature is kept below the critical value.

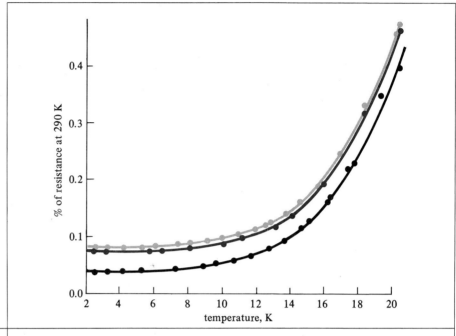

FIGURE 17.28

Electrical resistance of three specimens of sodium metal at low temperatures. [SOURCE: *D. K. C. MacDonald and K. Mendelssohn,* Proceedings of the Royal Society (London), **A202**, *111 (1950)*.]

highest filled band and the lowest empty band is rather small. Consequently, at low temperatures the substance behaves as an insulator. As the temperature is increased the thermal energy becomes sufficient to excite electrons into the empty band, causing the electric resistance to decrease with increasing temperature, just the opposite of metallic behavior. In diamond the energy gap is 5.33 eV, in silicon it is 1.14 eV, and in germanium it is only 0.67 eV. Consequently, diamond is a good insulator even at fairly high temperatures, whereas silicon and germanium are semiconductors at ordinary temperatures.

17.13 MAGNETIC PROPERTIES

The magnetic balance, shown in Figure 17.30, provides a simple method for investigating the magnetic properties of various substances. The sample is suspended from a balance arm so that it extends partly into the coil of an electromagnet or solenoid. When the electromagnet is turned on, most substances are repelled away from the magnetic field, and the repulsive force decreases the weight required on the right-hand pan to balance the sample; such substances are called **diamagnetic.** Some substances are attracted into the magnetic field, and the attractive force increases the weight required on the right-hand pan; these sub-

17.13 MAGNETIC PROPERTIES

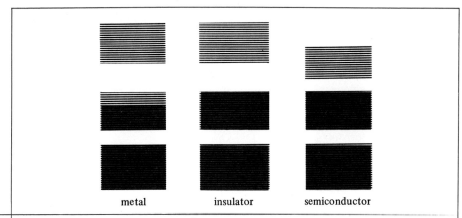

FIGURE 17.29 Energy bands in a metal, an insulator, and a semiconductor. Colored shading indicates the extent to which the bands are filled with electrons at low temperatures.

stances are called **paramagnetic.** A few metals, alloys, and other substances, such as magnetite or lodestone (Fe_3O_4), exhibit an intense form of paramagnetism called **ferromagnetism,** in which the attractive force is very large compared to most paramagnetic substances. In addition, a ferromagnetic substance retains its magnetic properties to some extent (residual magnetism) when removed from the

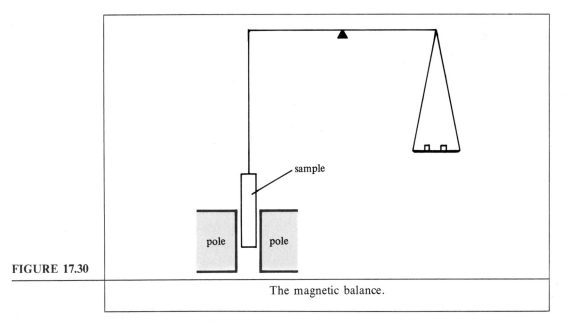

FIGURE 17.30 The magnetic balance.

magnetic field. For example, an iron or steel needle acquires a north-seeking magnetic pole at one end, and a south-seeking magnetic pole at the other end.

Paramagnetic and ferromagnetic substances owe these properties to the presence of one or more unpaired electrons. Closely associated with its spin, each electron has a magnetic moment, that is, it behaves like a small compass needle. Unless another electron is in the same orbital with opposite spin, there is a net attraction into a magnetic field, causing the substance to be paramagnetic or ferromagnetic.

Diamagnetic substances have all of their electrons paired (↑ ↓), and the magnetic moment of each electron is canceled by the magnetic moment of an electron with opposite spin. Thus the magnetic moments of the individual electrons have no effect on the behavior of the substance. Why, then, is the substance repelled by a magnetic field? The theory of diamagnetism is rather complex. Briefly, this phenomenon is related to the effect of a magnetic field in causing electric currents to flow, as in an electric generator or dynamo. These electric currents themselves create new magnetic fields, always in a direction *opposite* to the original magnetic field. Thus, when a diamagnetic substance is placed in a magnetic field, the motions of the electrons in the atoms are affected. The tiny electric currents in the individual atoms set up new magnetic fields in the opposite direction, so that the substance is repelled.

This effect is present even in a paramagnetic or ferromagnetic substance. Although it is usually relatively small, it is necessary to correct the paramagnetism of a substance for the diamagnetic effect. The corrected paramagnetism can then be used to calculate the number of unpaired electrons in the atoms and molecules of the substance. These results have been particularly important in studies of transition metal compounds (Chapter 21).

Most metals are weakly paramagnetic owing to the relatively few unpaired electrons at high energies (in the region of the Fermi energy). This paramagnetism is often independent of temperature, a result that can be understood in terms of two opposing temperature effects. As the temperature is increased the number of singly occupied levels increases, and this should increase the paramagnetism. But at higher temperatures the greater random thermal motion makes it more difficult for the electrons to orient their magnetic moments in the direction of the magnetic field. The two effects are equal and opposite, and they effectively cancel each other.

The most important ferromagnetic metals are iron, cobalt, and nickel, which are adjacent to one another in the first long period of the periodic table. Certain compounds of chromium and manganese, which precede iron in the periodic table, are also ferromagnetic. The individual atoms of these elements have incomplete *d* subshells, in which there are two or more electrons with their spins aligned parallel. For example, an iron atom in its ground state has six electrons in the 3*d* subshell, four with unpaired spins:

17.14 CRYSTAL DEFECTS

In a ferromagnetic solid, there are small regions called domains, usually about 0.1 mm on a side. Within each domain the magnetic moments of the unpaired electrons are lined up in the same direction. Thus each domain has an intense magnetic effect. When placed in a strong magnetic field the various domains have their directions of magnetization all lined up in the direction of the field, producing a very intense magnetization of the entire sample.

All ferromagnetic substances lose their ferromagnetism at a sufficiently high temperature, called the **Curie temperature** (after Pierre Curie), becoming paramagnetic at still higher temperatures. For iron, cobalt, and nickel the Curie temperatures are 770°C, 1120°C, and 358°C, respectively.

17.14 CRYSTAL DEFECTS

Many important properties of crystalline solids cannot be accounted for on the assumption that there is a completely regular arrangement of atoms throughout the crystal. Instead there are usually **defects** in crystals, either atoms missing from various lattice sites, called **point defects,** or more extensive irregularities called **dislocations.** Crystal defects in metals and alloys are responsible for many physical properties and phenomena of technological importance—malleability, ductility, annealing, work-hardening, fatigue, and fracture. Electric properties of solids are also affected by crystal defects.

The simplest type of point defect (Schottky defect) consists of vacancies at various positions in the crystal lattice. A compound of formula MX may have as many missing atoms of one element as of the other, so point defects need not alter the stoichiometry. In another type of point defect (Frenkel defect) the atoms are still present but are out of position. For example, if, in the regular crystal structure, every other interstice between X atoms is occupied by an M atom, then a Frenkel point defect occurs when an M atom occupies the wrong interstice (Figure 17.31). Point defects have an important effect on the rate of diffusion in solids. An atom can move into a vacant lattice position much more easily than if it had to trade places with a neighbor in the crowded confines of the crystal lattice.

A dislocation is an extensive irregularity in a crystal. In an edge dislocation, there is an extra sheet of atoms extending partway into the crystal. Under stress, a dislocation may move progressively through the crystal, as shown in Figure 17.32. This is the principal mechanism for plastic deformation of a metal, as when it is pressed or beaten into a sheet or drawn into a wire.

Nonstoichiometric Compounds

A number of important compounds violate the law of definite proportions—their composition varies depending on the method and conditions of preparation. For example, iron sulfide, nominally FeS, has compositions ranging from 50 to 53.3 atomic % sulfur. Studies of the densities and crystal structures of samples of various compositions show that, in the "sulfur-rich" samples, there are actually iron atoms missing from their lattice positions. Substances having variable composition are called **nonstoichiometric compounds** or **berthollides,** after Berthollet, who

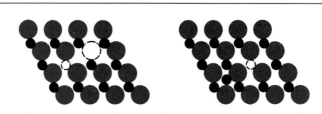

FIGURE 17.31 Schottky (left) and Frenkel (right) point defects in an ionic crystal MX. (SOURCE: A. F. Wells, *Structural Inorganic Chemistry,* 3rd ed., The Clarendon Press, Oxford, 1962, p. 163.)

unsuccessfully maintained that the composition of a compound was variable (see Section 1.8). Titanium monoxide, which has the NaCl structure, exhibits a particularly wide range of composition, and there are 15% vacancies in the lattice at room temperature even for the composition TiO. Table 17.2 shows the proportion of defects at various compositions.

Solid Solutions

Many elements and compounds having similar crystal structures form solid solutions of the substitutional type in which atoms of one element are found at sites normally occupied by atoms of the other element. Gold and silver form a continuous range of solid solutions from pure gold to pure silver, with gold and silver atoms randomly placed in the lattice positions of a face-centered cubic lattice.

Addition of a small amount of a group IIIA or a group VA element to the semiconducting elements of group IVA produces substitutional solid solutions having useful electric characteristics. An arsenic atom occupying a lattice site in a germanium crystal brings with it five valence electrons. As only four are needed to

TABLE 17.2 **COMPOSITION AND DEFECTS IN "TITANIUM MONOXIDE"**

O:Ti ratio	% Ti sites occupied	% O sites occupied
1.33	74	98
1.25	77	96
1.12	81	91
1.00	85	85
0.69	96	66

Source: A. F. Wells, *Structural Inorganic Chemistry,* 3rd ed., The Clarendon Press, Oxford, 1962, p. 167.

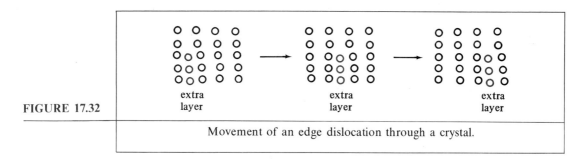

FIGURE 17.32 Movement of an edge dislocation through a crystal.

form bonds to the four neighboring germanium atoms in the diamondlike lattice of the host germanium crystal, the fifth electron is rather loosely held.

$$\begin{array}{c} |\\ -\text{Ge}- \\ | \quad \textcircled{e^-} \longrightarrow \\ -\text{Ge}-\text{As}^+-\text{Ge}- \\ | \quad | \\ -\text{Ge}- \\ | \end{array}$$

Thus the arsenic atoms provide conduction electrons. Such semiconductors are referred to as *n* type, because the current-carrying species is negatively charged.

Addition of gallium to a germanium crystal produces a *p*-type semiconductor, with the current carried by positive "holes." A gallium atom has only three valence electrons, so that, to form four bonds, it must borrow an electron from a germanium atom. The resulting deficiency of one electron migrates through the crystal as a positive hole:

$$\begin{array}{c} |\\ -\text{Ge}- \; \textcircled{+} \longrightarrow \\ | \quad | \\ -\text{Ge}-\text{Ga}^--\text{Ge}- \\ | \quad | \\ -\text{Ge}- \\ | \end{array}$$

As little as 0.001% of a group IIIA or group VA element added to germanium can decrease its electric resistance by a factor as large as 10^3.

PROBLEMS

17.1 Using the data in Figure 17.3, classify each of the following spatial arrangements in one of the seven systems of crystals.
   (a) a building divided into cubical rooms all of the same size
   (b) a building divided into rooms all having length = width = 10 ft, height = 8 ft, and 90° angles

(c) the building in part (a) following an earthquake that changes the angles between the walls to $\alpha = \beta = \gamma = 89°$ but does not otherwise damage the building

(d) a honeycomb

17.2 Classify each of the following substances in one of the seven systems of crystals.
   (a) Fluorite, CaF_2, has a unit cell in which the three edges have equal lengths and all angles are 90°.
   (b) Tetraborane, B_4H_{10}, has a unit cell in which the edges have different lengths, two angles are 90°, and the third angle is 105.9°.
   (c) Potassium nitrate, KNO_3, has a unit cell in which the edges have different lengths, and all angles are 90°.

17.3 A sodium chloride crystal ($d = 2.82$ Å) reflects X rays of unknown wavelength. The smallest angle that gives an intense reflection is $\theta = 8°31'$, for which $\sin \theta = 0.148$. Calculate λ from the Bragg equation.

17.4 When a potassium chloride crystal reflects X rays of the same wavelength as in the preceding problem, the smallest angle that gives an intense reflection is $\theta = 7°38'$, for which $\sin \theta = 0.133$. Calculate d for KCl.

17.5 (a) The unit cell of a metal crystal having the body-centered cubic structure contains two atoms: one at the center, and one-eighth of each of the eight corner atoms. (A corner atom is shared equally with the seven other unit cells that adjoin each corner.) Show that the density of the unit cell (hence the density of the crystal) expressed in terms of the length of one edge of the unit cell, l, and the mass of one atom, m, is $2m/l^3$.
   (b) Calculate l for metallic sodium (which has the body-centered cubic structure) from its atomic mass, density (0.97 g cm^{-3}), and Avogadro's number.
   (c) Calculate from l the shortest internuclear distance in metallic sodium.

17.6 (a) Using the type of reasoning illustrated in the preceding problem, calculate the number of atoms in a unit cell of a metal crystal having the face-centered cubic structure.
   (b) Derive a formula for the density of a face-centered cubic crystal in terms of the length of one edge of the unit cell, l, and the mass of one atom, m.
   (c) Calculate l for metallic aluminum (which has the face-centered cubic structure). Its density is 2.70 g cm^{-3}.
   (d) Calculate from l the shortest internuclear distance in aluminum.

17.7 Lithium iodide has the NaCl crystal structure. The length of one edge of the unit cell (Figure 9.2) is 6.05 Å. On the assumption that there is anion–anion contact in this crystal, calculate the ionic radius of I⁻. Compare your result with the I⁻ radius listed in Table 9.1.

17.8 Show that anion–anion contact in the NaCl structure occurs only when $R_+/R_- < 0.414$.

17.9 (a) From the following data calculate the crystal lattice energy of KCl.

	ΔE (kJ mol^{-1})
$K(g) \longrightarrow K^+(g) + e^-(g)$	414
$Cl^-(g) \longrightarrow Cl(g) + e^-(g)$	347
$K(c) \longrightarrow K(g)$	88
$Cl_2(g) \longrightarrow 2Cl(g)$	226
$K(c) + \tfrac{1}{2}Cl_2(g) \longrightarrow KCl(c)$	-435

(b) Compare your result from part (a) with the value calculated using Eq. 17.1. [The distance between adjacent K⁺ and Cl⁻ ions in KCl(c) is 3.14 Å.]

17.10 Explain the similarity of boron nitride and elemental carbon in their allotropic crystalline forms by comparing their electronic structures.

17.11 Using the electron-pair repulsion theory (Section 9.8), explain why the sheet structure of arsenic is puckered rather than planar as in graphite.

17.12 Suggest an explanation for the fact that the biphenyl molecule is twisted in the gas phase.

17.13 Nitrogen forms much stronger hydrogen bonds than chlorine even though nitrogen and chlorine have the same electronegativity. Suggest an explanation.

17.14 Because oxygen and sulfur are neighboring elements in group VI, one might expect a similarity between the structures of H₂O and H₂S, but actually the two structures are quite different (Sections 17.10 and 17.9). Why?

17.15 In NH₄F each ammonium ion has only four neighboring fluoride ions, whereas in NH₄Cl each ammonium ion has six neighboring chloride ions. Suggest an explanation.

17.16 Explain the Tolman effect.

17.17 (a) The density of metallic sodium is 0.97 g cm^{-3}. Assuming that each sodium atom has one valence electron, calculate the total number of valence electrons in a cube of sodium 1.00 cm on each side.
(b) Using this result and Eq. 17.3, calculate the Fermi energy of sodium.
(c) From the number of occupied energy levels and the Fermi energy, calculate the average energy difference between adjacent occupied energy levels.

17.18 Using the electronic band theory, explain the difference between a metal, an insulator, and a semiconductor.

17.19 (a) Which of the following kinds of substance is attracted into a magnetic field?
diamagnetic substance, paramagnetic substance, ferromagnetic substance
(b) How does a ferromagnetic substance differ from a paramagnetic substance in its behavior in a magnetic field?

17.20 Explain the difference between a Schottky defect and a Frenkel defect. How does each affect the density of a substance?

17.21 A sample of iron sulfide containing 53.3 atomic % sulfur has the same crystal structure as FeS, except that some of the iron atoms are missing from their positions in the crystal lattice. (All of the sulfur positions are occupied.) What percentage of the Fe sites are occupied?

17.22 (a) How does a small amount of a group IIIA or group VA element affect the electric resistance of a group IVA element?
(b) Explain the difference between an n-type and a p-type semiconductor.

17.23 Consider a set of hard spheres, all of the same diameter, that are packed together. Show that they are packed more closely in a face-centered cubic structure than in a body-centered cubic structure by calculating the fraction of empty space in each structure. (The formula for the volume of a sphere is $4\pi R^3/3$, where R is the radius of the sphere.)

17.24 If the attraction between adjacent ions of opposite sign were the only important term in the crystal lattice energy, what would the Madelung constant be for the NaCl structure? For the CsCl structure?

CHAPTER 18
THE ELEMENTS OF GROUPS IA–IVA

18

The next five chapters present a systematic review of the chemical elements and their compounds in considerably more detail than was possible in the brief introductory chapter to this subject (Chapter 11). At that time it was necessary to omit some important topics either for lack of space or because, at that early stage, the theoretical background was not sufficiently well developed. We are now in a position to consider the chemistry of the elements with the aid of the fundamental principles of chemistry relating to energy, structure, equilibria, and kinetics.

The present chapter covers the chemistry of the first four main groups of the periodic table, and Chapter 19 treats the last four groups. The transition metals are discussed in Chapters 20 and 21. Throughout these chapters the emphasis will be on comparative trends in physical and chemical properties within each group or from one group to the next. Particular emphasis will be placed on the reactions of substances in aqueous solutions because such solutions are by far the most important in man and his environment. Chapter 22 focuses special attention on organic chemistry, the chemistry of the element carbon.

At the outset a few generalizations are in order. The chemical and physical properties of the first four groups of elements exhibit less uniformity within a group of elements as we go from group IA to group IVA. The chemical properties of the group IA elements below hydrogen are remarkably uniform as all the elements are metals and their compounds are soluble strong electrolytes. The group IIA elements are all metals and their compounds are principally ionic except for those of beryllium. Beginning with group IIIA, we find that boron is a metalloid and that the elements become more metallic as we go down the table. This behavior is repeated in group IVA, with carbon being a nonmetal, silicon and germanium being metalloids, and tin and lead being metals. Associated with this change in behavior from nonmetal to metal, as the size of the atom increases, are changes in chemical behavior: acidic oxides to amphoteric oxides to basic oxides; decreasing volatility of the halides; and increasing ionic character of the compounds. The tendency for the elements to change from metallic to nonmetallic character as we go across the periodic table, in conjunction with the increase in metallic behavior as we go down a group, leads to diagonal similarities between elements, which are most marked in the second and third periods:

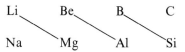

(The same generalizations may be carried over to the other main groups of elements, which are considered in Chapter 19.)

18.1 THE GROUP IA ELEMENTS

IA
H 1
Li 3
Na 11
K 19
Rb 37
Cs 55
Fr 87

The atoms of a group IA element have a single valence electron each. For the first element, hydrogen, the similarity to the other group IA elements stops there. Hydrogen is a gas with an extremely low boiling point. The other group IA elements are highly reactive metals called the **alkali metals.**

Hydrogen not only has just one valence electron, it lacks just one electron from having a complete outer electronic shell. In this respect, it is similar to the group VIIA elements. In some of its properties hydrogen resembles the first group VIIA element, fluorine: Both hydrogen and fluorine are nonmetals, both are diatomic gases, and both form an extensive series of compounds with carbon (**hydrocarbons** and **fluorocarbons,** respectively). However, hydrogen does differ from fluorine in many important respects; for example, hydrogen is much less reactive and has a much lower electronegativity.

Some of the properties, methods of preparation, and reactions of hydrogen are summarized in Section 11.2; the role of hydrogen in acid–base chemistry is discussed in Sections 11.3 and 13.1; and hydrogen bonding is described in Section 10.4. At this point we shall consider some additional aspects of acids in aqueous media.

Protonic acids dissociate in aqueous solution to give aquated ions. They may be divided into two groups: (1) oxyacids in which the proton is bound to oxygen (H—O—X) and (2) binary acids in which the proton is bound to an element other than oxygen (H—X).

The oxyacids vary widely in strength in aqueous solution, as shown in Table 18.1. In this table the different acids are grouped together according to the value of m when the molecular formula is written in the form $(HO)_n XO_m$, where X is the element other than hydrogen or oxygen. For example, the formula of phosphoric acid, usually written H_3PO_4, becomes $(HO)_3PO$. Therefore $m = 1$ for phosphoric acid. (When there are equal numbers of hydrogen and oxygen atoms, the formula can be written $(HO)_n X$, and $m = 0$.)

As shown in Table 18.1, acids having the same value of m show a certain similarity in their values of K_1, the first dissociation constant. Furthermore, successive dissociation constants of the polyprotic acids change by about the same factor in each case. These observed regularities are the basis of the following two empirical rules, which may be used to estimate the dissociation constants of an oxyacid.

RULE 1

The first dissociation constant of an oxyacid with formula $(HO)_n XO_m$ is determined by the value of m:

If $m = 0$, then $K_1 \approx 1 \times 10^{-8}$ or less   (a very weak acid)

If $m = 1$, then $K_1 \approx 1 \times 10^{-3}$ (a weak acid)

If $m \geq 2$, then K_1 is large   (a strong acid)

TABLE 18.1 DISSOCIATION CONSTANTS OF OXYACIDS $(HO)_nXO_m$ IN WATER AT 25°C

Acid	K_1	K_2	K_3
$(HO)_nX$ ($m = 0$)			
HClO, hypochlorous acid	2.9×10^{-8}	—	—
HBrO, hypobromous acid	2.2×10^{-9}	—	—
HIO, hypoiodous acid	2.3×10^{-11}	—	—
H_3AsO_3, arsenious acid	5.1×10^{-10}	3×10^{-14}	—
H_3BO_3, boric acid	5.8×10^{-10}	—	—
H_3SbO_3, antimonous acid	1×10^{-11}	—	—
H_4GeO_4, germanic acid	3×10^{-9}	1×10^{-13}	—
H_4SiO_4, silicic acid	3.1×10^{-10}	—	—
$(HO)_nXO$ ($m = 1$)			
$HClO_2$, chlorous acid	1.1×10^{-2}	—	—
HNO_2, nitrous acid	4.5×10^{-4}	—	—
H_2SeO_3, selenious acid	2.4×10^{-3}	4.8×10^{-9}	—
H_3AsO_4, arsenic acid	6.0×10^{-3}	1.1×10^{-7}	3.0×10^{-12}
H_3PO_4, phosphoric acid	7.1×10^{-3}	6.3×10^{-8}	4.5×10^{-13}
$(HO)_nXO_2$ ($m = 2$)			
$HClO_3$, chloric acid	Large	—	—
HNO_3, nitric acid	Large	—	—
H_2SO_4, sulfuric acid	Large	1.0×10^{-2}	—
H_2SeO_4, selenic acid	Large	1×10^{-2}	—
$(HO)_nXO_3$ ($m = 3$)			
$HClO_4$, perchloric acid	Very large	—	—
$HMnO_4$, permanganic acid	Very large	—	—

RULE 2

The ratios of successive dissociation constants, K_2/K_1, K_3/K_2, and so on, are about 1×10^{-5}.

If we apply these rules to phosphoric acid, then, as $m = 1$, we would expect $K_1 \approx 1 \times 10^{-3}$ from rule 1. From rule 2 we would then have $K_2 \approx 1 \times 10^{-8}$ and $K_3 \approx 1 \times 10^{-13}$. As can be seen from the actual values listed in Table 18.1, these estimates are roughly correct. (Note that the value of n is not used in estimating the dissociation constants by these rules.)

The first rule can be explained qualitatively in the following manner. When $m = 0$, the proton is attracted to the oxygen from which it has dissociated by a full negative charge:

$$(HO)_nX \rightleftharpoons (HO)_{n-1}XO^- + H^+$$

However, if $m = 1$, we have

$$(HO)_nXO \rightleftharpoons (HO)_{n-1}XO_2^- + H^+$$

There are now two equivalent oxygen atoms to which the proton is attracted.[1] On dividing the negative charge of the anion between these two oxygen atoms, we see that the average charge on each oxygen atom is only $-\frac{1}{2}$, and therefore the proton is subjected to a smaller attractive force. For this reason oxyacids with $m = 1$ are stronger acids than those with $m = 0$. (This argument may be extended to larger values of m.) Carbonic acid (H_2CO_3) and phosphorous acid (H_3PO_3), which apparently do not agree with rule 1 above, will be discussed in Sections 18.4 and 19.1, respectively. Trends in the acid strengths of binary acids (HX) of the elements of groups VIA and VIIA will be discussed in Sections 19.2 and 19.3, respectively.

The group IA elements below hydrogen (the alkali metals) are soft, silvery white metals that react readily with nonmetals and hence are never found in the elemental state in nature. Compounds of sodium and potassium are relatively abundant in the earth's crust; those of lithium, rubidium, and cesium are less common; and compounds of francium, which is radioactive, occur in only trace amounts. Sodium and potassium, which are essential to life processes, are found in large amounts in plant ashes as the carbonates, Na_2CO_3 (soda ash) and K_2CO_3 (potash). The term *alkali* is derived from the Arabic word for the ashes of the saltwort, a thistlelike plant.

The metallic elements may be prepared by electrolysis of their fused chlorides (see Figure 14.2). Once prepared, they must be stored in an inert atmosphere or under oil because of their high reactivity with water and air. Sodium reacts vigorously with water to produce hydrogen gas, whereas the heavier elements of the group react explosively with water. The elements are all metallic and are good conductors of electricity as they have specific conductances greater than 5×10^4 ohm^{-1} cm^{-1} at 0°C. (The specific conductance is simply the reciprocal of the electrical resistance in ohms between opposite faces of a cube of the metal that measures 1 cm on a side.)

Of all the groups in the periodic table, the alkali metals show the most uniform trends in chemical and physical properties with increasing atomic number. As shown in Table 18.2, the ionization potential, the melting point, and the electronegativity decrease with increasing atomic number. With the exception of

---

[1] The equivalence of the two oxygens is apparent from the Lewis structure of the anion when both oxygens are singly bonded to the central atom X.

$$\left[(HO)_{n-1}X \begin{matrix} \ddot{O}: \\ \diagdown \\ \ddot{O}: \end{matrix} \right]^{-}$$

Even if one oxygen atom is bonded to X by a double bond, in the anion the two oxygens are again equivalent because of resonance, the resonance forms being

$$\left[(HO)_{n-1}X \begin{matrix} \ddot{O}: \\ \diagup\diagup \\ \ddot{O}: \end{matrix} \right]^{-} \longleftrightarrow \left[(HO)_{n-1}X \begin{matrix} \ddot{O}: \\ \diagdown \\ \ddot{O}: \end{matrix} \right]^{-}$$

TABLE 18.2 | **PROPERTIES OF THE ALKALI METALS**

Property	Li	Na	K	Rb	Cs	Fr
Atomic number	3	11	19	37	55	87
Electron configuration	[He]2s	[Ne]3s	[Ar]4s	[Kr]5s	[Xe]6s	[Rn]7s
Ionization potentials (eV):						
I_1	5.39	5.14	4.34	4.18	3.89	—
I_2	75.6	47.3	31.8	27.5	25.1	—
Melting point (°C)	181	98	64	39	28.6	—
$\mathcal{E}°$ (V)[a]	−3.02	−2.71	−2.92	−2.99	−3.02	—
Pauling electronegativity	1.0	0.9	0.8	0.8	0.7	0.7
Density (g cm^{-3})	0.53	0.97	0.86	1.53	1.88	—

[a] Standard electrode potential for M^+(aq) $+ e^- = M$(s).

lithium, the standard electrode potential becomes more negative with increasing atomic number of the element. As we shall notice for other groups, the first element of the group often exhibits more marked deviations from the uniform trend in properties than do other elements of the group.

The chemistry of the alkali metals is extremely simple because it consists primarily of the properties of the M^+ ion. This is readily explained by the low energy required to remove the s electron from the valence shell (first ionization potential) and the relatively large energy required to remove a second electron from the noble gas configuration underlying the valence shell (second ionization potential). Indeed, no oxidation state in compounds other than $+1$ is known, nor is one expected, in view of the large second ionization potentials.

The alkali metal hydroxides form salts with nearly all acids. In most cases the salts are colorless, crystalline, ionic solids characterized by high melting points, by high electrical conductivity when molten, and by ready solubility in water. The few colored salts owe their color to the properties of the anion (e.g., $KMnO_4$ is purple because of the characteristic color of the permanganate ion).

The melting points (mp), solubility in water, and the enthalpies of solution of the alkali metal chlorides, together with the specific conductance of the molten chloride at its melting point, are given in Table 18.3. Again we note the anomalous behavior of lithium in that its chloride has the lowest melting point; the melting points of the other chlorides decrease with increasing atomic number from a maximum value at NaCl. It is also interesting to note that the most soluble chloride is LiCl. Because aqueous solutions are so common, we shall consider the various factors that influence the solubility of a salt in more detail.

We note that NaCl is soluble in water to the extent of 6.2 mol L^{-1} and that the enthalpy of solution (ΔH_s) is only 3.8 kJ mol^{-1}. Such a small enthalpy change associated with the formation of the solvated ions is somewhat surprising in view of the extremely large enthalpy change (777 kJ mol^{-1}) required to convert 1 mol of NaCl to the gaseous ions (the crystal lattice enthalpy, ΔH_c). The difference between the enthalpy changes for formation of the ions in solution and in the gas

18.1 THE GROUP IA ELEMENTS

TABLE 18.3 — SOME PROPERTIES OF THE ALKALI METAL CHLORIDES

Property	LiCl	NaCl	KCl	RbCl	CsCl
Melting point (°C)	610	801	770	722	645
Solubility in water at 25°C (mol L^{-1})	20	6.2	4.8	7.8	11.4
Enthalpy of solution, ΔH_s (kJ mol^{-1})	−37	3.8	17	17	18
Crystal lattice enthalpy, ΔH_c (kJ mol^{-1})	845	777	706	679	663
Hydration enthalpy, ΔH_h (kJ mol^{-1})	−882	−773	−689	−662	−645
Ionic radius of M$^+$ (Å)	0.60	0.95	1.33	1.48	1.69
Specific conductance of MCl(l) at mp (ohm^{-1} cm^{-1})	5.67	3.58	2.25	1.52	1.12

phase is due to solvation effects, as shown by the following thermodynamic cycle:

$$\text{NaCl(s)} \xrightarrow{\Delta H_c} \text{Na}^+(g) + \text{Cl}^-(g) \xrightarrow{\Delta H_h} \text{Na}^+(aq) + \text{Cl}^-(aq)$$

$$\text{NaCl(s)} \xrightarrow{\Delta H_s} \text{Na}^+(aq) + \text{Cl}^-(aq)$$

The enthalpy of hydration of the gaseous ions (ΔH_h) is given by

$$\Delta H_h = \Delta H_s - \Delta H_c = 3.8 - 777 = -773 \text{ kJ mol}^{-1}$$

Thus the small enthalpy of solution of NaCl may be attributed to the large enthalpy of hydration of its ions, indicating that considerable solvation of the ions must occur.

The ability of water to dissolve ionic compounds is due primarily to its dipolar character, which enables it to solvate both positive and negative ions. For example, the Na$^+$ ion is strongly bound to four water molecules in a "primary hydration sphere," as shown in Figure 18.1. Surrounding the primary hydration sphere are layers of additional water molecules bound with decreasing strength. Hydration of a Cl$^-$ ion is similar, except that the orientation of the water molecules is in the opposite direction, as Figure 18.1 illustrates.

As shown by the data in the fifth and sixth rows of Table 18.3, the magnitude of the hydration enthalpy decreases with increasing size of the positive ion. This decrease in hydration enthalpy might be expected from the increase in the distance between the center of the cation and the center of negative charge in the water molecule as the size of the cation increases.

Even though the fundamental properties of crystal lattice enthalpy and hydration enthalpy show definite trends as we go down the group (see Table 18.3), it is not surprising that the small difference in these values, the enthalpy of solution,

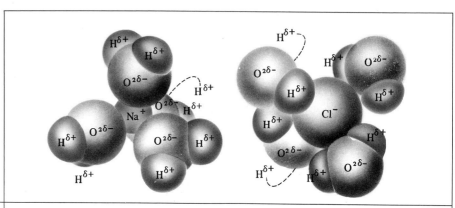

FIGURE 18.1

Hydration of Na^+ and Cl^- ions. The four water molecules are located at the corners of a tetrahedron, with the ion located at the center of the tetrahedron. If we let δ represent the fractional charge transferred from each hydrogen to the oxygen in water, then we can represent the dipolar character of the water molecule by $O^{2\delta-}(H^{\delta+})_2$. Water molecules hydrating Na^+ would be oriented so that the negative end of the dipole (oxygen) would be closest to the Na^+ ion. The orientation is reversed for Cl^- where the positive ends of the water dipole (two hydrogens) are closest to the Cl^- ion.

does not show a definite trend. Because many chemical properties depend on several distinct processes, one should use caution in attributing an observed trend in chemical behavior of a group of elements to a single property of those elements.

The alkali metals react readily with oxygen, but only lithium gives the simple monoxide (Li_2O) under all conditions. The other metals form the monoxides when excess metal is present; with excess oxygen, sodium gives the peroxide,

$$2Na(s) + O_2(g) = Na_2O_2(s)$$

whereas the heavier alkali metals give superoxides of formula MO_2 that contain the superoxide ion, O_2^-.

The alkali metals have recently been used in the construction of a new type of battery for applications requiring high electric currents. A sodium/sulfur battery, which operates at about 300°C, uses liquid sodium metal as the anode and liquid sulfur as the cathode. A lithium/iron sulfide battery, which operates at about 400°C, uses liquid lithium metal as the anode and an iron sulfide (FeS or FeS_2) as the cathode.

The normal oxides and sulfides of the alkali metals dissolve readily in water to produce OH^- or SH^- by hydrolysis:

$$Na_2O(s) + H_2O = 2Na^+ + 2OH^-$$

$$Na_2S(s) + H_2O = 2Na^+ + 2SH^-$$

The hydrides of the alkali metals can be prepared by heating the alkali metal in the presence of hydrogen:

$$2\text{Na}(s) + \text{H}_2(g) = 2\text{NaH}(s)$$

The hydrides are white solids that conduct electricity when molten, indicating their ionic character (Na^+H^-). These hydrides react readily with water to produce gaseous hydrogen:

$$\text{NaH}(s) + \text{H}_2\text{O} = \text{Na}^+ + \text{OH}^- + \text{H}_2(g)$$

18.2 THE GROUP IIA ELEMENTS

IIA
Be 4
Mg 12
Ca 20
Sr 38
Ba 56
Ra 88

The group IIA elements resemble the alkali metals in that the elements are all metals and their compounds are mainly ionic in character. Their oxides have been called **alkaline earths** since the early days of chemistry, and the group IIA elements are accordingly known as the **alkaline earth metals.**

Like the alkali metals, the alkaline earth metals are highly reactive and are never found in the metallic form in nature. The most abundant of the group IIA elements are calcium and magnesium (3.6% and 2.1% of the earth's crust, respectively); the other elements are present in amounts less than 0.1%. Radium, which is radioactive, occurs in small amounts in uranium ores as it is produced by nuclear disintegration of uranium.

The metals are prepared by electrolysis of the molten chlorides. Their specific conductance is similar to that of the alkali metals, ranging from 2×10^4 ohm^{-1} cm^{-1} for barium to 25×10^4 ohm^{-1} cm^{-1} for magnesium at 0°C. Thus all of the metals are good conductors of electricity. Magnesium metal is used in lightweight alloys for structural purposes such as aircraft frames.

The chemistry of the group IIA elements is primarily that of the +2 oxidation state, which is somewhat surprising in view of the relatively large values of the second ionization potentials (see Table 18.4). For example, we may use the

TABLE 18.4 PROPERTIES OF THE GROUP IIA ELEMENTS

Property	Be	Mg	Ca	Sr	Ba	Ra
Atomic number	4	12	20	38	56	88
Electron configuration	[He]$2s^2$	[Ne]$3s^2$	[Ar]$4s^2$	[Kr]$5s^2$	[Xe]$6s^2$	[Rn]$7s^2$
Ionization potentials (eV):						
I_1	9.32	7.65	6.11	5.70	5.21	5.28
I_2	18.2	15.0	11.87	11.0	10.0	10.1
I_3	154	80	51	—	—	—
Melting point (°C)	1280	650	810	800	850	—
$\mathcal{E}°$ (V)^a	—	−2.37	−2.87	−2.89	−2.90	−2.92
Pauling electronegativity	1.5	1.2	1.0	1.0	0.9	0.9
Ionic radius of M^{2+} (Å)	0.31	0.65	0.99	1.13	1.35	—

^{a} Standard electrode potential for M^{2+}(aq) + 2e^- = M(s).

ionization potentials to show that gaseous Ca^+ ions are stable with respect to gaseous Ca atoms and gaseous Ca^{2+} ions. The first ionization potential of calcium is 6.11 eV per atom or 589 kJ mol^{-1}; its second ionization potential is 11.87 eV per atom or 1145 kJ mol^{-1}. Thus we have

$$\begin{aligned} Ca(g) &= Ca^+(g) + e^- & \Delta H^\circ &= 589 \text{ kJ} \\ \underline{Ca^{2+}(g) + e^- = Ca^+(g)} & & \underline{\Delta H^\circ = -1145 \text{ kJ}} \\ Ca(g) + Ca^{2+}(g) &= 2Ca^+(g) & \Delta H^\circ &= -556 \text{ kJ} \end{aligned}$$

Because calcium exists as the metal at room temperature, we can combine the above equation with that for the sublimation of calcium,

$$Ca(s) = Ca(g) \qquad \Delta H^\circ = 176 \text{ kJ}$$

to obtain

$$Ca(s) + Ca^{2+}(g) = 2Ca^+(g) \qquad \Delta H^\circ = -380 \text{ kJ} \qquad (18.1)$$

However, in aqueous solution the situation is reversed owing to the large value of the hydration enthalpy of Ca^{2+}. For example,

$$Ca^{2+}(g) + 2Cl^-(g) = Ca^{2+}(aq) + 2Cl^-(aq) \qquad \Delta H_h^\circ = -2414 \text{ kJ} \qquad (18.2)$$

Although the hydration enthalpy of gaseous CaCl is not known, we may approximate its value by that of KCl, as the radii of Ca^+ and K^+ would be nearly the same:

$$2Ca^+(g) + 2Cl^-(g) = 2Ca^+(aq) + 2Cl^-(aq) \qquad \Delta H_h^\circ \approx -1377 \text{ kJ} \qquad (18.3)$$

If we subtract Eq. 18.2 from the sum of Eqs. 18.1 and 18.3, we obtain

$$Ca^{2+}(aq) + Ca(s) = 2Ca^+(aq) \qquad \Delta H^\circ \approx 657 \text{ kJ}$$

Thus we see that the enthalpy change favors Ca^+ in the gas phase but that Ca^{2+} is favored in aqueous solution owing to the large enthalpy of hydration of Ca^{2+}. A similar calculation may be used to show that the +2 oxidation state is favored in ionic crystals because of the large crystal lattice energy.

Magnesium salts and, to a lesser extent, calcium salts are effective drying agents owing to their high hydration enthalpies. In fact the halides of magnesium and calcium must be protected from the moisture of the air so that they do not absorb sufficient water to form solutions.

The lightest element of group IIA, beryllium, differs markedly from the other alkaline earth metals. Beryllium and its compounds are so poisonous that extreme precautions must be taken in handling them. (In contrast, magnesium and calcium are essential elements in living organisms.) Unlike the other elements of group IIA, beryllium tends to form covalent bonds. This is exemplified by the fact that the specific conductance of molten $BeCl_2$ (0.0032 ohm^{-1} cm^{-1}) is less than 1% that of the molten chlorides of the other alkaline earth metals.

Because beryllium oxide and beryllium hydroxide have both acidic and basic properties, they are said to be amphoteric (see Section 11.3). They dissolve in strongly acid solutions to form hydrated Be^{2+} or in strongly basic solutions to form an anion postulated to be $Be(OH)_4^{2-}$. For beryllium hydroxide, the reactions are

$$Be(OH)_2(s) + 2H^+ = Be^{2+} + 2H_2O$$

$$Be(OH)_2(s) + 2OH^- = Be(OH)_4^{2-}$$

Thus $Be(OH)_2$ acts as a base by donating its OH^- to H^+, and it acts as an acid by accepting OH^- ions to form the complex ion $Be(OH)_4^{2-}$. In general, the acidic properties of a metal oxide or hydroxide are most pronounced when the cation is small and has a high positive charge, as in the case of Be^{2+}. The larger group IIA cations such as Mg^{2+} and Ca^{2+} have basic oxides and hydroxides.

The alkaline earth nitrates and chlorides are very soluble in water. Slightly soluble compounds include most of the hydroxides, fluorides, sulfates, carbonates, and phosphates. Variations in solubility with increasing cation size are not uniform. (As indicated in the preceding section, these trends depend on a delicate balance between hydration and crystal lattice enthalpies.) The solubilities of the sulfates, carbonates, and phosphates show a general decrease with increasing cation size. Thus $BeSO_4$ and $MgSO_4$ are soluble in water, but $BaSO_4$ is only slightly soluble. The latter property is used in qualitative analysis in the test for sulfate ion:

$$Ba^{2+} + SO_4^{2-} = BaSO_4(s)$$

The solubility trend of the hydroxides is just the opposite: $Be(OH)_2$ and $Mg(OH)_2$ are only slightly soluble in water, whereas $Ba(OH)_2$ is quite soluble, and its solution is used in qualitative analysis to detect carbon dioxide:

$$CO_2(g) + Ba^{2+} + 2OH^- = BaCO_3(s) + H_2O$$

The fluorides show still another variation in solubility: The solubility in water decreases from BeF_2 (which is very soluble) to a minimum at CaF_2, and then increases.

In qualitative analysis, Ca^{2+} and Ba^{2+} are separated from solutions of the alkali and alkaline earth salts by addition of sodium carbonate to precipitate $CaCO_3$ and $BaCO_3$:

$$M^{2+} + CO_3^{2-} = MCO_3(s) \qquad (M = Ca \text{ or } Ba)$$

The mixture of solid carbonates is then dissolved in acetic acid and tested for Ba^{2+} by addition of ammonium acetate and sodium chromate to form a yellow precipitate of $BaCrO_4$, with the calcium remaining in solution as Ca^{2+}:

$$MCO_3(s) + CH_3CO_2H = M^{2+} + HCO_3^- + CH_3CO_2^-$$

$$Ba^{2+} + CrO_4^{2-} = BaCrO_4(s)$$

After removal of the $BaCrO_4$ by filtration, the filtrate may be tested for Ca^{2+} by neutralizing the acid and adding sodium carbonate to reprecipitate the white $CaCO_3$.

The solubility of calcium carbonate in water is enhanced by the presence of dissolved carbon dioxide. The dissolved CO_2 reduces the carbonate ion concentration by reacting with it to form bicarbonate ion. The equilibria are

$$\begin{aligned} CaCO_3(s) &\rightleftharpoons Ca^{2+} + CO_3^{2-} \\ CO_2(aq) + H_2O &\rightleftharpoons H^+ + HCO_3^- \\ H^+ + CO_3^{2-} &\rightleftharpoons HCO_3^- \\ \hline CaCO_3(s) + CO_2(aq) + H_2O &\rightleftharpoons Ca^{2+} + 2HCO_3^- \end{aligned} \quad (18.4)$$

Thus the amount of calcium carbonate that can be dissolved in ground waters depends on the concentration of dissolved carbon dioxide, which, in turn, depends on the carbon dioxide content of the atmosphere in contact with the water. Ground water percolating through decaying organic matter dissolves appreciable amounts of CO_2, and such water can dissolve natural deposits of limestone, producing large caves such as the Carlsbad Caverns in New Mexico.

The weird limestone formations within such caves (Figure 18.2) can also be explained using this equilibrium (Eq. 18.4). As water saturated with calcium carbonate slowly seeps into the cave and drips from the roof, carbon dioxide is lost to the atmosphere. This shifts the equilibrium of Eq. 18.4 to the left, and solid calcium carbonate is deposited on the roof and the floor of the cave. The formations of calcium carbonate so produced are called **stalactites** (deposits hanging from the roof like icicles) and **stalagmites** (deposits standing on the floor). They may eventually grow together to form columns.

Water containing appreciable amounts of Ca^{2+}, Mg^{2+}, or Fe^{2+} is known as **hard water** because it produces precipitates on the walls of boilers or water heaters and objectionable scums when treated with ordinary soaps. A typical soap is sodium stearate, $C_{17}H_{35}CO_2Na$; the reaction of the stearate anions with calcium ion is

$$Ca^{2+} + 2C_{17}H_{35}CO_2^- = Ca(C_{17}H_{35}CO_2)_2(s)$$

18.2 THE GROUP IIA ELEMENTS

FIGURE 18.2

Stalactites and stalagmites in a limestone cave. (*Courtesy of National Park Service. Photo by Fred Mang, Jr.*)

A general method for removal of the divalent ions is the use of ion exchange, which replaces the divalent ions with Na^+. If water containing Ca^{2+} is allowed to flow slowly through a bed of the mineral zeolite ($Na_2H_4Al_2Si_3O_{12}$), the Ca^{2+} ions replace the Na^+ ions in the zeolite. With zeolite represented as Na_2Z, the reaction is

$$Ca^{2+} + Na_2Z(s) \rightleftharpoons CaZ(s) + 2Na^+ \qquad (18.5)$$

When the available Na^+ is used up, the Na_2Z can be regenerated by passing a strong solution of rock salt (NaCl) through the bed, thus shifting the equilibrium to the left.

Another method of reducing the Ca^{2+} concentration to low levels is formation of a soluble complex ion. For example, hexametaphosphate ion reacts with Ca^{2+} to produce a complex ion:

$$2Ca^{2+} + (PO_3)_6^{6-} = Ca_2(PO_3)_6^{2-}$$

For household use it is also possible to use detergents that do not contain ions capable of precipitating the divalent ions.

18.3 THE GROUP IIIA ELEMENTS

IIIA
B 5
Al 13
Ga 31
In 49
Tl 81

All the elements of group IIIA except boron are metals, and they have specific conductances similar to those of the group IA and IIA metals, ranging from 5×10^4 to 4×10^5 ohm^{-1} cm^{-1} at 0°C. Boron is a metalloid. Its electrical characteristics are those of a semiconductor (Section 17.12)—a very low specific conductance (5×10^{-7} ohm^{-1} cm^{-1} at 0°C) that increases with increasing temperature. Boron and aluminum compounds show many similarities, but in general boron tends to form bonds that are more covalent and less ionic. This tendency may be correlated with the higher ionization potentials and higher electronegativity of boron (see Table 18.5).

Boron, which constitutes only $3 \times 10^{-4}\%$ of the earth's crust, occurs principally as borates near regions of former volcanic activity in southern Asia and in California. Boric acid (H_3BO_3) has been found in volcanic steam jets in Italy. Aluminum is the most common metallic element (8% of the earth's crust), whereas gallium, indium, and thallium are found only in trace amounts.

Elemental boron can be prepared by reducing B_2O_3 with magnesium:

$$B_2O_3(s) + 3Mg(s) = 2B(s) + 3MgO(s)$$

Aluminum metal is obtained by electrolytic reduction of the oxide, Al_2O_3, dissolved in molten cryolite, Na_3AlF_6:

$$2Al_2O_3(l) = 4Al(l) + 3O_2(g)$$

TABLE 18.5 **PROPERTIES OF THE GROUP IIIA ELEMENTS**

Property	B	Al	Ga	In	Tl
Atomic number	5	13	31	49	81
Electron configuration	[He]$2s^22p$	[Ne]$3s^23p$	[Ar]$3d^{10}4s^24p$	[Kr]$4d^{10}5s^25p$	[Xe]$4f^{14}5d^{10}6s^26p$
Ionization potentials (eV):					
I_1	8.30	5.99	6.00	5.79	6.11
I_2	25.2	18.8	20.5	18.8	20.4
I_3	37.9	28.4	30.6	27.9	29.7
I_4	259	120	64.2	54.4	50.7
Melting point (°C)	~2250	659	29.8	157	304
$\mathcal{E}°$ (V)[a]	—	−1.66	−0.53	−0.34	+0.72[b]
Pauling electronegativity	2.0	1.5	1.6	1.7	1.8

[a] Standard electrode potential for $M^{3+} + 3e^- = M(s)$.
[b] For $Tl^+ + e^- = Tl(s)$, $\mathcal{E}° = -0.34$ V.

Gallium, indium, and thallium may be prepared by electrolytic reduction of the ions in aqueous solution.

The melting points of the group IIIA elements show a minimum at gallium. It melts at 29.8 °C, and therefore it is a liquid on a hot summer day. Because the boiling point of gallium is quite high (2070 °C), it has the longest liquid range of all the elements. The metals are quite reactive. Aluminum does not ordinarily appear to be so because it is usually protected by a thin oxide coating that is impermeable to oxygen. When this coating is removed by aqueous NaOH or HCl, aluminum metal is quite reactive. Aluminum metal is soft, light, and malleable, but it forms alloys that are hard with other metals. Owing to their low densities, these alloys have many structural uses, as in airplane construction.

In binary compounds with other elements, the common oxidation state of the group IIIA elements is $+3$, which is to be expected if all three valence electrons enter into bond formation with more electronegative atoms. All the elements except boron also exhibit the $+1$ oxidation state. The tendency toward the $+1$ oxidation state increases in the order $Al < Ga < In \ll Tl$. The reason for the $+1$ state is discussed below under the chemistry of thallium.

The simple boron hydride, BH_3, has not been isolated, but the dimer (diborane, B_2H_6) and larger aggregates, such as B_4H_{10}, constitute an important class of boron compounds called the boron hydrides, or **boranes.** They are all extremely reactive and inflame in the presence of air or oxygen. Diborane is a toxic gas with a vile odor. It is prepared by addition of boron trifluoride to a sodium borohydride solution:

$$4BF_3 + 3NaBH_4 = 3NaBF_4 + 2B_2H_6$$

At elevated temperatures, B_2H_6 undergoes a series of reactions to form other boron hydrides, such as tetraborane, B_4H_{10},

$$2B_2H_6 = B_4H_{10} + H_2$$

Along with a few other group IIIA and beryllium compounds, the boranes have peculiar bridge structures. That of diborane is the simplest (see Figure 18.3). There is a planar B_2H_4 unit that closely resembles ethylene (C_2H_4). The four hydrogens in this unit are each bonded to only one boron atom, and they are said to be in **terminal** positions. The other two hydrogens, each bonded to two boron atoms, are in **bridge** positions above and below the B_2H_4 plane.

A simple approach to the electronic structure of diborane is through a bent-bond model. Of the 12 valence electrons of B_2H_6, 8 electrons form 4 single bonds to the terminal hydrogen atoms. The other 4 electrons form 2 bent bonds between the boron atoms, with a bridging proton embedded in each. Thus each of these 2 electrons pairs binds the 2 boron nuclei and a bridge proton in a **three-center bond.**

Structures of some of the higher boranes are shown in Figure 18.4. Hydrogens are found in both bridge and terminal positions. Other molecules with bridge

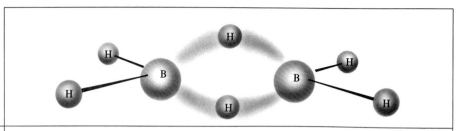

FIGURE 18.3

Bent-bond model for diborane.

structures are $Al_2(CH_3)_6$ (see Figure 18.5) and many of the group IIIA halides, such as Al_2Cl_6, Al_2Br_6, and Ga_2Cl_6. $Be(CH_3)_2$ forms a polymer $[Be(CH_3)_2]_n$, in which the methyl groups occupy bridge positions (Figure 18.5).

The halides of the group IIIA elements show considerable differences in their behavior; for example, BCl_3 (bp 12.5°C) is a gas at ordinary temperatures, whereas the remaining trichlorides are solids. Liquid BCl_3 is a nonconductor of electricity; the molten trichlorides of the other elements are conductors of electricity with specific conductances increasing from 5.6×10^{-7} ohm^{-1} cm^{-1} for $AlCl_3$ to 0.42 ohm^{-1} cm^{-1} for $InCl_3$. The tendency for boron to form covalent bonds may be explained by the large values of its first three ionization potentials when compared with the other elements of the group.

In the boron trihalides the three halogen atoms lie at the corners of an equilateral triangle with the boron atom at the center. The XBX bond angles are all 120°, and hybridization of the boron orbitals is of sp^2 type (Section 9.20). Possible resonance structures includes an all single-bond structure and three structures with double bonds:

The first structure violates the octet rule; the other structures are deficient in that they place a positive formal charge on a halogen atom, although the halogens are more electronegative than boron. It is, then, not surprising that the boron halides are excellent Lewis acids, readily combining with groups that can furnish a pair of electrons. Thus BF_3 accepts a pair of electrons from F^- to form BF_4^-, or from NH_3 to form F_3BNH_3:

$$BF_3 + :\!\ddot{F}\!:^- = [F_3B:\ddot{F}:]^-$$
$$BF_3 + :\!NH_3 = F_3B:NH_3$$

Aluminum complexes generally have coordination number 6 rather than 4; with F^-, they include not only $[AlF_6]^{3-}$ but also all of the complexes in the series

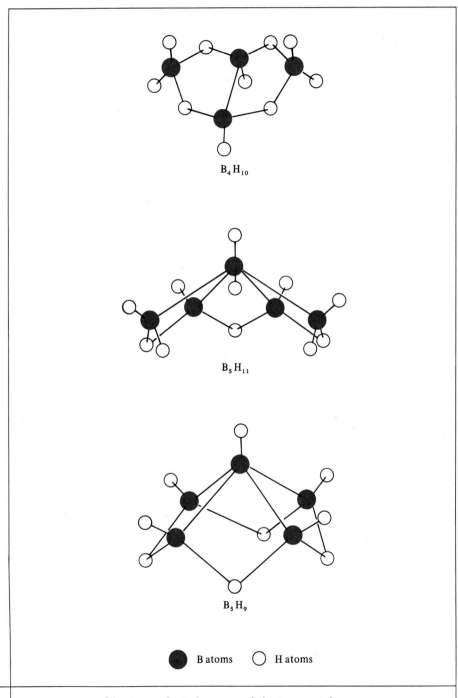

FIGURE 18.4 Structures of tetraborane and the two pentaboranes.

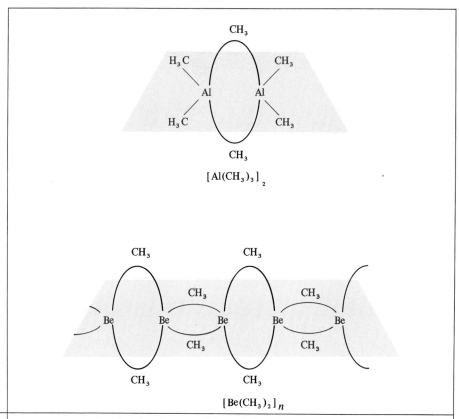

FIGURE 18.5 Structures of trimethylaluminum and dimethylberyllium. The colored portions depict atoms in the same plane.

$[AlF_n(H_2O)_{6-n}]^{+3-n}$, where n varies from 1 to 6. A more detailed discussion of complexes of this type is given in Chapter 21.

The sesquioxides of the group IIIA elements (M_2O_3) are all known. They increase in basicity as we go down the group, an effect previously noted for the group IIA elements. B_2O_3 is the anhydride of boric acid (H_3BO_3), which is a very weak monobasic acid. Boric acid does not act directly as a proton donor in aqueous solutions, but rather as a Lewis acid by accepting OH^- from water:

$$B(OH)_3 + H_2O = B(OH)_4^- + H^+ \qquad K = 5.8 \times 10^{-10}$$

The sesquioxides and hydroxides of aluminum and gallium are very slightly soluble and amphoteric, whereas those of indium and thallium are very slightly soluble and basic. The amphoteric behavior of $Al(OH)_3$ is illustrated by the following equations that show its ability to dissolve in acidic or basic solutions:

$$Al(OH)_3(s) + 3H^+ = Al^{3+} + 3H_2O$$

$$Al(OH)_3(s) + OH^- = Al(OH)_4^-$$

Aqueous solutions of aluminum salts are acidic, which may be explained by dissociation of the aquated Al^{3+} ion:

$$[Al(H_2O)_6]^{3+} = [Al(H_2O)_5OH]^{2+} + H^+ \quad K = 1.1 \times 10^{-5}$$

Thus, aquated Al^{3+} ion is about as strong an acid as is acetic acid.

Thallium also forms an important series of compounds in which it exhibits an oxidation state of +1. Often when an element exhibits more than one oxidation state, Roman numerals are used after the common name of the element to distinguish the different oxidation states. Thus Tl_2O is thallium(I) oxide and Tl_2O_3 is thallium(III) oxide. Thallium(I) compounds resemble those of the alkali metals in the solubility of the oxide, hydroxide, nitrate, fluoride, and sulfate. However, the sulfide and halides (except the fluoride) are only slightly soluble in water, showing a similarity between thallium(I) and silver(I). Soluble thallium(I) compounds are poisonous when ingested and should be used with suitable precautions.

In the thallium(I) compounds, the two $6s$ electrons do not participate in bond formation, and they are called the **inert pair.** Their inertness is due to the energy required to promote one of the s electrons to a vacant p orbital. As outlined in Section 9.20, promotion is necessary if all three valence electrons are to form chemical bonds, but promotion can occur only if the atom can form three strong bonds. As we go down the group the bonds become progressively weaker, the pair of s electrons becomes more inert, and the importance of oxidation states higher than +1 decreases. The inert-pair effect is also shown by the group IVA elements.

18.4 THE GROUP IVA ELEMENTS

IVA
C 6
Si 14
Ge 32
Sn 50
Pb 82

The elements of group IVA all exhibit quadrivalence, with all four valence electrons participating in bond formation. As we go down the group, bivalence becomes increasingly important because of the inert-pair effect noted in the last section. Carbon is a nonmetal; silicon and germanium are metalloids; and tin and lead are metals. As previously observed for the group IIIA elements, the compounds of the group IVA elements show increasing ionic character as we go down the group. Although carbon constitutes only about 0.1% of the earth's crust, it is the second most abundant element (after oxygen) in the human body, and it is the basis of all plant and animal tissues. Silicon is second only to oxygen in the earth's crust (28%). It occurs mainly as SiO_2 (sand, quartz, flint) or as complex silicates that are found in many rocks, clays, and soils. Germanium, tin, and lead are relatively rare (about 0.1%) but are of commercial importance.

The most notable chemical property of carbon is its ability to form covalent bonds, especially carbon–carbon bonds. This ability to form bonds between like

TABLE 18.6 **PROPERTIES OF THE GROUP IVA ELEMENTS**

Property	C	Si	Ge	Sn	Pb
Atomic number	6	14	32	50	82
Electron configuration	[He]$2s^2 2p^2$	[Ne]$3s^2 3p^2$	[Ar]$3d^{10} 4s^2 4p^2$	[Kr]$4d^{10} 5s^2 5p^2$	[Xe]$4f^{14} 5d^{10} 6s^2 6p^2$
Ionization potentials (eV):					
I_1	11.26	8.15	7.90	7.34	7.42
I_2	24.4	16.3	15.9	14.6	15.0
I_3	47.9	33.5	34.2	30.6	32.0
I_4	64.5	45.1	45.7	39.6	42.3
I_5	392	167	93	81	69.4
Melting point (°C)	~3600[b]	1414	958	232	328
$\mathcal{E}°$ (V)[a]	—	—	—	−0.14	−0.13
Pauling electronegativity	2.5	1.8	1.8	1.8	1.8
M—M bond energy (kJ mol^{-1})	356	238	188	151	—

[a] Standard electrode potential for $M^{2+}(aq) + 2e^- = M(s)$.
[b] Sublimation point.

atoms (**catenation**) decreases markedly in the order C $\gg$ Si > Ge $\approx$ Sn $\gg$ Pb and may be attributed in part to the marked decrease in the bond energy between like atoms as we go down the group (see Table 18.6). Some of the chemistry of carbon has been discussed in Sections 11.8 and 11.9, and organic compounds are the subject of Chapter 22. In this chapter we shall consider some of the carbon-containing substances that occur in aqueous solutions.

At 25°C, carbon dioxide is soluble in water to the extent of 0.0337 M when the partial pressure of CO_2 above the solution is 1 atm. The resulting solution is slightly acidic ([H$^+$] = 1.2 × 10^{-4} M). The acidity of aqueous solutions of carbon dioxide may be explained by assuming that the dissolved carbon dioxide reacts with water to produce carbonic acid (H_2CO_3), for which the Lewis structure is

$$H-\ddot{\underset{..}{O}}-\underset{\underset{\ddot{O}:}{\|}}{C}-\ddot{\underset{..}{O}}-H$$

The H_2CO_3 dissociates to produce H$^+$ and HCO_3^-:

$$H_2O + CO_2 \rightleftharpoons H_2CO_3 \qquad K_\alpha = \frac{[H_2CO_3]}{[CO_2]} \qquad (18.6)$$

$$H_2CO_3 \rightleftharpoons H^+ + HCO_3^- \qquad K_{H_2CO_3} = \frac{[H^+][HCO_3^-]}{[H_2CO_3]} \qquad (18.7)$$

At one time it was assumed that the first equilibrium was very far to the right, and that all the dissolved CO_2 was present as H_2CO_3. On this basis the value of $K_{H_2CO_3}$ would be

$$K_{H_2CO_3} = \frac{(1.2 \times 10^{-4})(1.2 \times 10^{-4})}{0.0337 - (1.2 \times 10^{-4})} = 4.3 \times 10^{-7} \qquad (18.8)$$

However, this value is much less than we would predict for an oxyacid of this formula (Section 19.1). According to rule 1 for oxyacids, we would estimate $K_{H_2CO_3} \approx 1 \times 10^{-3}$ because $m = 1$ for $(HO)_2CO$.

Additional evidence that all the dissolved carbon dioxide is not present as H_2CO_3 is provided by the slow rate of neutralization of aqueous solutions containing dissolved carbon dioxide. Qualitatively this may be demonstrated by comparing the time required for neutralization of sodium hydroxide solutions containing phenolphthalein indicator by (a) an aqueous CO_2 solution or (b) a solution of acetic acid of the same molarity. The acetic acid solution decolorizes the indicator as soon as the solutions are mixed, whereas the aqueous solution of CO_2 requires from 10 to 15 s after mixing for the indicator color to disappear. A kinetic interpretation of the results is

(1) $CH_3CO_2H + OH^- \xrightarrow{\text{fast}} CH_3CO_2^- + H_2O$

(2) $\begin{cases} CO_2(aq) + H_2O \xrightarrow{\text{slow}} H_2CO_3 \\ H_2CO_3 + OH^- \xrightarrow{\text{fast}} HCO_3^- + H_2O \end{cases}$

The results of several different experimental methods indicate that only about 0.2% of the dissolved CO_2 is present as H_2CO_3 molecules. The other 99.8% of the dissolved CO_2 is present as hydrated linear CO_2 molecules. Thus a saturated solution of carbon dioxide at 25°C contains $0.002 \times 0.0337 = 7 \times 10^{-5} M\ H_2CO_3$, and $K_{H_2CO_3}$ is therefore

$$K_{H_2CO_3} = \frac{(1.2 \times 10^{-4})^2}{7 \times 10^{-5}} = 2 \times 10^{-4}$$

This result is within the range of values expected for an oxyacid of this formula.

Although Eqs. 18.6 and 18.7 could be used in equilibrium calculations, it is more convenient to use a thermodynamic convention that ignores the distinction between the two kinds of hydrated molecules. Both species are denoted by the term "carbonic acid" and by the symbol $CO_2(aq)$ and their *total* concentration is $[CO_2(aq)]$. Then a single equation expresses the acidity of aqueous solutions of CO_2:

$$CO_2(aq) + H_2O = H^+ + HCO_3^- \qquad K_1 = \frac{[H^+][HCO_3^-]}{[CO_2(aq)]} \qquad (18.9)$$

Called the first dissociation constant of carbonic acid, K_1 has a numerical value of 4.3×10^{-7} (calculated by the same procedure as in Eq. 18.8). The second dissociation

$$HCO_3^- = H^+ + CO_3^{2-} \qquad K_2 = \frac{[H^+][CO_3^{2-}]}{[HCO_3^-]} \qquad (18.10)$$

has a dissociation constant K_2 of 4.7×10^{-11}.

The neutralization of carbonic acid can be carried out in two stages. One mol of $CO_2(aq)$ will neutralize 1 mol of OH^- to produce HCO_3^-:

$$OH^- + CO_2(aq) = HCO_3^-$$

Then further addition of OH^- removes a proton from HCO_3^- to produce CO_3^{2-}:

$$HCO_3^- + OH^- = CO_3^{2-}$$

The sodium salts of these anions are important industrial chemicals. Sodium carbonate (Na_2CO_3) is used in the manufacture of glass and in making soap. The decahydrate ($Na_2CO_3 \cdot 10H_2O$), known as washing soda, is obtained by crystallization from water. Redissolved in water, it produces a mildly basic solution owing to hydrolysis of the carbonate ion, which assists the action of soaps in laundering:

$$CO_3^{2-} + H_2O = HCO_3^- + OH^-$$

Sodium bicarbonate ($NaHCO_3$) is also known as baking soda. It is mixed with acidic substances, such as cream of tartar ($KHC_4H_4O_6$), to form baking powders. When a baking powder is wet, small bubbles of CO_2 gas are released by the following reaction, causing cakes and quick breads to rise:

$$HCO_3^- + HC_4H_4O_6^- = C_4H_4O_6^{2-} + H_2O + CO_2(g)$$

In baking bread the same effect is produced by using yeast (a microorganism), which produces an enzyme that converts sugar into alcohol and carbon dioxide:

$$C_6H_{12}O_6 = 2C_2H_5OH + 2CO_2(g)$$

Other important acids of carbon include the carboxylic acids (RCO_2H) such as acetic acid (CH_3CO_2H), the substance that gives vinegar its characteristic odor and sour taste. Acetic acid is a weak acid,

$$CH_3CO_2H = CH_3CO_2^- + H^+ \qquad K = 1.8 \times 10^{-5}$$

Carbon forms several important compounds with nitrogen, the simplest of which is cyanogen (C_2N_2), a colorless, extremely poisonous gas. The chemistry of

18.4 THE GROUP IVA ELEMENTS

cyanogen is very similar to that of the heavier halogens. In basic solutions it disproportionates to produce cyanide ion (CN^-) and cyanate ion (OCN^-):

$$C_2N_2(g) + 2OH^- = CN^- + OCN^- + H_2O$$

The chemistry of CN^- resembles that of Cl^-, the principal exception being that HCN is a weak acid ($K = 6.1 \times 10^{-10}$). Cyanide ion forms very stable complex ions with many transition metals. Extreme precautions must be taken when working with cyanide solutions or hydrogen cyanide because they are lethal.

The dioxides of silicon, germanium, tin, and lead are all solids at room temperature. Although SiO_2 is not soluble in water, it dissolves in hydrofluoric acid, producing the SiF_6^{2-} ion, and therefore HF may be used to etch glass:

$$SiO_2(s) + 6HF = SiF_6^{2-} + 2H_2O + 2H^+$$

Glass is prepared by fusing sand (SiO_2) with smaller quantities of basic substances, such as CaO and Na_2CO_3. Borosilicate glass also contains about 12% B_2O_3, which, by imparting a low coefficient of thermal expansion to the glass, makes it suitable for articles subjected to sudden changes in temperature.

Elemental silicon and germanium used for the manufacture of transistors must be very pure because their electrical properties are markedly affected by traces of impurities. To obtain silicon (or germanium) of the desired purity the oxide is first reduced by carbon to form the impure element:

$$SiO_2 + 2C = Si + 2CO$$

Reaction with chlorine produces the volatile tetrachloride (bp 58°C):

$$Si + 2Cl_2 = SiCl_4$$

After purifying the tetrachloride by distillation, it is reduced by hydrogen to produce purified silicon:

$$SiCl_4 + 2H_2 = Si + 4HCl$$

Finally, the pure silicon is made "ultrapure" by a process known as **zone refining.** In this process a rod of silicon is heated at one end until it melts. Then the molten cross section is moved slowly down the rod by moving the heat source. As the silicon cools, it crystallizes in a state of higher purity—the impurities are more soluble in the liquid than in the solid, and they stay in the molten section as it moves to the end of the rod. The process is repeated several times to obtain the desired purity, and the impure end is then removed.

Silicon will dissolve in concentrated solutions of a strong base to give silicates and hydrogen:

$$Si(s) + 2OH^- + H_2O = SiO_3^{2-} + 2H_2$$

The aqueous chemistry of silicon is primarily that of its complex halides in acidic solution and that of its silicates in strongly basic solution.

There are many examples of polymeric units containing Si—O—Si bonds. Many minerals contain alternating —Si—O—Si—O chains (Section 17.6). Siloxanes are polymeric substances based on chains or networks containing Si—O—Si units. Methyl silicone is a linear polymer containing CH_3 groups:

$$\begin{array}{c} \quad\quad CH_3 \quad CH_3 \\ \quad\quad | \quad\quad\; | \\ \diagdown O \diagup Si \diagdown O \diagup Si \diagdown O \diagup \\ \quad\quad | \quad\quad\; | \\ \quad\quad CH_3 \quad CH_3 \end{array}$$

It is used industrially as a lubricant because it has greater chemical and thermal stability than the hydrocarbon oils. The chemistry of germanium is somewhat similar to that of silicon. As germanium is extremely rare, its chemistry will not be discussed in more detail.

Tin and lead are metals, and they differ from the other members of group IVA in that many of their compounds contain the elements in the +2 oxidation state. In tin(II) and lead(II) compounds, only the two valence electrons in p orbitals take part in bond formation. For reasons discussed at the end of the preceding section, the pair of s electrons in the outer shell ($5s$ for tin, and $6s$ for lead) is relatively inert in these elements.

The standard electrode potentials for tin and lead are both slightly negative (Table 18.6), indicating that these metals should dissolve in dilute acids; the rate of solution is, however, quite slow. Like aluminum, ordinary tin forms a thin oxide coating that makes it quite inert. Because of this inertness tin is used as a protective coating over steel in the production of "tin" cans.

For metals having two important oxidation states, suffixes are often used to distinguish between them. The ending -ous is used for the cation having the lower oxidation state, -ic for the higher. Thus, Sn^{2+} is stannous ion and Sn^{4+} is stannic ion. Similar terms are used for the oxidation states of lead (+2, plumbous; +4, plumbic). When anionic species are formed in basic solution, the ending -ite is used for the lower oxidation state, -ate for the higher. For example, in basic solution +2 tin exists as stannite ion, $Sn(OH)_3^-$, and +4 tin as stannate ion, $Sn(OH)_6^{2-}$.

In aqueous solutions of soluble tin(II) compounds, the stannous ion hydrolyzes to such an extent that it is as strong an acid as HSO_4^-:

$$Sn^{2+} + H_2O = SnOH^+ + H^+ \quad K = 1 \times 10^{-2}$$

With Cl^-, Sn^{2+} forms a series of complexes such as $SnCl^+$, $SnCl_2$, and $SnCl_3^-$. Acidic solutions of Sn^{2+} are often used as mild reducing agents, with the tin being converted to tin(IV). The standard electrode potential is

$$Sn^{4+} + 2e^- = Sn^{2+} \quad \mathcal{E}° = +0.15 \text{ V}$$

18.4 THE GROUP IVA ELEMENTS

The hydroxide $Sn(OH)_2$ is very slightly soluble and is amphoteric, dissolving in basic solutions to produce $Sn(OH)_3^-$. In basic solutions tin(II) slowly disproportionates to form metallic tin and tin (IV):

$$2Sn(OH)_3^- = Sn(s) + Sn(OH)_6^{2-}$$

Like tin(II), tin(IV) undergoes extensive hydrolysis, forms complexes with anions such as Cl^- (e.g., $SnCl_6^{2-}$), and has a slightly soluble amphoteric hydroxide. Both tin(II) and tin(IV) are precipitated from acidic solution by H_2S as the sulfides, SnS and SnS_2, respectively. Stannous sulfide varies from brown to almost black; stannic sulfide is yellow. Stannic sulfide dissolves in solutions containing sulfide ion to produce thiostannate ion, SnS_3^{2-}:

$$SnS_2(s) + S^{2-} = SnS_3^{2-}$$

As might be predicted from equilibrium considerations, SnS_2 may be reprecipitated by acidifying the solution. Although stannous sulfide does not dissolve in sulfide ion solutions, it is soluble in polysulfide solutions, where it is oxidized to thiostannate ion:

$$SnS(s) + S_2^{2-} = SnS_3^{2-}$$

The most common oxidation state of lead is the +2 state. Only a few lead(II) salts, such as the nitrate and acetate, are readily soluble in water. Plumbous ion forms complexes with halide ions similar to those of stannous ion. It hydrolyzes to a much smaller extent than stannous ion (in keeping with the general decrease in acidity going down the group):

$$Pb^{2+} + H_2O = PbOH^+ + H^+ \qquad K = 1 \times 10^{-8}$$

The hydroxide is slightly soluble and is amphoteric, dissolving in basic solutions to give $Pb(OH)_3^-$. Other important slightly soluble lead(II) compounds are $PbCl_2$, $PbSO_4$, $PbCO_3$, and PbS. Lead compounds are extremely poisonous.

The most important lead(IV) compound is PbO_2, which is a very strong oxidizing agent in acidic solutions:

$$PbO_2(s) + 4H^+ + 2e^- = Pb^{2+} + 2H_2O \qquad \mathcal{E}° = +1.46 \text{ V}$$

It is the oxidizing agent in the lead storage battery (Section 14.9). Although PbO_2 is amphoteric, it is relatively inert toward dilute acids and bases. On dissolving in concentrated acids, it oxidizes water to oxygen (or Cl^- to Cl_2 in the case of hydrochloric acid). On dissolving in concentrated bases, it forms plumbate ion, $Pb(OH)_6^{2-}$.

PROBLEMS

18.1 Which elements of groups IA, IIA, IIIA, and IVA are nonmetals? Which are metalloids? Which are metals?

18.2 Draw Lewis structures for HClO, $HClO_2$, $HClO_3$, $HClO_4$, and their anions. (In each case the hydrogen atom is bonded to an oxygen atom.) What are the dissociation constants of the acids (see Table 18.1)? Explain the variation in dissociation constants.

18.3 Using an approach similar to that given in Section 18.1 to explain rule 1, explain rule 2 as it applies to phosphoric acid (H_3PO_4).

18.4 Using rules 1 and 2, estimate K_1 and K_2 for a diprotic acid of formula $(HO)_2XO$.

18.5 Using the data in Table 18.3, draw a graph showing crystal lattice enthalpy (ΔH_c) and the negative of the hydration enthalpy ($-\Delta H_h$) versus period number for the alkali metal chlorides. Connect the experimental points for each property by a smooth curve. Show that the enthalpy of solution, ΔH_s, changes sign where the two curves cross.

18.6 Explain why water molecules in the primary hydration sphere are oriented differently toward Na^+ and Cl^- (Figure 18.1).

18.7 In the electrolysis of liquid NaH, hydrogen is produced at the anode. Explain. What volume of hydrogen at standard conditions would be produced if 0.10 F were used?

18.8 What are the formulas of sodium oxide, sodium peroxide, and potassium superoxide? What are the molecular orbital electron configurations of the O_2^- and O_2^{2-} ions? What is the net number of bonding electrons in each? Which should have the shortest bond length, O_2, O_2^-, or O_2^{2-}? Which should have the longest bond length?

18.9 Compare sodium and magnesium qualitatively with respect to each of the following properties: metallic character, melting point, first and second ionization potentials, oxidation state in compounds, ionic character of compounds, basic character of hydroxide, standard electrode potential, solubility of chloride and sulfate in water, solubility of carbonate and hydroxide in water.

18.10 A sample of solid calcium chloride is left in an open bottle. At a later date it is noticed that the bottle contains a liquid solution. Suggest an explanation.

18.11 (a) Calculate the equilibrium constant for Eq. 18.4 from the dissociation constants of carbonic acid ($K_1 = 4.3 \times 10^{-7}$ and $K_2 = 4.7 \times 10^{-11}$) and the solubility product of calcium carbonate ($K_{sp} = 4.5 \times 10^{-9}$).
(b) Calculate the solubility of calcium carbonate in $0.0337 M$ CO_2(aq). (Assume that contact with CO_2(g) at 1 atm pressure maintains [CO_2(aq)] constant.)

18.12 After dissolving a small amount of solid calcium carbonate in a solution of carbonic acid, the solution is heated to boiling for a short time and then cooled. At this point it is noted that the calcium carbonate has precipitated from the solution. Suggest an explanation.

18.13 "Hard water" containing Ca^{2+}, Mg^{2+}, HCO_3^-, and SO_4^{2-} ions is passed through a zeolite ion exchanger, producing "soft water." What ions are present in the soft water?

18.14 A $1M$ NaCl solution is passed through an ion exchange resin that replaces the Na^+ ions by H^+ ions, and then through another ion exchange resin that replaces the Cl^- ions by OH^- ions. What are [H^+] and [OH^-] in the final solution?

18.15 Which substances are Lewis acids and which are Lewis bases in the following reactions?
$BF_3 + F^- = BF_4^-$
$(C_2H_5)_2O + BF_3 = (C_2H_5)_2OBF_3$
$B_2H_6 + 2CO = 2H_3BCO$
$Al_2Cl_6 + 6Cl^- = 2Al(Cl)_6^{3-}$

18.16 Using the standard electrode potentials for thallium given in Table 18.5, calculate the equilibrium constant for the reaction
$3Tl^+ = 2Tl(s) + Tl^{3+}$

18.17 As an alternative explanation of the inert-pair effect, consider the possibility that this effect is caused by changes in the second and third ionization potentials of the group IIIA elements with increasing atomic number.
(a) What trend in ionization potentials would be required to explain the inert-pair effect?
(b) Do the group IIIA elements exhibit such a trend? (See Table 18.5.)
(c) Apply a similar line of reasoning to the inert-pair effect in the group IVA elements.

18.18 Draw Lewis structures for F_3BNH_3, BF_4^-, HCN, OCN^-, C_2N_2 (a linear molecule with structure NCCN), and CO_3^{2-} (three resonance structures). Indicate all formal charges.

18.19 Assuming the mechanisms described in Section 18.4, what rate law would you expect for the neutralization of a solution of NaOH by
  (a) CH_3CO_2H?
  (b) $CO_2(aq)$?

18.20 Compare boron and silicon qualitatively with respect to each of the following properties: metallic character, melting point, first ionization potential, oxidation state in compounds, electronegativity, ionic character of compounds, basic character of oxides, volatility of chlorides, structure of hydride M_2H_6.

18.21 Write balanced equations for the net reactions that occur when
  (a) sodium metal reacts with water to form gaseous hydrogen and a solution of sodium hydroxide
  (b) potassium metal reacts with oxygen to form potassium superoxide
  (c) magnesium metal reacts with aqueous HCl to form gaseous hydrogen and a solution of magnesium chloride
  (d) solid calcium carbonate is deposited in a cave by evaporation of water containing Ca^{2+} and HCO_3^- ions
  (e) aluminum metal dissolves in aqueous HCl forming gaseous hydrogen
  (f) aluminum metal dissolves in aqueous NaOH forming gaseous hydrogen
  (g) bicarbonate ion, HCO_3^-, acts as an acid, neutralizing OH^-
  (h) bicarbonate ion acts as a base, neutralizing H^+
  (i) stannic sulfide, SnS_2, is precipitated by adding dilute HCl to a solution containing thiostannate ion, SnS_3^{2-}
  (j) salt water accidentally enters a lead storage battery, and gaseous chlorine is produced

18.22 As noted in Section 18.2, the acidic properties of a metal hydroxide are most pronounced when the cation is small. Suggest an explanation.

CHAPTER 19

THE ELEMENTS OF GROUPS VA-VIIA AND THE NOBLE GASES

19

The elements of groups VA, VIA, VIIA, and the noble gases include all of the nonmetals except hydrogen and carbon. Their chemical behavior is characterized by a greater tendency for electron-pair bonding than is that of the metallic elements. There is usually a pronounced change in character between the first two elements of a group (as in groups IIIA and IVA) and a general trend toward more metallic properties within each group with increasing atomic number.

19.1 THE GROUP VA ELEMENTS

VA
N 7
P 15
As 33
Sb 51
Bi 83

At room temperature nitrogen is a stable unreactive gas, whereas phosphorus is a nonmetallic solid that is very reactive. Arsenic and antimony are metalloids; bismuth is usually considered to be a metal although it has a low electric conductivity. As a group the elements show a considerable range in chemical behavior. Nitrogen and phosphorus form mainly covalent bonds in their compounds, whereas arsenic, antimony, and bismuth show increasing tendencies toward cationic behavior although no simple cations of these elements occur in aqueous solutions. The relatively large ionization potentials (Table 19.1) provide an explanation for the failure of these elements to form simple cations. The steady decrease in electronegativity as the atoms increase in size accounts for the increase in cationic behavior from N to Bi. The characteristic oxidation states for the group VA elements are -3, $+3$ and $+5$; however, nitrogen exhibits every integral oxidation state from -3 to $+5$.

Nitrogen occurs primarily as N_2 in the atmosphere (78%), as nitrate deposits such as $NaNO_3$ (Chile saltpeter), and in all animal and vegetable proteins. The chemistry of nitrogen is characterized by the extreme stability of the nitrogen molecule (because of the very strong nitrogen–nitrogen triple bond), and the high electronegativity of nitrogen (3.0). Only fluorine and oxygen have higher electronegativities than nitrogen. Nitrogen exhibits all oxidation states from $+5$ to -3 in its compounds with hydrogen and oxygen, as shown in Table 19.2.

Ammonia is a gas at room temperature, and it is quite soluble in water, producing a basic solution:

$$NH_3(g) = NH_3(aq)$$
$$NH_3(aq) + H_2O = NH_4^+ + OH^- \quad (19.1)$$

19.1 THE GROUP VA ELEMENTS

TABLE 19.1 **PROPERTIES OF THE GROUP VA ELEMENTS**

Property	N	P	As	Sb	Bi
Atomic number	7	15	33	51	83
Electron configuration	[He]$2s^2 2p^3$	[Ne]$3s^2 3p^3$	[Ar]$3d^{10}4s^2 4p^3$	[Kr]$4d^{10}5s^2 5p^3$	[Xe]$4f^{14}5d^{10}6s^2 6p^3$
Ionization potentials (eV):					
I_1	14.53	10.49	9.81	8.64	7.29
I_2	29.6	19.7	18.6	16.5	16.7
I_3	47.4	30.2	28.3	25.3	25.6
I_4	77.5	51.4	50.1	44.1	45.3
I_5	97.9	65.0	62.6	56	56.0
Melting point (°C)	−210	44[a]	814[b]	630	271
Pauling electronegativity	3.0	2.1	2.0	1.9	1.9

[a] White phosphorus.
[b] At 36 atm.

Although solutions of ammonia in water are often called **ammonium hydroxide** (NH$_4$OH) solutions, there is no evidence for the existence of NH$_4$OH molecules. Therefore aqueous ammonia solutions are best described by the equilibrium shown in Eq. 19.1 with

$$K = \frac{[NH_4^+][OH^-]}{[NH_3]} = 1.77 \times 10^{-5}$$

where [NH$_3$] is the total amount of ammonia in solution, whether as NH$_3$ or as a hydrated species.

There are many salts of ammonium ion (NH$_4^+$) and their chemistry resembles that of potassium salts, with the exception that ammonium salts are much more volatile, decomposing to form gases on heating. For example, at about

TABLE 19.2 **OXIDATION STATES OF NITROGEN**

Oxidation state	Compound or ion	Name
+5	N$_2$O$_5$, HNO$_3$	Dinitrogen pentoxide, nitric acid
+4	NO$_2$, N$_2$O$_4$	Nitrogen dioxide, dinitrogen tetroxide
+3	N$_2$O$_3$, HNO$_2$	Dinitrogen trioxide, nitrous acid
+2	NO	Nitric oxide
+1	N$_2$O, H$_2$N$_2$O$_2$	Nitrous oxide, hyponitrous acid
0	N$_2$	Nitrogen
−1	NH$_2$OH	Hydroxylamine
−2	N$_2$H$_4$	Hydrazine
−3	NH$_3$, NH$_4^+$	Ammonia, ammonium ion

300°C ammonium chloride decomposes to form ammonia and hydrogen chloride:

$$NH_4Cl(s) = NH_3(g) + HCl(g)$$

Ammonia acts as an electron donor (Lewis base) in many of its reactions owing to its unshared pair of electrons. In addition to its reaction to form NH_4^+, ammonia forms stable complex ions with many transition metal ions such as silver ion in aqueous solutions:

$$NH_3 + H^+ = NH_4^+$$
$$2NH_3 + Ag^+ = Ag(NH_3)_2^+$$

Liquid ammonia is a surprisingly good solvent for the alkali and alkaline earth metals. Dilute solutions of the metals in liquid ammonia exhibit a blue color, but the more concentrated solutions look like molten bronze. The dissolved metal in the dilute solutions exists as the solvated positive ions [M^+(am)] and solvated electrons [e^-(am)]:

$$Na(s) = Na^+(am) + e^-(am)$$

Metal ammonia solutions slowly decompose to yield gaseous hydrogen and an amide:

$$2e^-(am) + 2NH_3 = 2NH_2^- + H_2(g)$$

Nitrous oxide is sometimes called *laughing gas* because small doses are mildly intoxicating. Larger doses may be used to produce general anesthesia. It is also used to make self-whipping cream: The N_2O is dissolved in cream under pressure and when the pressure is released the N_2O escapes in the form of small gas bubbles, producing whipped cream.

The aqueous chemistry of nitrogen compounds other than ammonia is primarily that of nitrous and nitric acid and their salts. Nitric acid is a strong acid and nearly all the metallic nitrates are soluble strong electrolytes. Nitrous acid is a weak acid,

$$HNO_2 = H^+ + NO_2^- \qquad K = 7.1 \times 10^{-4} \tag{19.2}$$

Aqueous solutions of nitrous acid are unstable and decompose when heated to give nitric acid and nitric oxide:

$$3HNO_2 = H^+ + NO_3^- + H_2O + 2NO(g)$$

The possible electron transfer reactions involving nitrogen compounds are so numerous that only the more important reactions will be considered. However,

19.1 THE GROUP VA ELEMENTS

many of the possibilities may be shown by using Latimer reduction-potential diagrams (originated by Wendell M. Latimer) as follows:

acidic solution [H⁺] = 1M

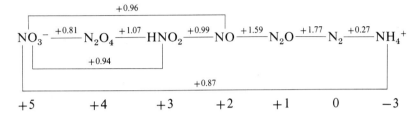

At the bottom of the diagram the oxidation state of the nitrogen atom in each ion or molecule is shown. The numbers between any two oxidation states of nitrogen represent the $\mathcal{E}°$ voltage for the balanced reduction half-reaction. For example,

$$NO_3^- \xrightarrow{+0.96} NO$$

is a shorthand notation for the half-reaction

$$NO_3^- + 4H^+ + 3e^- = NO + 2H_2O \qquad \mathcal{E}° = +0.96 \text{ V} \qquad (19.3)$$

To illustrate further the information obtainable from Latimer reduction diagrams, let us obtain the NO_3^-—NO half-reaction voltage from the series of reductions:

$$NO_3^- \xrightarrow{+0.81} N_2O_4 \xrightarrow{+1.07} HNO_2 \xrightarrow{+0.99} NO$$

The individual half-reactions would be as follows:

$$HNO_2 + H^+ + e^- = NO + H_2O \qquad \mathcal{E}° = +0.99 \text{ V}$$
$$N_2O_4 + 2H^+ + 2e^- = 2HNO_2 \qquad \mathcal{E}° = +1.07 \text{ V}$$
$$2NO_3^- + 4H^+ + 2e^- = N_2O_4 + 2H_2O \qquad \mathcal{E}° = +0.81 \text{ V}$$

We note that, if we multiply the first equation in the above set by 2, we may then add the resulting three equations to eliminate all nitrogen species except NO_3^- and NO. As indicated previously, when reactions are added to obtain an overall reaction, the free-energy change for the overall reaction is equal to the sum of the individual free-energy changes. Using $\Delta G° = -nF\mathcal{E}°$ for each reaction, we obtain

$$\begin{array}{ll} 2HNO_2 + 2H^+ + 2e^- = 2NO + 2H_2O & \Delta G_1^\circ = -2F(0.99) \\ N_2O_4 + 2H^+ + 2e^- = 2HNO_2 & \Delta G_2^\circ = -2F(1.07) \\ 2NO_3^- + 4H^+ + 2e^- = N_2O_4 + 2H_2O & \Delta G_3^\circ = -2F(0.81) \\ \hline 2NO_3^- + 8H^+ + 6e^- = 2NO + 4H_2O & \Delta G_T^\circ = \Delta G_1^\circ + \Delta G_2^\circ + \Delta G_3^\circ \\ & = -5.74F \end{array}$$

We then divide by 2 to obtain Eq. 19.3:

$$NO_3^- + 4H^+ + 3e^- = NO + 2H_2O \qquad \Delta G^\circ = \frac{\Delta G_T^\circ}{2} = -2.87F$$

For this reaction $\Delta G^\circ = -3F\mathcal{E}^\circ$, and therefore $\mathcal{E}^\circ = (-2.87F)/(-3F) = +0.96$ V, in agreement with Eq. 19.3.

The Latimer diagrams are quite useful in predicting possible disproportionation reactions, that is, the simultaneous oxidation and reduction of a substance in an intermediate oxidation state to produce a species in a higher oxidation state and one in a lower oxidation state. As an example, consider the disproportionation of the +3 oxidation state (HNO_2) to the +2 oxidation state (NO) and the +5 oxidation state (NO_3^-). The appropriate half-reactions are

$$HNO_2 + H^+ + e^- = NO + H_2O \qquad \mathcal{E}^\circ = +0.99 \text{ V}$$
$$NO_3^- + 3H^+ + 2e^- = HNO_2 + H_2O \qquad \mathcal{E}^\circ = +0.94 \text{ V}$$

We may multiply the first equation by 2 and then add the reverse of the second equation to obtain the equation for the reaction:

$$\begin{array}{ll} 2HNO_2 + 2H^+ + 2e^- = 2NO + 2H_2O & \mathcal{E}^\circ = +0.99 \text{ V} \\ HNO_2 + H_2O = NO_3^- + 3H^+ + 2e^- & \mathcal{E}^\circ = -0.94 \text{ V} \\ \hline 3HNO_2 = NO_3^- + H^+ + 2NO + H_2O & \Delta\mathcal{E}^\circ = +0.05 \text{ V} \end{array}$$

Because the $\Delta\mathcal{E}^\circ$ value is positive, we predict that the above reaction can occur, as it indeed does. We may generalize the above result to state that, for Latimer reduction-potential diagrams, *disproportionation of a substance occurs spontaneously whenever the voltage to the right of the substance is larger than the voltage to the left*. (If the voltage to the right is smaller than the voltage to the left, disproportionation does *not* occur spontaneously.) Thus nitrous acid is unstable with respect to NO and NO_3^- in acidic solution but is stable with respect to NO and N_2O_4. The above rule also indicates that N_2O is unstable with respect to N_2 and NO. (However, the rate at which favorable disproportionation reactions occur cannot be predicted and must be determined experimentally.)

The large positive values of $\mathcal{E}^\circ$ for the reduction of NO_3^- to lower oxidation states indicate that NO_3^- is a very powerful oxidizing agent. In fact nitric acid is capable of dissolving nearly all metals. There are many possible lower oxidation states available, and the composition of the reduction products of NO_3^- depends on the concentration of H^+ and NO_3^- as well as on the temperature and the reducing agent. For example, copper (which is not soluble in $2\,M\,H^+$) is readily dissolved by hot $2\,M\,HNO_3$ with NO as the principal reduction product. The half-reactions are:

$$Cu(s) = Cu^{2+} + 2e^- \qquad \mathcal{E}° = -0.34\text{ V}$$
$$NO_3^- + 4H^+ + 3e^- = NO(g) + 4H_2O \qquad \mathcal{E}° = +0.96\text{ V}$$

If the first equation is multiplied by 3 and the second equation by 2, the sum is

$$3Cu(s) + 2NO_3^- + 8H^+ = 3Cu^{2+} + 2NO(g) + 8H_2O$$
$$\Delta\mathcal{E}° = +0.62\text{ V}$$

With concentrated nitric acid the principal reduction product is NO_2.

Gold and platinum are not affected by nitric acid, but they may be dissolved by a mixture of concentrated nitric and hydrochloric acids called **aqua regia.** The dissolving power of aqua regia is due to the strong oxidizing power of nitrate ion together with the complexing ability of chloride ion for gold and platinum.

More reactive metals, such as zinc, may reduce nitric acid to ammonium ion under appropriate conditions:

$$4Zn(s) + 10H^+ + NO_3^- = 4Zn^{2+} + NH_4^+ + 3H_2O$$

The effectiveness of NO_3^- as an oxidizing agent is strongly influenced by the concentration of hydrogen ion. In fact in basic solution, NO_3^- is a very weak oxidant, as is evident from the corresponding Latimer reduction diagram:

basic solution $[OH^-] = 1\,M$

$$NO_3^- \xrightarrow{+0.01} NO_2^- \xrightarrow{-0.46} NO \xrightarrow{+0.76} N_2O \xrightarrow{+0.94} N_2 \xrightarrow{-0.74} NH_3$$

$$+5 \qquad\qquad +3 \qquad\qquad +2 \qquad\quad +1 \qquad\quad 0 \qquad\quad -3$$

Note that in contrast to the instability of HNO_2 in acid solution, nitrite ion is stable with respect to disproportionation to NO and NO_3^- in basic solution.

The nitrogen halides include NF_3, N_2F_2, N_2F_4, and NCl_3. The trihalides provide a marked contrast in chemical reactivity: Whereas NF_3 is an unreactive gas at room temperature, NCl_3 is a reactive liquid (bp 71°C) that explodes on heating, when exposed to light, or in the presence of organic compounds. The fluoride N_2F_2 exists in separable cis and trans forms:

FIGURE 19.1

The structure of P_4 and As_4 molecules.

Phosphorus occurs mainly in the form of phosphates (PO_4^{3-}), with the principal minerals being apatite [$Ca_5(PO_4)_3F$], hydroxyapatite [$Ca_5(PO_4)_3OH$], and tricalcium phosphate [$Ca_3(PO_4)_2$]. Hydroxyapatite is the main constituent of bones and teeth. Organic compounds containing phosphorus are essential components of nerve and brain tissue as well as some proteins. Arsenic, antimony, and bismuth are occasionally found uncombined, but their most important minerals are the sulfides. In contrast to the essential nature of some nitrogen and phosphorus compounds, the water-soluble compounds of arsenic, antimony, and bismuth are toxic to animals. The soluble compounds of arsenic are quite toxic to humans (fatal dose $\approx$ 0.06 to 0.20 g), but it is reported that a tolerance has been developed by some individuals by gradually increasing the dose up to gram amounts.

At ordinary temperatures phosphorus is a solid with several allotropic forms, including white phosphorus, black phosphorus, and at least one red form. White phosphorus (mp 44°C, bp 280°C) is a molecular solid containing P_4 molecules. Phosphorus vapor also contains P_4 molecules, which dissociate to P_2 molecules at very high temperatures (above $\sim$ 800°C). When cooled quickly, phosphorus vapor condenses to the white solid, which changes on heating to a more stable red form. Red phosphorus does not appear to have an ordered structure. Its color and properties depend on the conditions of preparation, and its structure appears to consist of random chains of phosphorus atoms. Black phosphorus forms under high pressure or with the aid of a mercury catalyst.

The molecule P_4 is a regular tetrahedron, with each phosphorus atom singly bonded to the other three atoms (Figure 19.1). The resulting PPP bond angle is 60°. The instability of white phosphorus no doubt results from its small bond angle.

As noted earlier, arsenic, antimony, and bismuth crystals have a hexagonal double layer or "puckered sheet" structure (see Figure 17.21) somewhat similar to graphite. Each atom is singly bonded to three atoms in the other half of the double layer, giving bond angles of 97° (As), 96° (Sb), and 94° (Bi), as compared to 120° in graphite. Black phosphorus has a similar structure, but each atom is bonded to two atoms in the same half of the double layer and one atom in the other half.

The common oxidation states and some typical compounds of the group VA elements (excluding nitrogen) are given in Table 19.3.

19.1 THE GROUP VA ELEMENTS

TABLE 19.3

OXIDATION STATES OF GROUP VA ELEMENTS (EXCLUDING NITROGEN)

Oxidation state	Compounds of			
	P	As	Sb	Bi
+5	PCl_5 P_4O_{10} H_3PO_4	$AsCl_5$ As_2O_5 H_3AsO_4	$SbCl_5$ Sb_2O_5 $HSb(OH)_6$	—
+3	PCl_3 P_4O_6 H_3PO_3	$AsCl_3$ As_4O_6 H_3AsO_3	$SbCl_3$ Sb_4O_6 $Sb(OH)_3$	$BiCl_3$ Bi_2O_3 BiO^+
−3	PH_3	AsH_3	SbH_3	BiH_3

The hydrides phosphine (PH_3), arsine (AsH_3), and stibine (SbH_3) are gases at room temperature and decrease in stability in the above order. In fact, bismuthine (BiH_3) is so unstable that only trace amounts have been prepared. These hydrides are all poisonous, and the gases are not appreciably soluble in water. Compared to ammonia they are relatively poor Lewis bases.

Phosphorus forms two oxides. When phosphorus is in excess, it burns in the presence of oxygen to yield a white oxide (P_4O_6) commonly called **phosphorus trioxide** in accordance with the empirical formula P_2O_3. With excess oxygen, a different white solid, **phosphorus pentoxide** (P_4O_{10}), is formed and again the common name is based on the empirical formula P_2O_5. However, present evidence indicates that in both the solid and the gaseous states the molecular formulas of the oxides are P_4O_6 and P_4O_{10}. The molecular structure of P_4O_6 consists of a P_4 tetrahedron with an oxygen atom inserted between each pair of P atoms. The P_4O_{10} molecule has a similar structure except that one additional O is attached to each P atom as shown in Figure 19.2.

Phosphorus trioxide reacts with water to produce phosphorous acid (H_3PO_3). This acid may also be prepared by the reaction of phosphorus trichloride and water:

$$PCl_3(l) + 3H_2O = H_3PO_3 + 3H^+ + 3Cl^-$$

Although phosphorous acid contains three hydrogens, only two of them are acidic. This is in accord with the structure of phosphorous acid in which one hydrogen is attached to phosphorus—only the protons bonded to the oxygens dissociate:

H—Ö—P—Ö: H_3PO_3 or $(HO)_2HPO$

(with H above P and :O: with H below)

$$H_3PO_3 = H^+ + H_2PO_3^- \qquad K_1 = 1.6 \times 10^{-2}$$
$$H_2PO_3^- = H^+ + HPO_3^{2-} \qquad K_2 = 7 \times 10^{-7}$$

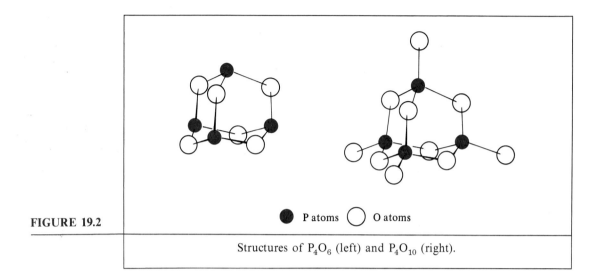

FIGURE 19.2 Structures of P_4O_6 (left) and P_4O_{10} (right).

The value of K_1 is in accord with the dissociation constants of other oxyacids with $m = 1$ (Table 18.1).

Phosphorus trichloride is much more stable than nitrogen trichloride. Liquid PCl_3 does not conduct electricity, but molten $BiCl_3$ is a good conductor. This trend reflects the change from covalent to ionic character of the M—Cl bond as we go down the group of elements. Phosphorus reacts with excess chlorine to form the pentachloride, PCl_5. In the gaseous state PCl_5 has a trigonal bipyramid structure (Figure 9.6), which was discussed in Section 9.8.

Phosphorus pentoxide has a great affinity for water and is one of the most efficient drying agents known. It even removes water from sulfuric acid to produce sulfur trioxide. Phosphorus pentoxide reacts violently with excess water to produce phosphoric acid (H_3PO_4). This triprotic acid dissociates stepwise:

$$H_3PO_4 = H_2PO_4^- + H^+ \quad K_1 = 7.1 \times 10^{-3}$$
$$H_2PO_4^- = HPO_4^{2-} + H^+ \quad K_2 = 6.3 \times 10^{-8}$$
$$HPO_4^{2-} = PO_4^{3-} + H^+ \quad K_3 = 4.5 \times 10^{-13}$$

Thus there are three series of salts possible. The sodium salts are sodium dihydrogen phosphate (NaH_2PO_4), disodium monohydrogen phosphate (Na_2HPO_4), and trisodium phosphate (Na_3PO_4). The aqueous chemistry of phosphorus is primarily that of phosphoric and phosphorous acids and their anions. Except for the sodium, potassium, and ammonium salts, most phosphates are only slightly soluble in water, although many of the dihydrogen phosphates are quite soluble in water. Phosphorus resembles silicon in its tendency to form polymeric species (involving tetrahedral P—O—P bonds). The phosphate linkage is present in many biologically active molecules, such as ribonucleic acid (RNA), deoxyribonucleic acid (DNA), and adenosine triphosphate (ATP) (see Chapter 23).

In contrast to those of nitrogen, the higher oxidation states of phosphorus are very poor oxidizing agents. This is evident from the negative potentials in the following Latimer reduction-potential diagrams:

acidic solution [H$^+$] = 1M

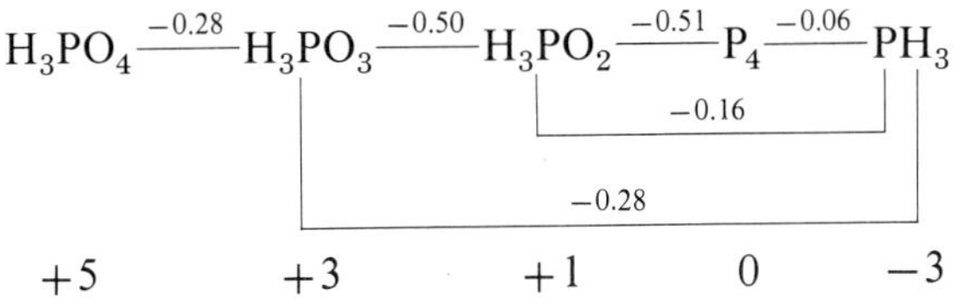

$$H_3PO_4 \xrightarrow{-0.28} H_3PO_3 \xrightarrow{-0.50} H_3PO_2 \xrightarrow{-0.51} P_4 \xrightarrow{-0.06} PH_3$$

with -0.16 bridging H_3PO_2 to P_4 region and -0.28 bridging further.

+5   +3   +1   0   −3

basic solution [OH$^-$] = 1M

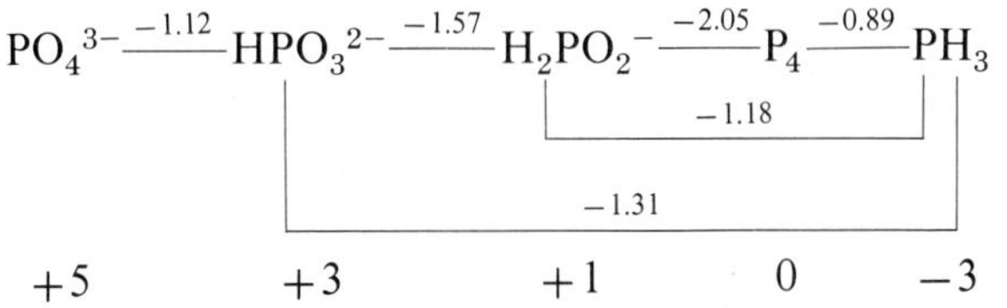

$$PO_4^{3-} \xrightarrow{-1.12} HPO_3^{2-} \xrightarrow{-1.57} H_2PO_2^- \xrightarrow{-2.05} P_4 \xrightarrow{-0.89} PH_3$$

with -1.18 and -1.31 bridges.

+5   +3   +1   0   −3

From these diagrams we predict that elemental phosphorus will disproportionate in either acidic or basic solutions. White phosphorus dissolves fairly rapidly in hot basic solutions to produce hypophosphite ion (H$_2$PO$_2^-$) and phosphine:

$$P_4(s) + 3OH^- + 3H_2O = 3H_2PO_2^- + PH_3(g)$$

$$\Delta\mathcal{E}° = +1.16 \text{ V}$$

Hypophosphite ion is the anion of hypophosphorous acid (H$_3$PO$_2$), a monoprotic acid with the following structure:

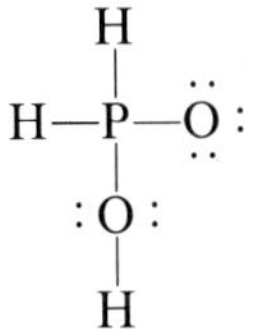

$$\begin{array}{c} H \\ | \\ H-P-\ddot{O}: \\ | \\ :\ddot{O}: \\ | \\ H \end{array} \quad H_3PO_2 \text{ or } (HO)H_2PO$$

The principal oxidation states of arsenic are the +3 and +5 states. Like phosphorus, arsenic forms two oxides. The structure of the trioxide, As$_4$O$_6$, is similar to that of P$_4$O$_6$. It is slightly acidic, whereas the corresponding oxide of antimony (Sb$_4$O$_6$) is amphoteric, dissolving in both acidic and basic solutions:

$$Sb_4O_6 + 6H_2O + 4OH^- = 4Sb(OH)_4^-$$

$$Sb_4O_6 + 4H^+ = 4SbO^+ + 2H_2O$$

The oxide of bismuth(III), Bi_2O_3, is basic and not very soluble in water. It dissolves in strongly acid solutions to give various cations, including polymers of the BiO^+ cation such as $[Bi_6O_6]^{6+}$.

The higher oxidation states of arsenic are much better oxidizing agents in acidic solution than the corresponding phosphorus compounds but are not as good as the corresponding nitrogen compounds. The reduction diagrams for acidic and basic solutions are

acidic solution $[H^+] = 1M$

$$H_3AsO_4 \xrightarrow{+0.56} H_3AsO_3 \xrightarrow{+0.25} As \xrightarrow{-0.60} AsH_3$$
$$+5 +3 0 -3$$

basic solution $[OH^-] = 1M$

$$AsO_4^{3-} \xrightarrow{-0.67} H_2AsO_3^- \xrightarrow{-0.68} As \xrightarrow{-1.43} AsH_3$$
$$\phantom{AsO_4^{3-}}+5 +3 0 -3$$

Unlike phosphorus, elemental arsenic in contact with acidic or basic solutions does not disproportionate.

19.2 THE GROUP VIA ELEMENTS

VIA
O 8
S 16
Se 34
Te 52
Po 84

As in the preceding groups, the properties of the first element of group VIA (oxygen) differ markedly from those of other members of the group (see Table 19.4). Oxygen is a nonmetal and a gas; sulfur is a nonmetallic solid; selenium and tellurium are metalloids; and polonium has the electric conductivity of a metal. The ionization potentials are large (see I_1 and I_2 in Table 19.4) and no simple cationic species exist. The -2 oxidation state is common to all the elements of the group; sulfur, selenium, tellurium, and polonium also form compounds with oxidation states of $+4$ and $+6$.

About one-half of the earth's crust is composed of oxygen, principally as water and the oxides of silicon and aluminum. The properties of elemental oxygen and the electronic structure of the O_2 molecule were considered in Sections 11.12 and 9.8, respectively. The most important compounds of oxygen are the oxides. The acidic, basic, or amphoteric behavior of the oxides has been discussed for each element and is summarized in Figure 19.3.

The alkali metals, alkaline earth metals, zinc, and cadmium form peroxides (containing O_2^{2-} ions) as well as the normal oxides. Hydrogen peroxide is a weak acid whose first dissociation constant is 2.4×10^{-12}. It is a good oxidizing agent in both acidic and basic solutions, as is evident from the Latimer diagrams:

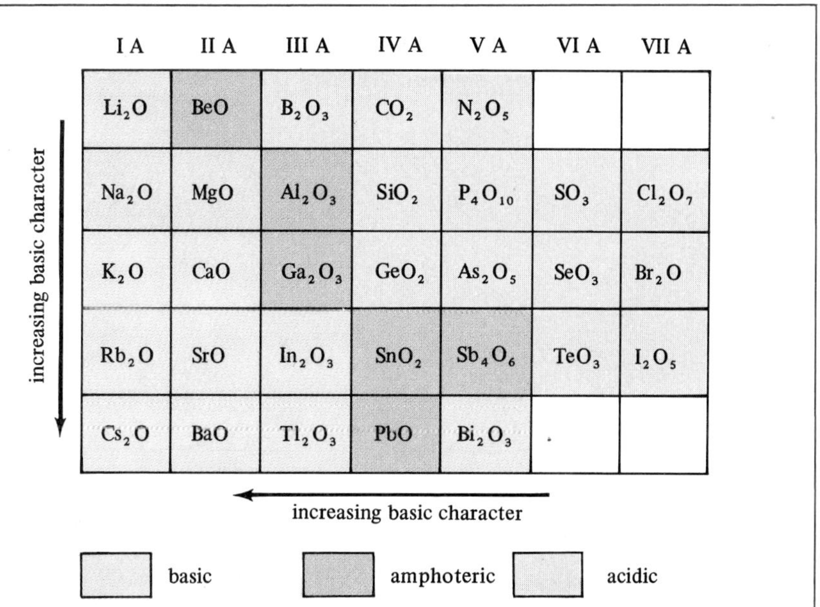

19.2 THE GROUP VIA ELEMENTS

	I A	II A	III A	IV A	V A	VI A	VII A
	Li_2O	BeO	B_2O_3	CO_2	N_2O_5		
	Na_2O	MgO	Al_2O_3	SiO_2	P_4O_{10}	SO_3	Cl_2O_7
	K_2O	CaO	Ga_2O_3	GeO_2	As_2O_5	SeO_3	Br_2O
	Rb_2O	SrO	In_2O_3	SnO_2	Sb_4O_6	TeO_3	I_2O_5
	Cs_2O	BaO	Tl_2O_3	PbO	Bi_2O_3		

(increasing basic character ↓)

← increasing basic character

□ basic   ▨ amphoteric   □ acidic

FIGURE 19.3 Acidic and basic properties of oxides of elements in groups IA–VIIA.

acidic solution $[H^+] = 1M$

$$O_2 \xrightarrow{+0.68} H_2O_2 \xrightarrow{+1.78} H_2O$$
$$\xrightarrow{+1.23}$$

0   −1   −2

basic solution $[OH^-] = 1M$

$$O_2 \xrightarrow{-0.08} HO_2^- \xrightarrow{+0.87} OH^-$$
$$\xrightarrow{+0.40}$$

0   −1   −2

Note that hydrogen peroxide is capable of disproportionation to form oxygen and water:

$$2H_2O_2 = 2H_2O + O_2$$

The rate of decomposition is slow unless catalysts are present. Most of the homogeneous catalysts are substances whose oxidized form can oxidize H_2O_2 and whose reduced form can reduce H_2O_2 (i.e., oxidation-reduction half-reactions

TABLE 19.4 **PROPERTIES OF THE GROUP VIA ELEMENTS**

Property	O	S	Se	Te	Po
Atomic number	8	16	34	52	84
Electron configuration	[He]$2s^22p^4$	[Ne]$3s^23p^4$	[Ar]$3d^{10}4s^24p^4$	[Kr]$4d^{10}5s^25p^4$	[Xe]$4f^{14}5d^{10}6s^26p^4$
Ionization potentials (eV):					
I_1	13.62	10.36	9.75	9.01	8.43
I_2	35.1	23.4	21.5	18.6	—
Electron affinity (eV)	1.47	2.07	—	—	—
Melting point (°C)	−219	119	217	450	—
Pauling electronegativity	3.5	2.5	2.4	2.1	2.0

with reduction potentials in the range of +0.68 to +1.78 V in acidic solution). The Br⁻—Br$_2$ half-reaction is in this range and, if Br⁻ is added to hydrogen peroxide solutions, the reddish brown color of bromine appears and then fades away.

$$H_2O_2 + 2H^+ + 2e^- = 2H_2O \qquad \mathcal{E}° = +1.78 \text{ V}$$
$$2Br^- = Br_2 + 2e^- \qquad \mathcal{E}° = -1.065 \text{ V}$$
$$\overline{2Br^- + H_2O_2 + 2H^+ = Br_2 + 2H_2O \qquad \Delta\mathcal{E}° = +0.72 \text{ V}}$$

$$Br_2 + 2e^- = 2Br^- \qquad \mathcal{E}° = +1.065 \text{ V}$$
$$H_2O_2 = O_2 + 2H^+ + 2e^- \qquad \mathcal{E}° = -0.68 \text{ V}$$
$$\overline{Br_2 + H_2O_2 = 2Br^- + O_2 + 2H^+ \qquad \Delta\mathcal{E}° = +0.39 \text{ V}}$$

Thus Br⁻ can reduce H$_2$O$_2$, and Br$_2$ can oxidize H$_2$O$_2$. The sum of the two reactions is the decomposition of H$_2$O$_2$:

$$2H_2O_2 = 2H_2O + O_2$$

Oxygen is also a good oxidizing agent in acidic solutions, but the rate of reaction is often slow. An extremely interesting and important reaction of molecular oxygen is its rapid equilibration with the protein hemoglobin, which accounts for the oxygen-carrying capacity of blood. Each molecule of hemoglobin contains four iron(II)–porphyrin complexes (see Figure 19.4) in addition to the globin protein. The protein is believed to be attached to the iron(II) through an Fe—N linkage. When hemoglobin is exposed to high partial pressures of oxygen in the lungs, the oxygen molecule binds to the iron atom at the position shown in Figure 19.4. The reverse reaction occurs at low partial pressures of oxygen in the body tissues. Thus the hemoglobin in blood acts as an oxygen carrier. Because both

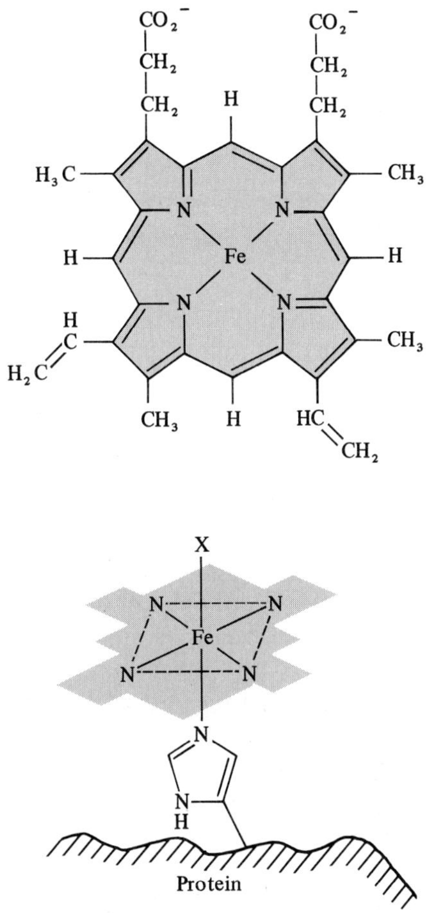

FIGURE 19.4 Top: Heme, an iron(II)–porphyrin complex. Bottom: Binding of heme in hemoglobin and oxyhemoglobin. Only the iron atom and four nitrogen atoms of the heme are shown. The position marked X is occupied by an O_2 molecule in oxyhemoglobin. The iron atom is bound to the protein by a nitrogen atom of a histidine side chain, as shown at the bottom of the figure.

carbon monoxide and cyanide ion coordinate with the iron(II) in hemoglobin more strongly than oxygen, they are extremely toxic, preventing oxygen from entering the bloodstream in sufficient amounts to maintain normal body processes.

Only fluorine is more electronegative than oxygen so that the only positive oxidation states of oxygen are in the oxygen fluorides, such as OF_2 and O_2F_2, in which oxygen is in the +2 and +1 states, respectively.

Sulfur constitutes about 0.05% of the earth's crust, occurring as the free

FIGURE 19.5 The S_8 molecule—plan view and side view.

element and in the form of sulfides and sulfates. Elemental sulfur is obtained from underground deposits by melting the sulfur with superheated steam at about 170°C, and then forcing the mixture of liquid sulfur and water to the surface with compressed air. The product obtained, after drying, is 99.5% sulfur.

The more important allotropic forms of sulfur are described in Sections 11.12 and 17.1. Rhombic sulfur, the stable form at room temperature, contains S_8 molecules, which have a puckered ring or crown form (Figure 19.5).

The most common oxidation states of sulfur are -2, 0, $+4$, and $+6$. These states together with examples of important compounds are shown in Table 19.5.

The electronic structure of hydrogen sulfide is similar to that of water. It is a vile-smelling (odor of rotten eggs), poisonous gas that is appreciably soluble in water (0.1M at 1 atm pressure of H_2S), producing a slightly acidic solution:

$$H_2S(aq) = H^+ + HS^- \quad K_1 = 1.0 \times 10^{-7}$$
$$HS^- = H^+ + S^{2-} \quad K_2 = 1.3 \times 10^{-14}$$

The sulfides of the alkali and alkaline earth metals are colorless and are readily soluble in water, producing solutions that are basic because of the extensive hydrolysis of the sulfide ion:

$$S^{2-} + H_2O = HS^- + OH^-$$

The sulfides of other metals vary in color and in their solubility in acidic and basic solutions. By taking advantage of these differences, we can separate and identify the various metallic ions (Section 13.12).

In addition to the simple sulfides there are many polysulfides (S_n^{2-}) in which n ranges from 2 to 6. This tendency toward catenation is greater for sulfur than for any other element except carbon. Acidification of polysulfide solutions yields sulfanes (H_2S_n) in which n ranges from 2 to 6. The sulfanes readily decompose to give hydrogen sulfide and sulfur.

Sulfur burns readily in air to produce sulfur dioxide (SO_2),

$$S(s) + O_2(g) = SO_2(g)$$

TABLE 19.5 — OXIDATION STATES OF SULFUR

Oxidation state	Compound or ion	Name
+6	H_2SO_4	Sulfuric acid
	SO_4^{2-}	Sulfate ion
	SO_3	Sulfur trioxide
	SF_6	Sulfur hexafluoride
+4	SO_3^{2-}	Sulfite ion
	SO_2	Sulfur dioxide
	SF_4	Sulfur tetrafluoride
0	S_8	Sulfur
−1	Na_2S_2	Sodium disulfide
−2	H_2S	Hydrogen sulfide

Sulfur dioxide causes a choking tendency and is somewhat poisonous. It dissolves in water to produce acidic solutions. Although one might expect formation of the monohydrate, H_2SO_3, called sulfurous acid (and in fact this name is often applied to aqueous solutions of SO_2), this species has never been detected, and it is not considered to be present in significant amounts. An aqueous solution of SO_2 behaves like a diprotic acid:

$$SO_2(aq) + H_2O = H^+ + HSO_3^- \qquad K_1 = 1.3 \times 10^{-2}$$
$$HSO_3^- = H^+ + SO_3^{2-} \qquad K_2 = 6.2 \times 10^{-8}$$

Sulfur dioxide is used as a preservative in the preparation of dried fruits, as it destroys fungi and bacteria. A solution of calcium hydrogen sulfite [$Ca(HSO_3)_2$] is used in the manufacture of paper pulp from wood. Sulfur dioxide is used in large quantities in the manufacture of sulfuric acid.

Sulfuric acid is produced by oxidizing sulfur dioxide to the trioxide and then hydrating the trioxide. The first step is

$$2SO_2(g) + O_2(g) = 2SO_3(g)$$

Although the reaction of sulfur dioxide with oxygen to produce sulfur trioxide has a negative free-energy change, the rate of reaction is extremely slow. However, the reaction is catalyzed by vanadium pentoxide or platinum, and nearly all sulfur trioxide is catalytically produced. Direct hydration of the sulfur trioxide thus produced is quite slow, and therefore SO_3 is first dissolved in sulfuric acid to produce pyrosulfuric acid ($H_2S_2O_7$):

$$SO_3 + H_2SO_4 = H_2S_2O_7$$

The required amount of water is then added to convert the pyrosulfuric acid to sulfuric acid:

$$H_2S_2O_7 + H_2O = 2H_2SO_4$$

Sulfuric acid is a diprotic acid in aqueous solutions. In dilute aqueous solutions the first dissociation is complete, but HSO_4^- is only partially dissociated:

$$HSO_4^- = H^+ + SO_4^{2-} \qquad K_2 = 1.0 \times 10^{-2}$$

Sulfuric acid is a viscous liquid that conducts electricity owing to a slight dissociation to form ions:

$$2H_2SO_4 = H_3SO_4^+ + HSO_4^-$$

The acid has a great affinity for water and is extensively used as a drying agent.

Most metal sulfates are fairly soluble in water, with the notable exception of some of the alkaline earth sulfates and lead sulfate. The solubility of the alkaline earth sulfates decreases in the order $CaSO_4 > SrSO_4 > BaSO_4$, with $BaSO_4$ soluble to the extent of about $1 \times 10^{-5} M$.

Sulfur will dissolve in a boiling solution of sulfite ion to produce thiosulfate ion ($S_2O_3^{2-}$):

$$S(s) + SO_3^{2-} = S_2O_3^{2-}$$

The $S_2O_3^{2-}$ ion differs from the sulfate ion (SO_4^{2-}) in that one oxygen atom has been replaced by a sulfur atom, and this fact is indicated by the name *thio*sulfate. (The prefix is from the Greek *theion* meaning brimstone or sulfur.)

When thiosulfate is prepared from radioactive sulfur and nonradioactive sulfite and then decomposed by addition of acid, all the radioactivity is found in the precipitated sulfur. If the radioactive sulfur atoms are designated as S*, the reactions are

$$S^*(s) + SO_3^{2-} = S^*SO_3^{2-}$$
$$S^*SO_3^{2-} + 2H^+ = S^*(s) + H_2O + SO_2(aq)$$

This experiment shows that the sulfur atoms in thiosulfate are not equivalent. Therefore the following structure is proposed for thiosulfate ion:

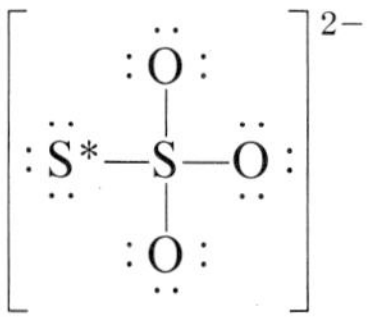

$$\begin{bmatrix} & \ddot{\underset{..}{O}}: & \\ & | & \\ :\overset{..}{\underset{..}{S}}{}^*-\overset{|}{\underset{|}{S}}-\overset{..}{\underset{..}{O}}: \\ & | & \\ & :\underset{..}{\overset{..}{O}}: & \end{bmatrix}^{2-}$$

This structure has been confirmed by an X-ray diffraction study of $Na_2S_2O_3 \cdot 5H_2O$ crystals. Both SO_4^{2-} and $S_2O_3^{2-}$ have tetrahedral bond angles as would be expected from electron-pair repulsion theory (Section 9.8). Thiosulfate is easily oxidized to tetrathionate ($S_4O_6^{2-}$) and is often used in the quantitative determination of iodine:

$$2S_2O_3^{2-} + I_2 = S_4O_6^{2-} + 2I^-$$

The thiosulfate ion also forms stable complex ions with metallic ions, such as silver ion,

$$Ag^+ + 2S_2O_3^{2-} = Ag(S_2O_3)_2^{3-}$$

In fact solutions of sodium thiosulfate (called "hypo" in photography) are used to remove silver bromide from photographic films.

The oxidation-reduction behavior of sulfur compounds in acidic and basic solutions is summarized in the following diagrams:

acidic solution $[H^+] = 1M$

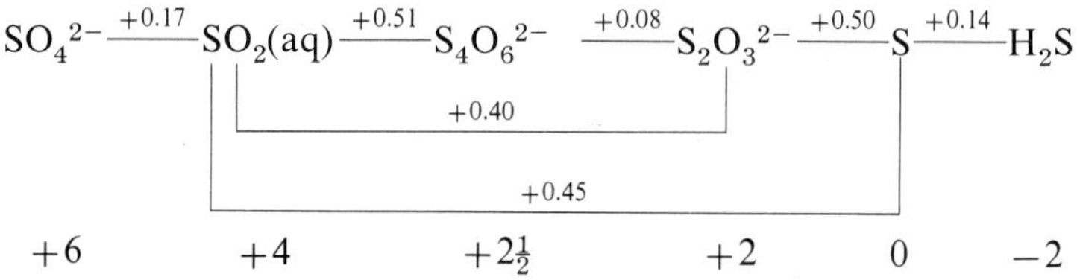

$$SO_4^{2-} \xrightarrow{+0.17} SO_2(aq) \xrightarrow{+0.51} S_4O_6^{2-} \xrightarrow{+0.08} S_2O_3^{2-} \xrightarrow{+0.50} S \xrightarrow{+0.14} H_2S$$

with $+0.40$ bridging $SO_2(aq)$ to $S_4O_6^{2-}$, and $+0.45$ bridging $SO_2(aq)$ to $S_2O_3^{2-}$.

Oxidation states: $+6 \quad +4 \quad +2\tfrac{1}{2} \quad +2 \quad 0 \quad -2$

basic solution $[OH^-] = 1M$

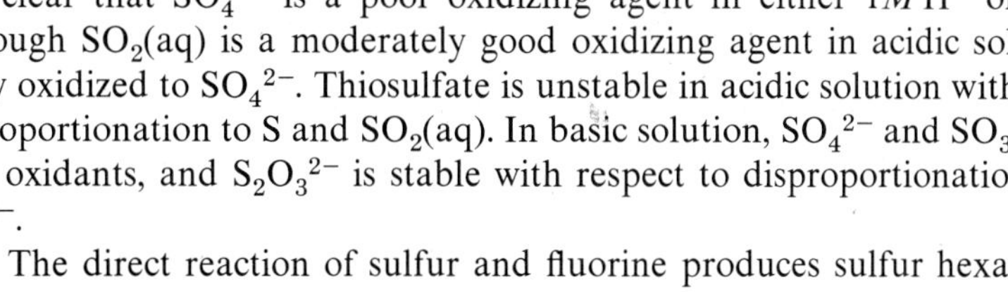

$$SO_4^{2-} \xrightarrow{-0.93} SO_3^{2-} \xrightarrow{-0.80} S_4O_6^{2-} \xrightarrow{+0.08} S_2O_3^{2-} \xrightarrow{-0.74} S \xrightarrow{-0.48} S^{2-}$$

with -0.58 bridging SO_3^{2-} to $S_4O_6^{2-}$, and -0.66 bridging SO_3^{2-} to $S_2O_3^{2-}$.

Oxidation states: $+6 \quad +4 \quad +2\tfrac{1}{2} \quad +2 \quad 0 \quad -2$

It is clear that SO_4^{2-} is a poor oxidizing agent in either $1M\,H^+$ or $1M\,OH^-$. Although $SO_2(aq)$ is a moderately good oxidizing agent in acidic solution, it is easily oxidized to SO_4^{2-}. Thiosulfate is unstable in acidic solution with respect to disproportionation to S and $SO_2(aq)$. In basic solution, SO_4^{2-} and SO_3^{2-} are very poor oxidants, and $S_2O_3^{2-}$ is stable with respect to disproportionation to S and SO_3^{2-}.

The direct reaction of sulfur and fluorine produces sulfur hexafluoride:

$$S(s) + 3F_2(g) = SF_6(g)$$

At room temperature SF_6 is an unreactive gas. It is widely used as an electric insulator in high-voltage generators. In the SF_6 molecule the central sulfur atom is surrounded by the six fluorine atoms at the corners of an octahedron (Figure 9.7). This structure was discussed in Section 9.8.

Sulfur tetrafluoride is a highly reactive gas with an unusual structure. The Lewis structure,

$$\begin{array}{c} \ddot{\text{F}} \quad \ddot{\text{F}} \\ \diagdown \diagup \\ \text{S} \\ \diagup \diagdown \\ \ddot{\text{F}} \quad \ddot{\text{F}} \end{array}$$

shows that there are five pairs of electrons around the sulfur kernel (one unshared pair and four shared pairs). According to electron-pair repulsion theory, these electrons should be arranged in the trigonal bipyramid structure previously described for PCl_5, with the unshared pair of electrons occupying an equatorial position, and this is the structure of SF_4 found experimentally. (As noted in Section 9.8, the most important repulsions involve unshared pairs. In an equatorial position an unshared pair has only two neighboring pairs at 90°, whereas in an axial position it would have three. The neighboring pairs farther away, at 120° and 180°, are relatively unimportant.) Figure 19.6 summarizes this and other structures of polyatomic molecules deduced from electron-pair repulsion theory. The central atom is designated A, surrounding atoms are X, and unshared pairs of valence electrons around the central atom are E. In this notation SF_4 is of AX_4E type.

Selenium and tellurium are much less abundant than sulfur, making up about 10^{-5} and $10^{-7}\%$, respectively, of the earth's crust. They are found mainly as trace constituents in metal sulfide ores. In the stable forms of elemental selenium and tellurium, the atoms are connected in long helical chains. Polonium is extremely rare and is found mainly in uranium ores such as pitchblende. It is extremely radioactive and decomposes by emitting alpha particles. Amounts of the order of 10^{-11} g are lethal to humans, so the chemistry of polonium compounds has not been thoroughly investigated. Selenium has the interesting property of increasing its electric conductivity when exposed to light, and thus its primary use is in the preparation of photoconductive cells to measure light intensity.

The chemical properties of selenium and tellurium resemble those of sulfur. The hydrides (H_2Se and H_2Te) are vile-smelling, poisonous gases that are moderately soluble in water to give acidic solutions. The acid strength of the hydrides increases in the order $H_2O < H_2S < H_2Se < H_2Te$:

$$H_2Se = H^+ + HSe^- \quad K_1 = 2 \times 10^{-4}$$

$$H_2Te = H^+ + HTe^- \quad K_1 = 2 \times 10^{-3}$$

As in groups IVA and VA, the stability of the hydrides of group VIA decreases as we go down the group, with H_2Se and H_2Te being unstable with respect to their elements. The decrease in bond energy with increase in size of the central atom is the main reason for these trends in acid strength and stability.

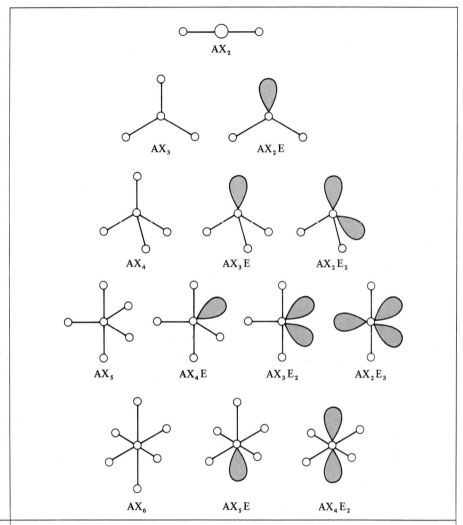

FIGURE 19.6 Common shapes of polyatomic molecules. The central atom (A) is connected by covalent or polar bonds to two or more other atoms (X). Unshared pairs of valence electrons (E) around the central atom are indicated by colored lobes. [SOURCE: R. J. Gillespie, Journal of Chemical Education, **40**, 297 (1963).]

Both SeO_2 and TeO_2 are solids at room temperature in contrast to SO_2, which is a gas. Whereas SeO_2 dissolves in water to give acidic solutions, and the acid H_2SeO_3 has been prepared, TeO_2 is not soluble in water, although it does dissolve in base to give tellurite ion:

$$TeO_2(s) + 2OH^- = TeO_3^{2-} + H_2O$$

Selenites (SeO_3^{2-}) may be oxidized to selenates (SeO_4^{2-}). Selenic acid (H_2SeO_4) is similar to sulfuric acid in acid strength. The formula of telluric acid is $Te(OH)_6$; it is a weak acid ($K_1 \approx 1 \times 10^{-7}$) in contrast to sulfuric and selenic acids.

19.3 THE GROUP VIIA ELEMENTS

VIIA

F 9
Cl 17
Br 35
I 53
At 85

The group VIIA elements, commonly known as the halogens, have similar chemical properties and show less variation in these properties with increasing atomic number than any other group with the exception of the alkali and alkaline earth metals and the noble gases. Some of the properties of the elements are summarized in Table 19.6. Astatine has been detected in extremely small amounts in the radioactive decay products of thorium and uranium but little is known about its chemistry.

Fluorine is the most electronegative element and hence exhibits only the -1 oxidation state in its compounds. The other halogens exhibit positive oxidation states as well as the -1 state, as shown in Table 19.7.

Elementary fluorine was first prepared in 1886 by the French chemist Henri Moissan by electrolysis of what he thought was anhydrous hydrogen fluoride. However, when asked to demonstrate his preparation before a special committee of the Academy of Sciences, he was unable to pass any electric current through his carefully purified hydrogen fluoride. A few days later he found that his original hydrogen fluoride contained some potassium fluoride, which provided the liquid HF with some electric conductivity and hence allowed fluorine to be produced. Fluorine is still prepared by electrolysis of solutions of potassium fluoride in anhydrous hydrogen fluoride:

$$2HF(l) = H_2(g) + F_2(g)$$

Chlorine may be prepared by electrolysis of seawater, with hydrogen and

TABLE 19.6 PROPERTIES OF THE GROUP VIIA ELEMENTS

Property	F	Cl	Br	I	At
Atomic number	9	17	35	53	85
Electron configuration	[He]$2s^22p^5$	[Ne]$3s^23p^5$	[Ar]$3d^{10}4s^24p^5$	[Kr]$4d^{10}5s^25p^5$	[Xe]$4f^{14}5d^{10}6s^26p^5$
Ionization potentials (eV):					
I_1	17.42	12.97	11.81	10.45	—
I_2	35.0	23.8	21.6	19.1	—
Electron affinity (eV)	3.45	3.61	3.36	3.06	—
Melting point (°C)	-223	-102	-7	114	—
Boiling point (°C)	-188	-35	59	184	—
Pauling electronegativity	4.0	3.0	2.8	2.5	2.2

TABLE 19.7 OXIDATION STATES OF HALOGENS

Oxidation state	Compounds of			
	F	Cl	Br	I
+7	—	Cl_2O_7, $HClO_4$	$HBrO_4$	H_5IO_6, HIO_4
+6	—	Cl_2O_6	—	—
+5	—	$HClO_3$	$HBrO_3$	I_2O_5, HIO_3
+4	—	ClO_2	BrO_2	—
+3	—	$HClO_2$	—	—
+1	—	Cl_2O, $HOCl$	Br_2O, $HOBr$	HOI
0	F_2	Cl_2	Br_2	I_2
−1	HF	HCl	HBr	HI

sodium hydroxide as by-products:

$$2Na^+ + 2Cl^- + 2H_2O = Cl_2(g) + H_2(g) + 2Na^+ + 2OH^-$$

Chlorine has many industrial uses because of its strong oxidizing power, including the conversion of bromide ion in seawater to bromine:

$$Cl_2 + 2Br^- = Br_2 + 2Cl^-$$

Although fluorine, chlorine, and bromine occur naturally mainly in the −1 oxidation state, the primary source of iodine is naturally occurring iodates (IO_3^-). Iodine is liberated by reduction with bisulfite:

$$2IO_3^- + 5HSO_3^- = I_2(s) + 2SO_4^{2-} + 3HSO_4^- + H_2O$$

The halogens react directly with many elements to form compounds, called **halides,** in which the halogen is in the −1 oxidation state. With metals in the +1 or +2 oxidation states the bonding is generally ionic, and this is particularly true of the fluorides. With metals in higher oxidation states and with nonmetals, the bonding is generally covalent, and this is particularly true of the chlorides, bromides, and iodides.

The hydrogen halides are gases at room temperature and are readily soluble in water to produce acidic solutions. The aqueous solutions are called hydrofluoric acid, hydrochloric acid, and so on. They are all strong acids with the exception of hydrofluoric acid:

$$HF = H^+ + F^- \qquad K = 6.7 \times 10^{-4}$$

To determine the relative strengths of HCl, HBr, and HI as acids they must be dissolved in a solvent (such as acetic acid) that is not as good a proton acceptor as water (Section 13.4). Under such conditions the acid strengths increase in the order HF < HCl < HBr < HI. This trend parallels that noted in the preceding section for the group VIA hydrides, and it occurs for the same reason. Because hydrofluoric acid dissolves glass, it is usually handled in plastic or steel containers.

Fluoride ion shows quite marked differences from the other halide ions in the solubility of its salts in water and in the stability of its complex ions. In general the least soluble salts and the most stable complex ions are formed (1) by small cations and F^- and (2) by large cations and Cl^-, Br^-, or I^-, with the solubility decreasing (and the stability of the complex ion increasing) in that order. Examples of type (1) are the salts LiF and BeF_2, and the complex ions BF_4^-, AlF_6^{3-}, and SiF_6^{2-}. Examples of type (2) are the halide salts and complex ions of Ag(I), Hg(I), Hg(II), Tl(I), and Pb(II). Although a number of factors must be considered in analyzing trends in solubility (or complex ion stability), as discussed in Section 18.2, the main factor here is the very strong bonding in both types (1) and (2). In type (1) the high electronegativity of fluorine results in highly ionic bonds. Because of the small ionic radii the interionic distances are short and the electrostatic forces of attraction are very strong. In type (2) the lower electronegativity of the other halogens causes the bonds to be less ionic and more covalent, but there is an important contribution from van der Waals forces of attraction, which are largest when the halide ion and the cation are large and have high polarizabilities.

Much of the chemistry of the halogens involves electron-transfer reactions, and many possible reactions may be predicted using the Latimer diagrams:

acidic solution $[H^+] = 1M$

$$F_2 \xrightarrow{+2.87} F^-$$

$$ClO_4^- \xrightarrow{+1.19} ClO_3^- \xrightarrow{+1.21} HClO_2 \xrightarrow{+1.64} HOCl \xrightarrow{+1.63} Cl_2 \xrightarrow{+1.36} Cl^-$$
$$\xrightarrow{+1.47}$$

$$BrO_4^- \xrightarrow{+1.76} BrO_3^- \xrightarrow{+1.49} HOBr \xrightarrow{+1.59} Br_2 \xrightarrow{+1.07} Br^-$$
$$\xrightarrow{+1.51}$$

$$H_5IO_6 \xrightarrow{+1.7} IO_3^- \xrightarrow{+1.14} HOI \xrightarrow{+1.45} I_2 \xrightarrow{+0.54} I^-$$
$$\xrightarrow{+1.20}$$

+7 +5 +3 +1 0 −1

19.3 THE GROUP VIIA ELEMENTS

basic solution $[OH^-] = 1M$

$$F_2 \xrightarrow{+2.87} F^-$$

$$ClO_4^- \xrightarrow{+0.36} ClO_3^- \xrightarrow{+0.33} ClO_2^- \xrightarrow{+0.66} ClO^- \xrightarrow{+0.42} Cl_2 \xrightarrow{+1.36} Cl^-$$
$$ClO_3^- \xrightarrow{+0.50} ClO_2^-$$
$$ClO^- \xrightarrow{+0.89} Cl^-$$

$$BrO_4^- \xrightarrow{+0.93} BrO_3^- \xrightarrow{+0.54} BrO^- \xrightarrow{+0.45} Br_2 \xrightarrow{+1.07} Br^-$$
$$BrO^- \xrightarrow{+0.76} Br^-$$

$$IO_3^- \xrightarrow{+0.14} IO^- \xrightarrow{+0.45} I_2 \xrightarrow{+0.54} I^-$$
$$IO^- \xrightarrow{+0.49} I^-$$

+7   +5   +3   +1   0   −1

We note from the above potentials that the halogens are good oxidizing agents, with the oxidizing strength decreasing in the order $F_2 > Cl_2 > Br_2 > I_2$ so that the lighter halogens readily oxidize the heavier halides to the corresponding halogen. For example,

$$Br_2 + 2I^- = I_2 + 2Br^- \qquad \Delta\mathcal{E}° = +0.53 \text{ V}$$

Elemental fluorine, chlorine, and several other oxidizing agents shown in the preceding diagrams are capable of oxidizing water to oxygen and are therefore unstable in aqueous solutions. In addition, several halogen species with oxidation states intermediate between −1 and +7 are unstable with respect to disproportionation, although many of these reactions are slow.

The halogen oxides are unstable reactive substances. With the exception of iodine pentoxide (I_2O_5), which is a white solid, the oxides are all gases or volatile liquids at room temperature. All the oxides are thermodynamically unstable with respect to their elements and thus cannot be prepared by direct reaction of the halogens with oxygen. The oxides Cl_2O and Br_2O may be prepared by the reaction of the halogen with solid mercuric oxide:

$$2X_2(g) + 2HgO = HgO \cdot HgX_2 + X_2O(g)$$

In the presence of water the hypohalous acid (HOX) is formed:

$$2X_2(g) + 2HgO + H_2O = HgO \cdot HgX_2 + 2HOX$$

The hypohalous acids are weak acids, good oxidizing agents, and unstable with respect to disproportionation.

The halate ions may readily be prepared by disproportionation of the hypohalites in basic solution:

$$3XO^- = 2X^- + XO_3^-$$

The corresponding halic acids [chloric acid (HClO$_3$), bromic acid (HBrO$_3$), and iodic acid (HIO$_3$)] are strong acids in aqueous solution. The halic acids have not been prepared in the pure state and attempts to concentrate chloric acid solutions by distillation result in violent explosions. The halate ions are powerful oxidizing agents in acidic solutions although reactions involving ClO$_3^-$ are usually slow.

Perchloric acid (HClO$_4$) is a very strong oxidizing agent. In concentrated solutions, it reacts explosively with many organic compounds. However, in dilute aqueous solutions, it is much less reactive and behaves simply like a strong acid. In fact, $1M$ perchloric acid solutions show no evidence of reacting with most reducing agents (in spite of favorable $\Delta\mathcal{E}°$ values) even over extended periods of time, owing to the extreme slowness of the reactions.

The halogens form a wide variety of interhalogen compounds, such as ClF, BrCl, ClF$_3$, CsICl$_2$, KICl$_4$, and BrF$_5$. The polyatomic molecules and anions have unusual structures illustrating particularly well the electron-pair repulsion theory. With five pairs of electrons around the central atom, the structures are derived from a trigonal bipyramid. The Lewis structures show that ClF$_3$ is of AX$_3$E$_2$ type, and ICl$_2^-$ is of AX$_2$E$_3$ type, in the nomenclature of Figure 19.6:

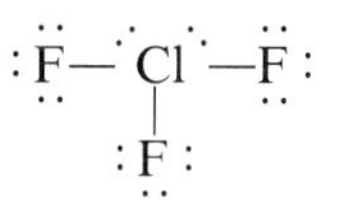

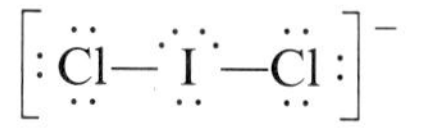

$$:\!\ddot{F}\!-\!\ddot{Cl}\!-\!\ddot{F}\!: \qquad \left[:\!\ddot{Cl}\!-\!\ddot{I}\!-\!\ddot{Cl}\!:\right]^-$$
$$\phantom{:\!\ddot{F}\!-\!}|\phantom{\ddot{Cl}\!-\!\ddot{F}\!:}$$
$$\phantom{:\!\ddot{F}\!-\!}:\!\ddot{F}\!:$$

As indicated in Figure 19.6, the unshared pairs occupy equatorial positions, with the result that the atoms in ClF$_3$ have a T-shaped structure and the atoms in ICl$_2^-$ have a linear structure. The structures of BrF$_5$ and ICl$_4^-$ are derived from an octahedron, as shown in the figure, with BrF$_5$ being of AX$_5$E type and ICl$_4^-$ of AX$_4$E$_2$ type:

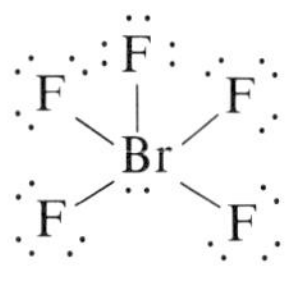

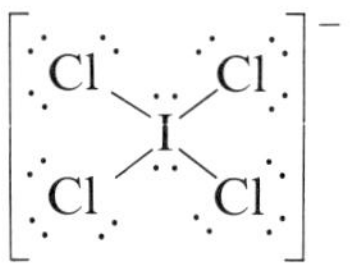

The five fluorine atoms in BrF$_5$ form a square pyramid with the bromine atom at the center of its base. In ICl$_4^-$ the unshared pairs occupy positions on opposite sides of the iodine atom, and the four chlorine atoms lie at the corners of a square with the iodine atom at its center.

19.4 THE NOBLE GASES

The occurrence and uses of the noble gases are treated in Section 11.4. Some properties of these elements are listed in Table 19.8. Until 1962 it was believed that the noble gases did not take part in ordinary chemical reactions although such species as He$_2^+$, Ar$_2^+$, ArH$^+$, and CH$_3$Xe$^+$ had been detected as transient gaseous ions. In 1962, Neil Bartlett found that the deep red-brown gaseous PtF$_6$

19.4 THE NOBLE GASES

TABLE 19.8 PROPERTIES OF THE GROUP 0 ELEMENTS

Property	He	Ne	Ar	Kr	Xe	Rn
Atomic number	2	10	18	36	54	86
Electron configuration	$1s^2$	$[He]2s^22p^6$	$[Ne]3s^23p^6$	$[Ar]3d^{10}4s^24p^6$	$[Kr]4d^{10}5s^25p^6$	$[Xe]4f^{14}5d^{10}6s^26p^6$
Ionization potentials (eV): I_1	24.59	21.57	15.76	14.00	12.13	10.75
Boiling point (°C)	−269	−246	−186	−153	−107	−65

0
He 2
Ne 10
Ar 18
Kr 36
Xe 54
Rn 86

reacted with oxygen at room temperature to produce an orange solid that he found to be $O_2^+[PtF_6]^-$. Because the first ionization potential of molecular oxygen (12.08 eV) is nearly the same as that of xenon, Bartlett reasoned that xenon should also form a compound with PtF_6. He verified his prediction by mixing xenon and platinum hexafluoride—they reacted immediately to form a yellow solid compound. Shortly thereafter Howard H. Claassen, Henry Selig, and John G. Malm prepared xenon tetrafluoride by the direct reaction of xenon and fluorine at 400°C. It is now known that XeF_2 is formed first and that, in the presence of excess fluorine, XeF_4 and XeF_6 can be prepared. The average bond energy of the xenon–fluorine bond in these compounds is about 125 kJ mol⁻¹. The XeF_2 molecule is linear, and XeF_4 is square planar. Both structures illustrate electron-pair repulsion theory. The Lewis structures are

$$:\ddot{F}-\ddot{Xe}-\ddot{F}: \qquad \begin{array}{c} :\ddot{F} \quad \ddot{F}: \\ \diagdown\diagup \\ Xe \\ \diagup\diagdown \\ :\ddot{F} \quad \ddot{F}: \end{array}$$

Thus XeF_2 is of AX_2E_3 type, whereas XeF_4 is of AX_4E_2 type. The resulting structures are shown in Figure 19.6.

Xenon hexafluoride, a colorless solid at room temperature, turns yellow at 43°C and melts to form a yellow liquid at 48°C. It must be stored in containers made of an unreactive material, such as nickel, because it reacts with the silicon dioxide in glass to produce SiF_4:

$$2XeF_6(s) + SiO_2(s) = 2XeOF_4(l) + SiF_4(g)$$

The aqueous chemistry of the xenon fluorides has been investigated; XeF_2 reacts rapidly in aqueous base to yield xenon and oxygen:

$$2XeF_2(s) + 4OH^- = 2Xe(g) + O_2(g) + 4F^- + 2H_2O$$

Both XeF_4 and XeF_6 are hydrolyzed by water to produce a xenon(VI) species. Evaporation of the solutions yields the explosive solid XeO_3:

$$6XeF_4(s) + 12H_2O = 2XeO_3(s) + 4Xe(g) + 3O_2(g) + 24HF$$
$$XeF_6(s) + 3H_2O = XeO_3(s) + 6HF$$

Approximate values of the aqueous reduction potentials for xenon are

acidic solution: $H_4XeO_6 \xrightarrow{+2.36} XeO_3 \xrightarrow{+2.12} Xe$

basic solution: $HXeO_6^{3-} \xrightarrow{+0.94} HXeO_4^- \xrightarrow{+1.26} Xe$

$\qquad\qquad\qquad +8 \qquad\qquad\qquad +6 \qquad\qquad 0$

Thus the +6 oxidation state of xenon is stable with respect to disproportionation in acidic solution but it disproportionates in basic solution. XeO_3 may have interesting uses as an oxidizing agent in acidic solution because its reduction product is xenon gas, which escapes and does not complicate the chemistry in the resulting solution.

So far no chemical compounds of helium, neon, or argon have been produced even with fluorine. At low temperatures krypton does react with atomic fluorine to produce a fluoride (KrF_2), a white crystalline solid that slowly decomposes at room temperature. Like the XeF_2 molecule, the KrF_2 molecule is linear.

PROBLEMS

19.1 Compare ammonia, NH_3, and ammonium ion, NH_4^+, as Lewis bases.

19.2 Draw Lewis structures for nitrous oxide, N_2O (structure NNO), nitrosyl chloride, NOCl (structure ClNO), and nitrous acid, HNO_2 (structure HONO). What values would you expect for the XNO bond angles (X = N, Cl, or O, respectively)?

19.3 Hydrazine, N_2H_4, reacts with hydrochloric acid to form a crystalline salt, N_2H_5Cl. With excess acid a second crystalline salt, $N_2H_6Cl_2$, can be formed. Draw Lewis structures for N_2H_4, $N_2H_5^+$, and $N_2H_6^{2+}$, indicating any formal charges.

19.4 (a) Calculate $\Delta\mathcal{E}°$ for the following reactions in $1M\ OH^-$ solution:
$$3NO_2^- + H_2O = 2NO + NO_3^- + 2OH^-$$
$$Br_2 + 2OH^- = Br^- + BrO^- + H_2O$$
(b) Which species is unstable with respect to disproportionation in basic solution, NO_2^- or Br_2?

19.5 What is the coordination number of phosphorus in P_4, PH_3, P_4O_{10}, H_3PO_4, H_3PO_3, H_3PO_2, PCl_3, and PCl_5?

19.6 From the fact that N_2F_2 exists in separable cis and trans forms, what can you say regarding rotation about the nitrogen–nitrogen double bond?

19.7 There are three hydrogen atoms per molecule in each of the acids H_3PO_4, H_3PO_3, and H_3PO_2, but the first is a triprotic acid, the second a diprotic acid, and the third a monoprotic acid. Why?

19.8 Estimate the dissociation constant of hypophosphorous acid, H_3PO_2, on the basis of its structure (Section 19.1) and the rules given in Section 18.1.

19.9 Crystalline PCl_5 contains PCl_4^+ cations and PCl_6^- anions arranged in an ionic lattice of the CsCl type. Draw Lewis structures for PCl_4^+ and PCl_6^-. On the basis of electron-pair repulsion theory, what do you expect for the molecular geometry of PCl_4^+ and PCl_6^-?

19.10 Elemental arsenic reacts with hot nitric acid to produce arsenic acid, H_3AsO_4, and nitrogen dioxide. Write a balanced equation for the reaction.

19.11 (a) The standard electrode potential is +0.77 V for the half-reaction
$$Fe^{3+} + e^- = Fe^{2+}$$
Is this within the range that would catalyze the decomposition of hydrogen peroxide?
(b) Write balanced equations for the oxidation of H_2O_2 by Fe^{3+}, and the reduction of H_2O_2 by Fe^{2+}, and calculate $\Delta\mathcal{E}°$ for each.

19.12 Is hydrogen peroxide a sufficiently good oxidizing agent to oxidize H_3AsO_3 to H_3AsO_4? Is hydrogen peroxide a sufficiently good reducing agent to reduce H_3AsO_4 to H_3AsO_3?

19.13 How many oxygen molecules can be bound by each hemoglobin molecule? At what point in the body is oxyhemoglobin formed? Where does oxyhemoglobin lose oxygen?

19.14 Compare phosphorus and sulfur qualitatively with respect to each of the following properties: electronegativity, first ionization potential, number of electrons in the valence shell of an atom, highest common oxidation state, formulas of oxides, affinity of oxide of highest oxidation state for water, tendency of the element to disproportionate in contact with acidic or basic solutions, and tendency to expand the outer octet of valence electrons in molecule formation.

19.15 Draw Lewis structures, showing formal charges on each atom, for H_2SO_4, HSO_4^-, and $H_3SO_4^+$.

19.16 Draw Lewis structures for thiosulfate ion ($S_2O_3^{2-}$) and tetrathionate ion ($S_4O_6^{2-}$). (The latter contains an S—S—S—S chain, with the oxygen atoms bonded to the end sulfur atoms.) What is the average oxidation state of sulfur in each?

19.17 In xM H_2SO_4, the HSO_4^- ions are 50% dissociated. Calculate x.

19.18 Why is H_2S a stronger acid than H_2O?

19.19 Write balanced equations for the reaction that occurs for each of the following:
(a) Bromine is dissolved in $1M$ NaOH.
(b) Bromine is added to an acidic solution of NaI.

19.20 Draw the Lewis structure of triiodide ion, I_3^-. On the basis of electron-pair repulsion theory, do you expect this to be a linear or a nonlinear ion?

19.21 The equilibrium constant for the formation of triiodide ion from iodine and iodide ion is 1.40×10^3 at 25°C:
$I_2(aq) + I^-(aq) = I_3^-(aq)$ $K = 1.40 \times 10^3$
If 6.5 g of iodine and 12 g of potassium iodide are dissolved in enough water to form 500 mL of solution, what are the equilibrium concentrations of I_3^-, I^-, and I_2?

19.22 Because iodine pentoxide (I_2O_5) can react with the toxic gas carbon monoxide to form carbon dioxide and iodine, it is used for the detection and determination of CO, which is a major pollutant of the atmosphere. Write a balanced equation for the oxidation of CO by I_2O_5.

19.23 (a) Draw a Lewis structure for the $XeOF_4$ molecule, showing all unshared pairs of valence electrons. (The four fluorine atoms and the oxygen atom are all bonded to the xenon atom.)
(b) How many pairs of valence electrons (shared plus unshared) are there around the central xenon atom?
(c) According to the VSEPR theory, which of the following geometrical figures describes the structure of $XeOF_4$ (including unshared pairs of valence electrons around the xenon atom)?
   tetrahedron, square, trigonal bipyramid, octahedron

CHAPTER 20
THE TRANSITION METALS—GENERAL ASPECTS

20

Those elements having partly filled d and f subshells are commonly called the **transition elements** or, because they are all metals, the **transition metals.** The "transition" is from the group IIA metals, with an outer pair of ns electrons over a vacant $(n-1)d$ subshell, to the group IIB metals, with an outer pair of ns electrons over a completely filled $(n-1)d$ subshell. There is some variation in assigning a beginning and an end to the transition series, but it is now customary to include the group IIIB metal as the first of each series, and the group IIB metal as the last (see periodic table). Thus the first transition series consists of the 10 elements from scandium through zinc. With this definition there are 62 known transition metals: 10 each in periods 4 and 5; 24 (including the 15 lanthanides) in period 6; and 18 (the 15 actinides and the new elements 104–106) in period 7.

IIIB	IVB	VB	VIB	VIIB	VIII			IB	IIB
Sc 21	Ti 22	V 23	Cr 24	Mn 25	Fe 26	Co 27	Ni 28	Cu 29	Zn 30
Y 39	Zr 40	Nb 41	Mo 42	Tc 43	Ru 44	Rh 45	Pd 46	Ag 47	Cd 48
*	Hf 72	Ta 73	W 74	Re 75	Os 76	Ir 77	Pt 78	Au 79	Hg 80
†	Rf 104	Ha 105	106						

*	La 57	Ce 58	Pr 59	Nd 60	Pm 61	Sm 62	Eu 63	Gd 64	Tb 65	Dy 66	Ho 67	Er 68	Tm 69	Yb 70	Lu 71
†	Ac 89	Th 90	Pa 91	U 92	Np 93	Pu 94	Am 95	Cm 96	Bk 97	Cf 98	Es 99	Fm 100	Md 101	No 102	Lr 103

20.1 GENERAL FEATURES

The general sequence of electronic structures of the transition metal atoms is shown by Figure 8.18, and details are given in Appendix E. Because of the shielding effect of the inner electrons (Section 8.7), in a neutral atom the $3d$ subshell has higher energy than the $4s$ subshell (Figure 8.15). Thus, after the inner subshells are completed at element 18 (Ar), the next two electrons enter the $4s$ subshell in elements 19 (K) and 20 (Ca), and it is only at element 21 (Sc) that an electron enters the $3d$ subshell. One would then expect that the succeeding elements would each have one additional electron in the $3d$ subshell until it reaches its full complement of 10 electrons at element 30 (Zn). This indeed happens in all but two cases: element 24 (Cr) has an electron configuration [Ar] $3d^5 4s$ (the expected [Ar] $3d^4 4s^2$ is a low-lying excited state), and element 29 (Cu) is [Ar] $3d^{10} 4s$ (the expected [Ar] $3d^9 4s^2$ is also a low-lying excited state). The basic problem here is that the energy difference between the $3d$ and $4s$ subshells is quite small, so that minor variations in the shielding effect are able to upset the normal pattern. Similar anomalies occur in the other transition metal series. (Most of the anomalies conform to a simple rule stating that electronic structures with filled or half-filled subshells are especially stable. Thus, in Cr, the $3d$ and $4s$ subshells are each half-filled; in Cu, the $3d$ subshell is filled and the $4s$ subshell is half-filled.)

Because the $4s$ subshell is filled before the $3d$, one might expect that electrons removed from a transition metal atom would first come from the $3d$ subshell. Such is not the case. This paradox results from a change in the energy of the $4s$ subshell relative to the $3d$ subshell as the nuclear charge increases. For positive ions, the higher positive charge on the nucleus stabilizes the subshell of lower principal quantum number. This effect is illustrated in Figure 20.1, which shows the energy of the $4s$ subshell relative to the energy of the $3d$ subshell for an isoelectronic series with only one valence electron.[1] As we go from left to right, the increasing positive charge on the nucleus stabilizes the $3d$ orbital. In K and Cu^+, the $4s$ orbital has the lower energy; in Sc^{2+} and Ti^{3+}, the $3d$ orbital has the lower energy. As a result, the ground states of K and Ca^+ have the electron configuration [Ar] $4s$, whereas Sc^{2+} and Ti^{3+} have the configuration [Ar] $3d$ in their ground states.

The lower energy of the $3d$ subshell in all transition metal *ions* of any importance simplifies the counting of d electrons in these ions. The periodic table is used to count the number of electrons outside the argonlike kernel for the neutral atom, and the number of electrons that have been removed to form the positive ion are subtracted. For example, iron has atomic number 26, and thus it has $26 - 18 = 8$ electrons outside the argonlike kernel; therefore ferrous ion, Fe^{2+}, has six d electrons (configuration [Ar] $3d^6$, often abbreviated d^6). The configurations of ions in the other series are most easily obtained by analogy with the first series. Thus ruthenium, directly below iron in the periodic table, also has a $+2$ ion with a d^6 electron configuration. (In this case, the six d electrons are in the $4d$ subshell.)

It is, of course, difficult to make other generalizations about 62 different elements, but they do have a number of features in common that make it convenient to group them together. In the elemental form they all exhibit characteristic metallic properties. They are strong, hard, and dense; are good electric and ther-

---

[1] Both subshells change in energy as the nuclear charge increases, but we are only interested here in their energy difference.

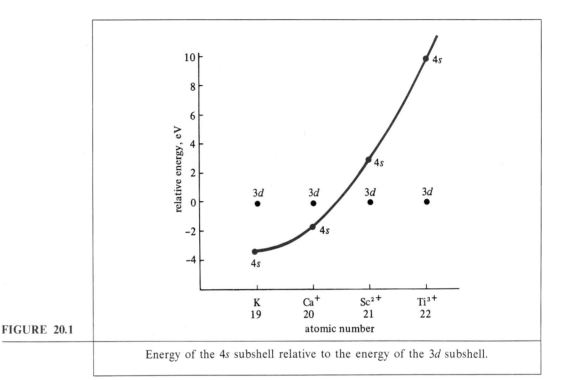

FIGURE 20.1 Energy of the 4s subshell relative to the energy of the 3d subshell.

mal conductors; have relatively high melting points (except for mercury) and boiling points; and form alloys with each other and with other metals. Like the other metals, they exhibit low electronegativities in their compounds. Most of them have two or more common oxidation states. They form coordination complexes in large numbers, most of them colored. The existence of several oxidation states and the variety of coordination complexes make transition metal chemistry particularly interesting.

Because of their periodic relations, transition metals are divided into vertical groups, as indicated in the periodic table. Rather than treat the different groups separately, in the remaining sections we shall make a comparative study of different aspects of transition metal chemistry.

20.2 NATURAL OCCURRENCE AND SEPARATION

Iron is the fourth most common element in the lithosphere, constituting 5.0% of the earth's crust, and titanium ranks ninth at 0.6%. Other fairly common transition metals, in the range from 0.1 to 0.01%, are manganese, chromium, zirconium, zinc, nickel, and vanadium. With the exception of zirconium, all of these more common metals are in the first transition series. Some naturally occurring minerals of the first transition series are listed in Table 20.1. The group VIII elements of the second and third transition series occur together in the elemental form, and

20.2 NATURAL OCCURRENCE AND SEPARATION

TABLE 20.1 SOME COMMON MINERALS OF THE FIRST TRANSITION SERIES

Element	Mineral	Formula
Scandium	Monazite	MPO_4 [a]
Titanium	Rutile	TiO_2
	Ilmenite	$FeTiO_3$
Vanadium	Vanadinite	$Pb_5(VO_4)_3Cl$
Chromium	Chromite	$FeCr_2O_4$
Manganese	Pyrolusite	MnO_2
Iron	Hematite	Fe_2O_3
	Magnetite	Fe_3O_4
Cobalt	Smaltite	$CoAs_2$
	Cobaltite	CoAsS
Nickel	Millerite	NiS
	Pentalandite	$NiFe_2S_3$
Copper	Native copper	Cu
	Chalcopyrite	$CuFeS_2$
Zinc	Sphalerite	ZnS
	Smithsonite	$ZnCO_3$

[a] M = a lanthanide, primarily La, Ce, Pr, and Nd. Sc is a minor constituent.

they are collectively known as the **platinum metals** after their most common member.

Scandium occurs in very small amounts with the other group IIIB elements in minerals such as monazite, MPO_4, where M is one of the lanthanides, primarily lanthanum, cerium, praseodymium, and neodymium. The lanthanides are so alike in their chemistry that they are especially difficult to separate from one another. An important modern method of separation is **ion-exchange chromatography**. A solution containing the lanthanides in their common oxidation state, +3, is poured through a column of an acidic ion-exchange resin. The metal ions are adsorbed on the solid resin, displacing hydrogen ions. With the acid resin represented by the symbol HR, the reaction is

$$M^{3+} + 3HR(s) = 3H^+ + MR_3(s)$$

Then the metal ions are washed through (eluted) with a solution containing a weak acid whose anion forms stable complex ions with the lanthanides. This second reaction is

$$MR_3(s) + 6HX = 3HR(s) + MX_6^{3-} + 3H^+$$

where the weak acid is represented as HX and the complex ion as MX_6^{3-}. The metal ions differ in the ease with which they are removed from the resin, and, under suitable conditions, a separation of the lanthanides can be achieved, as shown in the upper part of Figure 20.2. The order of removal is inversely related

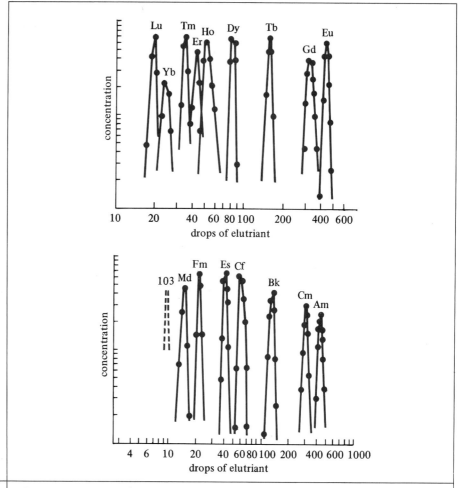

FIGURE 20.2 Separation of the lanthanides (upper figure) and actinides (lower figure) by ion-exchange chromatography. (SOURCE: *J. J. Katz* and *G. T. Seaborg*, The Chemistry of the Actinide Elements, *Wiley, New York, 1957, p. 435. By permission of the authors.*)

to position in the periodic table, with the last of the lanthanides, lutetium, eluted first, and so forth.

Ion-exchange chromatography played an important role in the discovery of several of the actinides. After attempts had been made to create new elements by nuclear reactions (Chapter 24), it was necessary to isolate and identify them. Their expected position in the periodic table allowed a prediction of the order in which the elements would be eluted from an ion-exchange column. This knowledge, combined with the radioactivity of the actinides, made possible the discovery of some elements on a fantastically small scale—with literally a few atoms. Results for the actinides are shown in the lower part of Figure 20.2. Note the inverse relation between order of removal and position in the periodic table.

20.3 METALS

Producing or "winning" a metal from a commercially valuable mineral deposit **(ore)** requires reduction, and one of the cheapest reducing agents is carbon. It is usually used in an impure form called coke, obtained from coal. Reduction with carbon is used to obtain iron, nickel, and zinc from their oxides. The oxide either occurs naturally or is obtained from the ore by air oxidation, a process called **roasting.** Carbon monoxide, from the incomplete combustion of coke in air or oxygen, is used to reduce the oxide in a high-temperature process:

$$Fe_2O_3 + 3CO = 2Fe + 3CO_2$$

Production of iron and its alloys **(steels)** is a major industry.

Metals that form particularly stable compounds with carbon require other reducing agents, and the light metals are commonly used. Titanium is obtained by first forming the tetrachloride, followed by reduction with magnesium:

$$TiO_2 + C + 2Cl_2 = TiCl_4 + CO_2$$
$$TiCl_4 + 2Mg = Ti + 2MgCl_2$$

Chromium and manganese oxides are reduced with aluminum:

$$Cr_2O_3 + 2Al = 2Cr + Al_2O_3$$

Hydrogen gas is particularly useful as a reducing agent to prepare metals of high purity. Another method of obtaining especially pure metals is electrolytic reduction from aqueous solutions of the cations (e.g., Fe^{2+}, Cu^{2+}, or Zn^{2+}).

Most transition metals crystallize in one or more of the three common metal structures (Section 17.4). Manganese is peculiar in that it has three crystalline forms under various conditions of temperature and pressure—none of them simple. Zinc has a distorted hexagonal close-packed structure. It is elongated along the axis perpendicular to the close-packed layers, as though the atoms were prolate spheroids (egg-shaped).

The transition series include the metals with the highest melting and boiling points. Figure 20.3 shows the melting points of the metals. In each series a peak is reached at or near group VIB, and a minimum at or near group IIB. The boiling points show similar trends. These properties reflect the particularly strong metal–metal bonding resulting from the large number of unpaired d electrons in atoms in the middle of a transition series. Another manifestation of these strong bonds is found in the interatomic distances, or atomic diameters (Figure 20.4). These reach minima at the group VIIB and VIII metals. Because the smallest diameters are about the same in each series and the elements of the third series have the highest

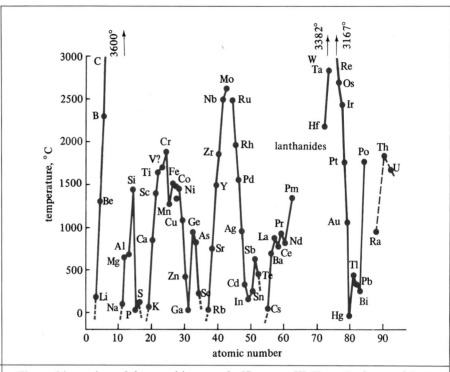

FIGURE 20.3 The melting points of the transition metals. [SOURCE: *W. Hume-Rothery* and *B. R. Coles*, Advances in Physics, **3**, *160 (1954)*.]

atomic masses, the elements in these groups in the third series have the highest densities of any elements: rhenium, 21.0; osmium, 22.6; iridium, 22.5; and platinum, 21.5 g cm^{-3}.

The transition metals, particularly the group IB metals, are good electric and thermal conductors. Magnetic properties of the transition metals were discussed in Section 17.13.

Resistance to corrosion is an important property of many transition metals. These include the group IB metals, the platinum metals, titanium, chromium, and nickel. Titanium has assumed considerable importance recently in the aerospace industry because of its strength, low density, and corrosion resistance. Iron and carbon steels corrode rapidly in the atmosphere, and it is necessary to protect them by covering them with paint, thin layers of zinc (galvanized iron), tin (food containers commonly referred to as tin cans), or chromium (automobile bumpers and ornaments), or by alloying the iron with other metals such as chromium and nickel (stainless steels).

Some of the transition metals are widely used as catalysts for various chemical reactions. Nickel and platinum are effective catalysts for hydrogenation reactions, and iron is used in the synthesis of ammonia from nitrogen and hydrogen.

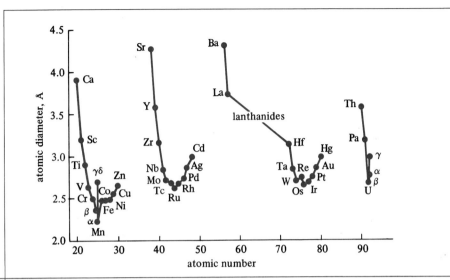

FIGURE 20.4 The atomic diameters of the transition metals. [SOURCE: *W. Hume-Rothery and B. R. Coles,* Advances in Physics, **3,** *156 (1954).*]

Many of the transition metals form interstitial compounds with the light elements, particularly boron, carbon, and nitrogen, in which the small atoms are located in the open spaces (interstices) between the larger transition metal atoms; for example, tungsten forms two carbides, WC and W_2C. Some of these interstitial compounds are stronger and harder and have higher melting points than the metals, giving rise to a variety of important uses such as drill tips and jet engine parts. They are sometimes called the **hard metals** although they are not true metals.

20.4 OXIDATION STATES

Table 20.2 shows the more important oxidation states for the first transition series. Beginning with vanadium, the +2 oxidation state becomes increasingly more important. The most common oxidation states are underlined in the table.

For the first five elements the highest oxidation state is the same as the number of valence electrons [ns and $(n-1)d$], that is, +3 for scandium, +4 for titanium, +5 for vanadium, +6 for chromium, and +7 for manganese. In these states, all of the valence electrons of the transition metal atom are involved in the formation of ionic or polar bonds with more electronegative atoms. In this respect there is a similarity between these transition metals and the corresponding main group elements of the periodic table. For example, sulfur in group VIA and chromium in group VIB have their highest oxidation state of +6 in substances such as sulfate ion, SO_4^{2-}, and chromate ion, CrO_4^{2-}, respectively.

TABLE 20.2   **IMPORTANT OXIDATION STATES OF THE FIRST TRANSITION SERIES**[a]

Oxidation state	Sc	Ti	V	Cr	Mn	Fe	Co	Ni	Cu	Zn
+7					MnO_4^-					
+6				CrO_4^{2-}	MnO_4^{2-}					
+5			VO_2^+							
+4		TiO^{2+}	VO^{2+}		$MnO_2(s)$					
+3	Sc^{3+}	Ti^{3+}	V^{3+}	Cr^{3+}	Mn^{3+}	Fe^{3+}	Co^{3+}			
+2			V^{2+}	Cr^{2+}	Mn^{2+}	Fe^{2+}	Co^{2+}	Ni^{2+}	Cu^{2+}	Zn^{2+}
+1									Cu^+	

Increasing atomic number →

[a] The most common oxidation states are underlined.

In general the transition metal ions in their lower oxidation states in acidic solution are hydrated cations [e.g., Cr^{3+} is really $Cr(H_2O)_6^{3+}$, the water ligands being usually omitted for brevity]. In their higher oxidation states, they are well-defined oxyanions (e.g., +6 manganese and +7 manganese exist as manganate ion, MnO_4^{2-}, and permanganate ion, MnO_4^-, respectively). In some cases the exact nature of the oxygen ligands is not clear; for example, +4 titanium is usually represented as the titanyl ion, TiO^{2+}, but there is evidence that it is really $Ti(H_2O)_4(OH)_2^{2+}$. The transition metal ions sometimes polymerize to form complex ions containing two or more metal ions. Of particular importance are mercurous ion, Hg_2^{2+}, and the red dichromate ion formed on acidification of solutions containing the yellow chromate ion:

$$2CrO_4^{2-} + 2H^+ = Cr_2O_7^{2-} + H_2O$$
   yellow   red

The energies required to remove electrons from the free atoms (ionization potentials) largely determine which oxidation states are important. The first four ionization potentials for the first transition series are shown in Table 20.3. Scandium has relatively low ionization potentials for removal of three electrons, and a very high fourth ionization potential, explaining why the +3 state is the only important oxidation state of scandium. The elements toward the end of the series have much higher third ionization potentials, and the +3 state is relatively unimportant for nickel and copper, and absent in zinc. Copper has a rather high second ionization potential, causing Cu^+ to be of some importance. Titanium and vana-

TABLE 20.3 — IONIZATION POTENTIALS OF THE FIRST TRANSITION SERIES

Element	Ionization potentials (eV)			
	1st	2nd	3rd	4th
Scandium	6.54	12.80	24.75	73.9
Titanium	6.82	13.57	27.47	43.2
Vanadium	6.74	14.65	29.31	48
Chromium	6.77	16.49	30.95	49.6
Manganese	7.44	15.64	33.69	52
Iron	7.87	16.18	30.64	57.1
Cobalt	7.86	17.05	33.49	53
Nickel	7.64	18.15	35.16	56
Copper	7.73	20.29	36.83	58.9
Zinc	9.39	17.96	39.7	62

dium have the lowest fourth ionization potentials, and the +4 state is important for both elements.

The nature of the species surrounding an ion in a solution or crystal also has an important effect on its oxidation state. In aqueous solutions the ions are complexed with water or with other ligands, depending on the particular species present and their concentrations. Thus the ionization potentials of the free atom can be taken as only a rough guide to the ease of removal of electrons from the ion in solution. For example, $Co(NH_3)_6^{2+}$ is much more easily oxidized than $Co(H_2O)_6^{2+}$.

In the second and third transition series the higher oxidation states are of greater importance and the lower oxidation states of lesser importance. In group IVB, zirconium and hafnium form the +4 state only. The chemistry of niobium and tantalum, elements in group VB, is largely that of the +5 state.

Because of the increasing positive charge on the nucleus, there is a general decrease in ionic radius with increasing atomic number for a given oxidation state of each transition series. The decrease is particularly striking for the lanthanides (Figure 20.5) and is called the **lanthanide contraction.** Atomic radii exhibit a similar trend, as discussed in Section 8.11.

Because of the lanthanide contraction, the ionic radius of an ion in the third transition series is about the same as that of the corresponding ion in the same group in the second transition series. For example, in group IVB the radii of Zr^{4+} and Hf^{4+} are 0.74 and 0.75 Å, respectively, compared to 0.68 Å for Ti^{4+} in the first transition series. These ions behave alike chemically because of their similarity in charge and size.

20.5 STANDARD ELECTRODE POTENTIALS

Table 20.4 lists the standard electrode potentials for the metals of the first transition series and their +2 cations (or the +3 cation in the case of scandium). Except

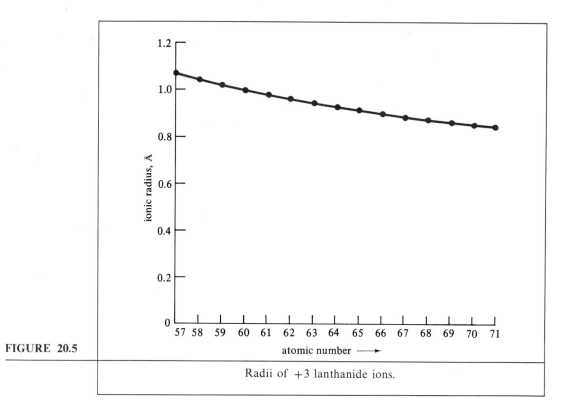

FIGURE 20.5 Radii of +3 lanthanide ions.

for copper, all the metals have standard potentials that are negative with respect to the standard hydrogen electrode, and therefore all but copper are base metals (dissolve in acid, such as hydrochloric acid, with the evolution of hydrogen gas). Copper is a noble metal, and more powerful oxidizing agents are required to dissolve it. It is soluble in hot nitric acid, forming nitrogen oxides (NO and NO_2). The other noble metals are in the second and third transition series: silver, gold, mercury, and the platinum metals.

Because transition metals exhibit such a wide variety of oxidation states, their electrochemistry is a subject of considerable complexity. Several of the transition metal ions are important oxidizing and reducing agents; for example, chromous ion (Cr^{2+}) is a good reducing agent. The standard electrode potential is

$$Cr^{3+} + e^- = Cr^{2+} \quad \mathcal{E}° = -0.41 \text{ V}$$

By contrast, +6 chromium in the form of dichromate ion ($Cr_2O_7^{2-}$) is a powerful oxidizing agent in acidic solution:

$$Cr_2O_7^{2-} + 14H^+ + 6e^- = 2Cr^{3+} + 7H_2O \quad \mathcal{E}° = +1.33 \text{ V}$$

Because this half-reaction requires $14H^+$ per dichromate ion, its potential is

TABLE 20.4 — STANDARD ELECTRODE POTENTIALS OF THE FIRST TRANSITION SERIES

Reaction	$\varepsilon°$
$Sc^{3+} + 3e^- = Sc(s)$	-2.08 V
$Ti^{2+} + 2e^- = Ti(s)$	-1.63 V
$V^{2+} + 2e^- = V(s)$	-1.18 V
$Cr^{2+} + 2e^- = Cr(s)$	-0.91 V
$Mn^{2+} + 2e^- = Mn(s)$	-1.18 V
$Fe^{2+} + 2e^- = Fe(s)$	-0.440 V
$Co^{2+} + 2e^- = Co(s)$	-0.277 V
$Ni^{2+} + 2e^- = Ni(s)$	-0.250 V
$Cu^{2+} + 2e^- = Cu(s)$	$+0.337$ V
$Zn^{2+} + 2e^- = Zn(s)$	-0.763 V

strongly dependent on acidity, and in basic solution +6 chromium is a much weaker oxidizing agent:

$$CrO_4^{2-} + 4H_2O + 3e^- = Cr(OH)_3(s) + 5OH^-$$

$$\varepsilon° = -0.13 \text{ V}$$

The standard electrode potential of a transition metal ion varies with the type of ligand complexed to the ion. For example, the aquo complex of +3 cobalt, $[Co(H_2O)_6]^{3+}$, is a powerful oxidizing agent, but the corresponding complex with ammonia, $[Co(NH_3)_6]^{3+}$, is a relatively poor one:

$$[Co(H_2O)_6]^{3+} + e^- = [Co(H_2O)_6]^{2+} \quad \varepsilon° = +1.82 \text{ V}$$

$$[Co(NH_3)_6]^{3+} + e^- = [Co(NH_3)_6]^{2+} \quad \varepsilon° = +0.1 \text{ V}$$

The familiar flashlight battery or dry cell is powered by transition metals. Manganese in the form of MnO_2 is the oxidizing agent, and zinc metal is the reducing agent (Section 14.9).

20.6 OXIDES, HYDROXIDES, HYDROUS OXIDES, AND OXYACIDS

The most important binary compounds of the transition metals are the oxides. Table 20.5 lists the oxides of the first transition series according to the oxidation state of the metal. As shown by the three sections in the table, the oxides of high oxidation state are acidic, those of low oxidation state are basic, and in the intermediate region the oxides are amphoteric.

The effects of oxidation state and atomic number on the relative acid-base character of the oxides may be correlated with covalent-ionic character and hence with electronegativity. In general the metals of lowest electronegativity (on the left side of the periodic table) form ionic bonds, basic oxides, and cationic species in solution. Moving to the right in a period, we find that the electronegativity

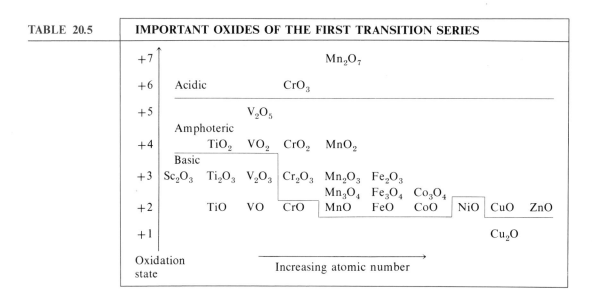

TABLE 20.5 IMPORTANT OXIDES OF THE FIRST TRANSITION SERIES

increases for a given oxidation state, the bonds become less ionic and more covalent, the oxides become amphoteric and then acidic, and anionic species are produced in solution. Electronegativity also increases with increasing oxidation state, and therefore, for a given element, the oxides of higher oxidation state are more acidic. For example, CrO is basic, dissolving in acids to give solutions that contain Cr^{2+} ions and forming the basic hydroxide $Cr(OH)_2$. Unless it has been heated to very high temperatures (when it changes to a form that reacts very slowly with either acid or base), Cr_2O_3 is amphoteric, dissolving in acids with the formation of Cr^{3+} cations and in bases with the formation of anions, such as $Cr(OH)_4^-$, and producing the amphoteric hydroxide $Cr(OH)_3$. The trioxide, CrO_3, is acidic, dissolving in bases to give solutions that contain CrO_4^{2-} ions and forming the strong oxyacids H_2CrO_4 and $H_2Cr_2O_7$.

The oxides Mn_3O_4, Fe_3O_4, and Co_3O_4 appear to have the metal atom in a fractional oxidation state, $+8/3$. Actually all three have structures related to that of spinel, $MgAl_2O_4$. Considering spinel as an ionic crystal, $Mg^{2+}[Al^{3+}]_2[O^{2-}]_4$, we note that the oxide ions are arranged in a cubic close-packed array, with the Mg^{2+} ions in tetrahedral interstices and the Al^{3+} ions in octahedral interstices (see Figure 20.6). Each Mg^{2+} ion is surrounded by four O^{2-} ions at the corners of a tetrahedron, and each Al^{3+} ion is surrounded by six O^{2-} ions at the corners of an octahedron.

In Mn_3O_4 there is one Mn^{2+} for every two Mn^{3+} ions, so that the average oxidation state is $+8/3$; the $+2$ ions occupy tetrahedral holes and the $+3$ ions occupy octahedral ones as in spinel. The cobalt oxide Co_3O_4 also has a normal spinel structure, but Fe_3O_4 has an inverted spinel structure. There is again one $+2$ ion for every two $+3$ ions, but half of the Fe^{3+} ions are in tetrahedral inter-

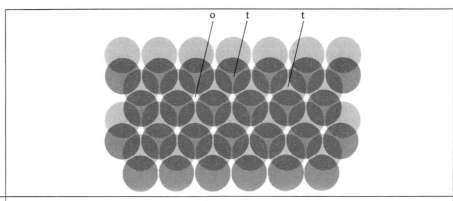

FIGURE 20.6 Two close-packed layers of spheres showing the octahedral (o) and tetrahedral (t) interstices.

stices. The other half of the Fe^{3+} ions, along with all the Fe^{2+} ions, occupy octahedral interstices. An explanation of this rather curious behavior will be found in Section 21.7.

In the second and third transition series the higher oxidation states are more stable. In contrast to the group VIII elements of the first series, several of the platinum metals form dioxides. Ruthenium and osmium also form highly volatile tetroxides: RuO_4, bp 100°C; OsO_4, bp 101°C.

The +2 hydroxides such as $Mn(OH)_2$ and $Fe(OH)_2$ are true hydroxides of the formulas indicated and have the same crystal structure as $Mg(OH)_2$. The +3 and +4 oxidation states form hydrous oxides of variable composition. Thus addition of a strong base to a solution containing Fe^{3+} ions produces a reddish brown precipitate of formula $Fe_2O_3 \cdot nH_2O$. The amount of water present depends on the conditions of precipitation. When the precipitate is dried, the oxide is produced. Although there is no evidence that a hydroxide of formula $Fe(OH)_3$ is ever produced, this symbol is often used for the hydrous oxide to simplify discussion of the precipitation reaction. The corresponding equation,

$$Fe^{3+} + 3OH^- = Fe(OH)_3(s)$$

is an abbreviated notation for

$$Fe^{3+} + 3OH^- + \tfrac{1}{2}(n-3)H_2O = \tfrac{1}{2}Fe_2O_3 \cdot nH_2O(s)$$

When amphoteric oxides and hydroxides dissolve in strong bases, anionic species are formed. These are often represented as oxyanions to emphasize the fact that the oxide or hydroxide is acting as an acid:

$$Zn(OH)_2 + 2OH^- = 2H_2O + ZnO_2^{2-}$$

However, the anions are hydrated, and a somewhat better representation of them is the hydroxo-complex ion:

$$Zn(OH)_2 + 2OH^- = Zn(OH)_4^{2-}$$

Formation of the hydroxo-complex ions can proceed stepwise by loss of protons from the hydrated cation. For example,

$$[Zn(H_2O)_4]^{2+} + OH^- = [Zn(H_2O)_3OH]^+ + H_2O$$
$$[Zn(H_2O)_3OH]^+ + OH^- = [Zn(H_2O)_2(OH)_2] + H_2O$$
$$[Zn(H_2O)_2(OH)_2] + OH^- = [Zn(H_2O)(OH)_3]^- + H_2O$$
$$[Zn(H_2O)(OH)_3]^- + OH^- = [Zn(OH)_4]^{2-} + H_2O$$

If only enough base is added to remove two protons, the neutral molecule $Zn(H_2O)_2(OH)_2$ polymerizes with loss of water to form a precipitate of zinc hydroxide:

$$Zn(H_2O)_2(OH)_2 = 2H_2O + Zn(OH)_2(s)$$

The remaining chemistry of the transition metals is largely that of coordination complexes, which are discussed in the next chapter.

PROBLEMS

20.1 In each of the following atoms and ions, how many $3d$ electrons are there? How many $4s$ electrons?
Sc, Sc^{3+}, Fe, Fe^{2+}, Fe^{3+}, Cu, Cu^+, Cu^{2+}, Zn, Zn^{2+}

20.2 The titanium atom has the electron configuration $3d^2 4s^2$ in its ground state, whereas V^+ and Cr^{2+} have the electron configuration $3d^4$ in their ground states. Suggest an explanation.

20.3 How many $4d$ electrons are there in each of the following ions?
Zr^{4+}, Mo^{3+}, Ru^{2+}, Ru^{3+}, Ag^+

20.4 Using the data in Appendix E, describe the filling of the $4d$ subshell in the transition metals of atomic numbers 39 to 48.

20.5 (a) Suggest an explanation for the high fourth ionization potential of scandium (see Table 20.3).
(b) As shown in Table 20.3, each of the first four ionization potentials of the first transition series shows a general increase across the series. [Disregard a few minor fluctuations and the high ionization potential mentioned in part (a).] Suggest an explanation.

PROBLEMS

20.6 The densities of the lanthanide oxides of formula M_2O_3 increase steadily from La_2O_3 to Lu_2O_3. Suggest an explanation.

20.7 The equilibrium constant for the reaction
$$Hg(l) + Hg^{2+} = Hg_2^{2+}$$
is 1.70×10^2. If excess liquid mercury metal is added to a $0.100\,M$ $Hg(NO_3)_2$ solution, what will be the concentration of Hg^{2+} after equilibrium is reached?

20.8 Assuming that all other substances are present at unit activity, calculate the electrode potential at $pH = 1.00$ of the dichromate half-reaction
$$Cr_2O_7^{2-} + 14H^+ + 6e^- = 2Cr^{3+} + 7H_2O$$
$$\mathcal{E}° = +1.33 \text{ V}$$

20.9 Draw Lewis structures obeying the octet rule for the CrO_4^{2-}, $Cr_2O_7^{2-}$, and MnO_4^- ions, indicating all formal charges. (The $Cr_2O_7^{2-}$ ion has a Cr—O—Cr linkage.) What O—Cr—O (or O—Mn—O) bond angles would you expect in these ions?

20.10 Write balanced equations for the net reactions that occur when
(a) zinc blende ore is roasted to form zinc oxide and sulfur dioxide
(b) ferric oxide is reduced to metallic iron by reaction with hydrogen gas at high temperature
(c) chromous hydroxide, $Cr(OH)_2$, is dissolved in dilute nitric acid
(d) chromic hydroxide, $Cr(OH)_3$, is dissolved in dilute nitric acid
(e) chromic hydroxide is dissolved in dilute NaOH solution
(f) chromium trioxide, CrO_3, is dissolved in dilute NaOH solution

20.11 Compare iron and cobalt qualitatively with respect to each of the following properties: electronic structures of the atoms, occurrence in the earth's crust, atomic diameters, magnetic properties, ionization potentials, common oxidation states, standard electrode potentials, formulas of oxides, acidic or basic character of oxides, and normal or inverted spinel structure of oxide of formula M_3O_4.

20.12 Explain why the lanthanide contraction causes the Hf^{4+} ion to have about the same ionic radius as the Zr^{4+} ion.

CHAPTER 21
THE TRANSITION METALS—COORDINATION CHEMISTRY

21

The transition metal ions form a great many complex ions and molecules. Typically the bonding is of the donor–acceptor type in which a **ligand** (an ion such as Cl^- or a neutral molecule such as NH_3) donates a pair of electrons to form an electron-pair bond with the metal ion. Formation of the complex is therefore a Lewis acid–base reaction (Section 13.1) in which the metal ion is the Lewis acid, and the ligand is the Lewis base. The donor–acceptor bonds are often called **coordinate covalent bonds,** and the complexes are called **coordination complexes.** Thus platinum in its +4 oxidation state combines with six chloride ions to form the hexachloroplatinate(IV) ion, $PtCl_6^{2-}$. This complex ion, as well as potassium ion, is present in crystals and aqueous solutions of K_2PtCl_6. The number of ligands attached to the metal ion is called its **coordination number;** in this example the coordination number of platinum is 6.

In the nineteenth century, coordination complexes were considered to be "double compounds." For example, K_2PtCl_6 was written as $PtCl_4 \cdot 2KCl$, a double compound of platinum tetrachloride and potassium chloride. Just why two compounds should combine in this way was not known. The modern theory of coordination compounds largely originated with the work of Alfred Werner who, in the early years of the twentieth century, synthesized literally hundreds of these coordination complexes. From his investigations of isomerism in these substances, he was able to reach an understanding of their structure, later confirmed by X-ray diffraction methods.

The most common number of ligands coordinated to a transition metal ion is six, and the geometrical arrangement of the six ligands is almost always octahedral. Complexes of this type are considered in the next sections. Some other important structures are discussed in Section 21.7.

The nontransition metal ions also form coordination complexes, such as the octahedral complex ions $Al(H_2O)_6^{3+}$ and AlF_6^{3-}, and the principles discussed in this chapter apply to them as well. But the number, variety, and unusual properties are greater for the coordination complexes of transition metals, primarily because of their incomplete shells of d electrons. As a result the terms "coordination chemistry" and "transition metal chemistry" have become almost synonymous.

21.1 OCTAHEDRAL COMPLEXES

The method by which Werner determined the coordination number of a metal ion is illustrated by his work on the isomers of chromium(III) chloride hexahydrate, $CrCl_3 \cdot 6H_2O$. Depending on the temperature and concentrations used, either violet or green crystals having this formula are obtained from aqueous solutions that contain Cr^{3+} and Cl^- ions. Freezing-point measurements show that when the violet form dissolves in water, four ions are formed per formula unit, and from this solution three chloride ions are readily precipitated by addition of a silver nitrate solution $[Cl^- + Ag^+ \rightarrow AgCl(s)]$. Thus the chloride ions cannot be strongly bonded to the chromium(III) ion. However, the water molecules are not removed when the violet crystals are dried over sulfuric acid, showing that the water molecules are firmly attached to the chromium(III) ion. Thus the violet form contains one hexaaquachromium(III) ion and three chloride ions per formula unit and may be represented as $[Cr(H_2O)_6]^{3+}(Cl^-)_3$.

When the green crystalline form dissolves in water, only two ions are formed per formula unit (as shown by the freezing-point depression), and only one chloride ion is readily precipitated by silver nitrate solution. Furthermore, drying easily removes two molecules of water. These results indicate that two of the chloride ions and four of the water molecules are strongly bonded to the chromic ion. Thus the green form contains one tetraaquadichlorochromium(III) ion, one chloride ion, and two molecules of water per formula unit and may be written as $[Cr(H_2O)_4Cl_2]^+Cl^- \cdot 2H_2O$.

By examining a series of platinum(IV) chlorides that contained varying amounts of ammonia (called **ammines**), Werner found that their properties were consistent with a coordination number of 6 for the platinum(IV) ion. The experimental data are summarized in Table 21.1. The first column gives the double compound designation of the solid substance; the second column lists the total number of ions formed when the solid dissolves in water (as determined by freezing-point measurements); and the third column gives the number of Cl^- ions precipitated by Ag^+. The last column shows the formula of the coordination complex deduced from the experimental data. We see that the number of Cl^- ligands varies from zero to six as the number of NH_3 ligands varies from six to zero. When

TABLE 21.1	THE PLATINUM(IV) AMMINE CHLORIDES			
	Substance	Total number of ions	Number of Cl^- ions	Complex
	$PtCl_4 \cdot 6NH_3$	5	4	$[Pt(NH_3)_6]^{4+}$
	$PtCl_4 \cdot 5NH_3$	4	3	$[Pt(NH_3)_5Cl]^{3+}$
	$PtCl_4 \cdot 4NH_3$	3	2	$[Pt(NH_3)_4Cl_2]^{2+}$
	$PtCl_4 \cdot 3NH_3$	2	1	$[Pt(NH_3)_3Cl_3]^+$
	$PtCl_4 \cdot 2NH_3$	0	0	$[Pt(NH_3)_2Cl_4]$
	$PtCl_4 \cdot NH_3 \cdot KCl$	2	0	$[Pt(NH_3)Cl_5]^-$
	$PtCl_4 \cdot 2KCl$	3	0	$[PtCl_6]^{2-}$

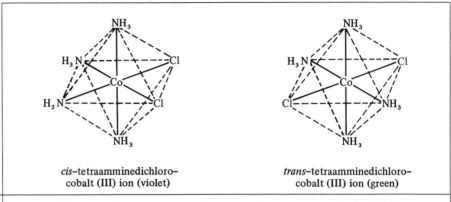

FIGURE 21.1

cis-tetraamminedichloro-cobalt (III) ion (violet)

trans-tetraamminedichloro-cobalt (III) ion (green)

The cis and trans isomers of an octahedral complex of cobalt(III).

the number of Cl^- ligands reaches four, the complex is an electrically neutral molecule. As solutions of $PtCl_4 \cdot 2NH_3$ contain no ions, the substance is a nonelectrolyte.

The spatial arrangement of the six ligands in a coordination complex is almost always octahedral (Figure 9.7). This may be understood as the result of the repulsion between the six pairs of electrons that bind the ligands to the metal ion, as discussed in Section 9.8. By studies of geometrical and optical isomerism in coordination complexes, Werner was able to determine their structures at a time when modern structural methods, such as X-ray diffraction, had not yet been developed.

In geometrical isomerism the same ligands are attached to the metal ion with different spatial arrangements. A good example is provided by the cis and trans isomers of any octahedral complex of general formula MX_2Y_4 (where M is a metal ion and X and Y are ligands). Figure 21.1 shows the cis and trans isomers of the ion $[Co(NH_3)_4Cl_2]^+$. In the cis isomer the Cl^- ligands are adjacent, with a Cl—Co—Cl angle of 90°; in the trans isomer the two Cl^- ligands are arranged on opposite sides of the Co^{3+} ion, with a Cl—Co—Cl angle of 180°. (No other isomers are possible.) These isomers are different chemical substances having different properties; for example, the cis isomer is violet, and the trans isomer is green. At the time Werner began his work only the latter was known; the discovery of the violet form by Werner provided experimental evidence supporting an octahedral structure.

One molecule or ion may bond to the metal ion at two or more positions, a phenomenon called **chelation** (from the Greek *chele,* claw). Ligands occupying two positions, called **bidentate** (from Latin *bidentatus,* "having two teeth"), include oxalate ion, $C_2O_4^{2-}$, and ethylenediamine, $H_2N—CH_2—CH_2—NH_2$. Figure 21.2 shows the green complex ion, which is similar to the trans isomer of Figure 21.1 except that, in Figure 21.2, two ethylenediamine molecules occupy the four positions that are occupied by four ammonia molecules in Figure 21.1. The

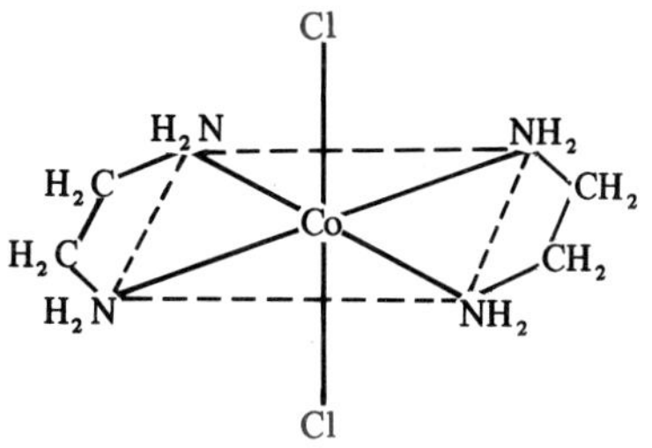

21.2 THE d ORBITALS

FIGURE 21.2

Chelation in a trans isomer of cobalt(III).

corresponding cis complex is shown in Figure 21.3. In this case there are two optical isomers, or **enantiomers,** that are mirror images of each other. Like the enantiomers of lactic acid (Section 11.9), they yield solutions that rotate plane-polarized light in opposite directions. Discovered by Werner, the existence of these enantiomers provided further evidence of an octahedral structure for these coordination complexes.

21.2 THE d ORBITALS

As indicated earlier, it is the filling of the inner electronic subshells that is the distinctive feature of the transition metals. In particular, the properties of the coordination complexes of the main transition series can be understood only with reference to the number and energies of the electrons in d orbitals. We recall from Chapter 8 that an orbital is a one-electron wave function, ψ, and that there are five

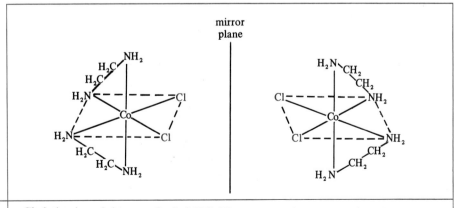

FIGURE 21.3

Chelation in a cis isomer of cobalt(III). The two enantiomers are mirror images and are nonsuperimposable.

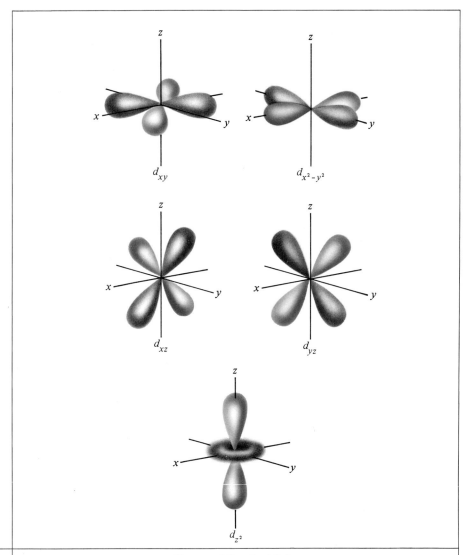

FIGURE 21.4 Shapes of the d orbitals.

3d orbitals (which have the same energy in a free atom or ion), five 4d orbitals, and five 5d orbitals.

Figure 21.4 shows the shapes of the five d orbitals (i.e., how the orbitals depend on direction at a given distance from the nucleus). These drawings apply to any set of d orbitals of the same principal quantum number—for example, the five 3d orbitals. Three of the d orbitals, labeled d_{xy}, d_{yz}, and d_{xz}, are quite similar. Each has four lobes, resembling a four-leaf clover, oriented at 45° angles to the x,

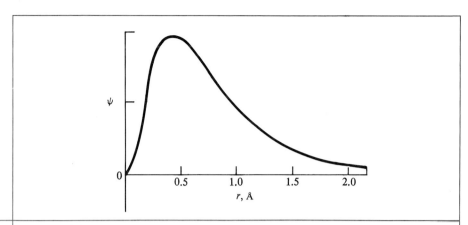

FIGURE 21.5 Effect of distance from the nucleus on a Hartree-Fock 3d orbital of the manganese atom.

y, and z axes. The d_{xy} orbital is concentrated in and about the xy plane, and its nodal surfaces (the surfaces where the orbital is zero) are the xz and yz planes. The d_{yz} and d_{xz} orbitals are similarly concentrated in and about the yz and xz planes, respectively. Because of symmetry properties, this set of three orbitals is often represented by the symbol t_{2g}. (It would take us too far afield to go into the origin of this symbol, but its use is so common that it is well worth remembering.)

The fourth d orbital, $d_{x^2-y^2}$, has four lobes directed along the x and y axes in either direction. (Note how it differs from d_{xy}, which is directed *between* the axes.) The remaining d orbital, d_{z^2}, has a rather different shape. There are two large lobes along the z axis in either direction, and a smaller toroidal or "doughnut" lobe in the middle. The symbol e_g is used for the $d_{x^2-y^2}$ and d_{z^2} orbitals because of their symmetry properties. (The similarity between d_{z^2} and $d_{x^2-y^2}$ is greater than appearances would indicate. It can be shown that the d_{z^2} orbital is a combination of two functions of the same shape as $d_{x^2-y^2}$—one directed along the y and z axes, the other along the x and z axes.)

To obtain a complete picture of a d orbital, one must consider not only its shape but also the way that the orbital varies with increasing distance from the nucleus. Figure 21.5 shows the radial dependence of a Hartree-Fock 3d orbital of the manganese atom. (All five 3d orbitals have the same radial dependence.) The amplitude of the orbital is zero at the nucleus, increases to a maximum at about 0.40 Å, and then decreases toward zero at larger distances. The radial dependence of the other d orbitals is more complicated, with one spherical node in a 4d orbital, two in a 5d orbital, and so on.

21.3 LIGAND-FIELD SPLITTING

In a completely free atom or ion (practically speaking, a gaseous atom or ion), the five 3d orbitals have exactly the same energy. But when an atom or ion is in the

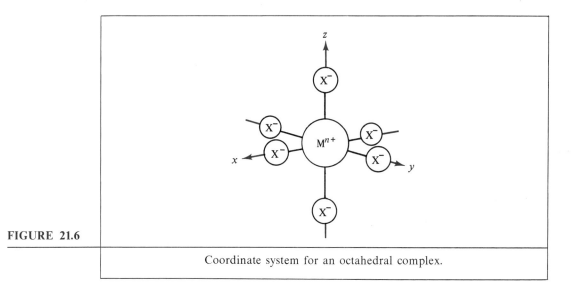

FIGURE 21.6 Coordinate system for an octahedral complex.

vicinity of other atoms, ions, or molecules, as in a crystal or coordination complex, the energies of the d orbitals change in a very important way. The theory of the origin of these changes in energy and their effect on the properties of coordination complexes is called **ligand-field theory.**

The simplest form of ligand-field theory, originated by Hans Bethe in 1929, is called **crystal-field theory.** It assumes that the metal ion–ligand bonds are of the ion–dipole type. As the ligand dipoles are oriented with their negative ends toward the positive metal ion, there is an electrostatic force of attraction between the ion and the ligands. It is further assumed that there is no covalent bonding, and thus the d orbitals of the metal ion do not participate in bonding. (Inclusion of covalent bonding does not significantly alter the most important results obtained from the theory, as we shall see in Section 21.9.) The only effect of the ligands on the d electrons is the coulombic repulsion by the negative ends of the ligand dipoles. The magnitude of this repulsion depends on the orientation of the d orbitals with respect to the ligands.

Figure 21.6 shows the coordinate system that is most convenient to describe the d electron–ligand interaction in an octahedral complex. Its origin is at the nucleus of the metal ion, and the x, y, and z axes pass through the centers of the six ligands. The orientation of the d orbitals with respect to these axes was described in the preceding section. (Although the choice of coordinate system is in principle an arbitrary one, this particular choice greatly simplifies the mathematical treatment.) Because the two e_g orbitals are directed along the axes and *toward* the ligands and the three t_{2g} orbitals are directed between the axes and *away from* the ligands, the d electrons in e_g orbitals are more strongly repelled by the negative ends of the ligand dipoles and have higher energies, as shown in Figure 21.7. The energy difference between the e_g and t_{2g} orbitals is called the **ligand-field**

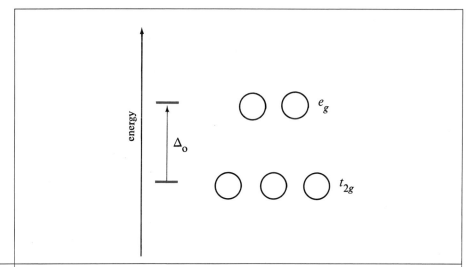

FIGURE 21.7 Energies of t_{2g} and e_g orbitals in an octahedral coordination complex or crystalline lattice site.

splitting, and in an octahedral complex it is designated by the symbol Δ_o. Its magnitude depends on both the central ion and its ligands.

To take a specific example, consider $[\text{Ti}(\text{H}_2\text{O})_6]^{3+}$, an octahedral coordination complex with one $3d$ electron. The water molecules are oriented with their negative ends (the oxygen atoms) nearest the positive Ti^{3+} ion (Figure 21.8).

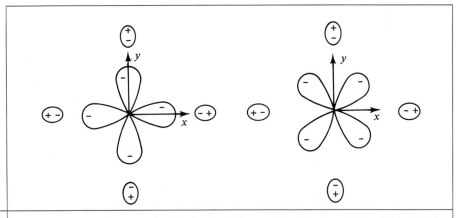

FIGURE 21.8 Distribution of electron density in the $3d_{x^2-y^2}$ orbital (left) and the $3d_{xy}$ orbital (right) surrounded by six dipoles. (The two dipoles above and below the plane of the paper are not important in these two cases and are not shown.) Plus and minus signs indicate the electric charge, not the amplitude of the wave function.

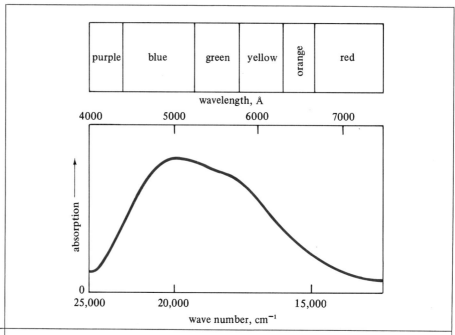

FIGURE 21.9 Absorption spectrum in the visible region of $[Ti(H_2O)_6]^{3+}$. [SOURCE: F. A. Cotton, *Journal of Chemical Education*, **41**, *474* (1964).]

Therefore the Ti^{3+} ion is at the center of an octahedron formed by the negative ends of six dipoles. The lobes of the e_g orbitals (such as $d_{x^2-y^2}$) point *directly toward* the surrounding water molecules; therefore if the $3d$ electron is in an e_g orbital, it will be strongly repelled by the negative ends of the water molecules. However, the lobes of the t_{2g} orbitals (such as d_{xy}) point *between* the water molecules; therefore the repulsion experienced by the $3d$ electron when it is in a t_{2g} orbital is not nearly so great. Consequently the e_g orbitals have higher energy than the t_{2g} orbitals, as shown in Figure 21.7. These same arguments apply to any octahedral coordination complex. (Some complexes are formed by anions such as Cl^-, rather than by polar molecules. In such a case, an electron in a $3d$ orbital is repelled by the anions, rather than by the negative ends of polar molecules, but the pattern of orbital energies is still the same.)

In a simple case, such as titanium(III), which has only one $3d$ electron, Δ_o can be measured directly from the absorption spectrum. Consider, for example, the $[Ti(H_2O)_6]^{3+}$ ion, which is a pale red-purple. Figure 21.9 shows its absorption spectrum (the fraction of light of various frequencies or wavelengths absorbed by this ion). Maximum absorption occurs for blue light of wavelength 5000 Å. The ion owes its color to the fact that light in the central part of the visible spectrum is absorbed to a greater extent than light at the ends of the spectrum, and therefore the visible transmitted light is mainly red and purple.

The interpretation of the spectrum is very simple. In the ground state the d electron is in a t_{2g} orbital. When light is absorbed, its energy promotes the d electron from a t_{2g} orbital to an e_g orbital. Therefore Δ_o for the $[\text{Ti}(\text{H}_2\text{O})_6]^{3+}$ ion is the same as the energy of a photon of wavelength 5000 Å, or 5.0×10^{-5} cm. (The absorption is spread over a rather wide frequency band because electronic excitation may be accompanied by either an increase or a decrease in the vibrational and rotational energy of the coordination complex. The magnitude of Δ_o also varies somewhat during the vibrational motion of the coordination complex as the ligands move in toward the metal ion, exerting a greater influence on the metal ion's d electrons, and then move away from the metal ion, exerting a smaller influence.)

The energy of a photon of wavelength λ may be calculated by combining Eqs. 7.2 and 7.5:

$$E = h\nu = \frac{hc}{\lambda}$$

Therefore the energy of a photon is proportional to the reciprocal of its wavelength, $1/\lambda$, called the **wave number.** It is conventional to report ligand-field splittings as the corresponding wave numbers in units of reciprocal centimeters. In this instance the ligand-field splitting is $1/(5.0 \times 10^{-5} \text{ cm}) = 20{,}000 \text{ cm}^{-1}$.

For a given central ion the effects due to various ligands exhibit a systematic trend. In almost all cases the magnitude of Δ_o induced by I^- is less than that due to Br^-, which, in turn, is smaller than the effect of Cl^-, and so on. These results are summarized in the **spectrochemical series,** discovered by Ryutaro Tsuchida in 1938:

$$\text{I}^- < \text{Br}^- < \text{Cl}^- < \text{F}^- < \text{OH}^- < \text{H}_2\text{O} < \text{NH}_3 < \text{en} < \text{CN}^- < \text{CO}$$

increasing field $\longrightarrow$

The ligands at the left end of the series are said to be **low-field ligands,** and those at the right, **high-field ligands.** Many other oxygen ligands occupy the same position as H_2O in the series, and many other nitrogen ligands occupy the same position as NH_3. The symbol en represents ethylenediamine.

Before we consider some important consequences of the spectrochemical series, a few other regularities should be noted. For a given ligand, the ions of a particular period and oxidation state have approximately the same values of Δ_o; for a higher oxidation state, Δ_o is larger. For example, hexahydrates of +2 ions of the first transition series have splittings of roughly 10,000 cm^{-1}, whereas the corresponding +3 ions have splittings about twice this. Some typical data are $\text{Ti}(\text{H}_2\text{O})_6^{3+}$, 20,000; $\text{V}(\text{H}_2\text{O})_6^{3+}$, 18,400; $\text{Cr}(\text{H}_2\text{O})_6^{3+}$, 17,400; $\text{Mn}(\text{H}_2\text{O})_6^{3+}$, 21,000; $\text{Fe}(\text{H}_2\text{O})_6^{3+}$, 14,300; $\text{Co}(\text{H}_2\text{O})_6^{3+}$, 18,100 cm^{-1}. Finally, the magnitude of Δ_o increases by approximately one-third from the first to the second transition series, and by another one-third from the second to the third transition series. In the next sections, we shall consider how the ligand-field splitting affects some of the properties of coordination complexes.

21.4 MAGNETIC PROPERTIES

The ligand-field splitting has an important effect on the number of unpaired electrons and hence on the magnetic properties of many transition metal compounds. Of course, for the d^1, d^2, and d^3 ions the electronic structure of octahedral complexes follows directly from Hund's rule (Section 8.9), with each electron going into a different t_{2g} orbital (Figure 21.10). But, beginning with the d^4 ions, there are two possibilities: Does the fourth electron go into an e_g orbital, even though it has a higher energy, or does it pair up with one of the electrons in a t_{2g} orbital? The answer turns out to depend on the magnitude of the ligand-field splitting, Δ_o, as compared to the pairing energy, P, required to overcome the interelectronic repulsion between two electrons in the same t_{2g} orbital. (The latter quantity can be determined from a study of the excited states of the transition metal ion.)

When the ligand field is low ($\Delta_o < P$), the fourth electron goes into an e_g orbital, as shown on the left side of Figure 21.11. Thus the coordination complex has four unpaired electrons. When the ligand field is high ($\Delta_o > P$), the fourth electron goes into a t_{2g} orbital, as shown on the right side of Figure 21.11. The two electrons in the same orbital have opposite spins, as required by the Pauli exclusion principle, and therefore the coordination complex has only two unpaired electrons. Thus the coordination complexes of a d^4 transition metal ion are of two types: (1) complexes with the low-field ligands having four unpaired electrons, called **high-spin complexes,** and (2) complexes with high-field ligands having two unpaired electrons, called **low-spin complexes.** Note that the terms *high-spin* and *low-spin* are used in a *relative sense for a particular transition metal ion*; that is, complexes with more unpaired electrons are called high-spin, and complexes with fewer unpaired electrons are called low-spin.

The electron configurations for the d^5 and d^6 coordination complexes are

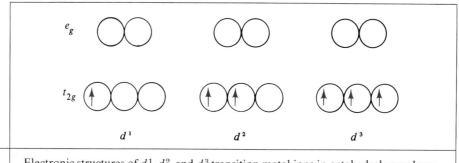

FIGURE 21.10

Electronic structures of d^1, d^2, and d^3 transition metal ions in octahedral complexes.

21.4 MAGNETIC PROPERTIES

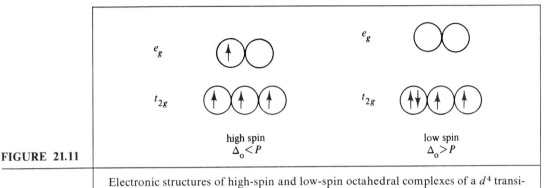

FIGURE 21.11 Electronic structures of high-spin and low-spin octahedral complexes of a d^4 transition metal ion.

shown in Figure 21.12. The high-spin d^5 complexes have five unpaired electrons, the low-spin complexes only one. The high- and low-spin complexes of a d^6 ion have four and zero unpaired electrons, respectively.

For any particular compound or coordination complex, the number of unpaired electrons affects its magnetic properties (Section 17.13). A substance having no unpaired electrons is weakly repelled by a magnetic field—it is said to be **diamagnetic.** A substance having one or more unpaired electrons is attracted by a magnetic field—it is said to be **paramagnetic.** Furthermore, the strength of the attractive force depends on the number of unpaired electrons. Therefore the number of unpaired electrons in a coordination complex can be determined by measuring its magnetic properties.

The experimental results support the interpretation given above. In particular, they show that low-field ligands such as the halide ions tend to form high-spin

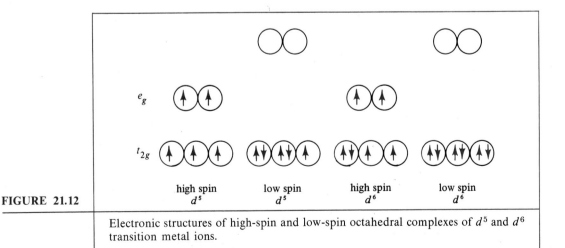

FIGURE 21.12 Electronic structures of high-spin and low-spin octahedral complexes of d^5 and d^6 transition metal ions.

complexes, whereas high-field ligands such as the cyanide ion tend to form low-spin complexes. For each transition metal ion, the spectrochemical series can be divided at some point: To the left, there are low-field ligands forming high-spin complexes and, to the right, high-field ligands forming low-spin complexes. For manganese(II) this point occurs near the right end of the spectrochemical series—only $[Mn(CN)_6]^{4-}$ is a low-spin complex. For cobalt(III) this point occurs near the left end of the series, and only the halide ions form high-spin complexes. Generally speaking, the $+2$ ions of the first transition series, which have relatively small ligand-field splittings, form the greatest number of high-spin complexes; the corresponding $+3$ ions, with larger splittings for a given ligand, have fewer high-spin complexes; and in the second and third transition series, the $+3$ ions seldom form high-spin complexes.

21.5 VISIBLE ABSORPTION SPECTRA

In general the coordination complexes of the transition metal ions have ligand-field splittings in the energy range corresponding to light visible to the human eye. Therefore they usually absorb light in the visible range and have various colors that depend on the positions and shapes of their absorption peaks. (The absorption spectrum and color of $[Ti(H_2O)_6]^{3+}$ have already been considered in Section 21.3.) If the ligand-field splitting is relatively small, absorption occurs mainly at the low-energy, or red, end of the spectrum. The light *not* absorbed, which gives the coordination complex its color, is therefore at the high-energy, or blue, end of the spectrum. Conversely, if the ligand-field splitting is relatively high, the absorption maximum is at the blue end of the spectrum, and the transmitted light is mainly at the red end. The exact color, of course, depends on the details of the absorption spectrum, that is, on just how much light is absorbed at each wavelength.

Table 21.2 shows the color of chromium(III) when it forms octahedral complexes with H_2O molecules, NH_3 molecules, and various mixtures of these two ligands. As indicated by their relative positions in the spectrochemical series, the ligand field of NH_3 is higher than that of H_2O. And, as H_2O is displaced by NH_3, the position of the absorption maximum shifts from yellow-green to blue, and the color of the complex ion changes from violet to yellow.

TABLE 21.2 COLORS OF SOME OCTAHEDRAL CHROMIUM(III) COMPLEX IONS

Ion	Absorbed light	Color (transmitted light)
$[Cr(H_2O)_6]^{3+}$	Yellow-green	Violet
$[Cr(NH_3)_2(H_2O)_4]^{3+}$	Green	Purple
$[Cr(NH_3)_3(H_2O)_3]^{3+}$	Blue-green	Red
$[Cr(NH_3)_4(H_2O)_2]^{3+}$	Green-blue	Orange-red
$[Cr(NH_3)_5(H_2O)]^{3+}$	Green-blue	Orange-yellow
$[Cr(NH_3)_6]^{3+}$	Blue	Yellow

21.5 VISIBLE ABSORPTION SPECTRA

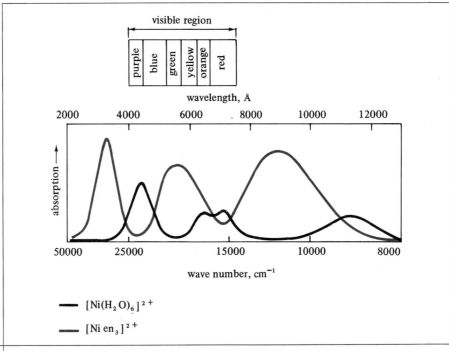

FIGURE 21.13 Absorption spectra of $[Ni(H_2O)_6]^{2+}$ and $[Ni\,en_3]^{2+}$. [SOURCE: F. A. Cotton, *Journal of Chemical Education*, **41**, *474 (1964)*.]

In some cases there is more than one absorption peak, each caused by a transition from the ground state to a different excited state. (Different excited states may occur, for instance, when different numbers of electrons are excited from t_{2g} to e_g orbitals.) Figure 21.13 shows the absorption spectrum of the octahedral complexes of nickel(II) with water and with ethylenediamine. In each case there are three absorption peaks. The hexaaqua complex ion absorbs light at each end of the visible region and transmits light in the middle, or green, region of the spectrum. This accounts for the green color of aqueous solutions of many nickel(II) salts such as $NiSO_4$. On addition of ethylenediamine (en), the six water molecules are displaced by three bidentate ethylenediamine molecules:

$$[Ni(H_2O)_6]^{2+} + 3en = [Ni\,en_3]^{2+} + 6H_2O$$

Because ethylenediamine has a higher ligand field than water, as shown by its position in the spectrochemical series, *each* absorption peak moves to higher energy and therefore to shorter wavelength. The red peak moves to the green, and the blue peak moves to the ultraviolet. The visible light transmitted by the solution is mainly blue and purple, along with some red light.

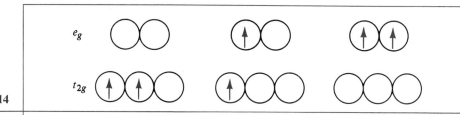

FIGURE 21.14 Ground state (left) and two excited states (right) of a d^2 ion.

When a coordination complex changes from its ground state to an excited state by absorption of light, there is usually no change in the total number of unpaired electrons. For example, if the ground state has two unpaired electrons, the excited state also has two unpaired electrons. Figure 21.14 shows the ground state and two of the excited states of a d^2 ion: Each has two unpaired electrons; the ground state has lowest energy, the next state has higher energy, and the last state has still higher energy. Transition from the ground state to the first excited state by absorption of light requires less energy than transition from the ground state to the second excited state, and therefore the first transition occurs at longer wavelength than the second. (Following absorption of light the metal ion quickly returns to the ground state. The excitation energy is emitted as light, or it is given up to neighboring molecules during collisions.)

A d^2 ion actually has three excited states with two unpaired electrons each, and therefore it has three absorption bands at different wavelengths (like the nickel complexes shown in Figure 21.13). An explanation of the differences between these excited states is rather involved and will be omitted here.

For a high-spin d^5 ion, only one state having five unpaired electrons is possible, and this state is of course the ground state. Because all excited states have different numbers of unpaired electrons, transitions to them occur with very low probability, and coordination complexes of d^5 ions such as Mn^{2+} are almost colorless. (Aqueous solutions of Mn^{2+} are very pale pink because of extremely weak transitions to these excited states.)

When the d subshell is completely empty or completely full, no d transitions are possible, and the coordination complexes are colorless (unless of course the ligands are colored). This is true of d^0 ions such as Sc^{3+} and Ti^{4+}, and of d^{10} ions such as Zn^{2+} and Ag^+. Their compounds are of course diamagnetic.

21.6 LIGAND-FIELD STABILIZATION ENERGIES

Some surprising features are observed when one examines the variation in a property along a series of transition metal complexes. These features are illustrated in Figures 21.15 and 21.16, which show the hydration energies and ionic radii, respectively, of the M^{2+} ions of the first transition series. (The hydration

21.6 LIGAND-FIELD STABILIZATION ENERGIES

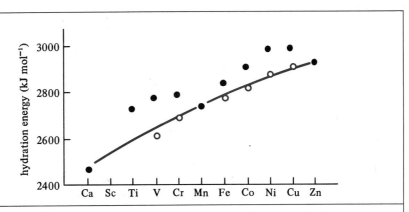

FIGURE 21.15

Hydration energies of the +2 ions of the first transition series. Solid circles are experimentally derived hydration energies. Open circles are energies corrected for the ligand-field stabilization energies. (SOURCE: *From F. A. Cotton and G. Wilkinson,* Advanced Inorganic Chemistry, *3rd ed. Copyright © 1972 by John Wiley and Sons, Inc. By permission of John Wiley and Sons, Inc.*)

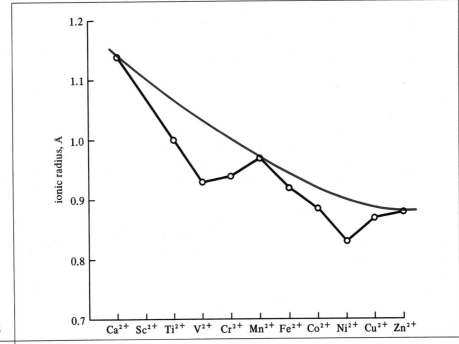

FIGURE 21.16

Ionic radii of +2 ions of the first transition series in high-spin complexes with coordination number six.

energy is the energy change, ΔE, for the reaction

$$[M(H_2O)_6]^{2+}(aq) = M^{2+}(g) + 6H_2O(l)$$

The ionic radius is that observed for an ionic crystal in which the M^{2+} ion is high-spin with coordination number six.) If a smooth curve is drawn through the ends and midpoint, then the other points are all to one side, giving rise to a pair of maxima in one case, and a pair of minima in the other case.

The explanation of these results starts from the realization that, in the d^0 and d^{10} ions at the ends and the d^5 ion at the midpoint, the d electrons are distributed uniformly among the d orbitals. In the d^0 ion, each d orbital is of course unoccupied; in the d^5 ion (high-spin complex), each d orbital has one electron, and, in the d^{10} ion, each d orbital has two electrons. The smooth curves drawn through the ends and midpoints may then be taken as the hydration energies and ionic radii that the other ions *would* have *if* their d electrons were uniformly distributed among the d orbitals.

We may compare the actual distribution with the hypothetical uniform distribution as follows. If nd electrons were distributed uniformly, then $n/5$ electrons would occupy each d orbital. Thus the three t_{2g} orbitals would have $3n/5$ electrons, and the two e_g orbitals would have $2n/5$ electrons. The resulting uniform distributions are compared in Table 21.3 with the actual distributions for the high-spin complexes.

Because of the ligand-field splitting Δ_o, the energy of a complex ion with a nonuniform distribution is lower than it would be if the d electrons were uniformly distributed among the d orbitals. For example, in going from the uniform to the actual distribution of the d^1 complex, two-fifths of an electron is transferred from an e_g orbital to a t_{2g} orbital whose energy is lower by an amount Δ_o per electron. Therefore the energy decreases by $\frac{2}{5}\Delta_o$. This energy difference is called the **ligand-field stabilization energy,** (LFSE). It is the energy by which the field of the ligands stabilizes the ion in its actual state, as compared to the hypothetical uniform distribution. The last column of Table 21.3 lists the LFSE for each ion. Because of the LFSE of an ion with a nonuniform distribution, the ion's hydration energy is higher than it would be if the distribution were uniform. When the LFSE is subtracted from the experimental value of the hydration energy, a corrected value is obtained, represented by an open circle in Figure 21.15. The open circles all lie on or near the curve drawn through the d^0, d^5, and d^{10} points, and therefore the explanation of the two maxima is in quantitative agreement with the experimental data.

Other properties associated with the stabilities of coordination complexes show similar characteristics. Of special interest are the equilibrium constants (called **stability constants**) for the formation of coordination complexes of the $+2$ ions with a given ligand. If the ligand is a neutral molecule L, the general reaction is

$$M^{2+} + 6L = ML_6^{2+}$$

21.6 LIGAND-FIELD STABILIZATION ENERGIES

TABLE 21.3 — DISTRIBUTION OF d ELECTRONS IN OCTAHEDRAL COMPLEXES

Total no. of d electrons	Uniform distribution		Actual distribution[a]		LFSE
	t_{2g}	e_g	t_{2g}	e_g	
0	0	0	0	0	0
1	$\frac{3}{5}$	$\frac{2}{5}$	1	0	$\frac{2}{5}\Delta_o$
2	$\frac{6}{5}$	$\frac{4}{5}$	2	0	$\frac{4}{5}\Delta_o$
3	$\frac{9}{5}$	$\frac{6}{5}$	3	0	$\frac{6}{5}\Delta_o$
4	$\frac{12}{5}$	$\frac{8}{5}$	3	1	$\frac{3}{5}\Delta_o$
5	3	2	3	2	0
6	$\frac{18}{5}$	$\frac{12}{5}$	4	2	$\frac{2}{5}\Delta_o$
7	$\frac{21}{5}$	$\frac{14}{5}$	5	2	$\frac{4}{5}\Delta_o$
8	$\frac{24}{5}$	$\frac{16}{5}$	6	2	$\frac{6}{5}\Delta_o$
9	$\frac{27}{5}$	$\frac{18}{5}$	6	3	$\frac{3}{5}\Delta_o$
10	6	4	6	4	0

[a] High-spin complexes.

In the second half of the first transition series, the stability constants for a given ligand increase from Mn^{2+} to a maximum at Cu^{2+}:

$$K_{Mn^{2+}} < K_{Fe^{2+}} < K_{Co^{2+}} < K_{Ni^{2+}} < K_{Cu^{2+}} > K_{Zn^{2+}}$$

This trend, which parallels that shown in Figure 21.14, is also explained by ligand-field stabilization.

The two minima in Figure 21.16 are also explained by the nonuniform distribution of d electrons. Opposing the close approach of a ligand and a metal ion is the repulsion between ligands and e_g electrons (previously discussed in connection with Figure 21.8). Consequently the larger the number of e_g electrons, the larger the ionic radius of the metal ion.

Table 21.3 shows that if the d electrons were distributed uniformly, the number of e_g electrons would increase by equal increments from the d^0 to the d^{10} ion. The colored curve shown in Figure 21.16, which is drawn through the experimental points for the three ions having a uniform distribution (Ca^{2+}, Mn^{2+}, and Zn^{2+}), would then represent the ionic radii for all of the ions. However, for an ion with a nonuniform distribution, the number of e_g electrons is smaller than it would be if the d electrons were distributed uniformly. Therefore the ligands can approach the metal ion more closely, and the ionic radius is smaller.

In Figure 21.16, the vertical distance between each experimental point and the colored curve represents the shortening of the ionic radius caused by the nonuniform distribution of d electrons. The effect is largest for the d^3 and d^8 ions, where for high-spin complexes the number of e_g electrons in the actual distribution is smaller by $\frac{6}{5}$, and it is here that the experimental values of the ionic radii show the greatest deviations from the colored curve.

21.7 LINEAR, TETRAHEDRAL, AND SQUARE PLANAR COMPLEXES

The coordination complexes of the d^0–d^6 transition metal ions generally have six ligands in an octahedral structure. For the d^7–d^{10} ions several other structures are found (although octahedral complexes are also formed by some of these ions).

Coordination with only two ligands in a linear structure is typical of the d^{10} ions having a +1 oxidation state [copper(I), silver(I), and gold(I)]. Examples include the silver ammonia complex ion, $[Ag(NH_3)_2]^+$,

$$\left[\begin{array}{c} H \\ H-N-Ag-N-H \\ H \quad\quad\quad H \end{array}\right]^+$$

The d^{10} ions in +2 oxidation states [zinc(II), cadmium(II), and mercury(II)] are usually coordinated with four ligands in a tetrahedral structure. (The tetrahedral structure is discussed in Section 9.8.) Examples include the zinc ammonia complex ion, $[Zn(NH_3)_4]^{2+}$. Both tetrahedral and octahedral complexes are important in the chemistry of the d^7 ions such as cobalt(II).

Square planar structures are most common for the d^8 ions. Nickel(II) has both square planar and octahedral complexes, whereas palladium(II), platinum(II), and gold(III) coordination complexes are usually square planar; $[Pt(NH_3)_4]^{2+}$ is an example:

$$\left[\begin{array}{c} \text{structure} \end{array}\right]^{2+}$$

Cis and trans isomers are formed in square planar complexes of the type MA_2B_2, where M is a metal ion, and A and B are ligands. An important example is diamminedichloroplatinum(II), $Pt(NH_3)_2Cl_2$. The isomers are

cis isomer   trans isomer

In the cis isomer the NH_3 ligands are in adjacent positions, whereas in the trans isomer they are on opposite sides of the platinum ion. (The same is true of the

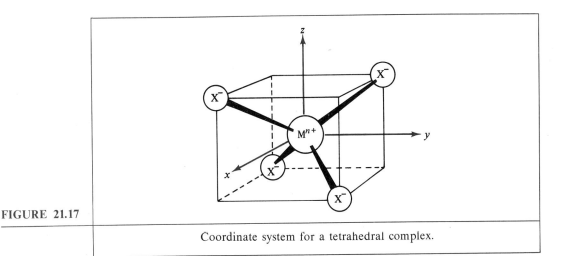

FIGURE 21.17 Coordinate system for a tetrahedral complex.

chloride ligands.) The cis isomer has been found to have significant biological activity, and it is now used in treating certain types of cancer. By contrast, the trans isomer has no therapeutic effect.

Unlike square planar complexes, tetrahedral complexes of formula MA_2B_2 do not have geometrical isomers, because in tetrahedral geometry each ligand is adjacent to every other ligand.

Because the positions of the ligands relative to the various d orbitals depend on the structure of the coordination complex, the pattern of ligand-field splittings differs for each structure. Figure 21.17 shows the coordinate system for a tetrahedral coordination complex. Here the interaction between the various d orbitals and the ligands is just the reverse of that previously encountered in an octahedral complex. Because the t_{2g} orbitals are directed between the axes and toward the ligands, electrons in these orbitals are more strongly repelled by the ligands and have higher energy. The e_g electrons are directed along the axes and away from the ligands, are repelled less, and have lower energy. The resulting splitting pattern is shown in Figure 21.18.

The energy difference for the tetrahedral complex, Δ_t, is not so large as the splitting in the octahedral structure, Δ_o. For the same central ion and the same ligands, crystal-field theory predicts that $\Delta_t = \frac{4}{9}\Delta_o$; experimental values of Δ_t are usually about $\frac{1}{2}\Delta_o$. (For different ligands, the spectrochemical series previously given for Δ_o applies to Δ_t as well.) Consequently, transition metal complexes with tetrahedral structures are generally of the high-spin type, and they absorb light of longer wavelengths than do comparable octahedral complexes. For example, cobalt(II) forms two important series of coordination complexes: octahedral complexes generally having a pale pink color due to a weak absorption band in the blue region of the spectrum, such as $[Co(H_2O)_6]^{2+}$, and tetrahedral complexes generally having a deep blue color due to a much more intense absorption band in the red, such as $[CoCl_4]^{2-}$ (see Figure 21.19).

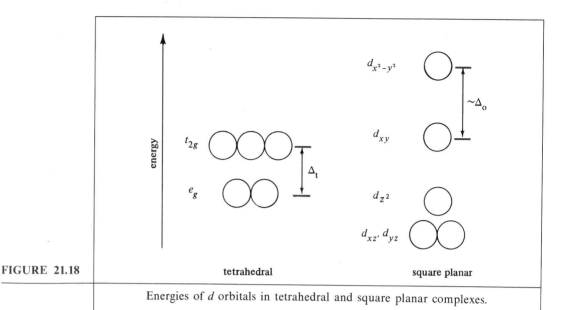

FIGURE 21.18 Energies of d orbitals in tetrahedral and square planar complexes.

In a square planar structure the ligand-field splitting is somewhat more complicated, as shown in Figure 21.18. If the axis perpendicular to the plane of the complex ion is designated the z axis (see Figure 21.20), then the electrons that are least repelled by the ligands (thus having the lowest energy) are those farthest from the ligands in orbitals on or near the z axis, that is, d_{z^2}, d_{xz}, and d_{yz}. The electrons that are most repelled by the ligands (thus having the highest energy) are those in the $d_{x^2-y^2}$ orbital, which is directed toward the four ligands. The d_{xy} orbital (directed between the ligands) has intermediate energy, usually about Δ_o below the energy of the $d_{x^2-y^2}$ orbital, as in an octahedral complex. Consequently, a d^8 ion has six electrons in the three orbitals of lowest energy and, depending on the ligand-field splitting, either two electrons in the d_{xy} orbital (the high-field low-spin complex) or one electron in d_{xy} and one electron in $d_{x^2-y^2}$ (the low-field high-spin complex). All known square planar complexes of the d^8 ions are of the low-spin type.

It is not fully understood just why a given coordination complex has a particular structure. Certainly an important factor is the LFSE. The LFSE's for octahedral and tetrahedral structures are compared in Table 21.4. (The LFSE's for the tetrahedral structure were calculated by the method used in the preceding section for the octahedral structure, taking into account the different splitting pattern.) As examples of the effect of LFSE's on structures, consider cobalt(II) and nickel(II), which have seven and eight d electrons, respectively. A d^7 ion has maximum LFSE for a tetrahedral structure, and only an average LFSE for an octahedral structure. The LFSE's of a d^8 ion are just the opposite—maximum for octahedral, and average for tetrahedral. These results help explain the very different stereo-

21.7 LINEAR, TETRAHEDRAL, AND SQUARE PLANAR COMPLEXES

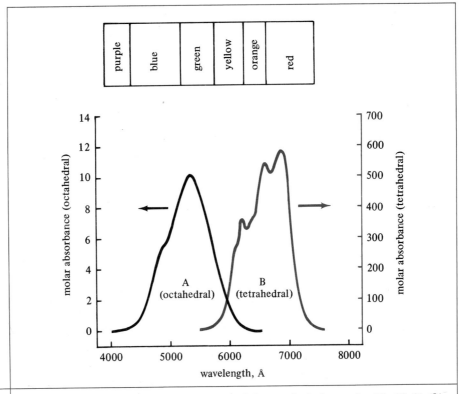

FIGURE 21.19

Absorption spectra in the visible region of the octahedral complex $[Co(H_2O)_6]^{2+}$ (curve A) and the tetrahedral complex $[CoCl_4]^{2-}$ (curve B). Note the difference in scale. (SOURCE: From F. A. Cotton and G. Wilkinson, *Advanced Inorganic Chemistry*, 2nd ed. Copyright © 1966 by John Wiley and Sons, Inc. By permission of John Wiley and Sons, Inc.)

chemistry of cobalt(II) and nickel(II): Cobalt(II) forms a great many tetrahedral complexes, whereas tetrahedral coordination is much less important for nickel(II). (Because $\Delta_o \approx 2\Delta_t$, even for cobalt(II) the LFSE is larger in an octahedral structure, and therefore the assistance of other factors is required in forming a tetrahedral structure.)

The LFSE's are also important factors in determining the location of transition metal ions in the spinels (Section 20.6). In Mn_3O_4 the d^5 Mn^{2+} ions have zero LFSE in either environment. Thus the LFSE's of the d^4 Mn^{3+} ions determine the structure. As the LFSE of a d^4 ion is much larger in an octahedral structure (assuming $\Delta_o \approx 2\Delta_t$), the Mn^{3+} ions occupy octahedral interstices, and Mn_3O_4 has a normal spinel structure. In Fe_3O_4 it is the $+3$ ions that are d^5 and have zero LFSE in either environment. Thus the LFSE's of the d^6 Fe^{2+} ions determine the structure. As a d^6 ion has a larger LFSE in an octahedral structure, the Fe^{2+} ions

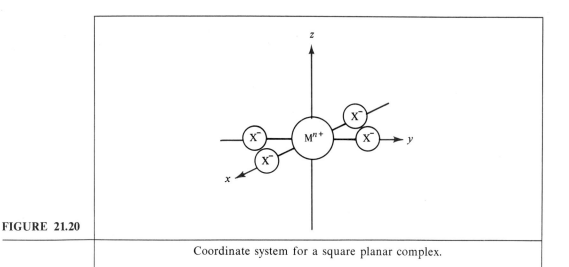

FIGURE 21.20 Coordinate system for a square planar complex.

occupy octahedral interstices, and Fe_3O_4 has an inverted spinel structure. (A cautionary note: There are other factors besides LFSE's that affect structure, and only when these other factors are fairly constant can one correctly predict a structure from the LFSE's. In the relatively few cases where the actual structure differs from that predicted by the method outlined above, it has been found that the experimental results can be explained by consideration of all factors involved.)

TABLE 21.4   **LIGAND-FIELD STABILIZATION ENERGIES OF OCTAHEDRAL AND TETRAHEDRAL STRUCTURES**

Total no. of d electrons	Octahedral LFSE[a]	Tetrahedral LFSE[a]
0	0	0
1	$\frac{2}{5}\Delta_o$	$\frac{3}{5}\Delta_t$
2	$\frac{4}{5}\Delta_o$	$\frac{6}{5}\Delta_t$
3	$\frac{6}{5}\Delta_o$	$\frac{4}{5}\Delta_t$
4	$\frac{3}{5}\Delta_o$	$\frac{2}{5}\Delta_t$
5	0	0
6	$\frac{2}{5}\Delta_o$	$\frac{3}{5}\Delta_t$
7	$\frac{4}{5}\Delta_o$	$\frac{6}{5}\Delta_t$
8	$\frac{6}{5}\Delta_o$	$\frac{4}{5}\Delta_t$
9	$\frac{3}{5}\Delta_o$	$\frac{2}{5}\Delta_t$
10	0	0

[a] High-spin complexes.

21.8 THE JAHN-TELLER EFFECT

The d^9 ions, such as copper(II), have octahedral structures that are distorted to a significant extent. In most cases bonds to two ligands on opposite sides of the metal ion are much longer and weaker than the other four bonds. (This type of distortion is called **elongation**.) It is not uncommon to disregard these weakly bound ligands and to emphasize the other four, which have a square planar structure. Thus the ammonia complex of copper(II) is often written $[Cu(NH_3)_4]^{2+}$, with a structure analogous to that of $[Pt(NH_3)_4]^{2+}$ shown in the last section.

In some cases the bonds to two ligands on opposite sides of the metal ion are shorter and stronger than the other four bonds. (This type of distortion is called **compression**.)

The theory of these distorted octahedra is based on a theorem derived in 1937 by H. A. Jahn and E. Teller from the principles of quantum mechanics. As it applies to the present case, the theorem states that distortion in an octahedral complex will occur whenever the orbitals of a subshell are occupied by *different numbers of electrons*. For example, consider a d^9 ion in an octahedral complex. Its e_g subshell has two electrons in one e_g orbital, but only one electron in the other e_g orbital, and therefore a Jahn-Teller distortion occurs.

Neither the type (elongation or compression) nor the extent of the distortion can be predicted from the theorem. Experimental measurements show that elongation occurs in most cases. The magnitude of the distortion is relatively small when it is the orbitals of the t_{2g} subshell that contain different numbers of electrons, and relatively large when it is the orbitals of the e_g subshell that contain them.

There is a simple physical explanation for the Jahn-Teller effect. A d^9 ion in an octahedral field either has two electrons in the d_{z^2} orbital and one electron in the $d_{x^2-y^2}$ orbital, or else has one electron in the d_{z^2} orbital and two electrons in the $d_{x^2-y^2}$ orbital. In the former case, the two ligands on the z axis are pushed away from the metal ion by the two d_{z^2} electrons, leading to elongation. (Of course the single electron in the $d_{x^2-y^2}$ orbital repels the ligands along the x and y axes, but the effect is smaller because there is only one electron involved.) In the latter case, the four ligands located on the x and y axes are pushed away from the metal ion by the two electrons in the $d_{x^2-y^2}$ orbital. The other two bonds are shorter than these four, so the distortion is of the compression type. Thus, regardless of how the electrons are distributed, a distortion must occur.

Other ions in octahedral complexes that should exhibit Jahn-Teller distortions include the d^1, d^2, d^4, low-spin d^5, high-spin d^6, and d^7 ions.

21.9 MOLECULAR ORBITALS OF AN OCTAHEDRAL COMPLEX

The application of molecular orbital theory to an octahedral coordination complex of a first transition series ion leads to the pattern of orbital energies shown in Figure 21.21. [Electrons are shown in the orbitals of lowest energy for a complex

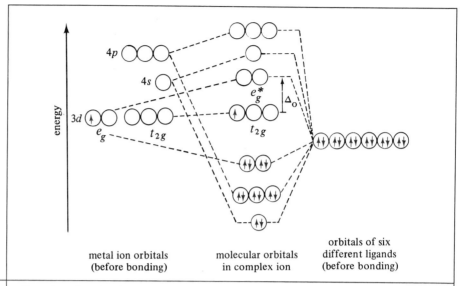

FIGURE 21.21

Energies of molecular orbitals in an octahedral coordination complex of a d^1 transition metal ion.

of a d^1 ion such as titanium(III), but other transition metal ions have the same pattern of orbital energies in their complexes.] The relevant atomic orbitals of the metal ion before bonding are shown along the left side, and the ligand orbitals before bonding along the right. These orbitals combine to form molecular orbitals. Just which orbitals combine depends on the symmetry of the atomic orbitals in relation to the octahedron of ligand orbitals, and the results are as follows.

The three $3d$ orbitals of t_{2g} symmetry have the wrong symmetry to combine with the ligand orbitals, and they remain unchanged as three *nonbonding* orbitals of intermediate energy, shown in the middle of Figure 21.21. The six molecular orbitals of lowest energy are *bonding* orbitals, and they are combinations of the six ligand orbitals and six of the metal ion orbitals—the two $3d$ orbitals of e_g symmetry, the $4s$ orbital, and the three $4p$ orbitals. The same orbitals also combine in a different way to form six *antibonding* molecular orbitals, shown at the top of Figure 21.21. The two antibonding molecular orbitals of lowest energy are of special importance: They have e_g symmetry and are antibonding; hence they are labeled e_g^*.

The 12 electrons in the bonding molecular orbitals may be considered to come entirely from the 6 ligands, and thus the $3d$ electrons of the central ion must populate the t_{2g} and e_g^* orbitals. Except for explicit recognition that the e_g^* orbitals are antibonding, this is exactly the same as the description of d electrons in

21.10 CHELATES

Coordination complexes of the chelate type, mentioned in Section 21.1, are especially important. In general, chelates are formed by ligands having two or more groups that can act as Lewis bases in forming donor–acceptor bonds with metal ions. The ethylenediamine molecule,

$$H_2\ddot{N}-CH_2-CH_2-\ddot{N}H_2$$

is one of the most important bidentate ligands. Two of its complexes with cobalt(III) are shown in Figures 21.2 and 21.3. Each nitrogen atom donates its unshared pair of electrons to form a bond to the Co^{3+} ion. Note that the cobalt ion, the two nitrogen atoms, and the two carbon atoms of each ethylenediamine molecule form a five-membered ring. The most stable chelate complexes have five- and six-membered rings.

Coordination complexes with chelating agents are more stable than comparable complexes lacking chelate rings. This chelate effect is illustrated by the following equilibrium constants:

$$[Ni(H_2O)_6]^{2+} + 6NH_3 = [Ni(NH_3)_6]^{2+} + 6H_2O \qquad K = 4.1 \times 10^8$$

$$[Ni(H_2O)_6]^{2+} + 3en = [Ni\,en_3]^{2+} + 6H_2O \qquad K = 1.9 \times 10^{18}$$

Even though the structures of the individual coordinating groups are quite similar, the ethylenediamine (en) complex is more stable by a factor of about 10^{10}. Studies of the enthalpy and entropy contributions to the free-energy changes of these reactions have shown that the chelate effect is caused by a more favorable entropy change. If one looks at the entropy changes in these two reactions from a probability point of view, the chance of a metal ion having three ethylenediamine molecules in its vicinity is greater than its chance of having six ammonia molecules nearby. In other words, if one end of an ethylenediamine molecule is near, the other end cannot be far behind.

The chelate effect is even greater for ligands with more than two coordinating groups. Diethylenetriamine is an example of a tridentate ligand:

$$:\!\ddot{N}H\!\!\begin{array}{l}\diagup CH_2-CH_2-\ddot{N}H_2\\ \diagdown CH_2-CH_2-\ddot{N}H_2\end{array}$$

Two molecules of this ligand are sufficient to form an octahedral complex. A hexadentate ligand, which is one of the best chelating agents known, is the ethylenediaminetetraacetate ion ($EDTA^{4-}$):

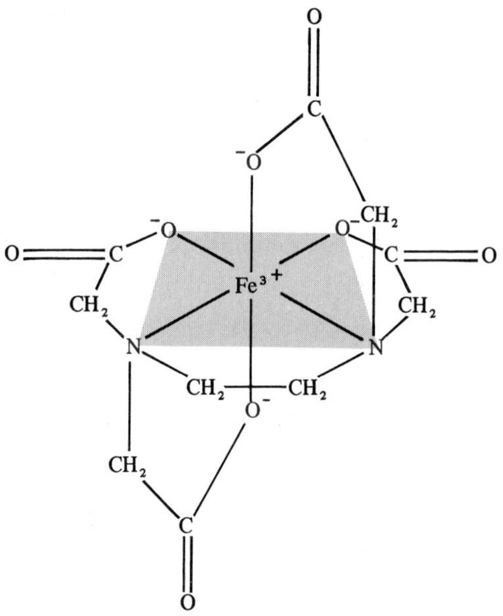

FIGURE 21.22 An octahedral complex of Fe^{3+} and $EDTA^{4-}$.

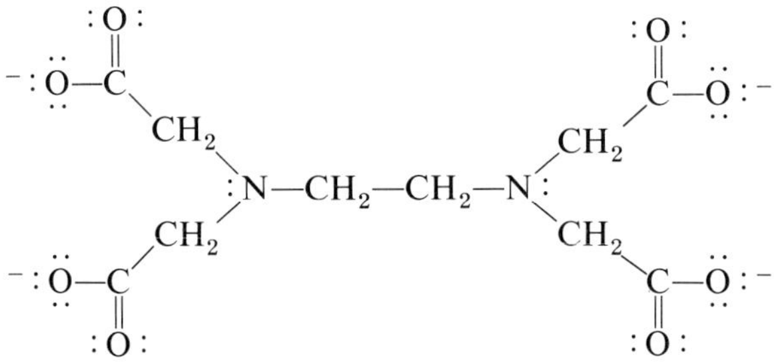

The two nitrogen atoms and the four COO^- groups occupy all six positions in an octahedral complex (see Figure 21.22). The protonated species $EDTAH^{3-}$ is pentadentate.

The porphyrin ring is a quadridentate ligand that occurs complexed with Mg^{2+} in chlorophyll, the green pigment in plants, and complexed with Fe^{2+} in heme, the red pigment in animals (Figure 19.4).

Practical applications of chelation include complexing Ca^{2+}, Mg^{2+}, and Fe^{3+} ions in hard water to prevent their interference with the cleaning action of soaps and detergents (Section 18.2). Iron deficiency in plants is remedied by application of the iron–EDTA complex. In analytical chemistry, nickel(II) is precipitated as the slightly soluble, square planar complex with the bidentate dimethylglyoxime anion. The complex is stabilized by hydrogen bonds between the ligands:

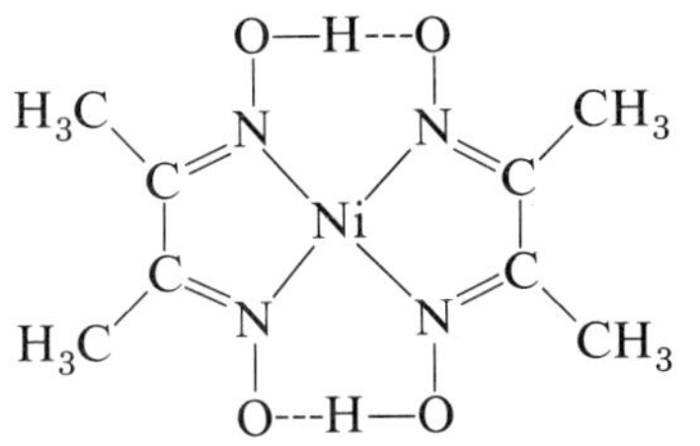

PROBLEMS

21.1 How many isomers does the $[Pt(NH_3)_3Cl_3]^+$ ion have? Sketch the structure of each isomer.

21.2 The NH_3 molecule forms coordination complexes with transition metal ions; the NH_4^+ ion does not. Why?

21.3 How many isomers are there of a square-planar transition metal complex of a metal ion M^{2+} with four different substituents A, B, C, and D? Draw a sketch of each isomer.

21.4 How many isomers are there of a transition metal complex $[M(H_2O)_2Cl_2]$ if it is
(a) square planar?
(b) tetrahedral?

21.5 Two compounds of cobalt(III) have the same empirical formula, $Co(NH_3)_5(SO_4)Br$. One gives a precipitate of AgBr with silver nitrate, but no precipitate with $BaCl_2$. The other gives a precipitate of $BaSO_4$ with barium chloride, but no precipitate with $AgNO_3$. What is the formula (including net electric charge) of the coordination complex for each isomer? (Assume a coordination number of six for the cobalt ion.)

21.6 How many d electrons are there in each of the following ions?
Ti^{3+}, Ti^{4+}, Cr^{2+}, Cr^{3+}, Co^{2+}, Co^{3+}, Zn^{2+}

21.7 How many d electrons are there in a coordination complex of iron(II)? Of iron(III)? Of nickel(II)?

21.8 Which d orbital has lobes concentrated *along* the x and y axes? Which d orbital has lobes concentrated *between* the x and y axes?

21.9 Compare the $2p_x$ orbital (Section 8.4) and the $3d_{x^2-y^2}$ orbital with respect to
(a) number of lobes

(b) axes along which the lobes are concentrated
(c) number and shape of nodal surfaces

21.10 Which d orbital(s) in a Ni^{2+} ion would have the highest energy if the Ni^{2+} ion were bonded to
(a) two ligands lying along the z axis on either side of the Ni^{2+} ion?
(b) six ligands arranged octahedrally?
(c) four ligands arranged tetrahedrally?
(d) four ligands arranged at the corners of a square?

21.11 A particular coordination complex has an absorption maximum in the green at 5680 Å. Calculate the wave number (in reciprocal centimeters).

21.12 Draw diagrams similar to Figure 21.12 for the high-spin and low-spin octahedral complexes of a d^7 transition metal ion. How many unpaired electrons are there in each?

21.13 Which of the following ions have low-spin octahedral complexes that are paramagnetic?
    Cr^{2+}, Fe^{2+}, Fe^{3+}, Co^{2+}, Co^{3+}

21.14 Which of the following transition metal complexes has two unpaired electrons? (Assume octahedral coordination.)
    high-spin d^4, low-spin d^7, d^8, d^{10}

21.15 Which of the following transition metal complexes is diamagnetic? (Assume octahedral coordination.)
    high-spin d^4, low-spin d^6, high-spin d^7, d^9

21.16 How many t_{2g} electrons are there in a d^5 octahedral complex in which the pairing energy is smaller than the ligand-field splitting, $P < \Delta_o$?

21.17 (a) In a Co^{3+} ion the pairing energy P is 209 kJ mol^{-1}. For the F^- ligand, Δ_o is 155 kJ mol^{-1}. Is the $[CoF_6]^{3-}$ complex of the low-spin or of the high-spin type?
(b) Given that Δ_o is 276 kJ mol^{-1}, is the $[Co(NH_3)_6]^{3+}$ complex of the low-spin or of the high-spin type?

21.18 When the water ligands in the $[Ni(H_2O)_6]^{2+}$ ion are replaced by ammonia ligands, does the wavelength of each absorption peak increase or decrease?

21.19 When a high-spin d^4 octahedral complex absorbs light, what is the electronic structure of the excited state? How many unpaired electrons are there in this state? (Ignore any Jahn-Teller distortions.)

PROBLEMS

21.20 (a) For each number of d electrons in Table 21.4, calculate the difference (in units of Δ_t) between the octahedral and tetrahedral LFSE, assuming $\Delta_o = 2\Delta_t$.
(b) For what numbers of d electrons is this difference a maximum?

21.21 Calculate the ligand-field stabilization energy of a low-spin d^5 octahedral complex, in terms of Δ_o.

21.22 Show that the ligand-field stabilization energy of a d^1 tetrahedral complex is $\frac{3}{5}\Delta_t$.

21.23 Which ion would you expect to have the larger ionic radius in an octahedral complex, a low-spin iron(II) ion or a high-spin iron(II) ion?

21.24 On diagrams similar to Figure 21.18 show the electronic structures of both the high-spin and the low-spin square planar complexes of a d^8 transition metal ion. Which is diamagnetic and which is paramagnetic?

21.25 The following compounds have the spinel structure. Predict for each compound whether the structure is normal or inverted.
(a) $CoFe_2O_4$ (containing Co^{2+} and Fe^{3+} ions)
(b) $ZnCr_2O_4$ (containing Zn^{2+} and Cr^{3+} ions)

21.26 Which of the following ions in an octahedral complex will undergo Jahn-Teller distortion?
d^3, low-spin d^6, low-spin d^7, d^8

21.27 In recent years nitrilotriacetate (NTA) has been used to replace polyphosphates in laundry soaps and detergents. How many coordination positions around a Ca^{2+} ion can one NTA ion occupy? The structure of NTA is

$$^-OOC-H_2C-\overset{..}{N} \begin{array}{c} CH_2-COO^- \\ \\ CH_2-COO^- \end{array}$$

21.28 Compounds A and B both contain the coordination complex ion $[Co(NH_3)_4(NO_2)_2]^+$. Compound A reacts with carbonate ion,
$[Co(NH_3)_4(NO_2)_2]^+ + CO_3^{2-} = [Co(NH_3)_4CO_3]^+ + 2NO_2^-$
Compound B does not react with carbonate ion. Which compound contains the cis isomer and which contains the trans?

21.29 Two coordination compounds X and Y have the same formula, $[MA_3B_3]$. When one A group is replaced by another B group,
$[MA_3B_3] + B = [MA_2B_4] + A$

compound X forms two products (cis and trans) while compound Y forms only one product (cis). Suggest an explanation.

21.30 Aqueous solutions of $CoCl_2$ are generally either light pink or intense blue, the pink form being found at low concentrations and low temperatures, and the blue form at high concentrations and high temperatures. Addition of HCl or KCl to a pink solution of $CoCl_2$ turns it blue; the pink color is restored by addition of $ZnCl_2$ or $HgCl_2$. Suggest an explanation for these observations.

CHAPTER 22
ORGANIC CHEMISTRY

22

Carbon is unique among the elements in that it has a much greater tendency to form bonds to itself than does any other element. Each carbon atom may form bonds with as many as four other atoms (usually C, H, N, O, or halogen). Thus there are an amazing number and variety of compounds containing carbon. Because many of the essential constituents of living organisms are chemical compounds containing carbon, the branch of chemistry devoted to the study of carbon compounds is known as **organic chemistry.**

Many of the more common organic materials, such as wood, paper, gasoline, oil, perfumes, and rubber, are mixtures that can be separated into their component compounds by chemical or physical methods. Many additional organic materials have been synthesized by chemists, including dyes, drugs, textile fibers, synthetic rubber, plastics, and pesticides. In this chapter, we shall consider the major types of organic compounds and some of the reactions they undergo. In general, organic molecules are composed of a relatively unreactive portion containing C and H, called the **hydrocarbon skeleton,** to which are attached one or more groups of other atoms that are more reactive. These reactive groups are called **functional groups.** For example, ethyl alcohol (C_2H_5OH) consists of the relatively unreactive ethyl group (C_2H_5) and the reactive hydroxyl group (OH). We shall confine this brief description of organic chemistry to a few of the more common functional groups.

The simplest organic compounds are the hydrocarbons, for they contain only carbon and hydrogen. We shall start our discussion of organic chemistry with the saturated hydrocarbons, which are the main components of natural gas and petroleum. A second reason for beginning with the saturated hydrocarbons is that the names of other organic compounds are based on the names of the saturated hydrocarbons.

22.1 SATURATED HYDROCARBONS (ALKANES)

In the **saturated hydrocarbons,** which are also called **alkanes,** each carbon atom forms a single bond to four neighboring atoms (carbon or hydrogen) located at the corners of a tetrahedron (see Figure 22.1) with the carbon atom in the center. The general formula for an alkane containing n carbon atoms is C_nH_{2n+2}. Thus

22.1 SATURATED HYDROCARBONS (ALKANES)

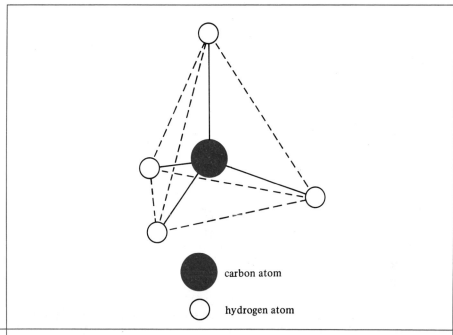

FIGURE 22.1 Methane. The hydrogen atoms are located at the corners of a tetrahedron (dashed lines), and the carbon atom is located at the center of the tetrahedron; the HCH bond angles are all 109°28′.

the first three saturated hydrocarbons are CH_4, C_2H_6, and C_3H_8. Their names and Lewis structures are

$$\underset{\text{methane}}{\text{H}-\overset{\overset{\text{H}}{|}}{\underset{\underset{\text{H}}{|}}{\text{C}}}-\text{H}} \qquad \underset{\text{ethane}}{\text{H}-\overset{\overset{\text{H}}{|}}{\underset{\underset{\text{H}}{|}}{\text{C}}}-\overset{\overset{\text{H}}{|}}{\underset{\underset{\text{H}}{|}}{\text{C}}}-\text{H}} \qquad \underset{\text{propane}}{\text{H}-\overset{\overset{\text{H}}{|}}{\underset{\underset{\text{H}}{|}}{\text{C}}}-\overset{\overset{\text{H}}{|}}{\underset{\underset{\text{H}}{|}}{\text{C}}}-\overset{\overset{\text{H}}{|}}{\underset{\underset{\text{H}}{|}}{\text{C}}}-\text{H}}$$

Often it is more convenient to use condensed structural formulas, such as

$$\underset{\text{methane}}{CH_4} \qquad \underset{\text{ethane}}{CH_3CH_3} \qquad \underset{\text{propane}}{CH_3CH_2CH_3}$$

The name of the group formed by removing one hydrogen atom from an alkane is obtained by replacing the suffix *-ane* by *-yl*. Thus CH_3 is **methyl**, CH_3CH_2 (or C_2H_5) is **ethyl**, and in general C_nH_{2n+1} is **alkyl**. The letter R is commonly used to represent the alkyl group. These names are the basis of the

TABLE 22.1	NAMES AND NUMBERS OF ISOMERS OF SOME ALKANES					
No. of carbon atoms	Formula	Name	No. of isomers	Boiling point of n-isomer (°C)
1	CH_4	Methane	1	−161
2	C_2H_6	Ethane	1	−89
3	C_3H_8	Propane	1	−42
4	C_4H_{10}	Butane	2	−1
5	C_5H_{12}	Pentane	3	36
6	C_6H_{14}	Hexane	5	69
7	C_7H_{16}	Heptane	9	98
8	C_8H_{18}	Octane	18	126
9	C_9H_{20}	Nonane	35	151
10	$C_{10}H_{22}$	Decane	75	174
11	$C_{11}H_{24}$	Undecane	159	196
12	$C_{12}H_{26}$	Dodecane	355	216

common names of many organic compounds; for example, the compounds in which a methyl group is bonded to the functional groups Cl, OH, and NH_2 are

CH_3Cl CH_3OH CH_3NH_2
methyl chloride   methyl alcohol   methyl amine

The properties of alkyl chlorides (RCl), alcohols (ROH), and amines (RNH_2) will be considered in later sections.

The names of some of the alkanes are listed in Table 22.1. When the alkane has five or more carbon atoms, its name is based on the Greek word for the number of carbon atoms, followed by the suffix *-ane*. Thus C_5H_{12} is pentane, and C_6H_{14} is hexane.

In the saturated hydrocarbons containing four or more carbon atoms, there are two or more structural isomers (Section 11.9). For example, there are two isomers with the general formula C_4H_{10}:

$CH_3CH_2CH_2CH_3$ and CH_3CHCH_3
                                                             CH_3

n-butane, bp −1°C   isobutane, bp −12°C

Both *n*-butane (the *n* stands for "normal") and isobutane occur in natural gas, and they may be separated by fractional distillation because they have different boiling points. Whenever the carbon atoms are linked together in a single chain, as in *n*-butane, the alkanes are called *normal* to distinguish them from alkanes in which the carbon chain is branched.

The numbers of isomers of some alkanes are given in Table 22.1 together

22.1 SATURATED HYDROCARBONS (ALKANES)

with the boiling points of the first 12 members of the normal alkane series. The number of structural isomers increases rapidly with the number of carbon atoms in a molecule, as shown in Table 22.1. Use of common names, such as isobutane, to designate isomers is inadequate to deal with the large number of possible isomers, and a systematic method of nomenclature has been devised by the International Union of Pure and Applied Chemistry (IUPAC). The main points of the IUPAC system as it applies to alkanes are as follows:

1. The name of the compound is based on the name of the longest carbon chain. This chain is numbered consecutively to indicate the points at which substituents are attached, starting from the end nearest the most substituents.
2. A substituent is indicated using the name of the alkyl group as a prefix, preceded where necessary by the number of the carbon atom to which the substituent is attached. For example, the name 3-ethylpentane shows that an ethyl group is bonded to carbon atom number 3 of a five-carbon chain.
3. When there is more than one substituent of a given type, the prefixes di, tri, tetra, and so on, are used, preceded where necessary by sufficient numbers to indicate all points of attachment. For example, the name 3,3,4-trimethylhexane indicates that three methyl groups are attached to a six-carbon chain. Two methyl groups are attached to carbon atom number 3 (note the repetition of the 3), and one methyl group is attached to carbon atom number 4.

Further illustrations are provided by the five isomers of hexane:

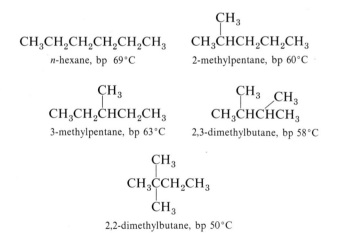

Cycloalkanes are saturated hydrocarbons in which the carbon atoms are connected in a ring. Cyclopropane and cyclobutane are

TABLE 22.2	**DISTILLATION FRACTIONS OF PETROLEUM**		
Fraction	Distillation temperature range (°C)	Alkanes (no. of carbon atoms)	
Natural gas	<20	C_1–C_4	
Petroleum ether	20–70	C_5–C_6	
Ligroin	70–100	C_6–C_7	
Gasoline	85–200	C_6–C_{12} and cycloalkanes	
Kerosene	200–275	C_{12}–C_{15} and aromatics	
Fuel oil	>275	C_{15}–C_{18}	
Lubricating oils, greases, paraffin wax, asphalt, tar	Nonvolatile	C_{16}–C_{24} and cyclic structures	

$$\underset{CH_2—CH_2}{\overset{CH_2}{\triangle}} \quad \text{and} \quad \begin{array}{c} CH_2—CH_2 \\ |\quad\quad| \\ CH_2—CH_2 \end{array}$$

The general formula of a cycloalkane with one carbon ring is C_nH_{2n}.

Natural gas and petroleum are the chief sources of the alkanes. Large deposits of these materials have been formed over long periods of time by the gradual decomposition of the remains of living material. These deposits, trapped by rock structures, are found near the surface of the earth and can be obtained by drilling a well through the protective cap. The natural gas, which is under pressure, flows freely, carrying with it some of the more volatile liquids. The gas is then passed through an oil to remove the less volatile components. The resulting product, called **natural gas,** is primarily methane with small amounts of ethane, propane, butane, carbon dioxide, nitrogen, and hydrogen sulfide. After the flow of gas ceases, the liquid petroleum is removed by pumping. The crude oil or petroleum is a complex mixture of hydrocarbons, primarily alkanes. It is separated by distillation into fractions with different boiling ranges, as shown in Table 22.2.

The principal use of gasoline is as an energy source in the internal combustion engine of motor vehicles. When a mixture of gasoline vapor and air is ignited by an electric spark, the alkanes and oxygen react to form water vapor and carbon dioxide. For a heptane, the reaction is

$$C_7H_{16} + 11O_2 \longrightarrow 8H_2O + 7CO_2 \tag{22.1}$$

The reaction is highly exothermic, and part of the energy released is harnessed by the engine to propel the vehicle. Other products in the exhaust gases include carbon monoxide, unreacted alkanes, and partially oxidized alkanes. Oxides of nitrogen are also produced by the reaction of the nitrogen and oxygen from air at the high temperatures in the engine cylinder:

$$N_2 + O_2 \longrightarrow 2NO$$

$$2NO + O_2 \longrightarrow 2NO_2$$

These various undesirable by-products often accumulate in the atmosphere over metropolitan areas to the point where they pose a major problem in the all-too-familiar form of "smog."

At room temperature the alkanes are unreactive. At higher temperatures or under the influence of light, they react with oxygen (Eq. 22.1) and with the halogens. For example, chlorine reacts with methane at temperatures over 100°C or when the gaseous mixture is illuminated with ultraviolet light:

$$CH_4 + Cl_2 \longrightarrow CH_3Cl + HCl \tag{22.2}$$

(With excess chlorine the other hydrogen atoms may be replaced to form CH_2Cl_2, $CHCl_3$, and CCl_4.) The proposed mechanism for the reaction is

$$Cl_2 \xrightarrow{h\nu \text{ or heat}} 2Cl\cdot \tag{22.2a}$$

$$Cl\cdot + CH_4 \longrightarrow CH_3\cdot + HCl \tag{22.2b}$$

$$CH_3\cdot + Cl_2 \longrightarrow CH_3Cl + Cl\cdot \tag{22.2c}$$

$$2Cl\cdot \longrightarrow Cl_2 \tag{22.2d}$$

This is an example of a free-radical chain mechanism (Section 16.10). In this mechanism the free radicals are chlorine atom, $Cl\cdot$, and methyl radical, $CH_3\cdot$. (Each of these highly reactive chemical species has an unpaired valence electron, which is explicitly indicated in the symbols $Cl\cdot$ and $CH_3\cdot$.) The second and third steps form a chain, because each generates the free radical required by the other. Furthermore, the net result of Eqs. 22.2b and 22.2c is the overall reaction, Eq. 22.2. The first step, Eq. 22.2a, initiates the chain, and the last step, Eq. 22.2d, terminates the chain. Because the last step is relatively slow, a pair of chlorine atoms formed in the first step may cause the production of as many as 10,000 molecules of methyl chloride before two chlorine atoms combine to form a chlorine molecule.

22.2 UNSATURATED HYDROCARBONS

The **unsaturated hydrocarbons** contain one or more double or triple carbon–carbon bonds. The multiple carbon–carbon bond is a point of chemical reactivity, and unsaturated hydrocarbons are much more reactive than the saturated hydrocarbons.

Hydrocarbons with double bonds are called **alkenes,** and those with triple bonds **alkynes.** The general formula for an alkene with one double bond is C_nH_{2n}, whereas that for an alkyne with one triple bond is C_nH_{2n-2}. Although the simpler molecules are usually known by common names, the systematic nomenclature described earlier is extended to unsaturated hydrocarbons by indicating the presence of a double bond with the suffix -*ene*, and a triple bond by -*yne*. Its position is indicated by the number of the doubly or triply bonded carbon atom nearest the beginning of the chain. (The chain is numbered from the end closest to the multiple bond, even if this is not the end closest to the point of branching.)

$$CH_2{=}CH_2$$
ethylene or ethene

$$HC{\equiv}CH$$
acetylene or ethyne

$$CH_3CH{=}CHCH_2CH_3$$
2-pentene

$$CH_3C{\equiv}CCH_2\overset{\overset{\displaystyle CH_3}{|}}{C}HCH_3$$
5-methyl-2-hexyne

Multiple bonds between carbon atoms lead to stronger bonds and to a decrease in bond length. For example, in ethane the C—C bond energy is 347 kJ mol^{-1} and the bond length is 1.54 Å. In ethylene the C=C bond energy is 615 kJ mol^{-1} and the bond length is 1.33 Å, whereas in acetylene the C≡C bond energy is 812 kJ mol^{-1} and the bond length decreases to 1.20 Å. The interpretation of the multiple bonds in ethylene and acetylene in terms of σ and π bonds has already been discussed (Sections 9.19 and 9.20), as has geometrical isomerism in molecules containing carbon–carbon double bonds (Section 11.9).

The presence of multiple bonds in molecules greatly increases the number of possible isomers. For example, there are four known isomers of C_4H_8 that contain a double bond:

2-methyl-1-propene, bp $-7°C$ cis-2-butene, bp $4°C$

trans-2-butene, bp $1°C$ 1-butene, bp $-6°C$

The existence of geometrical cis and trans isomers shows that rotation about the carbon–carbon double bond does not occur, at least at room temperature.

The principal source of alkenes is through decomposition of alkanes at high temperatures:

$$CH_3CH_2CH_3 \longrightarrow CH_3CH{=}CH_2 + H_2$$

Acetylene is produced by reaction of lime and coke, followed by hydrolysis of the resulting calcium carbide:

$$CaO + 3C \longrightarrow CaC_2 + CO$$
$$CaC_2 + 2H_2O \longrightarrow Ca(OH)_2 + HC{\equiv}CH$$

The principal source of reactivity in an unsaturated hydrocarbon is the multiple bond, which is often represented in an alkene by the functional group

$$\diagdown_{\!\!\!\!}C{=}C\diagup_{\!\!\!\!}$$

or in an alkyne by the functional group —C≡C—. Alkenes and alkynes readily undergo addition reactions with hydrogen, halogens, hydrogen halides, and water. If we use XY to represent H_2, Cl_2, HCl, or H—OH, the addition reactions can be generalized as follows:

$$\diagdown_{\!\!\!\!}C{=}C\diagup_{\!\!\!\!} + XY \longrightarrow -\underset{X}{\overset{|}{C}}-\underset{Y}{\overset{|}{C}}- \tag{22.3}$$

$$-C{\equiv}C- + XY \longrightarrow \underset{X}{\diagdown}C{=}C\underset{Y}{\diagup} \tag{22.4}$$

Note that addition of hydrogen converts an alkene to an alkane.

Unsaturated hydrocarbons also add to themselves in polymerization reactions. For example, in the presence of a small amount of oxygen, ethylene under a pressure of 1000 atm and at temperatures over 100°C polymerizes to form a very long saturated hydrocarbon chain known as polyethylene, a familiar plastic material:

$$nCH_2{=}CH_2 \longrightarrow -[CH_2CH_2]_n-$$

22.3 AROMATIC HYDROCARBONS

Aromatic hydrocarbons include benzene, C_6H_6, hydrocarbons containing benzene rings, and a few other hydrocarbons having properties similar to benzene. As we have seen in Section 9.10, benzene may be represented by the resonance structures

A simplification in notation is achieved by use of a hexagon with three double sides to represent benzene (omitting the ring carbons and the hydrogens attached to them):

Other examples of aromatic hydrocarbons are

naphthalene   anthracene   pyrene   toluene or methylbenzene

(Only one resonance structure is shown for each.) The group obtained by removing a hydrogen atom from benzene, C_6H_5, is called the **phenyl group**. The general name for a group obtained by removing a hydrogen atom from a ring of any aromatic hydrocarbon is the **aryl group**. The primary source of aromatic compounds is the petroleum industry although some are obtained from coal.

In marked contrast to the alkenes, aromatic substances do not readily undergo addition reactions. In valence bond theory this decreased reactivity is explained by the fact that each carbon–carbon bond has only partial double-bond character owing to the several resonance forms. In molecular orbital theory, the description of benzene starts from a framework in which each carbon atom is attached to two other carbon atoms and a hydrogen atom by σ bonds (sp^2 hybridization) [Figure 22.2(a)]. The remaining electron on each carbon atom is in a p orbital perpendicular to the plane of the ring [Figure 22.2(b)]. All the p orbitals overlap each other to form π bonds, which can be represented by doughnut-shaped π-electron clouds above and below the plane of the benzene ring (Figure 22.3). Often this structure is indicated by writing the symbol

to indicate that the π-electron density is distributed evenly around the ring. The relative inertness toward addition reactions is then explained by delocalization of the six π electrons over six carbon–carbon bonds, as compared to two π electrons per carbon–carbon bond in an alkene such as ethylene.

Aromatic substances undergo substitution reactions more readily than do alkanes. Chlorine or bromine reacts readily with benzene in the presence of a Lewis acid catalyst to produce the halogenated benzene:

The catalyst in this case is ferric bromide.

22.3 AROMATIC HYDROCARBONS

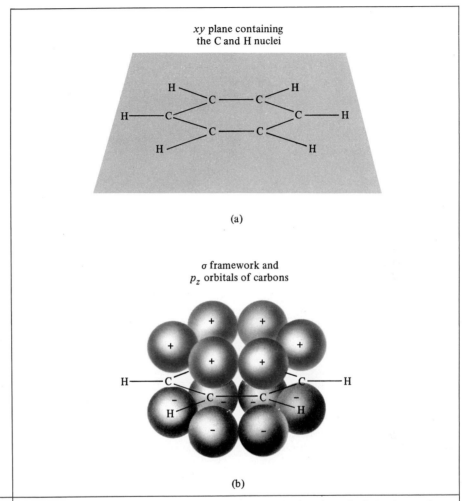

FIGURE 22.2 Bonding in benzene. (a) Sigma bonding. Each C—C σ bond results from the overlap of the sp^2 hybrid orbitals of two carbon atoms as in ethylene, and each C—H bond results from the overlap of an sp^2 orbital of carbon and the $1s$ orbital of hydrogen. (b) The p_z orbitals of the carbons are shown perpendicular to the plane containing the carbon and hydrogen nuclei.

Alkyl groups may be attached to the benzene ring by treating benzene with an alkyl halide in the presence of an aluminum chloride catalyst, a reaction discovered by Charles Friedel and James Crafts in 1877:

$$\text{C}_6\text{H}_6 + \text{CH}_3\text{Cl} \xrightarrow{\text{AlCl}_3} \text{C}_6\text{H}_5\text{CH}_3 + \text{HCl}$$

FIGURE 22.3

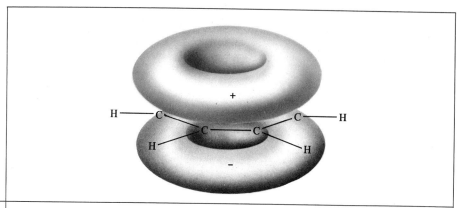

Sigma and pi bonding in benzene. The σ bonds are represented by the straight lines connecting the carbon and hydrogen nuclei, and the π bonding is represented by the doughnut-shaped clouds above and below the plane of the benzene ring.

Further reaction leads to dialkyl and polyalkylbenzenes. There are three isomers of dimethylbenzene:

| *o*-xylene or 1,2-dimethylbenzene, bp 144°C | *m*-xylene or 1,3-dimethylbenzene, bp 139°C | *p*-xylene or 1,4-dimethylbenzene, bp 138°C |

They are designated systematically by numbering the carbon atoms consecutively around the ring so as to obtain the smallest integers indicating the positions of substitution. The common name of dimethylbenzene is xylene; the three isomers are distinguished as *ortho* (*o*-) to indicate substitution at adjacent carbon atoms, *meta* (*m*-) to indicate that the positions of substitution are separated by one carbon atom, and *para* (*p*-) to indicate substitution at opposite vertices.

22.4 ALKYL HALIDES

The general formula of the **alkyl halides** is RX, where R may be any alkyl group and X is any halogen. Because the alkyl halides undergo substitution reactions with many reagents, allowing the introduction of different functional groups into the molecule, they are extremely important in organic chemistry.

22.4 ALKYL HALIDES

The alkyl chlorides and bromides may be formed from alkanes by substitution reactions (Eq. 22.2) and from alkenes and alkynes by addition reactions (Eqs. 22.3 and 22.4). In the laboratory they are often prepared by reacting hydrogen chloride or phosphorus trichloride with the corresponding alcohol:

$$ROH + HCl \longrightarrow RCl + H_2O$$

$$3ROH + PCl_3 \longrightarrow 3RCl + H_3PO_3$$

(The alkyl bromides are prepared in a similar way from HBr or PBr_3.)

Some of the alkyl chlorides are of commercial importance. Chloroform, also known as trichloromethane, $CHCl_3$, was one of the earliest anesthetics, but now its main use is as a solvent in the manufacture of penicillin. Carbon tetrachloride or tetrachloromethane, CCl_4, is also an important solvent. Dichlorodifluoromethane, CCl_2F_2, is widely used as the working fluid of refrigerators, freezers, and air conditioners. Many halogen derivatives are highly toxic, and several chlorinated hydrocarbons are important insecticides. The best known is DDT, an acronym of three letters in the common name *d*ichloro*d*iphenyl*t*richloroethane.

Following its discovery in 1939 by Paul Müller, it greatly advanced world health by controlling insect-borne diseases, and Müller received a Nobel prize for his work. More recently, insects have developed a resistance to DDT, and it has been found that DDT persists for long periods in the environment, where it accumulates to the detriment of wildlife.

The alkyl halides contain a polar $C^{\delta+}-X^{\delta-}$ bond and the halogen can be readily displaced by bases stronger than the halide ion. Bases possess an unshared pair of electrons and hence are attracted to positive sites such as the carbon atom bonded to a halogen atom in an alkyl halide. The general reaction is

$$R-X + Z^- \longrightarrow R-Z + X^-$$

As shown in Table 22.3, a variety of functional groups may be introduced into the molecule by reaction of an alkyl halide with different bases. Aryl halides (in which the halogen atom is attached to a ring carbon) do not readily undergo the substitution reactions listed above.

TABLE 22.3 **FUNCTIONAL GROUPS INTRODUCED BY DIFFERENT BASES**

Base	Product	Name
OH^- or H_2O	ROH	Alcohol
OR'^-	ROR'	Ether
NH_2^- or NH_3	RNH_2	Amine
SH^-	RSH	Mercaptan or thiol

22.5 OXYGEN DERIVATIVES— ALCOHOLS, PHENOLS, AND ETHERS

Oxygen bonded to two other atoms is a characteristic feature of several classes of organic compounds. In the alcohols, a hydroxyl group, —OH, is bonded to an alkyl group. In the phenols an —OH group is bonded to an aryl group. Ethers contain the C—O—C linkage with oxygen bonded to two groups, either alkyl or aryl. Examples of each class are

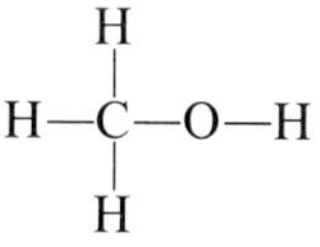

methyl alcohol or methanol

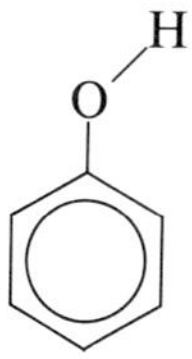

phenol or hydroxybenzene

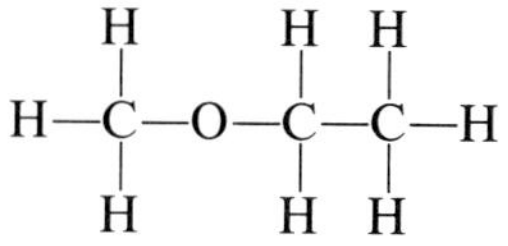

methyl ethyl ether

(Unshared electron pairs are omitted.) The corresponding condensed structural formulas are

$$CH_3OH \qquad C_6H_5OH \qquad CH_3OCH_2CH_3$$

The —OH group is one of the most important functional groups in organic chemistry.

Alcohols are subdivided into primary, secondary, and tertiary alcohols depending on whether the carbon atom bonded to the oxygen atom is bonded to one, two, or three other carbon atoms. In their systematic names the terminal -e of the corresponding alkane is replaced by -ol, and a numerical prefix indicates the number of the carbon atom to which the OH group is attached. Three isomers of molecular formula C_4H_9OH are

$$CH_3CH_2CH_2CH_2OH \qquad CH_3\underset{\underset{OH}{|}}{C}HCH_2CH_3 \qquad CH_3\underset{\underset{CH_3}{|}}{\overset{\overset{OH}{|}}{C}}CH_3$$

n-butyl alcohol or n-butanol (a primary alcohol), bp 118°C

sec-butyl alcohol or 2-butanol (a secondary alcohol), bp 100°C

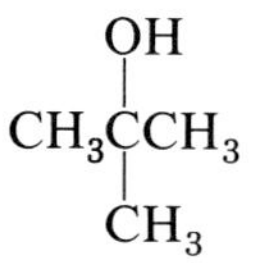

tert-butyl alcohol or 2-methyl-2-propanol (a tertiary alcohol), bp 83°C

TABLE 22.4 PHYSICAL PROPERTIES OF WATER AND SOME ALCOHOLS

Formula	Common name	bp (°C)	Dielectric constant	Solubility in water[a]
HOH	Water	100	78.5	—
CH_3OH	Methyl alcohol	65	32.6	∞
CH_3CH_2OH	Ethyl alcohol	78	24.3	∞
$CH_3CH_2CH_2OH$	n-Propyl alcohol	97	20.1	∞
$CH_3(CH_2)_2CH_2OH$	n-Butyl alcohol	118	17.1	7.9
$CH_3(CH_2)_4CH_2OH$	n-Hexyl alcohol	157	13.3	0.6
$CH_3(CH_2)_6CH_2OH$	n-Octyl alcohol	195	10.3	0.05
$HOCH_2CH_2OH$	Ethylene glycol	198	37.7	∞
$HOCH_2CHOHCH_2OH$	Glycerol	291	42.5	∞

[a] g (100 g of $H_2O)^{-1}$. The symbol ∞ indicates that the substance is miscible with water in all proportions.

Alcohols are organic analogs of water in which a hydrogen atom is replaced by an alkyl group. Through hydrogen bonding the OH group imparts waterlike properties to the alcohols. Thus, compared to an alkane with similar molecular mass, alcohols have high boiling points and high dielectric constants. The extent to which the alcohols have waterlike properties decreases with increasing size of the alkyl group (see Table 22.4). Ethylene glycol, a *diol* with two hydroxyl groups, and glycerol, a *triol* with three hydroxyl groups, have very high boiling points and are miscible with water in all proportions:

$$\begin{array}{cc} CH_2CH_2 & CH_2CHCH_2 \\ |\ \ | & |\ \ |\ \ | \\ OH\ OH & OH\ OH\ OH \end{array}$$

ethylene glycol or 1,2-ethanediol   glycerol or 1,2,3-propanetriol

Ethylene glycol is commonly used as a permanent antifreeze in automobile radiators.

Alcohols are formed from alkyl halides by substitution reactions and from alkenes by addition reactions:

$$CH_3CH_2Br + OH^- \longrightarrow CH_3CH_2OH + Br^-$$

$$CH_2{=}CH_2 + H_2O \longrightarrow CH_3CH_2OH$$

Alcohols can be dehydrated by sulfuric acid above 150°C to form the corresponding alkenes:

$$CH_3CH_2OH \longrightarrow CH_2{=}CH_2 + H_2O$$

Below 150°C, in the presence of sulfuric acid, an intermolecular dehydration occurs, with two molecules of alcohol losing only one molecule of water and forming an ether:

$$2CH_3CH_2OH \longrightarrow \underset{\text{ethyl ether}}{CH_3CH_2OCH_2CH_3} + H_2O$$

Phenols are much more acidic than alcohols, and it is mainly for this reason that they are considered as a separate class. (An old name for phenol itself is "carbolic acid.") The dissociation constants of phenol and ethanol are 1.0×10^{-10} and 1×10^{-16}, respectively. It takes an extremely strong base to remove the proton from an alcohol. The usual method is reaction with an alkali metal:

$$2CH_3CH_2OH + 2Na \longrightarrow \underset{\substack{\text{sodium} \\ \text{ethoxide}}}{2CH_3CH_2ONa} + H_2$$

The products of this reaction, called **alkoxides,** are useful in synthesizing ethers with different alkyl or aryl groups bonded to the oxygen atom:

$$CH_3CH_2ONa + CH_3Br \longrightarrow CH_3CH_2OCH_3 + NaBr$$

Because they lack OH groups, the ethers are unable to engage in hydrogen bonding. They have relatively low boiling points, about the same as the alkanes of similar molecular mass:

$$\underset{\text{ethyl ether, bp 35°C}}{CH_3CH_2OCH_2CH_3} \qquad \underset{n\text{-pentane, bp 36°C}}{CH_3CH_2CH_2CH_2CH_3}$$

22.6 OXYGEN DERIVATIVES—ALDEHYDES, KETONES, CARBOXYLIC ACIDS, AND ESTERS

Oxygen doubly bonded to carbon forms the **carbonyl group** $\left(\begin{array}{c}\diagdown \\ \diagup\end{array} C{=}O\right)$. It is present in four important classes of organic compounds which differ in the other groups bonded to the carbonyl carbon atom. If we designate an alkyl or aryl group by the letter R, the characteristic structure of each class is

$$\underset{\text{aldehyde}}{\underset{\text{O}}{\overset{\|}{R-C-H}}} \qquad \underset{\text{ketone}}{\underset{\text{O}}{\overset{\|}{R-C-R'}}} \qquad \underset{\text{carboxylic acid}}{\underset{\text{O}}{\overset{\|}{R-C-O-H}}} \qquad \underset{\text{ester}}{\underset{\text{O}}{\overset{\|}{R-C-O-R'}}}$$

(In ketones and esters the two R groups may differ, hence R' is used for one of them.) The condensed structural formulas are written RCHO, RCOR', RCOOH, and RCOOR', respectively.

22.6 OXYGEN DERIVATIVES—ALDEHYDES, KETONES, CARBOXYLIC ACIDS, AND ESTERS

Aldehydes may be formed by oxidation of a primary alcohol:

$$CH_3CH_2OH = CH_3CHO + 2H^+ + 2e^- \qquad \mathcal{E}° = -0.18 \text{ V}$$
$$\text{ethanol} \qquad \text{acetaldehyde}$$

Further oxidation of an aldehyde yields the corresponding acid:

$$CH_3CHO + H_2O = CH_3COOH + 2H^+ + 2e^- \qquad \mathcal{E}° = +0.13 \text{ V}$$
$$\text{acetaldehyde} \qquad \text{acetic acid}$$

Indeed, the conversion of wine to vinegar is due to the oxidation of ethanol by oxygen to acetic acid:

$$CH_3CH_2OH + O_2 \longrightarrow CH_3COOH + H_2O$$

Because aldehydes are oxidized to acids more easily than alcohols are oxidized to aldehydes, it is necessary to use special methods to obtain the aldehydes from the alcohols. (Ordinarily the oxidation of a primary alcohol results in the formation of the corresponding carboxylic acid and, conversely, the reduction of a carboxylic acid yields the corresponding primary alcohol.) In the laboratory, an aldehyde can be prepared by oxidation of the corresponding alcohol if one removes the aldehyde from the reaction mixture before it is oxidized to the acid. For example, acetaldehyde may be prepared by dripping a solution of sulfuric acid and potassium dichromate (a strong oxidizing agent) into boiling ethanol (bp 78°C). The acetaldehyde (bp 20°C) vaporizes as fast as it is formed and is thereby removed from the reaction mixture before it undergoes further oxidation. Another method of preparing aldehydes from alcohols is by dehydrogenation. For example, acetaldehyde may be formed by passing ethanol vapor over metallic copper at 200 to 300°C:

$$CH_3CH_2OH \xrightarrow{Cu,\ 200-300°C} CH_3CHO + H_2$$

Ketones are formed by oxidation of secondary alcohols:

$$CH_3CHOHCH_3 \longrightarrow CH_3COCH_3 + 2H^+ + 2e^-$$
$$\text{isopropyl alcohol or 2-propanol} \qquad \text{acetone or propanone}$$

Conversely, reduction of a ketone produces a secondary alcohol. Neither ketones nor tertiary alcohols can be oxidized without breaking down the carbon skeleton.

Systematic names of aldehydes and ketones are derived from the name of the alkane by replacing the terminal *-e* with either *-al* or *-one*, respectively, the position of the carbonyl group being indicated for ketones by the usual numbering system. (The aldehyde functional group is necessarily at the end of the carbon chain. Its carbon atom is numbered 1.) Examples are:

TABLE 22.5 PHYSICAL PROPERTIES OF SOME CARBOXYLIC ACIDS

Formula	Common name	bp (°C)	Solubility in water[a]	Dissociation constant
HCOOH	Formic acid	101	∞	1.8×10^{-4}
CH_3COOH	Acetic acid	118	∞	1.8×10^{-5}
CH_3CH_2COOH	Propionic acid	141	∞	1.3×10^{-5}
$CH_3(CH_2)_2COOH$	Butyric acid	163	∞	1.5×10^{-5}
$CH_3(CH_2)_3COOH$	Valeric acid	187	3.7	1.4×10^{-5}
$CH_3(CH_2)_4COOH$	Caproic acid	205	1.0	1.4×10^{-5}
C_6H_5COOH	Benzoic acid	249	0.34	6.1×10^{-5}

[a] g (100 g of $H_2O)^{-1}$. The symbol ∞ indicates that the substance is miscible with water in all proportions.

$$CH_3\overset{H}{\underset{|}{C}}=O$$
acetaldehyde or ethanal

$$CH_3\overset{CH_3}{\underset{|}{C}}HCH_2\overset{H}{\underset{|}{C}}=O$$
isovaleraldehyde or 3-methylbutanal

$$CH_3\overset{O}{\underset{\|}{C}}CH_3$$
acetone or propanone

$$CH_3\overset{O}{\underset{\|}{C}}CH_2\overset{CH_3}{\underset{|}{C}}HCH_3$$
methyl isobutyl ketone or 4-methyl-2-pentanone

Like the double bond in alkenes, the carbonyl double bond in aldehydes and ketones readily undergoes addition reactions with a variety of substances. These reactions are of special importance in synthetic organic chemistry (Section 22.8). Several ketones are also widely used as solvents.

The common names and some properties of several carboxylic acids are shown in Table 22.5. Hydrogen bonding is even more important in the carboxylic acids than in the alcohols. (Hydrogen bonds in crystalline formic acid were described in Section 17.10.) Carboxylic acids have higher boiling points than alcohols of similar molecular mass, and the first four members of the series are completely miscible with water. As with the alcohols, the solubility decreases as the size of the alkyl group increases.

To form the systematic name of a carboxylic acid, the terminal -e of the name of the corresponding alkane is replaced with -oic acid. When numbering the carbon atom skeleton, the carbon atom of the carboxyl group is numbered 1. Examples are:

HCOOH
methanoic acid

$$CH_3\overset{CH_3}{\underset{|}{C}}HCH_2CH_2COOH$$
4-methylpentanoic acid

Carboxylic acids are all moderately weak acids with dissociation constants of about 1×10^{-4} to 1×10^{-5} (Table 22.4). Their anions are named by replacing

22.6 OXYGEN DERIVATIVES—ALDEHYDES, KETONES, CARBOXYLIC ACIDS, AND ESTERS

the ending *-ic* with *-ate*. Thus the anions of formic acid (HCOOH) and acetic acid (CH$_3$COOH) are formate ion (HCOO$^-$) and acetate ion (CH$_3$COO$^-$).

There is an enormous difference between the acidity of a carboxylic acid and that of the corresponding alcohol, even though each contains an —OH group. Clearly the adjacent carbonyl group has a major effect on the acid–anion equilibrium. This effect is attributed to the role of the carbonyl group in stabilizing the anion by resonance. The resonance structures of acetate ion, including formal charges (Section 9.7), are

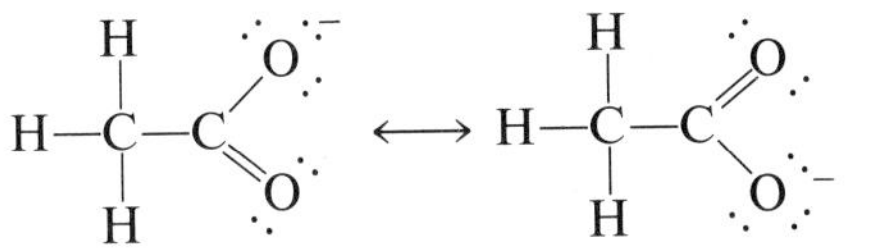

The negative formal charge is thus delocalized between the two oxygen atoms, weakening the attraction of either oxygen for a proton.

The explanation of the acidity of phenols, noted in the preceding section, is similar. Resonance structures of the phenolate anion include

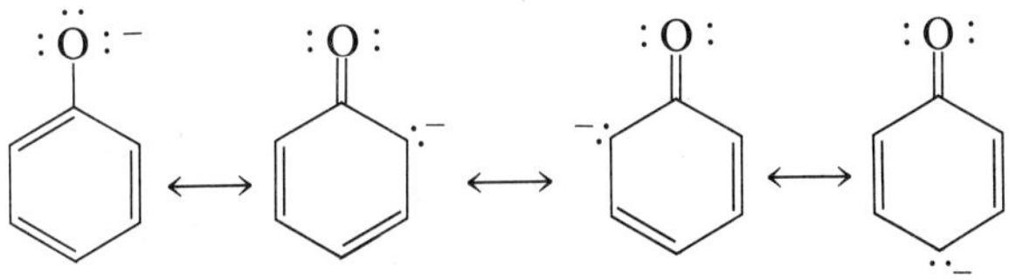

Thus the negative formal charge is delocalized between the oxygen atom and the carbon atoms in the ortho and para positions. Because the electronegativity of carbon is much lower than that of oxygen, even three carbon atoms are not so effective as one oxygen atom in withdrawing electrons, and phenol is not nearly so strong an acid as the carboxylic acids.

Esters are formed from an acid and an alcohol with the elimination of water. For example, acetic acid and methyl alcohol form methyl acetate and water.

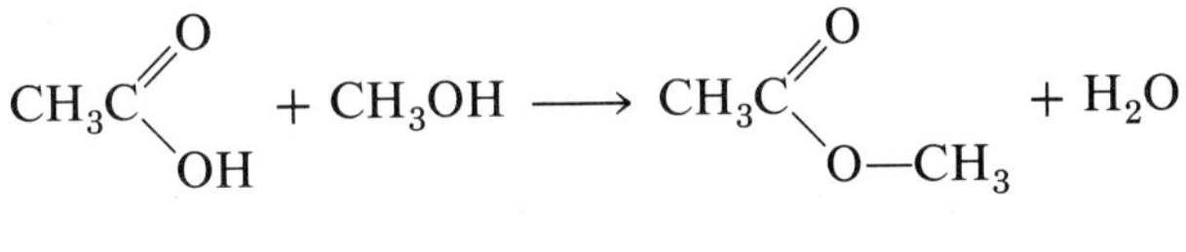

$$CH_3C\begin{matrix}O\\OH\end{matrix} + CH_3OH \longrightarrow CH_3C\begin{matrix}O\\O-CH_3\end{matrix} + H_2O$$

The name of an ester is derived from the alcohol name followed by the name of the acid anion. Note how methyl acetate (shown above) differs from ethyl formate,

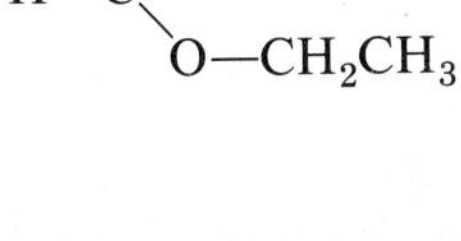

$$H-C\begin{matrix}O\\O-CH_2CH_3\end{matrix}$$

Fats and vegetable oils are esters of glycerol and long-chain carboxylic acids. Saturated acids such as stearic acid, $CH_3(CH_2)_{16}COOH$, predominate in fats, whereas vegetable oils contain mainly unsaturated acids such as linoleic acid, $CH_3(CH_2)_4CH=CHCH_2CH=CH(CH_2)_7COOH$. Aqueous base hydrolyzes the esters to form glycerol and the anion of the carboxylic acid:

$$\begin{array}{l} RCOOCH_2 \\ | \\ RCOOCH \\ | \\ RCOOCH_2 \end{array} + 3OH^- \longrightarrow 3RCOO^- + \begin{array}{l} HOCH_2 \\ | \\ HOCH \\ | \\ HOCH_2 \end{array}$$

The reaction is called **saponification**, and the sodium salts of the long-chain carboxylic acids constitute ordinary soap.

22.7 NITROGEN DERIVATIVES

Nitric acid mixed with concentrated sulfuric acid reacts with hydrocarbons to form substances with the functional group $-NO_2$, called **nitro compounds**:

toluene + $3HNO_3 \longrightarrow$ 2,4,6-trinitrotoluene + $3H_2O$

Trinitrotoluene (TNT) is an important high explosive. Nitric acid also reacts with alcohols to form nitrates in which the functional group is $-ONO_2$:

$$C_2H_5OH + HNO_3 \longrightarrow C_2H_5ONO_2 + H_2O$$
$$\text{ethyl nitrate}$$

Glycerol forms a trinitrate, usually called **nitroglycerin** although it is not a true nitro compound. It is also a common high explosive.

$$\begin{array}{l} CH_2OH \\ | \\ CHOH \\ | \\ CH_2OH \end{array} + 3HNO_3 \longrightarrow \begin{array}{l} CH_2ONO_2 \\ | \\ CHONO_2 \\ | \\ CH_2ONO_2 \end{array} + 3H_2O$$

Nitro compounds are easily reduced to primary **amines**, in which the functional group is $-NH_2$:

22.7 NITROGEN DERIVATIVES

$$\text{nitrobenzene} + 6H^+ + 6e^- \longrightarrow \text{aniline or phenylamine} + 2H_2O$$

Amines are derivatives of ammonia in which one or more of the hydrogens are replaced by alkyl or aryl groups. They are classified as primary (RNH_2), secondary (R_2NH), or tertiary (R_3N) amines depending on the number of hydrogens that are replaced. They are all weak bases.

Amines may be prepared using the substitution reaction of an alkyl halide with ammonia. With an alkyl bromide, the reaction is

$$RBr + 2NH_3 \longrightarrow RNH_2 + NH_4Br$$

Further reaction produces secondary amines and tertiary amines:

$$RNH_2 + RBr + NH_3 \longrightarrow R_2NH + NH_4Br$$

$$R_2NH + RBr + NH_3 \longrightarrow R_3N + NH_4Br$$

A tertiary amine and an alkyl halide form a salt called a quaternary ammonium halide, $R_4N^+X^-$, which is analogous to the ammonium halide $NH_4^+X^-$:

$$R_3N + RBr \longrightarrow R_4N^+Br^-$$

Amines are named by adding the suffix *-amine* to the name of the alkyl group. Quaternary ammonium halides are named as derivatives of the ammonium salt.

$CH_3CH_2NH_2$ $\quad$ CH_3NHCH_3 $\quad$ $(CH_3)_3N$ $\quad$ $(CH_3)_4N^+Br^-$
ethylamine $\quad$ dimethylamine $\quad$ trimethylamine $\quad$ tetramethylammonium bromide

Secondary and tertiary amines are often designated by using the letter N to indicate that substitution has occurred on the nitrogen atom:

N,N-dimethylaniline $\qquad$ $CH_3CH_2CH_2CH_2NHCH_3$
$\qquad\qquad\qquad\qquad\qquad$ N-methyl-*n*-butylamine

Substances having both the amine functional group and the carboxyl functional group are called **amino acids.** Some representative amino acids are

$$H_2NCH_2COOH \qquad CH_3\underset{\underset{NH_2}{|}}{C}HCOOH \qquad C_6H_5CH_2\underset{\underset{NH_2}{|}}{C}HCOOH$$
$$\text{glycine} \qquad\qquad \text{alanine} \qquad\qquad\qquad \text{phenylalanine}$$

Proteins are important naturally occurring high polymers of amino acids. Amino acids and proteins are discussed in more detail in the next chapter.

22.8 FORMATION OF CARBON–CARBON BONDS

To verify the structure of complicated organic molecules the organic chemist synthesizes them from simple, readily available substances of known structure. He or she isolates and verifies the structure of each intermediate obtained in the sequence of reactions used to prepare the desired product. The general procedure is to form a carbon skeleton of the desired dimensions, with functional groups attached at the appropriate points. Thus the formation of carbon–carbon bonds is a process of central importance in organic chemistry.

We have already seen in Section 22.3 how the Friedel-Crafts reaction is used to form carbon–carbon bonds in aromatic hydrocarbons. Another important type of reaction utilizes alkyl magnesium halides, discovered by Philippe Barbier in 1899, and extended by his student Victor Grignard in 1901 to a wide variety of reactions. These substances are called **Grignard reagents.**

A Grignard reagent is prepared from magnesium metal and an alkyl halide in ether solution. A typical reaction is

$$CH_3CH_2Br + Mg \longrightarrow CH_3CH_2MgBr$$

Grignard reagents react by addition with substances containing the carbonyl group. The alkyl group bonds to the carbon atom, and the magnesium atom bonds to the oxygen atom, from which it may be removed by hydrolysis. Depending on the carbonyl compound used, a single Grignard reagent can form a variety of products. Formaldehyde forms primary alcohols, other aldehydes form secondary alcohols, and ketones form tertiary alcohols:

$$RMgBr + H\underset{\underset{O}{\|}}{C}H \longrightarrow R\underset{\underset{OMgBr}{|}}{C}H_2 \xrightarrow{H_2O} RCH_2OH + Mg(OH)Br$$

$$RMgBr + R'\underset{\underset{O}{\|}}{C}H \longrightarrow R\underset{\underset{OMgBr}{|}}{C}HR' \xrightarrow{H_2O} R\underset{\underset{OH}{|}}{C}HR' + Mg(OH)Br$$

$$RMgBr + R'\underset{\underset{O}{\|}}{C}R'' \longrightarrow \underset{\underset{OMgBr}{|}}{\overset{\overset{R'}{|}}{R}CR''} \xrightarrow{H_2O} \underset{\underset{OH}{|}}{\overset{\overset{R'}{|}}{R}CR''} + Mg(OH)Br$$

22.8 FORMATION OF CARBON–CARBON BONDS

As examples of the synthesis of larger molecules from smaller ones, consider the formation of the three butyl alcohols mentioned in Section 22.5.

1. *n*-Butyl alcohol. This is a four-carbon primary alcohol, and it can therefore be synthesized from a three-carbon Grignard reagent and formaldehyde. Starting with *n*-propyl bromide, the synthetic scheme is

$$CH_3CH_2CH_2Br \xrightarrow{Mg} CH_3CH_2CH_2MgBr$$

$$\downarrow HCHO$$

$$CH_3CH_2CH_2CH_2OMgBr$$

$$\downarrow H_2O$$

$$CH_3CH_2CH_2CH_2OH + Mg(OH)Br$$

2. *sec*-Butyl alcohol. This is a four-carbon secondary alcohol, and it can therefore be synthesized from a two-carbon Grignard reagent and the two-carbon aldehyde (acetaldehyde). Starting with ethyl bromide, the synthetic scheme is

$$CH_3CH_2Br \xrightarrow{Mg} CH_3CH_2MgBr \xrightarrow{CH_3CHO} CH_3CH_2\underset{OMgBr}{\overset{|}{C}}HCH_3$$

$$\downarrow H_2O$$

$$CH_3CH_2\underset{OH}{\overset{|}{C}}HCH_3 + Mg(OH)Br$$

3. *tert*-Butyl alcohol. This is a four-carbon tertiary alcohol, and it can therefore be synthesized from a one-carbon Grignard reagent and the three-carbon ketone (acetone). Starting with methyl bromide, the synthetic scheme is

$$CH_3Br \xrightarrow{Mg} CH_3MgBr \xrightarrow{CH_3COCH_3} CH_3\underset{OMgBr}{\overset{\overset{\displaystyle CH_3}{|}}{\underset{|}{C}}}CH_3$$

$$\downarrow H_2O$$

$$CH_3\underset{OH}{\overset{\overset{\displaystyle CH_3}{|}}{\underset{|}{C}}}CH_3 + Mg(OH)Br$$

If the two-carbon and three-carbon substances used in these synthetic schemes were not available, they could be synthesized from one-carbon substances by these same methods. Once the four-carbon alcohols are synthesized they can be used to prepare a variety of other four-carbon compounds by reactions described previously. For example, *n*-butyl alcohol can be dehydrated to form 1-butene, or oxidized to *n*-butyraldehyde or butyric acid. By reaction with suitable Grignard reagents, the *n*-butyraldehyde can be used to prepare compounds having even longer carbon chains.

The syntheses described above illustrate a method by which synthetic organic chemists, starting from relatively simple substances, have been able to build complicated carbon skeletons with various functional groups at desired points.

22.9 STRUCTURE DETERMINATION

One of the major problems in organic chemistry is the determination of the structure of new compounds isolated from natural sources or synthesized in the laboratory. The methods used by organic chemists to determine the structures of large molecules are impressive feats of deductive logic.

The first step in any such problem is the purification of the compound, followed by its elemental analysis and a determination of its molecular mass. From these data one computes the numbers of atoms of each element in a molecule of the substance, that is, the molecular formula. In addition, the C:H ratio indicates the extent of ring or multiple-bond formation.

At one time chemical reactions were used to determine the various functional groups, and the molecule was then broken down into simpler fragments that were isolated and identified. The structures of many complicated molecules, such as cholesterol and morphine,

cholesterol

morphine

were established by such methods.[1]

In recent years, physical methods that greatly simplify the task of structure proof have been developed. The mass spectrometer (Section 6.6) may be used for

---

[1] In these diagrams there is a carbon atom, and hydrogen atoms as required to satisfy the quadrivalence of carbon, at each vertex and at the end of each line-segment not marked with HO, O, or N. In this type of diagram ethylcyclopentane, for example, would be represented as

a rapid and accurate determination of the molecular mass. With the proper equipment, the molecular mass of a microgram sample of a substance can be determined to 1 part in 10^8. Other spectrometers measure the energy of quanta of radiation that are absorbed by the molecule. Such spectrometers all operate in a manner similar to that indicated in the following block diagram:

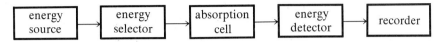

The energy source produces radiation of the frequency range desired. This radiation then passes into the energy selector, which selects a narrow band of frequencies to pass into the absorption cell containing the substance (or a solution of the substance) to be analyzed. The energy detector measures the energy that is not absorbed by the sample, and the recorder presents the results in chart form with frequency as the x coordinate and some function of the energy reaching the detector as the y coordinate. Thus spectrometers of this type give us information on the frequency and amount of radiation that is absorbed by the substance being investigated.

The ultraviolet and visible absorption spectrometer uses radiation of wavelength 2×10^{-7} to 2×10^{-6} m (Figure 7.6). The energy of photons in this range corresponds to the energy differences between molecular orbitals, and hence absorption of these photons results in transitions of electrons from occupied to vacant molecular orbitals. Ultraviolet and visible spectroscopy is very useful in the study of carbon compounds that contain alternate double and single bonds, such as certain unsaturated hydrocarbons, unsaturated carbonyl compounds, and aromatic compounds.

Infrared spectroscopy covers the longer wavelength region from 2×10^{-5} to 2×10^{-6} m (or wave number region from 500 to 5000 cm^{-1}), and hence the photon energy is lower than in the visible or ultraviolet region. The energy of photons in the infrared region corresponds to differences between the vibrational energy levels of molecules. As a first approximation the absorption frequency of a particular bond type (such as C—H, C=O, and C—Cl) is independent of the other groups in the molecule. Thus the infrared spectrum of a substance gives valuable information on the possible functional groups in a molecule.

In nuclear magnetic resonance (nmr) spectroscopy, the sample is placed between the poles of a powerful magnet. Then it is exposed to radiation in the radio-frequency region, at 60 or 100 MHz (λ = 5 or 3 m). The spectra obtained yield information about the chemical environment in the vicinity of atoms with nuclei that have magnetic moments such as ^{1}H, ^{11}B, ^{13}C, and ^{14}N. We shall restrict our discussion to protons. The proton nmr spectrum of an organic molecule is very valuable for structure proof because it provides evidence about the carbon skeleton of the molecule that is not available from the ultraviolet or infrared spectra. Like electrons, protons have both spin angular momentum and a magnetic moment. Thus they act as tiny magnets, and they can occupy one of two energy states in a magnetic field. In the lower-energy state the nuclear magnet is

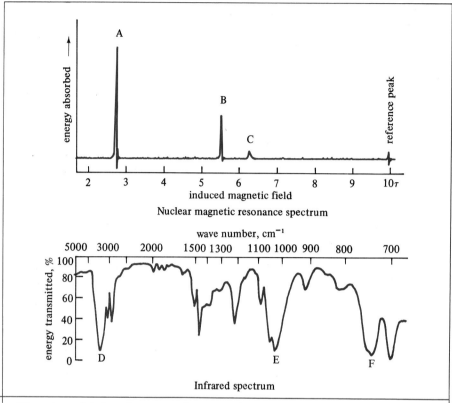

FIGURE 22.4

The nmr and infrared spectra of a pure substance of molecular formula C_7H_8O. Note that the vertical scale of the infrared spectrum is energy transmitted, and therefore the main absorption "peaks" (D, E, and F) appear as the deepest minima.

aligned in the same direction as the magnetic field; the higher-energy state results when the nuclear magnet is opposed to the field. The energy required to effect a transition between these two states is proportional to the strength of the magnetic field in the neighborhood of the proton.

If a molecule is placed in a magnetic field, a small but measurable magnetic field is *induced* in the molecule in the direction opposing the external field. (It is this induced magnetic field that is responsible for diamagnetism, as noted in Section 17.13.) The extent to which the induced field diminishes the effect of the external field varies from point to point in the molecule, so that protons in different chemical environments are subjected to different magnetic fields. In general terms the nmr spectrum of a molecule shows the amount of energy absorbed by protons in each chemical environment (on the vertical scale) as a function of the induced magnetic field due to that chemical environment (on the horizontal scale). The nmr spectra of known compounds show that it is possible to distinguish between H bonded to C, O, N, or some other atom, and to distinguish between C—H bonds in groups such as CH_3, CH_2, CH, and $C\!\!\begin{array}{c}\nearrow O\\ \searrow H\end{array}$. Furthermore, it is possible to determine how many hydrogen atoms there are of each type by measuring the area under each absorption peak.

Although the above descriptions of spectra are greatly simplified, they are sufficient background for an illustration of the use of spectra in determining the structure of a simple molecule. (A more detailed treatment of spectra is necessary when determining the structures of more complicated molecules.) The molecular formula of a certain substance, obtained from a molecular mass determination and an elemental analysis of the compound, is C_7H_8O. Its infrared and nmr spectra are shown in Figure 22.4. The nmr spectrum shows that there are three types of protons in the molecule (A, B, C). The ratios of the areas under the absorption peaks are 5:2:1. As there are eight protons per molecule, we conclude that there are five protons of type A, two of type B, and one of type C. By comparison with the nmr spectra of known compounds, we find that aromatic ring protons absorb at the left peak (A), CH_2 protons at the middle peak (B), and OH protons at the right peak (C). We therefore propose the structure

CH_2OH
⎯⟨benzene ring⟩

benzyl alcohol

The infrared spectrum is in agreement with this structure assignment because the absorption at 3200 to 3600 cm^{-1} (D) is characteristic of O—H and that near 1020 cm^{-1} (E) is characteristic of C—O. The strong absorption near 750 cm^{-1} (F) is similar to that of monosubstituted benzenes.

PROBLEMS

22.1 Draw the condensed structural formulas for the nine isomers of formula C_7H_{16}. Give the IUPAC name for each isomer.

22.2 Give the IUPAC names of the following compounds

$$CH_3CH_2\underset{\underset{CH_3}{|}}{C}HCH_2\underset{\underset{CH_3}{|}}{C}HCH_3 \qquad CH_3CH_2\underset{\underset{\underset{\underset{CH_3}{|}}{CH_2}}{|}}{\overset{\overset{CH_3}{|}}{C}}CH_3$$

22.3 The heat of combustion of a typical gasoline is 52 kJ g^{-1}. One horsepower represents an energy production of 745.7 J s^{-1}. How many grams of gasoline per minute will an engine consume if it operates at 30% efficiency and develops 100 horsepower? How many liters per hour is this? (Use 0.68 g mL^{-1} for the density of gasoline.)

22.4 In addition to Eq. 22.2d, there are two other chain-terminating steps of lesser importance in the mechanism for the reaction of CH_4 with Cl_2. Write an equation for each.

22.5 Draw Lewis structures and condensed structural formulas for propyne, propene, propane, and cyclopropane.

22.6 Draw condensed structural formulas for all of the isomers of pentene, C_5H_{10}, having one double bond. Give the IUPAC name of each. (Indicate any cis and trans isomers.)

22.7 Draw condensed structural formulas for all of the isomers of pentyne, C_5H_8, having one triple bond. Give the IUPAC name of each.

22.8 Draw a condensed structural formula for
(a) 2-methyl-2-butene
(b) 3-ethyl-1-pentyne

22.9 By using the sigma-pi classification, describe the bonding in each of the following molecules. Your answer should include the atomic or hybrid orbitals used to form each bond and the classification of the bond formed (σ or π).
(a) ethane
(b) propylene
(c) propyne
(d) cyclopentane

22.10 Alkenes and cycloalkanes both have the general formula C_nH_{2n}. Why?

22.11 Starting with the appropriate alkene, give a method of preparing each of the following:
(a) ethane
(b) 1,2-dichloropropane
(c) ethyl alcohol
(d) *sec*-butyl alcohol
(e) acetic acid
Write equations for all reactions.

22.12 Write the condensed structural formulas and names of all of the isomers of trimethylbenzene.

22.13 Draw three resonance structures of naphthalene. (One structure is shown in Section 22.3.)

22.14 What alkyl chloride should be used in the Friedel-Crafts reaction with benzene to form cumene? The structure of cumene is

$$CH_3CHCH_3$$

(attached to benzene ring)

22.15 How many isomers are there of $C_2H_4Br_2$? Draw the Lewis structure of each isomer.

22.16 Draw Lewis structures for all of the isomers of $C_2H_3Cl_2F$.

22.17 Draw condensed structural formulas for all of the isomers formed by the substitution reaction of chlorine with 2,3-dimethylbutane:
$$C_6H_{14} + Cl_2 = C_6H_{13}Cl + HCl$$

22.18 Suggest an explanation for the following statements that, on the surface, appear to be contradictory.
 (a) Alkyl halides may be prepared from the corresponding alcohols by reaction with a hydrogen halide, that is, ROH + HX ⟶ RX + H_2O.
 (b) Alcohols may be prepared from the corresponding alkyl halides by reaction with a strong base such as NaOH, that is, RX + OH⁻ ⟶ ROH + X⁻.

22.19 How could you prepare ethylene glycol from 1,2-dichloroethane?

22.20 Draw a condensed structural formula for each of the following compounds:
 (a) 3-methyl-2-butanol
 (b) methyl ether
 (c) methanal (or formaldehyde)
 (d) 2-butanone (or methyl ethyl ketone)
 (e) propanoic acid

22.21 Give an explanation for the fact that most oxygen-containing organic compounds are fairly soluble in concentrated sulfuric acid.

22.22 Using the symbol R to represent any alkyl group, write a general formula for
 (a) a primary alcohol
 (b) a secondary alcohol

(c) a tertiary alcohol
(d) a ketone
(e) a secondary amine

22.23 What functional group is represented by each of the following structures?

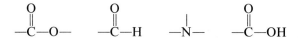

22.24 In wine production the enzymes of the yeast convert sugars into ethyl alcohol. The net reaction is
$$C_6H_{12}O_6 \xrightarrow{yeast} 2C_2H_5OH + 2CO_2$$
The fermented product is then aged in oak barrels or casks. Why is it necessary to keep the barrels full of wine at all times or to store the wine under nitrogen? Write a chemical equation for the reaction that occurs if the above precautions are not taken.

22.25 Write a half-reaction for the oxidation of
(a) a primary alcohol to an acid
(b) a secondary alcohol to a ketone
(c) an aldehyde to an acid

22.26 Identify the functional groups in each of the following compounds:
(a) $CH_3CHOHCH_3$
(b) CH_3CHCH_2
(c) $CH_3CHOHCH_2NH_2$
(d) H_2NCH_2COOH
(e) $CH_3CHOHCOOH$
(f) CH_2CHCl

22.27 Draw condensed structural formulas for the following:
(a) 1,2-dichloropropene
(b) *o*-nitrobromobenzene
(c) *m*-nitrophenol
(d) lactic acid (or 2-hydroxypropanoic acid)
(e) methyl ethyl ether
(f) ethyl acetate
(g) the four isomeric butylamines

22.28 Write Lewis structures (including formal charges) for NH_4^+ and $(CH_3)_4N^+$.

22.29 An analysis of papaverine, an alkaloid found in opium, gave 70.8% carbon, 6.2% hydrogen, 4.1% nitrogen, and 18.9% oxygen. The molecular mass of papaverine was found to be 339.4. What is the molecular formula of papaverine?

22.30 All of the substances listed below have about the same molecular mass, but the last four have much higher boiling points than the first five. Suggest an explanation.

Class	Substance	Molecular mass	bp (°C)
Alkane	n-Pentane	72	36
Alkene	1-Pentene	70	30
Alkyne	1-Pentyne	68	39
Alkyl halide	n-Butyl fluoride	76	32
Ether	Ethyl ether	74	35
Alcohol	n-Butyl alcohol	74	118
Carboxylic acid	Propionic acid	74	141
Diol	1,3-Propanediol	76	214
Amino acid	Glycine	75	290

22.31 Which alcohol can be synthesized starting from a one-carbon Grignard reagent and the three-carbon aldehyde?

22.32 Suggest a method of synthesizing $(CH_3)_2CHCH_2OH$ (isobutyl alcohol) from starting materials that have fewer than four carbon atoms each.

22.33 Although the carbon–carbon double bond is substantially stronger than the carbon–carbon single bond, alkenes readily undergo exothermic addition reactions (Eq. 22.3). Suggest an explanation.

CHAPTER 23
BIOLOGICAL CHEMISTRY

23

As with most of the topics treated in this book, the subject of this chapter has been developed in great detail by many workers, and many introductory texts in it alone are larger than the present volume. The chemistry of living systems involves a vast array of substances, most of which are quite complex. We shall, however, confine our attention mainly to three important classes of substance found in biological systems: proteins, carbohydrates, and nucleic acids, with particular emphasis on their molecular structure. We shall also examine the kinetics and the thermodynamics of chemical reactions that occur in biological systems. The rate at which these reactions occur involves biological catalysts called enzymes. An analysis of the thermodynamic aspects of these reactions shows that organisms possess a remarkable ability for utilizing available free energy.

23.1 AMINO ACIDS, POLYPEPTIDES, AND PROTEINS

The **proteins** constitute a class of chemical substances that is of major importance in biological chemistry. They are present in all living organisms. Apart from water, they are the major constituent of the soft tissues of animals and of the interior regions of plant cells. They have a wide variety of functions essential to life processes.

All the thousands of different naturally occurring proteins are built up from **amino acids.** These amino acids have the common structure

$$NH_2-\underset{H}{\overset{R}{\underset{|}{\overset{|}{C}}}}-COOH$$

Table 23.1 lists the common naturally occurring amino acids, brief notations for each, and the structures of their R side chains. (The complete molecular structures are shown for the last three. Cystine is a double amino acid derived from cysteine, and the last two are actually imino acids, since the R group in each is also connected to the nitrogen atom to form a secondary amine group.)

In all these amino acids except glycine, there are four different groups attached to the central carbon atom. Therefore each of these 23 amino acids has two possible enantiomeric forms (see Section 11.9). The two configurations (which are mirror images of each other) are

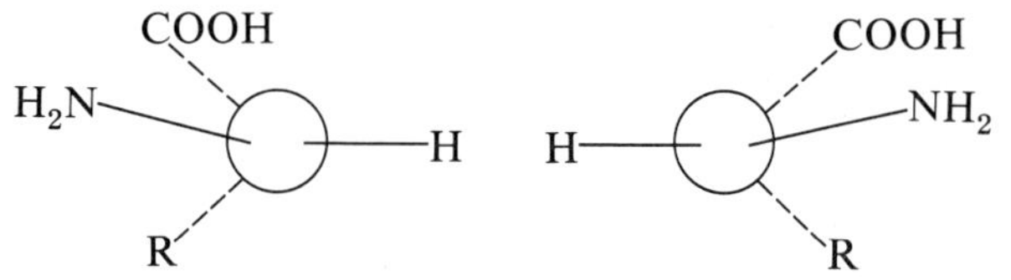

$$
\begin{array}{cc}
\text{L-configuration} & \text{D-configuration}
\end{array}
$$

(structures showing L- and D-configurations with COOH, NH_2, H, and R groups around a central carbon)

In these drawings the solid lines represent bonds to groups above the plane of the paper, and the dashed lines represent bonds to groups below the plane of the paper. It is a remarkable fact that *all* the amino acids obtained from naturally occurring proteins have the L-configuration.

The simpler living organisms are able to synthesize all the amino acids necessary to their survival, but the higher animals do not have this ability. Eight of the amino acids, called the **essential** amino acids, must be supplied to man either directly or combined in foods such as proteins. (In either case only the L-configurations are essential.) The average minimum daily nutritional requirements for man are listed in Table 23.1.

A molecule formed from only a few amino acids is called a **peptide**; molecules of a size intermediate between peptides and proteins are called **polypeptides**. It is convenient to take a molecular mass of 5000 as the dividing line between the polypeptides and the proteins.

A peptide may be formed from two amino acids by the reaction of the carboxyl group of one and the amino group of the other. Water is also formed in the process, as shown below for alanine and glycine:

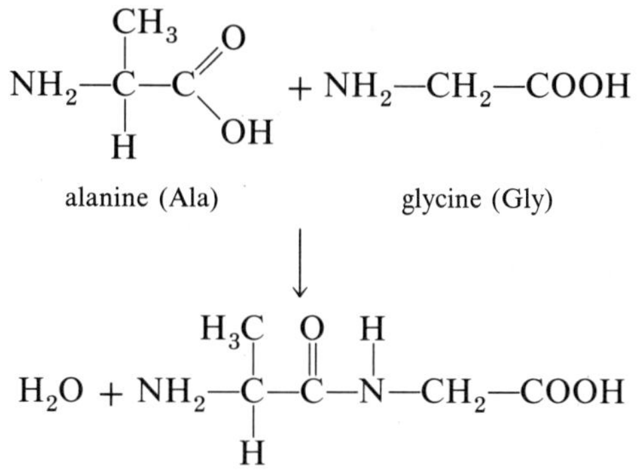

$$NH_2-\underset{\underset{H}{|}}{\overset{\overset{CH_3}{|}}{C}}-C\overset{O}{\underset{OH}{\diagup\!\!\!\diagdown}} + NH_2-CH_2-COOH$$

alanine (Ala) glycine (Gly)

↓

$$H_2O + NH_2-\underset{\underset{H}{|}}{\overset{\overset{H_3C}{|}}{C}}-\overset{O}{\underset{|}{\overset{\|}{C}}}-\underset{}{\overset{H}{\overset{|}{N}}}-CH_2-COOH$$

alanylglycine (Ala · Gly)

The peptide is written with the terminal NH_2 group to the left and the terminal COOH group to the right. The name of the peptide is formed from the amino acid names (taken from left to right) with the suffix *-yl* for all but the last (e.g., alanylglycine). A brief notation is formed from the amino acid symbols taken in the same order (e.g., Ala · Gly).

TABLE 23.1 THE COMMON, NATURALLY OCCURRING AMINO ACIDS

Name	R group	Minimum daily requirement (g)
Glycine (Gly)	—H	—
Alanine (Ala)	—CH_3	—
Valine (Val)	—$CH(CH_3)_2$	0.8
Leucine (Leu)	—$CH_2CH(CH_3)_2$	1.1
Isoleucine (Ile)	—$CH(CH_3)CH_2CH_3$	0.7
Serine (Ser)	—CH_2OH	—
Threonine (Thr)	—$CH(OH)CH_3$	0.5
Aspartic acid (Asp)	—CH_2COOH	—
Asparagine (Asp NH_2)	—CH_2CONH_2	—
Glutamic acid (Glu)	—CH_2CH_2COOH	—
Glutamine (Glu NH_2)	—$CH_2CH_2CONH_2$	—
Lysine (Lys)	—$CH_2CH_2CH_2CH_2NH_2$	0.8
Hydroxylysine (Hylys)	—$CH_2CH_2CH(OH)CH_2NH_2$	—
Cysteine (Cy SH)	—CH_2SH	—
Methionine (Met)	—$CH_2CH_2SCH_3$	1.1
Arginine (Arg)	—$CH_2CH_2CH_2NHC(=NH)NH_2$	—
Phenylalanine (Phe)	—CH_2—C$_6$H$_5$	1.1
Tyrosine (Tyr)	—CH_2—C$_6$H$_4$—OH	—
Histidine (His)	—CH_2—(imidazole)	—
Tryptophan (Try)	—CH_2—(indole)	0.25
Thyroxine	—CH_2—C$_6$H$_2$I$_2$—O—C$_6$H$_2$I$_2$—OH	—
Cystine (CyS-SCy)	NH_2—CH(COOH)—CH_2S—SCH_2—CH(COOH)—NH_2	—

23.1 AMINO ACIDS, POLYPEPTIDES, AND PROTEINS

Name	R group	Minimum daily requirement (g)
Proline (Pro)	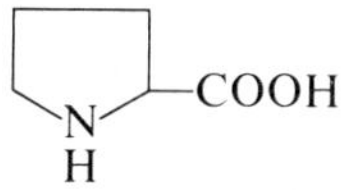 (pyrrolidine ring with N–H)—COOH	—
Hydroxyproline (Hypro)	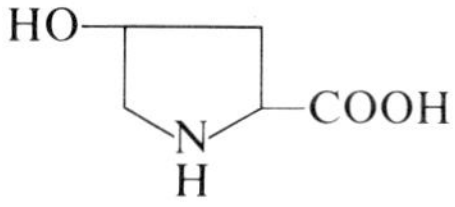 HO–(pyrrolidine ring with N–H)—COOH	—

An isomer of alanylglycine is glycylalanine:

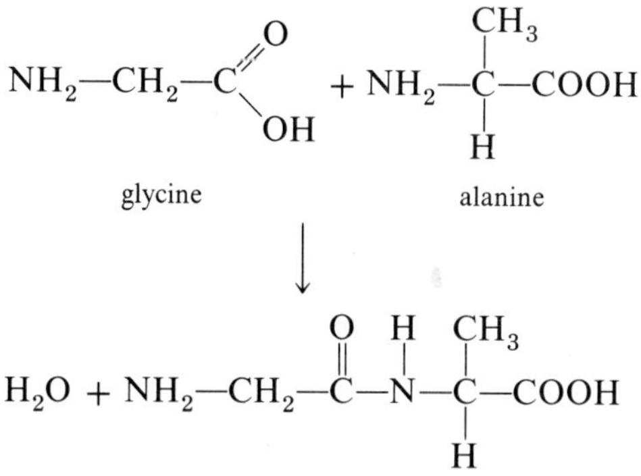

$$NH_2\text{—}CH_2\text{—}C(\!=\!O)OH \;+\; NH_2\text{—}\underset{H}{\underset{|}{C}}(CH_3)\text{—}COOH$$

glycine   alanine

↓

$$H_2O \;+\; NH_2\text{—}CH_2\text{—}\underset{}{\overset{O}{\overset{\|}{C}}}\text{—}\underset{H}{\overset{H}{\overset{|}{N}}}\text{—}\underset{H}{\overset{CH_3}{\overset{|}{C}}}\text{—}COOH$$

glycylalanine (Gly · Ala)

Because it is the connecting link between amino acids in all peptides, polypeptides, and proteins, the linkage $-\overset{O}{\overset{\|}{C}}-\overset{H}{\overset{|}{N}}-$ is of special importance. It is called the **peptide bond.** The hydrolysis of the peptide bond to form the constituent amino acids may be represented by the reverse of the reactions shown above.

The two main classes of proteins are the simple proteins and the conjugated proteins. The simple proteins are built up from hundreds or thousands of amino acid molecules. In the conjugated proteins, a simple protein is bonded to a nonprotein group called a **prosthetic group.** For example, the prosthetic group of the protein hemoglobin, which carries oxygen from the lungs to the tissues of higher vertebrates, is the iron chelate heme (Figure 19.5). In the following discussion we shall be primarily concerned with the simple proteins.

The statements made in this section have an extensive basis of experimental data. For example, the following lines of experimental evidence support the fundamental concept that polypeptides and proteins consist of amino acids linked by peptide bonds.

1. Complete hydrolysis of a polypeptide or simple protein yields only amino acids.
2. Acid–base titrations show that a polypeptide or protein has relatively few

free amino and carboxyl groups. These are equal in number to the terminal amino group and the amino groups of the R side chains of basic amino acids (such as lysine; see Table 23.1) and to the terminal carboxyl group and the carboxyl groups of the R side chains of acidic amino acids (e.g., aspartic acid), respectively. On hydrolysis, the numbers of free amino and carboxyl groups increase rapidly, and they appear in equal numbers.

3. Several polypeptides and one of the smaller proteins, insulin (molecular mass 5733), have been synthesized from amino acids under controlled conditions that yield products of known structure. The synthetic and natural products have identical properties.

4. The structures of several crystalline proteins have been determined by X-ray crystal structure methods.

23.2 STRUCTURES OF POLYPEPTIDES AND PROTEINS

Polypeptides and proteins consist of chains of amino acids linked by peptide bonds. The number of amino acids of each kind and the sequence in which they occur along a chain form the **primary structure** of the polypeptide or protein. Although determination of the primary structure is a complicated matter, the following types of experimental evidence are used.

1. Analysis of the amino acids formed by hydrolysis, together with a molecular mass determination, show the number of amino acids of each kind in a molecule of the polypeptide or protein.

2. Reagents that combine with the terminal NH_2 groups are used to determine the number of terminal NH_2 groups and hence the number of polypeptide chains in a molecule. An example is 1-fluoro-2,4-dinitrobenzene:

$$O_2N\text{-C}_6H_3(NO_2)\text{-F} + NH_2\text{-R} \longrightarrow O_2N\text{-C}_6H_3(NO_2)\text{-NH-R} + HF$$

Subsequent hydrolysis yields the 2,4-dinitrobenzene derivatives of those amino acids having terminal NH_2 groups, which may be isolated and identified. In this way the protein insulin has been shown to contain two polypeptide chains—one terminating in glycine and one in phenylalanine. (Amino acids with side-chain NH_2 groups also react with this reagent, but the position of substitution shows that these amino acids are not located at the ends of polypeptide chains.) Analogous methods may be used to determine which amino acids have free carboxyl groups and hence are located at the other ends of the polypeptide chains.

3. Partial hydrolysis with various enzymes yields lower peptides that are

simple enough for positive identification by comparison with peptides of known structure. Each of these lower peptides establishes the amino acid sequence in one part of a polypeptide chain. Further, lower peptides having amino acids in common indicate amino acid sequences over longer sections. For example, identification of the tripeptides Ser · Leu · Tyr and Tyr · Glu · Leu suggests the possibility that the polypeptide chain has Ser · Leu · Tyr · Glu · Leu as part of its amino acid sequence. With enough information from overlapping peptides a unique structure may be assigned to each chain.

4. Cysteine side chains frequently form sulfur–sulfur bonds as in the double amino acid, cystine (see Table 23.1). Such bonds often link the various polypeptide chains in a protein. Prior to hydrolysis the S—S bonds are broken with a reagent such as performic acid:

$$R_1-S-S-R_2 \xrightarrow{H\overset{O}{\overset{\|}{C}}-O_2H} R_1-SO_3^- + R_2-SO_3^-$$

where R_1 and R_2 are different polypeptide chains.

Figure 23.1 shows the primary structure of the protein bovine insulin, determined by the methods outlined above. It is composed of two polypeptide chains linked together by two S—S bonds.

Different regions of a polypeptide chain are linked by hydrogen bonds between oxygen and nitrogen atoms:

$$\overset{\diagdown}{\underset{\diagup}{C}}=O\cdots H-\overset{\diagup}{\underset{\diagdown}{N}}$$

In this way hydrogen bonding determines the spatial relationships between various parts of the chain—an aspect of the configuration of proteins called **secondary structure**. A common configuration is the **alpha helix** (Figure 23.2), in which the polypeptide chain forms a left-handed or right-handed spiral or helix. Note that the R side chains, which may be quite bulky (Table 23.1), project out away from the helical axis. Every amino acid is linked to the fourth amino acid in either direction, and thus the repeating unit may be represented in two dimensions as

$$-\overset{H}{\underset{}{N}}-\left[\overset{}{\underset{\overset{\|}{O}}{C}}-CHR-\overset{}{\underset{H}{N}}\right]_3 \overset{\overset{O}{\|}}{C}-CHR-\overset{H}{\underset{}{N}}-\left[\overset{}{\underset{\overset{\|}{O}}{C}}-CHR-\overset{}{\underset{H}{N}}\right]_3 \overset{\overset{O}{\|}}{C}-$$

In most proteins the polypeptide chain is folded back and forth to form a highly compact molecule (in some cases roughly spherical or ellipsoidal). This aspect of the configuration of proteins is called **tertiary structure.** Figure 23.3 shows the manner in which the polypeptide chain is arranged in space in the

FIGURE 23.1

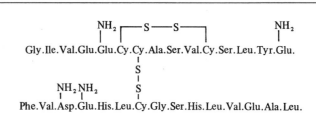

Primary structure of bovine insulin. [SOURCE: *Reprinted with permission from J. I. Harris and V. M. Ingram, in P. Alexander and R. J. Block (eds.), A Laboratory Manual of Analytical Methods of Protein Chemistry, vol. 2, 1960, Pergamon Press Ltd., Oxford, p. 445.*]

protein myoglobin (molecular mass 17,500), whose structure has been determined by X-ray diffraction. Regions having the alpha-helix structure are indicated by double lines. The forces responsible for tertiary structure are thought to involve interactions between the various R side chains, primarily van der Waals interactions between the relatively nonpolar hydrocarbon side chains. Another protein whose structure has been determined by X-ray diffraction is lysozyme, an enzyme of molecular mass 14,400. A photograph of a model of the lysozyme molecule is shown in Figure 23.4.

Finally, some proteins contain two or more polypeptide chains that are not covalently linked by sulfur–sulfur bonds as in insulin. An example is hemoglobin, which has four polypeptide chains (each with a prosthetic heme group). The spatial configuration of these chains relative to one another is called **quaternary structure.**

Under adverse conditions the higher-order structure of a protein is altered, a process called **denaturation.** As a consequence the physical and chemical properties may change markedly, and the protein loses its ability to perform its biological function. Conditions causing denaturation include high temperatures, ultraviolet radiation, or various reagents such as urea or certain salts. A familiar example is the effect of cooking an egg white. Denaturation may be viewed as essentially an unfolding of the polypeptide chain, without altering its primary structure. Under some conditions denaturation is reversible; for instance, removal of urea generally restores a protein to its native state (Figure 23.5). Under other conditions (e.g., high temperatures) the process is irreversible.

23.3 CARBOHYDRATES

The **carbohydrates** form one of the major components of living matter. Their general formula can be written as a combination of carbon and water, $C_x(H_2O)_y$, hence the name carbohydrate. Usually x is 5 or 6, or some multiple of 5 or 6. Carbohydrates are formed in plants from carbon dioxide, water, and sunlight by photosynthesis:

23.3 CARBOHYDRATES

$$\text{Leu.Glu.Asp.Tyr.Cy.Asp} \begin{matrix} \text{NH}_2 & \text{NH}_2 \\ | & | \\ & \end{matrix}$$

$$\overset{|}{\underset{|}{\text{S}}}$$

$$\text{Tyr.Leu.Val.Cy.Gly.Glu.Arg.Gly.Phe.Phe.Tyr.Thr.Pro.Lys.Ala.}$$

$$x\text{CO}_2 + y\text{H}_2\text{O} \xrightarrow{\text{light}} x\text{O}_2 + \text{C}_x(\text{H}_2\text{O})_y$$

The simpler carbohydrates are called **sugars;** examples include ribose, $C_5H_{10}O_5$, and glucose, $C_6H_{12}O_6$. Sucrose, the common table sugar extracted from sugar beets and sugar cane, is $C_{12}H_{22}O_{11}$.

Polymeric carbohydrates are called **polysaccharides.** Starch and cellulose, both with the general formula $(C_6H_{10}O_5)_n$, are important examples. Cellulose is the main constituent of plant cell walls, and starch is a major component of foods such as wheat, rice, corn, and potatoes.

The carbohydrates contain carbon chains to which oxygen atoms are attached as hydroxyl and carbonyl groups. A typical six-carbon sugar has the structure

$$\text{HOCH}_2-\overset{\text{OH}}{\underset{|}{\text{CH}}}-\overset{\text{OH}}{\underset{|}{\text{CH}}}-\overset{\text{OH}}{\underset{|}{\text{CH}}}-\overset{\text{OH}}{\underset{|}{\text{CH}}}-\overset{\text{O}}{\underset{||}{\text{CH}}}$$

The four central carbon atoms are asymmetric, and thus there are a number of different optically active isomers (16 in all). The most important is glucose. Its structure is

$$\begin{array}{c} \text{HC}=\text{O} \\ | \\ \text{HCOH} \\ | \\ \text{HOCH} \\ | \\ \text{HCOH} \\ | \\ \text{HCOH} \\ | \\ \text{CH}_2\text{OH} \end{array}$$

As for the amino acid enantiomers shown in Section 23.1, the convention followed in this drawing is that groups to the left and right of each asymmetric carbon atom

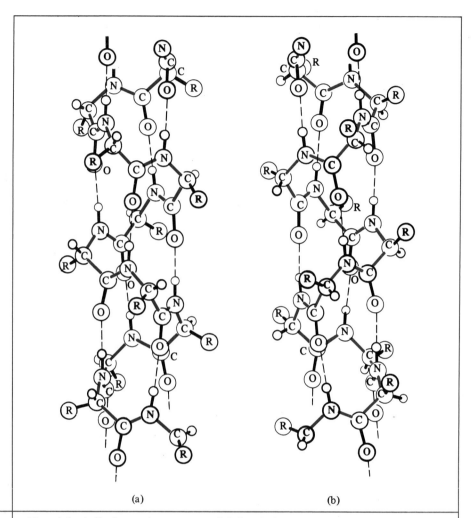

FIGURE 23.2   (a)   (b)

The alpha-helical protein structure—left-handed (a) and right-handed (b) forms. The helices are shown in color. In both forms the amino acids have the L-configuration. (SOURCE: *From Linus Pauling:* The Nature of the Chemical Bond, *Third Edition.* © *1960 by Cornell University. Used by permission of the publisher, Cornell University Press.*)

23.3 CARBOHYDRATES

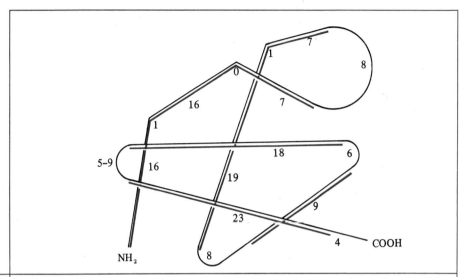

FIGURE 23.3 Tertiary structure of sperm whale myoglobin. Numbers indicate the number of amino acids in each region of the primary structure. Regions having the alpha-helix structure are indicated by double lines. [SOURCE: D. W. Green, "*Spacial Configuration in Proteins,*" in M. Florkin and E. H. Stotz (eds.), Comprehensive Biochemistry, *vol. 7,* Elsevier, Amsterdam, 1963, p. 249.]

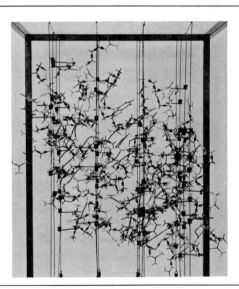

FIGURE 23.4 A model of the lysozyme molecule. (SOURCE: D. C. Phillips, Laboratory of Molecular Biophysics, Oxford University, England.)

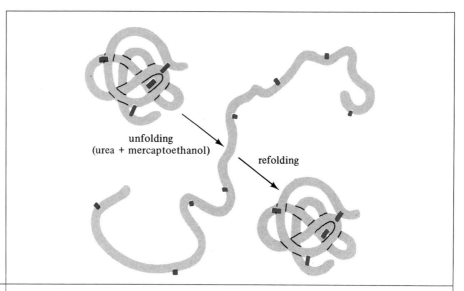

FIGURE 23.5

Reversible denaturation of a protein. Double-line segments indicate disulfide bonds broken during denaturation. [SOURCE: C. J. Epstein, R. F. Goldberger, and C. B. Anfinsen, Cold Spring Harbor Symposia on Quantitative Biology, **28**, *440 (1963)*.]

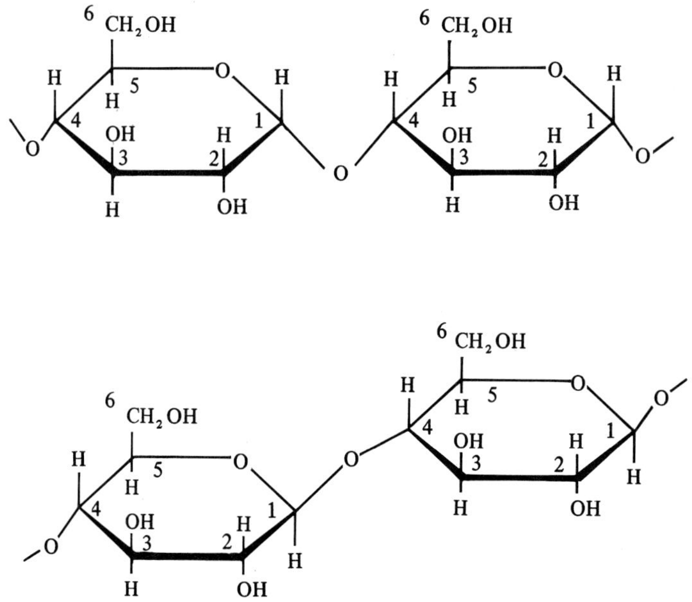

FIGURE 23.6

Structures of starch (top) and cellulose (bottom).

23.3 CARBOHYDRATES

(taken one at a time) are above the plane of the paper, and the upper and lower groups are below.

An organic reaction of special importance to carbohydrate structure is **hemiacetal** formation, in which an alcohol and an aldehyde combine:

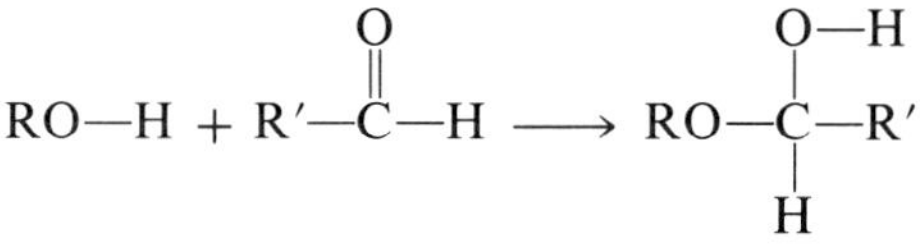

$$RO-H + R'-\overset{O}{\underset{\|}{C}}-H \longrightarrow RO-\underset{H}{\overset{O-H}{\underset{|}{C}}}-R'$$

(This reaction may be regarded as an addition to the carbonyl double bond, with the alcohol hydrogen adding to the carbonyl oxygen, and the RO group adding to the carbonyl carbon.) When a hemiacetal is formed by the aldehyde group at one end of a sugar molecule and an alcohol group at or near the other end, a cyclic structure is formed. Glucose forms the following products:

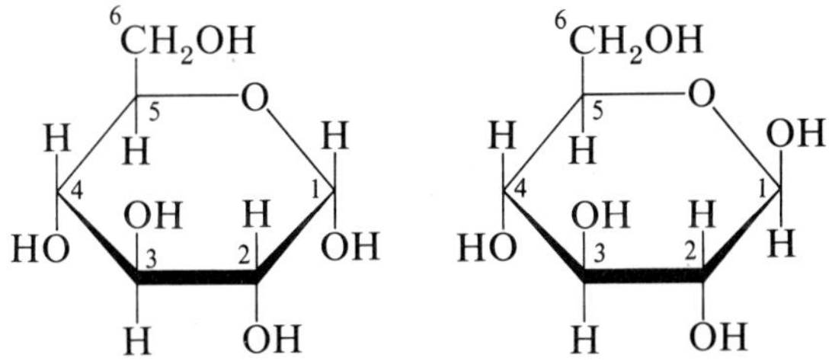

α-glucose   β-glucose

They differ in the relative orientation (trans or cis) of the groups attached to the carbon atoms on either side of the ring oxygen. In an aqueous solution of glucose at room temperature, about one-third is in the alpha form, two-thirds is in the beta form, and a very small amount is in the open-chain form.

Starch is a polymer of α-glucose, whereas cellulose is a polymer of β-glucose (Figure 23.6). In either polymer the C_1 carbon atom of one molecule is linked to the C_4 carbon atom of the next, with the elimination of water:

$$nC_6H_{12}O_6 \rightarrow (C_6H_{10}O_5)_n + nH_2O$$

Ribose is a C_5 carbohydrate of special importance in the nucleic acids (to be described in the next section). In its cyclic form ribose has the structure

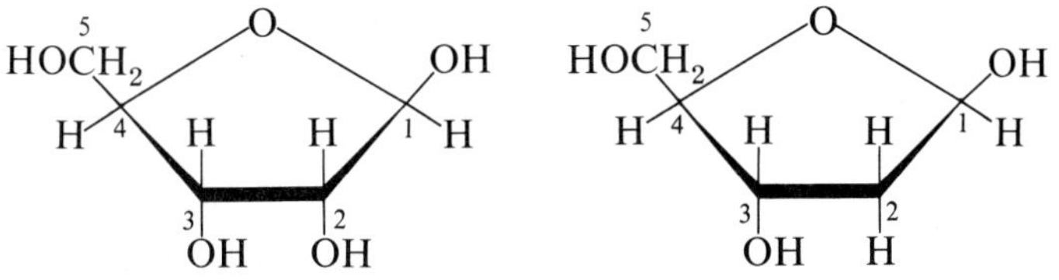

ribose   deoxyribose

(The carbon atoms are numbered for later reference.) A closely related C_5 molecule is deoxyribose, which differs from ribose in that it has H rather than OH attached to the 2 position.

23.4 NUCLEIC ACIDS

Two types of nucleic acid are found in living cells: **ribonucleic acid (RNA)** and **deoxyribonucleic acid (DNA)**. Both are polymers of high molecular mass, DNA's generally ranging from about 10^6 to 10^9, and RNA's from 10^4 to 10^6. Deoxyribonucleic acids are usually found concentrated in the cell nuclei. Ribonucleic acids are distributed more widely throughout the cell, but they are mainly found in particles called **ribosomes.** It has been shown that genes—the substances that store the information controlling hereditary characteristics, and transmit it to succeeding generations—are DNA molecules. Ribonucleic acid molecules control the synthesis of proteins in living cells.

Both RNA's and DNA's are highly acidic—at the pH of the cell the acids are dissociated, so the polymers are anionic. They are usually associated with cations such as Mg^{2+} and protonated amines or proteins.

On hydrolysis, RNA's form ribose, phosphoric acid (H_3PO_4), and four organic nitrogen compounds: adenine, guanine, cytosine, and uracil. Deoxyribonucleic acids yield deoxyribose, phosphoric acid, and the same set of nitrogen compounds except that thymine is formed instead of uracil:

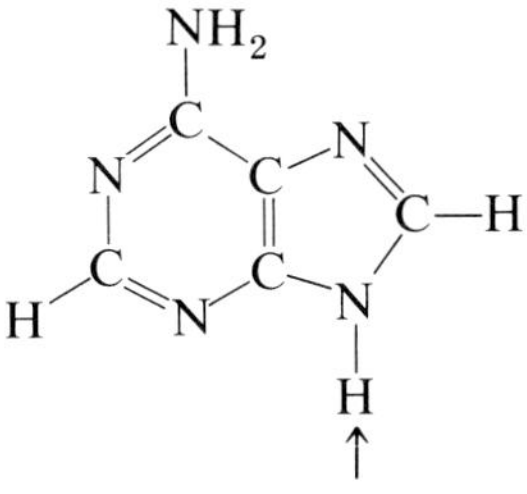

adenine 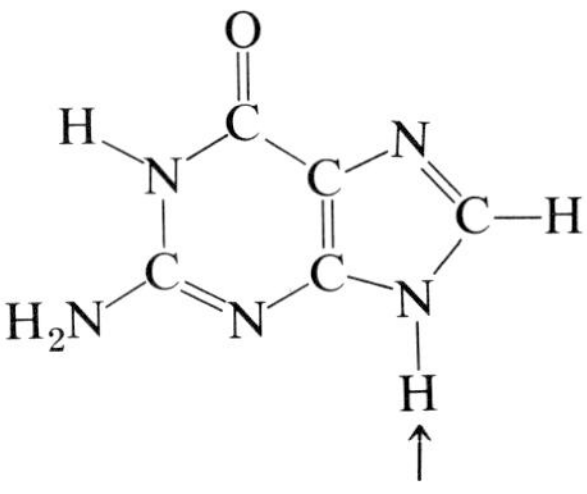 guanine

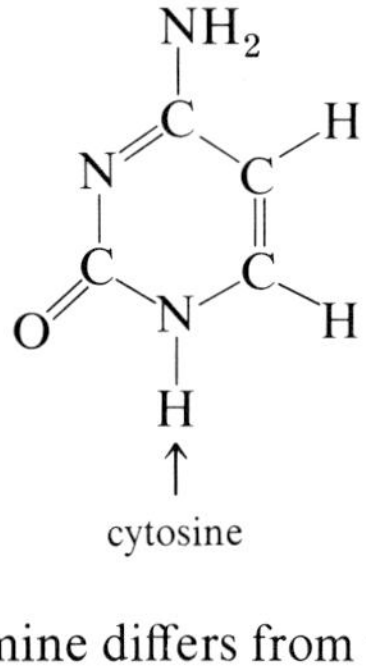

cytosine 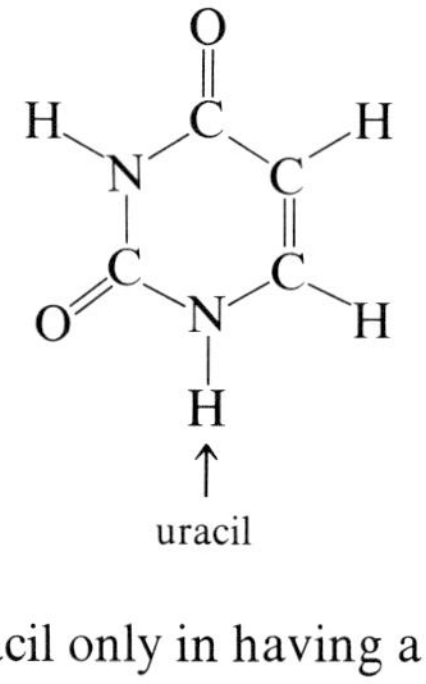 uracil 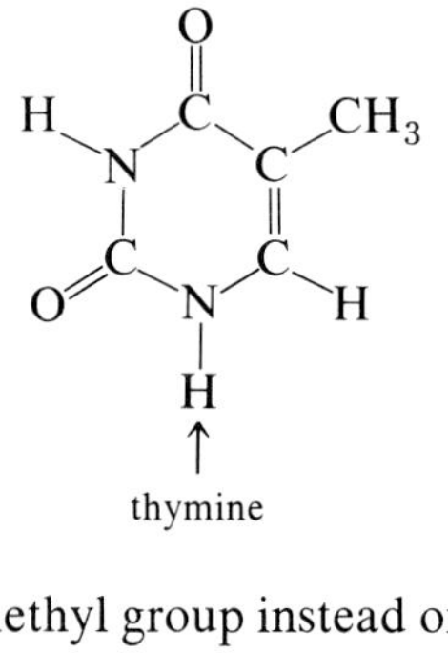 thymine

(Thymine differs from uracil only in having a methyl group instead of a hydrogen atom attached to one of the ring carbons.) The nitrogen compounds have weak basic properties, and they are referred to as nitrogen bases.

23.4 NUCLEIC ACIDS

In both RNA and DNA the basic structure is an alternating sequence of phosphoric acid and sugar molecules (ribose for RNA and deoxyribose for DNA). In addition, each sugar molecule is bonded to a nitrogen base:

```
—phosphoric—sugar—phosphoric—sugar—phosphoric—sugar—
    acid      |         acid      |         acid      |
        nitrogen base      nitrogen base      nitrogen base
```

The phosphoric acid and sugar molecules are connected through ester linkages at the 3 and 5 positions of the sugar molecule. Thus for each phosphoric acid molecule, the polymerization reaction is

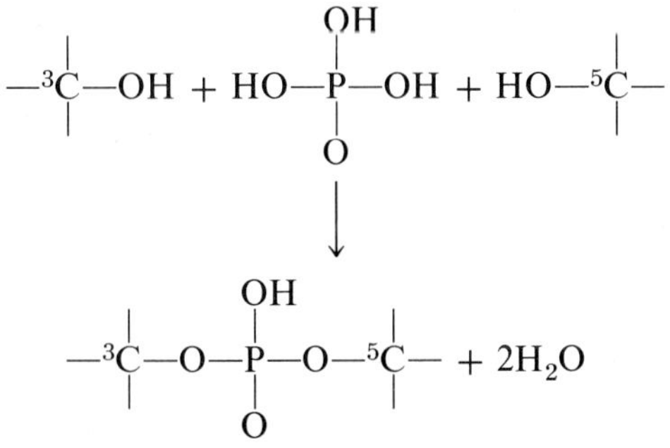

$$-{}^3\text{C}-\text{OH} + \text{HO}-\overset{\overset{\text{OH}}{|}}{\underset{\underset{\text{O}}{|}}{\text{P}}}-\text{OH} + \text{HO}-{}^5\text{C}- $$

$$\downarrow$$

$$-{}^3\text{C}-\text{O}-\overset{\overset{\text{OH}}{|}}{\underset{\underset{\text{O}}{|}}{\text{P}}}-\text{O}-{}^5\text{C}- + 2\text{H}_2\text{O}$$

Linkage between the sugar and nitrogen base occurs in a similar way at the 1 position of the sugar molecule and at the points indicated by arrows in the structures shown above for the nitrogen bases. A portion of a DNA chain, with cytosine and thymine as the nitrogen bases, is shown in Figure 23.7.

Deoxyribonucleic acid typically forms a secondary structure that is a **double helix.** In this structure two DNA chains are intertwined with the nitrogen bases projecting in toward the helical axis. The molecular planes of the nitrogen bases are perpendicular to the axis in a stacking arrangement (Figure 23.8). Furthermore, the nitrogen bases in the two chains are linked by hydrogen bonds in a highly specific manner: Cytosine is always hydrogen-bonded to guanine, and adenine is always hydrogen-bonded to thymine. Only these two pairs have the right dimensions to fit inside the double helix (Figure 23.9). Guanine and cytosine form three hydrogen bonds, adenine and thymine form two. The distance between the deoxyribose carbon atoms to which the nitrogen compounds are attached (designated C′ in Figure 23.9) is 10.85 Å for each pair.

As a consequence of the specificity in hydrogen bonding, each chain of the double helix must have a sequence of nitrogen bases complementary to the other. Thus, designating each nitrogen base by its first letter, if the sequence in one chain is

A G T T C G T A C

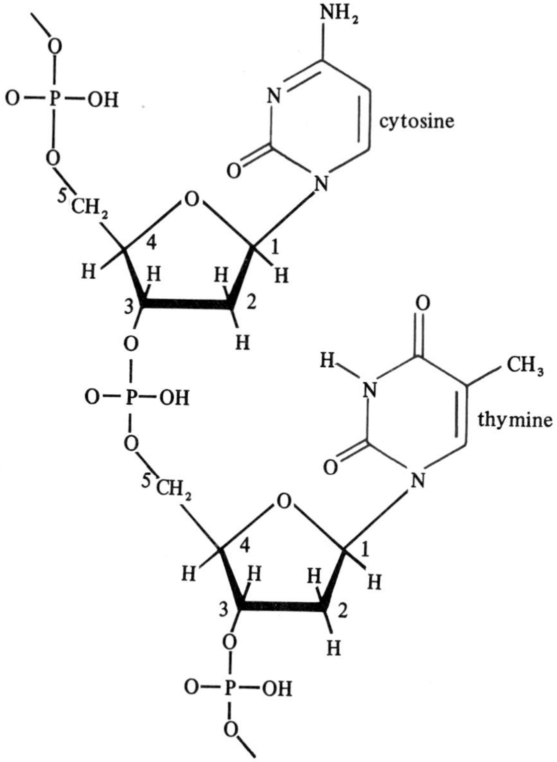

FIGURE 23.7

Portion of a DNA chain. The chain is read in the direction $-{}^3C-O-P-O-{}^5C-$, or from top to bottom in this figure. A shorthand notation for the cytosine-thymine sequence shown here is CT.

then the corresponding sequence in the complementary chain must be

T C A A G C A T G

Each of these sequences represents information coded in a particular way. Deoxyribonucleic acid transmits this information to succeeding generations by forming duplicate molecules, that is, it brings about the synthesis of molecules identical to itself—a process called **replication.** As each strand in the double helix contains the same information, either strand may be used to replicate the other.

Replication occurs through an uncoiling of the double helix followed by formation of a new chain using each of the original strands as a template, provided, of course, that the sugar–phosphate–nitrogen base monomers (called **deoxyribonucleotides**) are available. The following diagram illustrates replication

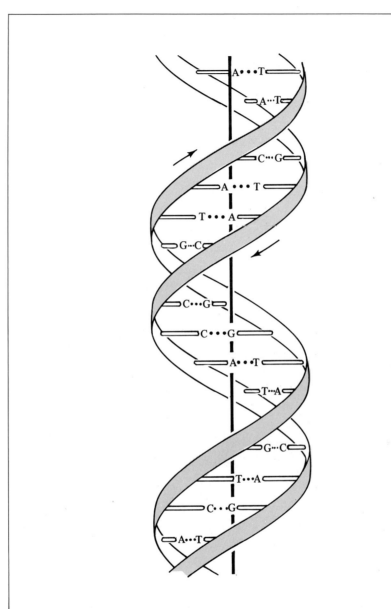

FIGURE 23.8

A schematic illustration of the double helix structure of DNA. The vertical line indicates the axis of the double helix. Nitrogen bases are indicated by first letters. The two sugar-phosphate backbones (in color) twist about on the outside, with the flat hydrogen-bonded base pairs forming the core. Seen this way, the structure resembles a spiral staircase with the base pairs forming the steps. (SOURCE: *From The Double Helix by James D. Watson. Copyright © 1968 by James D. Watson. Reprinted by permission of Atheneum Publishers.*)

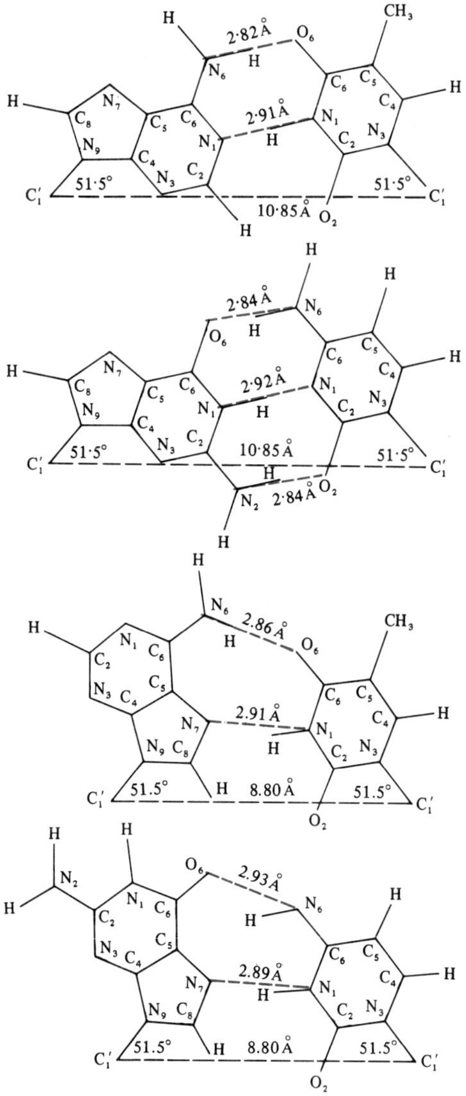

FIGURE 23.9 Structures of hydrogen-bonded base pairs. Top, adenine-thymine base pair in DNA. Second, guanine-cytosine base pair in DNA. Note that the distance across the double helix between the sugar-phosphate chains (C_1'-C_1') is the same for each base pair (10.85 Å). Hydrogen bond distances are also indicated. Third, adenine-thymine base pair in a molecular complex (a 1:1 complex of 9-N-methyladenine and 3-N-methylthymine). Bottom, guanine-cytosine base pair similarly oriented. These latter structures have been shown to be incompatible with X-ray diffraction data on DNA. [SOURCE: S. Arnott, M. H. F. Wilkins, L. D. Hamilton, and R. Langridge, *Journal of Molecular Biology*, **11**, *392 (1965). Copyright by Academic Press Inc., (London) Ltd.*]

at a point where a fourth monomer is being added to each of two new strands:

```
                                  C G
                                  A T
                                  T A
                                  G ← C
        A G T T C G T A C        A G T T C
                          ⟶
        T C A A G C A T G        T C A A G
                                  C ← G
                                  A T
                                  T A
                                  G C
```

Evidence supporting this mechanism for DNA replication has been obtained by isotope studies using ^{15}N. Replication of a DNA containing only ^{15}N in a medium containing only ^{14}N produces new DNA molecules that are half ^{14}N and half ^{15}N. After two replications, half the DNA molecules are ^{14}N–^{15}N hybrids, and the other half contain only ^{14}N (see Figure 23.10).

The information coded in DNA molecules controls the synthesis of proteins that are characteristic of each kind of organism. Ribonucleic acid molecules act as intermediaries in this process. The basic mechanism by which information is utilized in living organisms is

$$\text{DNA} \longrightarrow \text{RNA} \longrightarrow \text{proteins} \longrightarrow \text{other substances}$$

Ribonucleic acid molecules complementary to specific regions of a DNA molecule are formed using a DNA strand as a template, in a process analogous to that described above for DNA replication. Then the RNA acts, in turn, as a template on which protein synthesis occurs. (Many important but complicating features, such as different kinds of RNA and the locations in the living cell where various functions are carried out, are omitted here.) Thus the sequence of nitrogen bases in DNA determines the sequence of amino acids in proteins. Each successive group of three nitrogen bases in the DNA sequence (called a **triplet**) leads to one amino acid in the protein sequence. The correspondence between triplets and amino acids is called the **genetic code.** For example, the triplet adenine–adenine–adenine causes the addition of lysine to the growing polypeptide chain, whereas guanine–cytosine–adenine causes the addition of alanine. (The direction along the chain is that indicated in the caption to Figure 23.7.) In this way the sequence in which the nitrogen bases are arranged in DNA is believed to contain all the information needed to direct the formation of a human being or other living organism.

23.5 ENZYMES

A wide variety of chemical reactions occur in biological systems. Most are quite slow under the conditions usually found in the living cell unless a catalyst is

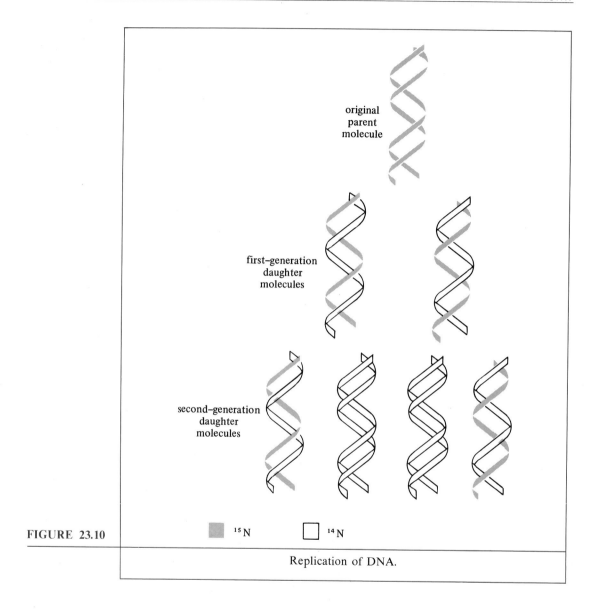

FIGURE 23.10 Replication of DNA.

present. For example, an aqueous solution of glucose can be maintained in contact with air for a long time without appreciable air oxidation of the glucose. Yet organisms use the oxidation of glucose to form carbon dioxide and water as an important energy source.

The catalysts found in biological systems are called **enzymes.** All enzymes have been found to be proteins. Their molecular masses range from about 10^4 to 10^7. Many thousands of enzymes have been isolated and studied. The human body contains upward of 10^4 different enzymes. Examples include ptyalin, or salivary amylase (found in the saliva), that catalyzes the hydrolysis of starch, and pepsin (found in the gastric juice) that catalyzes the hydrolysis of proteins.

An enzyme is usually quite specific in its catalytic activity. It catalyzes only one type of reaction and it does this for only one chemical substance or group of closely related substances. For example, amylase, which catalyzes the hydrolysis of starch in animals, is unable to catalyze the hydrolysis of cellulose despite its similarity to starch in chemical nature and structure (Figure 23.6).

Chemical substances whose reactions are catalyzed by enzymes are called **substrates.** Like other catalysts, an enzyme does not itself undergo any net change: After catalyzing the reaction of one substrate molecule, the enzyme is regenerated in its original form, ready to catalyze the reaction of another substrate molecule. One enzyme molecule is thus able to catalyze the reaction of many substrate molecules in a brief interval. Catalytic activity of an enzyme is often reported as the **turnover number,** that is, the number of molecules of substrate that react per minute for each molecule of enzyme present. Turnover numbers generally range from 10^2 to as high as 10^8.

Studies of enzyme-catalyzed reactions show that usually the substrate becomes attached to the enzyme, forming a substrate–enzyme complex in equilibrium with the enzyme and substrate. For simple reactions the complex then decomposes in a first-order step to form the products of the reaction. If we designate an enzyme molecule by E, a substrate molecule by S, and the reaction products by P, the mechanism is

$$E + S \xrightleftharpoons{K} ES \xrightarrow{k} E + P$$

where K is the equilibrium constant for complex formation, and k is the first-order rate constant (Section 16.2) for decomposition of the complex to form the products. The law of chemical equilibrium requires that

$$\frac{[ES]}{[E][S]} = K$$

The rate of formation of products is

$$\text{rate} = k[ES]$$
$$= kK[E][S] \qquad (23.1)$$

Of the total enzyme present, $[E]^{\text{total}}$, part is present in the enzyme–substrate complex and part as free enzyme:

$$[E]^{\text{total}} = [ES] + [E]$$
$$= K[E][S] + [E]$$
$$= [E](K[S] + 1)$$

Thus

$$[E] = \frac{[E]^{\text{total}}}{K[S] + 1} \qquad (23.2)$$

Substituting this last expression into Eq. 23.1 gives

$$\text{rate} = \frac{kK[\text{E}]^{\text{total}}[\text{S}]}{K[\text{S}] + 1} \tag{23.3}$$

By dividing numerator and denominator by K, we obtain

$$\text{rate} = \frac{k[\text{E}]^{\text{total}}[\text{S}]}{[\text{S}] + (1/K)} \tag{23.4}$$

This result is referred to as the Michaelis-Menten equation for enzyme kinetics. Figure 23.11 shows the relation between rate and substrate concentration given by this equation.

The Michaelis-Menten equation reduces to simpler forms at both very low and very high substrate concentrations. At very high substrate concentrations, where $[\text{S}] \gg (1/K)$, it reduces to

$$\text{rate} = \frac{k[\text{E}]^{\text{total}}[\text{S}]}{[\text{S}]}$$

$$= k[\text{E}]^{\text{total}} \tag{23.5}$$

Under these conditions the reaction is first-order in total enzyme concentration and independent of substrate concentration. (This is the flat region at the right in Figure 23.11.)

At very low substrate concentrations, where $[\text{S}] \ll (1/K)$, Eq. 23.4 reduces to

$$\text{rate} = kK[\text{E}]^{\text{total}}[\text{S}] \tag{23.6}$$

and the rate is first-order in both total enzyme and substrate concentrations. (This is the steep rise at the left in Figure 23.11.)

Enzymes, then, are proteins that owe their catalytic activity to the formation of an enzyme–substrate complex that decomposes to products and enzyme. In most cases the substrate molecule is much smaller than the enzyme molecule, and this fact has given rise to the concept that the substrate is complexed at (and later reacts at) an **active site** on the enzyme surface. Because of the extreme specificity of enzyme catalysis, it has been proposed that only certain substrate structures will "fit" at the active site; in effect, the active site in the complex three-dimensional structure of an enzyme will accommodate and bind particular molecules but not others. Figure 23.12 shows this relationship schematically. In support of this mechanism of enzyme action is the correlation between enzyme structure and enzyme activity. For example, the same conditions of pH and temperature that denature an enzyme also destroy its catalytic activity. When originally proposed, this mechanism was thought to involve a rigid enzyme structure in the neighborhood of the active site, giving rise to the name **lock-key** hypothesis. Recent experi-

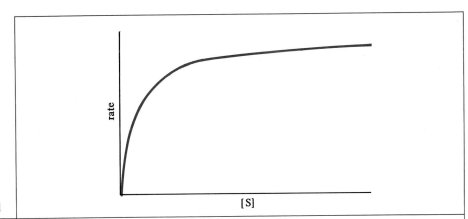

FIGURE 23.11 Effect of substrate concentration on the rate of a reaction following the Michaelis-Menten equation.

mental evidence has led to a newer **induced-fit** hypothesis, according to which the substrate molecule induces substantial changes in the active-site structure as it fits into the active site—changes necessary to give just the right geometrical relationship between the various functional groups of the enzyme and substrate for catalysis.

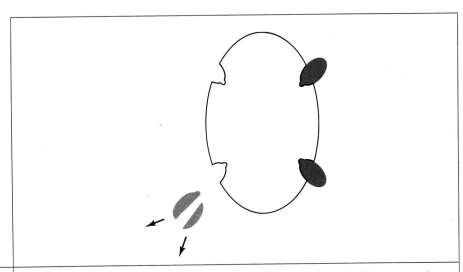

FIGURE 23.12 A schematic representation of the theory of enzyme specificity. This enzyme has four active sites. Substrate molecules are shown in color at the two active sites on the right; the products of the enzyme-catalyzed reaction are leaving the active site on the left.

The higher-order structure of a protein brings together in space various points along the peptide chain that may be far apart in the primary structure (Figure 23.4). The active site may therefore be composed of widely separated groups of the peptide chain. In several cases it has been possible to determine the structure of the enzyme in the region of the active site. A typical example is the enzyme acetylcholinesterase that catalyzes the hydrolysis of acetylcholine (an ester):

$$(CH_3)_3N^+CH_2CH_2O\overset{O}{\underset{\|}{C}}CH_3 + H_2O$$
acetylcholine

$$\downarrow$$

$$(CH_3)_3N^+CH_2CH_2OH + CH_3\overset{O}{\underset{\|}{C}}OH$$
choline   acetic acid

The enzyme has a molecular mass of about 3×10^6, and it contains about 100 active sites that catalyze the above hydrolysis. The active sites have been studied by different methods including (1) determining the effect of the pH of the medium on the rate of hydrolysis, and (2) adding an inhibitor, partially hydrolyzing the enzyme into peptides, and determining the groups that are attached to the inhibitor. The results of the above investigations can be interpreted as showing that each active site consists of two adjacent subsites: an anionic site and an esteratic site. The esteratic site contains an acidic group (a phenol), a basic group (imidazole nitrogen =N—), and a serine OH group, as shown in Figure 23.13(a). Figure 23.13(b) gives a suggested structure for the complex between the enzyme and acetylcholine. Note that the anionic site attracts the positively charged nitrogen atom of the acetylcholine, whereas the esteratic site attracts the ester group.

23.6 THERMODYNAMICS OF CHEMICAL REACTIONS IN BIOLOGICAL SYSTEMS

Some of the reactions necessary to the function of a biological system have negative free-energy changes; others have positive free-energy changes. Because those reactions with a negative free-energy change are spontaneous and those with a positive free-energy change are not (Section 15.7), the obvious question arises as to how nonspontaneous reactions can occur in living organisms. Detailed studies of these reactions show that there is a *coupling* between the reactions with negative and positive free-energy changes in such a way that the former reactions drive the latter.

To present in symbolic form the problem and how it is solved by organisms,

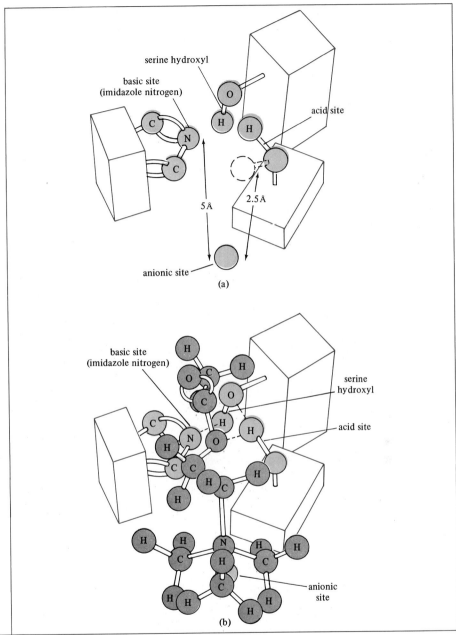

FIGURE 23.13 (a) Proposed structure of an active site of the enzyme acetylcholinesterase. (b) A molecule of acetylcholine (gray) complexed at an active site. [SOURCE: *R. M. Krupka and K. J. Laidler,* Journal of the American Chemical Society, **83,** *1459* (1961). Copyright 1961 by the American Chemical Society. Reprinted by permission of the copyright owner.]

consider the following two reactions:

$$A \longrightarrow X \quad \Delta G = -10 \text{ kJ mol}^{-1}$$
$$B \longrightarrow Y \quad \Delta G = +8 \text{ kJ mol}^{-1}$$

Although the conversion of B to Y is not spontaneous, if the reactions can be coupled, then

$$A + B \longrightarrow X + Y \quad \Delta G = -10 + 8 = -2 \text{ kJ mol}^{-1}$$

and the combined reaction is indeed spontaneous.

Although this solution is a clever one, it imposes the restriction that the two reactions must occur at the same place at the same time. In cells, however, it may be more efficient to have them occur in different regions of the system and, perhaps, even at different times. This is accomplished by means of a set of general intermediates that act as a sort of cash currency in the system's energy economy. To illustrate, designate the intermediates as I_1 and I_2, and suppose that the free-energy change is

$$I_1 \longrightarrow I_2 \quad \Delta G = +9 \text{ kJ mol}^{-1}$$

Then, if the intermediates are first coupled with the spontaneous reaction $A \rightarrow X$,

$$A + I_1 \longrightarrow X + I_2 \quad \Delta G = -10 + 9 = -1 \text{ kJ mol}^{-1}$$

The intermediate I_2 can later drive the nonspontaneous reaction $B \rightarrow Y$ at a different time and place:

$$B + I_2 \longrightarrow Y + I_1 \quad \Delta G = +8 - 9 = -1 \text{ kJ mol}^{-1}$$

The intermediate commonly found in biological systems is adenosine triphosphate (ATP), which contains ribose, adenine (linked to ribose at the 1 position), and triphosphate (linked to ribose at the 5 position):

adenosine triphosphate (ATP)

23.6 THERMODYNAMICS OF CHEMICAL REACTIONS IN BIOLOGICAL SYSTEMS

Hydrolysis of ATP to form adenosine diphosphate (ADP) and phosphoric acid yields about 54 kJ mol^{-1} of free energy under typical conditions:

$$\text{ATP} + \text{H}_2\text{O} \longrightarrow \text{ADP} + \text{H}_3\text{PO}_4 \qquad \Delta G = -54 \text{ kJ mol}^{-1}$$

(These symbols denote the typical mixture that includes other protonated species, such as H_2PO_4^- and HPO_4^{2-}, and ion pairs with ions such as Mg^{2+}, in equilibrium at the prevailing concentrations of H^+, Mg^{2+}, etc.)

This reaction is used to drive a variety of chemical reactions that would otherwise be nonspontaneous. They include the synthesis of proteins and nucleic acids as well as the chemical reactions resulting in muscle contraction.

An important energy-releasing reaction that leads to ATP formation is the oxidation of glucose:

$$\text{C}_6\text{H}_{12}\text{O}_6(aq) + 6\text{O}_2(g) \longrightarrow 6\text{CO}_2(g) + 6\text{H}_2\text{O}(l)$$
$$\Delta G = -2820 \text{ kJ mol}^{-1}$$

The mechanism by which this reaction occurs in living cells is quite complex, involving many different steps. One of these steps is the reaction of ADP with phosphoenolpyruvic acid (an intermediate in the oxidation of glucose) to form ATP and pyruvic acid:

$$\text{ADP} + \text{CH}_2\!\!=\!\!\underset{\underset{\text{OPO}_3\text{H}_2}{|}}{\text{C}}\!\!-\!\!\text{COOH} \longrightarrow \text{ATP} + \text{CH}_3\!\!-\!\!\underset{\underset{\text{O}}{\|}}{\text{C}}\!\!-\!\!\text{COOH}$$

Other steps also produce ATP. In all, the oxidation of 1 molecule of glucose forms about 36 to 38 molecules of ATP. Because each of these is capable of generating about -54 kJ mol^{-1} of free energy, about $(-54) \times (37) = -1998$ kJ mol^{-1} of free energy is then available to the system in the form of ATP. By comparing this figure with the -2820 kJ mol^{-1} released in glucose oxidation, we see that approximately $(-1998)/(-2820) \approx 70\%$ of the available free energy has been stored for later use—a remarkably high efficiency.

In conclusion it should be emphasized that, although progress to date in understanding biological chemistry is most impressive, relatively little is yet known about many important aspects of biological systems. Because this is a rapidly developing field, the rate at which ideas become obsolete is rather high. Nevertheless, the basic concepts described here appear to be sufficiently well grounded in experimental observations that the interpretation of all biological phenomena in terms of atomic and molecular processes now seems a realistic objective.

PROBLEMS

23.1 Name each of the following amino acids. Does it have the D- or L-configuration? Is it an essential amino acid?

$$\begin{array}{cc}
\text{COOH} & \text{NH}_2 \\
| & | \\
\text{H}_2\text{N}—\text{C}—\text{H} \quad & \text{H}—\text{C}—\text{CH}_2\text{CH}_2\text{SCH}_3 \\
| & | \\
\text{HCOH} & \text{COOH} \\
| & \\
\text{CH}_3 &
\end{array}$$

23.2 The substance glutathione, commonly found in living cells, has the structure

$$\begin{array}{c}
\text{COOH} \quad\ \text{O} \ \ \text{H} \ \ \text{O} \ \ \text{H} \\
| \qquad\quad \| \ \ \ | \ \ \ \| \ \ \ | \\
\text{H}_2\text{NCHCH}_2\text{CH}_2\text{C—NCHC—NCH}_2\text{COOH} \\
| \\
\text{CH}_2\text{SH}
\end{array}$$

What amino acids are formed when glutathione is hydrolyzed?

23.3 Draw condensed structural formulas for the following peptides:
(a) serylmethionine
(b) methionylserine

23.4 How many isomers are there of a peptide formed from glycine, alanine, and serine? Name each, using brief notation.

23.5 Complete hydrolysis of a certain peptide yields two molecules of glycine and one molecule each of alanine, threonine, lysine, and tryptophan per molecule of peptide. Partial hydrolysis yields the following peptides: Ala·Gly, Try·Thr·Ala, and Lys·Gly·Try. What is the structure of the original peptide?

23.6 If the S—S bonds in insulin (Figure 23.1) are broken by a reagent such as performic acid, how many polypeptides will be formed?

23.7 How many free COOH groups are there in an insulin molecule (Figure 23.1)? How many free COOH groups are there after complete hydrolysis of the insulin?

23.8 Approximately how many amino acids are there per turn of the alpha-helix?

23.9 Draw the Lewis structure of the hemiacetal formed by methanol, CH_3OH, and acetaldehyde, CH_3CHO.

PROBLEMS

23.10 Explain how the structure of starch differs from that of cellulose.

23.11 Lactose ("milk sugar") constitutes about 5% of human milk and cow's milk. Its molecular formula is $C_{12}H_{22}O_{11}$. On hydrolysis glucose and galactose are formed in equal amounts:

$$C_{12}H_{22}O_{11} + H_2O \longrightarrow C_6H_{12}O_6 + C_6H_{12}O_6$$
$$\text{lactose} \qquad\qquad\qquad \text{glucose} \qquad \text{galactose}$$

The structure of lactose is

[structure of lactose showing galactose half and glucose half linked by an oxygen bridge]

How does the structure of galactose differ from that of glucose? Write the structures of glucose and galactose in their open-chain forms.

23.12 Given the following sequence of nitrogen bases in one DNA chain of a double helix, what is the corresponding sequence in the complementary chain? G A C C T A

23.13 Certain characteristics of the composition of most DNA's are summarized by Chargaff's rules, one of which is as follows: The number of moles of adenine is equal to the number of moles of thymine; the number of moles of guanine is equal to the number of moles of cytosine. (These rules were important aids to the discovery of the double helix model.) Explain why DNA's having the double helix structure obey this rule.

23.14 Draw a figure similar to Figure 23.7 for the sequence G A.

23.15 (a) In the DNA base pairs shown in Figure 23.9, classify the different hydrogen bonds as N—H···N, N—H···O, O—H···N, or O—H···O.
(b) In the adenine-thymine base pair, why doesn't the hydrogen bonded to the number C_2 carbon atom of the adenine molecule form a hydrogen bond to the number O_2 oxygen atom of the thymine molecule?

23.16 If the replication shown in Figure 23.10 is extended, what proportion of the third generation of daughter molecules will contain only ^{14}N, and what proportion will contain half ^{14}N–half ^{15}N?

23.17 In the rate curve shown in Figure 23.11, which region is zero-order, and which is first-order, in substrate concentration?

23.18 What proportions of free enzyme, E, and enzyme-substrate complex, ES, would you expect at very high substrate concentrations? On the basis of these proportions, suggest an explanation in physical terms for the fact that the rate of the enzyme-catalyzed reaction is independent of [S] at very high [S].

23.19 A solution contains 0.50 g L^{-1} of an enzyme having a molecular mass of 6.2×10^4, and a substrate at a concentration of $0.50M$. At 70°C the rate of the enzyme-catalyzed reaction is $1.25 \times 10^{-3} M$ min^{-1}. Calculate the turnover number under these conditions.

23.20 Some substances markedly inhibit (decrease the rate of) enzyme-catalyzed reactions. In general, enzyme inhibition is highly specific, just as enzyme catalysis is highly specific. Suggest a mechanism for the action of an enzyme inhibitor.

23.21 Outline a possible sequence by which the energy of sunlight is converted to the energy expended when a person lifts an object.

23.22 (a) Rearrange Eq. 23.4 in the form of a linear equation $y = mx + b$, where $y = (1/\text{rate})$, and $x = (1/[S])$. How are the slope m and the y-intercept b related to k, K, and $[E]^{\text{total}}$?

(b) How could you use this form of the Michaelis-Menten equation (called the Lineweaver-Burk form) to calculate k and K from experimental data for the rate of the reaction at known values of [S] and $[E]^{\text{total}}$?

CHAPTER 24
NUCLEAR CHEMISTRY

24

Up to this point we have largely ignored the nucleus, even though all of the positive electric charge and almost all of the mass of the atom are concentrated there. That we have ignored the nucleus is reasonable from a chemical viewpoint, because the nucleus does not change during a chemical reaction. Nuclear reactions, in which the nucleus does undergo a transformation, are important for a number of reasons, and these reactions are the subject of this chapter. Discovery of the naturally occurring radioactive substances, in which nuclei spontaneously disintegrate (see Section 6.4), was a major event leading to our present understanding of the nature of matter. Artificial means of creating new nuclei led to new chemical elements. The energy changes involved in nuclear reactions are immense, and nuclear energy is a source of energy that dwarfs all others. Finally, the various isotopes created by nuclear reactions have a number of important applications in chemistry.

24.1 NATURAL RADIOACTIVITY

The naturally occurring elements that spontaneously undergo nuclear transformation are found mainly at the end of the periodic table, starting with element 84 (polonium). They emit three types of particles: alpha rays, beta rays, and gamma rays. [These names were taken from the first three letters of the Greek alphabet, α (alpha), β (beta), and γ (gamma), at a time when the nature of the rays was not understood.] **Alpha rays** are ^{4}He nuclei and hence carry a charge of $+2$. (Natural radioactivity is considered to be the source of the large underground helium deposits in Kansas and Texas.) **Beta rays** are very fast-moving electrons. **Gamma rays** are photons of very high energy, and they usually accompany alpha or beta rays, carrying away part of the large amounts of energy evolved in radioactive disintegration.

Two important types of apparatus for the study of radioactive substances are the **cloud chamber** and the **Geiger counter.**

A cloud chamber makes use of the fact that positive and negative ions, which are created by alpha and beta rays, catalyze the condensation of a supercooled gas to a liquid. Thus an alpha or beta ray leaves a trail of tiny droplets, similar to the familiar vapor trails left by high-altitude airplanes. A diagram of a cloud chamber is shown in Figure 24.1. The chamber contains an easily condensable gas such as water vapor or ethyl alcohol vapor. If one lowers the piston, the gas

24.1 NATURAL RADIOACTIVITY

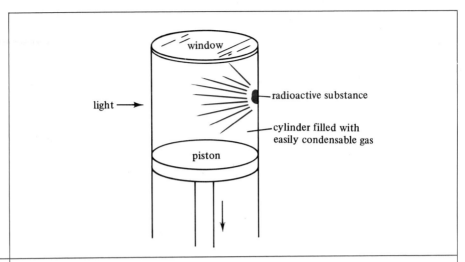

FIGURE 24.1

A cloud chamber. When the piston is suddenly lowered, the gas in the cylinder is cooled below its condensation temperature. Droplets of liquid condense on the ions formed by alpha rays or beta rays, making their tracks visible through the window at the top.

suddenly expands and is cooled to a temperature below its condensation point. Then, if a high-energy charged particle enters the chamber, it collides with the gas molecules along its path, knocking off electrons, and creating numerous positive and negative ions. Liquid drops condense on the ions, marking the path of the high-energy particle. A photograph of cloud chamber tracks made by alpha rays is shown in Figure 24.2.

A diagram of a Geiger counter is shown in Figure 24.3. The positive and negative ions created by a high-energy charged particle are attracted to negative and positive electrodes, respectively, so that when an alpha or a beta ray passes through a Geiger counter, a pulse of electric current flows through the circuit. By counting the current pulses, one may determine the number of alpha and beta rays passing through the counter.

The isotope ^{238}U disintegrates by alpha-particle emission. Loss of an alpha particle decreases the atomic number by 2 (for the two protons of the alpha particle) and decreases the mass number by 4 (for the two protons and two neutrons of the alpha particle). The remaining nucleus has an atomic number of 90 (hence it is an isotope of thorium) and a mass number of 234. Thus the equation for the decomposition of ^{238}U is

$$^{238}_{92}\text{U} \longrightarrow {}^{234}_{90}\text{Th} + {}^{4}_{2}\text{He}$$

(In writing equations for reactions that involve a change in the nucleus, we must

FIGURE 24.2

Photograph of cloud chamber tracks of alpha particles from a deposit of $^{212}_{83}\text{Bi}$ and $^{212}_{84}\text{Po}$. (SOURCE: *E. Rutherford, J. Chadwick, and C. D. Ellis,* Radiations from Radioactive Substances, *Cambridge University Press, London, 1930, Plate II, Figure 3.*)

indicate that electric charge is conserved as in any chemical reaction. In the above equation, charge is conserved as each atom is electrically neutral.)

The energies involved in these reactions are so high that ionization commonly occurs, and one could write the equation as, for example,

$$^{238}_{92}\text{U} \longrightarrow {}^{234}_{90}\text{Th} + {}^{4}_{2}\text{He}^{2+} + 2e^{-}$$

where the alpha particle is ${}^{4}_{2}\text{He}^{2+}$. But the total number of electrons (92) is unchanged in this reaction. (The neutral $_{92}\text{U}$ and $_{90}\text{Th}$ atoms have 92 and 90 electrons, respectively.) Eventually the alpha particle slows down and picks up two electrons to form ${}^{4}_{2}\text{He}$, justifying the simpler equation given before.

The thorium isotope thus produced is itself radioactive, disintegrating by beta emission. This is most easily understood as the transformation of one of the neutrons in the thorium nucleus into a proton and an electron:

$$^{1}_{0}n \longrightarrow {}^{1}_{1}\text{H}^{+} + e^{-}$$

(Note that total electric charge is conserved.) The electron is ejected from the nucleus as a beta ray, while the proton remains behind in the nucleus, so that the

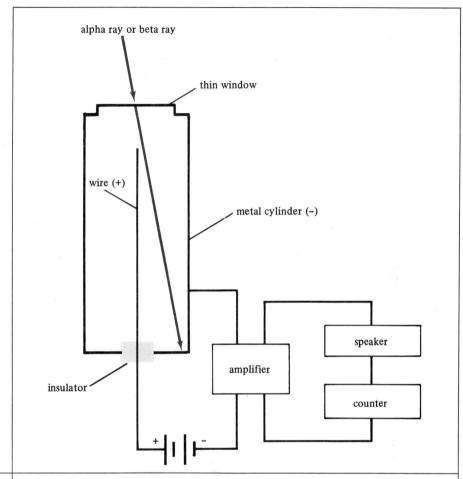

FIGURE 24.3 A Geiger counter. A battery maintains a difference of electric potential between the metal cylinder ($-$) and the central wire ($+$). When an alpha ray or beta ray enters through the thin window of mica, glass, or aluminum at the top, it forms positive and negative ions that move to the cylinder and the wire, respectively, creating a surge of electric current through the circuit. The amplified signal is heard as a clicking noise on a speaker, and it is registered on a counter.

atomic number of the nucleus increases by 1 and its mass number is unchanged. Thus the reaction is

$$^{234}_{90}\text{Th} \longrightarrow {}^{234}_{91}\text{Pa}^+ + e^- \longrightarrow {}^{234}_{91}\text{Pa}$$

where we have indicated that the protactinium ion eventually attracts an electron and becomes a neutral atom.

(One should not assume that beta decay implies that a neutron is "composed" of a proton and an electron. In beta decay one fundamental particle, a neutron, is transformed into two other fundamental particles, a proton and an electron.)

The reactants and products of nuclear disintegrations are often referred to in familiar terms as mothers and daughters, respectively. Thus, ^{234}Th is the daughter, and ^{234}Pa the granddaughter, of ^{238}U. The isotope ^{234}Pa is itself radioactive by beta emission, forming ^{234}U, the great-granddaughter of ^{238}U. Radioactive disintegrations continue until a nonradioactive (stable) isotope is reached. For the uranium series or uranium family, as this one is called, the end product is the lead isotope $^{206}_{82}$Pb. The overall reaction, involving eight alphas and six betas, is

$$^{238}_{92}\text{U} \longrightarrow {}^{206}_{82}\text{Pb} + 8\,{}^{4}_{2}\text{He}$$

[Note that the number of electrons comes out even. The 92 electrons of uranium and the 6 beta emissions, a total of 98 electrons, are just what are needed by the lead atom (82 electrons) and eight helium atoms (16 electrons).]

Because the descendants in a radioactive family are formed by loss of an alpha particle or a beta particle, the mass numbers of all members of a family differ by a multiple of 4. The uranium family ranges from ^{238}U to ^{206}Pb and is known as the $4n + 2$ series, where n is an integer. [For ^{238}U, $4n + 2 = 238$, and $n = (238 - 2)/4 = 59$.] The thorium family is the $4n$ family (its mother is ^{232}Th and the last descendant is ^{208}Pb), and the actinium family is the $4n + 3$ family (its mother is ^{235}U and the last descendant is ^{207}Pb). The remaining possibility, $4n + 1$, begins with the artificially prepared ^{237}Np and ends with ^{209}Bi.

Among the elements of lower atomic number, natural radioactivity is comparatively rare. An interesting example is $^{40}_{19}$K, which undergoes a nuclear transformation called **electron capture.** One of the electrons (usually one in the innermost, or 1s orbital) is absorbed by the nucleus. The process is the reverse of beta decay: A proton in the nucleus combines with the electron to form a neutron,

$$^{1}_{1}\text{H}^+ + e^- \longrightarrow {}^{1}_{0}n$$

Accordingly, the atomic number decreases by 1, the mass number remains the same, and the transformation produces ^{40}Ar, which is a stable isotope:

$$^{40}_{19}\text{K} \longrightarrow {}^{40}_{18}\text{Ar}$$

Following capture of an electron from an inner shell, an electron in one of the outer shells takes its place; this process is accompanied by the emission of light in the X-ray region. Thus naturally occurring potassium, which contains 0.012% ^{40}K, spontaneously emits X rays characteristic of argon.

All the elements whose atomic numbers are smaller than 84 have at least one stable isotope except for technetium ($Z = 43$) and promethium ($Z = 61$). Most have several stable isotopes. (The mass spectrum of mercury, which has seven, is shown in Figure 6.12.) The number of neutrons in the stable isotopes is

24.1 NATURAL RADIOACTIVITY

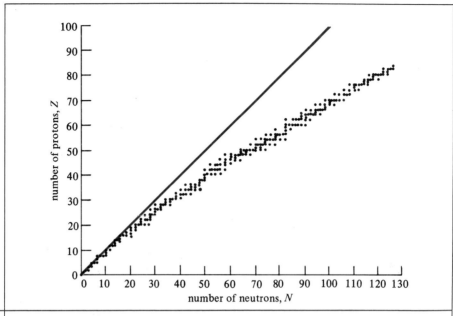

FIGURE 24.4 The known stable nuclei on a plot of Z versus N. The line represents equal numbers of protons and neutrons. (SOURCE: G. Friedlander, J. W. Kennedy, and J. M. Miller, *Nuclear and Radiochemistry*, 2nd ed. Copyright © 1964 by John Wiley and Sons, Inc. Reprinted by permission of John Wiley and Sons, Inc.)

plotted against the number of protons in Figure 24.4. From this diagram one may conclude that, *for the lighter elements, stability depends on having about as many neutrons as protons, whereas, for the heavier elements, stability depends on having somewhat more neutrons than protons.*

The data plotted in Figure 24.4 define a **region of stability.** If the numbers of neutrons and protons of an isotope lie outside this region, then the isotope undergoes radioactive decay in a manner leading toward the stability region. The three modes of disintegration considered so far lead toward the stability region from different directions. Table 24.1 summarizes the effects of radioactive decay of

TABLE 24.1

CHANGES IN NUMBERS OF PROTONS AND NEUTRONS IN THE NUCLEUS FOR VARIOUS TYPES OF RADIOACTIVE DECAY		
Type of radioactivity	Change in no. of protons, Z	Change in no. of neutrons, N
Alpha	-2	-2
Beta	$+1$	-1
Electron capture	-1	$+1$
Positron	-1	$+1$

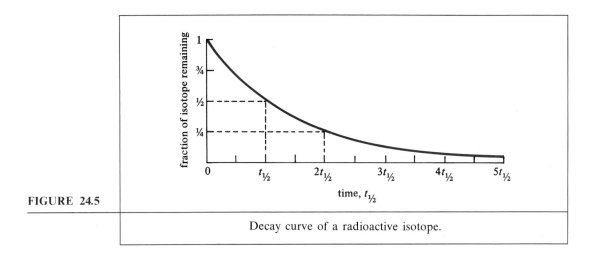

FIGURE 24.5 Decay curve of a radioactive isotope.

various types on numbers of protons and neutrons in the nucleus. In general, when the nucleus is too large, alpha emission occurs, reducing the number of protons and the number of neutrons. When the neutron:proton ratio is too large (below the stability region), beta emission occurs. When the neutron:proton ratio is too small (above the stability region), one finds either electron capture or positron emission (described in the next section).

The rate at which a radioactive substance disintegrates can be studied by using, for example, a Geiger counter to measure the number of disintegrations per second. A typical curve showing how the amount of a radioactive substance changes with time is shown in Figure 24.5. It is invariably found that radioactive transformations are first-order reactions, that is, the number of nuclei disintegrating per second is proportional to the number present (see Section 16.2). For radioactive decay, the rate constant (usually denoted by k for a chemical reaction) is called the **decay constant,** λ (lambda). Thus the rate at which isotope A undergoes radioactive decay, A $\rightarrow$ B, is given by Eq. 16.9a,

$$\text{rate} = -\frac{d[A]}{dt} = \lambda[A] \tag{24.1}$$

where λ has been substituted for k.

It is often more convenient to use the integrated form of the first-order rate law, Eq. 16.9b or 16.9c. From the latter equation we obtain the relation

$$\log \frac{[A]_0}{[A]} = \frac{\lambda t}{2.303} \tag{24.2}$$

where $[A]_0$ is the initial concentration of A, and $[A]$ is the concentration of A at time t. The decay constant is independent of both temperature and the state of chemical combination of the isotope (i.e., the metal and the various compounds all decay at the same rate).

24.1 NATURAL RADIOACTIVITY

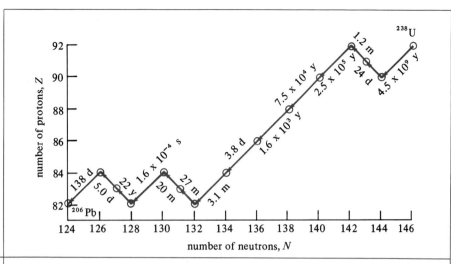

FIGURE 24.6

The uranium family on a plot of Z versus N. Abbreviations are y, year; d, day; m, minute; s, second.

As the rate law is first-order, each radioactive isotope may be characterized by a half-life, $t_{1/2}$, the time required for half of a particular sample to disintegrate. For ^{234}Th, $t_{1/2}$ is 24 d. Thus, after 24 d, only one-half of the nuclei in a sample of ^{234}Th remain undecomposed; after 48 d one-half of that, or one-quarter, are still present; after 72 d, $\frac{1}{2} \times \frac{1}{4} = \frac{1}{8}$ remain; and so on.

The half-life is inversely related to the decay constant: The more rapidly the substance decays, the higher the decay constant, and the shorter the half-life. The relation between half-life and decay constant is given by an equation analogous to Eq. 16.10:

$$t_{1/2} = \frac{0.693}{\lambda} \tag{24.3}$$

Half-lives of the uranium family are shown in Figure 24.6. Each isotope intermediate between ^{238}U and ^{206}Pb is present in a uranium ore at a steady-state concentration, in which it is decaying to its daughter just as fast as it is produced by its mother. The more abundant species are those that decay slowly.

Besides the decay path shown in Figure 24.6, several isotopes of the uranium family exhibit the phenomenon of branching, in which small amounts decay by a different route. For example, 0.04% of ^{214}Bi reaches ^{210}Pb not by the route shown, but by alpha decay to ^{210}Tl, which, in turn, undergoes beta decay to ^{210}Pb. Branching also occurs in the radioactive decay of the $^{40}_{19}$K isotope. Only 11% of the nuclei form $^{40}_{18}$Ar by electron capture (as described above); the other 89% undergo beta decay with the formation of $^{40}_{20}$Ca.

Finally, an important natural source of radiation is cosmic radiation. Primary cosmic rays are positively charged particles (mostly protons) of fantastically high energies that enter the earth's atmosphere from outer space. On colliding with air molecules they undergo a variety of nuclear reactions, creating various secondary cosmic rays, some of which eventually strike the earth. Because of cosmic rays, there is always present a rather high radiation level, called **background radiation** to distinguish it from radiation emanating from the substance under investigation. To gain some idea of the general level of background radiation, consider a Geiger counter 5 cm in diameter and 20 cm long: It registers a background of about 500 counts per minute at sea level, and about 6000 counts per minute at an elevation of 9000 m.

24.2 ARTIFICIAL NUCLEAR REACTIONS AND RADIOACTIVITY

The spontaneous transmutation of one element into another observed in natural radioactivity attracted much attention, for it raised the possibility that one element could be converted to another by artificial means—a prospect that had fascinated mankind since the days of alchemy. Apparently what was needed was a high-speed projectile that could penetrate an atom and alter its nucleus. Ernest Rutherford had used alpha particles from natural radioactivity in his discovery of the nucleus (see Section 6.4); 8 years later, in 1919, he used the same particles to effect the first artificial nuclear reaction:

$$^{14}_{7}N + ^{4}_{2}He \longrightarrow ^{17}_{8}O + ^{1}_{1}H$$

A shorthand notation is $^{14}N(\alpha, p)^{17}O$, where α and p designate the incident alpha particle and the proton produced by the reaction, respectively. In this case the products of the reaction are stable isotopes. A photograph of this reaction occurring in a cloud chamber is shown in Figure 24.7.

The first artificial radioactivity was induced in several elements in 1934 by Frédéric and Irène Joliot-Curie. (Mme. Joliot-Curie was the daughter of Pierre and Marie Curie.) Positrons, which are similar to electrons except for the sign of their electric charge, had been discovered 2 years earlier in cosmic rays by Carl D. Anderson. It was soon discovered that several light elements bombarded with alpha particles produce positrons, and the Joliot-Curies showed that radioactive positron-emitting isotopes were formed. For example, with aluminum the reaction is

$$^{27}_{13}Al + ^{4}_{2}He \longrightarrow ^{30}_{15}P + ^{1}_{0}n$$

The shorthand notation is $^{27}Al(\alpha, n)^{30}P$. The product of the reaction, ^{30}P, emits positrons with a half-life of 2.5 min:

$$^{30}_{15}P \longrightarrow ^{30}_{14}Si^{-} + e^{+}$$

FIGURE 24.7

Photograph of the ^{14}N(α,p)^{17}O reaction in a cloud chamber containing nitrogen gas. The diverging lines are alpha-particle tracks. In the upper center an alpha particle has effected the transmutation of a nitrogen nucleus into an oxygen nucleus. The proton has been ejected toward the left edge of the photograph, while the oxygen nucleus has recoiled a short distance to the upper right. [SOURCE: *P. M. S. Blackett and D. S. Lees*, Proceedings of the Royal Society of London, **A136,** *338 (1932).*]

In this process a proton decays to form a neutron and a positron, and thus the atomic number decreases by one unit and the mass number is constant. [Positron emission is sometimes called beta-plus (β^+) emission, and ordinary beta emission is then designated beta-minus (β^-).]

Positrons are quickly annihilated by reaction with electrons to produce gamma rays:

$$e^+ + e^- \longrightarrow \text{energy}$$

Thus, the overall reaction is similar to electron capture:

$$^{30}_{15}\text{P} \longrightarrow\ ^{30}_{14}\text{Si}$$

Like electron capture, positron emission is a decay mode of isotopes whose neutron:proton ratio is too low. Although no positron-emitting isotopes occur naturally, many have been created by artificial means.

Following this pioneering work on radioactivity, a wide variety of nuclear reactions were discovered and studied. Elaborate machines were designed to accelerate charged particles such as protons, deuterons ($^2_1$H nuclei), alpha particles, and electrons to very high energies. In the cyclotron, invented by Ernest O. Lawrence, the particles are held in a spiral path between huge magnets, while their energy is increased by electric pulses (see Figure 24.8). When the particles reach a sufficiently high energy, they are released to strike the target nuclei.

Neutrons proved to be one of the most interesting particles for nuclear reactions. Unlike protons, deuterons, and alpha particles, they were not repelled by the positive charge of the nucleus, and many nuclei absorbed neutrons to form new radioactive isotopes. As the neutron:proton ratio was then usually too high, the new isotopes were often beta emitters, decaying to form the element of next higher atomic number. This finding was particularly exciting because it offered a method of creating the **transuranium elements**—elements of atomic number greater than that of uranium, the highest then known.

Nuclear reactions, such as the α,n reaction described above, are the only sources of neutrons. In general, α,n reactions produce neutrons of very high energy. An important discovery was made in 1934 by Enrico Fermi and his co-workers, who found that neutrons slowed by passage through hydrogen-containing materials such as paraffin were more readily absorbed by silver nuclei. They applied this technique to uranium, expecting that the uranium nuclei would absorb neutrons and then decay by beta emission to nuclei of atomic number 93, but a bewildering array of radioactive products of various half-lives were formed. Some of them strongly resembled elements of lower atomic number in their chemical properties; for example, one product had properties similar to barium and radium, and it was naturally thought to be a new element of group IIA. In 1938, Otto Hahn and Fritz Strassmann made the startling discovery that it *was* barium. The uranium nucleus had split in two—a process that was named **fission.**

It was soon found that ^{235}U, which constitutes only 0.72% of ordinary uranium, was the isotope that led to the fission reaction. It absorbs a slow neutron to form ^{236}U, which then splits into two large fragments plus several neutrons. The fission process is not unique. It leads to a rather wide variety of products, but usually one nuclear fragment is somewhat larger than the other, and two or three neutrons are formed at the same time. A typical fission reaction is

$$^{235}_{92}\text{U} + ^1_0n \longrightarrow ^{236}_{92}\text{U} \longrightarrow ^{90}_{36}\text{Kr} + ^{143}_{56}\text{Ba} + 3^1_0n$$

Because of the way the neutron:proton ratio required for stability varies with atomic number (Figure 24.4), the neutron:proton ratio of the fission fragments is too large, and all undergo a chain of beta emissions. For example:

$$^{90}_{36}\text{Kr} \xrightarrow[33\text{ s}]{\beta} {}^{90}_{37}\text{Rb} \xrightarrow[2.8\text{ min}]{\beta} {}^{90}_{38}\text{Sr} \xrightarrow[29\text{ y}]{\beta} {}^{90}_{39}\text{Y} \xrightarrow[64\text{ h}]{\beta} {}^{90}_{40}\text{Zr (stable)}$$

The reaction sought by Fermi was discovered in 1940 by Edwin M. McMillan and Philip H. Abelson, who showed that one of the products of the

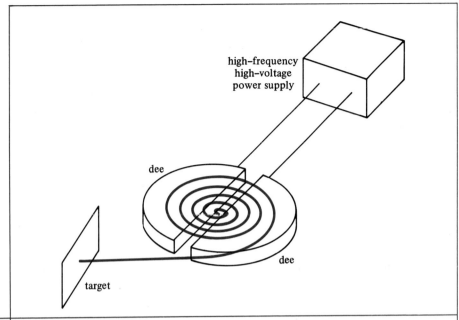

FIGURE 24.8 A cyclotron. Two semicircular hollow electrodes (each shaped like the letter D and called a dee) are placed in a high-vacuum chamber between the pole pieces of a powerful magnet (not shown). Ions are injected at the center. Each time the ions pass from one dee to the other they are accelerated by the potential difference between the dees, which changes sign each half-revolution. In the magnetic field the ions follow curved paths (color), which increase their radius of curvature each time the ions are accelerated. Thus the ions follow a spiral path, eventually emerging with very high energies and striking the target.

uranium–neutron reaction was indeed an isotope of atomic number 93:

$$^{238}_{92}\text{U} + ^{1}_{0}n \longrightarrow {}^{239}_{92}\text{U} \xrightarrow[23.5 \text{ min}]{\beta} {}^{239}_{93}\text{Np}$$

They named the element neptunium after Neptune, the next planet beyond Uranus in the solar system. Neptunium decays by beta emission to element 94, plutonium (named after the next planet, Pluto):

$$^{239}_{93}\text{Np} \xrightarrow[2.3 \text{ d}]{\beta} {}^{239}_{94}\text{Pu}$$

Like ^{235}U, the ^{239}Pu isotope undergoes fission following absorption of a neutron.
  Since then a number of other transuranium elements have been synthesized, mainly by Glenn T. Seaborg and his co-workers. In general these elements have been formed by reactions of uranium or one of the transuranium elements with

neutrons or small nuclei. For example, rutherfordium (atomic number 104) was produced by bombarding a $^{249}_{98}$Cf target with high-energy $^{12}_{6}$C ions. Hahnium (atomic number 105) was produced by a similar reaction with the use of high-energy $^{15}_{7}$N ions.

24.3 NUCLEAR ENERGY

Although we ordinarily think of mass and energy as different properties, Albert Einstein discovered, through his theory of relativity, that they are really different manifestations of the same fundamental property, which might be called "mass-energy" or "energy-mass." He found that the relation between energy and mass is expressed by the equation

$$E = mc^2 \tag{24.4}$$

where E is energy, m is mass, and c is the velocity of light (3.00×10^8 m s^{-1}). If 1.00 kg of mass is converted to energy, there would be 9.00×10^{16} J of energy produced:

$$E = (1.00 \text{ kg})(3.00 \times 10^8 \text{ m s}^{-1})^2$$
$$= 9.00 \times 10^{16} \text{ J}$$

In chemical reactions the energy changes are sufficiently small that the corresponding changes in mass are too small to detect. In nuclear reactions the energy changes are much larger, and one must reorient one's thinking accordingly.

As a typical example, consider alpha emission by ^{238}U. The reaction is

$$^{238}_{92}\text{U} \longrightarrow {}^{234}_{90}\text{Th} + {}^{4}_{2}\text{He}$$

The masses of the isotopes are 238.0508, 234.0436, and 4.0026, respectively. Thus when 1 mol of uranium atoms decays by this process, the original mass of 238.0508 g is replaced by products having a total mass of only $234.0436 + 4.0026 = 238.0462$ g. This is a loss of 0.0046 g, or almost 5 mg, which has been converted to energy. Equation 24.3 may be used to calculate the energy released:

$$E = (4.6 \times 10^{-6} \text{ kg})(3.00 \times 10^8 \text{ m s}^{-1})^2$$
$$= 4.1 \times 10^{11} \text{ J}$$

Thus the energy released is indeed an enormous quantity. For comparison, the energy liberated by the combustion of the same mass (238 g) of carbon is only 7.8×10^6 J, which is smaller by a factor of about 50,000.

The unit of energy usually used in nuclear chemistry to describe events on an atomic scale is the million electron volt, or MeV. Since 1 atomic mass unit (u) has a mass of 1.661×10^{-24} g (see Section 2.2), the loss of 1 u would be accompanied by the production of 1.493×10^{-10} J of energy:

$$E = (1.661 \times 10^{-27} \text{ kg})(2.998 \times 10^8 \text{ m s}^{-1})^2$$
$$= 1.493 \times 10^{-10} \text{ J}$$

Since 1 eV is 1.602×10^{-19} J (see Appendix B), the loss of 1 u results in the production of 932 MeV of energy:

$$[1.493 \times 10^{-10} \text{ J u}^{-1}]/[1.602 \times 10^{-19} \text{ J (eV)}^{-1}] = 932 \text{ MeV u}^{-1}$$

Thus the decay of one ^{238}U atom releases $(0.0046 \text{ u})(932 \text{ MeV u}^{-1}) = 4.3$ MeV of energy. Part of this energy is released as a gamma ray, but most of it appears as the kinetic energy of the alpha particle.

One way to compare the energies of different nuclei is to compute for each nucleus the binding energy per nucleon, where a *nucleon is either a proton or a neutron*. The mass of an isotope with Z protons, Z electrons, and $A - Z = N$ neutrons is compared with the total mass of Z hydrogen atoms and N neutrons. The mass difference, converted to million electron volts, represents the binding energy of that nucleus. Dividing by the number of nucleons gives the binding energy per nucleon.

For example, $^{12}_{6}$C has six electrons and a nucleus with six protons and six neutrons. At least in principle it could be formed by the nuclear reaction

$$6\,^{1}_{1}\text{H} + 6\,^{1}_{0}n \longrightarrow \,^{12}_{6}\text{C} \tag{24.5}$$

The masses of $^{1}_{1}$H, $^{1}_{0}n$, and $^{12}_{6}$C are 1.007825, 1.008665, and 12.000000, respectively. Thus, when one atom of carbon-12 forms by Eq. 24.5, the mass decreases by $(6 \times 1.007825) + (6 \times 1.008665) - 12.000000 = 0.098940$ u. This corresponds to an energy release of $(0.098940 \text{ u})(932 \text{ MeV u}^{-1}) = 92.2$ MeV. The energy released in this reaction is a measure of the extent to which the protons and neutrons are bound to one another in the ^{12}C nucleus. As there are 12 nucleons in ^{12}C, the binding energy per nucleon is 92.2 MeV/12 = 7.68 MeV.

Figure 24.9 shows the binding energy per nucleon for the different isotopes. The highest binding energies are found in the middle range of mass numbers, at iron and nickel. Thus the energy released in the fission process can be understood as an increase in binding energy per nucleon. The fission of a ^{235}U nucleus releases about 200 MeV of energy.

Following the discovery of fission, it was found that two or three neutrons are formed during each fission. Because neutron bombardment initiated the fission reaction, the production of more neutrons in this reaction allows a branched chain reaction that causes an explosive release of energy. (The neutrons from one ^{235}U fission would cause fission in two or three other ^{235}U nuclei. These, in turn, would each release two or three neutrons, causing fission in a total of four to nine other ^{235}U nuclei, etc.) This discovery came during the early stages of World War

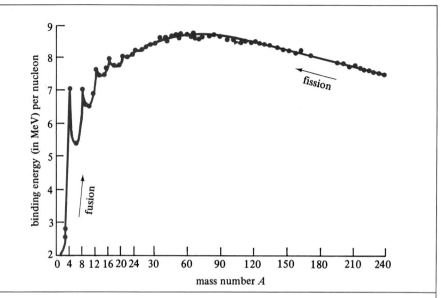

FIGURE 24.9

Average binding energy per nucleon of various isotopes. Note the change in the mass number scale at $A = 30$. (SOURCE: W. J. Moore, *Physical Chemistry*, 3rd ed., © 1962, p. 796. Reprinted by permission of Prentice-Hall, Inc., Englewood Cliffs, New Jersey.)

II. A team of scientists in the United States, led by J. Robert Oppenheimer, developed a bomb based on this principle. Three of these bombs were exploded in 1945: one (a test) at Alamogordo, New Mexico, in July, and two in Japan in August, one at Hiroshima and one at Nagasaki. Since then the controlled release of energy by the fission process has been achieved, and a number of nuclear power plants fueled by uranium are in operation.

From Figure 24.9 it can be inferred that the fusion of small nuclei to form larger ones would be accompanied by an increase in binding energy per nucleon. The isotope ^{4}He has a particularly high binding energy per nucleon compared to the other light isotopes. There is evidence that the fusion of hydrogen to form helium,

$$4\,^1_1\text{H} \longrightarrow \,^4_2\text{He} + 26.7 \text{ MeV}$$

is the energy source of the sun and other stars. The fusion of lithium deuteride,

$$^6_3\text{Li}\,^2_1\text{H} \longrightarrow 2\,^4_2\text{He} + 22.4 \text{ MeV}$$

$$^7_3\text{Li}\,^2_1\text{H} \longrightarrow 2\,^4_2\text{He} + \,^1_0n + 15.1 \text{ MeV}$$

is the basis of a fusion bomb or hydrogen bomb. This type of bomb, which is far more destructive than the fission bombs used at Hiroshima and Nagasaki, has been successfully tested by several nations. The controlled release of fusion energy for the production of electric power presents a number of difficult technical problems, but it offers the possibility of a cheap and plentiful source of useful energy.

24.4 NUCLEAR STRUCTURE

The simplest and one of the oldest theories of the nucleus, called the **liquid-drop model,** conceives of the nucleus as consisting of Z protons and $A - Z$ neutrons tightly packed together in a tiny sphere. The radius of the sphere varies from about 1×10^{-13} cm for the smallest nuclei to about 1×10^{-12} cm for the largest. This model predicts that all nuclei should have about the same density and that the density falls off sharply to zero at a particular distance, the nuclear radius.

Electron-scattering experiments support the view that the density of different nuclei is approximately the same. Such experiments also indicate that in each nucleus a "core" of constant density is surrounded by a "skin," in which the density falls off rather gradually to zero. Figure 24.10 shows the variation of density with distance from the center of the gold nucleus.

A central problem in nuclear theory is to explain how protons can be held together in such a small volume in spite of the coulombic force of repulsion between charged particles of like sign. Both to explain this fact and to explain the enormous energies involved in nuclear reactions, it is clear that there must be very strong forces of attraction between nucleons. These nuclear forces are not apparent between nucleons at distances that are large compared to nuclear dimensions, hence they are called **short-range forces.**

As a result of the long-range coulombic force, a positively charged particle (proton, deuteron, or alpha particle) approaching a nucleus experiences a force of repulsion: Its potential energy increases with decreasing distance, as shown in Figure 24.11. But as the particle nears the nuclear surface the short-range nuclear forces predominate, it is attracted to the nucleus, and its potential energy drops sharply. The potential energy therefore has a shape roughly as indicated in Figure 24.11, called the **square-well** potential energy diagram because of the vertical sides of the potential energy depression. Because the surface of the nucleus is a thick skin rather than a thin surface, the square-well potential is no doubt an oversimplification, but it is nevertheless a useful concept.

A variety of experimental data indicate that, within the potential energy well, the different nucleons occupy various energy levels, similar to the energy levels of electrons in atoms. In the shell model of the nucleus, proposed independently by Maria Goeppert Mayer and J. Hans D. Jensen in 1948, the energy levels are arranged in shells. These nucleon shells are analogous to the electronic shells of atoms, where the numbers of electrons in especially stable filled shells and subshells are those of the noble gases: 2, 10, 18, 36, 54, and 86. For protons the "magic numbers" are 2, 8, 20, 28, 50, and 82, and for neutrons they are the same set of numbers followed by 126. The shell model explains a variety of nuclear

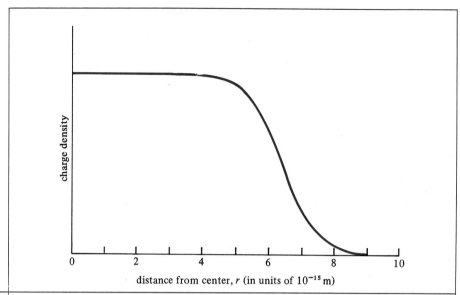

FIGURE 24.10

Charge density in the gold (^{197}Au) nucleus. [SOURCE: R. *Hofstadter,* Annual Review of Nuclear Science, **7,** *296 (1957)*.]

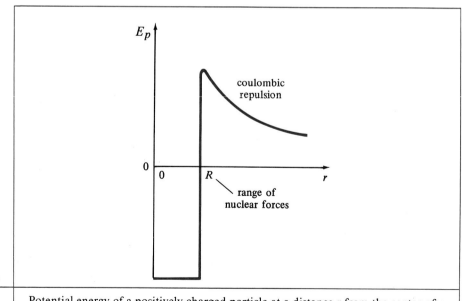

FIGURE 24.11

Potential energy of a positively charged particle at a distance r from the center of a nucleus.

phenomena: the particularly high binding energies of $^4_2$He and $^{16}_8$O shown in Figure 24.9; the very large number of stable isotopes (10) of tin ($Z = 50$); and the stable isotopes of lead ($Z = 82$) at the end of the natural radioactive series discussed in Section 24.1. Additional evidence supporting the shell model is provided by other nuclear properties that have not been considered here, particularly nuclear spin.

24.5 APPLICATIONS OF ISOTOPES

Because of the ease with which very small quantities of a radioactive substance may be detected, radioactive isotopes are widely used as tracers to follow the course of chemical reactions. By using carbon dioxide containing the radioactive isotope ^{14}C, Melvin Calvin and his co-workers were able to study the path followed by carbon in photosynthesis. For some elements there are no isotopes of convenient half-life, and stable isotopes must be used as tracers. The mass spectrograph (see Section 6.6) is used to analyze the products. With ^{18}O (a stable isotope) as a tracer, it was shown that, in ester formation, it is the carbon–oxygen bond in the acid that breaks:

$$\underset{\underset{\text{RC—OH}}{\|}}{\text{O}} + \text{R}'^{18}\text{O—H} \longrightarrow \underset{\underset{\text{RC—}^{18}\text{OR}'}{\|}}{\text{O}} + \text{H}_2\text{O}$$

Although the isotopes of an element are similar in their chemical and physical properties, there are small differences that can be used to separate the isotopes. These differences depend on the ratio of masses and therefore are greatest for hydrogen isotopes. Only the hydrogen isotopes have separate names and symbols: ^{2}H is deuterium (D) and ^{3}H is tritium (T). (Tritium is radioactive by beta decay, with a half-life of 12.3 y.) **Heavy water,** D_2O, has a density of 1.105 at 25°C; its melting and boiling points are 3.8° and 101.4°C, respectively, and its dissociation constant is only 0.16×10^{-14}.

Naturally occurring radioactive isotopes of a particular half-life serve as "clocks" for determining events in the history of the earth over the same time scale. The isotopes of longest half-life, particularly the mothers of the three natural radioactive series, have provided information on the age of the earth and the ages of rocks and meteorites. For example, it is possible to calculate the age of a rock from the ratio of ^{238}U to ^{206}Pb. Independent results are obtained from other isotopic ratios (^{238}U : ^{4}He, ^{235}U : ^{207}Pb, ^{232}Th : ^{208}Pb, ^{87}Rb : ^{87}Sr, and ^{40}K : ^{40}Ar). A correction for lead originally present in the rock may be obtained from the content of ^{204}Pb, which is not a product of the natural radioactive series. In this way the age of the earth since its formation as a planet, and the age of the oldest meteorites (time since solidification), have been found to be about the same: (4.5 to 4.6) $\times 10^9$ y. The oldest terrestrial rocks have an age (time since solidification) of about 3.0×10^9 y. Lunar rocks, obtained during the Apollo program, are somewhat older, ranging from 3.1×10^9 to 4.1×10^9 y; the age of lunar soil is 4.6×10^9 y.

Another interesting application of radioactivity is the determination of the age of carbonaceous materials by measurement of their ^{14}C content. Naturally occurring carbon is about 98.89% ^{12}C and 1.11% ^{13}C. Depending on its source, the carbon may also contain a very small amount of the ^{14}C isotope, which is radioactive:

$$^{14}_{6}C \xrightarrow{\beta} {}^{14}_{7}N \qquad t_{1/2} = 5720 \text{ y}$$

Carbon-14 is apparently formed continuously in the upper atmosphere by neutrons from cosmic rays reacting with nitrogen:

$$^{14}_{7}N + {}^{1}_{0}n \longrightarrow {}^{14}_{6}C + {}^{1}_{1}H$$

After oxidation to carbon dioxide in the atmosphere, the carbon dioxide (containing both radioactive and nonradioactive carbon) is assimilated by living plants. Animals that eat the plants or other animals also incorporate ^{14}C in their tissues. Through a natural balance of uptake of ^{14}C and radioactive decay, living organisms reach a steady-state ratio of ^{14}C to ^{12}C equivalent to that in the atmosphere and corresponding to about 15.3 disintegrations per minute per gram of carbon. When the plant or animal dies, the uptake of carbon ceases and the radioactive carbon content begins to decrease owing to decay. After 5720 years, it would decrease to half its original value so that the disintegration rate would drop to (15.3)/2, or 7.65, disintegrations per minute per gram of carbon. Thus, by measuring the decay rate of the carbon in samples of wood, paper, fibers, fossils, and the like, it is possible to determine when the living organism died. The radiocarbon dating method, developed by Willard F. Libby, has been used to determine the dates of occurrence of a number of historical, archaeological, and geological events, as shown in Figure 24.12.

High-energy radiations from radioactive substances and from particle accelerators have important biological effects. Large doses of radiation can cause serious illness or death. Even low levels of radiation can cause genetic damage. Just what constitutes tolerable levels of radiation is currently a subject of intense controversy. The problem is complicated by the fact that radioactive isotopes are concentrated by biological processes. For example, fission yields substantial amounts of ^{90}Sr both directly and by decay of other radioactive isotopes, as shown in Section 24.2, and it became widely distributed in the environment during the 1950s. Because strontium is similar in its chemical properties to calcium (the preceding group IIA element), when strontium enters the body it is concentrated in high-calcium areas such as bones and teeth. It can then do more damage than one might expect simply on the basis of the average level of ^{90}Sr in the environment. A group of scientists led by Linus Pauling were successful in convincing most of the major governments of the world that for this and other reasons they

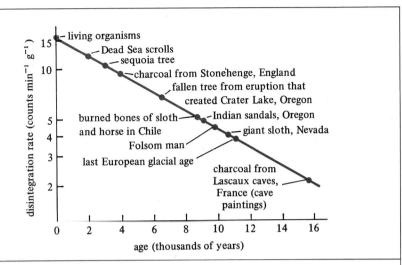

FIGURE 24.12

Carbon-14 dates of selected objects. A plot of carbon-14 disintegration rate (counts per minute per gram of carbon) against the age of the sample in thousands of years. (The vertical axis has a logarithmic scale.)

should stop nuclear bomb tests in which radioactive material was released into the atmosphere.

Under controlled conditions, high-energy radiations have highly beneficial results and they are widely used in medicine. The use of X rays in the diagnosis of tuberculosis, bone fractures, and tooth decay is familiar. Radioactive isotopes are now used to form images of abnormalities such as brain tumors. The destruction of harmful cells as in the treatment of cancer was one of the first applications of radioactive isotopes. In recent years, new methods and new radioactive isotopes such as ^{60}Co have improved the treatment and prognosis. Various physiological processes, such as blood flow, oxygen utilization, and thyroid activity, are studied with the aid of radioactive isotopes that provide a way to monitor the movement of various substances within the body.

PROBLEMS

24.1 A radioactive source emits alpha rays, beta rays, and gamma rays. If these rays pass between two metal plates, one positive and the other negative, how will the rays be deflected?

24.2 The longer alpha-particle tracks shown in Figure 24.2 all have about the same length. One can also see a set of shorter tracks, and they too have about the same length. Suggest an explanation.

24.3 The $4n + 3$ family's mother is $^{235}_{92}$U and the last descendant is $^{207}_{82}$Pb. How many alpha particles are emitted in between? How many beta particles?

24.4 The mother of the thorium (4n) series is $^{232}_{90}$Th. The nucleus undergoes six alpha emissions and four beta emissions, usually in the following order: α, β, β, α, α, α, α, β, β, α.
(a) Identify each descendant by name, atomic number, and mass number.
(b) Depict the thorium series on a graph similar to Figure 24.6.
(c) About one-third of the nuclei have the reverse order for the last two steps (α, β rather than β, α). Show this branching on your diagram.

24.5 (a) From the half-life of ^{90}Sr of 29 y, calculate the decay constant λ (in y^{-1}).
(b) Calculate the time required for 90.0% of a sample of ^{90}Sr to decay.

24.6 The half-life of tritium, $^{3}_{1}$H, is 12 y. What percentage of a sample of tritium will decay in 36 y?

24.7 The half-life of ^{234}Th is 24 d. What length of time is required for 99.9% of a sample of this isotope to decay?

24.8 The isotope $^{15}_{8}$O is radioactive. What type of radioactivity would you expect? What isotope is formed?

24.9 In which region of Figure 24.4 would you expect to find radioactive isotopes having each of the following decay modes: alpha-particle emission, beta-particle emission, positron emission, electron capture?

24.10 What type of particle reacts with a positron to produce gamma rays?

24.11 In each of the following reactions, what is the missing product? (Give the chemical symbol, atomic number, and mass number.)

$$^{237}_{93}\text{Np} \longrightarrow {}^{4}_{2}\text{He} + ?$$
$$^{60}_{27}\text{Co} \longrightarrow e^- + ? \quad \text{[beta decay]}$$
$$^{41}_{20}\text{Ca} \longrightarrow ? \quad \text{[electron capture]}$$
$$^{11}_{6}\text{C} \longrightarrow e^+ + ?$$
$$^{24}_{12}\text{Mg} + {}^{2}_{1}\text{H} \longrightarrow {}^{25}_{12}\text{Mg} + ?$$
$$^{6}_{3}\text{Li} + {}^{1}_{0}n \longrightarrow {}^{4}_{2}\text{He} + ?$$

24.12 In each of the following reactions, what kind of particle is produced?

$$^{9}_{4}\text{Be} + {}^{4}_{2}\text{He} \longrightarrow {}^{12}_{6}\text{C} + ?$$
$$^{226}_{88}\text{Ra} \longrightarrow {}^{222}_{86}\text{Rn} + ?$$
$$^{14}_{6}\text{C} \longrightarrow {}^{14}_{7}\text{N}^+ + ?$$
$$^{18}_{9}\text{F} \longrightarrow {}^{18}_{8}\text{O}^- + ?$$
$$^{3}_{1}\text{H} + {}^{2}_{1}\text{H} \longrightarrow {}^{4}_{2}\text{He} + ?$$

24.13 Deuterons are accelerated in a cyclotron having a 250-kV power supply. How many revolutions of the deuterons are required to raise their energy to 21 MeV?

24.14 The atomic mass of deuterium ($^2_1$H) is 2.01410. Calculate the binding energy of deuterium in MeV per nucleon.

24.15 What type of particle reacts with $^{235}_{92}$U to produce fission?

24.16 In what part of the periodic table does one find isotopes with the smallest binding energy per nucleon? With the largest binding energy per nucleon?

24.17 The fission of one ^{233}U atom yields, on the average, 2.28 neutrons. To control the fission of this isotope for the steady production of energy (as in an "atomic power" plant), it is necessary that x of these neutrons be absorbed by control rods, lost to the surroundings, and so on, leaving 2.28 − x neutrons to carry on the chain reaction. What is x?

24.18 There are only two stable isotopes of potassium, $^{39}_{19}$K and $^{41}_{19}$K, the former being the more abundant (93.1%). Calcium has six stable isotopes, with mass numbers 40, 42, 43, 44, 46, and 48. The isotope $^{40}_{20}$Ca is the most abundant (97.0%). How do these data illustrate the "magic numbers" of the nuclear shell model?

24.19 (a) The half-life of ^{238}U is 4.51 × 10^9 y. In a rock containing N atoms of ^{238}U, how many helium atoms are produced by radioactive decay of ^{238}U to ^{206}Pb in 4.51 × 10^9 y?
(b) A rock is analyzed for its helium and uranium content, and found to contain 1.00 × 10^{-9} mol of ^{238}U atoms, and 2.00 × 10^{-9} mol of ^{4}He atoms. Assuming that all the helium was formed by radioactive decay of ^{238}U to ^{206}Pb and that it all remained trapped in the rock, how many moles of ^{238}U atoms were there when the rock was formed?
(c) Calculate the age of the rock.

24.20 Carbon constitutes about 18% (by mass) of the human body. Calculate the number of radioactive disintegrations of ^{14}C in your body per minute.

24.21 Coal, petroleum, and natural gas contain practically no radioactive carbon, although they are generally believed to have originated from plants and animals that lived millions of years ago, and that presumably contained about as much radioactive carbon as present-day plants and animals. Suggest an explanation.

24.22 (a) In excavating an ancient Egyptian tomb, believed on historical evidence to have been erected about 3000 B.C., an archaeologist finds a well-preserved cigarette butt. To determine between two possible explanations (early use of tobacco by the Egyptians or carelessness by a fellow worker), he determines its radioactivity, and obtains a result of 15.3 disintegrations per minute per gram of carbon. Which explanation is supported by this evidence?

(b) A papyrus scroll from the same site is found to have a radioactivity of 8.24 disintegrations per minute per gram of carbon. Estimate the age of the scroll, using Figure 24.12. Calculate a more accurate value, using Eqs. 24.2 and 24.3 and the half-life of ^{14}C of 5720 y.

APPENDIXES

A THE INTERNATIONAL SYSTEM OF UNITS (SI)

In 1960 the General Conference on Weights and Measures established a practical system of units known as the **International System of Units,** with the international abbreviation **SI** (after "Le Système International d'Unités"). In this system there are seven base units. There are also units derived from these base units, decimal multiples and submultiples, and other units that may be used for a limited time.

The seven base units are shown in Table A-1. The significance and use of these quantities are discussed in the main body of the text at appropriate points.

TABLE A-1

THE SEVEN BASE UNITS

Quantity	Name	Symbol
Length	meter	m
Mass	kilogram	kg
Time	second	s
Thermodynamic temperature	kelvin	K
Amount of substance	mole	mol
Electric current	ampere	A
Luminous intensity	candela	cd

The approved multiples and submultiples of SI units are given in Table A-2.

TABLE A-2

MULTIPLES AND SUBMULTIPLES OF BASE UNITS

Factor	Prefix	Symbol	Factor	Prefix	Symbol
10^1	deka	da	10^{-1}	deci	d
10^2	hecto	h	10^{-2}	centi	c
10^3	kilo	k	10^{-3}	milli	m
10^6	mega	M	10^{-6}	micro	μ
10^9	giga	G	10^{-9}	nano	n
10^{12}	tera	T	10^{-12}	pico	p
10^{15}	peta	P	10^{-15}	femto	f
10^{18}	exa	E	10^{-18}	atto	a

It shows, for instance, that 1 millimeter (mm) is 1×10^{-3} meter (m) and 1 kilometer (km) is 1×10^3 m. Since 1 centimeter (cm) is 1×10^{-2} m, then 1 cubic centimeter (cm^3) is 1×10^{-6} m^3.

There are many units that can be derived from the base units. Some of the common derived units often used in chemistry are shown in Table A-3.

TABLE A-3

SI DERIVED UNITS

Quantity	SI unit Name	Symbol	Expression in base units
Volume	cubic meter	m^3	m^3
Pressure	pascal	Pa	$m^{-1}\,kg\,s^{-2}$
Force	newton	N	$m\,kg\,s^{-2}$
Energy	joule	J	$m^2\,kg\,s^{-2}$
Frequency	hertz	Hz	s^{-1}
Electric charge	coulomb	C	$s\,A$
Electric potential	volt	V	$m^2\,kg\,s^{-3}\,A^{-1}$

Some other units that are outside the International System, but that are currently accepted by the International Committee for Weights and Measures for use with the SI system, are shown in Table A-4.

TABLE A-4

UNITS OUTSIDE THE INTERNATIONAL SYSTEM

Quantity	Name	Symbol	Value in SI units
Length	ångström	Å	$1\,\text{Å} = 0.1\,\text{nm} = 10^{-10}\,\text{m}$
Volume	litre[a]	L	$1\,\text{L} = 1\,\text{dm}^3 = 10^{-3}\,\text{m}^3$
Pressure	standard atmosphere	atm	$1\,\text{atm} = 101{,}325\,\text{Pa}$
Time	minute	min	$1\,\text{min} = 60\,\text{s}$
	hour	h	$1\,\text{h} = 60\,\text{min} = 3600\,\text{s}$
	day	d	$1\,\text{d} = 24\,\text{h} = 86{,}400\,\text{s}$

[a] In the United States, the common spelling is "liter."

Still other units used in chemistry are listed in Appendix B, together with conversion factors to SI units.

Additional derived units and information on SI units are given in the National Bureau of Standards Special Publication 330, 1977 edition, entitled *The International System of Units (SI)*, published by the U.S. Government Printing Office, Washington, D.C. 20402.

B FUNDAMENTAL CONSTANTS AND CONVERSION FACTORS

B-1 FUNDAMENTAL CONSTANTS TO FOUR SIGNIFICANT FIGURES

Atomic mass unit	$u = 1.661 \times 10^{-27}$ kg
Avogadro's number	$N_0 = 6.022 \times 10^{23}$ mol^{-1}
Boltzmann constant	$k_B = 1.381 \times 10^{-23}$ J K^{-1}
Electron charge	$e = 1.602 \times 10^{-19}$ C
Electron mass	$m_e = 9.110 \times 10^{-31}$ kg
Faraday constant	$F = 9.648 \times 10^4$ C mol^{-1}
Gas constant	$R = 0.08206$ L atm K^{-1} mol^{-1}
	$R = 8.314$ J K^{-1} mol^{-1}
Neutron mass	$m_n = 1.675 \times 10^{-27}$ kg
Planck's constant	$h = 6.626 \times 10^{-34}$ J s
Proton mass	$m_p = 1.673 \times 10^{-27}$ kg
Speed of light in vacuum	$c = 2.998 \times 10^8$ m s^{-1}
Standard acceleration of free fall	$g_n = 9.807$ m s^{-2}
Standard molar volume of ideal gas	$V_0 = 22.41$ L mol^{-1}

B-2 CONVERSION FACTORS

Energy	1 calorie (cal) = 4.184 joules (J)
	1 electron volt (eV) = 1.602×10^{-19} J
	1 liter atmosphere (L atm) = 101.3 J
Length	1 inch (in.) = 2.540 centimeters (cm)
	1 mile (mi) = 1.609 kilometers (km)
	1 ångström (Å) = 0.1 nanometer (nm)
	= 10^{-10} meter (m)
	1 bohr (b) = 0.5292×10^{-10} m
Mass	1 avoirdupois pound (lb)
	= 0.4536 kilogram (kg)
Pressure	1 standard atmosphere (atm) =
	1.01325×10^5 pascals (Pa)
	1 atm = 760 torr (Torr)
	= 760 millimeters of mercury (mm of Hg) at 0°C
	= 14.70 pounds per square inch (lb in.$^{-2}$)
	1 Torr = 1 mm of Hg = 133.3 Pa

Standard temperature and pressure (STP)	0°C and 1 atm
Temperature	273.15 kelvins (K) = 0° Celsius (C)
	= 32° Fahrenheit (F)
	$t(°C) = T(K) - 273.15$
Time	1 hour (h) = 60 minutes (min)
	= 3600 seconds (s)
Velocity	1 mile per hour (mi h^{-1}) =
	0.4470 meters per second (m s^{-1})
Volume	1 liter (L) = 10^{-3} cubic meter (m^3)
	1 quart (qt; U.S., liquid) = 0.9463 L
	1 gallon (gal; U.S., liquid) = 3.785 L
	1 quart (Imperial) = 1.1365 L
	1 gallon (Imperial) = 4.546 L

MATHEMATICAL OPERATIONS

C-1 POWERS AND EXPONENTS

The product of repeated multiplication of a number by itself is called a **power.** Thus $4 \times 4 \times 4 = 64$ is the third power of 4. The number of times a factor appears is indicated by a right superscript called an **exponent.** In this example $4 \times 4 \times 4 = 4^3 = 64$, and the exponent is 3. Reciprocals are indicated by negative exponents. Thus

$$\frac{1}{5 \times 5} = \frac{1}{25} = 5^{-2}$$

Powers of 10 are commonly used in scientific work to express very large and very small numbers. For example, $157 = 1.57 \times 10 \times 10 = 1.57 \times 10^2$. (Note that when the decimal point is moved n places to the left, the exponent is n.) Very small numbers are expressed by using negative exponents. Thus

$$0.0049 = \frac{4.9}{10 \times 10 \times 10} = 4.9 \times 10^{-3}$$

(Note that when the decimal point is moved n places to the right, the exponent is $-n$.) Examples showing the reverse process, that is, converting numbers from exponential form to ordinary form, are

$8.3155 \times 10^3 = 8315.5$

$1.95 \times 10^{-5} = 0.0000195$

To add or subtract numbers expressed in exponential notation one must express them to the same power of 10. Thus

$$(1.3 \times 10^2) + (1.51 \times 10^3) = (0.13 \times 10^3) + (1.51 \times 10^3)$$
$$= 1.64 \times 10^3$$

When numbers are *multiplied,* the exponents are *added.* For example,

$(2.1 \times 10^2) \times (4.0 \times 10^3) = 8.4 \times 10^5$

Raising a number to a power is, of course, just a special case of multiplication:

$$(3.0 \times 10^3)^2 = 3.0 \times 10^3 \times 3.0 \times 10^3 = 9.0 \times 10^6$$

When numbers are *divided*, the exponents are *subtracted*. Thus

$$\frac{2.16 \times 10^2}{2.00 \times 10^3} = 1.08 \times 10^{-1}$$

To take the square root of a number, first change the exponent (if necessary) to a multiple of 2, then divide the exponent by 2:

$$\sqrt{3.60 \times 10^{-5}} = \sqrt{36.0 \times 10^{-6}} = 6.00 \times 10^{-3}$$

In taking the cube root, the exponent is first changed to a multiple of 3, and then divided by 3:

$$\sqrt[3]{1.25 \times 10^8} = \sqrt[3]{125 \times 10^6} = 5.00 \times 10^2$$

PROBLEMS

C-1.1 Convert each of the following numbers to exponential form:
(a) 327.5
(b) 60,271
(c) 0.00109
(d) 0.00086

C-1.2 Convert each of the following numbers to ordinary form:
(a) 7.2×10^{-1}
(b) 6.0×10^{-3}
(c) 4.9×10^1
(d) 6.022×10^{23}

C-1.3 Carry out the operations indicated:
(a) $(1.6 \times 10^{-3}) + (8.12 \times 10^{-2}) =$
(b) $(4.02 \times 10^3) - (2.6 \times 10^2) =$
(c) $(7.5 \times 10^{-4}) \times (6.0 \times 10^3) =$
(d) $(8.2 \times 10^5) \div (2.0 \times 10^{-2}) =$
(e) $\sqrt{8.1 \times 10^{-3}} =$

C-2 SIGNIFICANT FIGURES

In reporting scientific data, it is customary to express numbers to include all digits known with a high degree of certainty, plus the first doubtful digit. All such digits are called **significant figures**. For example, if we measure the length of a cylinder with a ruler having a scale divided into centimeters and millimeters, we may find

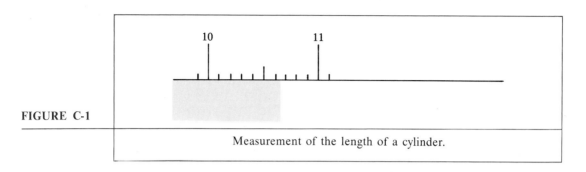

FIGURE C-1

Measurement of the length of a cylinder.

that the length is between 10.6 and 10.7 cm (Figure C-1). Thus there are three known digits (10.6). To obtain the first doubtful digit we estimate the distance between 10.6 and 10.7. In this case we estimate that the end of the cylinder is $\frac{5}{10}$ of the way between 10.6 and 10.7, so that the first doubtful digit is 5. Therefore the length is reported to four significant figures as 10.65 cm.

If three measurements yield 10.65, 10.63, and 10.66, the value 10.65 ± 0.01 is reported. The 10.65 is the average [(10.65 + 10.63 + 10.66)/3], and the ± 0.01 is an estimate of the uncertainty in the last digit. The uncertainty is here taken as the average deviation from the average (without regard to sign). Thus the deviations of the three measurements from the average are 0.00, 0.02, and 0.01, respectively, and their average value is $(0.00 + 0.02 + 0.01)/3 = 0.01$.

Zeros used to indicate the position of the decimal point are not significant figures. For example, 2.568 g = 0.002568 kg has only four significant figures in either form. However, the zeros in the following examples *are* significant figures: 2.207, 20.3, 26.0, 39.100. Expressions such as 25,000 are ambiguous and are usually avoided in scientific work; exponential notation is used instead. Thus the same number is written 2.5×10^4 or 2.50×10^4, and so on, depending on the correct number of significant figures.

When adding or subtracting numbers, one determines which number has the *smaller* number of *digits to the right of the decimal point* and expresses the sum or difference to the *same* number of *digits to the right of the decimal point*. Thus

$$
\begin{array}{ll}
\quad 2.80 & \quad 27.504 \\
+1.0815 & -16. \\
\hline
3.8815 \text{ or } \underline{3.88} & 11.504 \text{ or } \underline{12}
\end{array}
$$

In the first example the first number has two digits to the right of the decimal point, the second number has four, so the sum is expressed to two digits to the right of the decimal point. In the second example the first number has three digits to the right of the decimal point, the second has none, so the difference has no digits to the right of the decimal point. [When dropping extra digits (called **rounding off**), the preceding digit is increased by 1 if the extra digit is 5 or more.]

When multiplying or dividing numbers, one determines which number has the smallest *total number of significant figures*, and expresses the product or quotient to the *same total number of significant figures*. Thus

$$0.163 \times 2.0 = 0.3260 \text{ or } 0.33$$

$$1.80 \div 2.7000 = 0.66666 \text{ or } 0.667$$

In the first example, the total numbers of significant figures are three and two, respectively, and therefore the product is expressed to two significant figures. In the second example, the total numbers of significant figures are three and five, respectively, and therefore the quotient is expressed to three significant figures.

Note that the rule for addition or subtraction is based on the numbers of *digits to the right of the decimal point*, whereas the rule for multiplication or division is based on the total numbers of *significant figures*. (For squares, cubes, square roots, cube roots, and the like, use the rule for multiplication or division.)

The reason for these rules can be seen by examining the effect of changes of 1 in the last significant figure in each example. In the addition example, increasing 2.80 to 2.81 would increase the sum to 3.8915, whereas increasing 1.0815 to 1.0816 would increase the sum to 3.8816. As the second digit to the right of the decimal point is the first doubtful digit, it is the last significant figure. In the multiplication example, increasing 0.163 to 0.164 increases the product to 0.3280, whereas increasing 2.0 to 2.1 increases the product to 0.3423. Therefore the product has only two significant figures.

Exact numbers obtained from definitions or by counting small numbers of objects can be considered to have an infinite number of significant figures. Thus the diameter of a circle is twice the radius, and if the radius is 1.53 cm, then the diameter = (2)(1.53) = 3.06 cm, the 2 being actually 2.0000... by definition. In taking the average of three measurements in the first paragraph of this section, $(10.35 + 10.33 + 10.36)/3 = (31.04)/3 = 10.35$, the 3 being actually 3.0000... by counting.

PROBLEMS

C-2.1 How many significant figures are there in each of the following?
(a) 464.26
(b) 2.00×10^5
(c) 0.04064

C-2.2 The length of an object is given as 40,000 cm. To remove any doubt about the precision to which the length has been measured, use the exponential notation to express the length to
(a) two significant figures
(b) four significant figures

C-2.3 Carry out the operations indicated below and express each answer to the correct number of significant figures.

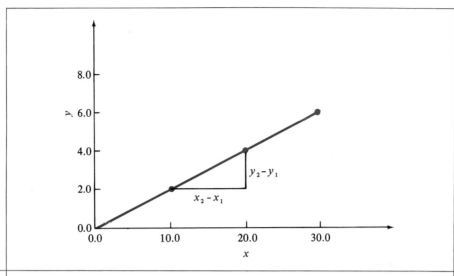

FIGURE C-2

Graph of three points lying on a straight line of positive slope and zero y intercept.

(a) $5.68 + 0.064 + 200.0 =$
(b) $6.8 - 4.447 =$
(c) $(62.78 \times 10^{10}) - (4.834 \times 10^8) =$
(d) $0.030 \times 4.835 \times 10^3 =$
(e) $(4.56 \times 10^4) \div (2.4 \times 10^{-2}) =$
(f) $(6.25 \times 10^3)^2 =$
(g) $\sqrt{1.600 \times 10^3} =$
(h) Convert 8465 s to minutes.

C-3 FUNCTIONS AND GRAPHS

When there is a relationship between two properties of a substance, we say that one property is a function of the other. This functional relationship can be described by tabulating the properties as in the example below, which shows the values of property y for three different values of property x.

When x is	Then y is
10.0	2.0
20.0	4.0
30.0	6.0

A function is most easily visualized by means of a graph such as Figure C-2. Property x is measured in the horizontal direction (called the x coordinate or **abscissa**) and property y in the vertical direction (called the y coordinate or **ordinate**). Each set of values of x and y is represented by a point on the graph. Thus

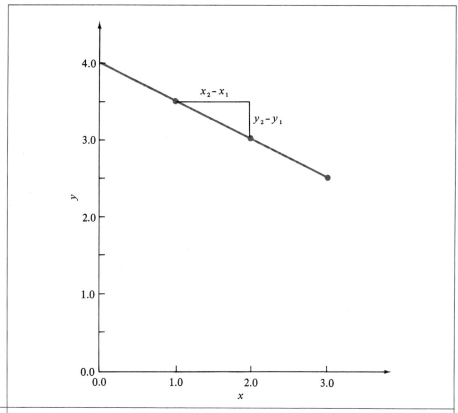

FIGURE C-3

Graph of three points lying on a straight line of negative slope and positive y intercept.

the lowest point has an abscissa of 10.0 and an ordinate of 2.0. The points can be connected by a smooth curve as shown. In this case the points lie on a straight line, and y is said to be a **linear function** of x.

Any linear function can be represented by a **linear equation** of the form $y = mx + b$, where y is the ordinate and x is the abscissa. The point at which the line (if extended sufficiently) strikes the y axis ($x = 0$) is called the y intercept, and in the above equation its value is b. In Figure C-2, $b = 0$. Another linear function is graphed in Figure C-3. In this case the y intercept $b = 4.0$. The quantity m in the equation above, called the **slope,** is a measure of how steeply the line is tilted. If any two points on the line have coordinates x_1, y_1, and x_2, y_2, respectively, the slope $m = (y_2 - y_1)/(x_2 - x_1)$. The quantities $y_2 - y_1$ and $x_2 - x_1$ are illustrated in Figure C-2. It may be seen that the slope is the ratio of the increase in y coordinate to the increase in x coordinate. In this case $m = (4.0 - 2.0)/(20.0 - 10.0) = 0.20$. In Figure C-3 the quantity $y_2 - y_1$ is negative (y decreases with increasing x) and therefore the slope is negative (-0.5). Thus the equations for the functions shown in the two figures are

$$y = 0.20\, x \qquad \text{(Figure C-2)}$$
$$y = -0.5x + 4.0 \qquad \text{(Figure C-3)}$$

PROBLEMS

C-3.1 One point on a straight line has the coordinates $x_1 = 1.0$, $y_1 = -1.0$. Another point on the same line has coordinates $x_2 = 3.0$, $y_2 = 3.0$. Calculate the slope of the line.

C-3.2 Draw a graph of the line described in the preceding problem. What is its y intercept? What is the equation of the line?

C-3.3 Two of the points in Figure C-3 have coordinates $x_1 = 1.0$, $y_1 = 3.5$, and $x_2 = 2.0$, $y_2 = 3.0$. Show that the slope of the line is -0.5.

C-4 SOLVING PROBLEMS

One of the simplest types of problem is that involving a single equation where all but one factor is known. For example, given $PV = nRT$, and $P = 2.0$, $n = 0.40$, $R = 0.082$, $T = 315$, $V = ?$, one can first solve the equation for V by dividing both sides of the equation by P. Then one can obtain the numerical value of V by substituting the known values of P, n, R, and T:

$$V = \frac{nRT}{P} = \frac{(0.40)(0.082)(315)}{2.0} = 5.2$$

Problems in chemistry usually involve properties that have dimensions such as length, which are expressed in various units such as centimeters. By explicitly showing the units for each term throughout the calculation, the correct units for the answer are automatically obtained. Furthermore, the dimensions of the answer provide an extra check on the method used. When units are included, the above problem is stated: Given $PV = nRT$, and $P = 2.0$ atm, $n = 0.40$ mol, $R = 0.082$ L atm mol^{-1} K^{-1}, $T = 315$ K, $V = ?$

$$V = \frac{nRT}{P} = \frac{(0.40\ \text{mol})(0.082\ \text{L atm mol}^{-1}\ \text{K}^{-1})(315\ \text{K})}{2.0\ \text{atm}}$$
$$= 5.2\ \text{L}$$

All the units cancel except liters, and the fact that liters have the dimension of volume is reassuring.

If the different terms had not been expressed in the same units, it would have been necessary to apply the appropriate conversion factors (see Section B-2).

The more interesting problems one encounters are also more complex, and there are no general rules that will automatically lead to solutions. Rather, one can approach problems following certain guiding principles:

1. Try to visualize the situation described in the problem. Draw a sketch showing the essential features of the apparatus, or the paths followed by the particles, or whatever is appropriate to the problem. If some change in conditions occurs, draw and label two sketches, one "before" and one "after." Make sure that you have noted all relevant information. To the maximum extent possible think of the problem as a physical, concrete situation, rather than as a matter of manipulating numbers.
2. Focus attention on the beginning (what is known) and on the end (what is required). Then look for relationships that will provide a means of linking the end to the beginning. Such relationships are generally found in the properties of the system one is dealing with, for example, the way that the volume of a gas depends on its pressure and temperature. The more easily one can conceptualize the appropriate relationships, the more easily one can solve problems depending on these relationships. Conversely, as one sees the fundamental concepts of chemistry illustrated by specific situations, the more fully will one understand these concepts, and that is the main reason why solving problems is an aid to learning chemistry.

C-5 LOGARITHMS

If a positive number N is expressed as a power of 10, or 10^a, the exponent a is called the **common logarithm** (log) of N. Conversely, the number N is called the **antilogarithm** of a. Some examples of numbers and logarithms are

Number	Logarithm
$100 = 10^2$	2
$10 = 10^1$	1
$1 = 10^0$	0
$0.1 = 10^{-1}$	-1

Note that only positive numbers have logarithms.

One of the chief uses of logarithms depends on the addition of exponents when multiplying numbers in exponential form (Section C-1). Thus, if $C = A \times B$, and $a = \log A$, $b = \log B$, and $c = \log C$, then $10^c = 10^a \times 10^b = 10^{a+b}$, and $c = a + b$. Therefore we know that the sum of the logarithms of two numbers is equal to the logarithm of their product: $\log C = \log (A \times B) = \log A + \log B$.

By definition, all integral powers of 10 have logarithms that are integers. Other numbers have logarithms that are not integers. In particular, numbers between 1 and 10 have logarithms between 0 and 1. To obtain the logarithm of a number, it is first written in exponential form as a number between 1 and 10, multiplied by the appropriate power of 10. The logarithm of the number between 1 and 10, written as a decimal fraction, is called the **mantissa**. (Table C-1 gives

C-5 LOGARITHMS

mantissas to four places.) To the mantissa one adds the logarithm of the power of 10, called the **characteristic**. For example,

$$\log (23.4) = \log (2.34 \times 10^1)$$
$$= \log (2.34) + \log (10^1) = 0.3692 + 1 = 1.3692$$

(The mantissa of 2.34 is found in row 2.3 and column 4 of Table C-1.)

To find an antilogarithm the above process is reversed. For example,

$$\text{antilog } (2.8048) = \text{antilog } (0.8048) \times \text{antilog } (2) = 6.38 \times 10^2$$

If a number is less than 1, its characteristic is negative. Thus

$$\log (0.0729) = \log (7.29 \times 10^{-2})$$
$$= \log (7.29) + \log (10^{-2}) = 0.8627 - 2$$
$$= -1.1373$$

$$\text{antilog } (-3.7878) = \text{antilog } (0.2122 - 4)$$
$$= \text{antilog } (0.2122) \times \text{antilog } (-4)$$
$$= 1.63 \times 10^{-4}$$

In the last example note that because the mantissa is always a positive number, it is necessary to write -3.7878 as the sum of 0.2122 and a negative characteristic (-4).

The **natural logarithm** (ln) is similar to the common logarithm except that the base is $e = 2.71828...$ instead of 10. Thus $\ln (e^3) = 3$, and so on. The relationship between natural and common logarithms is

$$\ln N = 2.3026 \log N$$

Unless otherwise specified, the term *logarithm* will denote the common logarithm.

PROBLEMS

C-5.1 What is the logarithm of each of the following numbers?
 (a) 10,000
 (b) 0.0001
 (c) 6.64
 (d) 1.07×10^3
 (e) 0.0347
 (f) 5.25×10^{-4}

C-5.2 What is the antilogarithm of each of the following logarithms?
 (a) 3
 (b) -5
 (c) 0.7505
 (d) 2.9175
 (e) -2.2140
 (f) -4.4034

TABLE C-1

LOGARITHMS

	0	1	2	3	4	5	6	7	8	9
1.0	.0000	.0043	.0086	.0128	.0170	.0212	.0253	.0294	.0334	.0374
1.1	.0414	.0453	.0492	.0531	.0569	.0607	.0645	.0682	.0719	.0755
1.2	.0792	.0828	.0864	.0899	.0934	.0969	.1004	.1038	.1072	.1106
1.3	.1139	.1173	.1206	.1239	.1271	.1303	.1335	.1367	.1399	.1430
1.4	.1461	.1492	.1523	.1553	.1584	.1614	.1644	.1673	.1703	.1732
1.5	.1761	.1790	.1818	.1847	.1875	.1903	.1931	.1959	.1987	.2014
1.6	.2041	.2068	.2095	.2122	.2148	.2175	.2201	.2227	.2253	.2279
1.7	.2304	.2330	.2355	.2380	.2405	.2430	.2455	.2480	.2504	.2529
1.8	.2553	.2577	.2601	.2625	.2648	.2672	.2695	.2718	.2742	.2765
1.9	.2788	.2810	.2833	.2856	.2878	.2900	.2923	.2945	.2967	.2989
2.0	.3010	.3032	.3054	.3075	.3096	.3118	.3139	.3160	.3181	.3201
2.1	.3222	.3243	.3263	.3284	.3304	.3324	.3345	.3365	.3385	.3404
2.2	.3424	.3444	.3464	.3483	.3502	.3522	.3541	.3560	.3579	.3598
2.3	.3617	.3636	.3655	.3674	.3692	.3711	.3729	.3747	.3766	.3784
2.4	.3802	.3820	.3838	.3856	.3874	.3892	.3909	.3927	.3945	.3962
2.5	.3979	.3997	.4014	.4031	.4048	.4065	.4082	.4099	.4116	.4133
2.6	.4150	.4166	.4183	.4200	.4216	.4232	.4249	.4265	.4281	.4298
2.7	.4314	.4330	.4346	.4362	.4378	.4393	.4409	.4425	.4440	.4456
2.8	.4472	.4487	.4502	.4518	.4533	.4548	.4564	.4579	.4594	.4609
2.9	.4624	.4639	.4654	.4669	.4683	.4698	.4713	.4728	.4742	.4757
3.0	.4771	.4786	.4800	.4814	.4829	.4843	.4857	.4871	.4886	.4900
3.1	.4914	.4928	.4942	.4955	.4969	.4983	.4997	.5011	.5024	.5038
3.2	.5051	.5065	.5079	.5092	.5105	.5119	.5132	.5145	.5159	.5172
3.3	.5185	.5198	.5211	.5224	.5237	.5250	.5263	.5276	.5289	.5302
3.4	.5315	.5328	.5340	.5353	.5366	.5378	.5391	.5403	.5416	.5428
3.5	.5441	.5453	.5465	.5478	.5490	.5502	.5514	.5527	.5539	.5551
3.6	.5563	.5575	.5587	.5599	.5611	.5623	.5635	.5647	.5658	.5670
3.7	.5682	.5694	.5705	.5717	.5729	.5740	.5752	.5763	.5775	.5786
3.8	.5798	.5809	.5821	.5832	.5843	.5855	.5866	.5877	.5888	.5899
3.9	.5911	.5922	.5933	.5944	.5955	.5966	.5977	.5988	.5999	.6010
4.0	.6021	.6031	.6042	.6053	.6064	.6075	.6085	.6096	.6107	.6117
4.1	.6128	.6138	.6149	.6160	.6170	.6180	.6191	.6201	.6212	.6222
4.2	.6232	.6243	.6253	.6263	.6274	.6284	.6294	.6304	.6314	.6325
4.3	.6335	.6345	.6355	.6365	.6375	.6385	.6395	.6405	.6415	.6425
4.4	.6435	.6444	.6454	.6464	.6474	.6484	.6493	.6503	.6513	.6522
4.5	.6532	.6542	.6551	.6561	.6571	.6580	.6590	.6599	.6609	.6618
4.6	.6628	.6637	.6646	.6656	.6665	.6675	.6684	.6693	.6702	.6712
4.7	.6721	.6730	.6739	.6749	.6758	.6767	.6776	.6785	.6794	.6803
4.8	.6812	.6821	.6830	.6839	.6848	.6857	.6866	.6875	.6884	.6893
4.9	.6902	.6911	.6920	.6928	.6937	.6946	.6955	.6964	.6972	.6981
5.0	.6990	.6998	.7007	.7016	.7024	.7033	.7042	.7050	.7059	.7067
5.1	.7076	.7084	.7093	.7101	.7110	.7118	.7126	.7135	.7143	.7152
5.2	.7160	.7168	.7177	.7185	.7193	.7202	.7210	.7218	.7226	.7235
5.3	.7243	.7251	.7259	.7267	.7275	.7284	.7292	.7300	.7308	.7316
5.4	.7324	.7332	.7340	.7348	.7356	.7364	.7372	.7380	.7388	.7396
5.5	.7404	.7412	.7419	.7427	.7435	.7443	.7451	.7459	.7466	.7474
5.6	.7482	.7490	.7497	.7505	.7513	.7520	.7528	.7536	.7543	.7551
5.7	.7559	.7566	.7574	.7582	.7589	.7597	.7604	.7612	.7619	.7627
5.8	.7634	.7642	.7649	.7657	.7664	.7672	.7679	.7686	.7694	.7701
5.9	.7709	.7716	.7723	.7731	.7738	.7745	.7752	.7760	.7767	.7774
6.0	.7782	.7789	.7796	.7803	.7810	.7818	.7825	.7832	.7839	.7846
6.1	.7853	.7860	.7868	.7875	.7882	.7889	.7896	.7903	.7910	.7917

C-5 LOGARITHMS

	0	1	2	3	4	5	6	7	8	9
6.2	.7924	.7931	.7938	.7945	.7952	.7959	.7966	.7973	.7980	.7987
6.3	.7993	.8000	.8007	.8014	.8021	.8028	.8035	.8041	.8048	.8055
6.4	.8062	.8069	.8075	.8082	.8089	.8096	.8102	.8109	.8116	.8122
6.5	.8129	.8136	.8142	.8149	.8156	.8162	.8169	.8176	.8182	.8189
6.6	.8195	.8202	.8209	.8215	.8222	.8228	.8235	.8241	.8248	.8254
6.7	.8261	.8267	.8274	.8280	.8287	.8293	.8299	.8306	.8312	.8319
6.8	.8325	.8331	.8338	.8344	.8351	.8357	.8363	.8370	.8376	.8382
6.9	.8388	.8395	.8401	.8407	.8414	.8420	.8426	.8432	.8439	.8445
7.0	.8451	.8457	.8463	.8470	.8476	.8482	.8488	.8494	.8500	.8506
7.1	.8513	.8519	.8525	.8531	.8537	.8543	.8549	.8555	.8561	.8567
7.2	.8573	.8579	.8585	.8591	.8597	.8603	.8609	.8615	.8621	.8627
7.3	.8633	.8639	.8645	.8651	.8657	.8663	.8669	.8675	.8681	.8686
7.4	.8692	.8698	.8704	.8710	.8716	.8722	.8727	.8733	.8739	.8745
7.5	.8751	.8756	.8762	.8768	.8774	.8779	.8785	.8791	.8797	.8802
7.6	.8808	.8814	.8820	.8825	.8831	.8837	.8842	.8848	.8854	.8859
7.7	.8865	.8871	.8876	.8882	.8887	.8893	.8899	.8904	.8910	.8915
7.8	.8921	.8927	.8932	.8938	.8943	.8949	.8954	.8960	.8965	.8971
7.9	.8976	.8982	.8987	.8993	.8998	.9004	.9009	.9015	.9020	.9026
8.0	.9031	.9036	.9042	.9047	.9053	.9058	.9063	.9069	.9074	.9079
8.1	.9085	.9090	.9096	.9101	.9106	.9112	.9117	.9122	.9128	.9133
8.2	.9138	.9143	.9149	.9154	.9159	.9165	.9170	.9175	.9180	.9186
8.3	.9191	.9196	.9201	.9206	.9212	.9217	.9222	.9227	.9232	.9238
8.4	.9243	.9248	.9253	.9258	.9263	.9269	.9274	.9279	.9284	.9289
8.5	.9294	.9299	.9304	.9309	.9315	.9320	.9325	.9330	.9335	.9340
8.6	.9345	.9350	.9355	.9360	.9365	.9370	.9375	.9380	.9385	.9390
8.7	.9395	.9400	.9405	.9410	.9415	.9420	.9425	.9430	.9435	.9440
8.8	.9445	.9450	.9455	.9460	.9465	.9469	.9474	.9479	.9484	.9489
8.9	.9494	.9499	.9504	.9509	.9513	.9518	.9523	.9528	.9533	.9538
9.0	.9542	.9547	.9552	.9557	.9562	.9566	.9571	.9576	.9581	.9586
9.1	.9590	.9595	.9600	.9605	.9609	.9614	.9619	.9624	.9628	.9633
9.2	.9638	.9643	.9647	.9652	.9657	.9661	.9666	.9671	.9675	.9680
9.3	.9685	.9689	.9694	.9699	.9703	.9708	.9713	.9717	.9722	.9727
9.4	.9731	.9736	.9741	.9745	.9750	.9754	.9759	.9763	.9768	.9773
9.5	.9777	.9782	.9786	.9791	.9795	.9800	.9805	.9809	.9814	.9818
9.6	.9823	.9827	.9832	.9836	.9841	.9845	.9850	.9854	.9859	.9863
9.7	.9868	.9872	.9877	.9881	.9886	.9890	.9894	.9899	.9903	.9908
9.8	.9912	.9917	.9921	.9926	.9930	.9934	.9939	.9943	.9948	.9952
9.9	.9956	.9961	.9965	.9969	.9974	.9978	.9983	.9987	.9991	.9996

D CONCEPTS FROM PHYSICS

D-1 VELOCITY AND ACCELERATION

When an object moves from one place to another, its motion is described by its **velocity,** which is a measure of its **speed** (distance traveled in unit time) and the direction of motion. Familiar units are miles per hour or kilometers per hour, as indicated by the speedometer of a motor vehicle. In scientific work the common units are centimeters per second or meters per second. Velocity may be symbolized by an arrow or vector, the direction of the vector indicating the direction of motion, and the length of the vector indicating the speed. The component of velocity in a given direction is the distance traveled in that direction per unit time. Figure D-1 illustrates the components of velocity in two directions.

Change in velocity per unit time is called **acceleration.** Thus if a stationary object acquires a uniform acceleration of 5.0 m s^{-2}, its velocity after 1 s is 5.0 m s^{-1}; after 2 s, 2×5.0 m s^{-1}; and after t s, $t \times 5.0$ m s^{-1}.

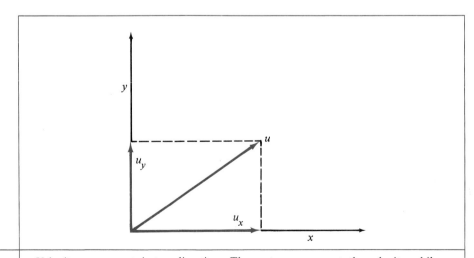

FIGURE D-1 Velocity components in two directions. The vector u represents the velocity, while u_x and u_y represent the components of velocity along the x and y axes, respectively. (The dashed lines are drawn parallel to the x and y axes and through the tip of the vector u.)

D-2 MASS, FORCE, AND WEIGHT

All objects have a tendency to continue in the same state of motion, a tendency called **inertia.** If at rest, they tend to stay at rest; if moving, to continue moving at the same velocity. **Mass** is a quantitative measure of inertia. In chemistry the masses of different materials are usually measured in grams.

To change the state of motion of an object requires a **force,** that is, a push or a pull. An object with a larger mass requires a greater force to slow it down or speed it up. The quantitative relation between force (F) and mass (m) is given by Newton's second law of motion, $F = ma$, where a is the acceleration of the object. In SI units, force is measured in units called **newtons** (symbol N). It takes just 1 s for a force of 1 N to accelerate a stationary object with a mass of 1 kg to a velocity of 1 m s^{-1}.

One of the most important kinds of force is **weight,** the downward force exerted on every object by gravitational attraction, moderated slightly by the effect of the earth's rotation. Whereas the mass of an object is the same at any place, the weight of an object varies with its position on the earth's surface, mainly because the earth is flattened slightly toward the poles and because the effect of the earth's rotation varies with latitude and elevation. For example, a mass of 1 kg has a weight of 9.795 N at San Diego, California, and 9.807 N at Seattle, Washington.

The ratio of weight to mass is the acceleration of free fall (g). As g varies from one place to another, it is convenient to have an arbitrary standard value for various calculations, and for this purpose the standard acceleration of free fall has been defined as 9.80665 m s^{-2}.

To determine the mass of an object, chemists use some form of analytical balance (see Figure D-2). Standard masses of various sizes are added to one pan until their weight balances the weight of the object on the other pan. The effect of gravity is essentially the same at the two pans because they are not very far apart. Therefore if the two arms of the balance have the same length, the mass of the object must be the same as the sum of the standard masses on the other pan. [In very accurate work, a small correction (buoyancy correction) must be made for the difference between the masses of air displaced by the object on one side and by the standard masses on the other.]

This experiment determines the mass of the object in grams, and not its weight. Thus if this experiment is repeated at some other place, the same result will be obtained, even though the object has a different weight. Nevertheless, it is common to refer to this procedure as weighing, to speak of the standard masses as standard weights, and rather generally to use the term weight for mass.

D-3 LINEAR MOMENTUM AND ANGULAR MOMENTUM

If an object of mass m is moving with a velocity u, the product mu is called its **linear momentum** (or simply momentum). A force is required to change the momentum. Because acceleration is change in velocity per unit time, Newton's sec-

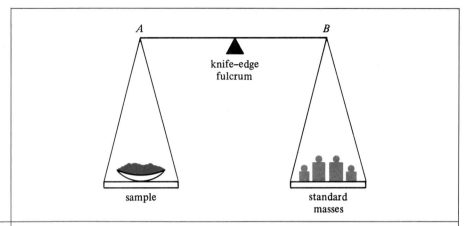

FIGURE D-2

Schematic diagram of an analytical balance. Points A and B from which the sample and the standard masses are suspended are equidistant from the fulcrum.

ond law of motion can be written to show that force equals change of momentum per unit time:

$$F = ma = \frac{m\,\Delta u}{\Delta t} = \frac{\Delta(mu)}{\Delta t}$$

where Δu is the change in velocity, $\Delta(mu)$ is the change in momentum, and Δt is the elapsed time. In any collision between two or more particles, the total momentum of the system is the same after the collision as before—the **law of conservation of momentum.**

If an object of mass m is moving in a circular path of radius r with velocity u, the product mur is called its **angular momentum.** If the path is not circular, the definition is generalized as follows: The object has angular momentum $mu_\perp r$ about a point P, where r is its distance from P, and $u_\perp$ is the component of velocity perpendicular to the line between P and the object.

D-4 ENERGY, WORK, AND HEAT

Energy is a quantity that appears in a variety of forms, such as kinetic energy, potential energy, and thermal energy, and that can be changed from one form to another, subject to the restriction that the total amount of energy is constant. By determining the relationships between the different forms of energy, it has been possible to develop a formula for each. Translational kinetic energy, the energy an object has by virtue of its motion from one place to another, is given by the formula $\frac{1}{2}mu^2$, where m is the mass and u is the velocity of the object. Gravitational potential energy is the energy an object has by virtue of its position relative to the earth. When an object of mass m and weight mg is raised to a height h, the gravitational potential energy increases by an amount *mgh*. Thermal energy is the energy an object has by virtue of its temperature.

The energy of an object can be increased by adding energy as either work or heat. **Work** is energy transferred by means of a force (F) moving through a distance (d); the amount of work is the product Fd. **Heat** is energy transferred from a hotter to a colder object because of the temperature difference.

The SI unit of energy is the joule (symbol J). It is equal to the work done when a force of 1 N moves through a distance of 1 m.

D-5 ELECTRIC CHARGE AND CURRENT

There are two kinds of electricity, labeled *positive* and *negative*. **Electric charge** is a measure of the kind and quantity of electricity. If an object contains a total positive charge $+Q_a$ and a total negative charge $-Q_b$, its net charge is $Q = +Q_a - Q_b$. In SI units, the unit of electric charge is the **coulomb** (symbol C).

The force between two objects having net charges Q_1 and Q_2, a distance r apart, is given by **Coulomb's law:**

$$F = \frac{Q_1 Q_2}{4\pi\epsilon_0 \epsilon r^2}$$

If Q_1 and Q_2 are of like sign, the force is positive, indicating repulsion; if they are of unlike sign, the force is negative, indicating attraction.

The quantity ϵ_0 appearing in Coulomb's law is a constant called the "permittivity of vacuum" or the "permittivity of free space." Its numerical value is 8.854×10^{-12} C^2 N^{-1} m^{-2}. Thus if the electric charges Q_1 and Q_2 are measured in coulombs, and the distance r is in meters, then Coulomb's law gives the force in newtons. (This, in fact, is the sole function of the quantity ϵ_0—it allows the force to be calculated directly in newtons.)

The quantity ϵ appearing in Coulomb's law is a dimensionless number called the **dielectric constant** of the material between Q_1 and Q_2. For a vacuum, the dielectric constant is precisely equal to 1. For any substance, the dielectric constant is a number greater than 1. For example, the dielectric constant of liquid water at 25°C and 1 atm pressure is 78.5.

Coulomb's law may also be written as the potential energy of two charges:

$$E = \frac{Q_1 Q_2}{4\pi\epsilon_0 \epsilon r}$$

As 1 newton–meter (N m) is equal to 1 J, this form of Coulomb's law gives the energy in joules when Q_1 and Q_2 are expressed in coulombs and r is expressed in meters.

Electric current (I) is the flow of electricity in a particular direction, as along a metal wire. The fundamental unit of current is the ampere. A current of 1 A carries a charge of 1 C past a fixed point in 1 s, or, in general,

$$Q \text{ (coulombs)} = I \text{ (amperes)} \times t \text{ (seconds)}$$

Thus 1 C is equal to 1 ampere-second (A s).

A direct current (as from an electric battery) flows in one direction only. An alternating current (as from most home and laboratory electric outlets) oscillates from one direction to the other, usually at a frequency of 60 cycles per second, or 60 hertz (Hz).

D-6 ELECTRIC POTENTIAL, FIELD STRENGTH, AND RESISTANCE

An electric charge has potential energy by virtue of its position relative to other charges. If the potential energy increases when a *positive* charge is moved from point A to point B, then point B is said to be at a higher *electric potential* than A. One is usually concerned with the difference of electric potential between two points. Electric potential difference (symbol V) is measured in volts, defined so that the potential energy changes by 1 J when 1 C of charge moves through a potential difference of 1 V. In general,

$$E \text{ (joules)} = V \text{ (volts)} \times Q \text{ (coulombs)}$$

Thus 1 J is 1 volt-coulomb (V C).

A rather common unit of energy is the electron volt (symbol eV), defined as the change in potential energy when 1 electron moves through a potential difference of 1 V. As the electric charge of an electron is 1.602×10^{-19} C,

$$1 \text{ eV} = (1 \text{ V})(1.602 \times 10^{-19} \text{ C}) = 1.602 \times 10^{-19} \text{ J}$$

A region in which an electric charge is attracted or repelled by other charges is called an **electric field.** The *strength* of the electric field is numerically equal to the force on a charge of unit size. In the region between two parallel metal plates having a potential difference V, the force on a charge Q is QV/d, where d is the distance between the plates, and therefore the electric field strength is V/d.

If two points differing in potential are connected by a metal wire, a current flows through the wire. The ratio of potential difference to current is called the **resistance** (R) of the wire, and it is measured in ohms:

$$R \text{ (ohms)} = \frac{V \text{ (volts)}}{I \text{ (amperes)}}$$

The reciprocal of resistance is called **conductance.**

ELECTRON CONFIGURATIONS AND FIRST IONIZATION POTENTIALS OF ATOMS

Atomic number	Symbol	Electron configuration	First ionization potential (eV)	Atomic number	Symbol	Electron configuration	First ionization potential (eV)	Atomic number	Symbol	Electron configuration	First ionization potential (eV)
1	H	$1s$	13.599	33	As	$-3d^{10}4s^24p^3$	9.81	65	Tb	$-4f^96s^2$	5.98
2	He	$1s^2$	24.588	34	Se	$-3d^{10}4s^24p^4$	9.75	66	Dy	$-4f^{10}6s^2$	5.93
3	Li	$[He]2s$	5.392	35	Br	$-3d^{10}4s^24p^5$	11.814	67	Ho	$-4f^{11}6s^2$	6.02
4	Be	$-2s^2$	9.323	36	Kr	$-3d^{10}4s^24p^6$	14.000	68	Er	$-4f^{12}6s^2$	6.10
5	B	$-2s^22p$	8.298	37	Rb	$-[Kr]5s$	4.177	69	Tm	$-4f^{13}6s^2$	6.18
6	C	$-2s^22p^2$	11.266	38	Sr	$-5s^2$	5.696	70	Yb	$-4f^{14}6s^2$	6.25
7	N	$-2s^22p^3$	14.53	39	Y	$-4d5s^2$	6.379	71	Lu	$-4f^{14}5d6s^2$	6.15
8	O	$-2s^22p^4$	13.618	40	Zr	$-4d^25s^2$	6.837	72	Hf	$-4f^{14}5d^26s^2$	7.0
9	F	$-2s^22p^5$	17.423	41	Nb	$-4d^45s$	6.883	73	Ta	$-4f^{14}5d^36s^2$	7.88
10	Ne	$-2s^22p^6$	21.565	42	Mo	$-4d^55s$	7.10	74	W	$-4f^{14}5d^46s^2$	7.98
11	Na	$[Ne]3s$	5.139	43	Tc	$-4d^55s^2$	7.28	75	Re	$-4f^{14}5d^56s^2$	7.87
12	Mg	$-3s^2$	7.646	44	Ru	$-4d^75s$	7.366	76	Os	$-4f^{14}5d^66s^2$	8.7
13	Al	$-3s^23p$	5.986	45	Rh	$-4d^85s$	7.464	77	Ir	$-4f^{14}5d^76s^2$	9.2
14	Si	$-3s^23p^2$	8.152	46	Pd	$-4d^{10}$	8.33	78	Pt	$-4f^{14}5d^96s$	9.0
15	P	$-3s^23p^3$	10.487	47	Ag	$-4d^{10}5s$	7.576	79	Au	$6s^a$	9.22
16	S	$-3s^23p^4$	10.360	48	Cd	$-4d^{10}5s^2$	8.994	80	Hg	$-6s^2$	10.437
17	Cl	$-3s^23p^5$	12.967	49	In	$-4d^{10}5s^25p$	5.786	81	Tl	$-6s^26p$	6.108
18	Ar	$-3s^23p^6$	15.760	50	Sn	$-4d^{10}5s^25p^2$	7.344	82	Pb	$-6s^26p^2$	7.415
19	K	$[Ar]4s$	4.341	51	Sb	$-4d^{10}5s^25p^3$	8.642	83	Bi	$-6s^26p^3$	7.287
20	Ca	$-4s^2$	6.113	52	Te	$-4d^{10}5s^25p^4$	9.01	84	Po	$-6s^26p^4$	8.43
21	Sc	$-3d4s^2$	6.54	53	I	$-4d^{10}5s^25p^5$	10.451	85	At	$-6s^26p^5$	—
22	Ti	$-3d^24s^2$	6.82	54	Xe	$-4d^{10}5s^25p^6$	12.130	86	Rn	$-6s^26p^6$	10.745
23	V	$-3d^34s^2$	6.74	55	Cs	$[Xe]6s$	3.894	87	Fr	$[Rn]7s$	—
24	Cr	$-3d^54s$	6.765	56	Ba	$-6s^2$	5.212	88	Ra	$-7s^2$	5.277
25	Mn	$-3d^54s^2$	7.435	57	La	$-5d6s^2$	5.61	89	Ac	$-6d7s^2$	6.9
26	Fe	$-3d^64s^2$	7.87	58	Ce	$-4f5d6s^2$	5.65	90	Th	$-6d^27s^2$	—
27	Co	$-3d^74s^2$	7.864	59	Pr	$-4f^36s^2$	5.42	91	Pa	$-5f^26d7s^2$	—
28	Ni	$-3d^84s^2$	7.635	60	Nd	$-4f^46s^2$	5.49	92	U	$-5f^36d7s^2$	6.08
29	Cu	$-3d^{10}4s$	7.726	61	Pm	$-4f^56s^2$	5.55	93	Np	$-5f^46d7s^2$	5.8
30	Zn	$-3d^{10}4s^2$	9.394	62	Sm	$-4f^66s^2$	5.63	94	Pu	$-5f^67s^2$	5.8
31	Ga	$-3d^{10}4s^24p$	5.999	63	Eu	$-4f^76s^2$	5.68	95	Am	$-5f^77s^2$	6.05
32	Ge	$-3d^{10}4s^24p^2$	7.900	64	Gd	$-4f^75d6s^2$	6.16	96	Cm	$-5f^76d7s^2$	—

^{a} Structure of closed shells $[Xe]4f^{14}5d^{10}$.

Source: Adapted from D. E. Gray (Coordinating Editor), *American Institute of Physics Handbook*, 3rd ed., McGraw-Hill, New York, 1972, pp 7–10 and 7–11.

THERMODYNAMIC PROPERTIES OF SUBSTANCES AT 1 atm PRESSURE AND 25°C

Table F-1 lists the standard enthalpies of formation (ΔH_f°) and the standard Gibbs free energies of formation (ΔG_f°) of substances in the specified state at 1 atm pressure and 25°C in kilojoules per mole (kJ mol^{-1}). The entropies (S°) and heat capacities (C_p°) are given in joules per kelvin per mole (J K^{-1} mol^{-1}).

The state of the substance is indicated by one of the following subscripts: (s) for solid, (l) for liquid, and (g) for gas.

TABLE F-1 THERMODYNAMIC PROPERTIES OF SUBSTANCES AT 1 atm PRESSURE AND 25°C

Substance	ΔH_f° kJ mol^{-1}	ΔG_f° kJ mol^{-1}	S° J K^{-1} mol^{-1}	C_p° J K^{-1} mol^{-1}
Hydrogen				
H$_2$(g)	0	0	130.57	28.8
H(g)	218.0	203.3	114.6	20.8
Group IA				
Lithium				
Li(s)	0	0	29.1	24.6
Li(g)	160.7	128.0	138.7	20.8
LiF(s)	−616.9	−588.7	35.7	41.8
LiCl(s)	−408.3	−384.0	59.3	48.0
Li$_2$O(s)	−599	−562	37.9	54.1
LiOH(s)	−485	−439	42.8	49.6
Sodium				
Na(s)	0	0	51.4	28.2
Na(g)	107.8	77.3	153.6	20.8
NaF(s)	−575.4	−545	51.2	46.9
NaCl(s)	−411.1	−384.0	72.1	50.5
NaBr(s)	−361.4	−349.3	86.8	51.4
NaI(s)	−288	−284.6	98.5	52.2
Na$_2$O(s)	−418	−379.1	75.0	69.1
NaOH(s)	−426	−380	64.4	59.5
Na$_2$SO$_4$(s)	−1381.1	−1265.7	157.9	163.1
NaNO$_3$(s)	−466.7	−365.9	116.3	93.1
Na$_2$CO$_3$(s)	−1130.8	−1048.1	138.8	111.0
NaH(s)	−56.4	−33.6	40.0	36.4
Potassium				
K(s)	0	0	64.7	29.5
K(g)	89.2	60.7	160.2	20.8
KCl(s)	−436.7	−408.8	82.6	51.3

THERMODYNAMIC PROPERTIES OF SUBSTANCES AT 1 atm PRESSURE AND 25°C

Substance	ΔH_f° kJ mol^{-1}	ΔG_f° kJ mol^{-1}	S° J K^{-1} mol^{-1}	C_P° J K^{-1} mol^{-1}
KBr(s)	−393.8	−380.4	95.9	52.3
K$_2$O(s)	−363	−322.1	94.1	83.7
KOH(s)	−425	−378.9	78.9	64.9
K$_2$SO$_4$(s)	−1437.4	−1319.4	175.6	131.2
K$_2$CO$_3$(s)	−1150	−1065	155.5	114.4
Group IIA				
Beryllium				
Be(s)	0	0	9.54	16.5
Be(g)	327.4	289.7	136.2	20.8
BeCl$_2$(s)	−490.9	−446.3	82.7	64.9
BeO(s)	−608	−579.0	13.77	25.6
Magnesium				
Mg(s)	0	0	32.7	25.0
Mg(g)	147.6	113.1	148.5	20.8
MgF$_2$(s)	−1124	−1071	57.2	61.6
MgCl$_2$(s)	−641.6	−592.1	89.6	71.4
MgO(s)	−601	−569	26.9	37.1
Mg(OH)$_2$(s)	−925	−833.7	63.2	77.2
MgCO$_3$(s)	−1112	−1028.2	65.9	76.2
Calcium				
Ca(s)	0	0	41.6	26.3
Ca(g)	179.3	145.4	154.8	20.8
CaF$_2$(s)	−1225	−1173.5	68.6	68.6
CaCl$_2$(s)	−796	−748.1	104.6	72.9
CaO(s)	−635.1	−603.5	38.2	42.1
Ca(OH)$_2$(s)	−986.1	−898.5	83.4	87.5
CaSO$_4$(s)	−1432.7	−1320.3	106	100
CaSO$_4 \cdot \frac{1}{2}$H$_2$O(s)	−1575.2	−1435.2	131	120
CaSO$_4 \cdot$ 2H$_2$O(s)	−2021.1	−1795.7	194	186
Ca$_3$(PO$_4$)$_2$(s)	−4126	−3890	241	231.6
CaCO$_3$(s) calcite	−1206.9	−1128.8	93	81.9
Barium				
Ba(s)	0	0	67	26.4
Ba(g)	179.1	147.0	170.1	20.8
BaCl$_2$(s)	−858.6	−810.3	123.7	75.1
BaO(s)	−548	−520.4	72.1	47
Ba(OH)$_2$(s)	−946.3	−859.5	107	101.6
BaSO$_4$(s)	−1465	−1353	132	101.8
BaCO$_3$(s)	−1219	−1139	112	85
Group IIIA				
Boron				
B(s)	0	0	5.87	11.1
B(g)	555.6	511.7	153.3	20.8
BF$_3$(g)	−1135.6	−1119.0	254.2	50.4
BCl$_3$(g)	−403.0	−388.0	290.1	62.4
B$_2$O$_3$(s)	−1272	−1192.8	54.0	62.6
H$_3$BO$_3$(s)	−1094.0	−968.6	88.7	81.3

TABLE F-1 (Continued)

Substance	ΔH_f° kJ mol^{-1}	ΔG_f° kJ mol^{-1}	S° J K^{-1} mol^{-1}	C_P° J K^{-1} mol^{-1}
$BH_3(g)$	107	110.9	187.8	36.2
$B_2H_6(g)$	41	91.8	233.1	58.1
Aluminum				
$Al(s)$	0	0	28.3	24.3
$Al(g)$	326	286	164.4	21.4
$AlF_3(s)$	-1510	-1431	66.5	75.1
$AlCl_3(s)$	-705.6	-630.1	109.3	91.1
$AlBr_3(s)$	-527	-504.4	180.2	100.6
$Al_2O_3(s)$	-1676	-1582.3	50.9	79.0
Group IVA				
Carbon				
$C(s)$ graphite	0	0	5.70	8.5
$C(s)$ diamond	1.90	2.89	2.38	6.3
$C(g)$	715.0	669.6	158.0	20.8
$CCl_4(l)$	-139.3	-68.6	214.4	131.8
$CO(g)$	-110.5	-137.2	197.5	29.1
$CO_2(g)$	-393.5	-394.4	213.7	37.1
$CS_2(l)$	87.9	63.6	151	75.7
$CH_4(g)$	-74.8	-50.8	186.2	35.7
$C_2H_2(g)$	226.7	209.2	200.8	44.1
$C_2H_4(g)$	52.5	68.4	219.2	42.9
$C_2H_6(g)$	-84.7	-32.9	229.5	52.7
$C_3H_8(g)$	-103.8	-23.5	269.9	73.6
$C_6H_6(l)$	49.0	124.5	172.8	—
$CH_3Cl(g)$	-83.7	-60.2	234.3	40.7
$HCN(g)$	131	120	201.8	35.9
$HCHO(g)$	-116	-110	218.7	35.4
$HCOOH(l)$	-409	-346	129.0	99.0
$CH_3OH(l)$	-238.6	-166.2	127	82
$CH_3CHO(g)$	-166.4	-133.7	266	63
$CH_3COOH(l)$	-487	-392	160	123
$C_2H_5OH(l)$	-277.6	-174.8	161	111.5
Silicon				
$Si(s)$	0	0	18.8	20.0
$Si(g)$	450.6	406.2	167.9	22.3
$SiCl_4(g)$	-663	-622.8	330.8	90.3
$SiO_2(s)$ quartz	-911	-856.5	41.5	44.6
$SiH_4(g)$	34	57	204.5	42.8
Tin				
$Sn(s)$ white	0	0	51.5	26.4
$Sn(g)$	301	268	168.4	21.3
$SnCl_2(s)$	-350	—	—	—
$SnCl_4(l)$	-545	-474	259	165
$SnO(s)$	-286	-257	56	44.4
$SnO_2(s)$	-581	-520	52	52.6
Lead				
$Pb(s)$	0	0	64.8	26.8

THERMODYNAMIC PROPERTIES OF SUBSTANCES AT 1 atm PRESSURE AND 25°C

Substance	ΔH_f° kJ mol^{-1}	ΔG_f° kJ mol^{-1}	S° J K^{-1} mol^{-1}	C_P° J K^{-1} mol^{-1}
Pb(g)	195.6	162.6	175.3	20.7
PbCl$_2$(s)	−359	−314.2	136	77.1
PbO(s) red	−219.4	−189.3	66.3	45.8
PbO(s) yellow	−218.1	−188.7	68.7	45.8
PbO$_2$(s)	−274	−215.4	71.8	61.2
PbS(s)	−98	−96.7	91.3	49.4
PbSO$_4$(s)	−918	−811.2	147	104
Group VA				
Nitrogen				
N$_2$(g)	0	0	191.5	29.1
N(g)	472.6	455.5	153.2	20.8
NO(g)	90.3	86.6	210.7	29.8
NO$_2$(g)	33.1	51.2	239.9	37.0
N$_2$O(g)	82.0	104.2	219.9	38.6
N$_2$O$_5$(g)	11.3	117.9	346	96.3
HNO$_3$(g)	−134	−74.0	266.4	53.3
NH$_3$(g)	−45.9	−16.4	192.6	35.6
NH$_4$Cl(s)	−314.6	−203.2	94.9	86.4
(NH$_4$)$_2$SO$_4$(s)	−1179.3	−900.4	220.3	187.5
Phosphorus				
P(s)	0	0	22.8	21.2
P(g)	333.9	292.0	163.1	20.8
PCl$_3$(g)	−271	−257.5	311.6	71.6
PCl$_5$(g)	−342.7	−278.3	364	111.9
P$_4$O$_{10}$(s)	−2940.1	−2675.5	222.8	212
H$_3$PO$_4$(s)	−1266.9	−1112.4	110.5	106.1
PH$_3$(g)	9.2	18.2	210	37.2
Group VIA				
Oxygen				
O$_2$(g)	0	0	205.0	29.4
O(g)	249.2	231.7	160.9	21.9
O$_3$(g)	143	163.2	238.8	39.2
H$_2$O(l)	−285.84	−237.19	69.94	75.3
H$_2$O(g)	−241.8	−228.6	188.7	33.6
Sulfur				
S(s) rhombic	0	0	31.9	22.6
S(g)	277.3	236.9	167.7	23.7
S$_8$(g)	101	49.1	430.2	156.0
SF$_6$(g)	−1220	−1116.5	291.4	97.0
SO$_2$(g)	−296.8	−300.2	248.1	39.9
SO$_3$(g)	−395.8	−371.1	256.7	50.7
H$_2$S(g)	−20.4	−33.3	205.6	34.2
H$_2$SO$_4$(l)	−814.0	−690.1	156.9	139
Group VIIIA				
Fluorine				
F$_2$(g)	0	0	202.7	31.3

TABLE F-1
(Continued)

Substance	ΔH_f° kJ mol^{-1}	ΔG_f° kJ mol^{-1}	S° J K^{-1} mol^{-1}	C_P° J K^{-1} mol^{-1}
F(g)	78.9	61.8	158.6	22.7
F_2O(g)	23	41	247	—
HF(g)	−272.5	−274.6	173.7	29.1
Chlorine				
Cl_2(g)	0	0	223.0	33.9
Cl(g)	121.3	105.3	165.1	21.8
Cl_2O(g)	88	105.0	267.9	47.8
HCl(g)	−92.3	−95.3	186.8	29.1
Bromine				
Br_2(l)	0	0	152.2	75.7
Br(g)	111.9	82.4	174.9	20.8
HBr(g)	−36.4	−53.5	198.6	29.1
Iodine				
I_2(s)	0	0	116.1	54.4
I(g)	106.8	70.2	180.7	20.8
HI(g)	26	1.57	206.5	29.2
ICl(s)	−35.4	−14.1	97.9	55.2
Transition elements				
Titanium				
Ti(s)	0	0	30.6	25.0
Ti(g)	473	428.2	180.2	24.4
$TiCl_4$(l)	−804.2	−737.3	252.4	145.2
TiO_2(s)	−945	−889.5	50.3	55.1
Vanadium				
V(s)	0	0	28.9	24.9
V(g)	515	469.8	182.2	26.0
VCl_2(s)	−452	−406	97	72.2
VCl_3(s)	−573	−502	131	93.2
V_2O_3(s)	−1219	−1139.1	98.1	105.0
V_2O_5(s)	−1551	−1419.4	130.5	130.6
Chromium				
Cr(s)	0	0	23.6	23.4
Cr(g)	397	352.6	174.2	20.8
$CrCl_2$(s)	−395.6	−356.3	115	70.6
$CrCl_3$(s)	−563	−494	126	90.1
Cr_2O_3(s)	−1135	−1053.1	81.2	120.4
Manganese				
Mn(s)	0	0	31.8	26.3
Mn(g)	285.9	243.6	173.6	20.8
$MnCl_2$(s)	−482	−441	117	72.9
MnO(s)	−385	−363	60	43.0
MnO_2(s)	−521	−466	53	54.0
Iron				
Fe(s)	0	0	27.3	25.1
Fe(g)	416.3	370.7	180.4	25.7
$FeCl_2$(s)	−342	−302.4	117.9	76.7
$FeCl_3$(s)	−399.4	−334.0	142.3	96.7

THERMODYNAMIC PROPERTIES OF SUBSTANCES AT 1 atm PRESSURE AND 25°C

Substance	ΔH_f° kJ mol^{-1}	ΔG_f° kJ mol^{-1}	S° J K^{-1} mol^{-1}	C_P° J K^{-1} mol^{-1}
FeO(s)	−272.0	−251.4	60.8	49.9
Fe$_2$O$_3$(s)	−822	−741.0	90	105
Fe$_3$O$_4$(s)	−1117	−1014	146	—
Fe(OH)$_3$(s)	−833	−706	105	102
FeS(s)	−95.1	−97.6	67	55
Cobalt				
Co(s)	0	0	30.0	24.8
Co(g)	424.7	380.1	179.4	23.0
CoCl$_2$(s)	−313	−270.0	109.3	78.5
CoO(s)	−237.7	−214.0	53.0	55.3
Co$_3$O$_4$(s)	−910	−795.0	114.3	123.1
Nickel				
Ni(s)	0	0	30.1	26.0
Ni(g)	425.1	379.8	182.4	23.4
NiCl$_2$(s)	−316	−272	107	78
NiO(s)	−244	−216	38.6	44.4
Copper				
Cu(s)	0	0	33.1	24.4
Cu(g)	339	299.2	166.3	20.8
CuCl(s)	−138	−120.9	87	49
CuCl$_2$(s)	−206	−162	108.1	71.9
Cu$_2$O(s)	−170	−147.7	92.9	63.6
CuO(s)	−155.9	−128.1	42.6	42.1
Cu(OH)$_2$(s)	−450.4	−372.7	108	95.2
Cu$_2$S(s)	−79	−86	121	76.3
CuS(s)	−49	−49	67	47.8
CuSO$_4$(s)	−770.0	−660.9	109.3	98.9
Silver				
Ag(s)	0	0	42.7	25.5
Ag(g)	289.2	250.4	172.9	20.8
AgCl(s)	−127.0	−109.7	96.1	50.8
AgBr(s)	−99.5	−95.9	107	52.4
Ag$_2$O(s)	−30.6	−10.8	121.7	65.6
Ag$_2$S(s)	−31.8	−40.3	146	—
AgNO$_3$(s)	−123.1	−32.2	140.9	93.1
Zinc				
Zn(s)	0	0	41.6	25.1
Zn(g)	130.5	94.9	160.9	20.8
ZnCl$_2$(s)	−416	−369.3	108	77
ZnO(s)	−348.0	−318.2	44	40.3
ZnS(s)	−202.9	−198.3	58	45
ZnSO$_4$(s)	−978.6	−871.6	125	117
Mercury				
Hg(l)	0	0	76.0	28.0
Hg(g)	61.3	31.8	174.9	20.8
Hg$_2$Cl$_2$(s)	−264.9	−210.5	192.5	102.0
HgCl$_2$(s)	−230	−184.1	144.5	73.9
HgO(s)	−90.8	−58.5	70.3	44.1
HgS(s)	−54	−46.2	83	—

G SUPPLEMENTAL DERIVATIONS

G-1 KINETIC THEORY INCLUDING MOLECULAR MOTION IN ANY DIRECTION

At the beginning of Section 4.1, it was assumed that a molecule moves parallel to the length of the box in which it is contained. Although this assumption simplifies the derivation of Eq. 4.4, it is in fact unnecessary if one is willing to accept the complications that follow from considering motion in any direction.

Instead of assuming that the molecule is moving parallel to the length of the box with velocity u, consider that it is moving in *any* direction with velocity u. Then, if the x direction is taken parallel to the length of the box, the *component* of the molecule's velocity in that direction may be represented by u_x. (For a discussion of velocity components, see Appendix D-1.)

One may then substitute u_x for u in the derivation in Section 4.1 all the way to Eq. 4.3a, so that we have the force caused by a single molecule as

$$F = \frac{mu_x^2}{l} \tag{G.1}$$

If now the number of molecules in the box is increased from one to N, then the number of collisions with the wall and hence the force on the wall will also increase by a factor of N:

$$F = N\frac{mu_x^2}{l} \tag{G.2}$$

As there is no reason why the magnitude of one component of the velocity should be larger or smaller than the magnitude of another component, we can say that *on the average* u_x^2 should be the same as either u_y^2 or u_z^2:

$$u_x^2 = u_y^2 = u_z^2 \tag{G.3}$$

(Of course, at any given instant these components will differ, but over a long period of time, as the molecules bounce from one wall to another, the differences will average out.)

One can now make use of the theorem of Pythagoras, which states that, for any right triangle, the square of the hypotenuse is equal to the sum of the squares

G-1 KINETIC THEORY INCLUDING MOLECULAR MOTION IN ANY DIRECTION

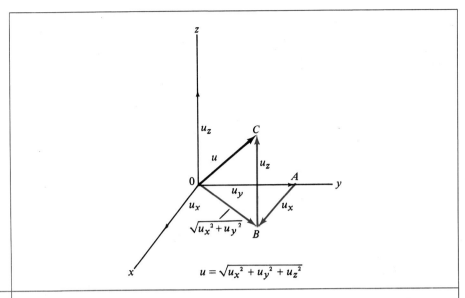

FIGURE G-1 Three perpendicular components of the velocity. The components u_x, u_y, and u_z are taken in the same directions as the coordinate axes x, y, and z, respectively.

of the other two sides. When applied to motion in only two directions, the result is $u^2 = u_x^2 + u_y^2$. For motion in three directions, application of the theorem is more complicated, though the result is rather simple. As shown in Figure G-1, two sides of right triangle OAB are u_x and u_y, and therefore the hypotenuse is $\sqrt{u_x^2 + u_y^2}$. Two sides of right triangle OBC are $\sqrt{u_x^2 + u_y^2}$ and u_z, and therefore its hypotenuse (which is equal to the velocity u) is $\sqrt{u_x^2 + u_y^2 + u_z^2}$. Thus

$$u^2 = u_x^2 + u_y^2 + u_z^2 \tag{G.4}$$

Substituting from Eq. G.3 into this last equation, one obtains

$$u^2 = u_x^2 + u_x^2 + u_x^2 = 3u_x^2$$

or

$$u_x^2 = u^2/3 \tag{G.5}$$

When this result is substituted into Eq. G.2, the result is

$$F = \frac{Nmu^2}{3l}$$

which is identical with Eq. 4.3.

To summarize, Eq. 4.3 has been derived while allowing for molecular motion in any direction.

G-2 RELATION BETWEEN COULOMBIC ENERGY OF ATTRACTION AND INTERNUCLEAR DISTANCE

In Section 9.4, we need to calculate the energy required to dissociate a positive ion and a negative ion in an alkali metal halide molecule. For example, what is the energy required to dissociate an Na$^+$ ion and a Cl$^-$ ion in an NaCl molecule? A simple formula is given there (Eq. 9.5) relating this energy to the internuclear distance R. We shall now derive this formula from Coulomb's law.

The energy of a pair of electric charges Q_1 and Q_2 is given by Coulomb's law (see Appendix D-5):

$$E = \frac{Q_1 Q_2}{4\pi\epsilon_0 \epsilon r}$$

where r is the distance between the charges. When using Coulomb's law, a spherical distribution of charge may be considered to have its charge concentrated at the center of the sphere, and therefore the net charge of an ion may be considered to be concentrated at the nucleus of the ion. Thus the appropriate distance between the charges in Coulomb's law is the internuclear distance, usually denoted by R for a diatomic molecule.

The quantity ϵ_0 is 8.854×10^{-12} C^2 N^{-1} m^{-2}, and the dielectric constant, ϵ, is exactly 1 for a vacuum. For singly charged ions such as Na$^+$ and Cl$^-$, Q_1 is $+e$ and Q_2 is $-e$, where e is the magnitude of the electronic charge, or 1.602×10^{-19} C. The quantity π is of course 3.1416. Substituting these various quantities into Coulomb's law, we obtain

$$E = \frac{-(1.602 \times 10^{-19} \text{ C})^2}{(4)(3.1416)(8.854 \times 10^{-12} \text{ C}^2 \text{ N}^{-1} \text{ m}^{-2})R}$$

$$= -\frac{2.307 \times 10^{-28} \text{ N m}^2}{R}$$

$$= -\frac{2.307 \times 10^{-28} \text{ J m}}{R}$$

Thus E would be in joules and R in meters. (To obtain the last expression we used the fact that 1 N m is equal to 1 J.)

If R is expressed in ångström units, then, as 1 m = 10^{10} Å,

$$E = -\frac{(2.307 \times 10^{-28} \text{ J m})(10^{10} \text{ Å m}^{-1})}{R} = -\frac{2.307 \times 10^{-18} \text{ J Å}}{R}$$

G-2 RELATION BETWEEN COULOMBIC ENERGY OF ATTRACTION AND INTERNUCLEAR DISTANCE

To obtain the energy in electron volts, we use the relation $1 \text{ eV} = 1.602 \times 10^{-19} \text{ J}$, or $1 \text{ J} = 6.242 \times 10^{18} \text{ eV}$.

$$E = -\frac{(2.307 \times 10^{-18} \text{ J Å})(6.242 \times 10^{18} \text{ eV J}^{-1})}{R}$$

$$= -\frac{14.40 \text{ eV Å}}{R}$$

Note that R must be in Å to obtain the energy in electron volts.

Finally, when the pair of ions is separated, Coulomb's law shows that the attraction energy goes to zero as the distance between the ions goes to infinity. Thus, for the process described by Eq. 9.6,

$$\Delta E = E_{\text{final}} - E_{\text{initial}} = 0 - [(-14.40 \text{ eV Å})/R]$$

$$= \frac{14.40 \text{ eV Å}}{R}$$

and this is identical to Eq. 9.5.

To summarize, we have derived Eq. 9.5 from Coulomb's law.

BRIEF ANSWERS TO PROBLEMS

CHAPTER 1

1.2 homogeneous, solution
1.4 (a) homogeneous
   (b) heterogeneous
   (c) homogeneous
1.6 (a) solution containing two or more volatile substances
   (b) heterogeneous mixture of a liquid and a solid
   (c) heterogeneous mixture of two liquids
1.7 eutectic; the solid formed on freezing is heterogeneous
1.9 0.572 g of O (g of N)$^{-1}$ and 2.29 g of O (g of N)$^{-1}$, ratio = 4.00, yes
1.11 7.5×10^{-2} m, 7.5×10^{-5} km, 0.75 dm, 75 mm, 7.5×10^{7} nm
1.12 (a) 1.6 m
   (b) 9.8 in.
   (c) 56.7 kg
   (d) 3.8 L
   (e) 0.57 L
1.13 (a) 35 mi (U.S. gal)$^{-1}$
   (b) 15 km L^{-1}
   (c) 42 mi (Imp. gal)$^{-1}$
1.14 19.3 g cm^{-3}, 1.93×10^{4} kg m^{-3}
1.15 7.68 cm
1.16 (a) 1.84 g of Y (g of X)$^{-1}$ and 3.07 g of Y (g of X)$^{-1}$, ratio = 1.67 = $\frac{5}{3}$
   (b) 5

CHAPTER 2

2.2 15.999
2.3 24.31
2.4 (a) 64.12
   (b) 256.5
   (c) 78.11
   (d) 63.02
   (e) 98.08
   (f) 84.01
2.5 (a) 1.50×10^{23} atoms
   (b) 120 atoms
   (c) 2.41×10^{23} atoms

CHAPTER 3

2.6 (a) 2.33×10^{-23} g
  (b) 0.50 mol of N atoms, 0.25 mol of N_2 molecules
  (c) 3.0×10^{23} atoms, 1.5×10^{23} molecules
2.7 (a) 0.202 mol
  (b) 1.27 mol
  (c) 1.66×10^{-22} mol
  (d) 6.70 mol
  (e) 1.7×10^{-23} mol
2.8 42.10% C, 6.480% H, 51.42% O
2.9 1.599% H, 22.23% N, 76.17% O
2.10 NaCl
2.11 SO_2
2.12 C_5H_7N
2.13 48
2.14 31
2.15 5.7
2.16 (a) $2K + Cl_2 \longrightarrow 2KCl$
  (b) $2SO_2 + O_2 \longrightarrow 2SO_3$
  (c) $2KClO_3 \longrightarrow 2KCl + 3O_2$
  (d) $2C_2H_2 + 5O_2 \longrightarrow 4CO_2 + 2H_2O$
  (e) $CH_4 + 4Cl_2 \longrightarrow CCl_4 + 4HCl$
2.17 (a) 2.27 mol
  (b) 0.999 mol
  (c) 2×10^{-18} mol
  (d) 0.35 mol
2.18 (a) $C_2H_3O_3$
  (b) $C_4H_6O_6$
2.19 (a) $\frac{1}{3}$ mol
  (b) 0.40 mol of CO_2, 0.60 mol of H_2O, 0.20 mol of O_2
  (c) 2.1 g
  (d) 1.2 g
2.20 (a) 0.25 mol
  (b) 0.0299 mol
2.21 2.3 g
2.22 2.6 mol
2.23 0.0126 mol
2.24 $1.2 M$
2.25 $0.576 M$
2.26 $12.0 M$
2.27 (a) 1.67×10^4
  (b) 6.68×10^4

CHAPTER 3

3.1 22.414 L
  (a) 19.6 g

(b) 4×10^{-17} L
(c) 4 L
3.2 10.4 m
3.3 3.0×10^3 Torr
3.4 1.4 atm
3.5 6 atm
3.6 $P_T = 2.54$ atm, $P_{O_2} = 0.0769$ atm, $P_{H_2} = 2.44$ atm, $P_{CO_2} = 0.028$ atm
3.7 278 mL
3.8 $-123\,°C$
3.9 $-23\,°C$
3.10 4.0 L, 2.0 L
3.12 C_2H_4O
3.13 19; 2, 4, 2, 1
3.14 2, 3, 1, 2
3.15 0.692 g L^{-1}
3.16 CH_4
3.17 (a) 77.8
 (b) C_6H_6
3.18 2×10^{13} molecules
3.19 (a) 6.10×10^{20} molecules
 (b) 69 g
3.20 1.327 atm
3.21 34 lb in.$^{-2}$
3.22 (a) 0.921 atm
 (b) 0.480 atm
 (c) 4.31×10^{-3} atm
 (d) 1.4 atm
 (e) 3.1 atm
3.23 16.7%
3.24 1.2×10^3 L
3.25 yes, 22.414 L atm^{-1} mol^{-1}
3.26 10.5 atm, 11.2 atm

CHAPTER 4

4.1 O_2
4.2 (a) same
 (b) $H_2O > N_2 > O_2 > Ar > CO_2$
4.3 (a) 5.66×10^{-21} J molecule^{-1}, 7.73×10^{-21} J molecule^{-1}
 (b) 3.40 kJ mol^{-1}, 4.65 kJ mol^{-1}
4.4 4×10^2 m s^{-1}, 5×10^2 m s^{-1}, u_{rms}
4.5 18.8 K or $-254\,°C$
4.6 (a) 3.04×10^3 mi h^{-1}
 (b) 9.18×10^2 mi h^{-1}

4.7 2
4.8 81 K or $-192\,°C$
4.9 gas B, 1.7
4.10 (a) 1.463
(b) 0.594 m from NH_3 end
4.11 increase T by a factor of 4, increase T by a factor of 2
4.12 (a) same
(b) same
(c) $u_{\text{rms},H_2} = 3.984\ u_{\text{rms},O_2}$
4.13 (a) same
(b) $E_{k,SO_2} = 2E_{k,CH_4}$
(c) $u_{\text{rms},CH_4} = 1.41\ u_{\text{rms},SO_2}$
4.15 2.76×10^{-10} m
4.16 (a) =
(b) <
(c) >
(d) ?
(e) >
4.17 (a) 475 m s^{-1}
(b) u_{rms}
4.18 (a) 2.92×10^{27} collisions s^{-1}
(b) $z_{H_2} = 3.984\ z_{O_2}$

CHAPTER 5

5.1 (a) decrease
(b) increase
(c) increase
(d) decrease
5.2 200 J, -50 J, 500 J, 650 J
5.3 (a) 0.981 J or 0.234 cal
(b) $0.235\,°C$
5.4 $25.1\,°C$
5.5 (a) 1.0×10^3 J
(b) 8.35×10^3 J
(c) 1.15×10^4 J
(d) 5.65×10^4 J
5.6 (a) 36.3 J, 8.31 J, 28.0 J, 36.3 J
(b) 12.5 J
(c) 15.5 J, rotational and vibrational energy
5.7 1.8×10^2 g
5.8 122.5 kJ
5.9 (a) 178.3 kJ, 175.8 kJ
(b) -850 kJ, -850 kJ
(c) -311.42 kJ, -306.46 kJ

(d) -1559.87 kJ, -1553.67 kJ
(e) -631.22 kJ, -623.79 kJ
5.10 (a) -20.2 kJ
(b) -3267.5 kJ
(c) -234.1 kJ
(d) -104.8 kJ
(e) 35 kJ
(f) -433.7 kJ
5.11 (a) -1366.91 kJ, -1364.43 kJ
(b) 43 kJ and 41 kJ
5.12 (a) 6.78 kJ
(b) $C_{12}H_{22}O_{11}(s) + 12O_2(g) = 12CO_2(g) + 11H_2O(l)$
(c) -5.65×10^3 kJ, -5.65×10^3 kJ
(d) -2.22×10^3 kJ mol^{-1}
5.14 Path 1: $A \longrightarrow C \quad q = 11.3$ kJ, $w = -4.54$ kJ, $\Delta E = 6.8$ kJ
$C \longrightarrow B \quad q = -6.81$ kJ, $w = 0$, $\Delta E = -6.81$ kJ
$A \longrightarrow B \quad q = 4.5$ kJ, $w = -4.54$ kJ, $\Delta E = 0$
Path 2: $A \longrightarrow D \quad q = -3.40$ kJ, $w = 0$, $\Delta E = -3.40$ kJ
$D \longrightarrow B \quad q = 5.67$ kJ, $w = -2.27$ kJ, $\Delta E = 3.40$ kJ
$A \longrightarrow B \quad q = 2.27$ kJ, $w = -2.27$ kJ, $\Delta E = 0$
5.15 (a) 50
(b) MO_2
(c) 48.0
5.16 $q = -0.63$ kJ, $w = 0$, $\Delta E = -0.63$ kJ

CHAPTER 6

6.1 (a) $8n$, $8p$, $8e^-$
(b) $10n$, $8p$, $10e^-$
(c) $18n$, $17p$, $18e^-$
(d) $2n$, $1p$, $0e^-$
(e) $138n$, $88p$, $86e^-$
6.2 (a) $^{31}_{15}P$
(b) $^{56}_{26}Fe^{3+}$
(c) $^{79}_{35}Br^-$
6.3 2, 2, 2; 11, 12, 10; $^{48}_{22}Ti$; $^{50}_{22}Ti$; $^{48}_{22}Ti^{4+}$
6.5 (a) increase by a factor of 2
(b) decrease by a factor of 2
(c) increase by a factor of 2
(d) decrease by a factor of 2
6.6 (a) increase by a factor of 2
(b) decrease by a factor of 2
6.7 2, ?, 3, 5
6.9 44.96 (44), 69.72 (68), 72.59 (72)
6.10 4.4×10^{18} Hz

6.11 (a) decrease by a factor of 2
  (b) increase by a factor of $\sqrt{2}$
  (c) increase by a factor of $\sqrt{2}$
  (d) decrease by a factor of $\sqrt{2}$
  (e) no change
6.12 $^6\text{Li}^{2+} < {}^7\text{Li}^{2+} < {}^6\text{Li}^+ = {}^{12}\text{C}^{2+} < {}^7\text{Li}^+ < {}^{12}\text{C}^+$
6.13 (a) 3
  (b) 2
  (c) 4
6.14 24.31
6.15 200.6
6.16 294, $C_{18}H_{30}O_3$
6.17 (a) 5×10^4
  (b) 1×10^{14}
6.18 (a) 11.996708
  (b) 99.972567%
  (c) 1×10^{14}

CHAPTER 7

7.1 0.050 Hz, 40 m, 2.0 m s^{-1}, 2 m
7.2 out of phase
7.4 5.090×10^{14} Hz, 3.373×10^{-19} J
7.5 $(2.114 \text{ to } 6.341) \times 10^{-18}$ J
7.7 (a) infrared region
  (b) 8.66×10^{13} Hz, 2.87×10^{-20} J molecule^{-1}, 17.3 kJ mol^{-1}
7.8 4.774×10^{14} Hz, 3.163×10^{-19} J
7.9 6.69×10^{-34} J s
7.10 1.440×10^{-27} kg m s^{-1}, 1.581×10^3 m s^{-1}
7.11 (a) 6.6×10^{-31} m
  (b) 1.03×10^{-10} m
  (c) 6.0×10^{-12} m
7.12 (a) 1.6×10^{-60} J
  (b) 1.8×10^{-17} J, ratio = 1.1×10^{43}
7.13 maxima at $\frac{1}{8}a, \frac{3}{8}a, \frac{5}{8}a, \frac{7}{8}a$; minima at $0, \frac{1}{4}a, \frac{1}{2}a, \frac{3}{4}a, a$
7.14 (a) threefold degenerate
  (b) $n_x = 2, n_y = 1, n_z = 1$; $n_x = 1, n_y = 2, n_z = 1$; $n_x = 1, n_y = 1, n_z = 2$
  (c) $14\, h^2/8\, m\, a^2$

CHAPTER 8

8.2 6.56465×10^{-7} m or 6564.65 Å
  4.86273×10^{-7} m or 4862.73 Å
  1.21568×10^{-7} m or 1215.68 Å
  1.02573×10^{-7} m or 1025.73 Å

8.3 (a) 0.8500 eV atom^{-1}
   (b) 82.01 kJ mol^{-1}
8.4 2d, 1p
8.5 $n = 4$ or any larger integer; $l = 3$ or any larger integer
8.6 $m_l = \pm 3, \pm 2, \pm 1,$ or 0
8.7 5, 7
8.8 (a) same
   (b) shell farther from nucleus has larger volume
8.9 (a) 0.529 Å
   (b) larger
8.10 shorter, longer
8.11 (a) [He] $2s^2 2p$
    (b) [Ne] $3s^2 3p$
    (c) [Ar] $3d^{10} 4s^2 4p$
8.12 [He] $2s^2 2p^6$ or [Ne]; [Ne] $3s^2 3p^6$ or [Ar]; [Ne]; [Ar]; [Ne]; [Ne]; Ne, O^{2-}, F^-, and Na^+ are isoelectronic; K^+ and Cl^- are isoelectronic
8.13 [Ne] $3p$
8.14 (a) [Ar] $3d^3 4s^2$
    (b) [Ar] $3d^6 4s^2$
    (c) [Ar] $3d^{10} 4s^2 4p^3$
    (d) [Kr] $5s^2$
    (e) [Kr] $4d^{10} 5s^2 5p^5$
8.15 0, 1, 2, 1, 0
8.16 Ne, Ca
8.18 He, Fr

CHAPTER 9

9.1 (a) 1, 3, 2, 1; F^-, N^{3-}, S^{2-}, Br^-
   (b) 1, 2, 3, 1; Li^+, Ca^{2+}, Al^{3+}, K^+
9.2 (a) K_2S
   (b) CaS
   (c) $CaBr_2$
9.3 (a) AlF_3, Al_2O_3, AlN, Mg_3N_2
   (b) +3 for Al, −1 for F, −2 for O, −3 for N, +2 for Mg
   (c) N^{3-}
9.4 (a) +2
   (b) +2
   (c) +3
   (d) −1
   (e) −2
9.5 (a) 6
   (b) 4
9.6 (a) 2.69 Å
   (b) NaCl

9.7 Cl^-, Na^+, K^+
9.8 5.45 eV, 3.67 eV
9.9 (b) 180°, ~109.5°, 180°, ~120°
9.10 (b) ~109.5°, ~120°, ~109.5°, ~120°, 180°
(c) F_2CO and HNCO
9.12 (a) +1
(b) nonplanar
9.13 N—B, N—O, S—O, one P—O (to O not bonded to H); donors are N, S, and P; B and O are acceptors
9.14 120°, 90°, 180°
9.17 3.1×10^2 kJ mol^{-1}
9.18 H—P and C—S; H, Br, O, N
9.19 32%
9.20 BF_3, CO_2
9.22 (a) constructively
(b) constructively in one region, destructively in another
(c) destructively
9.24 (a) σ^*1s
(b) $\pi 2p$
9.26 N_2^+ BeBe$(\pi_x 2p)^2(\pi_y 2p)^2(\sigma 2p)$; O_2^+ BeBe$(\pi_x 2p)^2(\pi_y 2p)^2(\sigma 2p)^2(\pi^*2p)$; 5 in each; N_2; O_2^+
9.27 $3s$, $4s$, $4s$ with $3s$, $3p$, $4p$, $4p$ with $3p$
9.28 (a) [Ne] $3s3p$
(b) [Ne] $3s3p_x3p_y$
(c) [Ne] $3s3p_x3p_y3p_z$
9.29 180°, 120°, 109.5°
9.30 (b) 4
(c) sp
9.31 (a) 2 in each
(b) sp
(c) 180°
(d) −1, +1, 0; 0, +1, −1
9.32 390.8 kJ
9.33 (a) a^3
(b) molecular volume = $2a^3$
(c) 2.82×10^{-10} m or 2.82 Å
9.34 NaCl, 1.16
9.35 (a) 4.70 eV
(b) 3.06 Å
9.39 (c) $2p_y$

CHAPTER 10

10.1 $(C_2H_5)_2O \sim 35°C$, $CCl_4 \sim 75°C$, $H_2O \sim 100°C$, $C_6H_5Br \sim 155°C$
10.2 10°C

10.3 it will boil
10.4 0.85
10.5 $V_l = 18.02$ mL mol^{-1}, $V_s = 19.7$ mL mol^{-1}, $V_g = 3.72 \times 10^3$ L mol^{-1}
10.6 (a) above
   (b) below
10.9 (a) ice
   (b) steam
10.11 H_2O, NH_3
10.12 (a) 2.77×10^{-3}, $0.154m$, $0.149M$
   (b) mole fraction and molality
10.13 4.0 g
10.14 125 mL
10.15 1.04, 0.95, 0.93, 0.92
10.16 62 Torr, 0.76
10.17 (a) 0.84
   (b) 1.3×10^2
10.19 water < acetic acid < benzene < carbon tetrachloride < camphor; water < benzene < acetic acid < carbon tetrachloride
10.20 40
10.21 8.06 mol
10.23 1.9×10^{-4} K, 5.1×10^{-5} K, 2.4 cm
10.24 (a) $0.008m$
   (b) 0.08
10.25 (a) Na^+ and Cl^-, Ba^{2+} and Cl^-
   (b) $H^+ + CH_3CO_2^- = CH_3CO_2H$
   $H^+ + OH^- = H_2O$
   $Ba^{2+} + SO_4^{2-} = BaSO_4(s)$
10.26 $0.32M$ and $0.16M$; $Na_2SO_4(s) = 2Na^+ + SO_4^{2-}$
10.27 (a) $Ba^{2+} + SO_4^{2-} = BaSO_4(s)$
   (b) 0.012 mol
   (c) $0.24M$, $0.08M$, $0.40M$
10.28 $0.060M$
10.29 (a) 0.24
   (b) 45 Torr
10.31 (a) 267
10.32 (a) 241
   (b) 2

CHAPTER 11

11.3 Li (a) and (f); F (b), (c), (d), (e)
11.4 0.729 g
11.5 256 mL
11.6 acids: H_2SO_4, $HClO_4$, CH_3CO_2H, HNO_3, H_3PO_4

bases: $Ca(OH)_2$, $NaOH$, KOH, NH_3
salts: $NaCl$, $NaNO_3$, $Ca(NO_3)_2$, Na_2SO_4, $KClO_3$, $(NH_4)_2SO_4$

11.7 initial: $[H^+] = [Cl^-] = 0.400 M$ and $[Na^+] = [OH^-] = 0.200 M$
final: $[H^+] = 0.040 M$, $[Cl^-] = 0.16 M$, $[Na^+] = 0.12 M$

11.8 4.00×10^{-3} mol, 2.00×10^{-3} mol

11.9 $0.320 M$

11.10 $0.254 M$

11.11 sample from method (1)

11.12 17 g for H_2 and 15 g for He

11.15 (a) $C_2H_4 + H_2 = C_2H_6$
(b) $C_2H_2 + 2H_2 = C_2H_6$

11.16 $C_2H_6 + Cl_2 = C_2H_5Cl + HCl$
$C_2H_5Cl + OH^- = C_2H_5OH + Cl^-$

11.18 same, one

11.20 (a) and (c)

11.21 (a) 5.63 tons
(b) $2NH_3 + H_2SO_4 = (NH_4)_2SO_4$, 21.8 tons

11.23 $P_2O_5 + 3H_2O = 2H_3PO_4$
$SeO_3 + H_2O = H_2SeO_4$
$TeO_3 + 3H_2O = H_6TeO_6$

11.25 SSF_2

11.28 87 J mol^{-1} K^{-1}

11.29 $0.15 M$

11.30 first is cis, second is trans

CHAPTER 12

12.1 (a) increase
(b) decrease
(c) increase

12.2 7.38×10^{-3} mol L^{-1}

12.3 (a) $K = P_{CO_2}^2 / P_{CO}^2 P_{O_2}$
(b) $K = P_{NO_2}^4 P_{H_2O}^6 / P_{NH_3}^4 P_{O_2}^7$
(c) $K = P_{N_2O_4} / P_{NO_2}^2$
(d) $K = P_{H_2O} / P_{H_2} P_{O_2}^{1/2}$

12.5 $[H_2] = [I_2] = 0.0320 M$, $[HI] = 0.236 M$

12.6 $P_{H_2O} = P_{CO} = 0.339$ atm, $P_{H_2} = P_{CO_2} = 0.161$ atm

12.7 (a) $K = P_{CO_2}$
(b) $K = P_{CO_2} / P_{CH_4} P_{O_2}^2$
(c) $K = P_{CO_2}$
(d) $K = P_{HCl}^2 / P_{H_2}$

12.8 2.6×10^2

12.9 (a) shift to right, shift to right, shift to left
(b) no effect, shift to right, shift to left

12.10 (a) larger
  (b) smaller
  (c) larger
12.11 (a) right, left
  (b) left, right
12.12 (a) right
  (b) left
  (c) unaffected
  (d) left
12.13 (a) about $2 \times 10^{-3} M$
  (b) $1.6 \times 10^{-3} M$
12.14 yes
12.15 2.0×10^{-7}, weaker
12.16 1.0×10^{-8}
12.17 5.6×10^{4}
12.18 $[H^+] = [CH_3CO_2^-] = 6.0 \times 10^{-4} M$, $[CH_3CO_2H] = 0.0194 M$
12.19 remain the same
12.20 (a) smaller
  (b) larger
  (c) same
  (d) larger
12.21 (a) smaller
  (b) larger
  (c) same
  (d) larger
12.22 (a) $1.0 \times 10^{-5} M$
  (b) $1 \times 10^{-2} M$
12.23 1.4×10^{-11}
12.24 6.3×10^{-6}
12.25 $1.1 \times 10^{-8} M$
12.26 (a) $1.6 \times 10^{-2} M$
  (b) $1.5 \times 10^{-3} M$, decrease
12.27 precipitation will occur in (a) and (b) but not in (c)
12.28 solubility of AgCl is $1.3 \times 10^{-5} M$
  solubility of La(IO$_3$)$_3$ is $6.9 \times 10^{-4} M$
12.29 strong acid
12.30 $[H_2] = [I_2] = -0.075 M$

CHAPTER 13

13.1 H_3O^+, OH^-
13.2 H_3O^+, H_2O; H_2O, OH^-; OH^-, O^{2-}; NH_4^+, NH_3; NH_3, NH_2^-; $H_3SO_4^+$, H_2SO_4; H_2SO_4, HSO_4^-; HSO_4^-, SO_4^{2-}; H_2S, HS^-; HS^-, S^{2-}
13.3 first reaction: H_3O^+ and CH_3CO_2H are acids, $CH_3CO_2^-$ and H_2O are bases
  second reaction: HCO_3^- and H_2O are acids, OH^- and CO_3^{2-} are bases

13.4 (a) base
    (b) acid
13.5 SO_4^{2-}
13.6 H_2O
13.7 BF_3 and S are acids, F^- and SO_3^{2-} are bases
13.8 →, ←, →, ←
13.9 between H_3O^+/H_2O and $CH_3CO_2H/CH_3CO_2^-$
13.10 (a) $1.0M$, $1.0 \times 10^{-14}M$, 0.00
    (b) $2.0 \times 10^{-4}M$, $5.0 \times 10^{-11}M$, 3.70
    (c) $1.0 \times 10^{-7}M$, $1.0 \times 10^{-7}M$, 7.00
    (d) $5.0 \times 10^{-15}M$, $2.0M$, 14.30
    (e) $6.0M$, $1.7 \times 10^{-15}M$, -0.78
    (f) $1.0 \times 10^{-10}M$, $1.0 \times 10^{-4}M$, 10.00
13.11 (a) $1.0 \times 10^{-7}M$, $1.0 \times 10^{-7}M$
    (b) $2.0 \times 10^{-4}M$, $5.0 \times 10^{-11}M$
    (c) $4.0 \times 10^{-9}M$, $2.5 \times 10^{-6}M$
    (d) $5.6 \times 10^{-5}M$, $1.8 \times 10^{-10}M$
13.12 3.26
13.13 2.4×10^{-6} mol
13.14 →, →, ←, ←
13.15 (a) $[H^+] + [Na^+] = [Cl^-] + [OH^-]$
    (b) $[NH_4^+] + [H^+] = [OH^-] + [Cl^-]$
    (c) $[NH_4^+] + [H^+] = [OH^-] + 2[SO_4^{2-}] + [HSO_4^-]$
13.16 (a) $[NH_3] + [NH_4^+] = 0.100M$
    (b) $[NH_3] + [NH_4^+] = 0.0900M$
13.17 (a) Na^+, $CH_3CO_2^-$, H^+, Cl^-, OH^-, CH_3CO_2H
    (b) $H_2O \rightleftharpoons H^+ + OH^-$
        $CH_3CO_2H \rightleftharpoons CH_3CO_2^- + H^+$
    (c) $[Na^+] = 0.010M$, $[Cl^-] = 0.0020M$
    (d) 4, 4
        (1) $[H^+][OH^-] = 1.0 \times 10^{-14}$
        (2) $\dfrac{[H^+][CH_3CO_2^-]}{[CH_3CO_2H]} = 1.8 \times 10^{-5}$
        (3) $[CH_3CO_2H] + [CH_3CO_2^-] = 0.010M$
        (4) $[H^+] + 0.008M = [OH^-] + [CH_3CO_2^-]$
    (e) $[CH_3CO_2^-] = 0.008M$, $[CH_3CO_2H] = 0.002M$
        $[H^+] = 5 \times 10^{-6}M$, $[OH^-] = 2 \times 10^{-9}M$
13.18 $1.1 \times 10^{-5}M$
13.19 $3.3 \times 10^{-12}M$
13.20 basic, acidic, neutral, neutral, basic
13.21 $CN^- + H_2O \rightleftharpoons HCN + OH^-$, $K_h = 1.6 \times 10^{-5}$
    $[Na^+] = 0.10M$, $[CN^-] = 0.10M$, $[OH^-] = [HCN] = 1.3 \times 10^{-3}M$
    $[H^+] = 7.7 \times 10^{-12}M$
13.23 $1M\,NH_3$, $1M\,CH_3CO_2H$, $1M\,NaCl$

13.24 $1M$ NaZ
13.25 $[Na^+] \approx [CH_3CO_2^-]$, $[CH_3CO_2H] \approx [CH_3CO_2^-]$
13.26 4.74
13.27 $[NH_4^+] = [Cl^-] = 0.020M$, $[NH_3] = 0.010M$
       $[H^+] = 1.1 \times 10^{-9}M$, $[OH^-] = 9.0 \times 10^{-6}M$
13.28 (a) 9.25
       (b) 8.77
       (c) 9.72
13.29 180 mL
13.30 $[NH_4^+] = 0.16M$, $[OH^-] = 9 \times 10^{-6}M$
13.31 1.4×10^{-6}
13.32 red, orange, yellow
13.33 3
13.34 between 6 and 8
13.35 (a) 50.00 mL, 50.01 mL
       (b) 0.01 mL
       (c) 0.02%
13.36 (b) methyl red
13.37 $[Na^+] \approx [CH_3CO_2^-]$, $[OH^-] \approx [CH_3CO_2H]$
13.38 8.38
13.39 (a) $HCO_3^- + OH^- = CO_3^{2-} + H_2O$
       (c) $CO_3^{2-} + H_2CO_3 = 2HCO_3^-$
       (d) $HCO_3^- + H^+ = H_2CO_3$
       (f) $H_2CO_3 + OH^- = H_2O + HCO_3^-$
13.40 (a) $1.0M$ NaHCO$_3$
       (b) $0.034M$ H$_2$CO$_3$
       (c) $1.0M$ Na$_2$CO$_3$
13.41 increase
13.42 $4.0 \times 10^{-8}M$, 8.5% H$_2$CO$_3$, 91.4% HCO$_3^-$, 0.11% CO$_3^{2-}$
13.43 (a) $H_2PO_4^-$
       (b) $H_2PO_4^-$
       (c) $H_2PO_4^-$
       (d) HPO_4^{2-}
       (e) HPO_4^{2-}
       (f) PO_4^{3-}
13.44 $[Na^+] = 0.100M$, $[H_2CO_3] = [CO_3^{2-}] = 9.8 \times 10^{-4}M$, $[HCO_3^-] = 0.098M$,
       $[OH^-] = 2.2 \times 10^{-6}M$
13.45 $1M$ Na$_2$CO$_3$, $0.034M$ H$_2$CO$_3$, $0.5M$ NaHCO$_3$
13.46 HCO$_3^-$, H$_2$CO$_3$, 1.3×10^2
13.47 (a) $1 \times 10^{-6}M$
       (b) $2 \times 10^{-32}M$
       (c) $3 \times 10^{-2}M$
       (d) $1 \times 10^{-5}M$
13.48 $3 \times 10^{-15}M$, $8 \times 10^{-11}M$
13.49 4.4

13.50 $5.9 \times 10^{-9} M$
13.51 $1.8 \times 10^{-7} M$
13.52 $Mg(OH)_2(s) + 2H^+ = Mg^{2+} + 2H_2O$
$CaCO_3(s) + 2H^+ = Ca^{2+} + H_2CO_3$
$CH_3CO_2Ag(s) + H^+ = Ag^+ + CH_3CO_2H$
13.53 (a) 10
(b) 45 mL
13.55 1.85

CHAPTER 14

14.1 (a) in H_2S, +1 for H and −2 for S; in SO_4^{2-}, +6 for S and −2 for O; in CH_3CO_2H, +1 for H, −2 for O, −3 for left C, and +3 for right C
(b) 0
14.2 (a) +3, −2
(b) +1, +5, −2
(c) +2, −1
(d) +4, −1
(e) +2, +6, −2
(f) +4, −2
(g) +1, +5, −2
(h) +7, −2
14.3 (a) reducing agent
(b) oxidizing agent
(c) oxidizing agent
(d) other type of reagent
14.4 5.7 g
14.5 (a) anode
(b) hydrogen
(c) $2H_2O(l) = 2H_2(g) + O_2(g)$
(d) 0.24 F, 2.3×10^4 C
14.6 (a) zinc
(b) carbon
(c) carbon
(d) carbon
(e) $Cl_2(g)$, $Zn(s)$
(f) 0.24 mol
14.7 $Fe(s) = Fe^{2+} + 2e^-$ at the anode
$Ag^+ + e^- = Ag(s)$ at the cathode
14.8 (a) $IO_3^- + 6H^+ + 6e^- = I^- + 3H_2O$
(b) $NO_3^- + 4H^+ + 3e^- = NO(g) + 2H_2O$
(c) $NO_3^- + 10H^+ + 8e^- = NH_4^+ + 3H_2O$
(d) $H_2S + 4H_2O = HSO_4^- + 9H^+ + 8e^-$
(e) $BrO^- + H_2O + 2e^- = Br^- + 2OH^-$

(f) $HO_2^- + OH^- = O_2(g) + H_2O + 2e^-$
(g) $Zn(s) + 3OH^- = Zn(OH)_3^- + 2e^-$

14.9 (a) $MnO_4^- + 5Fe^{2+} + 8H^+ = Mn^{2+} + 5Fe^{3+} + 4H_2O$
$MnO_4^-, Fe^{2+}, Fe^{2+}, MnO_4^-$
(b) $I_2(s) + 2OH^- = I^- + IO^- + H_2O$
$I_2(s), I_2(s), I_2(s), I_2(s)$

14.10 $2Al(s) + 6H^+ = 3H_2(g) + 2Al^{3+}$
$I_2(s) + H_2S = 2I^- + 2H^+ + S(s)$
$3Cu(s) + 2NO_3^- + 8H^+ = 3Cu^{2+} + 2NO(g) + 4H_2O$
$2Br^- + H_2O_2 + 2H^+ = Br_2(l) + 2H_2O$
$Cr_2O_7^{2-} + 6Fe^{2+} + 14H^+ = 2Cr^{3+} + 6Fe^{3+} + 7H_2O$
$2S_2O_3^{2-} + I_3^- = 3I^- + S_4O_6^{2-}$
$NH_4^+ + NO_3^- = N_2O(g) + 2H_2O$
$Fe(s) + 2Fe^{3+} = 3Fe^{2+}$
$3HNO_2 = 2NO(g) + H^+ + NO_3^- + H_2O$

14.11 $2Al(s) + 2OH^- + 6H_2O = 2Al(OH)_4^- + 3H_2(g)$
$3SO_3^{2-} + ClO_3^- = 3SO_4^{2-} + Cl^-$
$3Cl_2(g) + 6OH^- = 5Cl^- + ClO_3^- + 3H_2O$
$N_2H_4 + 2Cu(OH)_2(s) = N_2(g) + 2Cu(s) + 4H_2O$
$P_4(s) + 3H_2O + 3OH^- = PH_3(g) + 3H_2PO_2^-$

14.12 as written: (a), (d), (e)
reverse direction: (b), (c), (f)

14.13 0.59 V

14.14 $Cd^{2+} + 2e^- = Cd(s)$
$2H^+ + 2e^- = H_2(g)$
$Hg^{2+} + 2e^- = Hg(l)$

14.15 $O_2(g) + 2H^+ + 2e^- = H_2O_2$
$Br_2(l) + 2e^- = 2Br^-$
$Cl_2(g) + 2e^- = 2Cl^-$

14.16 (a) Fe^{3+}
(b) Cu^+

14.17 (a) 1.562 V
(b) 1.37 V
(c) 1×10^{53}

14.18 (a) 2.12 V
(b) 2.27 V
(c) 1×10^{72}

14.19 $8 \times 10^{-5} M$

14.20 (a) 0.13 V, 1.6×10^2
(b) 1.72 V, 1×10^{175}

14.21 (a) $MnO_2(s)$, $Zn(s)$; 0 to +2, +4 to +3
(b) $PbO_2(s)$, $Pb(s)$; 0 to +2, +4 to +2

14.22 (a) 3.60×10^5 C
(b) 387 g
(c) 4.3×10^6 J

14.23 (a) $4Al(s) + 3O_2(g) + 6H_2O = 4Al(OH)_3(s)$
(b) 2.70 V

14.24 $3CuS(s) + 2NO_3^- + 8H^+ = 3Cu^{2+} + 3S(s) + 2NO(g) + 4H_2O$
$4Fe(OH)_2(s) + O_2(g) + 2H_2O = 4Fe(OH)_3(s)$

14.25 (a) $Ag(s) + Br^- = AgBr(s) + e^-$
$Ag^+ + e^- = Ag(s)$
(b) 0.03 V
(c) 1×10^{-13}

14.26 (a) 1.562 V, Zn(s)
(b) 0.60 V, Zn(s)

CHAPTER 15

15.1 29.1 kJ mol^{-1}, 94.6 J K^{-1} mol^{-1}

15.2 (a) 1.3×10^{-3} mol, 4.7%; 2.77×10^{-2} mol, 100%
(b) yes

15.3 4.7 kJ mol^{-1}

15.4 0.091 atm, 69 Torr

15.5 water vapor

15.6 entropy

15.7 23.5 J K^{-1}

15.8 zero

15.9 4.40 J K^{-1}

15.10 it increases

15.12 (a) 20%
(b) 56.7%

15.13 (a) 245 J
(b) 17%; no

15.14 0, 0, -11.5 J K^{-1}, 3.45 kJ

15.15 12.0 kJ, 44.0 J K^{-1}, 0

15.16 (a) zero
(b) same

15.17 (a) zero
(b) negative
(c) positive (considering only water and evolved gases), negative (system also includes source of electric current)
(d) negative
(e) positive (considering only leaf and atmosphere), negative (system also includes sun and sunlight)

15.18 $\Delta H°$ and $\Delta S°$ are both negative

15.19 (a) reaction C
(b) reactions A and D
(c) reaction B

15.20 (a) -184.62 kJ, 20.0 J K^{-1}, -190.58 kJ
(b) -890.34 kJ, -242.7 J K^{-1}, -817.98 kJ

15.21 -89 kJ, -1.8×10^2 kJ
15.22 -56.8 kJ, 79.9 kJ
15.23 56 kJ
15.24 $+0.763$ V
15.25 173 kJ, 5×10^{-31}
130.9 kJ, 1.2×10^{-23}
-841.3 kJ, 3×10^{147}
-503.1 kJ, 1.4×10^{88}
-399.0 kJ, 7.9×10^{69}
15.26 (a) -137.2 kJ, -120.3 J K^{-1}, -101.3 kJ
(b) $\Delta H°$, $-T\Delta S°$
15.27 277.0 J K^{-1} mol^{-1}
15.28 (a) -41.2 J K^{-1}
(b) 1.8 J K^{-1}

CHAPTER 16

16.1 $-2.5 \times 10^{-4} M$ min^{-1}, $-4.2 \times 10^{-6} M$ s^{-1}, stoichiometry of reaction
16.2 $-9.1 \times 10^{-4} M$ s^{-1}, $1.8 \times 10^{-3} M$ s^{-1}
16.3 $\text{rate} = -\dfrac{d[Cr_2O_7^{2-}]}{dt} = -\dfrac{1}{14}\dfrac{d[H^+]}{dt} = -\dfrac{1}{6}\dfrac{d[Fe^{2+}]}{dt}$
$= \dfrac{1}{2}\dfrac{d[Cr^{3+}]}{dt} = \dfrac{1}{6}\dfrac{d[Fe^{3+}]}{dt}$

16.4 (a) rate doubles
(b) rate is halved
(c) rate is halved
16.5 (a) $2.7 \times 10^{-8} M$ s^{-1}
(b) $2.73 \times 10^{-7} M$ s^{-1}
(c) $2.7 \times 10^{-9} M$ s^{-1}
(d) $2.7 \times 10^{-8} M$ s^{-1}
(e) $[CH_3COCH_3] = 0.095 M$, $[H^+] = 0.015 M$,
$[I_2] = [CH_3COCH_2I] = [I^-] = 0.005 M$;
rate $= 3.9 \times 10^{-8} M$ s^{-1}
16.6 (a) rate increases by a factor of 9
(b) rate increases by a factor of 3
(c) $\dfrac{d[N_2]}{dt} = 2.9 \times 10^{-6} M$ s^{-1}
16.7 (a) s^{-1}
(b) M^{-1} s^{-1}
(c) M^{-2} s^{-1}
16.8 (a) second-order, first-order
(b) rate $= k[NO]^2[H_2]$, $k = 40 M^{-2}$ s^{-1}

16.9 (a) first order
(b) first order
16.10 (a) rate $= k[S_2O_8^{2-}][I^-]$
(b) $2.8 \times 10^{-3} M^{-1}\,s^{-1}$
16.11 (a) 1.1×10^3 s
(b) 89%
16.12 $3.3 \times 10^{-4}\,s^{-1}$
16.13 (a) first-order
(b) second-order
16.14 (a) 7.7×10^{-9} s
(b) 5.3×10^{-12} s
16.15 (a) 52 s
(b) $1.9 \times 10^2 M^{-1}\,s^{-1}$
16.16 (a) 38 s
(b) 19 s
(c) $0.18 M^{-1}\,s^{-1}$
16.17 rate $= k_r \dfrac{[CH_3COCH_2I][H^+]^2[I^-]}{[I_2]}$
16.19 54 kJ mol^{-1}
16.20 1.0×10^2 kJ mol^{-1}
16.21 none
16.22 50 kJ mol^{-1}
16.23 15 kJ mol^{-1}
16.24 (a) $3.0 \times 10^{-2} M^{-1}\,s^{-1}$
(b) $6 \times 10^5 M^{-1}\,s^{-1}$
16.25 (a) and (g)
16.28 more than one step
16.29 (a) $2B \longrightarrow D + E$
(b) A
(c) C
16.30 $NO_2 + O_3 \longrightarrow NO_3 + O_2$ (slow)
$NO_3 + NO_2 \longrightarrow N_2O_5$ (fast)
16.31 rate $= k[A][X]$
16.32 (a) $a = 1,\ b = \tfrac{1}{2}$
(b) $k_{obs} = k'K^{1/2}$
16.33 (c) first-order plot, first order
(d) 6.36×10^{-4} min^{-1}
16.34 (a) 0.26
(b) 8.4×10^{-13}

CHAPTER 17

17.1 (a) cubic
(b) tetragonal

(c) rhombohedral
(d) hexagonal
17.2 (a) cubic
(b) monoclinic
(c) orthorhombic
17.3 0.835 Å
17.4 3.14 Å
17.5 (b) 4.3 Å
(c) 3.7 Å
17.6 (a) 4 atoms
(b) $d = 4m/l^3$
(c) 4.05 Å
(d) 2.86 Å
17.7 2.14 Å, 2.16 Å
17.9 (a) 703 kJ mol^{-1}
(b) 774 kJ mol^{-1}
17.17 (a) 2.5×10^{22} electrons
(b) 5.0×10^{-19} J
(c) 3.8×10^{-41} J
17.19 (a) paramagnetic and ferromagnetic substances
17.21 87.6%
17.22 (a) decreases resistance
17.23 face-centered cubic, 26%
body-centered cubic, 32%
17.24 6, 8

CHAPTER 18

18.1 nonmetals: H, C
metalloids: B, Si, Ge
metals: other elements
18.4 $K_1 \approx 10^{-3}$, $K_2 \approx 10^{-8}$
18.7 1.1 L
18.8 Na_2O, Na_2O_2, KO_2
BeBe$(\pi_x 2p)^2(\pi_y 2p)^2(\sigma 2p)^2(\pi_x^* 2p)^2(\pi_y^* 2p)$
BeBe$(\pi_x 2p)^2(\pi_y 2p)^2(\sigma 2p)^2(\pi_x^* 2p)^2(\pi_y^* 2p)^2$
3, 2; O_2, O_2^{2-}
18.11 4.1×10^{-5}, $7.0 \times 10^{-3} M$
18.13 Na^+, HCO_3^-, and SO_4^{2-} ions
18.14 $[H^+] = [OH^-] = 1 \times 10^{-7} M$
18.15 Lewis acids: BF_3, BF_3, B_2H_6, Al_2Cl_6
Lewis bases: F^-, $(C_2H_5)_2O$, CO, Cl^-
18.16 1×10^{-54}
18.19 (a) rate = $k[CH_3CO_2H][OH^-]$
(b) rate = $k[CO_2(aq)]$

18.21 (a) $2Na(s) + 2H_2O = H_2(g) + 2Na^+ + 2OH^-$
(b) $K(s) + O_2(g) = KO_2(s)$
(c) $Mg(s) + 2H^+ = Mg^{2+} + H_2(g)$
(d) $Ca^{2+} + 2HCO_3^- = CaCO_3(s) + CO_2(g) + H_2O(g)$
(e) $2Al(s) + 6H^+ = 2Al^{3+} + 3H_2(g)$
(f) $2Al(s) + 6H_2O + 2OH^- = 2Al(OH)_4^- + 3H_2(g)$
(g) $HCO_3^- + OH^- = CO_3^{2-} + H_2O$
(h) $HCO_3^- + H^+ = CO_2(aq) + H_2O$
(i) $SnS_3^{2-} + 2H^+ = SnS_2(s) + H_2S$
(j) $PbO_2(s) + 4H^+ + SO_4^{2-} + 2Cl^- = PbSO_4(s) + 2H_2O + Cl_2(g)$

CHAPTER 19

19.2 $180°, \sim 120°, \sim 120°$
19.4 (a) -0.47 V, $+0.62$ V
(b) Br_2
19.5 3, 3, 4, 4, 4, 4, 3, 5
19.8 $K \approx 1 \times 10^{-3}$
19.9 PCl_4^+, tetrahedral; PCl_6^-, octahedral
19.10 $As(s) + 5NO_3^- + 5H^+ = H_3AsO_4 + 5NO_2 + H_2O$
19.11 (a) yes
(b) $H_2O_2 + 2Fe^{3+} = O_2(g) + 2H^+ + 2Fe^{2+}$
$\Delta\mathcal{E}° = +0.09$ V
$H_2O_2 + 2H^+ + 2Fe^{2+} = 2H_2O + 2Fe^{3+}$
$\Delta\mathcal{E}° = +1.01$ V
19.12 yes, no
19.13 4, lungs, tissues
19.16 $+2, +2\frac{1}{2}$
19.17 $6.7 \times 10^{-3} M$
19.19 $Br_2 + 2OH^- = Br^- + BrO^- + H_2O$
$Br_2 + 2I^- = 2Br^- + I_2$
19.20 linear
19.21 $[I_3^-] = 5.1 \times 10^{-2} M$, $[I^-] = 9 \times 10^{-2} M$
$[I_2] = 4 \times 10^{-4} M$
19.22 $I_2O_5(s) + 5CO(g) = I_2(s) + 5CO_2(g)$
19.23 (b) 6
(c) octahedron

CHAPTER 20

20.1 $3d$: 1, 0, 6, 6, 5, 10, 10, 9, 10, 10
$4s$: 2, 0, 2, 0, 0, 1, 0, 0, 2, 0
20.3 0, 3, 6, 5, 10
20.7 $5.85 \times 10^{-4} M$
20.8 1.19 V

20.9 109.5°, 109.5°
20.10 (a) $2ZnS(s) + 3O_2(g) = 2ZnO(s) + 2SO_2(g)$
      (b) $Fe_2O_3(s) + 3H_2(g) = 2Fe(s) + 3H_2O(g)$
      (c) $Cr(OH)_2(s) + 2H^+ = Cr^{2+} + 2H_2O$
      (d) $Cr(OH)_3(s) + 3H^+ = Cr^{3+} + 3H_2O$
      (e) $Cr(OH)_3(s) + OH^- = Cr(OH)_4^-$
      (f) $CrO_3(s) + 2OH^- = CrO_4^{2-} + H_2O$

CHAPTER 21

21.1 2 isomers
21.3 3 isomers
21.4 (a) 2 isomers
      (b) 1 isomer
21.5 first, $[Co(NH_3)_5SO_4]^+$; second, $[Co(NH_3)_5Br]^{2+}$
21.6 1, 0, 4, 3, 7, 6, 10
21.7 6, 5, 8
21.8 $d_{x^2-y^2}$, d_{xy}
21.9 (a) 2 lobes, 4 lobes
      (b) x axis, x and y axes
      (c) 1 nodal plane, 2 nodal planes
21.10 (a) d_{z^2}
      (b) d_{z^2} and $d_{x^2-y^2}$
      (c) d_{xy}, d_{xz}, and d_{yz}
      (d) $d_{x^2-y^2}$
21.11 1.761×10^4 cm^{-1}
21.12 3, 1
21.13 Cr^{2+}, Fe^{3+}, Co^{2+}
21.14 d^8
21.15 low-spin d^6
21.16 5
21.17 (a) high-spin
      (b) low-spin
21.18 decrease
21.19 t_{2g}^2, e_g^2, 4
21.20 (a) $0, \frac{1}{5}, \frac{2}{5}, \frac{3}{5}, \frac{4}{5}, 0, \frac{1}{5}, \frac{2}{5}, \frac{3}{5}, \frac{4}{5}, 0$
      (b) 3 and 8
21.21 $2\Delta_o$
21.23 high-spin
21.24 diamagnetic, low-spin; paramagnetic, high-spin
21.25 (a) inverted
      (b) normal
21.26 low-spin d^7
21.27 4
21.28 compound A, cis; compound B, trans

CHAPTER 22

22.2 2,4-dimethylhexane
3,3-dimethylpentane
22.3 2.8×10^2 g min^{-1}, 25 L h^{-1}
22.4 $2CH_3\cdot \longrightarrow C_2H_6$
$CH_3\cdot + Cl\cdot \longrightarrow CH_3Cl$
22.14 2-chloropropane
22.15 2 isomers
22.23 ester, aldehyde, amine, carboxylic acid
22.24 $CH_3CH_2OH + O_2(g) = CH_3COOH + H_2O$
22.29 $C_{20}H_{21}NO_4$
22.31 2-butanol (*sec*-butyl alcohol)

CHAPTER 23

23.1 L-threonine, yes; D-methionine, no
23.2 glutamic acid, cysteine, and glycine
23.4 6, Gly·Ala·Ser, Gly·Ser·Ala, Ala·Gly·Ser, Ala·Ser·Gly, Ser·Gly·Ala, Ser·Ala·Gly
23.5 Lys·Gly·Try·Thr·Ala·Gly
23.6 2
23.7 6, 55
23.8 $3\tfrac{1}{2}$
23.12 C T G G A T
23.15 (a) N—H···N in A T (lower) and G C (middle)
N—H···O in A T (upper) and G C (upper and lower)
no O—H···N or O—H···O
23.16 0.75, 0.25
23.17 upper right is zero-order, lower left is first-order
23.18 almost all enzyme is complexed by substrate
23.19 1.5×10^2 min^{-1}
23.22 (a) $m = 1/kK[E]^{\text{total}}$, $b = 1/k[E]^{\text{total}}$
(b) $k = 1/b[E]^{\text{total}}$, $K = b/m$

CHAPTER 24

24.1 toward $-$, toward $+$, no deflection
24.3 7 alpha particles, 4 beta particles
24.4 (a) radium, 88, 228
actinium, 89, 228
thorium, 90, 228
radium, 88, 224
radon, 86, 220
polonium, 84, 216

lead, 82, 212
bismuth, 83, 212
polonium 84, 212
lead, 82, 208

24.5 (a) $2.4 \times 10^{-2}\,y^{-1}$
(b) 96 y

24.6 87.5% decays

24.7 2.4×10^2 d

24.8 positron emission or electron capture; $^{15}_{7}N$

24.9 upper right (beyond stability region), lower right, upper left, upper left

24.10 electron

24.11 $^{233}_{91}Pa$, $^{60}_{28}Ni^+$, $^{41}_{19}K$, $^{11}_{5}B^-$, $^{1}_{1}H$, $^{3}_{1}H$

24.12 $^{1}_{0}n$, $^{4}_{2}He$, e^-, e^+, $^{1}_{0}n$

24.13 42

24.14 1.11 MeV per nucleon

24.15 neutron

24.16 beginning, middle

24.17 1.28

24.19 (a) $4N$
(b) 1.25×10^{-9} mol
(c) 1.45×10^9 y

24.20 $1.2 \times 10^3 \times$ (weight in pounds) disintegrations per minute

24.22 (a) carelessness by a fellow worker
(b) 5×10^3 y, 5.12×10^3 y

APPENDIX C

C-1.1 (a) 3.275×10^2
(b) 6.0271×10^4
(c) 1.09×10^{-3}
(d) 8.6×10^{-4}

C-1.2 (a) 0.72
(b) 0.0060
(c) 49
(d) 602,200,000,000,000,000,000

C-1.3 (a) 8.28×10^{-2}
(b) 3.76×10^3
(c) 4.5
(d) 4.1×10^7
(e) 9.0×10^{-2}

C-2.1 (a) 5
(b) 3
(c) 4

C-2.2 (a) 4.0×10^4 cm
(b) 4.000×10^4 cm

C-2.3 (a) 205.7
(b) 2.4
(c) 6.273×10^{11}
(d) 1.5×10^2
(e) 1.9×10^6
(f) 3.91×10^7
(g) 4.000×10^1
(h) 141.1 min

C-3.1 2.0

C-3.2 $-3.0, y = 2.0x - 3.0$

C-5.1 (a) 4
(b) -4
(c) 0.8222
(d) 3.029
(e) -1.460
(f) -3.280

C-5.2 (a) 1×10^3
(b) 1×10^{-5}
(c) 5.63
(d) 8.27×10^2
(e) 6.11×10^{-3}
(f) 3.95×10^{-5}

INDEX

A

Abelson, Philip H., 730
Absolute temperature scale, 52
Absorption spectra, 166, 634–635, 638–640
Acceleration, 71–73, A-18
Acetic acid, 306, 308
   dissociation constant of, 345–347
   equilibrium problems using, 368–369
   properties of, 570
   standard enthalpy of formation for, 108
Acetylene, 108
Acid-base indicators, 306–309
Acid-base titrations, 381–386
Acids, 306–309, 396–397
   amino, 690–694
   in aqueous media, 551–558
   arrhenius, 307
   binary, 386–389
   Brønsted-Lowry, 360–363
   carboxylic, 672–676
   dissociation constants of, 367
   generalized systems of, 360–363
   Lewis, 362–363
   nucleic, 702–707
   polyprotic, 386–391
   relative strengths of, 366–368
   weak, 345–347, 570
Actinides, 612–613
Actinium, 326
Activated-complex theory, 493–494
Activation energy, 486, 502

Activity, 341
Addition reactions, 315
Adenosine triphosphate (ATP), 586
Adiabatic processes, 110, 454
Air
   composition of, 310
   as a solution, 7
Air pollution, 2
Alcohols, 670–676
Aldehydes, 672–676
Alkali metals, 204, 310–311
   peroxides formed by, 588
   properties of, 551–557
   sulfides of, 592
Alkaline earth metals, 205, 311–312, 326, 557–561
   peroxides formed by, 588
   sulfides of, 592
Alkanes, 316–317, 658–663
Alkenes, 663
Alkoxides, 672
Alkyl groups, 316, 659
Alkyl halides, 668–670
Alkynes, 663
Allotropes, 301
Alpha helix, 695
Alpha rays, 128, 720
Aluminum, 562–564, 566–567
   electron affinity of, 195
   electron configuration of, 188, 312–313, A-23
   nitrates of, 531
   oxides of, 588
   thermodynamic properties of, A-26

Aluminum oxide, 108
Amines, 316, 676–677
Amino acids, 690–694
Ammines, 627
Ammonia
   in aqueous solution, 306–307, 578–579
   boiling point of, 103
   chemical properties of, 321, 578–580
   as an electron donor, 580
   as ligand, 393–394
   liquid, 580
   melting point of, 103
   molar enthalpies of fusion and vaporization for, 103, 108
   molecular structure of, 222
   production of, 305
   salts of, 579–580
Ammonium fluoride, 533
Ammonium hydroxide, 579
Amorphous solids, 511–512
Amphoteric hydroxides, 308
Amplitude, 142
Anderson, Carl D., 728
Angular momentum, A-19–A-20
Anions, 411, 522–523
Anisotropy, 510
Anode, 411
Antimony, 301, 320–322
   crystal structure of, 531, 584
   properties of, 578, 584, 587
Aqua regia, 583
Argon, 532, 604

Argon (*Continued*)
  electron configuration of, 188, A-23
  van der Waals radius of, 193
Aristotle, 13n
Aromatic hydrocarbons, 665-668
Arrhenius, Svante, 287, 360, 363, 367, 485
Arsenic, 320-322, 324, 544-545
  crystal structure of, 531, 584
  properties of, 578, 584, 587-588
Aryl group, 666
Atmospheric pressure, 47
Atomic masses, A-5. *See also* Mass
  definition of, 20-23
  determination of, 30-32, 55-57
Atomic-molecular theory
  development of, 12-14
  experimental investigation of gases and, 44
Atomic number, 118, 130
  of the alkali metals, 554
  of group IIA elements, 557
  of group IIIA elements, 562
  of group IVA elements, 568
  of group VA elements, 579
  of group VIA elements, 590
  ionization potential and, 194-195
Atomic radii, 192-193
Atomic spectra, 166-171
Atomic weights, 22
Atoms, 12-14, 22-23. *See also* Hydrogen atom
  general features of, 118-137
  ionization potentials of, A-23
  in metallic crystal structures, 518-521
  structure and properties of, 166-197
Autocatalytic reaction, 497
Avogadro, Amedeo, 24, 54-55
Avogadro's law, 55-56, 59
Avogadro's number, 24-29, 74, 81
  fundamental constant for, A-5
Axial positions, 218
Azeotropes, 8
Azimuthal quantum number, 173, 183

B

Background radiation, 728-732
Balanced equations, 32-33
Balmer, Johann J., 169n
Balmer series, 168, 169n
Barbier, Philippe, 678
Barium, 730, A-25
Barometer, 47
Bartlett, Neil, 310, 602-603
Base metals, 425

Bases, 306-309. *See also* Acids
  arrhenius, 307
  generalized systems of, 360-363
  relative strengths of, 366-368
Batteries, 431, 573, 619
Becquerel, Henri, 127
Benzene, 108
Berthollet, Claude Louis, 13, 543
Berthollides, 13, 543
Beryllium, 308, 311-312, 519, 558-559
  electron configuration of, 188, A-23
  thermodynamic properties of, A-25
Beta rays, 128, 720
Bethe, Hans, 632
Bidentate ligands, 628
Binary acids, 551, 553
Biological chemistry, 690-715
Biphenyl, 532
Bismuth, 320
  crystal structure of, 531, 584
  properties of, 578, 584, 588
Blackbody radiation, 148, 166
Blood, 379, 590
Body-centered cubic structures, 519
Bohr, Niels, 167, 172
Bohr orbit, 175
Boiling points, 263, 310, 737
  elevation of, 279-282
  enthalpies of fusion and vaporization and, 103
  of noble gases, 603
  of transition metals, 327
Boltzmann, Ludwig, 78
Boltzmann constant, 74-76, A-5
Bond energy, 203
Bond enthalpy, 203
Bonding
  carbon-carbon, 305, 314, 567-568, 678-680
  covalent, 210-213, 222-223, 520-521, 531, 559, 567, 626-652
  donor-acceptor, 214-216
  of halogens, 599
  hybridization and, 246-254
  hydrogen, 269-271, 533-534
  ionic, 204-210
  metallic, 520-521
  molecular orbitals and, 235-245
  molecular properties and, 202-204
  polar, 224-230, 524
  reaction mechanisms and, 494-501
  three-center, 563
  valence bond theory of, 246
Bond length, 202
Bond moment, 235
Boranes, 563
Born, Max, 161, 529-530

Born-Haber cycle, 529-531
Boron, 312, 531, 562-564
  electron affinity of, 195
  electron configuration of, 188
  thermodynamic properties of, A-25–A-26
Boron nitride, 313n
Boyle, Robert, 48-50
Boyle's law, 49-50, 57, 57n, 61
Boyle temperature, 61
Bragg, William Henry, 513, 515-516
Bragg, William Lawrence, 513, 515
Bragg equation, 513
Braun, Ferdinand, 343n
Bromine, 324-326
  electron affinity of, 195
  thermodynamic properties of, A-28
Brønsted, Johannes N., 360-362, 366, 374, 389, 394, 407
Bronzes, 319
Buerger, Martin J., 515
Buffers, 377-380
Building-up principle, 185-189

C

Cadmium, 326, 588
Calcite, 528
Calcium, 520, 558-561
  electron configuration of, 188, A-23
  thermodynamic properties of, A-25
Calcium carbonate, 312, 560-561
  standard enthalpy of formation for, 108
Calcium hydroxide, 108, 306
Calcium oxide, 108
Calorie, 100
Calorimetry, 110-114
Calvin, Melvin, 737
Cannizzaro, Stanislao, 55-57, 130
Carbohydrates, 696-702
Carbon. *See also* Hydrocarbons
  abundance of, 567
  acids of, 386-389, 570
  allotropic forms of, 313-314
  bonding characteristics of, 252
  compounds of, 314-319, 658-683
  electron affinity of, 195
  electron configuration of, 188
  as a reducing agent, 613
  thermodynamic properties of, A-26
Carbon-carbon bonds, 305, 314, 567-568, 678-680
Carbon dioxide, 108
Carbon-14, 738-739
Carbonic acid, 386-389
Carbon monoxide, 100, 613
Carbon steels, 7, 327, 614
Carbon-12, 22

Carbonyl group, 672
Carboxylic acids, 672–676
Carnot, N. L. Sadi, 454
Carnot cycle, 454–456
Catalysts, 304, 497, 501–502
Catenation, 568
Cathode rays, 121–124
Cations, 522–523, 578
Cells, electrochemical. *See*
 Electrochemical cells
Celsius temperature scale, 52
Cerium, 611
Cesium, 553
Chain reactions, 499–500
Chalcogens, 205, 323
Charge. *See* Electric charge
Charge balance equation, 369
Charles, Jacques, 52, 59
Chelation, 628, 651–652
Chemical bonding. *See* Bonding
Chemical elements. *See* Elements
Chemical equation, 32
Chemical equilibria. *See* Equilibria
Chemical kinetics, 472–502
Chemical reactions, 32–35
 and net reactions, 290–293
Chemistry
 biological, 690–715
 coordination, 626–652
 dynamic character of, 15
 of the elements, 300–337
 mole concept of, 23–29
 nuclear, 720–739
 observation and interpretation
  in, 11
 organic, 313–319, 658–683
 study of, 2–3
 transition metal, 626–652
Chlorine, 324–326, 598–602
 boiling point of, 103
 electron affinity of, 195
 electron configuration of, 188
 melting point of, 103
 molar enthalpies of fusion and
  vaporization for, 103
 molar heat capacity of, 100–101
 thermodynamic properties of, A-28
Chromium, 610–611, 613–617
 compounds formed by, 327
 electron configuration of, 326
 ionization potentials of, 617, A-23
 oxidation states of, 615–616
 thermodynamic properties of, A-28
Claassen, Howard H., 603
Clausius, Rudolf, 454
Cloud chamber, 720–728
Cobalt, 542–543, 611
 coordination complexes of, 628–629
 ionization potentials of, 617

oxidation states of, 616
 thermodynamic properties of, A-29
Cohesive energy, 529–531
Coinage metals, 327
Colligative properties of solutions,
 273
Collision theory, 488–493
Common-ion effect, 352
Complex ions, 393–395
Compounds
 carbon, 314–319, 658–683
 coordination, 626–653
 defined, 9–10
 nonstoichiometric, 543–544
Compressibility factor, 61
Compression, 649
Compton, Arthur H., 154
Compton effect, 154
Concentration, 271
Conductance, A-22
Conduction band, 540
Conjugate acid-base pair, 360
Constructive interference, 143
Conversion factors, A-5–A-6
Coordinate covalent bonds, 626–652
Coordination chemistry, 626–652
Coordination number, 626
Copper, 327, 611, 617–618
 ionization potentials of, 617, A-23
 as a reducing agent, 425
 standard electrode potentials of,
  618
 thermodynamic properties of, A-29
Corundum, 327
Cosmic radiation, 728
Coulomb's law, A-21, A-32–A-33
Covalent bond angles, 216–221
Covalent bond radii, 222–223
Covalent bonds, 221–223, 531, 626–
 652
 crystal-field theory and, 632
 formed by beryllium, 559
 formed by carbon, 567
 Lewis structures and, 210–213
 in metallic crystals, 520–521
Covalent network crystals, 531
Crafts, James, 667
Cristobalite, 524–525, 528
Crookes, William, 121
Crystal-field theory, 632
Crystal lattice energy, 529–531
Crystals, 510–534
 Born-Haber cycle and, 529–531
 covalent network, 531
 defects in, 543–545
 hydrogen-bonded, 533–534
 ionic, 521–529
 metallic, 518–521
 molecular, 531–533

seven systems of, 513
 transition metal, 613
 X-ray diffraction and, 510, 512–517
Cubic close-packed structures, 519
Cupric oxide, 108
Curie, Marie, 127, 728
Curie, Pierre, 128, 543, 728
Curie temperature, 543
Cyanogen, 570–571
Cycle, 142
Cycloalkanes, 661–662
Cyclotron, 730–731

D

Dalton, John, 13–14, 22, 50, 54, 118
Davisson, Clinton J., 155
DDT, 669
de Broglie, Louis, 155
de Broglie wavelength, 155–161, 171
Debye, Peter J. W., 289
Decay constant, 726
Deduction, 15
Defects in crystals, 543–545
Democritus, 13n
Denaturation, 696
Density, 17–18
 of the alkali metals, 554
Deoxyribonucleic acid (DNA), 586,
 702–707
Deoxyribonucleotides, 704
Derivations, A-30–A-33
Derivative, the, 474
Destructive interference, 144
Devitrification, 512
Diamagnetism, 541–542, 637
Diamonds, 313, 315
 crystal structure of, 510, 524, 531
 energy gap in, 541
Diatomic molecules, 324
Diborane, 563–564
Dielectric constant, 230–231
Diffraction, 144
Digonal hybrids, 246–249
Dipole moments, 231–236, 269
Diprotic acid, 386
Dirac, Paul A. M., 538
Dislocations, 543
Disorder and entropy, 446–447
Dispersion force, 268
Disproportionation, 389
Dissociation, 291, 307, 345–348
 degree of, 288
 of an oxyacid, 551–552
 simultaneous equilibria and, 363–
  364
 of water, 363–365
Distillation, 5–9
Donor-acceptor bonds, 214–216
d orbitals, 629–631

Double helix, 703
Doublet, 182
Dulong, Pierre L., 98
Dynamic equilibrium, 335

E

Efficiency, 456
Einstein, Albert, 152, 154, 732
Electric charge, A-21–A-22
 formal, 214–216
 net, 407–410
 oxidation state and, 407–410
Electric conductivity, 301, 327, 411
 free-electron theory of metals and, 534–540
 of selenium, 596
Electric current, A-21–A-22
Electric field strength, A-22
Electric potential, A-22
Electrochemical cells, 411, 418–434
 standard potential of, 418–421
 voltage in, 424–430
Electrodes, 411
 standard potentials for, 421–425
Electrolysis, 412, 553
 Faraday's law of, 120
Electrolytes, 286–289, 396–397
 solubility in water of, 290
 strong, 287, 292
 weak, 288, 291–292, 345–348
Electrolytic conduction, 411
Electromotive force, 418
Electron affinities, 194–196, 529n, 590
Electron capture, 724
Electron charge, 308, A-5
Electron configurations, 184–189, 196–197, A-23
 of the alkali metals, 554
 of the group IIA elements, 557
 of the group IIIA elements, 562
 of the group IVA elements, 568
 of the group VA elements, 579
 of the group VIA elements, 590
 of noble gases, 603
Electron density, 174–175, 192–193
Electron diffraction, 155
Electron-dot symbol, 196
Electronegativities, 227–229. *See also*
 Pauling electronegativity
Electronic band theory, 540–541
Electrons, 15. *See also* Electron
 affinity; Electron configurations
 correlation of, 190–192
 discovery of, 118–124
 Pauli exclusion principle and, 181, 186–187, 190, 236, 536, 538
 shielding effect, 181–184
 spin quantum number, 178–180
 valence, 196

wave properties of, 154–156
Electron transfer reactions, 406–434, 580–582
Elementary reaction, 488
Elements, 300–337. *See also*
 Halogens; Transition metals;
 and individual elements
 defined, 9–10
 group IA, 551–557
 group IIA, 557–561
 group IIIA, 562–567
 group IVA, 567–573
 group VA, 578–588
 group VIA, 588–598
 group VIIA, 598–602
 periodic relationship among, 130, 189, 300, 302–304
 symbols for, 22
 transition, 608–622
 transuranium, 730–731
Elongation, 649
Emeralds, 327
Emission spectrum, 166–171
Empirical formula, 27
Enantiomers, 318–319, 500, 629
End point, 382
Energy, 88–114, A-5, A-20–A-21
 activation, 486, 502
 atomic structure and, 166–197
 cohesive, 529–531
 conservation of, 88
 crystal lattice, 529–531
 Fermi, 536–539, 542
 Gibbs free, 458–461
 ionic resonance, 224–226
 internal, 93
 kinetic, 75
 levels of, 149
 nuclear, 732–735
 potential, 88–89
 radiant, 147–152
 zero-point, 150
Enthalpies, 96, 445
 of formation, standard, 107–109, A-24–A-29
 of fusion, 103–104
 of reaction, 104–107
 of vaporization, 103–104
Entropies, 441–465, A-24–A-29
 defined, 446–447
 probability and, 456–457
 in spontaneous irreversible processes, 451–454
Enzymes, 690, 707–712
 as catalysts, 501–502
Equation of state
 defined, 44
 for ideal gases, 59–61
 for real gases, 61–64
 van der Waals, 62–64

Equilibria, 92, 334–353. *See also*
 Equilibrium constants
 heterogeneous, 348–353
 reaction rates and, 483–484
 simultaneous, 360–397
 thermodynamics and, 440–465
Equilibrium constants, 336, 426–430
 conventions used for, 340–342
 temperature dependence of, 440–445
Equilibrium vapor pressure. *See*
 Vapor pressure
Equivalence point, 382
Ethane, 108
Ethers, 670–676
Ethyl, 659
Ethyl alcohol, 108
Ethylene, 108
Eutectics, 8
Excited states, 149, 188
Exclusion principle, 181, 186–187, 190, 236, 536, 538
Exponents, A-7–A-8
Extensive properties, 45
Eyring, Henry, 493

F

Faraday, Michael, 119–120, 120n
Faraday constant, A-5
Faraday's law of electrolysis, 120
Fermi, Enrico, 538, 730
Fermi-Dirac distribution, 538
Fermi energy, 536–539, 542
Ferric oxide, 108
Ferromagnetism, 541–543
Fission, 730
Fluorine, 324–326, 551, 591, 598–602, 604
 electron affinity of, 195
 electron configuration of, 188, A-23
 as an oxidizing agent, 425
 thermodynamic properties of, A-27–A-28
Fluorite, 527–529
Fluorocarbons, 551
Fock, Vladimir A., 192
Force, A-19
Formal charges, 214–216
Formation constant, 393
Formic acid, 533
Fractional distillation, 7, 278
Fractional ionic character, 233–234
Francium, 553
Free-electron Fermi gas, 538
Free-electron theory of metals, 534–540
Free-energy changes, standard, 462–465, A-24–A-29

Freezing point depression, 279–282
Frequency, wave, 142
Friedel, Charles, 667
Friedel-Crafts reaction, 678
Friedrich, Walter, 513
Fuel cells, 430–434
Functional groups, 316, 658
Functions, A-11–A-13
Fundamental constants, A-5

G

Gallium, 312–313, 531, 545, 562–563, 566
Gamma rays, 128, 720
Gas constant, 59, A-5
Gases. *See also* Ideal gases; Noble gases; Real gases
  entropies of, 446–447, 450–461
  equilibrium in, 334–338
  ionization potential and, 194–196
  kinetic molecular theory of, 70–83
  vapor pressure and, 260–263, 267, 273–279, 348–349
Gay-Lussac, Joseph, 52–54, 59
Geiger, Hans, 128–129
Geiger counter, 720–728
Genetic code, 707
Geometrical isomerism, 317
Gerlach, Walter, 178
Germanium, 319–320, 544–545
  crystal structure of, 531
  energy gap in, 541
  properties of, 567, 571–572
Germer, Lester H., 155
Gibbs, J. Willard, 458
Gibbs free energy, 458–461
Glasses, 511–512, 571
Gold, 326–327, 618, 644
  aqua regia and, 583
  charge density in nucleus of, 736
  solid solutions formed by, 7, 544
Goldstein, Eugen, 134
Goudsmit, Samuel A., 178
Graham, Thomas, 78
Graham's law of diffusion, 76–78
Graphite, 313–314, 531
Graphs, A-11–A-13
Grignard, Victor, 678
Grignard reagents, 678–680
Group, 130
Guldberg, Cato M., 337

H

Haber, Fritz, 321, 529–530
Haber process, 501
Hahn, Otto, 730
Hahnium, 326, 732
Half-life, 479
Half-reactions, 410–413
Halides, 599–602
  alkyl, 668–670
Halogens, 196, 205, 301, 324–326, 598–602
Hard metals, 615
Hard water, 560
Hartree, Douglas R., 192
Hartree-Fock orbitals, 192
Heat, A-20–A-21
  as energy transfer mode, 90
Heat capacities, 96–100, A-24–A-29
Heat engines, 454–456
Heavy water, 737
Heisenberg, Werner, 160, 172
Heisenberg uncertainty principle, 158–161
Heitler, Walter, 245
Helium, 309–310, 604, 734
  electron configuration of, 188, A-23
  ionization potential for, 173, 194
  natural radioactivity as source of, 720
  van der Waals radius of, 193
Hemiacetal formation, 701
Hemoglobin, 590–591
Henry's law, 277–278
Hertz, Heinrich, 146
Hess, Germain H., 106
Hess's law, 106, 106n, 203
Heterogeneous equilibria, 348–353
Heterogeneous systems, 3–5, 11
Hexagonal close-packed crystal structures, 519
High-field ligands, 635
High-spin complexes, 636
Homogeneous systems, 3–9, 11
Hückel, Erich, 289
Hund, Friedrich, 188
Hund's rule, 188
Huygens, Christiaan, 144
Hybridization, 246–254
Hydrocarbons, 304, 551
  aromatic, 665–668
  saturated, 316–317, 658–663
  straight-chain, 314
  unsaturated, 314–315, 663–665
Hydrocarbon skeleton, 658
Hydrochloric acid, 306
Hydrofluoric acid, 599–600
Hydrogen, 551. *See also* Hydrogen atom
  in acid-base chemistry, 306–308
  bonding properties of, 267–272, 533–534
  crystal structure of, 532
  electron affinity of, 195
  electron configuration of, 188, A-23
  ionization potential of, 173, 194
  periodic law and, 301, 304–306
  production and reactions of, 301–306
  thermodynamic properties of, A-24
Hydrogen atom. *See also* Hydrogen
  electron spin and, 178–180
  emission spectrum of, 166–171
  formation of, 237
  orbitals of, 174–178
  quantum numbers for, 171–173
Hydrogen-bonded crystals, 533–534
Hydrogen chloride, 103, 108
Hydrogen halides, 599
Hydrogen peroxide, 588–590
Hydrogen sulfide, 532, 592
Hydrolysis, 374
Hydronium ion, 361
Hypohalous acids, 601

I

Ice, 534–535
Ideal gases
  defined, 58
  equation of state for, 59–61
  kinetic theory model of, 70–80
  law of chemical equilibrium and, 340–342
  standard molar volume of, A-5
Ideal solutions, 340–342
Indicators, 306–309, 380–381
Indium, 312–313, 531, 562–563, 566
Induced-fit hypothesis, 711
Induction, 14–15
Inert pair, 567
Insulator, 540–541
Intensity distribution, 147–148
Intensive properties, 45
Interference, 143–144
Intermediates, 497
Intermolecular forces, 267–271
Internal energy, 93
International Bureau of Weights and Measures, 16
International System of Units (SI), 15, A-3–A-4
International Union of Pure and Applied Chemistry (IUPAC), 661
Internuclear distance, A-32–A-33
Iodine, 324, 349, 532, 599, 602
  electron affinity of, 195
  thermodynamic properties of, A-28
Ion-exchange chromatography, 611–612
Ionic bonding theory, 204–210
Ionic crystal structures, 206–207, 520–529
Ionic radii, 206–209
Ionic resonance energy, 224–226
Ionic valence, 205

Ionization potentials, 173, 191, 194–196, A-23
  of the alkali metals, 554
  of the group IIA elements, 557
  of the group IIIA elements, 562
  of the group IVA elements, 568
  of the group VA elements, 578–579
  of the group VIA elements, 588, 590
  of the group VIIA elements, 598
  of the noble gases, 603
  of transition metals, 617
Ions
  complex, 393–395
  negative, 119
  positive, 118
Iridium, 614
Iron, 327, 610–611, 613–614
  ferromagnetic properties of, 543–544
  ionization potential of, 617
  as a reducing agent, 425
  thermodynamic properties of, A-28–A-29
Irreversible processes, spontaneous, 451–454
Isomers, 317–319
Isotopes, 10–11, 15, 737–739
Isotopic mass, 134–137

J
Jahn, H. A., 649
Jahn-Teller effect, 649
Jensen, J. Hans D., 735
Joliot-Curie, Frédéric, 728
Joliot-Curie, Irene, 728
Joule, James P., 90
Joule experiment, 451–453, 456

K
Kelvin units (K), 52
Kernel, 196
Kernel charge, 408n
Ketones, 672–676
Kilogram (kg), 16–17
Kinetic molecular theory
  of gases, 70–83, A-30–A-33
  of liquids, 265–267
Kinetics, 472–502
Knipping, Paul, 513
Kopp, Hermann, 99
Kopp's law, 98–99, 511
Krypton, 604
  molecular crystals of, 532
  van der Waals radius of, 193
Kusch, Polykarp, 79

L
Lactic acid, 318–319
Lanthanides, 326, 611–612, 617
Lanthanum, 611
Latimer, Wendell M., 581
Latimer reduction-potential diagrams, 581–583, 587
Laughing gas, 580
Lavoisier, Antoine Laurent, 12
Law of chemical equilibrium, 337
Law of conservation of energy, 88
Law of conservation of mass, 12
Law of constant heat summation, 106
Law of definite proportions, 12–13
Law of mass action, 337n
Law of multiple proportions, 13–14
Law of Petit and Dulong, 98, 511
Lawrence, Ernest O., 730
Lead, 319–320, 567, 571–573
  thermodynamic properties of, A-26–A-27
Le Bel, Joseph A., 318
Le Châtelier, Henri Louis, 343
Le Châtelier's principle, 342–345, 350, 352–353, 363, 374, 393, 396
Length measurements, 16, A-5
Leveling effect, 367
Lewis, Gilbert N., 211, 362–363
Lewis acid-base reaction, 626
Lewis structures, 211, 214, 235, 245–247
  covalent bonding and, 210–213
Libby, Willard F., 738
Ligand-field splitting, 631–640
Ligand-field stabilization energies, 640–643
Ligand-field theory, 631–632
Ligands, 393, 396–397
  low-field, 635
  transition metal ions coordinated to, 626–652
Light
  emission spectrum of, 166–171
  energy quanta and, 147–152
  intensity of, 147
  photon concept of, 152–154
  wave theory of, 144–147
Lime, 312
Limestone, 560–561
Linear equation, A-12
Linear momentum, A-19–A-20
Linear planar complexes, 644–648
Line spectrum, 166
Liquid-drop model, 735
Liquids, 260–293
  entropies of, 446–465
  kinetic molecular theory of, 265–267
  standard state for, 341
Liter (L), 17
Lithium, 310, 553–554, 556
  electron affinity of, 195
  electron configuration of, 188, A-23
  thermodynamic properties of, A-24
Lithosphere, composition of, 320
Lock-key hypothesis, 710
Logarithms, 365n, A-14–A-17
London, Fritz, 245
London force, 268
Low-field ligands, 635
Lowry, Thomas M., 360–362, 366, 374, 389, 394
Low-spin complexes, 636
Lutetium, 612
Lyman series, 168–169

M
McMillan, Edwin M., 730
Madelung, E., 530
Madelung constant, 530
Magnesium, 311–313, 519, 558
  electron configuration of, 188, A-23
  in lithosphere, 320
  thermodynamic properties of, A-25
Magnetic properties, 541–543, 614
  of transition metal compounds, 636–638
Magnetic quantum number, 173
Malm, John G., 603
Manganese, 610–611, 613, 615–617
  ionization potential of, 617, A-23
  thermodynamic properties of, A-28
Manometer, 46–48
Marcasite, 517
Marsden, Ernest, 128–129
Mass, A-5, A-19. See also Atomic mass
  energy and, 732–735
  law of conservation of, 12
  measurement of, 16–18
Mass law, 337n
Mass number, 23
Mass spectrograph, 737
Mass spectrometer, 134–135, 137
Mass spectrum, 135–137
Material balance equation, 369
Mathematical operations, A-7–A-17
Maxwell, James Clerk, 78, 146, 457–458
Maxwell-Boltzmann distribution, 78–79, 162, 266, 538
Mayer, Maria Goeppert, 735
Melting point, 311, 313, 511
  of the alkali metals, 554–555
  of the group IIA elements, 557
  of the group IIIA elements, 562
  of the group IVA elements, 568
  of the group VA elements, 579
  of the group VIA elements, 590

of the group VIIA elements, 598
  normal, 263
Mendeleev, Dmitriĭ I., 130, 189
Mercury, 326, 618, 737
  molar heat capacity of, 100
  thermodynamic properties of, A-29
Metalloids, 300–301, 308, 312, 319–320, 322, 562, 567
Metals, 300–301, 308, 312–313, 319–320, 322. *See also* Transition metals
  alkali, 204, 310–311, 551–557, 588, 592
  alkaline earth, 205, 311–312, 557–561, 588, 592
  base, 425
  binary compounds formed by, 391
  bonding by, 520–521
  coinage, 327
  crystal structures of, 518–521
  electronic band theory and, 540–541
  free-electron theory of, 534–540
  group IIIA, 562–567
  group IVA, 567–573
  magnetic properties of, 541–543
  noble, 425–426, 618
  platinum, 611, 614, 618, 621
  rare earth, 326
  sulfides of, 592, 594
Methane, 532
  molecular structure of, 222
  standard enthalpy of formation for, 108
Methyl silicone, 572
Meyer, J. Lothar, 130, 189
Miller, Robert C., 79
Millikan, Robert, 124–125, 127
Moissan, Henri, 598
Molality, 271–272
Molar heat capacity, 97
Molarity, 37–38, 272–273
Molecular crystals, 531–533
Molecular formulas, 55–57
Molecular models, 27
Molecular orbitals. *See also* Orbitals
  chemical bonding and, 235–245
  in heteronuclear diatomic molecules, 244–245
  in homonuclear diatomic molecules, 238–244
Molecular theory, kinetic. *See* Kinetic molecular theory
Molecules, 13–14
  chemical bonding and, 204–254
  in crystalline solids, 510–545
  kinetic theory of, 70–83, 265–267, A-30–A-33
  in liquids and solutions, 260–293
  moles and, 26–29

properties of, 8–9, 202–204
  structure of, 202, 680–683
Mole fraction, 271
Moles, 23–29, 37
Momentum, 71–73, A-19–A-20
Moseley, Henry G. J., 131–134, 516
Müller, Paul, 669
Mulliken, Robert S., 228
Mulliken electronegativities, 227–229

N

Natural gas, 662
Natural logarithm, 440n
Negative ion, 119
Neodymium, 611
Neon, 532, 604
  electron configuration of, 188, A-23
  van der Waals radius of, 193
Neptunium, 731
Nernst, Walther, 427
Net electric charge, 407–410
Net reactions, 290–293
Net useful work, 460
Neutralization, 306
Neutrons, 118, A-5
  in nuclear reactions, 730
Newlands, John A. R., 130n
Newton, Isaac, 18, 142, 144, 152, 154
Newton (N), 18
Nickel, 327, 542–543
  coordination complexes of, 610–611, 613–617
  ionization potentials of, 617, A-23
  thermodynamic properties of, A-29
Niobium, 617
Nitrates, 322
Nitric acid, 306, 580, 582
Nitric oxide, 344
Nitrides, 320
Nitrites, 322
Nitro compounds, 676
Nitrogen, 320–322, 501, 578–579
  compounds of, 580–582
  derivatives of, 676–678
  electron affinity of, 195
  electron configuration of, 188, A-23
  thermodynamic properties of, A-27
Nitrogen fixation, 501
Nitrogen tetroxide, 322
Nitroglycerin, 676
Nitrous acid, 580, 582
Nitrous oxide, 580
Noble gases, 185, 202, 309–310, 602–604
  molecular crystals of, 532
  van der Waals radii of, 192–193
Noble metals, 425–426, 618
Nonelectrolytes

defined, 287
  solubility of, 285–286
Nonmetals, 300–301, 308, 310, 320, 322, 324–326, 567
Nonstoichiometric compounds, 13, 543–544
Nuclear chemistry, 720–739
Nuclear energy, 732–735
Nuclear fission, 730–735
Nuclear fusion, 734–735
Nuclei, 15
  discovery of, 127–130
Nucleic acids, 702–707

O

Octahedral complexes, 626–629, 649–651
Octet rule, 205–206
  violations of, 213–214, 324, 564
Oppenheimer, J. Robert, 734
Optical activity, 318
Optical isomerism, 318, 500
Orbitals, 172–178, 629–631. *See also* Molecular orbitals
  chemical bonding and, 235–245
  Hartree-Fock, 192
  of octahedral complexes, 649–651
  self-consistent field, 192
Ore, 613
Organic chemistry, 313–319, 658–683
Osmium, 614, 621
Osmotic pressure, 282–285
Oxidation, 406–407, 422n
  relative strength of agents for, 425–426
Oxidation states, 407–410
  of group VA elements, 585
  of halogens, 598–599
  of nitrogen, 579
  of sulfur, 593
  of transition metals, 615–617, 619
Oxyacids, 551–553
Oxygen, 322–324, 588, 590–591
  derivatives of, 670–676
  electron affinity of, 195
  electron configuration of, 188, A-23
  in lithosphere, 320
  thermodynamic properties of, A-27
Ozone, 323

P

Paramagnetism, 541–542, 637
Pauli, Wolfgang, 181
Pauli exclusion principle, 181, 186–187, 190, 236, 536, 538
Pauling, Linus, 225–226, 520, 738
Pauling electronegativity, 226–229
  of the alkali metals, 554
  of group IIA elements, 557

Pauling electronegativity (*Continued*)
    of group IIIA elements, 562
    of group IVA elements, 568
    of group VA elements, 579
    of group VIA elements, 590
    of group VIIA elements, 598
Peptide bonds, 693
Perchloric acid, 602
Perfect gas. *See* Ideal gas
Period, 130
Periodic law, 130–131, 189, 300–337, 578
Periodic table, 189, 300, 302–304
Period of a wave, 142
Peroxides, 409
Petit, Alexis T., 98
Petit and Dulong, law of, 98, 511
pH, 365–366, 379, 389
Phase, 3
Phase diagrams, 263–265
Phenols, 670–676
Phenyl group, 666
Phosphoric acid, 306
Phosphorus, 320–322, 324, 571, 584–588
    electron affinity of, 195
    electron configurations of, 188, A-23
    thermodynamic properties of, A-27
Phosphorus pentoxide, 584, 586
Phosphorus trioxide, 584–585
Photoelectric effect, 152
Photons, 152–154
Physical processes, 8
Physics, concepts from, A-18–A-22
Pi (π) bonds, 238–245
Planar complexes, 644–648
Planck, Max, 142, 149, 152, 154
Planck's constant, 160, 172, A-5
Platinum, 583, 614, 626–628
    coordination complexes of, 627
    properties of, 611, 614, 618, 621
Plücker, Julius, 121
Plutonium, 731
Point defects, 543
Polar bonds, 224–230, 524
Polarization, 230–235
Pollution, 2
Polonium, 322–324, 588, 596
    radioactivity of, 720
Polymers, 326, 499
Polypeptides, 690–696
Polyprotic acids, 386–391
Polysaccharides, 697
Positive ion, 118
Positron emission, 729
Potassium, 310, 520, 553
    electron configuration of, 188, A-23
    in lithosphere, 320
    as an oxidizing agent, 425
    thermodynamic properties of, A-24–A-25
Potassium hydroxide, 306
Potential energy, 88–89
Powers, A-7–A-8
Praseodymium, 611
Preexponential factor, 486
Pressure. *See also* Vapor pressure
    atmospheric, 47
    conversion factors for, A-5
    critical, 264
    gases and, 45–53, 70–83
    law of partial, 50
    measurement of, 45–53
    osmotic, 282–285
Primary cells, 430–434
Principal quantum number, 169
Probability and entropy, 456–457
Probability density, 161–162, 174
Probability distribution function, 171
Products, 32
Prosthetic group, 693
Proteins, 690–696
Protonic acids, 551
Protons, 118, A-5
Proust, Joseph Louis, 12–13
Pseudo-first-order reactions, 479

Q

Quantum of energy, 149
Quantum number
    azimuthal, 173, 183
    for the hydrogen atom, 171–173
    magnetic, 173
    principal, 169
    spin, 178–180
    vibrational, 149–152
Quantum theory, 158
Quartz, 525–526, 567

R

Racemate, 318
Radial distribution function, 175
Radiation, 737–739
    background, 728–732
    blackbody, 148, 166
    cosmic, 728
Radioactivity, 553, 594, 598, 612
    natural, 720–728
    of polonium, 596
Radiocarbon dating method, 738
Radium, 730, A-23
Radius of curvature, 134
Radon, 193
Raoult, François Marie, 273
Raoult's law, 274, 276–279
Rare earth metals, 326
Rare gases. *See* Noble gases
Rate constant, 476, 484–488
Rate laws, 475–483
Reactants, 32
Reaction mechanisms, 494–501
Reaction order, 476
Reaction rates, 472–494
    collision theory of, 488–493
    equilibrium and, 483–484
    initial rate method for, 477–478
    temperature and, 484–488
    transition-state theory of, 493–494
Real gases
    equations of state for, 61–64
    kinetic molecular theory, 81–83
    law of chemical equilibrium and, 340–342
Real solutions, 340–342
Reduction, 406–407, 411, 613
    relative strength of agents for, 425–426
Relativity, theory of, 158
Replication, 704
Resistance, A-22
Resonance, 223–224
Resonating-valence-bond theory of metals, 520–521
Reversible process, 446, 452–453
Ribonucleic acid (RNA), 586, 702–707
Ribosomes, 702–707
Ritz-Paschen series, 168
Roasting, 613
Röntgen, Wilhelm, 127
Root mean square (rms) speed, 76
Rubidium, 553
Rubies, 327, 510
Ruthenium, 621
Rutherford, Ernest, 128–130, 728
Rutherfordium, 326, 732
Rutile, 527

S

Salt bridge, 410
Salts, 306–309
    ammonium, 579
    formed by alkali metal hydroxides, 554–555
    halide, 600
    potassium, 579, 586
    solutions of, 374–377
Sapphires, 327
Saturated hydrocarbons, 316–317, 658–663
Scandium, 611, 616
    ionization potentials of, 617, A-23
Schrödinger, Erwin, 171
Schrödinger equation, 171–174, 192
Seaborg, Glenn T., 731

Seawater, composition of, 325
Second (s), 17
Selenium, 301, 322–324, 588, 596
Self-consistent field method, 192
Selig, Henry, 603
Semiconductors, 301, 540–541, 545, 562
Shielding effect, 181–184
Sigma (σ), bonds, 238–245
Significant figures, A-8–A-10
Silicon, 319–320, 567, 571–572, 588
  crystal structure of, 524, 531
  electron affinity of, 195
  electron configuration of, 188
  energy gap in, 541
  in lithosphere, 320
  thermodynamic properties of, A-26
Silver, 326–327, 393–394, 618
  isotopes of, 10
  as a solid solution, 7, 544
  thermodynamic properties of, A-29
Silver chloride, 396–397
Simplest formulas, 27–32
Simultaneous equilibria, 360–397
Sodalite, 525–526
Sodium, 310–311, 553
  atomic structure of, 182–185
  electron affinity of, 195
  electron configuration of, 188, A-23
  isotopes of, 11
  thermodynamic properties of, A-24
Sodium bicarbonate, 570
Sodium carbonate, 570
Sodium chloride, 103, 517, 528
  Born-Haber cycle for, 529
  molar enthalpies of fusion and vaporization for, 103
  molar heat capacities of, 100
Sodium hydroxide, 306
Soft-sphere model, 522, 531
Solids, 510–545
  amorphous, 511–512
  macroscopic properties of, 510–511
  solutions in the form of, 544–545
  vapor pressure for, 511
Solubility product, 349–351
Solutes, 271, 340–342
Solutions, 5–9
  buffer, 377–380
  gaseous, 50–53
  ideal, 340–342
  molarity and, 37
  properties of, 260, 271–293
  salt, 374–377
  solid, 544–545
Solvents, 271
Sommerfeld, Arnold, 535
Sørensen, Søren P. L., 365

Spectrochemical series, 635
Spectrograph, 737
Speed, 71–74, 76, A-18
Sphalerite, 524
sp hybrids, 246–249
sp^2 hybrids, 249–251
sp^3 hybrids, 251–254
Spin quantum number, 178
Spontaneous irreversible processes, 451–454
Square planar complexes, 644–648
Square-well potential energy diagram, 735
Stability constants, 642
Stalactites, 560–561
Stalagmites, 560–561
Standard electrode potentials, 421–425, 617–619
Standard enthalpies of formation, 107–109, A-24–A-29
Standard entropy of vaporization, 442
Standard free-energy changes, 462–465, A-24–A-29
Standard heat of formation, 108
Standard states, 341
Standard voltage, 420–425
Standing de Broglie waves, 156–158
State functions, 92
Stationary states, 149
Steels, 327, 613
Stern, Otto, 178
Stoichiometry, 12
Stokes, George, 125
Stoney, G. Johnstone, 121
Storage cells, 430–434
Straight-chain hydrocarbons, 314
Strassmann, Fritz, 730
Strontium, 520, 738
Structural formulas, 26–27
Structural isomers, 26, 317
Sublimation, 313, 511
Substances, 5–9
Substitution reactions, 316
Substrates, 709
Sugars, 697
Sulfide precipitations, 391–392
Sulfur, 322–324, 512, 588, 591–596
  electron affinity of, 195
  electron configuration of, 188, A-23
  thermodynamic properties of, A-27
Sulfur dioxide, 592–593
Sulfur hexafluoride, 595–596
Sulfuric acid, 306, 308, 593–594
Sulfur tetrafluoride, 596
Surroundings, 3, 92
Systems, 3, 92
  heterogeneous, 3–5, 11

T

Tantalum, 617
Teller, E., 649
Telluric acid, 598
Tellurium, 322–324, 588, 596
Temperature, 52, A-6
  Boyle, 61
  critical, 264
  Curie, 543
  equilibrium constants and, 440–445
  molecular disorder and, 446–447
  reaction rate constant and, 484–488
  vapor pressure and, 440–445
Tetraethyllead, 320
Tetrahedral angle, 217
Tetrahedral hybrids, 251–254
Tetrahedral complexes, 644–648
Thallium, 312–313, 562–563, 566–567
Thermodynamics, 52, 100–103, A-24–A-29
  chemical equilibria and, 440–465
  of chemical reactions in biological systems, 712–715
  empirical basis of, 454
  first law of, 91–96, 529
  second law of, 452–454, 456–457
  third law of, 449
Thiosulfate, 594
Thomson, George P., 155
Thomson, J. J., 121–124, 134, 155, 534
Thorium, 598
Three-center bond, 563
Time measurements, 17, A-6
Tin, 301, 319–320, 567, 571–573, 614
  gray, 531
  isotopes of, 11
  molar heat capacities of, 100
  thermodynamics of, A-26
Titanium, 327, 610–611, 614–617
  ionization potentials of, 617, A-23
  thermodynamic properties of, A-28
Tolman, Richard C., 534
Tolman effect, 534
Torricelli, Evangelista, 47
Transition metals, 326–327, 501, 608–622, 626–652
  compounds formed by, 542
  enzymes and, 501
  ionization potentials of, 617
  oxidation states of, 615–617, 619
  thermodynamic properties of, A-28–A-29
Transition-state theory, 493–494
Translational kinetic energy, 75, 88
Transuranium elements, 730–731
Tridymite, 525
Trigonal hybrids, 250–251

Triplet, 707
Trouton, Frederick, 103
Trouton's rule, 103, 448
Turnover number, 709

U

Uhlenbeck, George E., 178
Ultraviolet catastrophe, 149, 152
Uncertainty principle, 158–161
Unit cell, 513
Units and conversion factors, 15–18, A-3–A-6
University of Leiden, 178
Unsaturated hydrocarbons, 314–315, 663–665
Uranium, 598, 726–727, 730–731

V

Valence bond theory, 246, 520
Valence electrons, 196
Valence-shell electron-pair repulsion theory (VSEPR theory), 216–221
Vanadium, 610–611, 615–617
  ionization potentials of, 617, A-23
  thermodynamic properties of, A-28
van der Broek, Antonius, 130–131
van der Waals, Johannes D., 62–64, 80–83
van der Waals equation of state, 63–64, 489
van der Waals radii, 192–193
van der Waals repulsion, 522, 531

van't Hoff, Jacobus, 318
Vapor pressure, 51, 260–263, 267, 348–349
  lowering of, 273–279
  for solids, 511
  temperature dependence of, 440–445
Velocity, A-6, A-18
  kinetic theory of gases and, 71–73
Voltage, standard, 420–425
von Laue, Max, 513

W

Waage, Peter, 337
Water, 103
  dipolar character of, 555
  dissociation of, 363–365
  hard, 560
  heavy, 737
  molar enthalpies of fusion and vaporization for, 103
  molar heat capacity of, 100
  molecular structure of, 222
  ocean, 55
  standard enthalpy of formation for, 108
Water pollution, 2
Wave function, 161, 171
  Heitler-London, 245
Wavelength, 142
Wave motion, 142–144
Wave number, 635
Wave-particle duality, 156

Waves, 154–161
Wave theory of light, 144–147
Weight, A-19
Werner, Alfred, 626–629
White graphite, 531
Work, 460, A-20–A-21
Wurtzite, 523, 525, 533–534

X

Xenon, 532, 603–604
  van der Waals radius of, 193
X-ray diffraction, 510, 512–517
X-rays, 127, 739

Y

Young, Thomas, 145

Z

Zeeman, Pieter, 182
Zeeman effect, 182
Zeolite, 561
Zero-point energy, 150
Zinc, 103, 326, 583, 588, 610–611, 613–617, 619
  ionization potential of, 617
  molar enthalpies of fusion and vaporization for, 103
  molar heat capacity of, 100
  thermodynamic properties of, A-29
Zirconium, 610
Zone refining, 571

Atomic Masses

Expressed to four significant figures

Based on the atomic mass of $^{12}C = 12$ exactly

Number in brackets denotes isotope of longest known half-life

Name	Symbol	Atomic Number	Atomic Mass
Actinium	Ac	89	[227]
Aluminum	Al	13	26.98
Americium	Am	95	[243]
Antimony	Sb	51	121.8
Argon	Ar	18	39.95
Arsenic	As	33	74.92
Astatine	At	85	[210]
Barium	Ba	56	137.3
Berkelium	Bk	97	[247]
Beryllium	Be	4	9.012
Bismuth	Bi	83	209.0
Boron	B	5	10.81
Bromine	Br	35	79.90
Cadmium	Cd	48	112.4
Calcium	Ca	20	40.08
Californium	Cf	98	[251]
Carbon	C	6	12.01
Cerium	Ce	58	140.1
Cesium	Cs	55	132.9
Chlorine	Cl	17	35.45
Chromium	Cr	24	52.00
Cobalt	Co	27	58.93
Copper	Cu	29	63.54
Curium	Cm	96	[247]
Dysprosium	Dy	66	162.5
Einsteinium	Es	99	[254]
Erbium	Er	68	167.3
Europium	Eu	63	152.0
Fermium	Fm	100	[257]
Fluorine	F	9	19.00
Francium	Fr	87	[223]
Gadolinium	Gd	64	157.3
Gallium	Ga	31	69.72
Germanium	Ge	32	72.59
Gold	Au	79	197.0
Hafnium	Hf	72	178.5
Hahnium	Ha	105	[262]
Helium	He	2	4.003
Holmium	Ho	67	164.9
Hydrogen	H	1	1.008
Indium	In	49	114.8
Iodine	I	53	126.9
Iridium	Ir	77	192.2
Iron	Fe	26	55.85
Krypton	Kr	36	83.80
Lanthanum	La	57	138.9
Lawrencium	Lr	103	[256]
Lead	Pb	82	207.2
Lithium	Li	3	6.941
Lutetium	Lu	71	175.0